Brownfields

A COMPREHENSIVE GUIDE TO REDEVELOPING CONTAMINATED PROPERTY

TODD S. DAVIS
KEVIN D. MARGOLIS
with a preface by Vice President Al Gore

Section of Natural Resources, Energy, and Environmental Law
American Bar Association

Cover design by Catherine Zaccarine.

The materials contained herein represent the opinions of the authors and editors and should not be construed to be the action of either the American Bar Association or the Section of Natural Resources, Energy, and Environmental Law unless adopted pursuant to the bylaws of the Association.

Nothing contained in this book is to be considered as the rendering of legal advice for specific cases, and readers are responsible for obtaining such advice from their own legal counsel. This book and any forms and agreements herein are intended for educational and informational purposes only.

© 1997 American Bar Association. All rights reserved.
Printed in the United States of America.

Library of Congress Catalog Card Number 97-71739
ISBN 1-57073-439-9

Discounts are available for books ordered in bulk. Special consideration is given to state bars, CLE programs, and other bar-related organizations. Inquire at ABA Publishing, Book Development and Marketing, American Bar Association, 750 North Lake Shore Drive, Chicago, Illinois 60611.

01 00 99 98 97   5 4 3 2 1

CONTENTS

Acknowledgments xvii

Preface xix
Vice President Al Gore

Introduction: How to Use This Book xxi

PART I: Background Information

Chapter 1
Defining the Brownfields Problem 3
Todd S. Davis and Kevin D. Margolis
    What Is a Brownfield? 5
    Why Are Brownfields Demanding Attention? 6
    How Did the Brownfields Issue Evolve to a Crisis State? 7
    Barriers to Brownfields Redevelopment 9
    Bringing Down the Barriers 12
    Food for Thought 13

Chapter 2
Overview of Federal and State Law Governing Brownfields Cleanups 15
Wendy E. Wagner
    The Legal Landscape 15
    Cleanup of Brownfields under Federal or Parallel State Laws 17
    Federal and State Reforms for Brownfields Redevelopment 25
    Conclusion 28

Chapter 3
The Clinton Administration's Brownfields Initiative 41
Jonathan D. Weiss
    Clarification of Liability and Cleanup Issues 41
    USEPA's Pilot Program and Community Partnerships 45
    Government Coordination 47
    Conclusion 49

PART II: Legal, Business, Financial, and Political Issues Associated with Redeveloping Contaminated Property

Chapter 4
Doing the Brownfields Deal 53
Kevin D. Margolis and Todd S. Davis
    Valuing the Target Site—Setting a Baseline for Negotiating
        the Purchase Price 53
    Evaluating the "Environment" Surrounding the Deal 55
    The Outline of the Deal—the Purchase Agreement 56
    Environmental Liability—Defining Exposure 59
    Investigating the Target Site 61
    Do the Deal! 62

Chapter 5
Acquisition Considerations for Brownfields Properties 65
Adam Fishman and B. Robert Amjad
    Types of Properties Ripe for Redevelopment 65
    Return Needed to Justify Investment in a Brownfield 69
    Brownfields Investment Strategy 73
    To Invest or Not to Invest 74

Chapter 6
Valuing Brownfields 76
Bill Mundy
    Estimating Value of Clean Property 77
    Understanding Contamination 77
    Stigma 80
    Conclusion 86

Chapter 7
Creative Financing Strategies for Redeveloping Brownfields 87
Donald T. Iannone
    The Starting Point: Liability and Investment Risk 87
    Types of Brownfields and Associated Financing Strategies 90
    Understanding the Financial Requirements 91
    Current Brownfields Financing Strategies 94
    Federal Funding for Brownfields Redevelopment 95
    Future Commercial Real Estate Finance Trends 96
    Conclusion 98

Chapter 8
Brownfields Sites: Removing Lender Concerns as a Barrier to Redevelopment 100
Margaret Murphy
- The Brownfields Problem 101
- Lender Concerns at Brownfields Sites 102
- State Approaches to Limiting Lender Liability 102
- Federal Initiatives That Limit Lender Liability 108
- Insuring Against Liability 113
- Conclusion 114

Chapter 9
The Deductibility of Environmental Remediation Costs 121
Bernard J. Jerlstrom and Alan S. Doris
- To Deduct or to Capitalize?—That Is the Question 121
- The Statutory Scheme 121
- The Case Law 122
- The Service's Rulings 126
- Observations 131
- Proposed Legislation 132
- State Tax Incentives 133
- Applicability of the Tax Law to Brownfields 134
- Calculation of Actual Tax Savings 136

Chapter 10
Building Consensus for the Project 138
Howard C. Landau
- The Cast of Characters 138
- The Role of Public Relations 139
- The Direct Approach 140
- The Steps 140
- The Media 142
- Building Consensus: An Ongoing Process 143

Chapter 11
Environmental Insurance in the Brownfields Transaction 144
William McElroy and Todd S. Davis
- Brief History of Environmental Insurance 144
- Applicability of Environmental Insurance to Brownfields Projects 145
- Insurance Underwriting Risks for Environmentally Impacted Properties 146

The Underwriting Process................................. 147
Appropriate Coverage and the Insurance Contract 148
The Cost and Value of Environmental Insurance 150
Limitations to Available Insurance 151
A Comment on the State of the
 Environmental-Insurance Market 152

Chapter 12
Using Old Insurance Policies as Weapons........................... 154
Diane R. Archangeli
Insurance Archaeology 154
Making a Claim under Old Commercial
 General-Liability Policies 155
Coverage under CGL Policies.................................. 156
Coverage under Automobile Policies 157
Coverage under Garage-Liability Policies 157
Coverage under Environmental-Impairment Liability Policies 157
Coverage under Personal-Injury Section of the CGL 158
Coverage under First-Party Property Insurance 158
Accident versus Occurrence Policies............................. 159
Defenses Insurance Companies Often Use to Avoid Liability
 under CGLs... 159
How Much Coverage Can You Get? 163
Stacking 163
Good Approaches in Negotiating Settlements
 with Insurance Companies 164

Chapter 13
Hiring the Right Laboratory 170
Christine DiCato-Thaxton
Analytical Expenses 170
Coordination between the Client, Environmental
 Consultant, and Laboratory 170
Selecting the Appropriate Laboratory............................ 171
Selecting the Appropriate Analytical Tests 173
Insurance, Contracts, and Confidentiality 174
Data Validation, Certifications, and Prices....................... 175
Comfort Levels and Responsiveness............................. 176

Chapter 14
Rebuilding Communities through Brownfields Redevelopment........ 177
Scott D. Garson
The Need for a Community Brownfields Strategy 178
The Need for Infrastructure Subsidies 178
Limited Resources 179
State Voluntary Cleanup Programs: Costs and Benefits............ 180
The Role of CDCs: Educator, Broker, Facilitator 181

Chapter 15
Community Participation in Brownfields Redevelopment 183
John C. Chambers and Michelle A. Meertens
    The Community Perspective: A Historical Grounding 184
    Community Concerns 185
    Current Methods of Participation 189
    Conclusion 191

PART III: Scientific Concepts Used to Address Contaminated Property

Chapter 16
The Science of Brownfields 197
Dan B. Brown
    Subsurface Geology and Hydrogeology 197
    Phases of Contamination 201
    Site Investigation Process 203
    Determining Remedial Alternatives for the Property 206
    Established Protocols in the Ohio VAP 209
    The Context of the Cleanup Standards 210
    Conclusion 211

Chapter 17
The Role of Risk Assessment in Redeveloping Brownfields Sites 214
Michael L. Gargas and Thomas F. Long
    What Is Risk? 215
    The Rise of Environmental Risk Assessment 216
    The Risk Assessment Process 217
    Risk Assessment and the Creation of Brownfields 220
    Risk Assessment and the Elimination of Brownfields 223
    Pitfalls and Promises in Risk Assessment for Brownfields 226
    Brownfields Case Studies 242
    Conclusion 245

Chapter 18
The Risk-Based Corrective-Action Process 250
James R. Rocco and Lesley Hay Wilson
    The Traditional Corrective-Action Process 250
    Lessons Learned in the Development
        of the Corrective-Action Process 254
    Defining Corrective Action 254
    The Move toward Risk-Based Corrective Action 257
    A Case Study of the RBCA Approach 260
    Conclusion 266

Chapter 19
Risk Assessment—A Physician's Introduction 268
Richard L. Cristea, M.D.
    The Jargon of Health Risk Assessment 268
    Epidemiology and Study Design 270
    Conclusion 271

Chapter 20
Remediation Strategies for Brownfields Redevelopment 273
Edward J. Cichon
    General Remedial Strategy 274
    Risk-Based Cleanups 276
    Soil Remediation 277
    Groundwater Remediation 280
    Nonaqueous Phase Liquid Removal 282
    Conclusion 283

PART IV: State Voluntary Cleanup Programs

Chapter 21
Arizona 287
Joel L. Herz
    Summary of Major Provisions of the Arizona Programs 288
    Memorandum of Agreement 290
    Liability Protection 290
    Cleanup Standards 291
    The Mechanics of Participating in the Arizona
        Voluntary Programs 294
    Cost Recovery 295
    Conclusion 296

Chapter 22
Arkansas 300
Allan Gates
    Participation in the Arkansas Brownfields Program 300
    Memorandum of Agreement 301
    Liability Protection 302
    Financial Incentives 303
    Cleanup Standards 304
    The Mechanics of Participating in the Arkansas
        Brownfields Program 306
    Remedy Failure or Ineffectiveness 308
    Cost Recovery 308
    Conclusion 309

Chapter 23
California 313
Jane B. Kroesche
    Voluntary Cleanup Program 314
    Expedited Remedial Action Program 317
    Additional Measures to Facilitate Brownfields Redevelopment 323
    Conclusion 332

Chapter 24
Colorado 344
David F. Goossen
    Summary of Major Provisions 344
    Participation in the Program 345
    Memorandum of Agreement 346
    Protections Granted by the Colorado Program 348
    Financial Incentives 348
    Cleanup Standards 349
    The Mechanics of Participating in the Program 349
    Enforceability of Voluntary Cleanup Plans 352
    Conclusion 353

Chapter 25
Connecticut 357
Margaret Murphy
    Summary of Connecticut's Brownfields Initiatives 357
    Eligibility for Connecticut's Brownfields Programs 358
    Memorandum of Agreement 360
    Liability Protection for Brownfields Volunteers 361
    Financial Incentives 362
    Cleanup Standards 363
    The Mechanics of Participating in Connecticut's
        Brownfields Initiatives 364
    Liability Protection for Lenders 367
    Recording 367
    Admission of Liability 368
    Civil and Criminal Penalties 368
    Cost Recovery 368
    Conclusion 369

Chapter 26
Delaware 374
R. Judson Scaggs, Jr.
    Summary of Major Provisions 374
    Brownfields Legislation 375
    VCP 377
    Cost Recovery and Financial Assistance 380
    Conclusion 381

Chapter 27
Illinois 385
David J. Engel
- Site Remediation Program Created by the 1995 Legislation 385
- The Mechanics of Program Participation 390
- Memorandum of Agreement 394
- Cost Recovery 394
- The 1989 Program... 396
- Conclusion... 397

Chapter 28
Indiana... 402
Anne Slaughter Andrew
- Summary of Major Provisions 402
- Participation in the Voluntary Remediation Program 402
- Memorandum of Agreement 403
- Lender and Fiduciary Liability Protection 403
- Financial Incentives 404
- Cleanup Standards... 404
- The Mechanics of the Voluntary Remediation Program 405
- Liability Protection... 406
- Costs and Cost Recovery....................................... 407
- Conclusion... 407

Chapter 29
Kentucky 410
Henry L. Stephens, Jr.
- Summary of Major Provisions 410
- Background Surrounding the Passage of Kentucky's Brownfields Legislation................................. 410
- Implementation of Kentucky's Brownfields Legislation 412
- NFR Letters 413
- Conclusion... 415

Chapter 30
Maine... 418
James T. Kilbreth and Juliet T. Browne
- Summary of Major Provisions 418
- Mechanics of Participating in the Maine Program 419
- Cleanup Standards.. 422
- Liability Protection under the Program 423
- Financial Incentives under the Program........................ 426
- Recovery of Cleanup Costs..................................... 426
- Conclusion... 428

Chapter 31
Maryland 433
Anthony M. Carey and Judith A. Armold
    Summary of Major Provisions 433
    The Cleanup Programs 434
    Liability Protection 438
    The Financial-Incentive Programs 440
    Other Provisions of the House and Senate Bills 441
    Memorandum of Agreement with USEPA 442
    Conclusion 442

Chapter 32
Massachusetts 443
Ned Abelson, William M. Seuch, and Maura McCaffery
    Summary of Major Provisions 443
    Understanding the Massachusetts Superfund Statute 444
    Understanding the Massachusetts Contingency Plan 447
    The Massachusetts Covenant-Not-to-Sue Program 452
    Memorandum of Understanding 454
    Economic Incentives for Brownfields Redevelopment 454
    Potential Changes to the Massachusetts Brownfields Program 455
    Conclusion 456

Chapter 33
Michigan 461
Grant R. Trigger
    Brownfields Financial Incentives 462
    Liability Reform 464
    Special Liability Protection for Lenders and Others 465
    Cleanup Criteria 468
    Remedial Actions 470
    Coordination of Other Legal Requirements 472
    Due Care, Reasonable Precautions, and Exacerbation
        of Existing Contamination 473
    The BEA 473
    Affirmative Obligation to Remediate 475
    Conclusion 476

Chapter 34
Minnesota 487
Paul S. Moe
    Major Provisions of the VIC Program 487
    Eligibility for the VIC Program 489
    Memorandum of Agreement 490

    Liability Protection................................. 491
    Financial Incentives 492
    Cleanup Standards.................................. 493
    Confidentiality, Consultants, and Fees 494
    Mechanics of Participating in the VIC Program 495
    Cost Recovery 496
    Conclusion... 496

Chapter 35
Missouri.. 501
James T. Price and Jennifer A. Downs
    Summary of Major Provisions 501
    Eligibility for Programs.............................. 502
    Memorandum of Agreement 503
    Liability Protection................................. 504
    Financial Incentives 505
    Cleanup Standards.................................. 506
    Participating in the Missouri Brownfields Programs 506
    Conclusion... 508

Chapter 36
Nebraska 511
James T. Price and Annette Kovar
    Summary of Major Provisions 511
    Participation in the Program......................... 512
    Sites Ineligible for the Program 513
    Memorandum of Agreement 513
    Liability Protection................................. 513
    Financial Incentives under the Program................ 514
    Cleanup Standards.................................. 514
    Mechanics of Participating in the Program 514
    Cost Recovery 515
    Conclusion... 515

Chapter 37
New Jersey 518
I. Leo Motiuk and Sean T. Monaghan
    Eligibility 518
    How Does the VCP Program Work?.................... 519
    What Happens If the VCP Is Not Completed?........... 519
    Confidentiality 520
    Liability Protection................................. 520
    Financial Incentives 521
    Technical Regulations 522
    Cleanup Standards.................................. 522
    No-Further-Action Letters........................... 523

Developer Incentives 524
Recovery from Third Parties..................................... 525
Examples of Brownfields Projects in New Jersey 525
Recommendations for Brownfields Redevelopment
    in New Jersey 526

Chapter 38
New York.. 532
Margaret Murphy
Summary of Major Provisions 532
Eligibility for the New York Voluntary Cleanup Program 533
Memorandum of Agreement 533
Liability Protection... 534
Financial Incentives 535
Cleanup Standards.. 535
Mechanics of Participating in the Voluntary Cleanup Program...... 536
Recording, Admission of Liability, and Penalties................. 537
Conclusion... 538

Chapter 39
Ohio... 542
Todd S. Davis and Kevin D. Margolis
Summary of Major Provisions 542
Participation in the Voluntary Action Program.................... 543
Memorandum of Agreement 544
Liability Protection... 545
Financial Incentives under the Voluntary Action Program.......... 545
Cleanup Standards.. 546
Certified Professionals and Laboratories 547
Mechanics of Participating in the Ohio
    Voluntary Action Program 548
Recording of the NFA Letter and Covenant....................... 549
Auditing by Ohio EPA... 549
Admission of Liability 550
Civil and Criminal Penalties..................................... 550
Cost Recovery 550
Conclusion... 552

Chapter 40
Oregon... 557
Richard M. Glick
Summary of Major Provisions 557
Participation in the Recycled Lands Act Program.................. 558
Cleanup Standards.. 558
Prospective-Purchaser Agreements............................... 560
Memorandum of Agreement 561

    Liability Protection...562
    Financial Incentives...562
    Cost Recovery..562
    Conclusion..563

Chapter 41
Pennsylvania 567
Robert L. Collings
    Summary of Major Provisions..568
    Elements of the Pennsylvania Program............................568
    Cleanup Standards..571
    Procedures for Performing Act 2 Cleanups.......................572
    Liability Protection...574
    Financial Assistance...574
    Guidance and Regulations/Buyer-Seller Agreements...............575
    Conclusion..575

Chapter 42
Rhode Island..581
Ned Abelson, William M. Seuch, and Maura McCaffery
    Major Provisions of the Rhode Island
        Brownfields Program..582
    Fees and Private Cost Recovery.....................................586
    Lender-Liability Protection..588
    Financial Incentives...589
    Conclusion..589

Chapter 43
Tennessee 592
Greer Tidwell
    Summary of Major Provisions..593
    Eligibility for the Voluntary Action Program......................594
    Memorandum of Agreement..594
    Liability Protection...594
    Lender-Liability Protection..595
    Benefits for Participants in the Program..........................596
    Cleanup Standards..596
    Participating in the Voluntary Action Program...................597
    Admission of Liability..597
    Cost Recovery..597
    Conclusion..598

Chapter 44
Texas..601
John Slavich
    Summary of Major Provisions..601
    Eligibility for the Program...601

Contents xv

    Memorandum of Agreement 602
    Liability Protection.. 602
    Financial Incentives 603
    Cleanup Standards.. 603
    Participation in the Program..................................... 604
    VCP Approval Process.. 605
    Cost Recovery 605
    Audit Privilege 605
    Developments... 607

Chapter 45
Utah 610
H. Michael Keller
    Summary of Major Provisions 610
    Eligibility for the Utah Voluntary Agreement Program.............. 611
    Memorandum of Agreement 612
    Liability Protection.. 612
    Financial Incentives 613
    Cleanup Standards.. 613
    Mechanics of Participating in the Voluntary
        Agreement Program...................................... 614
    Cost Recovery 615
    Conclusion.. 616

Chapter 46
Vermont... 620
A. Jay Kenlan and W. Andrew Hazelton
    Summary of Major Provisions 620
    Application and Eligibility Determination......................... 621
    Site Investigation and Site Remediation 623
    Liability Protection.. 625
    Additional Obligations of the Program Participant 627
    Cleanup Standards.. 627
    Conclusion.. 628

Chapter 47
Virginia 631
Mark D. Anderson
    Summary of Major Provisions of Chapter 622...................... 631
    Eligibility for the VRP 632
    Memorandum of Agreement 632
    Liability Protection.. 633
    Cleanup Standards.. 633
    Mechanics of Participating in the VRP 635
    Site Access 638
    Cost Recovery 638
    Conclusion.. 638

Chapter 48
Washington 642
Bradley M. Marten and Robin K. Rock
    Summary of Major Provisions 642
    Eligibility for the Prospective-Purchaser Program 643
    Memorandum of Agreement 643
    Liability Protection 644
    Financial Incentives 646
    Cleanup Standards 646
    Risk Assessments 649
    Institutional Controls 649
    Mechanics of Participating in Washington's
        Prospective-Purchaser Program 649
    Cost Recovery 651
    Conclusion 651

Chapter 49
West Virginia 655
Barbara D. Little
    Summary of Major Provisions of the Act 655
    Eligibility for the Voluntary Action Program 656
    Liability Protection 656
    Criminal Penalties 659
    Incentives under the Act 659
    Mechanics of Participating in the West Virginia
        Voluntary Action Program 660
    Special Provisions under the Act Applicable
        to Brownfields Sites 661
    Conclusion 662

Chapter 50
Wisconsin 665
Arthur J. Harrington
    Summary of Major Provisions 665
    Cleanup Liability—the Wisconsin Spill Statute 666
    Liability Protection 667
    Soil Cleanup Standards 673
    Memorandum of Agreement 676
    Conclusion 676

About the Editors 683

About the Contributors 685

ACKNOWLEDGMENTS

The authors wish to express their appreciation to everyone who played a role in the publication of this book. In particular, we give thanks to the contributors to this publication who withstood our tireless prodding requesting delivery of their chapters and to the staff at the American Bar Association for their editorial insights. We also must highlight the substantial contribution of Mrs. Mary Farage, whose patience, attention to detail, and management skills have made this book a reality. Additionally, we thank Michael A. Fixler, Jennifer A. Lesny Fleming, Ryan P. McBride, Thomas R. O'Donnell, Terence L. Thomas, and Anna K. Tucker for their legal research and editorial assistance.

Additional thanks must go to our families, friends, and colleagues for providing support and inspiration. Special thanks goes to Dr. Richard L. Cristea, whose vision and encouragement have made this book possible. A debt of gratitude also goes to our colleagues at Benesch, Friedlander, Coplan & Aronoff for their support throughout this process. In particular, we thank the Information Processing Center staff, who provided endless hours of support in perfecting this manuscript.

Last, but of course not least, Todd Davis wishes to thank Lisa, Jordan, Dakota, and Elle for their unending patience, love, and support. Also, special thanks goes to his mother, Carole; father, Stanley; and Jill, Brian, Stacey, and Rob, who would ban him from Sunday night dinner if their inspiration and constant encouragement were not specifically acknowledged. Finally, to his Grandmother Ruth and mother-in-law, Shirley, for their love and support. Kevin Margolis also wishes to thank his family, but in particular, his wife, Sheila, for her love and never-ending tolerant sustenance of him throughout the production of this book.

Todd S. Davis
Kevin D. Margolis

PREFACE

VICE PRESIDENT AL GORE

As the President has often stated, we are entering an "Age of Possibility"—with vast new opportunities. And Americans are leading the way in meeting the new global challenges. Over the last three years, our economy has created over 9 million new jobs and the nation has experienced the lowest combined rate of unemployment, inflation, and mortgage rates in twenty-seven years.

But as we embark on this exciting new age, we cannot forget the many people and places left behind; people and places that want to join in the progress we are making. The estimated hundreds of thousands of abandoned and contaminated properties that are littered across our poorest communities—known as "brownfields"—present a significant barrier to economic revitalization in our nation's cities. By encouraging the cleanup and redevelopment of these brownfields, the Clinton Administration is forging new ways to empower distressed communities and create jobs for their residents.

The Administration has embarked on a sweeping Brownfields Economic Redevelopment Initiative. This effort is an integral part of the President's Community Empowerment Agenda, a series of initiatives across the federal government—from Empowerment Zones to community policing—designed to work together to provide greater opportunity to our distressed urban and rural communities. The Clinton Administration recognizes that the federal government alone cannot solve the problems of distressed communities. But we also recognize that the government can be a catalyst in empowering communities with the tools to solve their own problems and in encouraging the private sector to join in those efforts.

The Brownfields Initiative, led by Environmental Protection Agency (EPA) Administrator Carol Browner with the assistance of other agencies, demonstrates the virtues of community empowerment. It is bringing new development, new jobs, and new hope to communities—while engaging citizens and showing that environmental protection and economic development can go hand-in-hand.

The Initiative includes a nationwide pilot program that challenges communities, with the input of residents and a broad range of stakeholders, to devise their own strategies and approaches to redevelopment. The first of sixty brownfields pilots, based in Cleveland, has resulted in $3.2 million in new private investment, a $1 million increase in the local tax base, and more than 170 jobs through the creation of several businesses on the former industrial property. Other innovative projects—from Buffalo, New York to Laredo, Texas—are underway.

The federal government is also enacting common-sense reforms to the Superfund program to remove the obstacles to brownfields redevelopment. The USEPA, for example, has removed 27,000 sites from the Superfund inventory, clearing the way for their redevelopment. And to encourage private-sector cleanup, the President has proposed a tax incentive for brownfields redevelopment that would target distressed communities—including federal Empowerment Zones. The plan provides $2 billion in tax incentives over seven years and is expected to spur $10 billion in private investment. The Administration also strongly believes we need a responsible new Superfund law to promote the cleanup and redevelopment of brownfields.

Through our community empowerment efforts—including the Brownfields Initiative—the Administration is encouraging new ways of getting things done on the local level—ensuring that everyone and every place has the opportunity to share in the "Age of Possibility."

June 1996

INTRODUCTION

How to Use This Book

Due to the complicated nature of brownfields issues, overcoming brownfields problems means different things to different people. This book was written as both a guide and a resource manual for stakeholders in the brownfields redevelopment process. The complexity of addressing barriers to brownfields redevelopment stems, in part, from the large number of professional disciplines necessary to tackle brownfields projects, as well as the lack of detailed information readily available about newly emerging state and federal programs. Therefore, this book was developed to provide both information and strategic advice to assist parties to hurdle the barriers precluding brownfields redevelopment. This book also offers an in-depth look at all recently enacted state Voluntary Cleanup Programs.

The book is divided into four parts. Part I explains critical background information to put the "brownfields issue" in context. Part II details the most important legal, business, financial, and political issues associated with redeveloping contaminated real estate. This section offers insight into key issues and strategic advice from experts in various disciplines regarding effective approaches to managing environmental liabilities on a proactive basis. It provides a template for getting the brownfields deal done. Part III discusses both the basic science and the emerging concepts involved in risk-based science used to address contaminated property appropriately and cost-effectively. Finally, Part IV offers a detailed look at the most important elements of each state Voluntary Cleanup Program enacted as of the time of this publication.

Parts I–III are designed to be read "cover-to-cover" and will provide the reader with not only essential information important to understanding the issues but also with strategic advice toward addressing the problems posed by contaminated real estate. The authors did not intend for the reader to read Part IV cover-to-cover. Rather, this section of the book should serve as critical reference material and offers tremendous insight into any particular state Voluntary Cleanup Program of interest to the reader. Additionally, Part

IV of the book may be used to compare one state's Voluntary Cleanup Program with another state's program.

The state Voluntary Cleanup Program chapters were drafted to highlight important issues particularly relevant to a developer of contaminated real estate, including the following:

- Citations to relevant statutes and regulations
- Properties/entities eligible for the program
- Properties/entities excluded from the program
- Existence of any memorandum of agreement for a particular program with the United States Environmental Protection Agency
- Liability protections afforded to lenders, trustees/fiduciaries, and other program participants
- Financial incentives available to facilitate redevelopment
- Articulated cleanup standards applicable to sites participating in the Voluntary Cleanup Program
- Availability of risk assessment, and institutional and engineering controls, as acceptable approaches to remediation
- Statutory protections against admissibility/discovery based on voluntary cleanup or voluntary investigation
- General description of criteria for participating in the Voluntary Cleanup Program and necessary fees and certifications
- Description of the general process for obtaining "covenants-not-to-sue" and other liability protections
- New causes of action created by Voluntary Cleanup Program statutes for cost recovery associated with participating in such programs

A flow chart depicting the general process for participating in any particular state's Voluntary Cleanup Program is included at the end of each state's chapter.

This book is intended to function as a practical reference for persons interested in participating in this quickly evolving and innovative approach to environmental liabilities. The authors also intend to update and supplement the book periodically, to keep readers abreast of the latest developments, new laws, and strategies for practically addressing environmental liabilities. *AS THIS AREA OF ENVIRONMENTAL LAW CONTINUES TO DEVELOP AT A RAPID PACE, THE READER SHOULD CAREFULLY CHECK THE STATUS OF ANY PARTICULAR PROGRAM, LAW, OR REGULATION DESCRIBED IN THE BOOK BEFORE RELYING ON THE ACCURACY OF ANY PARTICULAR REFERENCE.*

PART I
Background Information
~

CHAPTER 1

Defining the Brownfields Problem

TODD S. DAVIS
KEVIN D. MARGOLIS

Imagine that you are standing at the locked gates of an abandoned industrial site. What do you see? Economic opportunity? Or environmental contamination and financial disaster? For most, interpretation is a matter of perspective.

If you own this property—a site where manufacturing plants for decades operated unencumbered by environmental regulations—you may have to contend with "smoking guns" or "dead bodies" buried deep beneath the surface. Corporate real estate owners typically are advised by their lawyers to keep property like this under wraps, a permanent fixture in their real estate portfolios. Individual property owners take these white elephants to the grave, leaving their children to devise a solution for their final disposition.

If you are a regulator, you may view this site as a threat to human health, safety, and the environment. To you, it is a potential battleground for protracted litigation involving hordes of lawyers and technical consultants. Sites like this are exactly what motivated you to become a regulator—to save future generations from years of corporate abuse.

Are you a lender? Then you hope you weren't responsible for recommending the loan on this property. If you are the actual loan officer, you pray there was enough cash flow for a long enough period of time to pay off the note. Confronting another work-out, or worse yet, foreclosure, on a site plagued by environmental issues would do little to further your career path.

To the member of an environmental interest group, this property represents yet another example of why we need to tighten, rather than relax, environmental regulations. Who knows what environmental nightmare lurks behind those gates? Without vigilance, corporate America will continue to abuse the environment. The individual or corporation that owned this plant during its heyday probably racked up millions of dollars in profits—the former owners should be held responsible for cleaning up the mess, no matter what the cost.

If you are a neighboring homeowner, you may look at an abandoned industrial lot and reminisce about how your parent worked at that plant for thirty years. You recall the site as it was then, a thriving enterprise that supported your family and your friends' families. Or you may see it from an entirely different point of view. Today, it may look more like a potential hazard, a vacant lot where your children might encounter harm should they wander there to play.

But what if you are the mayor of the community in which that formerly productive plant has sat idle for years? Instead of problems, you may see opportunity—the opportunity to attract new business, create hundreds of new jobs, and add millions of dollars to the city's tax coffers. To you, this property embodies continuing economic development, not mere historical significance.

And finally, suppose you are a developer. Perhaps you also can see economic opportunity, but through a different lens. You believe the property could have potential, provided you could:

1. convince the corporate property owners that you can address and eliminate their environmental liabilities;
2. work your way through the maze of federal, state, and local laws and regulations governing potentially contaminated properties;
3. see eye-to-eye with the mayor's office;
4. assure regulators that the site is not a toxic time bomb;
5. appease skeptical citizens groups; and
6. prove beyond the shadow of a doubt to your lender that it is worth taking the risk to finance this project.

If you could accomplish all the above, the only remaining challenge would be to earn a sufficient rate of return to compensate you and your company for all the time, money, effort—and potential ulcers—associated with bringing such a project to fruition. So, you see, it is all a matter of perspective.

Given these many diverging viewpoints, individuals trying to create momentum for developing abandoned industrial sites, or brownfields, face a formidable task. Yet the topic of brownfields redevelopment is alive and well. The news media is writing about it. Local governments have created committees to study it. And seminars on the topic are springing up by the dozens. Even President Clinton mentioned the importance of resolving the "brownfields issue" in his 1996 State of the Union address.

What is fueling the interest in brownfields redevelopment? Pure economics. The brownfields issue is the anchor weighing down the ship of today's urban redevelopment movement. Although this certainly oversimplifies the problem, the fact remains that the redevelopment of brownfields must be regarded as an integral component of successful urban redevelopment; yet the numerous and complex issues associated with brownfields redevelopment are so daunting that they discourage otherwise interested parties.

Brownfields redevelopment requires extensive knowledge of the law, environmental assessment and remediation, finance, real estate, insurance, and economic development. Our purpose here is to clarify and demystify the issues surrounding the successful redevelopment of brownfields.

What Is a Brownfield?

The United States Environmental Protection Agency (USEPA) Region 5 defines brownfields as "abandoned, idled or underused industrial and commercial sites where expansion or redevelopment is complicated by real or perceived environmental contamination that can add cost, time or uncertainty to a redevelopment project."[1] The United States Office of Technology Assessment (OTA) provides a similar, albeit broader, definition. The OTA definition of a brownfield includes a site whose redevelopment may be hindered not only by potential contamination, but also by poor location, old or obsolete infrastructure, or other less tangible factors often linked to neighborhood decline.[2]

Brownfields routinely are associated with distressed urban areas, particularly central cities and inner suburbs that once were heavily industrialized, but since have been vacated. A brownfield may be as small as an abandoned gas station on a one-acre plot or as expansive as a steel-manufacturing operation sprawled out over several hundred acres. Brownfields sometimes are defined as the opposite of "greenfields"—property that has not previously been used for commercial or industrial activities and thus is presumed free of contamination.

Brownfields sites may be divided into four categories:

1. sites that—despite needed remediation—remain economically viable, due to sufficient market demand;
2. sites that have some development potential, provided financial assistance or other incentives are available;
3. sites that have extremely limited market potential even after remediation; and
4. currently operating sites that are in danger of becoming brownfields because historical contamination will ultimately discourage new investment and lending.[3]

Thus, from the developer's perspective, the focus of real estate professionals, corporations, government authorities, and other stakeholders should be on brownfields that are viable for economic development. "Viable brownfields" are defined as underutilized properties with actual or perceived environmental liabilities that, due to their inherently positive market attributes, may be economically redeveloped into productive assets. Properties that cannot be characterized in this manner are the least likely to be redeveloped with private resources and are the most likely to require either significant public subsidies or intervention to spur redevelopment efforts.

Although a small percentage of brownfields sites may have high contamination levels and be candidates for addition to the list of most heavily contaminated sites identified in the nation, the National Priorities List (NPL) under the Comprehensive Environmental Response, Compensation, and Liability Act of 1980 (CERCLA), or similar state priority lists, a large number of brownfields sites will likely never be listed. These sites will not be listed because (1) they have much lower contamination levels, or (2) the environmental condition of these sites will not be evaluated. Abandoned or underutilized industrial and commercial properties with no actual contamination also may suffer from the "brownfields stigma" until a site assessment proves the property is clean. Little information about the environmental condition of many brownfields sites is currently available.

Essential to the brownfields issue is distinguishing between NPL sites—the worst known contaminated sites with little prospect for economically viable reuse—and those sites characterized by low to medium levels of environmental contamination. Presently, the USEPA has identified nearly 1,250 high-priority sites that pose significant risks to human health and safety. These NPL or "Superfund" sites demand monumental effort and resources to restore and manage. The balance of contaminated sites generally are easier to clean and offer greater opportunities for reuse.

Why Are Brownfields Demanding Attention?

The sheer enormity of the brownfields dilemma has drawn it into the national spotlight, provoking the United States Conference of Mayors to declare the situation an emergency.[4] There are an estimated 130,000 to 450,000 contaminated commercial and industrial sites around the country, according to the United States Government Accounting Office.[5] No community is immune. Officials in Cook County, Illinois have identified 329 polluted industrial sites within county boundaries. A survey of Toledo, Ohio businesses found that 62 percent of the area's commercial and industrial real estate transactions are encumbered by environmental issues.[6]

Although these numbers are impressive, the real impact of brownfields is more dramatically summed up in dollars and cents. Current estimates place the cost of cleaning up the nation's brownfields at $650 billion. That is just the initial cleanup tab. Brownfields also represent millions of unrealized tax dollars and millions in lost wages.[7] Their presence contributes to reduced economic development and job creation in urban areas, particularly in central cities and older suburbs.[8]

According to a survey by the United States Conference of Mayors, 33 cities with brownfields sites conservatively estimated their cumulative annual loss of tax revenues at $121 million. Using more optimistic estimates, they projected losses at $386 million. This data suggests that more than 20,000 cities and other municipalities nationwide could be losing billions of dollars each year in local tax receipts resulting from their failure to restore brownfields to economic viability.[9]

Most of the nation's brownfields are caught in a vicious cycle of decline, which can be depicted as follows:[10]

1. A property owner, unwilling or unable to sell contaminated property, mothballs it, thus undermining the local tax base.
2. Vacant facilities deteriorate and invite arson, illegal dumping, and vandalism, including the stripping of parts and materials.

3. Unaddressed contamination may spread, further eroding the property value, escalating the cleanup cost, and threatening the economic viability of adjoining properties.
4. Potential investors, faced with uncertain costs and legal liabilities, seek development opportunities elsewhere.
5. Brownfields sites become unwanted legal, regulatory, and financial burdens on the community and its taxpayers.

The stigmatic impacts of brownfields on communities are manifold. Potential investors, concerned about liability, shy away from developing abandoned industrial sites. Real estate buyers are reluctant to invest in brownfields, which further diminishes their value. Communities lose out on property-tax revenues. Public services become less available and area unemployment rates soar.[11] The convergence of these economic development and environmental issues comes at a critical time for local officials struggling to craft community revitalization strategies targeting older industrial areas.

How Did the Brownfields Issue Evolve to a Crisis State?

The proliferation of brownfields and the failure to address their redevelopment effectively can be traced to a number of forces, including:

1. the unintended effect of environmental laws on brownfields redevelopment;
2. enforcement policies that target lenders; and
3. ignorance of the science of contaminated property.

The Unintended Effect of Environmental Laws on Brownfields Redevelopment

Environmental laws are a relatively recent phenomenon. The most significant statutes were not enacted and actively enforced until the mid- to late 1970s. Among the most widely publicized of United States environmental laws is CERCLA, also known as the federal Superfund law. Hastily passed in 1980, in the wake of the Love Canal scare, CERCLA established a federal program to identify and remediate chemical spills and abandoned hazardous waste sites believed to pose a significant threat to human health, safety, and the environment. It created a mechanism for assessing the environmental condition of those sites and placing the worst sites on the NPL, making them eligible for federal funds.

Only about 1,250 of the nation's hundreds of thousands of hazardous waste sites are listed on the NPL. To address those sites that do not meet the NPL criteria, states have enacted their own legislation, in the form of mini-CERCLA statutes.

CERCLA, its state equivalents, and the Resource Conservation and Recovery Act (RCRA) were intended to create a comprehensible system for correcting environmental damage that occurred in the past and for preventing future contamination. Instead, applied in the brownfields context, they produced almost the opposite effect. Deciphering these laws has not been easy. Environmental lawyers themselves bemoan the thousands of pages of intricate, complex, and often contradictory requirements that many environmental programs impose. And until recently, cost/benefit analysis has not played a significant role in the development of new laws. The end result is that the confusion engendered by environmental laws has inadvertently subverted progress toward redeveloping brownfields, rather than contributing to a positive solution, as originally intended.

Targeting Deep-Pocketed Lenders

In *United States v. Fleet Factors Corp.*, the court found that a lender could be held liable under CERCLA for cleanup if the lender participated "in the financial management of a facility to a degree indicating a capacity to influence the corporation's treatment of hazardous wastes."[12] The court's ruling led other private parties and the government to target deep-pocketed lending institutions in Superfund cases, a trend that further exacerbated the brownfields problem.

Regulators, realizing that the due diligence process required for finalizing a loan would likely uncover any contamination of a property, adopted the view that lenders should act as environmental police. If they failed to uncover environmental hazards, they could become responsible for the cleanup. Lenders reacted by refusing to loan money on projects associated with even a hint of environmental liability. This practice, sometimes referred to as "greenlining," in addition to creating more brownfields, also triggered a credit crunch for industrial financing during the late 1980s and early 1990s. Thus, lessons learned by lenders through tough litigation and unsympathetic court doctrine added to the creation of brownfields.

Ignorance of the Science of Contamination

The underlying fears regarding human health and the environment associated with contaminated property have been aggravated in part by a basic lack of understanding within the scientific community concerning the true risks posed by contaminated sites. The science supporting currently mandated risk goals is inconclusive and unrealistic. Yet current policy continues to be driven by inferior scientific evidence, resulting in the proliferation of brownfields and their excessive cleanup for limited returns to human health and the environment.

As an example, current regulations in many cases dictate that contaminated sites be returned to "background" or "naturally occurring" levels of hazardous substances. Such policy decisions regarding levels of "acceptable environmental risk" have little relation to the types of risks people confront daily. After all, the risk to the average commuter of being killed in a car accident is significantly greater than the risk of developing cancer from years of exposure to a mildly contaminated site.[13]

Barriers to Brownfields Redevelopment

Cleveland Mayor Mike White has cited contamination as the number-one obstacle to urban redevelopment. In large part, the frustration of Mayor White and other officials stems from the ambiguity surrounding brownfields—ambiguity related to legal issues, cleanup standards, liability, and the unavailability of financing.

Brownfields redevelopment is not a zero-sum game. It should result in economic growth for all parties involved. However, until recently, the many barriers to brownfields redevelopment have discouraged progress. These barriers include:

1. ambiguous legal liability;
2. lack of concentrated expertise;
3. potentially substantial capital costs;
4. insufficient financing;
5. clouded federal, state, and local environmental and legal policies;
6. entrenched attitudes among regulators;
7. absence of a consistent redevelopment framework;
8. public opposition;
9. limited demand for redeveloped sites; and
10. competition from greenfields.

Ambiguous Legal Liability

Fear and uncertainty about liability are the greatest obstacles to brownfields redevelopment. The daunting complexity, ambiguity, and overlapping nature of CERCLA and other environmental laws preclude an accurate appraisal of the actual risk of liability.[14] One court has referred to RCRA as "mind-numbing."[15] CERCLA has been called much worse.

Property owners potentially responsible for contamination of a site cannot completely shift their liability to buyers, including redevelopers. As a result, they often mothball property that might otherwise be redeveloped. And with no federal safe harbor for redevelopers who might otherwise see economic promise in brownfields, these potential saviors shy away from abandoned industrial sites, largely out of fear of becoming mired in Superfund's legal quagmire.

Most legislators, developers, and property owners, frustrated with the lack of progress on actually remediating Superfund sites, agree that the Superfund program is due for an overhaul. Parties calling for regulatory change often cite the disparity between the huge sums of money spent on litigating these sites compared with the much smaller dollar amounts put toward actually cleaning them up. Critics also point out that few Superfund sites have actually been addressed and removed from the NPL. Recently, the pleas of these interested parties have begun to be heard.

Lack of Concentrated Expertise

Key players involved in commercial and industrial site reuse—including property owners, lawyers, environmental consultants, real estate brokers and professionals, economic development representatives, insurance specialists, lenders, and regulators—have little or no experience in working collectively toward a common goal. In fact, they often engage in counterproductive behavior when it comes to brownfields redevelopment. They are only now realizing that cooperation must replace antagonism to advance each others' interests.

Potentially Substantial Capital Costs

Available data on actual brownfields cleanup costs is limited. However, the price tag can be substantial. Worse yet, potential liability issues make it difficult to determine up front what the final costs will be.

Assessment and remediation costs may range from a few thousand dollars to millions, depending on the site. A significant investment, usually for due-diligence purposes, may be required merely to estimate the anticipated cost of remediation and development. In many cases, potential due-diligence costs prohibit the assessment of smaller sites deemed unworthy of the investment.

Once developers arrive at an estimated cost for assessment and remediation, they cannot assume the cost is finite. In some cases, the process of remediation uncovers unanticipated areas of contamination, which then sends what was originally deemed an economically viable project deep into the red.

Public and private resources for brownfields assessment and remediation are limited—just one more deterrent for would-be developers.

Insufficient Financing

The effect of environmental liabilities on lenders has been dramatic. According to one study, more than 40 percent of commercial mortgage bankers polled said they had backed out of mortgage deals on potentially contaminated properties. About 87 percent of those bankers said that fear of environmental liabilities had delayed transactions. And approximately 70 percent of the survey respondents said environmental problems actually had materialized on properties for which they had arranged mortgages.[16] Ultimately, the prospect of foreclosing on contaminated collateral in the event of default dampens lender interest in brownfields loans.[17]

Clouded Policies

Historically, federal, state, and local policies have done little to spur industrial redevelopment. Rehabilitation tax credits offered during the mid-1970s provided incentives to invest in real estate and redevelopment. These tax incentives helped stem the exodus of businesses from long-established neighborhoods and made reuse more economically attractive. However, these tax advantages effectively vanished under the 1986 tax code revisions limiting passive losses. As a result, investors turned to potentially more lucrative sources of return, such as Wall Street, and many rehabilitation projects failed to materialize.[18]

Entrenched Attitudes among Regulators

The latest trend at the legislative level has been to adopt a more user-friendly approach to redeveloping brownfields sites, including attempts to be more flexible and creative in addressing historical environmental liabilities. Yet despite these efforts, significant differences of opinion and philosophy concerning redevelopment, environmental risk, and liability persist within state and federal environmental regulatory agencies. In many instances, the belief that the polluter must pay continues to reign supreme.

Absence of a Consistent Redevelopment Framework

The absence of clear and coordinated federal and state guidelines for redeveloping brownfields—a deficiency closely related to ambiguous legal liability issues—has hindered redevelopment.

Meanwhile, the failure to establish local brownfields redevelopment programs presents an often overlooked barrier. Although many local politicians have elevated the brownfields issue to the crisis level within their communities, few communities or cities have taken positive, concrete steps toward implementing a meaningful brownfields redevelopment strategy. As a result, developers attempting to work their way through the maze of city programs and permitting processes frequently abandon the process out of frustration.

Public Opposition

Although certain community groups voice an interest in promoting the cleanup and redevelopment of neighborhood brownfields, their members understandably expect some assurance that remediation will adequately protect their health and the environment. Some are intent on ensuring that traditional, heavy manufacturing-type industry is replaced with nontraditional industries perceived as less harmful to the environment. Unfortunately, this often creates conflict between potential developers and community groups

Limited Demand for Redeveloped Sites

There is no question about the inventory of brownfields for potential redevelopment—as previously noted, there are hundreds of thousands of these sites nationwide. However, even if all these sites were identified and completely remediated, the evidence suggests there is insufficient market demand for many of these properties due to other market forces (such as poor location, high crime, decaying infrastructure, and similar matters). Therefore, it is unlikely that investors would rush in to develop a large number of these brownfields even if the liability issues were resolved.

Competition from Greenfields

Fierce competition from greenfields communities intent on attracting new development has contributed to what we refer to today as urban sprawl—the practice of building on previously undeveloped land outside the city limits. Urban sprawl is costly. It allows a city's existing roads, bridges, water lines, sewer systems, and rail spurs to go unused while similar infrastructures are duplicated elsewhere. For the community populated by numerous brownfields, billions of dollars in previous public and private investment may go to waste.

Yet many developers choose urban sprawl over brownfields redevelopment, in part because greenfields communities can offer financial incentives, such as tax abatement and low-cost financing, equal to those available from cities where brownfields predominate. To counteract this trend, communities truly interested in meaningful brownfields redevelopment must go beyond leveling the playing field—they must tilt it significantly in favor of brownfields reuse.

Bringing Down the Barriers

Leaping the multiple hurdles to the successful redevelopment of brownfields can be an arduous process. Nonetheless, stakeholders across the nation are attempting to do just that. State Voluntary Cleanup Programs are emerging as one of the most innovative trends in this area. These programs have been designed specifically to address the obstacles to brownfields redevelopment. The goals of these programs include integrating issues involving legal liability, technical requirements, and economic incentives. Many of these programs provide technical assistance from regulators, liability assurances through covenants-not-to-sue, and financial incentives, including tax abatement, not available through other state regulatory programs.

Voluntary programs are gaining in popularity because they allow private parties to initiate cleanups and work cooperatively with state agencies, thus avoiding some of the costs and delay that would likely occur if the sites were subject to enforcement-driven programs.[19] They have set the stage for brownfields redevelopment. The next logical step is to add a motivated team of professionals to these programs—professionals who are savvy in the various areas of brownfields redevelopment—to facilitate the process.

Food for Thought

The environmental and liability issues surrounding brownfields have had the same chilling effect on real estate developers and lenders that the movie *Jaws* has had on swimmers. We know the sharks are out there. And as is the case with certain sharks, some environmental liabilities will eat you alive. Being ripped to shreds in the jaws of a ferocious beast is a gruesome way to die. It is not unlike the experience of the unsuspecting loan officer who extends credit on property that is subsequently identified as a Superfund site. Yet do we allow the knowledge that sharks exist intimidate us into staying out of the water? Shark experts tell us that few people actually die from shark attacks.[20]

Whether or not you decide to swim will depend on how much you know about a given situation. Are these waters typically shark-infested? If so, what kinds of sharks lurk beneath the surface? Man-eaters or those who dine on plankton? Can a steel cage be built to protect you from the jaws of death? Clearly, where you swim—or whether you swim—will depend in great part on your knowledge of both sharks and the waters in which you intend to swim. Hopefully, the information contained in this book will encourage interested parties to dive into the waters once more.

Notes

1. USEPA REGION 5 OFFICE OF PUB. AFFAIRS, BASIC BROWNFIELDS FACT SHEET, (1996).
2. U.S. OTA, STATE OF THE STATES ON BROWNFIELDS: PROGRAMS FOR CLEANUP AND REUSE OF CONTAMINATED SITES, at 8 (1995) [hereinafter STATE OF THE STATES ON BROWNFIELDS].
3. *Id.*
4. U.S. Conference of Mayors, Brownfields Redevelopment Action Agenda Initial Framework (Jan. 25, 1996) [hereinafter U.S. Mayors Report].
5. U.S. GOV. ACCOUNTING OFFICE, GAO/RCED-95-172, COMMUNITY DEVELOPMENT—REUSE OF URBAN INDUSTRIAL SITES (1995).
6. Charles Bartsch, *Restoring Contaminated Industrial Sites*, 10 ISSUES SCI. & TECH., 74 (1994).
7. NORTHEAST-MIDWEST INST., COMING CLEAN FOR ECONOMIC DEVELOPMENT 1-2 (1996) [hereinafter COMING CLEAN].
8. STATE OF THE STATES ON BROWNFIELDS, *supra* note 2, at 4.
9. U.S. Mayors Report, *supra* note 4, at 1-3.
10. COMING CLEAN, supra note 7, at 1-2.
11. Steven Lerner, *Brownfields of Dreams*, AMICUS JOURNAL, Winter 1996, at 17.
12. U.S. v. Fleet Factors Corp., 901 F.2d 1550, 1557 (11th Cir. 1990), *cert. denied*, 498 U.S. 1046 (1991).
13. See Chapter 17, Role of Risk Assessment in Redeveloping Brownfields Sites.

14. STATE OF THE STATES ON BROWNFIELDS, *supra* note 2, at 6.

15. *See American Mining Congress v. Environmental Protection Agency*, 824 F.2d 1177, 1189 (D.C. Cir. 1987).

16. ENVIRONMENTAL WARRANTY, INC., SURVEY OF 30 TOP COMMERCIAL MORTGAGE BANKERS NATIONWIDE (1995).

17. STATE OF THE STATES ON BROWNFIELDS, *supra* note 2, at 8–9.

18. *Id.*

19. *Id.* at 13.

20. Between 1984 and 1994, sharks killed just seven people in United States waters—three each in Hawaii and California, and one in Florida. To put this number in perspective, according to certain mathematicians, the odds of being hit by lightning are 600,000 to 1; the chances of winning the Florida Lottery are 13 million to 1; and the odds of being attacked by a shark are roughly 300 million to 1. Steve D'Oliveira, *Odds Against Shark Attacks*, SUN-SENTINEL (Fort Lauderdale), Sept. 27, 1995, at 1E.

CHAPTER 2

Overview of Federal and State Law Governing Brownfields Cleanups

WENDY E. WAGNER

Federal laws governing remediation of contaminated sites were passed initially to address toxic disasters like Love Canal, where hazardous wastes were discovered leaching into residential basements.[1] The resulting legislation provided the United States Environmental Protection Agency (USEPA) with a powerful grant of response authority and broad liability provisions to ensure expeditious cleanup of the nation's most serious hazardous waste sites. For the hundreds of thousands of mildly or moderately contaminated properties that are also caught within the jurisdictional reach of these aggressive federal statutes, however, the expansive liability and unpredictable cleanup costs have had the unintended side effect of imposing formidable impediments to profitable brownfields redevelopments.

In this chapter, the current federal and state laws governing liability for cleanup of brownfields sites will be outlined, both to provide a fuller appreciation of their role in impeding urban redevelopment and to highlight the extent and nature of the liability risks that different types of brownfields sites present. Though some federal and state programs provide a clear direction and sense of finality to the remediation of a brownfield, in many more cases a series of legal ambiguities surrounding the extent of liability and the appropriate method and level of cleanup continue to deter brownfields redevelopments.

The Legal Landscape

Although industrial waste-disposal practices were largely unregulated for nearly a century, once it became evident that federal government involvement was necessary, that federal presence came with a vengeance. Congress passed the Resource Conservation and Recovery Act (RCRA) in 1976[2] and the Comprehensive Environmental Response, Compensation, and Liability Act (CERCLA) in 1980,[3] which together provide an ambitious federal program for remediating hazardous-waste sites and ensuring that future waste-disposal operations are performed safely and in accordance with minimum federal standards.

The Resource Conservation and Recovery Act (RCRA)

The primary purpose of RCRA is regulation of the generation, transport, treatment, and disposal of hazardous wastes from "cradle to grave." Stringent statutory requirements direct manufacturers who produce hazardous wastes to identify themselves to the state and federal Environmental Protection Agencies[4] and ensure, through a manifest tracking system, that their wastes end up at licensed recycling, incineration, or disposal facilities.[5] "Grave sites" for the wastes must obtain a "treatment, storage, or disposal" (TSD) permit[6] and meet hundreds of pages of onerous bonding, safety, recordkeeping, training, and design requirements.[7]

RCRA's impact on redevelopment of brownfields typically takes two forms. The first is through its often costly regulation of underground storage tanks—fixtures that are commonplace on property that has had a history of industrial use. Subchapter I of the statute sets forth liability and compliance requirements for owners of land with underground storage tanks if the tanks contain petroleum or hazardous substances.[8]

Second, RCRA provides authority for governments or citizens to require cleanup at sites that "may present an imminent and substantial endangerment to health or the environment" or where there is a release of hazardous waste in violation of a permit or other requirement of RCRA.[9] The owners or operators of the land,[10] and in cases of an "imminent hazard" all other persons who contributed to the contamination,[11] are typically held responsible under RCRA for conducting the cleanup or reimbursing the state or federal EPA for cleanup expenses.

The Comprehensive Environmental Response, Compensation, and Liability Act (CERCLA)

In contrast to RCRA, CERCLA does not create a regulatory program, but instead provides an elaborate liability scheme for the remediation of virtually all contaminated sites. To ensure that private parties bear the brunt of cleanup costs, Congress imposed expansive liability on a diverse number of parties who contributed to the hazardous substances present at the site,[12] even if the disposal occurred before the passage of CERCLA.[13] Liability for cleanup is not limited to sites where there is an "imminent hazard,"[14] but extends to all sites contaminated with even modest amounts of one or more "hazardous substances." The term "hazardous substances" includes hundreds of listed substances as well as unlisted toxins,[15] but expressly excludes most petroleum products.[16] The courts, guided by Congress's desire to have those responsible for careless disposal practices pay for cleanup whenever possible, have interpreted CERCLA to impose full liability on covered parties regardless of fault[17] and regardless of the presence of other responsible parties.[18]

To ensure that the worst sites are addressed first, Congress tasked USEPA with the responsibility for ranking the most severely contaminated sites in the nation on the National Priorities List (NPL)[19] and provided USEPA with multiple authorities to ensure their expeditious cleanup.[20]

When private parties cannot be found to finance or perform these cleanups, USEPA is authorized to use money from the Superfund, a revolving trust fund funded initially by a tax on chemical feedstocks and reimbursed in part with recoveries from settlements and lawsuits brought by USEPA against potentially responsible parties.[21]

Private parties may also avail themselves of CERCLA's expansive liability scheme in their efforts to recover cleanup costs from other responsible parties.[22] Injunctive relief is not available to private party plaintiffs under CERCLA.

State Parallel Programs

Many states have developed their own "mini-CERCLA" statutes.[23] These statutes typically provide the state with authority to force potentially responsible parties to undertake cleanup at contaminated sites[24] and to establish a state fund to finance state-led cleanups when immediate action is necessary to protect the public health and environment or when solvent responsible parties cannot be located.[25] Although state statutes often parallel CERCLA, several states have different standards for liability[26] or provide enhanced measures to encourage voluntary site cleanups.[27]

Some states have also been delegated authority under RCRA to require corrective actions and closures and to regulate underground storage tanks.[28] State programs typically adopt federal regulations verbatim, but occasionally impose more stringent requirements on the regulated community.[29]

Cleanup of Brownfields under Federal or Parallel State Laws

The broad jurisdictional reach, comprehensive liability provisions, and uncertain and typically costly cleanup standards of RCRA and CERCLA and parallel state programs have been blamed for impeding the redevelopment of many brownfields sites. Prospective purchasers who desire quantification of cleanup costs before purchasing a contaminated site are dismayed to learn that government entities are often unable or unwilling to provide assistance in determining what constitutes an acceptable cleanup, even though the ultimate redevelopment of a contaminated property may serve the public interest.[30] Familiarity with the cleanup requirements of applicable federal and state laws is thus imperative for successful brownfields redevelopment.

Jurisdictional Reach

What Sites Are Covered?

Under federal and parallel state law, virtually any site with very mild contamination can engender responsibility for cleanup. A brownfield capable of cleanup pursuant to federal laws and regulations includes

1. any place where a non-naturally occurring[31] hazardous substance has "come to be located"[32] if that hazardous substance is entering or threatens to enter the environment[33] without a permit;[34]
2. any site where petroleum has been disposed if the petroleum presents an "imminent and substantial endangerment to health or the environment";[35]
3. an underground storage tank that is leaking a hazardous substance or petroleum product;[36] and
4. a building that contains asbestos and will be subject to renovation or demolition,[37] or that has loose asbestos posing a health hazard to workers within the building[38] or violating state or local regulations.

The remedial purpose of CERCLA has reinforced a generous reading of what constitutes a contaminated site capable of cleanup. Courts have held that a site where any detectable amount of a hazardous substance has been released[39] or where concentrations of hazardous substances exceed background levels[40] may be the subject of a cleanup action. Although one court has held that an open container containing hazardous substances is not sufficient to trigger the reporting requirements of CERCLA,[41] liability for cleanup may be imposed if the container "threatens" to enter the environment[42] or if the container is abandoned.[43]

When Is Cleanup Required?

Exacerbating the sweeping jurisdictional reach of what constitutes a brownfield capable of cleanup is the large number of possible plaintiffs who can insist that a cleanup be done. For many sites, RCRA and CERCLA provide legal channels for federal, state, and local governments and private citizens to file suit for cleanup of a brownfield or for reimbursement of their cleanup expenses. This multitiered enforcement makes it virtually impossible for a purchaser to obtain liability waivers from all possible plaintiffs before the transaction, and can create legal uncertainties so potentially explosive that the liability risks presented by a brownfield do not offset the profits to be gained from its redevelopment. Each of the parties capable of determining that cleanup is necessary will be considered in turn, along with the tools they have to effectuate cleanup.

The federal EPA typically requires and oversees cleanups only at those sites that it ranks on the NPL,[44] or at sites in states without an authorized hazardous-waste program where the contamination presents an "imminent hazard"[45] or has been released in violation of RCRA.[46] At all these sites, USEPA can force a potentially responsible party to conduct the cleanup or it can seek reimbursement for the cleanup costs it expended, including overhead costs and litigation expenses.

States may take similar action pursuant to CERCLA or their own mini-CERCLA statutes at inactive sites,[47] and pursuant to RCRA at sites that present an "imminent hazard" or are otherwise in violation of RCRA and involve the unpermitted release of a hazardous waste.[48] These state actions can create even more trepidation for developers, given the possibility that a moderately—or at times even mildly—contaminated site will rank among the state's enforcement priorities and lead to a costly cleanup.[49]

Finally, a private party may determine that a cleanup is necessary and attempt to force an owner or other potentially responsible party to conduct or pay for the cleanup under CERCLA, RCRA, or state law.[50] Private parties typically fall into two categories. The first includes citizens who are concerned about the risks posed by a contaminated site. Under federal law these citizens typically file only injunctive claims under RCRA for cleanup. They must prove either that the site presents an "imminent and substantial endangerment to health or the environment"[51] or that the disposal of hazardous wastes occurred in violation of RCRA.[52]

The second, more frequent type of private-party actions are contribution suits brought by the current owner of a contaminated site or other responsible party. These plaintiffs bring suit against other responsible parties either for reimbursement of their cleanup costs[53] or, less frequently, to force responsible parties to undertake the cleanup if an "imminent hazard" exists at the site.[54] Contribution actions have proven very popular over the past decade,[55] even though most private-party plaintiffs ultimately bear some of the cleanup costs themselves,[56] including attorney fees.[57]

Who Is Liable for Cleanup and What Will It Cost?

There are essentially two types of cleanups under federal law: simple and complex. Which type of cleanup will be necessary at any particular brownfield site depends in large part on the type of contamination present. If the contamination at a site is *only:* (1) petroleum from a leaking underground storage tank, (2) a spill of polychlorinated biphenyls (PCBs), or (3) equipment or structures containing hazardous materials, the applicable federal or state law will generally provide relatively good direction regarding what will be required for cleanup. If the contamination consists of other hazardous substances listed under CERCLA,[58] the extent of liability for remediation costs may be substantially more difficult to determine in advance. In this section, both the legally determinable and the legally horrifying cleanups will be discussed in turn.

"Simple" Cleanups

PETROLEUM FROM LEAKING UNDERGROUND STORAGE TANKS. Underground storage tanks are regulated under subchapter I of RCRA, although when hazardous substances leak from such tanks the liability provisions of CERCLA may also apply.[59] Liability for underground storage tanks leaking petroleum is tightly circumscribed unless the release causes an "imminent hazard."[60] Only the current owner or operator of the tank, provided the tank was in use at some time after November 8, 1984, or the last owner of a tank before its use was discontinued before November 8, 1984, is liable under RCRA for cleanup.[61] USEPA or the state may undertake the cleanup under specified circumstances[62] and either seek reimbursement of their costs from the responsible owner or operator[63] or finance the cleanup with monies from the Leaking Underground Storage Tank Trust Fund if a solvent owner or operator cannot be identified.[64]

USEPA has promulgated regulations and particularized guidelines regarding the remediation of underground storage tanks.[65] Many states have been authorized by USEPA to run the federal program. Even in the absence of express federal approval, however, states have developed underground storage tank programs that typically occupy the regulatory field.[66] Because state programs for underground storage tanks can add more stringent requirements to the existing federal regulations, familiarity with state laws and regulations is crucial for owners or prospective owners of underground storage tanks.[67]

A party who conducts a cleanup of an underground storage tank under the RCRA guidelines may confront difficulties in recovering its expenses. Although the availability of injunctive relief to private-party plaintiffs under RCRA is clear, the Supreme Court recently held that private-party plaintiffs may not use RCRA's citizen-suit provision for reimbursement of their cleanup expenses.[68] Therefore, parties remediating leaking underground petroleum[69] tanks may be left with state and common-law claims for recovery of their cleanup costs.

PCB SPILLS. USEPA regulates spills of PCBs under section 6(e) of the Toxic Substances Control Act (TSCA),[70] although concurrent remedial authority under CERCLA also exists.[71] If TSCA enforcement authority is used, only the current owner of the property or source of PCBs[72] is held legally responsible for cleanup of spills occurring after May 4, 1987.[73] If the cleanup is conducted in compliance with the remediation procedures of both CERCLA and TSCA, other parties potentially responsible for the contamination may be sued for cleanup or for reimbursement of cleanup expenses.[74]

Although stringent cleanup standards for PCB spills have been set by USEPA under TSCA,[75] compliance with USEPA's PCB Spill Policy creates a presumption against the need for further cleanup if the spill response requirements have been followed[76] and if the cleanup standards have been met.[77] Thus, cleanup procedures can be determined solely by a private party without government oversight, provided the private party makes a good-faith effort to comply with the regulations.

EQUIPMENT AND STRUCTURES CONTAINING HAZARDOUS MATERIALS. In most cases, use of equipment or structures containing hazardous materials is allowed with minimal or no federal oversight, provided the hazardous substances are contained.[78] More proactive federal regulations apply if the use of this equipment, such as PCB-containing transformers, is discontinued[79] or if a building containing asbestos is renovated or demolished.[80] Regulatory responsibility typically falls exclusively on the current owner. State statutory or common law provides the primary means for current owners to recover some or all of their compliance costs from prior owners or other responsible parties.[81]

A purchaser should also be wary of features at the site, such as air-venting systems, storm-water discharges, and underground storage tanks, which may require a permit or costly upgrading for their continued use, regardless of whether they contain a hazardous substance. Underground tanks, for example, are subject to extensive monitoring,[82] design,[83] and financial assurance[84] requirements, even when they are not leaking and remediation is not necessary. New owners of property containing these features may, in some cases, be held responsible for permitting violations. Thus, purchasers should ensure that discharges and equipment containing hazardous substances are appropriately permitted and in compliance with all applicable state and federal regulations before transfer of the property.

"Complex" Cleanups

Though remediation of leaking underground petroleum tanks or removal of asbestos from buildings is hardly a cakewalk under existing federal laws and regulations, the aggravations and complications typically encountered in conducting other types of cleanups under CERCLA and RCRA are considerably more agonizing. Although the legal ambiguities prove irritating to all potentially responsible parties, they are most devastating for prospective purchasers who require a thorough quantification of the costs of remediation to determine whether a brownfield is worthy of purchase.

BROAD LIABILITY. Under CERCLA and the "imminent hazard" authority of RCRA, virtually any party who had meaningful contact with the contamination at a site, either as an owner, operator, generator, or transporter,[85] is potentially liable for some or all of the costs of cleanup.[86]

1. *Current Owner:* With the exception of (1) state and local governments that acquire a property involuntarily,[87] and (2) purely "innocent" landowners who have made a very thorough, good-faith investigation of the site before purchase and mistakenly believe the site to be clean,[88] all current owners and operators of a brownfield where a hazardous substance is released or likely to be released can be held liable under CERCLA and RCRA[89] for the full cost of cleanup. Because liability of the owner attaches when the property is transferred, a prospective purchaser can avoid liability only by ensuring the cleanup has been completed before taking title.

A purchaser who conducts a thorough investigation of a site and mistakenly concludes that it is clean has, in theory, a potential statutory escape from liability for cleanup. In practice, however, the courts have proven very reluctant to grant an innocent-owner exception, either because a prepurchase investigation was deemed incomplete[90] or because an owner did not take sufficient action to mitigate the spread of contamination once it was discovered.[91] This harsh treatment of nonculpable purchasers is offset to some extent by the fact that these purchasers frequently pay a relatively small portion of cleanup costs if other solvent, potentially responsible parties have been identified.[92]

2. *Seller:* A seller of a parcel contaminated with hazardous substances is also in danger of liability, even when the cleanup occurs long after its ownership. The Ninth Circuit has held that even nonculpable past owners of a site are liable for cleanup costs under CERCLA, provided the contamination was present during their ownership and passively moved through the site.[93] Because "potential liability under CERCLA is . . . eternal,"[94] many sellers refuse to transfer property, even if it has been partially remediated.[95] Obtaining a comprehensive indemnification clause from the purchaser for all future environmental liabilities may protect the seller to the extent the purchaser remains solvent,[96] but the courts have consistently held that an indemnification clause does not protect a covered party from liability to the government or other claimants.[97]

3. *Lenders:* In limited instances, lenders who foreclose on a property have been held liable as owners under CERCLA,[98] or as operators regardless of foreclosure if they participated in the management of a facility that disposed of the hazardous wastes.[99] The courts have approached the level of lender involvement necessary to trigger liability somewhat differently, with the most stringent reading implying that a lender may be liable for cleanup costs if it "participat[ed] in the financial management of a facility to a degree *indicating a capacity to influence* the corporation's treatment of hazardous wastes."[100]

In an effort to provide more predictability to the lending community, USEPA promulgated a rule clarifying CERCLA liability regarding lenders by limiting lender liability only to circumstances in which the lender actively managed the facility.[101] Although the USEPA regulation was subsequently struck down as a usurpation of authority,[102] USEPA maintained its lender rule as an enforcement position.

As discussed in Chapter 8, shortly before this book went to press, Congress passed the Asset Conservation, Lender Liability, and Deposit Insurance Protection Act of 1996, which amends CERCLA and codifies the USEPA lender-liability rule.[103] This modification of CERCLA should afford significant comfort to lenders and open new avenues of financing for brownfields projects, although lenders may all have understandable reluctance to finance some brownfields purchases because of the possibility of losing all collateral value on a contaminated site that is later determined to require expensive cleanup.

4. *Other Parties:* Generators who have produced or "otherwise arranged" for one or more hazardous substances to be disposed of at the site[104] and transporters who have transported hazardous substances to the site[105] are also liable for cleanup costs under CERCLA, and possibly under RCRA if the site constitutes an "imminent hazard."[106]

EXTENT OF LIABILITY. Determining cleanup costs under CERCLA and the "imminent hazard" provision of RCRA can be quite challenging.[107] Nevertheless, sophisticated purchasers may be able to determine in advance a reasonable range of cleanup costs and may even find properties that are undervalued by those less familiar with federal and state cleanup requirements.

1. *Cleanup Process:* Depending on the scope and nature of the cleanup, several federal and state procedures governing cleanup may apply, in some cases concurrently. If a private party conducting a cleanup is doing so under an enforcement order, it will be conducting the cleanup pursuant to the cleanup methods prescribed by the enforcement authority. If the private party is conducting the cleanup voluntarily, it will have more choice in selecting a type of cleanup method, but may wish to follow federal or state procedures if it intends to later seek reimbursement of its response costs from other responsible parties. The four different types of cleanups, all of which overlap, are outlined below.

CERCLA: The CERCLA procedures governing cleanup, set forth in the National Contingency Plan (NCP),[108] are used by the federal EPA and are incorporated into many state laws and regulations.[109] Thus, a private party remediating an NPL site or an inactive site listed on a state priorities list pursuant to a state enforcement order will likely be directed to follow the NCP in conducting the cleanup.

For a private party who will conduct a cleanup without government oversight,[110] but who wishes to recover its cleanup costs from other responsible parties, the broad liability provisions of CERCLA offer a marked improvement over the common law.[111] To recover costs under CERCLA, however, the private-party cleanup must be conducted in "substantial compliance" with the NCP.[112] Thus, the NCP should also guide private-party cleanups to the extent cost-recovery actions are contemplated under CERCLA.

The costs of complying with the NCP can be significant. At a minimum, the NCP requires significant public participation in the selection of a cleanup,[113] necessitates that cleanup standards first be identified and then met,[114] and sets forth a cumbersome investigation and cleanup process.[115] In addition, because CERCLA does not provide private parties with claims for recovery of remediation costs for petroleum spills[116] or for recovery of attorney fees,[117] reimbursement of cleanup costs under CERCLA will likely be incomplete.

RCRA: When the site constitutes an "imminent and substantial endangerment"[118] and is subject to an enforcement suit or falls under the corrective action[119] or closure requirements,[120] the party conducting the cleanup must follow the requirements set forth under USEPA's or a state's RCRA regulations for corrective actions[121] or closures.[122] The Supreme Court, however, has held that a private party cannot recover reimbursement of its past cleanup expenses under RCRA.[123] Thus, if the private party wishes to preserve its cost-recovery claims, the enforcement authority may be persuaded to allow the cleanup to be conducted under CERCLA concurrently, even though the more extensive requirements of the NCP may cause additional delays in the remediation process.

State: State laws may provide yet another procedure for cleanup.[124] Those states with parallel CERCLA statutes often adopt the NCP.[125] Some states, however, may have different cleanup procedures.[126] For example, voluntary cleanup laws that have been passed by many states often provide more flexible cleanup procedures, which, when followed, provide the party conducting the cleanup with immunity from state enforcement action as well as a state claim for reimbursement of cleanup expenses against other potentially responsible parties.[127]

Free-Form: A private party conducting a voluntary cleanup is also free to conduct the cleanup in any way it sees fit, provided it does not violate federal or state worker-protection and certification laws, waste-disposal regulations, or other laws that require cleanups to be conducted safely. As the private party conducting a cleanup typically loses most, if not all, of its cost-recovery claims under CERCLA if the NCP is not followed, a free-form cleanup has obvious drawbacks. Additionally, unless federal or state cleanup standards have been met, the owner or other party conducting the cleanup will not be safe from subsequent cost-recovery actions if further cleanup is conducted.

2. *Cleanup Standards:* Cleanup standards under CERCLA are "highly variable" and make it very difficult for prospective purchasers and sellers to predict in advance the cost of cleanup.[128] Cleanup standards under RCRA and state law generally parallel CERCLA and are similarly difficult to determine.[129]

Cleanups conducted under CERCLA must meet all "applicable" and "relevant and appropriate" federal and state standards (ARARs), which typically vary from site to site and cleanup to cleanup.[130] Though the NCP provides little direction on how to identify ARARs for a cleanup,[131] USEPA has promulgated extensive guidances to assist its own staff, as well as state governments and private parties, in this tedious exercise.[132] Unfortunately, even with this guidance, the identification of ARARs is a source of confusion among even the most highly trained CERCLA personnel[133] and a frequent source of litigation.[134]

If an ARAR does not exist for a particular contaminant in a particular media, which is often the case,[135] the cleanup standard will be set at a level that ensures that a person with a "reasonable maximum exposure" will encounter no more than a one-in-ten-thousand to a one-in-one-million risk of cancer.[136] CERCLA risk assessments add even greater ambiguity and expense to cleanups. Not only are these risk assessments very costly to conduct, they also have been criticized for being far too cautious in their assumptions and producing cleanup standards that are unnecessarily risk averse and that result in exorbitantly expensive "cadillac cleanups."[137] Recently the USEPA parted with some of its more conservative assumptions by allowing future land uses to be taken into account in selecting a remedy.[138] Some states have modified their risk-assessment guidelines in a similar manner.[139]

3. *Additional Liabilities:* Sadly, even if an owner or prospective owner believes that she or he has mastered the federal and state procedures and com-

pleted a cleanup successfully, a brownfield may still harbor additional liabilities under federal or state law. Highly unlikely, yet potentially devastating, is the possibility of a natural-resource damage suit, which can be filed under CERCLA by federal natural-resource trustees, a state agency, or Indian tribes for restoration of the environment and for lost-use damages.[140] Although natural resource damage claims are likely to be filed only on large sites where damages have been extensive, fear of liability and the extended statute of limitations for natural resource damages[141] continue to cast a dark cloud over many contaminated sites.

Under state law, an owner may also be liable for damage to neighboring property or persons resulting from the release of hazardous substances from the site.[142] Though these claims tend to be unsuccessful, at sites where the contamination is particularly severe they remain a source of liability that should be fully investigated before a site is purchased.

Federal and State Reforms for Brownfields Redevelopment

A number of recent federal and state reforms may, to varying degrees, assist in counteracting the broad reach of liability and the indeterminate cleanup costs that currently deter brownfields redevelopment. The most important of these reforms, the state voluntary cleanup acts, will be discussed individually in the chapters that follow. The remainder of the reforms are discussed below.

USEPA's Brownfields Action Agenda

In 1995, USEPA released its multifaceted Brownfields Action Agenda.[143] Each of the individual reforms that comprise the agenda is intended to offset one or more legal obstacles to brownfields redevelopments.

Prospective-Purchaser Agreements

One of the most important components of the Brownfields Action Agenda is USEPA's effort to provide liability waivers—termed Prospective-Purchaser Agreements—to a larger number of prospective purchasers.[144] Over the past decade, there has been a growing demand by purchasers of brownfields sites for some governmental protection to ensure that the federal government will not take enforcement action against the new owner for cleanup. The federal government was traditionally very reluctant to provide a purchaser of a contaminated site with such a covenant-not-to-sue, even when the prospective purchaser agreed to take action to minimize contamination at the site.[145] Statutorily required public-participation requirements[146] and the possibility of intervention by citizen groups[147] exacerbated this reluctance.

In response to growing pressure for a more flexible policy, USEPA revised its prospective-purchaser guidelines in 1995 and expects to enter into a greater number of prospective-purchaser agreements in the future.[148] Despite improvements in the potential availability of prospective-purchaser agreements under USEPA's new policy, however, these agreements are still offered only to the purchasers of NPL sites or sites where USEPA has taken, or anticipates taking, a response action;[149] even then the agreements are not recommended for purchasers "in a hurry."[150] USEPA's prospective-purchaser agreements may also offer incomplete protection unless skilled counsel is hired by a prospective purchaser to negotiate them. For example, USEPA does not automatically provide (1) contribution protection from other potential plaintiffs, such as state governments, private parties, and natural resources trustees, (2) equivalent protection from suits under RCRA, or (3) liability protection that is transferable to subsequent purchasers and lenders.[151]

Other Protections for Purchasers

Other USEPA reforms that attempt to allay the concerns of prospective purchasers include the development of a guidance document pledging not to pursue innocent owners of land with contaminated aquifers[152] and the issuance of "comfort letters" to sellers and buyers conducting voluntary cleanups, in which USEPA provides assurance that it does not intend to take enforcement action at the remediated site.[153]

Memorandums of Understanding with States regarding Voluntary Cleanups

At least one USEPA Region has entered into Memorandums of Understanding (MOUs) with state environmental protection agencies under CERCLA. These memoranda provide assurance that USEPA will not take enforcement action at sites where private parties have conducted cleanups under the state's direction or pursuant to state voluntary cleanup acts.[154] Thus far, the MOUs do not appear to contain similar assurances under RCRA.[155]

Delisting Sites from the CERCLIS Database

In an effort to remove the "stigma" associated with listing a site on the Comprehensive Environmental Response, Compensation, and Liability Information System list site inventory (CERCLIS), a national list of CERCLA sites, USEPA has delisted 25,000 sites for which USEPA has no federal interest.[156] Although this is a positive step, the delisting is nonbinding and USEPA reserves the right to become involved in the future if "information so warrants."[157] The USEPA also recently issued guidance that defers sites from ranking while states oversee cleanups[158] and ultimately reduces the likelihood of such a site being placed on the NPL.

Land-Use Policy for CERCLA Cleanups

In response to criticisms that USEPA was requiring cleanups far in excess of what was necessary based on the anticipated future use of the site, USEPA recently promulgated a directive allowing future land uses to be considered in selecting the appropriate remedial action at NPL sites.[159]

Pilot Projects

By the end of 1996, USEPA had funded seventy-six economic-redevelopment pilot projects for contaminated brownfields sites.[160] Each project receives a grant of up to $200,000 to address problems caused by contamination.[161]

USEPA Region IX also recently undertook a "pilot demonstration" project in Oakland, California, to develop standardized cleanup standards for properties contaminated with leaking underground petroleum tanks.[162]

Legislative Developments

Eliminating many of the current legal obstacles to brownfields redevelopment continues to be one of the major agenda items for Congress in its reauthorization of CERCLA.[163] Legislative proposals include setting aside funds to encourage specific brownfields redevelopment projects,[164] providing tax incentives for cleaning up environmental hazards,[165] limiting liability of lenders, developers, and purchasers who reuse brownfields,[166] and providing more cost-effective cleanup standards.[167]

State Reforms

Although the most significant state response to some of the adverse impacts of CERCLA on purchasers of brownfields has been through voluntary cleanup bills (discussed in the chapters that follow), there are some other state reforms worthy of mention.

Transfer Restriction Statutes: New Jersey

Over the past decade, the state of New Jersey has developed a very aggressive land-transfer statute that requires industrial sites to become "clean" before transfer.[168] After a number of significant glitches, including an administrative backlog that caused several years of delay before transfers could be completed,[169] the amended Environmental Cleanup Responsibility Act (ECRA) now functions to allow certain transfers to occur without actual state inspection. Amendments to ECRA also allow buyers and sellers to defer remediation of a brownfield if the new owner plans to use it for substantially the same purpose and if exposure will not exceed levels set for industrial use.[170] Although some current owners may be dissuaded from attempting to transfer contaminated parcels if they believe cleanup costs will exceed the real value of the property,[171] ECRA has given great comfort to prospective purchasers of industrial property because it provides a state certification that the site is clean at the time of deed transfer. Interestingly, despite criticisms of ECRA, particularly by the real estate community, New Jersey officials boast that there are far more cleanups in New Jersey than the USEPA has conducted under CERCLA nationwide.[172] In 1990 alone, for example, the New Jersey Department of Environmental Protection approved 234 private cleanup plans costing over $178 million.[173]

Registry States

Several states maintain a statutorily established list of contaminated sites.[174] The list in some cases provides detailed information about why the site has been identified as a brownfield. The states may also require that state agencies be notified, or even approve transfers, of listed sites.[175]

Disclosure Laws

Other states have enacted disclosure laws. In Illinois and Indiana, for example, a seller must complete a detailed disclosure form with information regarding past ownership and uses of the site, past and current releases of hazardous substances, and permits or other regulatory conditions that apply to the site.[176] Noncompliance by a seller generally affords prospective purchasers or lenders with an opportunity to rescind from the transfer without penalty.[177] There are variations on this approach employed in other states. Depending on the state, owners of brownfields may be required to disclose known or reasonably suspected releases of hazardous substances to purchasers, lessees, or renters in writing;[178] record information regarding environmental contamination on their property with county registrars who maintain deed records;[179] or ensure that notices of contamination or past disposal activities become part of the recorded deed for the property.[180]

Liability and Cleanup Reforms

Many states are developing legislation that narrows the scope of liability set forth under CERCLA, particularly regarding lenders, purchasers, and residential homeowners.[181] Some states are also attempting to provide greater certainty and flexibility in the identification of cleanup standards, including allowing future land use to be considered in selecting a remedy.[182]

Conclusion

In sum, federal and state remedial programs offer some direction for cleanup of brownfields. Absent more comprehensive federal or state reforms, however, the broad sweep of potential liability, uncertain and cumbersome cleanup procedures, and the wide jurisdictional reach of the applicable statutes warrant close study before a purchaser or developer attempts to redevelop a brownfield site.

Notes

1. *See generally* ADELINE G. LEVINE, LOVE CANAL: SCIENCE, POLITICS, AND PEOPLE (1982).
2. *See* 42 U.S.C. § 6901 *et seq.* (1988).
3. *See* 42 U.S.C. § 9601 *et seq.* (1988).
4. *See* 40 C.F.R. § 262.12 (1994).
5. *See* 42 U.S.C. § 6922(a) (1988); 40 C.F.R. § 262.20–.23 (1994); *see also id.* pt. 262. Transporters of hazardous wastes are regulated similarly to generators. *See* 42 U.S.C. § 6923(a) (1988); 40 C.F.R. pt. 263 (1994).
6. *See generally* 42 U.S.C. § 6925 (1988); 40 C.F.R. pt. 270 (1994).
7. *See* 42 U.S.C. § 6924 (1988); 40 C.F.R. pt. 264 (1994).
8. *See generally* 42 U.S.C. §§ 6991, 6991a, 6991b (1988).

9. *See, e.g.,* 42 U.S.C. §§ 6971(a)(1), 6973 (1988). For suits alleging noncompliance with RCRA, citizens must allege that the defendant "is in violation of RCRA." *Id.* § 6972(a)(1)(A); *cf.* Gwaltney of Smithfield v. Chesapeake Bay Found., 484 U.S. 49 (1987) (holding that similar citizen suit provision under Clean Water Act requires allegation of continuing violation).

10. *See, e.g.,* 42 U.S.C. § 6924(u) (1988).

11. *See id.* §§ 6972(a)(1)(B), 6973(a).

12. *See id.* § 9607(a).

13. *See, e.g.,* United States v. Monsanto Co., 858 F.2d 160, 174 (4th Cir. 1988) (upholding challenge to CERCLA's retroactive liability on due process grounds), *cert. denied,* 490 U.S. 1106 (1989). One of the most controversial aspects of CERCLA is its retroactive imposition of liability. *See, e.g.,* Amy Blaymore, *Retroactive Application of Superfund: Can Old Dogs Be Taught New Tricks?*, 12 B.C. ENVTL. AFF. L. REV. 1, 49 (1985).

14. *See, e.g.,* 42 U.S.C. §§ 9601(9), 9607(a) (1988) (imposing liability on covered parties whenever there is a "release, or a threatened release" of a "hazardous substance" at "any site or area where a hazardous substance has . . . come to be located").

15. *See, e.g., id.* § 9601(14) (defining "hazardous substance"); 40 C.F.R. § 302.4 (1994) (providing list of hazardous substances and defining universe of additional, unlisted hazardous substances under CERCLA).

16. *See* 42 U.S.C. § 9601(14) (1988) (hazardous substance "does not include petroleum, including crude oil or any fraction thereof which is not otherwise specifically listed or designated as a hazardous substance"). The general rule of thumb is that CERCLA does not provide liability for the release of petroleum products that contain no more than their indigenous components and common additives. *See, e.g.,* Wilshire Westfield Assoc. v. Atlantic Richfield Corp., 881 F.2d 801, 805 (9th Cir. 1989); City of New York v. Exxon Corp., 766 F. Supp. 177, 186–87 (S.D.N.Y. 1991). CERCLA also exempts from liability "federally permitted" releases, 42 U.S.C. § 9607(j) (1988), the application of a registered pesticide product, *id.* § 9607(i), and actions taken to address an imminent hazard at the direction of an on-scene coordinator or by state and federal officers. *Id.* § 9607(d). Finally, CERCLA does not provide USEPA with authority to respond to releases for naturally occurring pollution, loose asbestos in the walls of buildings, or leaching lead in drinking-water systems. *See id.* § 9604(a)(3).

17. *See, e.g.,* New York v. Shore Realty Corp., 759 F.2d 1032, 1044 (2d Cir. 1985) (holding that Congress intended that responsible parties be held liable regardless of fault, even though explicit provision for strict liability not included in statute).

18. *See, e.g., United States v. Stringfellow,* 661 F. Supp. 1053, 1060 (D.C. Cal. 1987) (holding that if harm is indivisible, each defendant is jointly and severally liable for entire injury).

19. *See* 42 U.S.C. §§ 9605(a)(8), 9605(c) (1988).

20. *See, e.g., id.* §§ 9604(a), 9604(b), 9604(e), and 9606(a).

21. *See generally id.* § 9611.

22. *See, e.g., id.* § 9613(f)(1). Allocation among responsible parties is typically determined by the courts using the "Gore Factors," which allow the relative culpability of each party for the release at the site to be considered. *See, e.g.,* Amoco Oil Co. v. Dingwell, 690 F. Supp. 78, 86–87 (D. Me. 1988), *aff'd sub nom.* Travelers Indem. Co. v. Dingwell, 884 F.2d 629 (1st Cir. 1989); *see also infra* notes 110–15 and accompanying text.

23. *See* DANIEL P. SELMI & KENNETH A. MANASTER, STATE ENVIRONMENTAL LAW ch. 9 (1993) (providing comprehensive overview of state supplemental approaches to CERCLA).

24. *See* Robert B. McKinstry, Jr., *The Role of State "Little Superfunds" in Allocation and Indemnity Actions under the Comprehensive Environmental Response, Compensation and Liabili-*

ty Act, 5 VILL. ENVTL. L.J. 83, 83 & n.3 (1994). In CERCLA, Congress expressly provided for parallel state cleanup programs. 42 U.S.C. § 9614(a) (1988).

25. *See, e.g.,* KAN. STAT. ANN. § 65-3454a (1994) (creation of environmental defense fund and parameters of use); N.J. STAT. ANN. § 58:10-23.11i (West 1995) (description of New Jersey spill compensation fund).

26. *See, e.g.,* N.J. STAT. ANN. § 58:10-23.11b (West 1995) (defining covered hazardous substances more broadly than CERCLA by adding petroleum products and substances included in state list of "environmental hazardous substances"); 35 PA. CONS. STAT. ANN. §§ 6070.703(d)–(e) (1995) (providing defenses to residential homeowners, builders of residential housing, and transporters of municipal waste).

27. See subsequent chapters in this book on state voluntary cleanup statutes.

28. *See* 42 U.S.C. §§ 6926, 6991c (1988).

29. *See, e.g.,* Laura Nagle, *RCRA Subtitle I: The Federal Underground Storage Tank Program*, 24 Envtl. L. Rep. (Envtl. L. Inst.) 10,057, 10,059 & n.34 (1994) (identifying states with underground storage tank regulations that are more stringent than their federal counterparts).

30. *See, e.g.,* William Buzbee, *Remembering Repose: Voluntary Contamination Cleanup Approvals, Incentives, and the Costs of Indeterminable Liability*, 80 MINN. L. REV. 35, 47–48 (1995) ("For over a decade, . . . neither the federal political branches nor USEPA created any program or policy to review and approve voluntary cleanups.").

31. *See* 42 U.S.C. § 9604(a)(3)(A) (1988) (a naturally occurring substance for which no response action may be taken pursuant to CERCLA is one that is "in its unaltered form, or altered solely through naturally occurring processes or phenomena, from a location where it is naturally found").

32. *See id.* § 9601(9)(B).

33. *See id.* § 9607(a).

34. *See id.* §§ 9601(10), 9607(j).

35. *See id.* §§ 6972(a)(1)(B), 6973(a); *see also* Zands v. Nelson, 779 F. Supp. 1254, 1262–64 (S.D. Cal. 1991) (holding "imminent hazard" claims may be brought for gasoline leaking from underground storage tank).

36. Underground storage tanks are technically defined as those tanks that are at least 10 percent beneath the surface of the ground and contain a "regulated substance" as defined in the statute. *See* 42 U.S.C. § 6991(1) (1994). Exempt underground storage tanks include farm or residential tanks with a capacity of 1100 gallons or less, which are used for noncommercial purposes, and tanks used for storing heating oil for use on premises where oil is stored. 42 U.S.C. §§ 6991(1)(A)–(B) (1988). For a discussion of other exemptions, see Laura Nagle, *supra* note 29, at 10061–62.

37. *See* 40 C.F.R. §§ 61.141, 61.145 (1995) ("any person who owns, leases, operates, controls, or supervises the facility being demolished or renovated [which contains asbestos] or any person who owns, leases, operates, controls, or supervises the demolition or renovation operation, or both" must comply with regulations governing such activities promulgated by USEPA under the Clean Air Act). Though federal laws and regulations do not cover asbestos in buildings unless it is being removed or disturbed as part of renovation, USEPA has nevertheless issued various guidelines on the proper management of in-place asbestos and recommends that owners develop a management plan. *See generally* Anne C. Weisberg, *Asbestos, in* ENVIRONMENTAL LAW PRACTICE GUIDE: STATE AND FEDERAL LAW § 36.05[2][d] (Michael B. Gerrard, ed., 1996).

38. Under the Occupational Safety and Health Act, employers must protect employees against specified concentrations of asbestos in the workplace. *See* 29 C.F.R. § 1910.1001(b)–(f) (1995).

39. *See, e.g.*, HRW Sys., Inc. v. Washington Gas Light Co., 823 F. Supp. 318, 340–41 (D. Md. 1993) (holding that presence of any detectable amount of hazardous substance is sufficient for proof of "disposal" and "release" under CERCLA).

40. *See* United States v. Ottati & Goss, Inc., 900 F.2d 429, 438 (1st Cir. 1990); *see also* Amoco Oil Co. v. Borden, Inc., 889 F.2d 664, 670 (5th Cir. 1989) (liability for cleanup under CERCLA does not occur until concentrations of contaminants at site exceed CERCLA cleanup standards).

41. Fertilizer Inst. v. EPA, 935 F.2d 1303 (D.C. Cir. 1991) (rejecting USEPA's definition of "release" as including open container of radionuclides and holding hazardous substances must move into environment, not merely be exposed to environment, for CERCLA release to occur).

42. *See* 42 U.S.C. § 9607(a) (1988).

43. *See id.* § 9601(22).

44. USEPA typically limits its cleanup authorities to sites that rank on the NPL, although it has much broader enforcement authority under CERCLA. *See, e.g., id.* §§ 9604(a), 9606(a). The ranking of a site on the NPL is generally an indication that extraordinary cleanup expenses lie ahead. The average cost of cleanups at NPL sites ranges from $20–30 million. *See, e.g.*, James R. Deason, *Clear as Mud: The Function of the National Contingency Requirement in a CERCLA Private Cost-Recovery Action*, 28 GA. L. REV. 555, 572 (1994). The ranking of sites on the NPL is revised "no less often than annually." 42 U.S.C. § 9605(a)(8)(B) (1988). Thus, there is little finality for more severely contaminated sites that may be ranked at some point in the future.

45. *See* 42 U.S.C. § 6973(a) (1988).

46. *See id.* § 6928. USEPA or the states may also require the owner or operator of a facility at which hazardous wastes have been released to conduct a "corrective action" if the facility has applied for, or received, a RCRA treatment, storage, or disposal permit. *Id.* § 6924(u).

47. *See supra* notes 23–27 and accompanying text.

48. *See, e.g.*, 42 U.S.C. §§ 6929, 6972(a) (1988).

49. *See, e.g.*, Carolyn Scott Kortge, *Taken to the Cleaners*, NEWSWEEK, Oct. 23, 1995, at 16 (describing state enforcement taken against mother for cleanup of hazardous wastes released by defunct dry-cleaning business that had leased her mother's building in past).

50. For a general discussion of private-party actions under environmental statutes, see Adam Babich & Kent E. Hanson, *Opportunities for Environmental Enforcement and Cost Recovery by Local Governments and Citizen Organizations*, 18 Envtl. L. Rep. (Envtl. L. Inst.) 10,165 (1988).

51. 42 U.S.C. § 6972(a)(1)(B) (1988).

52. *Id.* § 6972(a)(1)(A).

53. Private parties may seek reimbursement of cleanup costs under CERCLA, provided the cleanup is conducted in substantial compliance with the National Contingency Plan. *See generally* 42 U.S.C. §§ 9607(a), 9613(f)(1) (1988). Recently, the Ninth Circuit ruled that private parties may also obtain restitution of the costs of cleanup under section 7002(a)(1)(B) of RCRA, which previously was interpreted to provide citizen plaintiffs only with injunctive authority. *See* KFC Western, Inc. v. Meghrig, 49 F.3d 518, 521 (9th Cir. 1995). The Supreme Court reversed this decision, however, and held that the owners of a site contaminated with petroleum cannot recover their cleanup costs from the former owners under RCRA's citizen suit provision. *See* Meghrig v. KFC Western, Inc., 116 S. Ct. 1251 (1996).

54. 42 U.S.C. § 6972(a) (1988) (providing suit for injunctive relief).

55. *See, e.g.*, Daniel M. Steinway, *Private Cost Recovery Actions under CERCLA: The Impact of the Consistency Requirements*, 4 Toxics L. Rep. (BNA) 1364 (May 2, 1990).

56. Although private parties typically recover only those cleanup costs they expended beyond what constitutes their "equitable share," some courts have allowed nonculpable owners to recover all their cleanup costs (minus attorney fees) under CERCLA. *See, e.g.,* United Technologies Corp. v. Browning-Ferris Indus., Inc., 33 F.3d 96, 99 (1st Cir. 1994), *cert. denied*, 115 S. Ct. 1176 (1995); United States v. Colorado & Eastern R.R. Co., 50 F.3d 1530, 1533 (10th Cir. 1995).

57. *See, e.g.,* Key Tronic Corp. v. United States, 114 S. Ct. 1960, 1965 (1994) (attorney fees not recoverable in private-party cost-recovery actions brought under CERCLA); *but see* 42 U.S.C. § 6972(e) ("reasonable" litigation costs may be awarded under citizen-suit provision of RCRA); *see also* 35 PA. CONS. STAT. ANN. § 6020.1115(b) (1995) (allowing recovery of attorney fees in litigation for cleanup of contaminated sites); WASH. REV. CODE ANN. § 70.105D.050(5)(a) (West 1995) (same provision); N.J. STAT. ANN. § 2A:35A-10 (West 1995) (same provision).

58. *See* 40 C.F.R. § 302.4 (1994).

59. 42 U.S.C. §§ 9607(a), 9607(j) (1988) (imposing liability for cleanup of all releases and threatened releases of hazardous substances, provided releases not "federally permitted releases").

60. *Id.* §§ 6972(a)(1)(B), 6973(a) (imposing liability for imminent hazards on any person "who has contributed or who is contributing" to release).

61. *See id.* § 6991(3). "Owner" also includes lenders, provided they "participat[ed] in the management of an underground storage tank." *Id.* § 6991b(h)(9). *Cf. infra* notes 98–102 and accompanying text. USEPA has recently promulgated regulations limiting lender liability for underground storage cleanups. *See* 60 Fed. Reg. 46,692 (1996).

62. *See* 42 U.S.C. §§ 6991b(h)(1)–(2) (1988).

63. *See id.* § 6991b(h)(6).

64. *See id.* § 6991b(h)(11).

65. For cleanup requirements, see 40 C.F.R. §§ 280.60–.66, 280.71 (1994). Cleanup standards are based on a site-specific, risk-based exposure assessment. If, following initial abatement efforts, contamination remains at concentrations sufficient to trigger corrective action requirements (*id.* § 280.66(a)), final remediation levels will be set based on five specific considerations. *Id.* § 260.66(b).

66. *See* Nagle, *supra* note 29, at 10,059. For an overview of state laws and regulations governing underground storage tanks, see Dale E. Hermeling & Joseph B. Pereles, *Storage Tanks, in* ENVIRONMENTAL LAW PRACTICE GUIDE §§ 39.11, 39.12, *supra* note 37.

67. *See* Nagle, *supra* note 29, at 10,068.

68. *See supra* note 53.

69. *See supra* note 16 and accompanying text.

70. 15 U.S.C. § 2605(e) (1988); *see also* 40 C.F.R. §§ 761.120–.135 (1995) (policy for cleanup of PCB spills occurring after May 4, 1987).

71. A number of PCBs are listed as hazardous substances under CERCLA. *See* 40 C.F.R. § 302.4(a) (1995). Even if a PCB is not listed, however, it still may be an unlisted hazardous substance. *Id.* § 302.4(b).

72. *See id.* § 761.123 ("*responsible party* means the owner of the PCB equipment, facility, or other source of PCBs or his/her designated agent (e.g., a facility manager or foreman)").

73. *See id.* § 761.120(a) (limiting scope of USEPA's PCB spill policy to more recent spills); *but see id.* § 761.120(b) (USEPA retains authority to require cleanup at PCB concentrations less than 50 ppm). Cleanup requirements for spills occurring before May 4, 1987 will be evaluated by USEPA on a case-by-case basis under its regulatory authority under the TSCA. *Id.* § 761.120(a)(1).

74. *See* 42 U.S.C. §§ 9607(a)(1)–(4) (1988) (listing broad number of covered parties liable under CERCLA); *see generally infra* notes 110–15 and accompanying text.

75. In its TSCA Spill Policy, USEPA requires that soils and surfaces generally be cleaned to a concentration of 10 to 50 ppm, depending on the surface. *See generally* 40 C.F.R. § 761.125 (1995).

76. The spill response requirements are set forth at 40 C.F.R. §§ 761.120–130 (1995).

77. *See id.* § 761.135(a) ("Cleanup in accordance with this policy means compliance with the procedural as well as the numerical requirements of this policy.").

78. *See, e.g., id.* § 761.30 (regulations governing use and servicing of PCB items); *id.* § 761.40 (warning labels must be placed on PCB-containing transformers and other listed equipment).

79. *See, e.g., id.* §§ 761.60–.80 (regulations governing storage and disposal of PCB items taken out of use).

80. Pursuant to Clean Air Act regulations, owners or operators of a demolition or renovation activity must conduct building asbestos surveys (40 C.F.R. § 61.145(a) (1995)), notify regulatory authorities when certain types of asbestos are identified (*id.* § 61.145(b)), remove certain types of asbestos in accordance with regulations (*id.* § 61.145(c)), and dispose of the asbestos in accordance with detailed regulations (*id.* § 61.150).

81. *See generally* Weisberg, *supra* note 37, § 36.03[4][c] (discussing inability of owners to recover asbestos abatement costs under CERCLA), and § 36.05[2][e] (discussing common-law theories for recovery of asbestos abatement by current owners).

82. *See generally* 40 C.F.R. §§ 280.40–.45 (1994).

83. *See generally id.* §§ 280.20–.21.

84. *See id.* §§ 280.90–.116.

85. Depending on the level of involvement, "owner," "operator," "generator," and "transporter" can also include the officers, directors, and employees of a corporation and its successor or parent corporation. *See, e.g.*, United States v. Northeastern Pharmaceutical & Chem. Co., Inc., 810 F.2d 726, 743–44 (8th Cir. 1986) (holding officers and employees liable as owners and operators under CERCLA), *cert. denied*, 484 U.S. 848 (1987); United States v. Kayser-Roth Corp., 724 F. Supp. 15, 22–23 (D.R.I. 1989) (holding parent corporation liable as owner under CERCLA), *aff'd*, 910 F.2d 24 (1st Cir. 1990), *cert. denied*, 111 S. Ct. 957 (1991); United States v. Carolina Transformer Co., 739 F. Supp. 1030, 1039 (E.D.N.C. 1989) (holding if there is "substantial continuity" between successor corporation and its predecessor, successor can be bound by acts of predecessor). Even lessees of contaminated property may be caught in CERCLA's liability net. *See, e.g.*, United States v. Mexico Feed and Seed Co., Inc., 980 F.2d 478 (8th Cir. 1992) (holding lessee liable under CERCLA as owner and operator of leaking tanks at PCB-contaminated site).

86. *See generally* 42 U.S.C. §§ 6972(a)(1)(B), 6973 (1988) (listing broad category of liable parties under RCRA imminent-hazard provisions, provided they "contributed" or are "contributing to" the "past or present handling, storage, treatment, transportation, or disposal of any solid or hazardous waste which may present an imminent and substantial endangerment to health or the environment"); *id.* §§ 9607(a)(1)–(4) (providing list of covered parties liable for cleanup under CERCLA).

87. *See id.* § 9601(20)(D) (exempting municipalities and state governments as owners if they acquire ownership involuntarily); *see also* USEPA Policy on CERCLA Enforcement against Lenders and Government Entities that Acquire Property Involuntarily (Sept. 22, 1995). Unfortunately, this exemption from liability does not transfer to the purchaser and may be invalidated by state law. *See, e.g.*, Robert S. Berger, et al., *Recycling Industrial Sites in*

Erie County: Meeting the Challenge of Brownfield Redevelopment, 3 BUFF. ENVTL. L.J. 69, 106–07, 110–11 (1995).

88. *See* 42 U.S.C. §§ 9601(35)(A), 9607(b)(3) (1988) (innocent-landowner defense under CERCLA provided to owners who "unknowingly acquired contaminated property . . . and who undertook all appropriate inquiry at the time of acquisition").

89. Although there are no specific exemptions from liability for these nonculpable activities under the imminent-hazard provisions of RCRA, only a party that "has contributed or who is contributing" to the contamination is liable, which presumably results in similar types of exceptions for "innocent" parties. *See, e.g., id.* §§ 6972(a)(1)(B), 6973(a).

90. *See, e.g.*, United States v. A & N Cleaners & Launderers, Inc., 854 F. Supp. 229 (S.D.N.Y. 1994) (rejecting innocent-landowner defense under CERCLA when purchaser failed to inquire about disposal practices of business that previously operated on property); *cf.* United States v. Broderick Inv. Co., 862 F. Supp. 272 (D. Colo. 1994) (rejecting innocent-landowner defense under CERCLA when owner had actual knowledge that lessor discharged waste on property).

91. *See* Kerr-McGee Chem. Corp. v. Lefton Iron & Metal Co., 14 F.3d 321 (7th Cir. 1994) (innocent-owner defense rejected when owner failed to take precautions to prevent damage from known hazardous substances); *see generally* Stephen M. Feldman, *CERCLA Liability, Where It Is and Where It Should Not Be Going: The Possibility of Liability Release for Environmental Beneficial Land Transfers*, 23 ENVTL. L. 295, 298–99 (1993); *see also* United States v. Price, 523 F. Supp 1055, 1073–74 (D.C.N.J. 1981) (denying current owner's motion for summary judgment and holding that current owner of landfill may be liable under imminent-hazard provision of RCRA because of studied indifference to hazardous conditions at site, even though current owner purchased property after disposal operations had ceased), *aff'd* 688 F.2d 204 (3d Cir. 1982). Though "innocent landowners" typically are recent purchasers, victim-owners of midnight dumping or similar activities are likely to be immunized from liability under this same provision, provided they took all reasonable precautions to prevent such dumping and to respond to the dumping once it occurred. 42 U.S.C. § 9607(b)(3) (1988). An Act of God and Act of War also exempt an owner from liability under CERCLA. *See id.* §§ 9607(b)(1)–(2).

92. *See, e.g.*, Jersey City Dev. Auths. v. PPG Indus., 18 Envtl. L. Rep. (Envtl. L. Inst.) 20,364, 20,368 (D.N.J. 1987) (0 percent share for current landowner who was liable under CERCLA but had no responsibility for dumping), *aff'd* 866 F.2d 1411 (3d Cir. 1988).

93. *See, e.g.*, Nurad, Inc. v. William E. Hopper & Sons Co., 966 F.2d 837, 846 (4th Cir. 1992) (holding past owners liable during periods of passive leaking of contaminants occurring during their ownership), *cert. denied*, 113 S. Ct. 377 (1992); *but see* Joslyn Mfg. Co. v. Koppers Co., Inc., 40 F.3d 750 (5th Cir. 1994) (declining to hold intermediate owner liable under CERCLA when intermediate owner did not operate facility or actively dispose of hazardous substances at site).

94. *See* Buzbee, *supra* note 30, at 44.

95. Robert Franz, General Electric Company, statement at ABA Satellite Seminar, Brownfields Redevelopment: Cleaning Up the Urban Environment (Mar. 7, 1996).

96. *See, e.g.*, AM Int'l v. International Forging Equip., 982 F.2d 989, 993–95 (6th Cir. 1993) (holding parties may allocate financial liability for CERCLA cleanup between themselves with contractual agreements). The success of the indemnification clause to provide protection will depend on the language of the clause. *See, e.g.*, John S. Boyd Co., Inc. v. Boston Gas Co., 992 F.2d 401, 406–07 (1st Cir. 1993).

97. *See, e.g., John S. Boyd Co.*, 992 F.2d at 407 (indemnity agreement will not protect responsible party from suit by government or private party under CERCLA 107(a)); *see also* 42 U.S.C. § 9607(e) (1988) (same rule).

98. *See* United States v. Maryland Bank & Trust Co., 632 F. Supp. 573, 579 (D. Md. 1986) (holding bank that foreclosed and purchased facility liable as owner under CERCLA).

99. *See* United States v. Fleet Factors Corp., 901 F.2d 1550, 1557 (11th Cir. 1990) (holding lender liable as operator for decisions that reflected capacity to control waste disposal of borrower who was directly responsible for contamination at site), *cert. denied*, 498 U.S. 1046 (1991); *see generally* Daniel Michel, *The CERCLA Paradox and Ohio's Response to the Brownfield Problem: Senate Bill 221*, 26 U. TOL. L. REV. 435, 444–49 (1995) (providing overview of lender liability under CERCLA, including discussion of cornerstone cases).

100. *Fleet Factors Corp.*, 901 F.2d at 1557 (emphasis added); *compare In re* Bergsoe Metal Corp., 910 F.2d 668, 671, 672 (9th Cir. 1990) (holding that "whatever the precise parameters of 'participation,' there must be *some* actual management of the facility before a secured creditor will fall outside the exception.").

101. *See* 57 Fed. Reg. 18,344 (1992) (to be codified at 40 C.F.R. § 300.1100). USEPA specified that lenders would be directly liable under CERCLA only if they participated in the management of a facility either by taking direct "responsibility for the borrower's hazardous substance handling or disposal practices" or by assuming "overall management responsibility encompassing the day-to-day decision making of the enterprise." *Id.* § 300.1100(c)(1). Actual participation in the management of the facility was necessary (*id.*), and a lender was also able to foreclose without becoming liable as an owner provided a series of requirements are satisfied. *Id.* § 300.1100(d).

102. *See* Kelley v. EPA, 15 F.3d 1100, 1107 (D.C. Cir. 1994), *cert. denied sub nom.* American Bankers Ass'n v. Kelley, 115 S. Ct. 900 (1995).

103. *See* Asset Conservation, Lender Liability, and Deposit Insurance Protection Act of 1996, Public Law No. 104-208 (1996).

104. *See* 42 U.S.C. § 9607(a)(3) (1988) (liability under CERCLA); *id.* §§ 6972(a)(1)(B), 6973(a) (liability under RCRA imminent-hazard provisions). For a broad reading of CERCLA's liability provisions for generators, see *United States v. Aceto Agricultural Chemicals Corp.*, 872 F.2d 1373 (8th Cir. 1989) (holding manufacturers of pesticide ingredients potentially liable under CERCLA for disposal of hazardous substances by formulator, when manufacturers owned work in process and hazardous waste was inherent in the formulation process).

105. *See* 42 U.S.C. § 9607(a)(4) (1988) (liability under CERCLA if transporter selects site); *id.* §§ 6972(a)(1)(B), 6973(a) (liability under RCRA imminent-hazard provisions); *see also* Tippins, Inc. v. USX Corp., 37 F.3d 87, 90 (3d Cir. 1994) (holding transporter liable under CERCLA "if the transporter's advice was a substantial contributing factor in the decision.").

106. *See supra* note 86.

107. Current owners or operators of brownfields must typically notify federal or state authorities regarding the source, location, and approximate extent of contamination. *See, e.g.*, 42 U.S.C. § 9603 (1988) (notice of reportable quantity of release of hazardous substance required under CERCLA); *id.* § 6991a(a) (notification of presence of underground storage tanks required under RCRA). Reporting requirements for owners once hazardous substances have been discovered at the site or underground storage tanks have been located should not be taken lightly. Federal officials, and often state environmental officials, may enforce civil penalties and criminal liability on noncompliers. *See, e.g., id.* § 9603(b). Prospective purchasers of brownfields are not required to comply with reporting requirements until the title has transferred, however.

108. 40 C.F.R. pt. 300 (1994).

109. *See infra* note 125 and accompanying text.

110. *See, e.g.*, Wickland Oil Terminals v. Asarco, Inc., 792 F.2d 887, 892–93 (9th Cir. 1986) (approving USEPA's regulatory revisions in NCP, that specify that no governmental approval is necessary before private-party cleanup).

111. *See, e.g.*, James R. Haisley, *Private Party Recovery of Environmental Response Costs*, 6 BYU J. PUB. L. 261 (1992); Patricia T. Barmeyer & C. Paul Chalmers, *Cleanup Cost Recovery or How to Get the Money Back*, 474 PLI/Lit 269 (1993) (available on Westlaw).

112. *See* 40 C.F.R. § 300.700(c)(3)(i) (1994); *see also* 42 U.S.C. § 9607(a)(4)(B) (1988); 55 Fed. Reg. 8,792-94 (1990) (providing guidance on what constitutes a "CERCLA-quality cleanup").

113. *See, e.g.*, 40 C.F.R. § 300.430(c) (1994) (outlining community-relations requirements for CERCLA cleanups); *id.* § 300.430(f)(3) (outlining public-participation requirements for selection of CERCLA remedy); *id.* § 300.700(c)(6) (requiring public participation for private-party cleanups).

114. *See infra* notes 128–39 and accompanying text.

115. *See, e.g.*, 40 C.F.R. §§ 300.420–.440 (1994); *see also* U.S. EPA, GUIDANCE FOR CONDUCTING REMEDIAL INVESTIGATION AND FEASIBILITY STUDIES UNDER CERCLA, INTERIM FINAL, OSWER 9355.3-01, EPA/540/G-89/004 (Oct. 1988).

116. *See supra* note 16 and accompanying text.

117. *See* Key Tronic Corp. v. United States, 114 S. Ct. 1960, 1965 (1994).

118. 42 U.S.C. §§ 6972(a)(1)(B), 6973(a) (1988); *see also* Dague v. City of Burlington, 732 F. Supp. 458, 469 (D. Vt. 1989) (finding imminent and substantial endangerment under RCRA existed at landfill; even though leachate collection 90 percent effective, some hazardous wastes being discharged into soil, groundwater, and surface water).

119. Any facility that has or must apply for a permit to store, treat, or dispose of hazardous wastes pursuant to 42 U.S.C. § 6925 must clean up any release of hazardous constituents at the facility through "corrective actions," regardless of when the wastes were placed in the facility and regardless of whether they are hazardous wastes under RCRA. 42 U.S.C. § 6924(u) (1988); *see also id.* § 6928(h) (USEPA may issue corrective-action orders against interim status facilities and other RCRA-covered facilities with no TSD permit). TSD Permits may contain schedules for compliance for corrective action. *Id.* § 6924(u).

120. Closure requirements apply only to facilities falling under the treatment, storage, and disposal facility (TSD) permit requirements of RCRA. *See* 42 U.S.C. § 6925 (1988). Before a TSD permit will be issued, the TSD facility must prepare a written plan for closing the facility that has been approved by USEPA or the state. *See* 40 C.F.R. §§ 264.112, 265.112 (1994). Regardless of the type of treatment, storage, and disposal facility, regulations promulgated under RCRA require that a TSD facility be closed in a way that minimizes the release of hazardous constituents in the future and also minimizes the need for further maintenance and controls after closure. *See id.* §§ 264.111(a)–(b), 265.111(a)–(b).

121. *See* 40 C.F.R. § 264.100 (1994); U.S. EPA, RCRA CORRECTIVE ACTION PLAN, OSWER DIRECTIVE 9902.3-2A (May 1994) (available from NTIS, order number 520-R94-004).

122. Specific requirements are provided for closing specific types of TSD facilities. *See* 40 C.F.R. §§ 264.111(c), 265.111(c) (1994). If a TSD facility is not able to "clean close," (*see, e.g., id.* § 264.228(a)(1) [describing clean-closure requirements for surface impoundments]), it must conduct postclosure care for years after closure. *Id.* §§ 264.117(a), 264.118(a), and 265.117(a), 265.118(a). This could include groundwater monitoring (*id.* §§ 264.117(a)(1)(i), 265.117(a)(1)(i)), and maintaining security at the site. *Id.* § 264.117(b).

123. *See* Meghrig v. KFC Western, Inc., 116 S. Ct. 1251 (1996).

124. *See* 42 U.S.C. § 9614(a) (1988) (CERCLA does not preempt state liability and cleanup programs).

125. *See, e.g.*, CAL. HEALTH & SAFETY CODE §§ 25,350, 25,356.1 (West 1995); N.J. STAT. ANN. § 58:10-23.11f(3) (West 1995).

126. *See, e.g., supra* note 29 and accompanying text.

127. See the chapters that follow.

128. *See* Buzbee, *supra* note 30, at 47, 59 & n.32 ("Without government feedback about the type and extent of necessary cleanup, efforts to determine liabilities associated with contaminated land are fraught with uncertainty."); *see also* James Boyd, The Impact of Uncertain Environmental Liability on Industrial Real Estate Development: Developing a Framework for Analysis (Oct. 1994) (RFF Discussion Paper 94-03 Rev. 21) (noting that cleanup standards depend heavily on unique perspectives of state and federal agencies involved in any particular cleanup); Lynn Grayson & Stephen A.K. Palmer, *Brownfields Phenomenon: An Analysis of Environmental, Economic, and Community Concerns*, 25 Envtl. L. Rep. (Envtl. L. Inst.) 10,337, 10,337–38 (1995) (observing great variation in numerical standards and risk-assessment procedures used for cleanups and also stating that developers' fears of "future changes in environmental laws or improvements in technology" exacerbate "how-clean-is-clean" uncertainty).

129. Cleanups conducted solely under the corrective action or closure requirements of RCRA may have slightly less or more stringent standards than CERCLA, respectively. USEPA has indicated that with regard to corrective action requirements "for the most part, RCRA cleanups will be less complex and less expensive than those under CERCLA . . . and therefore the need to pursue complete cleanup at such [RCRA] facilities will often be less urgent." 55 Fed. Reg. 30,798, 30,833 (1990); *see also* Richard G. Stoll, *The New RCRA Cleanup Regime: Comparisons and Contrasts with CERCLA*, 44 SW. L.J. 1299, 1310 (1991). USEPA guidance for closures is more extensive. *See supra* note 122. States have in instances required very stringent cleanup levels. *See, e.g.*, OHIO EPA, DIV. SOLID & HAZARDOUS WASTE MANAGEMENT, CLOSURE PLAN REVIEW GUIDANCE 28-56 (June 1, 1991) (requiring cleanup standards to ensure that hazardous wastes present at site do not exceed background levels or conservative risk assessment demonstrating site is "clean" for residential purposes and will not require further monitoring).

States have been endeavoring to provide greater certainty in the identification of appropriate cleanup standards (*see, e.g., infra* note 182 and accompanying text), although much work remains to be done.

130. *See* 42 U.S.C. § 9621(d)(2) (1988).

131. *See* 40 C.F.R. §§ 300.400(g), 300.430(e)(2) (1994).

132. *See* U.S. EPA, CERCLA COMPLIANCE WITH OTHER LAWS MANUAL, OSWER 9234.1-01 (Mar. 6, 1988); U.S. EPA, CERCLA COMPLIANCE WITH OTHER LAWS MANUAL, PART II, OSWER 9234.1-02 (Aug. 1989); *see also* Richard G. Stoll, *Comprehensive Environmental Response, Compensation & Liability Act, in* ENVIRONMENTAL LAW HANDBOOK 471, 485–94 (11th ed. 1991).

133. *See* Stephen M. Smith, *CERCLA Compliance with RCRA: The Labyrinth*, 18 Envtl. L. Rep. (Envtl. L. Inst.) 10,518, 10,524–25 (1988).

134. *See, e.g.*, James T. O'Reilly, *Environmental Racism, Site Cleanup and Inner City Jobs: Indiana's Urban In-fill Incentives*, 11 YALE J. ON REG. 43, 49 (1994).

135. A risk-based approach is also required when an ARAR will not be sufficiently protective (a cumulative risk of greater than 10^{-4}) due to multiple contaminants or pathways. *See* 40 C.F.R. § 300.430(e)(2)(i)(D) (1994).

136. *See id.* § 300.430(e)(2)(i)(A)(2). USEPA has prepared a series of risk-assessment guidelines that supplement the bare NCP requirements. *See generally* OFFICE OF EMERGENCY & REMEDIAL RESPONSE, U.S. EPA, RISK ASSESSMENT FOR SUPERFUND: HUMAN HEALTH EVALUATION MANUAL, SUPPLEMENTAL GUIDANCE, STANDARD DEFAULT EXPOSURE FACTORS, INTERIM FINAL 5–6 (1991); 1 U.S. EPA, RISK ASSESSMENT GUIDANCE FOR SUPERFUND, HUMAN HEALTH

EVALUATION MANUAL (PART A), INTERIM FINAL, EPA/540/1-89/002 (Dec. 1989); U.S. EPA, EXPOSURE FACTORS HANDBOOK, EPA/600/8-89/043 (July 1989).

137. For example, until recently, USEPA tended to assume in its risk assessment that the groundwater would be used as a drinking-water source. *See, e.g.,* Alex S. Karlin, *How Long is Clean? The Temporal Dimension to Protecting Human Health under Superfund,* 9 NATURAL RESOURCES & ENV'T 6 (1994); *see generally* W. Kip Viscusi & James T. Hamilton, *Human Health Risk Assessments for Superfund,* 21 ECOLOGY L.Q. 573, 587–88, 608 (1994) (discussing USEPA's costly and dubious assumption that sites will be used in future for residential development when there appears no likelihood of such future use).

138. *See* U.S. EPA, DIRECTIVE ON LAND USE IN THE CERCLA REMEDY SELECTION PROCESS, OSWER DIRECTIVE No. 9355.7-04 (May 25, 1995); Comment, *Environmental Justice and Industrial Redevelopment: Economics and Equality in Urban Revitalization,* 21 ECOLOGY L.Q. 705, 740 (1994) (discussing similar state initiatives that allow future land use to be considered in setting cleanup standards).

139. *See infra* note 182 and accompanying text.

140. *See* 42 U.S.C. §§ 9607(a)(4)(C), 9607(f) (1988).

141. *See id.* § 9613(g)(1).

142. *See, e.g.,* Kevin R. Duncan & B. Todd Bailey, *Innocence Amid "Lust": The Innocent Buyer and Leaking Underground Storage Tanks Containing Petroleum,* 7 BYU J. PUB. L. 245, 266–79; Note, *Toeing the Line: Compliance with the National Contingency Plan for Private Party Cost Recovery under CERCLA,* 32 WASHBURN L.J. 190, 231 (1993); Andrew N. Davis & Santo Longo, *Stigma Damages in Environmental Cases: Developing Issues and Implications for Industrial and Commercial Real Estate Transactions,* 25 Envtl. L. Rep. (Envtl. L. Inst.) 10,345 (1995).

143. U.S. EPA, BROWNFIELDS ACTION AGENDA (1995).

144. *See id.* at 2–3.

145. *See* Howard M. Shanker & Laurence R. Hourcle, *Prospective Purchaser Agreements,* 25 Envtl. L. Rep. (Envtl. L. Inst.) 10,035 (1995). Federal prospective-purchaser agreements were entered into by USEPA in only rare cases. As of January 1995, only sixteen USEPA prospective-purchaser agreements had been finalized. None of these agreements had been entered into for a non-NPL site. *See* Buzbee, *supra* note 30, at 80 & n.149. USEPA's 1989 guidance on such agreements confirmed this strong resistance to federal involvement in real estate transactions. *See generally* EPA Guidance on Prospective Purchaser Agreements, 54 Fed. Reg. 34,235, 34,241–42 (1989) (requiring number of conditions to be met before USEPA will enter into such agreement, including determination that agreement generates "substantial benefit" not otherwise available through site cleanup).

146. *See* 42 U.S.C. §§ 9622(d)(2), 9622(i) (1988).

147. *See, e.g., id.* § 9613(i) (providing right of intervention to affected persons in any action commenced under CERCLA).

148. *See* EPA, Announcement and Publication of Guidance on Agreements with Prospective Purchasers of Contaminated Property and Model Prospective Purchaser Agreement, 60 Fed. Reg. 34,792–98 (1995) (revised guidance providing USEPA with greater flexibility to consider agreements with prospective purchasers).

149. *See supra* note 143, at 4–5.

150. Even after USEPA's new guidance, as of March 7, 1996, only twenty-three prospective-purchaser agreements had been finalized and each agreement took from eight to eighteen months to complete. Michele B. Corash, Morrison & Foerster, statement at ABA Satellite Seminar, Brownfields Redevelopment: Cleaning Up the Urban Environment (Mar. 7, 1996).

151. *See* Buzbee, *supra* note 30, at 80; Shanker & Hourcle, *supra* note 145; Corash statement, *supra* note 150.

152. *See* U.S. EPA, POLICY TOWARD OWNERS OF PROPERTY CONTAINING CONTAMINATED AQUIFERS (Nov. 1995).

153. *See EPA Region I Announces Measures to Speed Cleanup of Waste Sites*, BNA DAILY REP. EXECUTIVES, Feb. 22, 1995, at 35.

154. *See* Superfund Memorandums of Agreement between USEPA Region V and the states of Wisconsin, Minnesota, Illinois, and Indiana, *reprinted in* ABA, BROWNFIELDS REDEVELOPMENT: CLEANING UP THE URBAN ENVIRONMENT 117–25 (Mar. 7, 1996).

155. *See id.*; James D. Bower, Office of Regional Counsel, Region V USEPA, statement at ABA Satellite Seminar, Brownfields Redevelopment: Cleaning Up the Urban Environment (Mar. 7, 1996).

156. BROWNFIELDS ACTION AGENDA, *supra* note 143, at 1; CERCLIS Definition Change, 60 Fed. Reg. 16,053, 16,053–54 (1995).

157. 60 Fed. Reg. 16,053, 16,054–55 (1995).

158. U.S. EPA, GUIDANCE ON THE DEFERRAL OF NPL LISTINGS, OSWER DIRECTIVE NO. 9375.6-11 (May 3, 1995).

159. *See* ELLIOTT P. LAWS, U.S. EPA, LAND USE IN THE CERCLA REMEDY SELECTION PROCESS, OSWER DIRECTIVE NO. 9355.7-04 (May 25, 1995).

160. *See, e.g., Superfund: Reports Cite Savings in Remedy Selection Resulting from Superfund Reform at EPA*, 27 Env't Rep. (BNA) 1874 (1997).

161. *Id.*

162. *See, e.g., Underground Tanks: Regional Office Takes Up Pilot Project to Facilitate Oakland Redevelopment Program*, 26 Env't Rep. (BNA) 280 (1995).

163. *See, e.g., Superfund: House Leaders Support Bipartisan Effort on Brownfields, Other CERCLA Reform Issues*, 26 Env't Rep. (BNA) 2036 (1996) (reporting on recent bipartisan efforts for Superfund reform on brownfields issues).

164. *See, e.g.*, H.R. 2742, 104th Cong., 1st Sess. (1995); H.R. 2178, 104th Cong., 1st Sess. (1995); *see also Superfund: Amendment to Pending House Superfund Bill Would Set Grants, Loans for 'Brownfields,'* 26 Env't Rep. (BNA) 1882 (1996) (discussing planned amendment to leading House Bill, H.R. 2500, which includes substantial grants to local governments for environmental assessment and cleanup of brownfields sites).

165. *See, e.g.*, S. 1542, 104th Cong., 2d Sess. (1996); H.R. 2846, 2847, 104th Cong., 2d Sess. (1996).

166. *See* H.R. 2500, 104th Cong., 1st Sess. (1995).

167. *See, e.g., Superfund: House Republican Conference Leader Sees Superfund Floor Action in Summer*, 25 Env't Rep. (BNA) 2450 (1995) (discussing congressional hearings on Superfund reforms that establish national risk-based cleanup goal, take cost and future land use into consideration when developing site cleanup plans, and allow for consideration of technical feasibility in selecting appropriate remedy).

168. *See* N.J. STAT. ANN. §§ 13:1K-6–35 (West 1995).

169. *See, e.g.*, I. Leo Motiuk & Daniel J. Sheridan, *Analysis and Perspective, New Jersey's ECRA: Problems, Policies*, 21 Env't Rep. (BNA) 549, 551 (1990).

170. *See* N.J. STAT. ANN. § 13:1K-11 (deferring remediation if industrial use of site will be substantially same as prior use); *id.* § 13:1K-11.2 (providing for expedited review of site remediations); *id.* § 13:1K-11.3 (providing for limited site review); *id.* § 13:1K-11.4 (providing area-of-concern waiver for sites undergoing remediation); *see generally* N.J. STAT. ANN. § 13:1K-1 *et seq.*; *New Jersey: ECRA Reforms Streamline Cleanup Process, Allow DEPE to Adopt Differential Standards*, 24 Env't Rep. (BNA) 364 (1993).

171. The amended ECRA does loosen cleanup standards to reduce the likelihood that industrial properties will be abandoned or "mothballed," however. *See infra* note 182 and accompanying text; *see also* Matthew L. Wald, *Trenton Acts to Weaken Industrial Cleanup Law*, N.Y. TIMES, June 7, 1993, at B1.

172. Bill Paul, *Required Environmental Inspections Become Common*, WALL ST. J., Oct. 13, 1987, at 6.

173. N.J. DEP'T OF ENVTL. PROTECTION, ECRA UPDATE 2 (July 1990).

174. *See, e.g.,* MO. ANN. STAT. § 260.440 (Vernon 1995); N.Y. ENVTL. CONSERV. LAW § 27-1305 (McKinney 1995).

175. *See, e.g.,* IOWA CODE ANN. § 455B.430 (West 1995).

176. *See generally* 30 ILCS 901 *et seq.* (1995); IND. CODE § 13-7-22.5 *et seq.* (1995). Somewhat similarly, in Connecticut, a seller must warrant that there has been no discharge of hazardous waste at the site, that any discharge has been cleaned up or does not constitute a threat to human health or the environment, and that any waste remaining on the site is being managed properly. *See generally* CONN. GEN. STAT. § 22a-134 (1995).

177. *See, e.g.,* 30 ILCS 904(c) (1995).

178. *See, e.g.,* CAL. HEALTH & SAFETY CODE § 25359.7 (West 1995).

179. *See, e.g.,* MASS. GEN. L. ch. 21c, § 7 (1995); MINN. STAT. ANN. § 115B.16(2) (West 1995).

180. *See, e.g.,* N.C. GEN. STAT. § 130A-310.8 (1995); 35 PA. CONS. STAT. ANN. § 6018.405 (1995); W. VA. CODE § 20-5E-20(a) (1989).

181. *See, e.g.,* DEL. CODE ANN. tit. 7, § 9105 (1995); 35 PA. CONS. STAT. ANN. § 6027.1 *et seq.* (1995).

182. *See generally* N.J. STAT. ANN. § 58:10B-12 (1996) (remediation standards determined in part by intended use of property); 35 PA. CONS. STAT. ANN. §§ 6026.301–.305 (1995) (providing three types of cleanup standards, including site-specific standards allowing future land use to affect selection of ultimate cleanup standards).

CHAPTER 3

The Clinton Administration's Brownfields Initiative

JONATHAN D. WEISS[1]

> [T]here are literally hundreds of thousands of old neglected industrial sites now popularly called brownfields that can be redeveloped.... Protecting our environment in the urban areas can go hand-in-hand with redevelopment. It can create jobs and at the same time make more people want to live in the cities of America again.
>
> President Clinton, February 22, 1996

The Clinton Administration's Brownfields Initiative is designed to empower states, localities, and other agents of economic redevelopment to work together in a timely manner to prevent, assess, safely remediate, and sustainably reuse brownfields. United States Environmental Protection Agency (USEPA) Administrator Carol Browner launched the Initiative in late 1993, and escalated its importance in January 1995, with the announcement of the Brownfields Action Agenda. The Action Agenda includes a range of ambitious activities and proposals, all designed to work together to help "jump start" brownfields redevelopment efforts. This chapter will discuss these activities by dividing them into the following areas: (1) clarification of liability and cleanup issues, (2) USEPA's Brownfields Pilot Program and community partnerships, and (3) focused efforts on governmental coordination in the brownfields area.

Clarification of Liability and Cleanup Issues

A major barrier to the cleanup and redevelopment of brownfields sites has been innocent parties' fears of inheriting cleanup liabilities associated with these sites. The Clinton Administration is attempting to address these concerns through common-sense reforms to, and clarification of liability issues involving, the Superfund program. A number of these initiatives are described below.

Archiving CERCLIS Sites

The first action Administrator Browner announced as part of the Brownfields Action Agenda was to archive 24,000 sites from "CERCLIS," the Superfund site database made up of 40,000 potential hazardous waste sites. The archived sites represent those that USEPA had already screened, sometimes years before, and had found to be of no further federal interest; states themselves may still have an interest in these sites. The mere fact that these sites had remained on the CERCLIS list—an accounting problem more than anything else—had caused potential developers to shy away from them. Many lending and real estate investment sectors deny loans for businesses in or near CERCLIS sites as a matter of policy. This action clarified for the lending and business communities the distinction between archived sites and those sites remaining on CERCLIS. In January 1996, Administrator Browner furthered this effort by archiving an additional 4,000 sites.

Guidance on Prospective-Purchaser Agreements

In 1989, USEPA issued its first guidance on prospective-purchaser agreements (PPAs).[2] PPAs, which require Department of Justice (DOJ) approval, generally require a purchaser of contaminated property to commit to a specific cleanup as well as to make a payment to USEPA in return for a covenant-not-to-sue and protection from claims by USEPA and other federal agencies under environmental laws. The original guidance stipulated that such agreements would be entered only when enforcement action was anticipated and the new owner provides the government with direct, substantial benefits.

The potential benefit to using PPAs was exhibited by USEPA Region X's Port of Seattle agreement. In 1994, USEPA Region X entered into an agreement with the Port of Seattle, under which the port purchased the Pacific Sound Resources Site as part of a proposed terminal-expansion project. The agreement included a USEPA covenant-not-to-sue, which protects the port from liability for additional cleanup costs. The port is now in the process of investigating, cleaning up, and redeveloping these sites. Though the PPA worked well in this instance, between 1989 and 1995 USEPA was able to enter into only seventeen PPAs.

In May 1995, USEPA issued new guidance on the use of PPAs, expanding and clarifying the circumstances under which the agency will consider such agreements.[3] This guidance noted that "EPA's experience has demonstrated that prospective purchaser agreements might be both appropriate and beneficial in more circumstances than contemplated by the 1989 guidance."[4] Specifically, this guidance establishes four new criteria to determine whether a PPA may be appropriate:

1. enforcement action is anticipated by USEPA;
2. a direct cleanup or indirect economic benefit will occur;

3. the operation of the site will not aggravate or contribute to contamination or pose a health risk to the community or persons likely to be present at the site; and
4. the prospective purchaser is a financially viable party.

The new guidance, which includes a model agreement, eliminates much of the "retroactive liability" concern associated with purchasing contaminated or previously contaminated property, when some evidence of federal environmental interest exists. It thereby encourages parties to purchase, assess, clean up, and redevelop brownfields they might otherwise avoid due to a reasonable fear of incurring federal liability. Though these agreements can potentially have broad applicability, they are most often used in association with Superfund sites. Since this guidance was issued, USEPA has entered into eight PPAs, and PPAs are currently under consideration at approximately twenty other sites.

Policy Regarding Underground Aquifers

Another guidance issued in May 1995 was a "Final Policy Towards Owners of Property Containing Contaminated Aquifers."[5] This policy is designed to give assurance to owners of uncontaminated properties above groundwater systems that have been contaminated by sources outside the property. In such circumstances, the guidance directs that USEPA not take action against owners, provided that the owner has not caused contamination or made the contamination worse, such as through the handling or disposition of the same or similar substances at the property. USEPA must also be satisfied that the person who caused the release of the hazardous substance that contaminated the groundwater system is not an agent, employee, or party connected through contractual relationship, or otherwise related to the owner whose liability is to be resolved. The guidance would permit USEPA to consider affording contribution protection to qualified owners by entering a "de minimis agreement."

This policy is helping to lower the barriers to transferring property by reducing the uncertainty regarding future liability. If a prospective purchaser knows of aquifer contamination on a piece of property at the time of purchase, he or she will not automatically be liable for cleanup costs. In such a case, the purchaser's liability depends on the seller's involvement in the aquifer contamination. If the seller would have qualified for protection under this policy, then the purchaser will be protected.

Lender Liability and Related Clarifications

In September 1995, USEPA issued a joint memorandum affirming that both the USEPA and DOJ intend to apply, as guidance, the provisions of the vacated "Lender Liability Rule"[6] under the Comprehensive Environmental Response, Compensation, and Liability Act (CERCLA). The memorandum clarifies the policy of not pursuing cleanup costs against those lenders that provide money to an owner or developer of a contaminated property, but do not "actively participate" in the day-to-day management of the property. USEPA first issued this as a rule in 1992.[7] In 1994, the Court of Appeals for the District of Columbia ruled that USEPA did not have statutory authority to issue this rule.[8] However, the Court of Appeals' decision to vacate the original rule did not preclude USEPA from following the rule's provisions as an enforcement policy. The issuance of the guidance, though no substitute for a legislative solution, serves to increase the availability of financing by assuring lenders that USEPA will not hold them liable for cleanup costs of land in which they held a security interest to protect their collateral.

The September 1995 memorandum also clarifies policy regarding government entities that acquire property involuntarily. USEPA clarified which actions would be considered "involuntary" and would, therefore, not subject the governmental unit to potential liability. USEPA expanded the examples listed in CERCLA that refer to involuntary acquisitions by government entities to include, as a nonexhaustive list,

1. acquisitions made by government entities acting as conservators or receivers pursuant to a clear and direct statutory mandate or regulatory authority (such as the acquisition of the security interests of properties of failed lending or depository institutions);
2. acquisitions by government entities through foreclosure and its equivalents while administering a governmental loan, loan guarantee, or loan-insurance program; and
3. acquisitions by government entities pursuant to seizure or forfeiture authority.

Also in September 1995, USEPA issued a final lender-liability rule for underground storage tank (UST) liability under RCRA.[9] This rule is similar to the invalidated Superfund lender-liability rule, as it specifies the conditions under which a lender may be exempted from RCRA liability. It allows a lender that forecloses on a UST to remain exempt from RCRA liability if it complies with certain minimum requirements, and limits the regulatory obligations of financial institutions and others who hold security interests in property on which petroleum USTs are located.

This rule is expected to result in the expansion of credit to gas-station owners and operators and other small businesses that have USTs on their property. Secured creditors have been reluctant to extend loans to these small businesses for fear of incurring cleanup liability associated with releases from USTs. USEPA hopes this reduction in the regulatory obligations of lenders will remove this potential barrier to extending loans to small businesses intending to operate USTs.

Land Use and Soil Screening

USEPA issued land-use guidance in May 1995 that will help ensure the consideration of reasonably anticipated future land use early in the remedial process at National Priority List (NPL) sites under CERCLA. This guidance encourages discussions among local land-use planning authorities, other officials, and the community, as early as possible in the site assessment process. The consideration of future land use in Superfund remedies should ensure proactive, practical, and cost-effective cleanups.

USEPA has also issued draft guidance on soil screening. This guidance is designed to assist decision makers in determining which portions of a site require further study and which areas pose little risk to human health and, therefore, may be ready for redevelopment even without extensive cleanup. This will streamline the study of toxic chemicals in soils at Superfund sites, remove barriers that currently hinder the redevelopment of sites that pose little risk to human health, and allow cleanup efforts and funding to target those areas truly requiring remediation.

New Superfund Law

There is still a limit to how much USEPA can do, administratively, to promote brownfields redevelopment. The Clinton Administration remains strongly committed to enacting a responsible new Superfund law—a law that will promote economic redevelopment and protect human health and the environment. A new Superfund law, the liability reforms, and the assurances for lenders and developers—all of these new policies—will enable brownfields cleanup and redevelopment to proceed more rapidly.

USEPA's Pilot Program and Community Partnerships

In announcing the Brownfields Action Agenda in January 1995, Administrator Browner committed USEPA to funding fifty Brownfields Pilots. The pilots are funded at up to $200,000 for two years, and are designed to test redevelopment models, direct special efforts toward removing regulatory barriers without sacrificing environmental protection, and facilitate community-based and coordinated input.

The national pilots were selected on the basis of (1) the applicant's problem statement and needs assessment, (2) evidence of community-based planning and involvement, (3) the applicant's implementation plan, and (4) the long-term benefits and sustainability of the project. Cities, counties, states, and Indian tribes were eligible to compete.

USEPA's Brownfields Initiative itself was launched in November 1993 with the awarding of a Brownfields Pilot to Cuyahoga County, Ohio. That pilot showed the potential for the program. To date, the $200,000 grant has been leveraged to approximately $3.2 million to support environmental cleanup and property improvements to the bankrupt and abandoned Hauserman business site; payroll taxes alone have netted over $1 million for the local government, and over 170 new jobs have been created by establishing a distribution center on the site.

An important part of the pilot process has been the encouragement of community-based participation in the pilot program. USEPA is committed to involving residents in every step of brownfields redevelopment—from applying for pilot programs to deciding how sites will be used in the future. For instance, in the Buffalo, New York, pilot, to ensure public involvement of those citizens who live near the brownfields sites, the city is instituting a bottom-up master investment plan. This included at least nine summit meetings in the community. Now, as a result of community surveys on how to reuse the sites, plans are being developed to build a badly needed new supermarket on ten acres of formerly industrial property. As another example of strong community participation, the New Orleans pilot is conducting a series of eleven community meetings on the brownfields issue.

USEPA is eager to learn from these pilot experiences, and for the pilot communities to learn from each other. Toward that end, USEPA held its first Brownfields Pilots National Workshop in February 1996. Over 300 representatives from Brownfields Pilots, the Association of State and Territorial Solid Waste Management Officials (ATSWMO), other federal agencies, the National Environmental Justice Advisory Council (NEJAC), and other key stakeholders, as well as USEPA coordinators, all met over a two-day period.

NEJAC—which USEPA established in 1994 to ensure that environmental justice issues are addressed in USEPA programs, including brownfields—has played a critical role in developing community participation. Through NEJAC, representatives from community groups, local governments, states, Indian tribes, and environmental organizations work together to recommend actions and policies. Of note, USEPA and NEJAC sponsored a series of one-day dialogues on brownfields across the country—in Boston, Philadelphia, Detroit, Atlanta, and Oakland—in an effort to involve community groups and environmental-justice advocates in dialogue with other stakeholders, including the business community. USEPA incorporated comments from the NEJAC members into its revised application guidelines for USEPA pilots.

In making sure that brownfields redevelopment follows an integrated approach on the local level, USEPA is also funding and participating in job training and education efforts linked to the Brownfields Initiative. It has formed a partnership with the Department of Labor to promote these efforts, while also forming partnerships with local organizations and community colleges to develop long-term plans for fostering workforce development in brownfields communities. For example, USEPA is working with the Hazardous Materials Training and Research Institute to expand environmental training and curriculum development to community colleges located near Brownfields Pilot communities. In November 1995, the Hazardous Materials Training and Research Institute, with USEPA support, hosted a workshop in

Baltimore, Maryland, to assist community colleges from seventeen brownfields cities in developing environmental job-training programs.

Meanwhile, through a cooperative agreement with USEPA, Rio Hondo Community College in California has established an environmental education and training center to provide comprehensive technician-level training, with an emphasis on Superfund-related subjects. The target audience is female heads of households in a nearby Hispanic community.

Government Coordination
Federal Government

A number of agencies across the government—including the Department of Housing and Urban Development (HUD), the Department of Labor, the Economic Development Administration of the Department of Commerce, and the Department of Treasury are working with USEPA to promote brownfields redevelopment efforts. President Clinton and Vice President Gore have both strongly embraced the Brownfields Initiative. The President, in fact, has asked Congress to adopt a brownfields tax incentive. Under the President's proposed brownfields tax incentive, first announced in his 1996 State of the Union address, environmental cleanup costs for properties in designated areas would be fully deductible in the year in which they are incurred. This incentive would reduce the capital costs for these types of investments by more than one half. The $2 billion tax incentive is expected to leverage $10 billion in private investment and return an estimated 30,000 brownfields to productive use.

According to the proposal, eligible property must be located in one of the following areas: (1) census tracts with a poverty rate of 20 percent or more, (2) census tracts that have a population under 2,000, have 75 percent or more land zoned for industrial or commercial use, and are adjacent to one or more census tracts with a poverty rate of 20 percent or more, (3) empowerment zones and enterprise communities (those existing and proposed), and (4) USEPA's current Brownfields Pilots. Sites on USEPA's NPL list would be excluded.

USEPA has also worked closely with the Department of Treasury to enhance use of the Community Reinvestment Act (CRA) in the brownfields area. Enacted in 1977, the CRA requires banks and other financial institutions to make loans in distressed communities.[10] The revised CRA regulations, issued by the banking regulatory agencies in the spring of 1995, for the first time allow banks to meet their CRA obligations through the cleanup or redevelopment of brownfields properties.[11] This change will significantly boost economic activity in urban areas by making the financing of industrial-property redevelopment more attractive to large lenders, as part of their effort to obtain CRA credit. It is estimated that the CRA, as a whole, prompts $4 billion to $6 billion of lending per year.

Brownfields also play an important role in the Clinton Administration's Empowerment Zone/Enterprise Community (EZ/EC) Initiative. In December 1994, 105 communities from forty-two states were designated EZ/ECs in reward for having developed a comprehensive, locally based revitalization strategy. In the EZ/EC application process, brownfields were the most frequently cited environmental impediment to redevelopment. To ensure that EZ/ECs receive support in addressing environmental issues that arise with redevelopment efforts, USEPA has designated brownfields redevelopment as the agency's EZ/EC Signature Initiative. After EZ/EC designations were made, many communities requested and received assistance on brownfields-related issues from USEPA. At the same time, in awarding its pilots, the agency has agreed to give preference to EZ/EC applicants. Of the forty Brownfields Pilots announced by January 1996, twenty-five were awarded to EZ/ECs.

The Community Development Block Grants (CDBG), administered by HUD, are now also a fruitful area for brownfields redevelopment. These grants are given to communities throughout the country to carry out locally designed community and economic development projects. The CDBG program can also be used to provide direct funding for activities that support the reuse of industrial sites. Distributed by HUD according to a formula, CDBG resources can be used for grants, loans, loan guarantees, and technical-assistance activities. Specifically, HUD has determined that the costs of environmental reviews, as well as the actual cleanup of identified hazards, are eligible CDBG expenses.

A related HUD program, Section 108 loan guarantees, enables local governments to finance physical and economic-development projects too large for front-end financing with single-year CDBG grants. Under Section 108, localities issue debentures to cover the cost of such projects, pledging their annual CDBG grants as collateral. At the same time, HUD and USEPA are undertaking a joint research study on the relationship between the environmental regulatory process and urban redevelopment, and exploring other ways to coordinate their efforts.

State Governments

USEPA recognizes that many states are moving forward with successful Voluntary Cleanup Programs, and is developing ways to foster and encourage these efforts. USEPA's partnership with ATSWMO helps to ensure that state Voluntary Cleanup Programs play an important role in the Brownfields Initiative. In terms of financial assistance, USEPA is providing more than a half a million dollars to certain states to build and develop Voluntary Cleanup Programs. USEPA has also provided expertise to states and localities through intergovernmental personnel assignments. The states of Illinois and Colorado have two USEPA staff members each, and the cities of Dallas, Detroit, East Chicago, Los Angeles, and East Palo Alto have one staff member each.

USEPA is moving toward the approval of certain designated state programs, and is developing guidance for its regional offices to use when determining whether to approve state programs. Presently, regional offices can negotiate Memorandums of Agreement (MOAs) with states possessing approved Voluntary Cleanup Programs. USEPA Region V has taken the lead, signing such memorandums with Michigan, Minnesota, Wisconsin, Indiana, and Illinois. (Ohio's MOA is currently being negotiated.) These MOAs generally provide that USEPA will not, absent extraordinary circumstances posing an imminent danger to human health and the environment, initiate any federal action under CERCLA for sites that have been successfully addressed under the state's Voluntary Cleanup Program. Region V is expected to award state and local capacity-building site-assessment grants to the Illinois USEPA, the Michigan Department of Natural Resources, the Minnesota Pollution Control Agency, and the Indiana Department of Environmental Management.

In those states that do not have a voluntary cleanup MOA with USEPA, an applicant that successfully completes a Voluntary Cleanup Program may attempt to gain assurances regarding federal Superfund liability by obtaining a USEPA "comfort letter." Comfort letters are issued on a site-by-site basis and generally state that USEPA intends to take no further action at the site. In states that have signed voluntary cleanup MOAs with USEPA, obtaining an individual comfort letter is unnecessary.

Other regions outside Region V are now moving ahead with MOAs. For instance, in April 1996, Region VIII signed an MOA with Colorado. One month later, USEPA's Region VI signed an MOA with Texas. The Texas agreement is the first in the nation in which the USEPA pledges it will not take enforcement action under RCRA or CERCLA. To be sure, all of USEPA's regional offices are doing much to further the Brownfields Initiative—from entering MOAs to sponsoring regional pilots.

Conclusion

Through the Brownfields Initiative, the Clinton Administration firmly believes that environmental cleanup can go hand-in-hand with economic development. To encourage brownfields cleanup and redevelopment, the federal government has made significant administrative changes to the Superfund program, has expanded its innovative pilot program, and has formed partnerships at all levels. As Administrator Browner stated in announcing the Brownfields Action Agenda before the United States Conference of Mayors, "The consensus approach has worked well for this Administration. Let's show this nation how the job should be done—how we can revitalize our cities, how we can create good jobs, how we can preserve the good areas outside our cities."

Notes

1. This chapter represents the personal views of Mr. Weiss and is not necessarily the policy of the United States government.
2. 54 Fed. Reg. 34,235 (1989).
3. 60 Fed. Reg. 34,690 (1995).

4. *Id.*
5. 60 Fed. Reg. 34,792 (1995).
6. Memorandum from Steven A. Herman, Assistant Administrator, USEPA Office of Enforcement and Compliance Assurance, Policy on CERCLA Enforcement against Lenders and Government Entities That Acquire Property Involuntarily (Sept. 22, 1995).
7. 57 Fed. Reg. 18,344 (1992).
8. Kelley v. Environmental Protection Agency, 15 F.3d 1100 (D.C. Cir. 1994), *cert. denied sub nom.* American Bankers Ass'n v. Kelley, 115 S. Ct. 900 (1995).
9. 60 Fed. Reg. 46,692 (1995).
10. 12 U.S.C.A. 2901 *et seq.*
11. 60 Fed. Reg. 22,156, 22,160 (1995).

PART II

Legal, Business, Financial, and Political Issues Associated with Redeveloping Contaminated Property

~

CHAPTER 4

Doing the Brownfields Deal

KEVIN D. MARGOLIS
TODD S. DAVIS

The brownfields deal is unlike the traditional real estate transaction. At first, it may look like any other real estate transaction—a simple acquisition, often of older industrial or commercial real estate. The usual rules apply: (1) identify a site, (2) identify a use for the target site, (3) establish a value for the property, (4) make an offer to purchase, (5) negotiate terms and conditions and a purchase price, (6) close the deal and take title, and (7) proceed with development. The brownfields deal makes all of these seemingly simple steps much more complicated and time consuming. As a result, the time involved from start to finish, the uniqueness of each deal, the number of individuals from various disciplines who must participate, and the overall complexity of the transaction requires a team approach to the project. Perhaps the most important part of the brownfields deal is identifying the team, pulling the team together, assigning tasks to the various team members, and then assuming the role of "grand marshal" over this puzzle-like parade of professionals.

Valuing the Target Site—Setting a Baseline for Negotiating the Purchase Price

The primary consideration in all brownfields redevelopment projects is not the level of contamination at a target site, nor is it the methods and costs of environmental remediation that may be required to bring the site to full utilization. Rather, the first and foremost concern is the projected underlying real estate value of the property in a "clean" state, once the presumably underutilized or out-of-use target site is returned to the normal stream of commerce.

Often, this simple concept is either ignored or addressed at a point in the brownfields transaction when significant site-investigation costs have already been incurred. *If a site has no intrinsic real estate value, no amount of environmental cleanup will redeem it.* The brownfields redevelopment industry has had a difficult time with the real estate value issue, usually as a result of confusing the goals of public and private interests. Governments, particularly local governments, are interested in spurring redevelopment of abandoned urban property in the central cores of their cities. As a result of this focus, governmental development incentives and administrative cooperation are biased toward sites of this type. However, a common problem with these sites is that they lack true economic value, even if they are pristine. The market demand is low, and generally future demand is predicated on some projected urban revitalization that is sometimes nothing more than fervent hope. In contrast, the private redeveloper is focused on real estate value that, if not tangible in reference to nearby sites, is readily predictable or calculable and can be measured like any other traditional business risk. In other words, governments tend either to hope for value or define value in noneconomic terms; private developers tend to predict value based on real facts and almost exclusively based on economic considerations.

To make the valuation determination, the brownfields redeveloper must analyze several important factors:

- Current market value of similarly situated real estate
- Projected market value of similarly situated real estate and the target site once environmental remediation and redevelopment is complete
- Projected cash flow of the target site during environmental remediation and redevelopment, and when the redevelopment is complete

Bearing in mind that it will not be unusual for many of the members of the brownfields team to fill multiple roles, the brownfields redeveloper, or the "team leader," must identify an individual with the appropriate skill set to make this projected valuation determination. If the team leader does not personally possess the skills to carry out this task, a professional real estate appraiser or a real estate consulting firm should be used.

Care must be taken in this valuation exercise to consider the stigma that may linger and remain associated with the target site once it is redeveloped. It is not unusual for the value of sites, once environmentally impacted but presently free from hazardous contamination due to a cleanup, to suffer as a result of premediation status.[1] Although the real estate market likely will continue to absorb and understand environmental remediation over time, at present there is enough misinformation and disinformation remaining in the market, particularly residing with traditional real estate lenders (discussed below), to make consideration of the environmental stigma effect an important part of any valuation analysis of a brownfield.[2]

The projected valuation exercise is undertaken in order to establish one element of the potential purchase price for the brownfields target, as well as to assist the developer in determining whether the target site is a project with sufficient economic value to pursue. Other elements or pieces of information must be developed before a realistic offer to purchase can be made for the target site. However, once the team leader has completed the initial valuation and concluded that it is likely the target site represents enough potential value to make it worthy of pursuit, it is the time in the process to begin taking steps to make a deal.

Evaluating the "Environment" Surrounding the Deal

The environment surrounding the potential brownfields deal is critical to its success. A troublesome seller or a hostile local community or regulator can make a deal that, on paper, seems like a great opportunity, more likely a deal to be passed by. Before approaching the owner of the target site, it is useful to consider the character and sophistication of the owner as the proposed acquisition is structured. In general, there are two types of brownfields owners, and each presents different obstacles and opportunities in structuring the brownfields deal:

- *Unsophisticated Owners.* These owners are usually afraid of the environmental issue, have little or no environmental data about their site, and want to get out of their property in a clean, unfettered break. Interestingly, these are also the owners most likely to overvalue their property and require "education" about the costs and liabilities involved in brownfields redevelopment.
- *Sophisticated Owners.* These owners come to the table ready to negotiate, often have data regarding the environmental condition of their site, and may be interested in structuring a deal that leaves them with some ownership interest or the right to participate in the economic benefit of a successful redevelopment.

The unsophisticated owner may seem to present a better opportunity for the negotiation of a favorable deal for the acquiring entity. However, the inverse is often true. The efforts needed to educate the unsophisticated owner of the true value of his or her property when considering the potential (and potentially open-ended) costs of environmental remediation, the risk of liability for contamination at the site, and the stigma associated with a formerly contaminated site are, at times, very significant. In addition, the unsophisticated owner is usually less likely to possess reliable environmental data regarding the target site and less likely to be willing to participate in the new development of such information for the purpose of a sale. Therefore, the time it may take to initiate and complete an acquisition from an unsophisticated owner will often be substantially longer than an acquisition from a sophisticated owner.

The other important factor to investigate before proceeding with a brownfields transaction is an evaluation of the regulatory and political environment associated with the target site. Certain states and certain local communities may be more or less favorable to the idea of brownfields redevelopment. At times, the "favorable" environments in which to pursue brownfields acquisitions may be jurisdictions that offer statutorily prescribed regulatory incentives (that is, statutory limited liability, no-further-action letters, governmental covenants-not-to-sue, prospective-purchaser agreements, and so forth). However, an examination of the specific locale in which the target site is located may expose a political or community environment that, notwithstanding statutory or administrative incentives, is or could be hostile to the planned brownfields redevelopment.

Political impediments may take the form of a community that will be satisfied only with an extremely conservative level of environmental remediation for the target site, which would be too costly to undertake given the economic value of the property. A target site may also be situated in a community where the local government officials are wary of outside developers with some "newfangled" ideas for the remediation and renovation of a brownfield in their city. The prospect of a meddlesome local official or community group hounding the brownfields redeveloper throughout the redevelopment process with items such as permit issues and requests for information adds a layer of problems and associated costs that may make the target site a potentially undesirable economic proposition.[3] Finally, the brownfields redeveloper should take stock of the level of cooperation he or she (or more likely, the environmental professional working with the developer) expects from environmental regulatory officials (usually at the state level) who are expected to be involved in the remediation of the target site. The deal will be much easier to complete if the brownfields developer has a team member who previously has established a level of respect from the environmental agencies with jurisdiction over the site and can quickly and efficiently develop a dialogue with these regulators, if necessary.

The Outline of the Deal—the Purchase Agreement

Once the target site and its owner have been identified and the site and area surrounding the site have been satisfactorily investigated on a preliminary basis, it is time to make contact and begin direct discussions between the parties. It is advisable to attempt to structure the approach and position the parties so that at the very best, sophisticated principals are speaking directly with each other or, at the very worst, the parties each have proxies that have the ability to digest and understand the complicated concepts associated with environmental liabilities. If a broker or real estate agent is involved, assure the broker that he or she will remain a part of the process, but because of the extraordinary or special nature of a brownfields deal, you would prefer that the broker did not act as a funnel for communications between the parties. Brokers are very helpful, but their focus is on quickly closing a deal and collecting a commission, whereas in the typical brownfields deal the "process" of the transaction (which can sometimes take a long time) is distinctly important and cannot be unnecessarily hurried.

In general, the team member responsible for these negotiations should be a lawyer or an environmental advisor skilled in the nuances of environmental regulation. Once the parties have outlined the basic terms of the deal, an agreement must be crafted to reflect the deal structure indicated by the underlying agreements of the parties. Both seller and buyer will require frequent counsel, not only regarding the language to be negotiated into the purchase agreement, but to explain how to maneuver through the swiftly changing maze of environmental regulations that affects brownfields transactions.

There usually are two immediate choices presented for structuring an acquisition of any real property, including brownfields sites: (1) acquiring an option to purchase the target site, or (2) entering a contract to purchase the target site immediately. Each choice requires relatively the same investment of effort to negotiate and conclude, especially if traditional real estate advice is followed and a copy of the proposed purchase contract is agreed on by the parties and appended to an option to purchase as an exhibit, or the terms and conditions of purchase are included in the option itself. Because the typical outright purchase contract for a brownfield includes a long and flexible due-diligence period for environmental investigation, an option, which itself must be paid for, often provides little benefit to a brownfields redeveloper. The important caveat to this general conclusion is the deal in which a prospective purchaser can quickly negotiate a very low option price, which is deductible from the ultimate purchase price for the target site. In this case, the prospective brownfields redeveloper may be able to tie up the site and develop superior knowledge regarding the costs to develop the site, to the exclusion of other potential purchasers. If an option to purchase is desirable, it is advisable that a "rolling option" be acquired, permitting the optionee the ability to extend the option repeatedly by making additional option payments to the site owner.[4]

The most important provisions in any contract for the acquisition of a brownfield are, obviously, the clauses that deal with the environmental aspects of the transaction. Although the other provisions of the purchase contract in a brownfields deal will usually be similar to those included in a traditional real estate transaction,[5] the environmental provisions likely will be new items on the negotiation checklist for those potential redevelopers familiar with real estate acquisitions, but not the acquisition of a brownfield. Potential purchasers should not expect a potential seller to have as good an understanding of the environmental issues as compared with the real estate issues. As a result, be prepared for lengthy discussions between the parties and the likelihood that a potential seller will need to be offered detailed explanations of the reasons for negotiating positions on these environmental issues. Expert legal advice may be invaluable for managing these negotiations.

As a helpful checklist, identified below are some of the most salient issues that must be negotiated by the parties and a brief explanation for each one.

1. *Access to the Target Site:* The potential buyer must guarantee itself unfettered access to the target site to complete its prepurchase environmental investigation.

2. *Permissible Forms of Environmental Investigation:* The parties must agree on whether intrusive environmental investigation will be permitted (for example, soil borings, wells, and so forth).
3. *Rights and Access to New Environmental Data:* This is an interesting issue. The immediate impulse of most site owners is to require potential purchasers to provide the seller with all environmental information developed in the course of the prepurchase investigation of the target site. Site owners should consider, with advice from legal counsel, whether they truly want this information at this time, or ever, or whether they merely want the right to access or acquire this data in the future in the event the proposed acquisition is not consummated. The legal ramifications, release reporting requirements, and other consequences of having actual knowledge of certain environmental information may warrant resisting the impulse to know everything the buyer knows.
4. *Length of Time Permitted for Environmental Investigation:* The parties must agree on how long a period of time the potential purchaser has to investigate the target site. The seller will uniformly want a short time frame, while the buyer will want a longer time frame; however, potential buyers should consider how a request for a long investigation period may affect the negotiations over purchase price. Instead, a shorter investigation period that includes a requirement for the conclusion of certain tasks (that is, Phase I site assessment, etc.) tied to the right to extend the investigation period may be a preferable alternative. Buyers should carefully consult with their environmental professionals regarding how much time may actually be necessary for the investigation, considering, for example, laboratory turnaround time, inclement weather, and the need for a phased approach to the investigation.
5. *Costs of Environmental Investigation:* In many cases, the costs of environmental investigation will not be strictly shouldered by the prospective purchaser in a brownfields acquisition. To facilitate an acquisition, a site owner may be willing to participate in these costs, particularly if he or she expects to have access to the data.
6. *Future Environmental Liabilities:* This is likely to be one of the most significant and contentious aspects of negotiating the terms of the brownfields deal. The resolution of this particular deal point is often a function of the type of site owner with which the prospective site purchaser is confronted: unsophisticated or sophisticated. In general, the unsophisticated site owner usually wants to terminate his or her relationship completely with the target site on the consummation of its sale, specifically including a release from future liability for historic environmental contamination at the site. In any case, a potential purchaser must be careful not to acquire the liabilities of the seller that relate to *how* the operations at the target site were conducted (that is,

relating to waste disposal and transport practices). The sophisticated owner may be willing to bargain for some of the risk of future liability if he or she understands how the purchaser/brownfields redeveloper intends to mitigate those risks. The prospective brownfields redeveloper must then consider if, as a matter of strategy (which generally translates into a matter of price negotiation), he or she should share, specifically or generally, its plans for mitigating environmental liability and redeveloping the target site with the site owner.

7. *Indemnities:* An indemnification clause provides for the contractual right of one party (the indemnitee) to require the other party (the indemnitor) to pay directly, or reimburse the indemnitee, for costs or expenses related to an item that is the subject of the indemnity provision. In the brownfields deal, this provision will routinely be a principal term to be negotiated as part of the original business deal. Specifically, the parties must come to agreement on the following points:

   - What will be the subject of indemnification, and which events will trigger the right of indemnification?[6]
   - Who will be the indemnitor?
   - Will there be a cap or maximum amount of the indemnification obligation?
   - Will there be a base or a minimum amount of claims, singly or in the aggregate, that must be incurred by the indemnitee before a claim under the indemnity can be made? Will this "base" be shared by the parties?
   - How long will the indemnity last or survive?

8. *Future Cooperation:* The brownfields redeveloper should determine whether it will be useful to negotiate for the future cooperation of the seller. This concern relates to the potential for the pursuit of potentially responsible parties (PRPs), who may have caused or contributed to the environmental contamination at the target site, for the recovery of the costs of environmental remediation. A seller that has agreed to cooperate with the new owner if this avenue of cost recovery is pursued may make later identification of PRPs an easier task.

Environmental Liability—Defining Exposure

In addition to all the time spent reviewing the terms of the proposed acquisition agreement, there is another issue that demands attention at this point in the process. As part of the evaluation of the proposed acquisition, you must determine your legal exposure to the government and third parties for environmental liabilities associated with the target site. Be certain that this assessment contemplates the availability of state[7] and federal[8] programs that may be useful in limiting or entirely eliminating future environmental liability, and that the terms and implications of participating in these programs have been explained.[9]

Environmental insurance should be part of the investigation into techniques or tools to manage future environmental liability relating to a target site. The insurance industry now offers a variety of products that can mitigate and potentially determine or fix the environmental liability of a brownfields developer (or the seller of the target site or the lender to the developer) for historic contamination. There are policies available that can insure against excessive environmental remediation costs, the cleanup of undiscovered contamination, and third-party claims for personal injury or property damage. Many of the terms of these policies often can be tailored to meet the needs of a particular project. The environmental-insurance industry is new and constantly changing to meet market demands. Policy forms are complex, and coverages and exclusions, as well as premiums and deductibles, vary widely from carrier to carrier. As a result, an insurance specialist familiar with real estate development and environmental-insurance policy terminology should be part of this investigation.[10]

Future environmental liability is not a concern only of the parties involved in the brownfields transaction, but is a top priority of any lender that may be considering providing capital to the brownfields developer. Though private capital has begun to enter the brownfields redevelopment market from specialized brownfields lending sources claiming familiarity with environmental liability issues, to date, the traditional commercial real estate lending community has been exceedingly cautious about entering this market, principally out of fear of lender liability.[11] Lender-liability worries arise directly out of concerns that the borrower/brownfields redeveloper may not adequately manage or mitigate environmental contamination at the target site, exposing the lender as the "deep pocket" to be saddled with the resolution of these problems at a later date.

The only solution to lender-liability concerns is information. The brownfields redeveloper must involve his or her potential lender(s) as early as possible in the life of the transaction. The lender must be made a partner in the review of data flowing out of the deal, including information about (1) the magnitude of environmental liabilities that must be resolved, (2) the tools that will be used to minimize the exposure to environmental liabilities, and (3) the plan for remediation. Inevitably, lenders walk away from brownfields deals because they are afraid of the unknown. Working with a lender that will take the extra effort to understand the project, and providing that lender with as much understandable information as is reasonably practicable, distinctly raises the odds that the lender will be there at closing, ready to fund the acquisition and redevelopment.

Finally, do not forget to identify the type of formal entity that should ultimately take title to the target site when the transaction closes. (This decision must be made before entering any purchase or option contract, or the contract must permit the buyer to specify a nominee or designee to take title to the property at closing.) Depending on the applicable state laws, you will want to limit individual exposure to environmental liability by taking title in the form of a corporation, limited partnership, or limited-liability company (LLC). In addition, the structure of the investment into the brownfields deal (for example, active or passive investors, lender preference, etc.) may also dictate the type of entity selected to take title to the target site.

Investigating the Target Site

Once the parties have reached an agreement and the site-acquisition contract has been executed, the "due diligence" period usually begins. Due diligence refers to the efforts expended by a prospective buyer investigating a target site's real estate and environmental characteristics before purchase. It is at this stage in the brownfields redevelopment process when the most technically oriented team member, the environmental consultant, will be engaged.

There are several attributes an environmental consultant for a brownfields redevelopment project must possess. First and foremost is a familiarity with the types of environmental issues that are likely to be investigated and managed throughout the redevelopment process. This is somewhat of a guessing game, because in many cases you do not know what kinds of environmental problems you must deal with until the consultant uncovers them. Nevertheless, make the best attempt to use an environmental consultant who knows how to handle your type of problem. Using a consultant who is unfamiliar with the issues likely will cost you money and probably time. For instance, an environmental consultant who is an expert in removing underground storage tanks or dealing with asbestos may have little or no experience in managing the complex cleanup of an old contaminated industrial building.

A typical solution to the problems associated with a narrowly focused environmental consultant is to consider and hire only those consultants who offer a broad range of services—usually large, full-service companies. Although this approach will probably resolve issues relating to the availability of technical skills, it also may create its own set of problems. Before you conclude that the biggest firm is the best firm, analyze the firm's credentials as they relate to your particular brownfields redevelopment project. In most cases, the firm is only as good as the particular individual consultant assigned to your project. Once you have answered the questions listed below, you may come to the conclusion that a small, local firm focused on brownfields development may be the best choice. Specifically, ask these questions of any prospective environmental consulting firm:

1. Besides excellent technical credentials and personnel, does the consultant have experience with a site that may be similar to the target site?

2. Can the consultant point to examples of projects where "cutting edge" environmental remediation or risk-assessment techniques were used? In a brownfields redevelopment, knowing the availability of—and the capacity to make use of—such technology can create a cost savings to the project that makes the difference between a deal that goes forward and a deal that dies in its tracks.
3. Will the consultant give you a supportable estimate of the costs of environmental remediation? It *is* unreasonable to expect the environmental consultant to give you a fixed cost estimate of remediation costs *before* the full environmental investigation of the site is complete. But will they give you a fixed price *afterwards*, with a reasonable range for error? If the consultant's answer to this question is yes, ask the next question: How? The answer to the second question should tell you something about the kind of consultant you may be selecting. A believable, reasoned explanation of how these costs can be predicted may give you, as the brownfields redeveloper, a "warm and fuzzy" feeling about the professional who is likely to be holding your hand throughout this complicated process.
4. Does the consultant have any practical experience with the brownfields cleanup program in your state, if any? If your state has a Voluntary Cleanup Program, does the consultant know the state regulators specifically responsible for the operation of the program? These kinds of contacts can be invaluable, and can sometimes cut costly weeks and months off the time it takes to remediate the target site.

Do the Deal!

The end of the road approaches: "Closing Date" is here. Hopefully you, as the brownfields developer and team leader, have finished your site investigation and reviewed the results with your team members, concluding that the deal can get done. You have presumably completed the calculus necessary to determine that the combination of site acquisition costs, transaction costs, environmental remediation costs, and development costs leave a sufficient margin of profit to compensate you for your time and risk once the redevelopment is complete. If your project is like most brownfields redevelopments, it is a project that requires vision and reaches outside the "box" of traditional real estate development. Inevitably, you have traveled a long and arduous path to get to closing, only to face the real work of the actual redevelopment ahead.

The environmental issues discussed above that are part of any brownfields redevelopment project are important and can be very complicated, legally dense and, at times, scientific and technical. The potential brownfields redeveloper who is not himself or herself (1) a real estate professional with special expertise in environmental matters, (2) an experienced environmental lawyer, (3) an environmental-insurance specialist, or (4) an environmental consultant/engineer/scientist (with a full complement of skills in geology, hydrogeology, and risk-assessment analysis) must seek the assistance of qualified professional advisors who can provide these skills. Without this expertise or assistance, a real possibility exists that an important environmental issue will be poorly negotiated, badly investigated, or missed entirely. Bringing this kind of specialized professional expertise to offer aid, comfort, and cogent advice can be a costly enterprise, but will inevitably prove to be a worthwhile investment. It will allow the potential brownfields developer the opportunity to either negotiate the best deal possible, driving out all the relevant issues or, and perhaps more importantly, arm him or her with enough information and counsel to know when to walk away from a bad or too-risky acquisition.

Notes

1. *See, e.g.,* Chance v. BP Chemicals, Inc., 670 N.E.2d 985 (Ohio 1996); Westling v. County of Mille Lacs, 543 N.W.2d (Minn. 1996); Desario v. Industrial Excess Landfill, 587 N.E.2d 454 (Ohio 1991).

2. For a more detailed examination and discussion of the issues involved in appraising contaminated real estate, see Chapter 6 of this book, Valuing Brownfields.

3. For a discussion of certain issues relating to a community's response to brownfields redevelopment, see Chapter 15 of this book, Community Participation in Brownfields Redevelopment.

4. See S. SAFT, COMMERCIAL REAL ESTATE TRANSACTIONS (2d ed. 1995); A. ARNOLD, REAL ESTATE TRANSACTIONS: STRUCTURE AND ANALYSIS (1992). Both sources contain detailed discussions of the specifics of options and contracts to acquire real property.

5. *See* SAFT, *supra* note 4; ARNOLD, *supra* note 4.

6. The "trigger" of an indemnity, or the event that gives rise to an indemnifiable claim, is often some of the most intensely negotiated language in a purchase contract involving environmentally contaminated real estate.

7. *See, e.g.,* OHIO REV. CODE ANN. § 3746 *et seq.* (Baldwin 1995) (Ohio Voluntary Action Program, providing for covenant-not-to-sue from state of Ohio for "volunteers" that remediate contaminated brownfields sites); *see also* 415 ILCS 5/58.1–.12 (1995) (Illinois voluntary remediation program, providing liability protection to volunteers in form of no-further-remediation letters). *See also* Part IV of this book.

8. Announcement and Publication of Guidance on Agreements with Prospective Purchasers of Contaminated Property and Model Prospective Purchaser Agreement, 60 Fed. Reg. 34,792 (1995) (publishing *Guidance on Settlements with Prospective Purchasers of Contaminated Property* (May 24, 1995)); Superfund Program: *De Minimis* Landowner Settlements, Prospective Purchaser Settlements, 54 Fed. Reg. 34,235 (1989) (publishing *Guidance on Landowner Liability under Section 107(a)(1) of CERCLA, De Minimis Settlements under Section 122(g)(1)(B) of CERCLA, and Settlements with Prospective Purchasers of Contaminated Property* (June 6, 1989)). *See also* The Effect of Superfund on Lenders That Hold Security Interests in Contaminated Property, 6 Fed. Reg. 63,517 (1995).

9. Terms and conditions may include property-use restrictions or fees or premiums paid to government agencies in exchange for agreements limiting liability. *See, e.g.,* 60 Fed. Reg. 34,792 (1995); 54 Fed. Reg. 34,235 (1989); 6 Fed. Reg. 63,517 (1995).

10. See Chapter 11 of this book, Environmental Insurance in the Brownfields Transaction, for more information regarding environmental insurance.

11. See Chapter 8 of this book, Brownfields Sites: Removing Lender Concerns as a Barrier to Redevelopment, for a detailed discussion of the issues relating to lender liability for environmental contamination.

CHAPTER 5

Acquisition Considerations for Brownfields Properties

ADAM FISHMAN
B. ROBERT AMJAD

Many opportunities exist for individuals or entities interested in brownfields redevelopment. These opportunities arise due to the decrease in value associated with properties labeled environmentally contaminated. This investment opportunity most closely resembles an "arbitrage" between the value of a contaminated site and the value of a greenfields site. As defined in the financial money markets, arbitrage exists whenever two things that are essentially the same are priced differently.[1] Brownfields redevelopment requires an extra step to reach the arbitrage opportunity, namely converting a brownfield into a greenfield. This chapter will refer to this scenario as "brownfields arbitrage." Thus, in brownfields arbitrage, the investor hopes that after remediation the fair market value of the sites will at least equal the original purchase price plus any costs associated with cleanup, including rezoning (if necessary) and other associated costs, and a reasonable return to the investors.

As with any investment, the investor's desired risk level will help determine which property is the most suitable venture. Thus, it is useful to study the different risks associated with brownfields investing. This chapter examines the types of brownfields properties available for redevelopment, identifies the risks associated with these properties, and explores the returns needed to justify the investor risk. Our goals are to help brownfields investors understand the intrinsic risks associated with investing in brownfields redevelopment projects and to build a framework from which investors can negotiate with owners of contaminated sites.

Types of Properties Ripe for Redevelopment

Brownfields sites can be found along the entire spectrum of real estate, with sites ranging from industrial and warehouse real estate to shopping centers and residential real estate. The successful investor in brownfields must understand the relative risks associated with investment in each of these properties.

Two components account for most of the risk associated with brownfields redevelopment. This section addresses the first component, the risks of remediation (particularly the costs of remediation), and a subsequent section addresses the second component, the risks of real estate speculation.

The environmental risk associated with properties *generally* falls along a continuum as shown in Figure 1. The risks associated with certain categories of these properties may be obvious. However, the table shown in Figure 2 briefly reviews the likely contamination risks of each property type. These risks arise from either the activities conducted on the property or careless disposal of contaminants throughout the property's historic use. This chart is meant only as a guide and does not summarize all potential contaminants (such as asbestos, for example) on the various types of properties, as these issues are covered more thoroughly elsewhere in this book.

The spectrum of potential contamination should help determine which properties are ripe for redevelopment. Typically, the largest arbitrage opportunities exist with two types of sites: those with marginal contamination and those with more extreme contamination—Superfund-type sites. Several factors are important in determining which sites are the most viable for redevelopment: the purchase price, rezoning costs, remediation costs, and marketing costs. Finally, the investor must add an appropriate return for the risk undertaken.

These factors are represented mathematically as follows:

(Formula 1.0) $\qquad S + C + R = FMV$

where S equals the speculator's costs, such as the purchase price, costs to rezone (as many of the redevelopment opportunities require land-use changes), taxes, insurance, and any other fees associated with purchasing and owning real estate; C equals the cleanup costs; R equals the investor's required return; and FMV equals the current fair market value of the site if not for the contamination and if the site has the appropriate zoning. Thus, the purchase price of the brownfields site plus all the costs associated with transforming the property into a viable, sellable, or leasable property, including investor return, must—at most—equal the fair market value of the property if it was not contaminated and if it had the appropriate zoning.

It is difficult to label any one factor as most important because the costs associated with any individual factor may be enough to make a project unprofitable. For example, suppose a contaminated property was selling at a 50 percent discount from market value. The purchase price is $500,000 for a site worth $1 million, if clean. Cleanup costs are $300,000 and the cleanup will take one year to complete. The investor's required rate of return is 20 percent. The site must be rezoned to meet the use the market is demanding. Assuming

FIGURE 1. Likelihood of Site Contamination Based on Past Use

More Contaminated								Less Contaminated
Heavy Industrial	Auto/Truck Fueling Facilities	Light Industrial	Retail	Warehouse	Agricultural	Medical Office	Commercial Office	Residential

FIGURE 2. Contamination Risks by Property Type

Property Type	Contamination Risks
Heavy Industrial	Heavy metals, chemical wastes
Auto/Truck Fueling Facilities	Hydrocarbons, chemical wastes
Light Industrial	Heavy metals, chemical wastes
Retail	Various chemical wastes (dry-cleaning fluids, for example)
Warehouse	Varies largely with what materials were stored in warehouse
Agricultural	Pesticides
Medical Office	Medical wastes
Commercial Office	Various chemical wastes
Residential	Various chemical wastes

a cash purchase (to simplify the example), the cost of rezoning the property cannot exceed $32,500, or the project will not meet the investor's goals. Similarly, if remediation costs exceed $300,000 by more than 2.5 percent and zoning costs equal $25,000, then the project will also not meet the investor's financial goals. This scenario also assumes rezoning can occur during the time frame of the remediation.

The foregoing discussion focuses on vacant land. However, the discussion remains valid for brownfields currently producing income. These properties add one more factor to the analysis, namely the income stream generated by the property. The above formula becomes

(Formula 2.0) $\qquad S + C + R - I = FMV$

where I equals the income stream from the property.

Whether a property currently generates income should not significantly affect the analysis. The same steps should be taken, but the investor should realize that income-producing property offers more flexibility. Whether the cash flow generated by the property is used to fund the remediation or to return equity to the investors, the income will increase the rate of return of the project, thus making the investment more attractive.

Two opportunities exist for income-producing properties: (1) real estate with lease rates below market rates due to the contamination, and (2) properties with lease rates at market rates, but with sale prices below fair market value due to contamination. In either case, the assumption is as above: contaminated properties should sell at a discount, creating a brownfields arbitrage opportunity for investors able to control certain critical costs and risks.

Additionally, federal, state, or local governments may offer incentives to brownfields investors. These programs range from tax abatements to remediation assistance and are discussed elsewhere in this book. Investors should aggressively pursue these incentives because they are analogous to income from income-producing brownfields. The incentives will lower costs (mainly property taxes), thus increasing return over the project's life.

This discussion highlights properties ripe for redevelopment. Ultimately, the larger the discount for contamination relative to the actual contamination level, the higher the return for the project, regardless of whether the property is currently income-producing. The underlying assumption is that "tainting" or "stigmatizing" a property as contaminated will automatically lead to a discount in price. Thus, within a historic-use class, discounts for contamination should be similar when no information about the level of contamination is available. As a result, marginally contaminated properties in any historic-use class may present the greatest brownfields arbitrage opportunities. This opportunity is created primarily from the stigma of being labeled contaminated. Market perceptions regarding the appropriate discount for contamination will determine the viability of brownfields redevelopment in a given market. For example, if all contaminated properties in the historic-use category of light industrial receive a 25 percent discount in value, this discount will disproportionately affect those properties with marginal contamination. This theory should apply to all historic-use classes.

Consequently, it is reasonable to assume that similarly situated contaminated properties within the same historic-use class should, before due diligence, have roughly the same market value per acre. Thus, theoretically, an opportunity exits if investors can locate properties within a historic-use class that have relatively little contamination but suffer a large decrease in value merely by being labeled as contaminated sites. Relative to other contaminated properties in that historic-use class and in that market, these properties should suffer a disproportionate decrease in price. Once the actual level of contamination is determined, the arbitrage opportunity within a class will decrease, as that knowledge begins to "perfect" market inequities.

In general, minimally contaminated properties within a historic-use class, with the appropriate zoning already in place, should offer the largest arbitrage opportunity. These properties will incur less cleanup costs and no rezoning costs. As a result, investors will be able to recover the difference between actual and perceived value with the smallest amount of risk.

Another arbitrage opportunity exists with severely contaminated properties. These properties are the most contaminated and will require the most complex remediation techniques. The opportunity exists here because very few investors will pursue these opportunities and very few owners wish to own these properties. Thus, the market becomes a buyer's market. Inherent in this market is substantially more risk because, presumably, the more complex the cleanup, the more likely that cost overruns, time delays, and other issues will occur that impact the return on the project. Additionally, properties at this end of the brownfields spectrum may actually have a negative value. This occurs either because remediation expenses exceed the fair market value of a clean site, or for whatever reason, a company is willing to pay the investors to take the property. Regardless, at this end of the market, investors *must be wary* of higher returns that do not compensate for the increased risk.

The chart shown in Figure 3 highlights the above discussion. Basically, brownfields arbitrage opportunities should decrease as the complexity of the remediation increases until a point at which the remediation becomes so complex that very few, if any, have the risk tolerance or the technical expertise to remediate the site satisfactorily. At this point, the arbitrage opportunity begins to increase. However, as remediation complexity increases, the risk involved in the project should also increase. The risk associated with highly complex remediation may, in practice, exceed even the relative estimates below.

Return Needed to Justify Investment in a Brownfield

Brownfields investors should evaluate the investment decision as they would any other investment. Specifically, investors should determine whether the projected return on the investment adequately compensates the investor for the risks of the investment.[2] An example will clarify this statement: Two bonds mature at the same time, one a thirty-year government bond and the other a thirty-year corporate bond. The government bond yields 5.5 percent while the corporate bond yields 9 percent. The 3.5 percent premium paid for the corporate bond is compensation for the additional risk associated with the corporate bond. In brownfields redevelopment, the same scenario should

FIGURE 3. Brownfields Arbitrage Opportunity

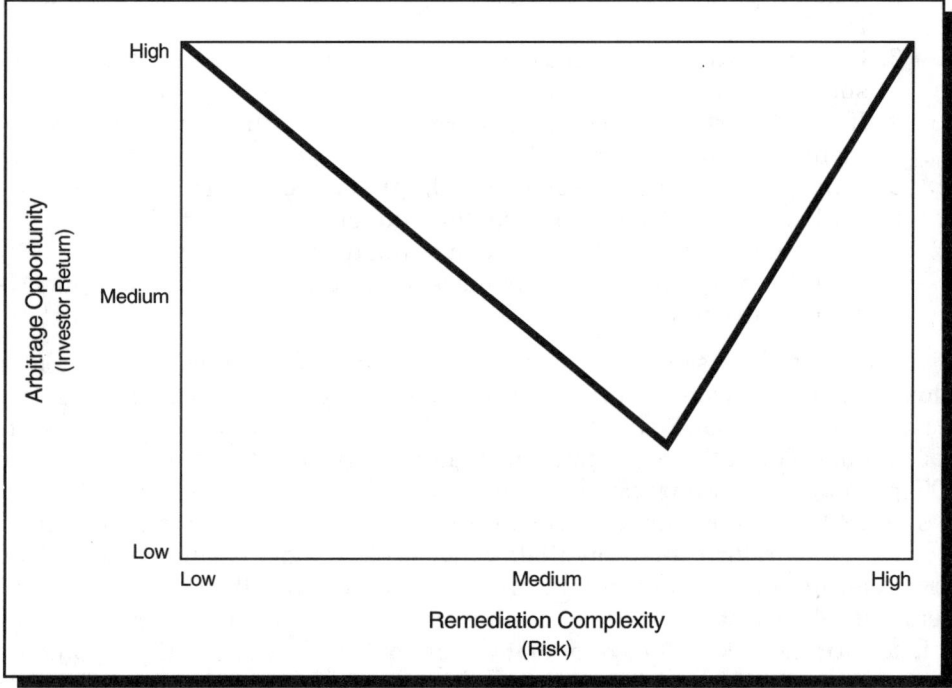

hold true. If a greenfields speculator requires a certain return, the brownfields speculator should require this return, plus an additional return to compensate for the risks associated with remediation, rezoning, and various other factors.

A baseline would be helpful to evaluate the return an investor should desire. Large-company stocks returned an average of 12.5 percent and small-company stocks averaged a 17.7 percent return over the seventy-year period from 1926 to 1995.[3] During this period, inflation averaged 3.2 percent.[4] Thus, if relatively liquid investments return 12.5 percent, illiquid investments, such as real estate, should expect commensurately higher returns due to illiquidity.[5] Further, brownfields investors should expect additional compensation for shouldering the risks associated with remediation and land speculation.

This said, the required rate of return for an investment in a brownfield must be determined by each investor individually. The more risk averse the investor, the higher the rate of return required to compensate for the project's risk. Investors should be aware that seemingly large returns could mean that the investor has merely underestimated the risk involved. As a result, investors must be careful to assess the risk of the investment accurately to ensure they are properly compensated for the risks being taken. Therefore, the following discussion focuses on elements of risk associated with investing in brownfields.

Generally, the major risks are as follows:

1. Will actual cleanup costs meet the budget?
2. Can the property be remediated in compliance with state and federal laws?
3. Once remediation is completed, will the state issue a covenant-not-to-sue?
4. If the state does issue a covenant-not-to-sue, will the federal government honor this determination?
5. If the property needs to be rezoned, can it be rezoned?
6. Will actual rezoning costs meet the budget?
7. Will the remediation take place in a timely manner?
8. Will the property regain fair market value after remediation and rezoning are completed?

Basically, the risks can be described as those of a real estate speculator plus the risks associated with environmental remediation. As a result, the return investors demand should be in excess of those required by real estate speculators. Typically, real estate speculators demand returns of 50 percent to 100 percent, depending on the perceived risk of the investment. Thus, a brownfields investor should start by evaluating the site as a greenfields site.

Required return for brownfields investors is most easily calculated by first determining, as a percentage, the return required if the site was a greenfields site. This factor, including zoning risks, will equal speculation risk (S_r). A factor for the riskiness of remediation, or environmental risk (E_r), should be included, as should a factor assessing the risk of the stigma or taint (T_r) continuing to affect the value of the property. A fourth factor, for the likelihood of obtaining a covenant-not-to-sue, should also be included (C_r).

Once risk factors are determined, the required rate of return for the project can be calculated. The required rate of return for the project is the weighted-average required rates of return for each component of risk. An example of a project shown in Figure 4 will illustrate this calculation. The total expected cost of the investment will be $1,600,000. The weighted-average required return for each component is calculated as follows:

(Formula 3.0)

(Required Return for Individual Risk Component) ×

[(Estimated Component Cost / Total Estimated Project Cost) × 100]

= Weighted-Average Required Return for the Individual Component

Using the above example, the premium required for greenfields speculation is equal to

S_{wr} = (50%) × [($500,000/$1,600,000) × 100]

= 15.625% (where S_{wr} equals the weighted-average required rate of return for S_r)

Figure 5 shows the weighted-average required return for each risk component. Thus, by adding the weighted-average individual risk components, the

FIGURE 4. Required Return by Risk Component

Risk Component	Estimated Cost	Required Return for Individual Risk Component
Speculation (S_r)	$500,000	50%
Environmental (E_r)	$1,000,000	90%
Taint (T_r)	$50,000	10%
Covenant-not-to-sue (C_r)	$50,000	10%

FIGURE 5. Required Return for Project

Component Risk	Weighted-Average Required Return
Speculation (S_{wr})	15.625%
Environmental (E_{wr})	56.25%
Taint (T_{wr})	.3125%
Covenant-not-to-sue (C_{wr})	.3125%
REQUIRED RATE OF RETURN FOR THE PROJECT ($S_{wr} + E_{wr} + T_{wr}) + C_{wr} = P_{RR}$)	72.50%

investor can calculate that the required rate of return for the entire project equals 72.50 percent.

Evaluating these risk components is the most challenging part of investing in brownfields. Several other components, less difficult to evaluate, will be addressed first. The risk of real estate speculation should be relatively easy to determine. Real estate speculation is pursued regularly in almost every market. Thus, recent speculatory transactions in the market should be a good indication of the return one should expect. Obviously, this factor will be greater for vacant land than for current income-producing properties. Zoning risks, which are part of greenfields speculation risk, can also be evaluated fairly easily through inquiry with local municipalities.

Remediation risks, however, are more difficult to determine. A qualified remediation firm should be able to provide an estimate of both the cost of remediation and probability of success. However, these estimates vary in accuracy depending upon the skill of the professional making the estimates and the complexity of the remediation. Investors would be wise to include a significant allowance for cost overruns, as these will likely make or break the investment. Alternatively, investors may be able to get the remediation firm to guarantee the estimate to eliminate cost-overrun risk completely. This "insurance policy" will cost extra, but it is probably money well spent.

The last two issues are more difficult to gauge. Will a site regain its entire fair market value upon successful remediation? Theoretically, it should. In reality, the answer will be determined by perceptions and attitudes in each local market. The risk that an investor will be able to resell the property at the presumed fair market value could be substantial. The final factor, the risk of not receiving a covenant-not-to-sue or of the federal government not recognizing a state's covenant-not-to-sue, is also difficult to assess at this time. Tightly written contracts with the environmental remediation firm can ensure that the firm is required to remediate to state standards, thus providing protection if a state will not issue a covenant-not-to-sue. Again, this "insurance policy" is money well spent if a firm offers it at a reasonable price. However, the question of whether the federal government will recognize a state-issued covenant-not-to-sue has not been fully answered. Pending legislation may eliminate this risk, but the timing and coverage of the legislation is not guaranteed until the legislation is passed.

Another crucial component of the investment decision-making process is timing. Basically, investors must estimate the duration of the holding period to arrive at an accurate project evaluation. Obviously, the length of the holding period will affect the investor's rate of return. Again, the remediation process will be the most likely determinant of the holding period's duration. The greater the remediation needed, the longer the holding period. Also, general inquiries into the difficulties of, or the possibility of, rezoning should be made during the due-diligence phase to determine the time needed to achieve rezoning.

Once the property is controlled, the rezoning process should begin immediately. Time savings may also occur by marketing the property as soon as a remediation completion date is reasonably certain. This approach will enable investors to recoup their investment more quickly, thus increasing the rate of return.

This discussion emphasizes the need for exit strategies from the investment. Because investors would like to return real estate to its highest and best use as quickly as possible, exit strategies must be contemplated before purchasing the asset. The typical decision will be whether to hold and lease the property or sell the property. This question will hinge primarily on the perspective of the investor. If the investor is, or desires to become, a real estate operating company that owns and manages real estate, then the hold-and-lease strategy should be the best choice. If the investors desire only to arbitrage brownfields, then the investment goals are best served by selling the clean properties and purchasing new brownfields properties.

Exit strategies must include contemplation of a failed remediation. This failure can occur in at least two ways: (1) the remediation does not satisfy the state authorities, or (2) the remediation does not satisfy the federal authorities. As stated above, if the state has brownfields redevelopment legislation in place, the investor should include a contractual obligation that requires remediation satisfactory to obtain a covenant-not-to-sue as a requirement of the contract with the environmental remediation firm. However, until the federal government promulgates its own brownfields redevelopment legislation, or embraces each state's regulations, the disposition of the federal government toward remediated sites is a risk that must be evaluated. Also, unsuccessful remediation, even with such a contractual clause, may significantly extend the holding period and turn a profitable project into a money loser. Failed remediation, for any of the above reasons, may eliminate any exit strategies from the investment.

Brownfields Investment Strategy

Investors must understand and estimate many risks before investing in a brownfields redevelopment project. Specifically, legal uncertainties, budget variances, timing, ability to obtain a covenant-not-to-sue, and the market for the property must be considered. Without an evaluation or an understanding of the risks associated with the project, investors will be unable to evaluate the project objectively against the risks and returns of other potential investments.

The required rate of return is determined by Formula 2.0. This rate of return can then be used to calculate the net present value (NPV) of the project.[6] If the NPV is zero or greater, the project is returning the required rate of return or greater and should be accepted. The above cost estimates can also be used to determine the internal rate of return (IRR), or yield, for the project.[7] These values can then serve as the basis for objective comparison with other investment opportunities.

An example will more clearly illustrate this approach. The NPV, using the required rate of return established above, will assist the investor in this decision. Using Figure 4, a property can be purchased for $500,000. The cost to remediate the property is estimated to be $1,000,000, and the cost is spread equally over the term of the remediation, which is three years. The cost to overcome any residual taint in the market is $50,000 and the cost to ensure obtaining a covenant-not-to-sue is $50,000. Both these costs will be incurred in the third year. The total project cost is $1,600,000. The project's required rate of return, as previously calculated, is 72.50 percent. Further, the holding period is equal to the length of the remediation. The fair market value of the property is $4 million. On the surface, this looks like a sound investment. The NPV confirms this assumption by resulting in a value of $99,744; thus the NPV is positive and the project should be undertaken. The IRR for this project is 88 percent, well above the project required rate of return of 72.50 percent. However, if remediation costs are estimated to be $1.35 million, then the NPV falls to negative $29,826 and the project would be rejected. Because the NPV is less than zero, the IRR is less than the required rate of return (72.50 percent). In this case, the IRR has fallen to 68 percent. Although this yield seems outstanding, the investors have not been fully compensated for the risks they have assigned to the project. Thus, investors should be wary of adjusting their required rate of returns once they know the IRR for the project. Similarly, investors should be wary of projects offering large returns with seemingly little risk, as it could be that the investors have merely underestimated the risk of the project.

To Invest or Not to Invest

In sum, investors must answer two questions before investing in brownfields. First, do the available brownfields have the same property uses forecasted to be in demand in the market once remediation is completed and, if not, can the property be rezoned to the appropriate use classification? Second, will the buyer's purchase price, remediation costs, rezoning costs, and a reasonable rate of return allow the property to be sold or leased at a competitive market rate? These questions must be answered with a thorough analysis of the risks associated with the project or the investor will not be able to evaluate the investment opportunity objectively against other opportunities in the marketplace.

Notes

1. Stephen A. Ross, *The Arbitrage Theory of Capital Asset Pricing*, 13 J. ECON. THEORY 341–60 (1976).

2. Franco Modigliani & Gerald A. Pogue, *An Introduction to Risk and Return*, 30 FIN. ANALYSTS J. 68–80 (1974).

3. IBOTSON ASSOCIATES, STOCKS, BONDS, BILLS, AND INFLATION 1996 YEARBOOK 33 & tab.2-1 (1990).

4. *Id.*

5. Yakov Amihud & Haim Mendelson, *Liquidity and Stock Returns*, 42 FIN. ANALYSTS J. 43–48 (1986).

6. NPV is a decision-making tool used to decide whether to undertake an investment. The methodology is relatively simple. The analysis consists of calculating the net cash flow for each year of a project. The results are then discounted back to today using the required rate of return as the discounting factor. If the NPV is zero, the project is achieving the required rate of return. If the NPV is greater than zero, the project is providing a return greater than the required rate of return. Thus, an NPV of zero or greater means the project should be accepted. Scott W. Bauman, *Investment Returns and Present Values*, 25 FIN. ANALYSTS J. 107–18 (1969).

7. The IRR is the discount rate at which the present value of the expected cash outflows equal the present value of the expected cash inflows. This is equal to the yield of the investment. In other words, the IRR of a project will equal the discount factor, which if used in the NPV calculation, would result in an NPV of zero. JAMES C. VAN HORNE, FINANCIAL MANAGEMENT AND POLICY, (9th ed. 1992).

CHAPTER 6

Valuing Brownfields[1]

BILL MUNDY

Valuing a property that is impacted by an environmental impairment is much like valuing a property in a takings case. The appraiser is, in fact, dealing with a taking—that is, a loss of some of the rights of use and enjoyment from the full bundle of rights.

The valuation process generally involves three steps:

1. estimating the value of the property as if there were no loss of value due to an environmental impairment;
2. valuing the property, taking into consideration the environmental impairment; and
3. calculating the difference between the two value estimates, which would yield the loss in value due to the impairment.[2]

There may be instances in which the appraiser can calculate the loss in value directly. This is the situation when the loss in value is equivalent to the cost to cure, or cost to control, the environmental problem.[3] This direct method in estimating damages is frequently possible when the contamination is simple, easily understood by the market, and can be quickly and inexpensively remediated. Cases include asbestos removal or containment, and underground storage tanks when leaking and spillage has been minimal and has not migrated beyond the property lines, and when complete remediation can be accomplished efficiently.

Often the direct method does not work for two reasons:

1. The cost to control is greater than the market value of the property. In certain jurisdictions (in takings cases), the just compensation cannot be greater than the market value of the property.
2. Frequently, the cost to control is not permanent. For example, hydrocarbon contamination in the soil might be remediated to meet United States Environmental Protection Agency (USEPA) and State Department of Ecology minimums, only to have the minimums changed sev-

eral years hence. In addition, even though a remediation program meets USEPA and state minimums, there is no assurance it will meet market prerequisites for a clean property.[4]

Estimating Value of Clean Property

To estimate the value of the property in its unimpaired state, the appraiser would use the standard approaches to value (cost, sales comparison, income) to estimate the property's value as if a normal appraisal is being done. It is important to make sure that when the unimpaired-value analysis is done, the market evidence used is not impacted by contamination. For example, in valuing an industrial property in a heavy-manufacturing area where there is a history of contamination, the appraiser needs to determine whether the transaction evidence is of property that is contaminated. Just because a market seems to be functioning normally does not mean that properties have not been adversely impacted because of contamination. There are situations, for example, where a broad geographical area might be affected from air emissions that have resulted in soil contamination. In this area, the soil contamination may be below USEPA standards, but above background standards. In the marketplace, sellers are selling and buyers are buying, with knowledge of the contamination. Although this might be "typical" market behavior for this particular geographical area, it does not mean that the properties transacted are not suffering a price discount due to contamination. One would need to compare these properties with similar properties selling in an area where the soil contamination is not present.

Understanding Contamination

First and foremost, the appraiser must realize that he or she is not an expert regarding environmental contaminants. Environmental consultants are professionals who have the training to understand the nature and extent of contamination. It is important for the appraiser to work closely with the client and the environmental consultant. An important component in analyzing the potential damage to a property from an environmental problem is the cost to control the problem. Environmental consultants can generally provide the appraiser with a number of different possible remediation programs and associated costs.

Another issue that the appraiser needs to understand, and about which the environmental consultant can provide insight, is the time that the impairment will last. That is, is the problem a temporary one or a permanent one? For example, consider the contamination of a piece of real estate by a petroleum product. That product might be gasoline, which is highly volatile, but the majority of it will vaporize and the nonvaporizing part will break down quickly, often through natural biodegradation. This would generally be classified as a temporary problem. On the other hand, the petroleum product might be crude oil, consisting of a high proportion of Polynuclear Aromatic

Hydrocarbons (PAHs)[5]—where one finds carcinogens such as benzene—and asphaltines that degrade very slowly, such as over a period of twenty to thirty years. Appraisers who are familiar with discounted-cash-flow analysis know that events happening twenty or more years in the future (significant changes in net cash flow or resale value) have little effect on present value. This would be considered permanent damage.

Different types of contamination have different effects on market behavior. Lenders are important intermediaries in the transaction process. There are frequently cases when a seller is willing to sell and a knowledgeable buyer is willing to buy, yet the transaction is not consummated because a lender, understanding the risks associated with contaminated real property, will not lend. A recent nationwide survey of lending institutions by Mundy & Associates found that lenders are much more concerned about radioactive material and groundwater contamination than they are with asbestos or electric transmission lines.[6] Figure 1 shows lenders' levels of concern for varying types of contamination.

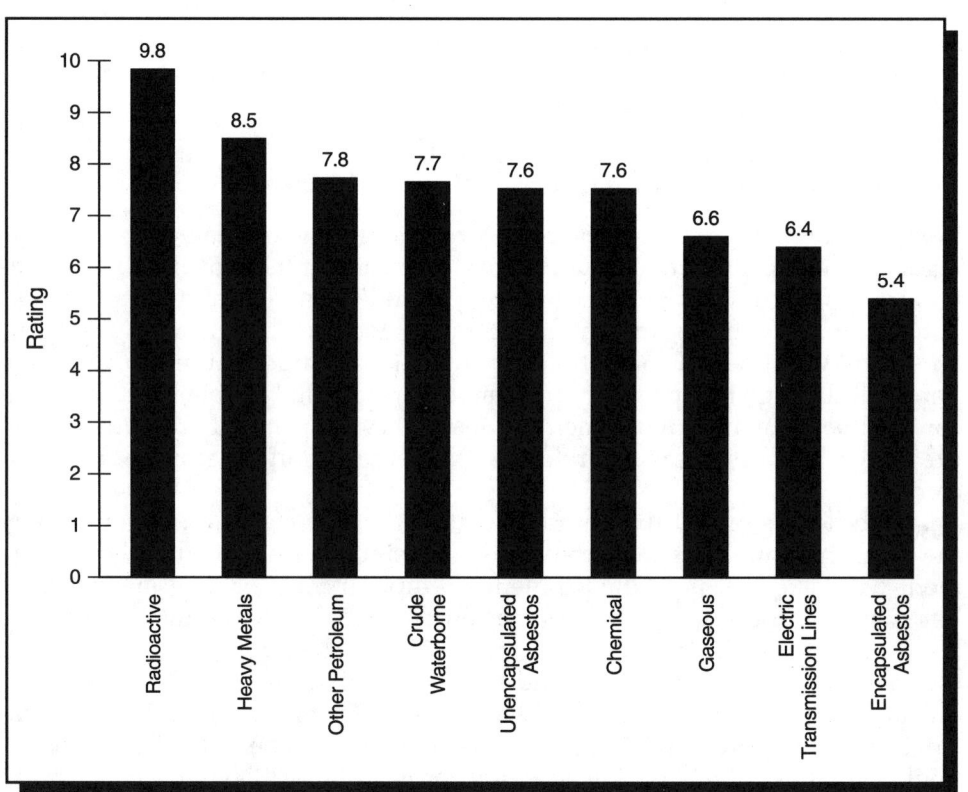

FIGURE 1. Average Ratings of Concern for Selected Property Contaminants

Nationwide research by psychologist Paul Slovic[7] shows that people perceive environmental problems along two discriminating continuums: unknown risk (not observable, unknown to those exposed, effect delayed, new risk, risks unknown to science), and dread risk (uncontrollable, involuntary, global catastrophic, consequences fatal, not equitable, high risk to future generations, not easily reduced, risk increasing). Slovic's research (see Figure 2, Contamination Risk) shows how eighty-one factors align themselves on these two risk scales. Environmental factors are shown as *. It is noteworthy that environmental contaminants rate high on both unknown and dread risk.

Disclosure

Market value is based on the notion of willing and knowledgeable sellers and buyers. Interviews with buyers of contaminated real estate frequently indicate the buyer has either no knowledge of a contamination problem or

FIGURE 2. Contamination Risk

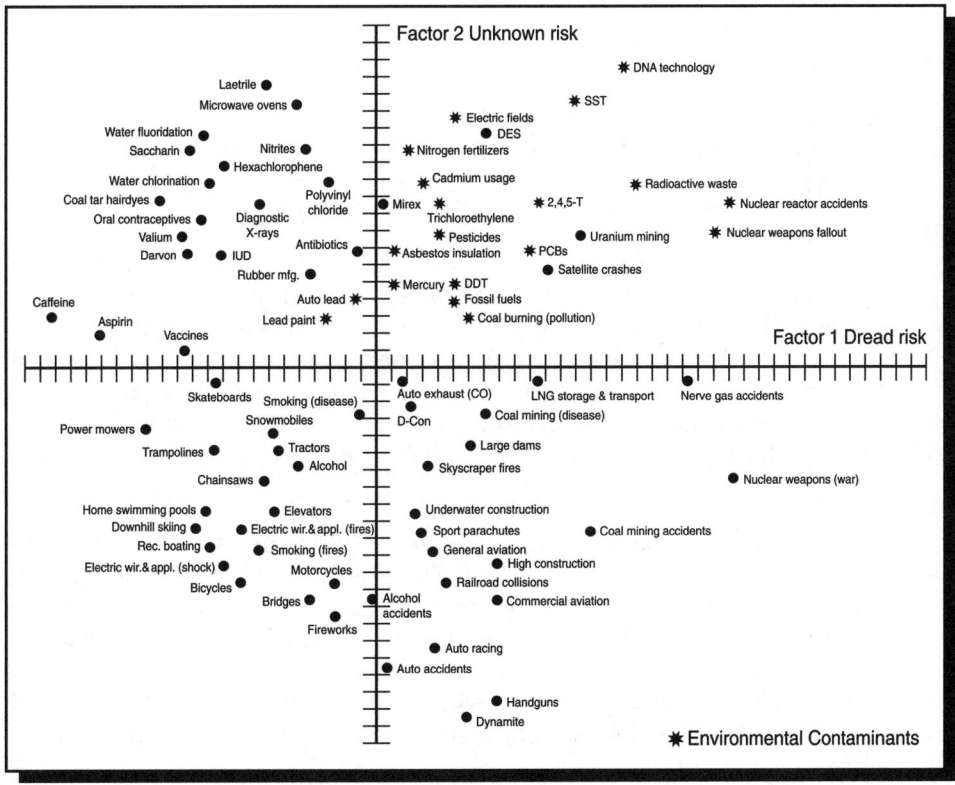

inadequate—or worse yet, inaccurate—information about it. In fact, to minimize liability risk from contamination, polluters will often carry out public-relations programs. However, this practice generally applies to residential property only.

Extensive interviewing in numerous contamination situations reveals that buyers would have behaved differently if full disclosure had been made. And interviews with a cross section of the population indicate that many would exhibit significantly different buyer behavior as more or less information about a contamination problem is provided.

Tenants also exhibit the same behavioral patterns. In fact, a recent survey of thirty-five tenants representing a cross section of tenants (small/large, national/regional/local) in the Sacramento, California, metropolitan area revealed that with proper disclosure (type of contamination, severity, potential cleanup liability, potential collateral impairment), 95 percent would either avoid the contaminated property or negotiate significantly different terms and conditions to protect themselves from liability (indemnification and hold-harmless clauses) or decrease risk (lower rent, leasing concessions).

Stigma

What Is Stigma?

The characteristics of social stigma may be summarized as follows:

> It is the essence of the stigmatizing process that a label marking the deviant status is applied, and this marking process typically has devastating consequences for emotions, thought, and behavior. Many words have been applied to the resulting status of the deviant person. He or she is flawed, blemished, discredited, spoiled, or stigmatized....
>
> The mark may or may not be physical: It may be embedded in behavior, biography, ancestry, or group membership. It may also be possible to conceal it. The mark is potentially discrediting and commonly becomes so when it is linked through attributional processes to causal dispositions, and these dispositions are seen as deviant. Furthermore, the discredit becomes more consequential when the deviant dispositions are judged to be persistent and central and, therefore, part of the marked person's identity.[8]

Environmental stigma is the result of an undesirable event that disrupts the balance of an environmental system.[9] This disruption may cause blame to be associated with it. When environmental features are viewed as repellent, upsetting, or disruptive, they are stigmatized as undesirable. This environmental stigma can result from a number of causes. One source of stigma is technologies, such as petroleum processing, nuclear power plants, and high voltage or transmission lines. A second source of environmental stigma is activities such as the transportation of hazardous materials, the development of hazardous-materials storage sites, or the underground storage of petroleum products. The third source of stigma is within the products themselves, including petroleum-based products and agricultural products associated with a health risk, such as the alar used by the Washington apple industry.

The consequences of an environmental stigma can be direct or indirect. Examples of direct consequences of various stigmas include an increasing incidence of cancer, lower work quality as a result of air or noise pollution, decreased occupancy in an apartment building, or lower market price or increased marketing time for single-family residences located adjacent to a sanitary landfill. However, consequences can also be indirect. The exodus of residents from an area affected by contamination, such as the Love Canal area of New York, and the negative impact on the apple industry in general caused by the few producers that used alar, are two examples.

An environmental stigma results from perceptions of uncertainty and risk. It may be relatively easy to quantify the cost to remedy a simple contamination problem, such as a leaking underground storage tank (LUST). However, as the complexity of the contamination increases, the level of uncertainty and perceived risk rises.

There are a number of variables that influence the degree to which a property may be stigmatized. Also, a stigma may apply to a property that is directly impacted by contamination. For instance, a buyer may refuse to buy or a lender may refuse to lend on a property where there is or has been a LUST, even if the cost to clean up the property is a small fraction of its value. Stigma may also be indirect. That is, a property may only be in close proximity to a contaminated property, and yet the ability to finance or market the property may be impaired.

The following are variables that can influence whether a property is stigmatized and, if so, to what extent.

1. *Responsibility:* This variable deals with the degree to which the responsibility for the stigmatizing event can be squarely placed on the shoulders of a person or entity (Exxon and its responsibility for the Exxon Valdez disaster, for example).
2. *Media exposure:* This is often called "risk amplification." It deals with the extent of media coverage for the stigmatizing event. The greater the level of media exposure, the greater the level of awareness, which has an influence on buyer behavior.
3. *Disruption:* This deals with the extent to which a contaminating event may alter a person's or firm's everyday activities. For example, in a North Tacoma arsenic soil-contamination matter, disruption was high in the area close to the old smelter site because of extensive neighborhood cleanup activities.
4. *Concealability:* This deals with the old bromide, "out of sight, out of mind." It used to be that if a contamination event was concealed, it

had little, if any, effect on market behavior. Today, it seems to be just the reverse. That is, if the contamination is concealable, it has a more dramatic impact on market behavior and, therefore, value.
5. *Aesthetic effect:* This is the degree to which the natural environment, or some other physical feature, is defaced by the contamination.
6. *Prognosis:* The degree to which the contamination problem can be remediated is influenced by two prognosis variables. The first is severity—that is, how bad is the problem? The second is persistence. This deals with how long the contamination problem will be around.
7. *Peril:* This deals with the degree to which human health, or the health of wildlife, might be impacted.
8. *Fear:* This deals with the stress that results because of the contamination. One's physiological health may not be affected; however, one's mental health can be affected by contamination.

In some cases, such as those involving asbestos that can be encapsulated, the level of stigma is minimal. This is a type of problem that can be dealt with knowledgeably. This risk can be quantified. In other cases, the level of stigma can be great. This occurs when the "loading" of the above factors is high.

Courts in more and more states are recognizing that fear can have an effect on property values. A recent study by lawyer Thomas Jaconetty found that in only three states (Alabama, Illinois, and West Virginia), fear was too speculative to be admissible as an agent affecting value.[10] Attention has recently been focused on electromagnetic fields, but there are also legal precedents involving crime (the value of a home where a murder was committed) and road rights-of-way where radioactive wastes are to be transported.

Approaches to Quantifying Impaired Values

The following section discusses the applicability of the three standard approaches to estimating value: the cost, sales comparison, and income approaches. In addition, two approaches that are not frequently employed by appraisers are also discussed: multiple regression and contingent valuation.

Cost Approach

In estimating the value of an environmentally impaired property, the appraiser faces the perennial problem with the cost approach—depreciation. Contamination can be viewed as a form of obsolescence. Whether it is functional obsolescence or external obsolescence will depend on the type of contamination. For example, asbestos might be viewed as a form of functional obsolescence. A migrating plume of hydrocarbons that originates on an adjacent property would be a form of external obsolescence. Regardless of the form of obsolescence, it would be extremely difficult to quantify the amount of obsolescence using standard valuation techniques. And, though it may be possible to quantify the capitalized value of the rent loss, it seems more straightforward to use the income approach, where a large set of variables that may influence value in a contamination situation can be addressed. Second, it is this author's opinion that using the capitalized value of rent loss really is employing the income approach, rather than the cost approach.

Sales-Comparison Approach

There are two techniques in which sales evidence can be used. The first is the direct-comparison method; that is, finding a property with similar contamination that has sold and comparing it directly to the subject property. A second method is to calculate price discounts on properties that have sold; that is, comparing what the property would have sold for without contamination with the amount it sold for because of the contamination, the difference between the two yielding a discount.

A disadvantage of the sales-comparison approach is that there are very few contaminated properties that sell. This often forces the valuer to extend the geographical coverage of the sale search, frequently to a nationwide basis. When this is done, it is much easier to use a percentage discount than a direct comparison, as the percentage discount internalizes locational differences and many other differences that would need to be addressed within the direct-sales method.

Because transaction evidence is so "thin," it is often difficult to use the sales-comparison approach properly. However, all is not lost. The sales-comparison approach can be very useful in showing that the market does discount the price of environmentally impaired property. It can also provide a range of discounts.

Income Approach

For valuing environmentally impaired property, the income approach can be a very useful tool because it is effective in modeling market behavior. This is due to the numerous variables (rent, vacancy, expenses, and rate) that the appraiser uses. The following is a discussion of what an appraiser might hypothesize regarding each of these variables when valuing contaminated property. The key is to test the hypothesis; that is, bring together market data to either confirm or refute the hypothesis.

RENT. Depending on the nature and extent of contamination, the rent stream may or may not be impaired. Unusual noise or smell may make it difficult to rent a property, thereby leaving the property owner with few options other than to decrease rent. Proper disclosure can decrease demand. To offset this problem, it may be necessary to lower rent.

OCCUPANCY (VACANCY). As with rent, occupancy levels may decrease because of environmental problems. At Times Beach, a short distance west of St. Louis, a dioxin-contamination case caused occupancies to drop to zero on a neighborhood-wide basis.

OPERATING EXPENSES. Expenses can increase due to environmental problems. Insurance costs can increase and professional fees, such as the cost of environmental engineers, can increase. Often there are direct costs associated with the remediation problem, such as asbestos encapsulation.

RATE. There are numerous ways that the capitalization or discount rate might be impacted. In situations of uncertainty, when there is not good-quality information on the nature of the contamination, the property may completely lose its marketability. In this case, no rate of return would be adequate to attract investors.[11] In other cases, when the nature of the contamination is known, the investment community can attach a risk rate to the unique character of the property. One method to develop such a rate would be to obtain lender and buyer information (debt and equity rates) that can be used in the band-of-investment model.[12]

RISK LADDER. Another method to approximate contamination is to use a risk ladder, comparing the relative contamination risk to risk associated with other forms of investments, such as corporate bonds, municipal bonds, and high-yield junk bonds. Liquidity differences between bonds and real estate must also be taken into account.

TEMPORARY VERSUS PERMANENT PROBLEMS. Direct capitalization is an inappropriate valuation technique when the contamination problem is temporary. This is the situation when the discounted-cash-flow method should be used to account for the dynamics of time. Figure 3 (Effect of Time on Value) shows how the inputs (rent, occupancy, expense, and rate) may each change over time as contamination effects (cleanup costs, uncertainty, media exposure, and lender concern) change.

If a contamination problem persists for a long period of time and there is little likelihood that market behavior toward the problem will change, then direct capitalization—dividing stabilized net operating income (NOI) by a direct cap rate—is an appropriate valuation method.

Multiple Regression

In situations when large areas are impacted (numerous single-family homes, for example) and when a large database is available for comparable sales evidence (from both the subject area and control areas), then multiple-regression analysis is often an effective tool to estimate damages. This is especially so if the contamination incident happened at a discrete period of time when there was substantial public knowledge of the incident. Three Mile Island would be an example. The drawback in using multiple-regression analysis is that it may not be useful in estimating current loss due to future problems.[13]

Contingent Valuation

Contingent valuation is a survey research technique developed in the 1960s. It is a method to estimate the value of, and especially damages to, natural re-

FIGURE 3. Effect of Time on Value

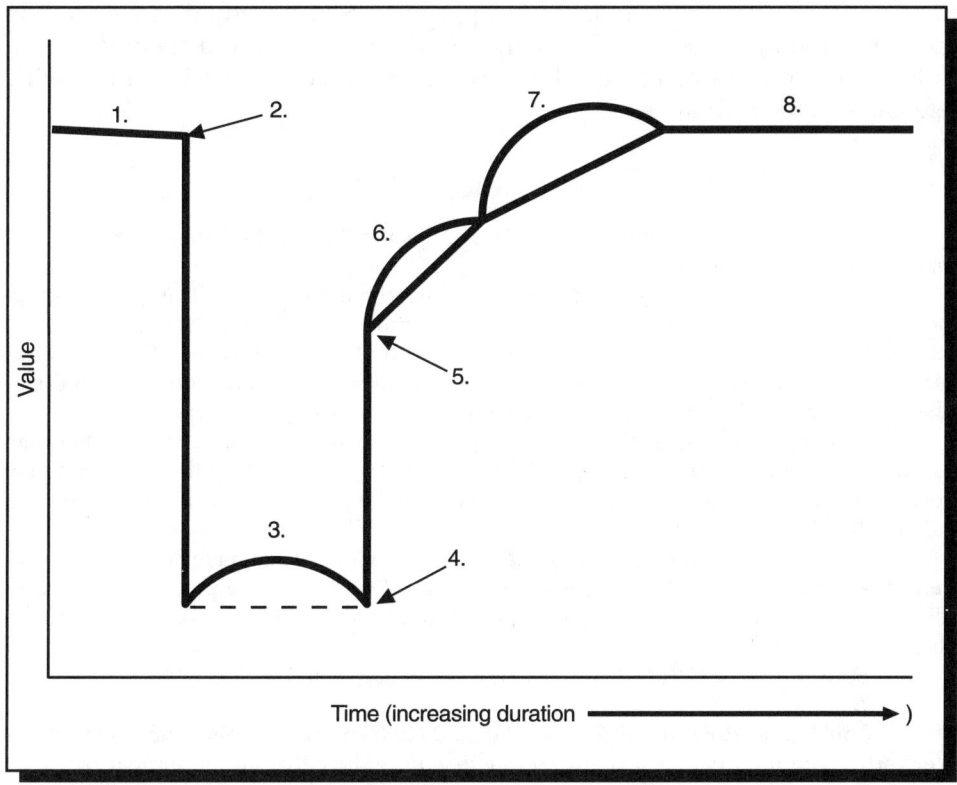

1. Value normal.
2. Contamination problem. Property loses marketability, value drops.
3. Period of uncertainty, unknown value.
4. Environmental investigation complete. Costs and risks documented.
5. Value recovers to point commensurate with remediation costs and risk.
6. Remediation program underway.
7. Temporary stigma decreases to zero.
8. Property value stabilizes, possible permanent stigma.

sources. It has recently received considerable attention, as it has been proposed as a method to quantify damages in oil-spill cases. It has received several years of intense scrutiny by a panel of experts, including several Nobel laureates, brought together by the National Oceanographic and Atmospheric Association. Contingent valuation uses two methods to estimate value, or damages. One is a person's willingness to accept compensation for the loss of a benefit. The second is a willingness to pay something (taxes, for example), to receive a benefit. Examples would be payments for creation of a park, or a payment used to create a cleaner environment.[14]

Conclusion

There are numerous methods available to the real estate appraiser to estimate the value of environmentally impaired property. Some methods are generally familiar to the appraiser, while others are more complex and relatively new. Multiple approaches can be used and when employed properly, can provide defensible value estimates.

Notes

1. This chapter is the modification of a paper submitted to the *Journal of Applied Real Estate Research* for publication.

2. There may be cases in which only the value of the property in its impaired condition might be estimated. The following discussion shows that although this value may be estimated, it is often easier and more defensible to first estimate the unimpaired value, then the impaired value. The Appraisal Foundation has issued Guide Note 8 which establishes some general parameters and pitfalls in valuing contaminated property.

3. Albert Wilson argues that it is more appropriate to use cost-to-control rather than cost-to-cure, as it is frequently not possible to say with certainty that the environmental problem has been cured. Albert Wilson, remarks at the Appraisal Institute Symposium, Philadelphia (Oct. 3–4, 1991).

4. There is considerable market evidence that properties that have been remediated, even when an environmental engineer's research indicates that the property is "clean," still carry a residual stigma. Therefore, what is clean to one is not necessarily clean to another.

5. Soc'y Indus. & Office Realtors, Professional Report (Spring 1995).

6. *Id.*

7. Paul Slovic, *Perception of Risk*, 236 Sci. 282 (1987). Slovic has also published several recent articles on Intuitive Toxicology, adding new insights to this earlier seminal article.

8. Edward E. Jones, et al., Social Stigma: The Psychology of Marked Relationships 4–7 (1984).

9. Michael R. Edelstein, Contaminated Communities: The Social and Psychological Impacts of Residential Toxic Exposure 6 (1988).

10. Thomas A. Jaconetty, *Stigma, Phobias and Fear: Their Effect on Valuation*, 3 Assessment J. 1, 61 (1996).

11. Bill Mundy, *The Impact of Hazardous and Toxic Materials on Property Value: Revisited*, Appraisal Journal, (Oct. 1992).

12. *Id.* at 466.

13. For a discussion of multiple-regression approaches to valuing contaminated property, see William H. Kinnard & Mary Beth Geckler, *The Effects on Residential Real Estate Prices from Proximity to Properties Contaminated with Radioactive Materials*, 16 Real Est. Issues 2 (special ed. Fall/Winter 1991).

14. For a more comprehensive explanation of contingent valuation, see Bill Mundy & David McLean, Valuing Market and Nonmarket Real Property Using the Contingent Value Approach: Natural Resource and Environmental Damage Applications (unpublished manuscript, submitted to *The Appraisal Journal* for publication).

CHAPTER 7

Creative Financing Strategies for Redeveloping Brownfields

DONALD T. IANNONE

The Starting Point: Liability and Investment Risk

More investment does not occur in brownfields because of real and perceived liability and investment risks. These risks are a problem for private investors, such as banks, developers, property owners, and institutional investors like insurance companies. They are also a problem for public-sector investors, including state and local governments, and various types of economic development and redevelopment organizations. Risk is present in any type of real estate investment. These risks fall along a spectrum from reasonable and acceptable to unreasonable and unacceptable. The presence of environmental contamination on a site increases both liability and investment risks. These risks must be lowered to levels that are reasonable and acceptable for increased public and private-sector investment to occur in brownfields.

Brownfields As Redevelopment Investments

Brownfields fall into the category of redevelopment investments, which often possess higher risk levels than new development projects. Redevelopment generally requires two types of activities. The first is renovation and construction, when existing buildings and other structures are preserved and enhanced. The second is demolition and clearance of existing buildings and other structures, and the construction of new improvements on a site. The redevelopment market covers all categories of development, including residential, commercial, industrial, and institutional. Though many brownfields were originally used for industrial purposes, their future-use potential may change as a result of shifting market conditions and other factors. Redevelopment projects are found in nearly all types of American communities, but are most common in larger, older cities that have limited clean, vacant land for development purposes.

In most cities, redevelopment is undertaken through the joint participation of public and private-sector entities. The public sector's participation is contingent upon a social and/or public purpose to be served by a redevelopment project. Job creation, the elimination of blight and poverty, and environmental protection are leading public purposes served by redevelopment in most cities. Businesses engage in redevelopment projects to locate or create production, office, and other facilities required to produce and distribute products and services demanded by society.

Environmental Contamination As a Source of Liability Risk

Government uses liability, rather than regulation per se, as a legal instrument to encourage the reduction of pollution. State and federal environmental regulators use liability to assign responsibility for cleanup costs to various parties directly and indirectly associated with a contaminated site. Liability is viewed by regulatory agencies as a way to (1) deter the mishandling of polluted materials, such as contaminated soil, (2) involve lenders and insurers in the regulatory process to encourage businesses and other entities responsible for polluted sites to comply with environmental laws, and (3) make liability joint and several in nature, which means the government can pursue reimbursement for cleanup costs from all parties directly and indirectly associated with the contamination.

The presence of environmental liability from any of these standpoints is a source of risk, which serves to reduce the probability of investment by lenders and other sources. By their financial participation in brownfields projects, banks are held liable for cleanup costs, which tends to discourage their involvement in these investments. In this sense, these liability risks create investment risks, jeopardizing financial returns on investments.

Lender-Liability Issues

Lenders are risk averse by nature. To minimize their liability related to brownfields, bankers have taken two major steps: (1) they have increased their scrutiny of real estate loans by requiring detailed environmental assessments and other site investigations, which in turn increase transaction costs, and (2) they have restricted their interaction with borrowers, to reduce liability exposure.

Phase I and II environmental site assessments are routinely required for sites with known or suspected contamination problems. These investigations increase the processing time for loans, and they add to transaction costs for borrowers. Both results have the effect of discouraging property owners and developers from redeveloping brownfields, if another site alternative is possible. Lenders also seek indemnification from sellers for any preexisting contamination. These indemnification agreements often address issues concerned with cleanup expenses and the costs of fines, third-party claims, the determination of "reasonable" costs, the definition of cleaning standards related to cleanup costs, limits on potential liability, and various other issues.

Banking Policies and the Lending Culture

Banks are in the process of increasing their environmental expertise to improve their judgment about investment projects with contamination problems. This goal is accomplished through hiring environmental specialists, retaining technical consultants for large-scale projects, and using other resources such as university researchers. This expertise is designed to help these institutions avoid high-risk investments while enabling them to accept reasonable risks in projects with manageable environmental problems.

In the short term, this lending culture is not expected to change substantially, even with increased incentives, such as the ability to gain Community Reinvestment Act (CRA) credits for brownfields loans. In the longer term, however, this reluctance may change as banks' roles in real estate lending become more focused, in light of the growing role of national and international capital markets in most areas of real estate finance. This trend is discussed later in this chapter.

Current policies limit lenders' ability to help brownfields redevelopers. It was common at one time for bankers to help borrowers avoid loan default through loan work-outs. Judicial rulings linking lenders' management assistance to liability have discouraged these institutions from playing this role in brownfields projects. Understandably, banks are now reluctant to foreclose on bad debts because they become responsible for cleanup costs. Lenders now prefer assets other than real estate (such as inventory, equipment, and machinery) as collateral instead of land and buildings, to reduce potential liability. In turn, this approach often makes it difficult for industrial borrowers to provide sufficient collateral to secure a financing deal. Finally, lenders have become more careful in their use of trusts as a source of equity for projects because of similar environmental liability risks associated with management involvement.

Another barrier precluding brownfields finance is the difficulty involved in accurately appraising contaminated property. Appraisal of contaminated property is a complex task, which requires a clear understanding of the pollution problems found on the site, and the impact of these problems on the property's value from a collateral standpoint for lending purposes. As a consequence, the banker experiences difficulty in determining collateral value, and the borrower experiences difficulty in providing sufficient collateral to support the loan request.

A Closer Look at Investment Risk

Four sets of issues must be examined in assessing investment risk related to any redevelopment project. These are relevant considerations for brownfields projects as well. These considerations are as follows: (1) transaction type (debt versus equity), (2) investment security (secured or unsecured), (3) investment liquidity (ability to sell on the secondary market), and (4) knowledge or information about the investment. Typically, the risk associated with an investment is offset by the interest rate charged on a loan. A higher-risk investment usually results in a higher interest-rate charge, unless the loan's security or liquidity can be improved.

Brownfields redevelopment projects offer less of an ability to improve security or liquidity given lender-liability concerns, reduced collateral value, and other problems. This explains in large part why public-sector entities are providing government loans, loan guarantees, grants, and tax credits to reduce the investment risks associated with these properties. The site assessments and investigations serve to improve knowledge about the project investment, and, therefore, reduce the uncertainty of undetected contamination problems and unknown cleanup costs.

Types of Brownfields and Associated Financing Strategies

Although brownfields vary considerably in terms of their levels of contamination, future-use potential, surrounding-area environments, and other characteristics, it is possible to categorize these sites according to two basic financial dimensions: (1) risk of environmental liability, and (2) expected rate of investment return.[1] Redevelopers should devise financial strategies that recognize the condition and reuse potential of these brownfields sites. Recognizing that financing options depend on the economic viability of the redevelopment project, three basic groups of brownfields opportunities may be identified:

1. *Viable sites:* economically viable sites where the private market is already working toward redevelopment without public-sector assistance. These sites have minor contamination problems, and possess favorable redevelopment potential. In general, these sites are seen as having greater advantages than risks, from an investment perspective.
2. *Threshold sites:* marginally viable sites that cannot be redeveloped without some public-sector assistance. Either greater environmental liabilities or lesser redevelopment potential and financial returns distinguish these properties from those projects considered viable.
3. *Nonviable sites:* sites with either overwhelming environmental liabilities or extremely limited economic advantage, from a redevelopment standpoint. Substantial public investment will be required to make these sites suitable for future use. In most cases, these sites should be left alone, unless they pose a major public-health threat.

It is important to recognize that in many cases, sites are only partially remediated before redevelopment. Partial remediation conserves capital resources, and also allows some use of the property to reduce blight conditions and avoid total abandonment of these sites. The partial-remediation strategy is common in both viable and threshold sites, where often some development activity or use of the site exists. Through proper site and area planning, the phased cleanup and reuse of the site is made possible. This approach is desirable to avoid the absolute loss of a business, residence, or other economic asset from the area. It also serves to reduce the potential for stigmatization of the real estate market in a neighborhood or community.

Public and private-sector organizations involved in financing brownfields deals are urged to approach the financing process in a strategic manner.

Not only is cleanup and redevelopment capital scarce, but considerable barriers hamper access to even available public and private finances for site cleanup and redevelopment. The financing of brownfields redevelopment projects is often complicated by complex legal regulatory, engineering and scientific, social, and economic issues, such as those discussed elsewhere in this book. These barriers can be reduced or overcome only through effective strategies addressing these issues as they relate to individual sites and larger-area redevelopment projects.

Understanding the Financial Requirements

In most cases, the successful redevelopment of brownfields requires considerable investment by government agencies and private businesses. These investments must be sufficient to correct site-contamination problems in a way that supports the property's planned future use. From a redevelopment perspective, they must also be sufficient to address surrounding area or neighborhood problems, such as poor transportation access and site acquisition and assembly, which also reduce the economic utility and value of these properties, even after environmental remediation.

Financial requirements for brownfields redevelopment fall into four broad categories: (1) site identification and assessment costs, (2) property cleanup costs, (3) site redevelopment costs, and (4) surrounding-area improvements central to the site's future use. Each of these four cost categories is discussed below.

Site Identification and Assessment Costs

Information and knowledge play a major role in most industries and businesses today. These are essential ingredients in financing and redeveloping brownfields. Phase I and II environmental site assessments are principal information tools used to determine the extent of site-contamination problems. These assessments form the foundation for reliable cleanup cost estimates, which can range from very minor improvements to major capital outlays. In the absence of reliable site-condition information, great uncertainty exists concerning the nature and extent of contamination identified at a site, and the feasibility of proposed future site uses. Beyond the environmental assessments, various financial and market analyses must be conducted to select the most appropriate redevelopment strategies for individual and area-based brownfields projects.

Site investigations require a cash outlay. Phase I investigations are based in large part upon public records and other readily available site and surrounding-area data. These investigations cost in the $3,000 to $5,000 range, depending upon the specific site's characteristics. The cost of Phase II investigations is generally much higher. Phase II costs also tend to be more variable because of differences in contamination levels and the intensity of site analysis required under different state laws. Environmental engineering studies, soil sampling and testing, and chemical analysis are common technical services that must be purchased during a Phase II investigation. Banks require these assessments for borrowers to receive financing, and state and federal agencies require them for the developer to receive development permits and approvals.

Private developers are looking increasingly to federal, state, and local government agencies for help in underwriting Phase I and II investigation costs. This approach is especially true when the redevelopment potential of a property is uncertain due to environmental issues. A number of proposals have been made for federal and state grants to help pay for these preliminary costs. States such as Connecticut, New Jersey, Michigan, and Pennsylvania now offer some assistance for site investigations and assessments. Proposals call for similar types of programs to be created in Ohio and several other states. Site-assessment grants should be targeted to those properties with high redevelopment potential.

Site Cleanup Planning and Execution

Brownfields redevelopers must be prepared to invest in plans to guide site cleanup efforts, as well as the actual costs to clean the site to future-use standards. Capital outlays for these items can be substantial, depending upon the contamination problem and the cleanup method required. Remediation cost estimates vary widely. It is not uncommon, however, for these costs to fall into the $150,000 to $250,000 per acre range. The cost to clean up a five-acre area could be $750,000 to $1,250,000. These are unrecoverable costs from the private redeveloper's standpoint, which suggests that some level of public-sector financing will be necessary to stimulate private-investor interest.

In many cases, costs will be a shared responsibility between the private and public sectors. For instance, Michigan, Minnesota, Wisconsin, New Jersey, and Pennsylvania offer such assistance through low-interest loans, matching grants, and other financial programs. New Jersey's Hazardous Discharge Site Grant Program provides grants and loans to communities and private developers for site cleanup and redevelopment. Massachusetts is developing a loan guarantee program and an Industrial Sites Recycling Program, which would provide low-interest loans for remediation and redevelopment. Ohio is embarking on a new loan program to help finance site cleanup, which will be capitalized by funds from the Ohio Environmental Protection Agency, the Ohio Department of Development, and the Ohio Water Development Authority.

Site Redevelopment Costs

Ultimately, communities and private developers invest in site cleanup to contribute to the reuse and redevelopment of the site in the future. Through redevelopment, both the community and the private investor receive returns on their investment in cleanup. The community's return comes in the form of new job creation, additional tax revenues, improved environmental quality, and other economic and community benefits. Returns to private investors come in the form of profit from their investments.

At one time, environmental expenditures were seen as separate from redevelopment investments. The fine line between the two has become more gray as cities attempt to grapple with both sets of issues through larger community and economic-development processes. This approach has increased the financial resources available to cities for brownfields cleanup and redevelopment. For example, the city of Cleveland has the ability to use a portion of a special community and neighborhood-development fund, called the Neighborhood Development Investment Fund, to help finance site cleanup in highly distressed city neighborhoods. Many cities across the country have committed a portion of their economic-development revolving loan funds, their tax-increment financing authority, and other development finance tools to the correction of environmental problems associated with inner-city sites. The good news is that more brownfields sites are able to receive the public financial support needed to spark private investment. The bad news is that brownfields financing demands have created an added burden for these already limited funding programs.[2]

Surrounding-Area Improvements

Brownfields redevelopers are encouraged to consider the costs of surrounding-area improvements needed to increase the redevelopment potential of brownfields targeted for cleanup and reuse. These costs typically are borne by the public sector, as they are designed to serve a larger public benefit. Cities should examine the current and future market demand for remediated land to help set future capital-budgeting priorities concerning public-purpose land acquisition, infrastructure improvements, and other area redevelopment investments.

Approaches similar to that adopted by the city of Birmingham for redeveloping the North Birmingham community should be considered.[3] Birmingham's approach to brownfields redevelopment is guided by market knowledge, which will help justify future public and private-sector investments in this severely distressed and environmentally challenged community. This market knowledge will also help establish realistic goals in terms of how much contaminated property should be cleaned during the next five to ten years. Moreover, this approach will ensure a proper balance in the financing roles played by city government and the private sector in the brownfields redevelopment process.

The Great Lakes Environmental Finance Center (GLEFC), based in Cleveland State University's Urban Center, is currently engaged in a general comparative study of industrial real estate market trends and the demand for remediated brownfields land in the cities of Cleveland, Chicago, Detroit, Milwaukee, Indianapolis, and Minneapolis. This study is financed by the United States Environmental Protection Agency (USEPA) and the

Economic Development Administration of the United States Department of Commerce. The results of the GLEFC study, once completed, will be used to advise local and state government officials about how much brownfields should be cleaned in the future.

Current Brownfields Financing Strategies

Brownfields financing strategies fall into two fundamental categories: (1) direct funding strategies, and (2) indirect financial strategies.[4] The first strategy is designed to provide money or tax benefits to finance the cleanup and redevelopment of a site. Indirect strategies, on the other hand, enhance the financing environment through improved regulations and administrative procedures. A few examples of each approach follow.

Direct Financing Strategies

These strategies are used to increase the rate of return on a project by either reducing capital costs or providing equity participation.

Equity Participation

Equity participation can be achieved by using lease arrangements, public ownership of a property, or a land reclamation bank. For instance, the Commerce Redevelopment Agency of Commerce City, California, acquired a contaminated site (formerly owned by Uniroyal Tire), cleaned the site, and leased the land to a private developer, who built a factory-outlet store on the site. The city of Detroit used public ownership to acquire and clean up a highly contaminated site, which was purchased by Chrysler for a new production facility in 1992. The cost of site cleanup was funded by the company, the city, and the state of Michigan. Finally, Minneapolis operates a land reclamation bank, which spends $5 million annually to acquire and redevelop contaminated industrial sites. The purchase and cleanup funds are generated by a tax-increment financing plan. Once cleaned up, the property is sold to private developers for redevelopment.

Fees

Various fees, including land-registration fees, inspection fees, and site-assessment fees can become a source of brownfields capital. Minnesota and various other states use this approach to generate revenue.

Taxes

Several tax incentives are available as inducements for private brownfields investment. These include special-assessment districts, tax-increment financing, and tax-abatements credits. Wisconsin and Minnesota have made considerable use of tax-increment financing[5] for this purpose. Ohio, Illinois, Delaware, and Pennsylvania are examples of states offering tax credits and abatements for brownfields investments.

Debt Financing

Many states and cities have created new, or tapped existing, revolving-loan funds and other economic-development loan funds to help fund brownfields cleanup. Connecticut is establishing a special insurance fund to foster brownfields redevelopment. Pennsylvania targets low-interest loans to companies for this purpose. A variety of other areas currently have, or plan to create, such public funding mechanisms for this purpose. Michigan's Environmental Bond Program has provided numerous grants and loans to municipalities for cleaning dirty sites for economic-development purposes.

Grant Programs

Federal Empowerment Zones and Enterprise Community grants are being used, in part, for brownfields cleanup. This is true in Cleveland, Chicago, and Detroit, according to a recent study by the GLEFC at Cleveland State University.[6]

Insurance Coverage

Though not a public-sector program, environmental insurance has become increasingly important to the brownfields redevelopment process. All parties involved in brownfields redevelopment—including the buyer, the seller, the financier, and the municipality—want protection against liability. For example, the buyer and financier want protection against unforeseen cost overruns.

Three types of environmental insurance exist: (1) errors and emission coverage, which protects against pollution caused by a contractor during the course of operation, (2) remediation stop-loss insurance, which caps the cost of remediation above initial cleanup cost estimates, and (3) remediation warranty insurance, which covers unknown expenses, third-party bodily damage, property damage, and other risks. The topic of environmental insurance is covered extensively in Chapter 11 of this book.

Indirect Financial Strategies

A variety of information bases, advisory opinions, liability assurance, legislative reforms, and other programs exist to create a more attractive environment for brownfields redevelopers to acquire public and private funding for cleanup and redevelopment. In many ways, state Voluntary Cleanup Programs generally fall into this category. No-further-action letters, covenants-not-to-sue, and liability releases and assurances, as well as other administrative actions, serve to give lenders and other private investors greater "comfort" in investing in brownfields. These mechanisms are often prerequisites to using direct-funding incentives by the private sector.

Federal Funding for Brownfields Redevelopment

To encourage brownfields cleanup and redevelopment, USEPA's Brownfields Pilot Program has provided $200,000 grants to forty United States cities and counties to develop planning processes for brownfields remediation. USEPA has plans to select another ten cities for planning grant awards in year end 1996.

Three funding programs under the United States Department of Housing and Urban Development (HUD) can help finance brownfields redevelopment. HUD's Community Development Block Grant funds may be used to pay for environmental assessments and to fund actual site cleanup. HUD's Section 108 Loan Guarantee Program funds may be used for land acquisition, clearance, demolition, and rehabilitation of structures, and to make other site-related improvements. HUD's Empowerment Zone funds can be used for site cleanup, environmental assessments, land acquisition, and other purposes. The Economic Development Administration, a part of the United States Department of Commerce, has a Public Works Program and a Revolving Loan Fund Program, which may be used in the future to help fund redevelopment costs associated with brownfields.

Various new federal initiatives are being explored to support brownfields redevelopment. These include President Clinton's recent proposal to provide a federal tax deduction for businesses investing in brownfields in highly distressed areas. Banks could have greater flexibility in using brownfields projects to help satisfy their CRA requirements. A number of proposals call for the use of a portion of the Superfund Trust Fund to finance brownfields cleanup. These and other proposals are expected in the near future to increase the federal role in financing brownfields redevelopment.

Future Commercial Real Estate Finance Trends

As mentioned earlier, real estate finance is changing dramatically, and these changes are expected to have major consequences for local lending institutions' future role in real estate in general, and brownfields specifically. Origination, pooling, and securitization are commonplace in the real estate industry today. Mortgage securities are being structured and sold with the backing of large, national commercial-loan pools. Wall Street is turning mortgage securities into commodity debt instruments, distinguished only by their rate of return.

The Shift from Local Banks to National Capital Markets

Banks and insurance companies, the traditional mainstream real estate capital providers, have been converted into intermediaries for mortgage capital. Loans processed by these entities are securitized, packaged, and sold to institutional investors, rather than being held in portfolios. Many predict that future real estate debt capital will come largely from Wall Street investment houses, institutional sources, and mutual funds. Commercial real estate loans will not be underwritten and funded by their original funding source. Instead, Wall Street investment houses will pool and secure these loans for sale to mutual funds, hedge funds, and other institutional investors.

Historically, real estate was financed by a regional bank providing a construction loan and an insurance company providing long-term financing. This has been the dominant model in cities like Cleveland, where banks historically played a significant role in real estate lending. Financing was project oriented, in the sense that each property was evaluated and financed as a separate and distinct transaction. After funding, the lender held the loan until payoff or maturity. Now the trend is toward lenders acting as conduits for ultimate sale or securitization. In this sense, commercial real estate finance parallels single-family home lending, where loans are sold in the secondary market to Wall Street firms and government-sponsored agencies, like the Federal National Mortgage Association. Increased securitization is causing the decoupling of the borrower from the original lender.

Many banks, thrift institutions, and insurance companies have curtailed or discontinued the financing of commercial-property investments and development. Risk-based rules for these entities have created disincentives for financing real estate, causing them to become more conservative in their underwriting standards. Environmental liabilities have had a major influence in this area. These changes have caused more developers and commercial property companies to turn to public equity markets for capital. This trend has led to a growth in mortgage-backed securities, the use of mortgage conduits, taking companies public, and the sponsoring of real estate investment trusts (REITs).

A Look Back at the Real Estate Boom of the 1980s

The real estate boom of the 1980s was fueled by capital supplied by pension funds, foreign investors, life-insurance companies, commercial banks, and savings and loan institutions. A large part of these investments was justified on the basis of discounted cash flow, which calculated a rising stream of future income based on rising rental rates and assumed future increases in the principal value of property, all inflation based. Huge amounts of capital were lent at a minimal or even spread, because it was assumed that the property would be sold at a major profit. Despite these predictions, the reverse occurred. As a result, lenders in many cases experienced tremendous losses, rather than expected huge profits.

Price inflation occurred in the commercial real estate industry in the 1980s because of the liberal use of tax benefits, attractive foreign exchange rates, and favorable financial terms. These advantages led investors and developers to overlook basic supply and demand relationships, causing an oversaturation of supply. This is another reason why future strategies to finance brownfields redevelopment must be based on market demand. Simple economic principles, such as constant-dollar valuation, were ignored, which could have reduced the number of bad investments made. Constant dollars account for real growth only, and are in direct contrast to projections based upon inflated dollars. Investors and lenders have come to learn that inflation does not create value, but rather has the opposite effect of decreasing the buying power of money.

Possible Future Brownfields Financing Strategies

What are the implications of these fundamental changes in real estate finance for brownfields redevelopment? A couple of possible developments are suggested. Wall Street and other national and international capital markets are more likely than local banking institutions to provide the financial resources needed for brownfields cleanup and redevelopment.

This suggests that new mechanisms like REITs for Environmentally Challenged Properties (REIT-ECPs) could become a reality, especially if linked to special federal-tax advantages, such as those recently proposed by President Clinton.[7] The REIT-ECP would, in principle, raise capital in much the same way as ordinary REITs, with the exception of its ability to accept public funds and donations to help fund site cleanup. Like other REITS, the REIT-ECP would specialize in contaminated-property recycling and reuse. Property held by the REIT-ECP would be improved through cleaning and redeveloped for a specified real estate use. These REIT-ECPs could specialize along industrial, commercial, and residential lines, which is common in the REIT industry now.

These developments could also increase the role of Business Development Corporations (BDCs) as brownfields financing entities. BDCs operate now in thirty states. These higher-risk-taking institutions are publicly chartered, private-development banks. BDCs generate their capital from private sources, such as banks, insurance companies, and other investors. By design, the BDC provides money to more risky projects, which is usually the case with brownfields redevelopment projects. National and international capital sources could provide the funds required by the BDCs to finance brownfields deals.

Conclusion

The financing of brownfields by public and private-sector organizations will grow in importance in the future, as each sector finds its appropriate niche in this area. In the short term, communities and states will continue to innovate with incremental improvements in environmental and economic-development funding programs. States will refine their Voluntary Cleanup Programs over time, leading the way for new tax credit and revolving-loan funds for site cleanup and redevelopment.

Cities will create special funds and use their taxing authority to provide greater incentives for investment in areas where brownfields sites are located. Landbanking, and larger area land-assembly projects will become more common as cities struggle to hold onto their population and job bases by providing more competitive real estate.

Long-term changes in real estate finance will open the door to national and international capital markets for brownfields. This should make capital-raising mechanisms like REIT-ECPs a reality.

Finally, brownfields redevelopers will approach their activities with greater awareness of market trends, which will help them pace their investments in cleanup and redevelopment. Continued improvements to the regulatory environment will concentrate on making compliance more flexible and cost-effective. These are but a few trends ahead that will give energy to future brownfields redevelopment.

Notes

1. USEPA, ENVTL. FIN. ADVISORY BD., FINANCING STRATEGIES FOR BROWNFIELDS REDEVELOPMENT, BROWNFIELDS REP. NO. 3 (1995).

2. Donald T. Iannone, *Sparking Investment Interest in Brownfields*, Urb. Land (Urb. Land Inst.) (June 1996).

3. Fluor Daniel Consulting Division and Donald T. Iannone & Associates, An Analysis of Industrial Land Redevelopment Potential in the North Birmingham Community, Report to City of Birmingham, Alabama (Nov. 1995).

4. USEPA, *supra* note 1.

5. Tax-increment financing is a public financing mechanism that uses anticipated growth in property taxes stemming from a redevelopment project to raise public-sector investment capital.

6. Robert A. Simons, Great Lakes Envtl. Fin. Ctr., Cleveland State Univ., Financing Environmentally Contaminated Land Redevelopment in the Great Lakes Urban Empowerment Zones (Mar. 1996).

7. President Clinton proposed a special federal-tax deduction for brownfields cleanup in highly distressed areas in March 1996. This deduction would require congressional support to become a reality. This author has been investigating the REIT-ECP concept and other innovative mechanisms to attract private cleanup and redevelopment capital to brownfields.

CHAPTER 8

Brownfields Sites: Removing Lender Concerns as a Barrier to Redevelopment

MARGARET MURPHY

Threatened by both economic-development pressures in declining manufacturing areas and a growing need to contain urban sprawl, state legislatures and the United States Environmental Protection Agency (USEPA) are developing new programs that seek to return properties contaminated with hazardous substances in urban areas to productive use. In general, these programs provide an opportunity for owners and developers of certain sites to receive a release from future liabilities at the site in return for voluntary cleanup to prescribed levels. Referred to as brownfields, these sites are commercial or industrial sites that are either inactive or have been abandoned. The problems created by brownfields take many forms, including unemployment in communities where manufacturing sites are inactive and foregone property taxes on abandoned sites.[1]

Often located in desirable locations near utilities, interstate highways, railways, water transport, and labor pools, brownfields scare owners, buyers, and lenders because of uncertainty about the liabilities that may attach to them. Barriers to brownfields redevelopment are several: prospective purchasers' fears about federal and state cleanup liability, the increased cost of redevelopment caused by cleanups, and the difficulties involved in identifying proper cleanup levels. A key additional barrier to brownfields redevelopment is the reluctance of banks to lend money to redevelopment projects out of fear of potential lender liability.

By providing financing, lenders can play a critical role in brownfields redevelopment. For brownfields to be redeveloped successfully, the environmental-

---

Margaret Murphy, a partner in the Environmental Practice Group of the New York law firm of Shearman & Sterling, was assisted in the preparation of this article by Kristopher D. Brown, a Shearman & Sterling associate, and Jeffrey Cross, a summer associate in the firm's 1995 summer program. This article was previously published in *The Banking Law Journal* in May 1996 and appears in this publication with the permission of *The Banking Law Journal*.

liability concerns of lenders must be addressed. Under current United States law, including in particular the federal Comprehensive Environmental Response, Compensation, and Liability Act (CERCLA, commonly known as Superfund),[2] a lender may be held liable for cleanup costs even when it holds only a security interest in contaminated property.[3] Because of CERCLA and similar state statutes, lenders are concerned that if they lend money to a brownfields redevelopment project, and take as collateral the land being redeveloped, they may be held liable for a cleanup if the borrower does not or cannot pay the cleanup costs. Clearly, to encourage banks to lend to brownfields redevelopment projects, the threat of lender liability must be minimized.

This chapter discusses state, federal, and private initiatives to limit lender liability associated with brownfields sites. Some brownfields initiatives have directly addressed the concerns of lenders;[4] many other programs have not. State initiatives are embodied primarily in recent state voluntary cleanup legislation. At the federal level, brownfields initiatives have taken the form of USEPA rules, proposed rules, guidance documents, and federal legislation. There are also Congressional bills to limit lender liability in general, including liability in connection with brownfields redevelopment projects. Shortly before this book went to press, a new federal law was enacted that limits lender liability under CERCLA and RCRA (1996 Lender Liability Law).[5] The 1996 Lender Liability Law codifies the 1992 lender-liability rule, which is discussed later in this chapter. Finally, insurance companies have pioneered new ways to work within the existing regulatory framework to limit cleanup costs at brownfields sites.

States have taken the lead in addressing brownfields problems, but Congress and many federal agencies have become more involved in recent years. In addition to federal agencies such as USEPA, the General Accounting Office, the Department of Housing and Urban Development, and the Department of Commerce's Economic Development Administration, interest groups such as the National Association for the Advancement of Colored People and the Mortgage Bankers Association of America are involved with the brownfields issue.[6]

The Brownfields Problem

In general, under applicable programs dealing with them, brownfields are not sufficiently contaminated to merit ranking on the National Priorities List (NPL)[7] or state lists of contaminated properties.[8] Recent estimates have put the number of brownfields sites nationwide at 500,000 or more.[9] Due to the absence of serious contamination, the cleanup of brownfields is not normally mandated by state or federal laws at the time of a transfer or sale.[10] Potential purchasers, however, are reluctant to buy brownfields because of strict federal and state cleanup liability laws such as Superfund, which can potentially impose cleanup liability on owners of brownfields regardless of whether they contributed to the contamination.[11] Many banks will also avoid lending money to projects involving brownfields sites for fear that they might assume the liabilities of the owner if a foreclosure is required.[12]

Lender Concerns at Brownfields Sites

According to a Bankers Roundtable survey of the 125 largest banking companies in the United States, 70 percent of banks responding indicated that they exercise greater caution for redevelopment lending versus lending for new development projects, particularly in urban, inner city areas, due to environmental concerns.[13] Forty percent of the responding banks stated that they automatically designate areas near Superfund sites as suspect in terms of environmental liability.[14] Some brownfields sites, although not listed on the NPL as Superfund sites, are listed on the Comprehensive Environmental Response, Compensation, and Liability Information System (CERCLIS) list of sites with suspected contamination. For some lenders, even properties located near CERCLIS sites are considered suspect.

Wariness by lenders adversely impacts the prospective redeveloper of a brownfields site. As one lender stated, "the specter of lender liability remains and dampens our enthusiasm [toward brownfields redevelopment] unless Congress, USEPA and the courts can offer greater certainty and protection."[15] A Pittsburgh bank official recently commented that the majority of Pittsburgh's former steel mills are now unused brownfields, and little redevelopment has occurred due to the fear of cleanup costs.[16] The official noted that his bank had refused to foreclose on several other former industrial sites because of similar concerns.[17]

State Approaches to Limiting Lender Liability

States have taken the lead in creating brownfields programs by passing new legislation to provide for voluntary cleanups. These legislative approaches vary considerably from state to state. Some states provide an exemption from liability for lenders concerning environmental cleanups in general, regardless of whether the cleanup is in connection with a brownfields project.[18] Other states provide for lender-liability exemptions only in the context of specific brownfields cleanups.

State Environmental Cleanup Statutes

When enacted in 1980, CERCLA provided an exception to the definition of the term "owner or operator" for purposes of liability under the statute to "a person who, without participating in the management of a vessel or facility, holds indicia of ownership primarily to protect his security interest in the vessel or facility."[19] This exception, as originally enacted, was commonly referred to as the CERCLA secured-creditor exemption. Almost every state has enacted legislation similar to the federal Superfund statute.[20] Some states adopted the language of the CERCLA secured-creditor exemption word for word;[21] other states created secured-creditor exemptions that employ slight variations on the language in the CERCLA secured-creditor exemption.[22] Because states differ on the exact language used within their CERCLA-type statutes, reference must be made to the individual state laws to determine the contours of the state exemptions for lenders.

As part of more recent lender-liability reform measures at the state level, some states have revised their secured-creditor exemptions while retaining the framework of a statute patterned after CERCLA. For example, some state exemptions contain CERCLA-like exemption language but add guidance language to provide that a high degree of control by the lender is required before liability will attach.[23] Other states reject CERCLA's approach entirely, exempting from liability all lenders involved with contaminated sites when the lender did not cause or contribute to the contamination.[24]

In any given situation, because state laws are changing at a rapid pace, reference should be made to an individual state's environmental laws and pending bills to ascertain the nature and current status of the state's secured-creditor exemption. State reform bills that are currently pending include legislation that would make the state's lender exemption conform to the USEPA's 1992 lender-liability rule;[25] some states have already passed such legislation.[26] In the meantime, certain states have enacted overall lender-liability reform through their state brownfields programs.

Brownfields Statutes That Eliminate All or Most Lender Liability under a State's Environmental Laws

A few states have chosen to reform lender liability through brownfields legislation that, in most cases, eliminates lender liability for environmental contamination under all or most of the state's existing environmental laws. Laws of this type exist in Pennsylvania, Ohio, and Illinois.

Pennsylvania

Pennsylvania's brownfields legislation reflects a dramatic reform of lender liability. The brownfields package, which was passed in the form of three separate acts, is intended to facilitate the development of brownfields. One of the acts in the package specifically addresses lender liability. Under the Pennsylvania Economic Development Agency, Fiduciary and Lender Environmental Liability Protection Act,[27] a lender is liable only if it directly causes or exacerbates a release or compels a borrower to cause or exacerbate a release.[28] The operative language states that

> A lender who engages in the practices of routine lending . . . shall not be liable under the environmental acts or common law . . . unless:
> (1) the lender . . . directly cause[d] an immediate release or directly exacerbate[d] a release of regulated substances on or from the property; or
> (2) [the lender] . . . knowingly and willfully compelled the borrower to:
>   (i) do an action which caused an immediate release of regulated substances; or
>   (ii) violate an environmental act.[29]

Even when liability attaches to a lender, however, Pennsylvania's law limits the potential liability of lenders: "[L]iability pursuant to this act [is] limited to the cost for a response action which may be directly attributable to the lender's activities."[30]

Ohio

Ohio passed its brownfields legislation in September 1994.[31] The Ohio law protects both lenders who lend to voluntary cleanup projects[32] and those who lend to any development project involving property subject to state environmental cleanup requirements.[33] The operative language provides that

> any person, who, without participating in the management of the property, holds indicia of ownership in a property primarily to protect a security interest in the property is not liable for:
> (a) the costs of conducting a voluntary action . . .
> (b) . . . the costs of investigating or remediating a release or threatened release of hazardous substances or petroleum at or upon any property . . . irrespective of whether such costs are incurred in connection with a voluntary action undertaken pursuant to any provision of this chapter of the Revised Code in a civil action brought under section 3746.23 of the Revised Code or otherwise brought under the Revised Code or common law of this state. . . .[34]

Ohio's statute defines the key phrase "participation in the management of property" as "actual participation in the management or operational affairs . . . not [the] mere capacity to influence, or ability to influence, or the unexercised right to control the operations of the property."[35] The statute lists the activities that constitute participation in management:

1. undertaking responsibility for hazardous-waste practices, or
2. exercising day-to-day control over the enterprise's environmental activities or substantially all its operational aspects.[36]

Under the Ohio scheme, however, a lender will lose its protection if it fails to take certain steps following a foreclosure. Specifically, after foreclosing on encumbered property, a lender must list the property with a broker or advertise the property for sale monthly within twelve months following the foreclosure; the lender must also accept any offer of fair consideration to purchase the property.[37]

Illinois

As part of a brownfields package enacted in December 1995, the Illinois legislature adopted several lender-liability reforms under the state's environmental laws.[38] The new law eliminates strict, joint and several liability against any party, lenders included, in suits by the Illinois Environmental Protection Agency to recover cleanup costs associated with voluntary cleanups.[39] Under this provision, the agency can recover costs from responsible parties based only on fault; environmental cleanup costs will be apportioned to responsible parties on the basis of their proportionate degree of responsibility for those costs.[40] This provision is favorable to lenders because in most instances, whether the site is a brownfields site or otherwise, the lender will likely bear no degree of responsibility for contamination at the site. Specifically, for brownfields, the Illinois law provides that "the No Further Remediation Letter [provided under Title 17] shall apply in favor of . . . [a]ny financial institution . . . that has acquired ownership, operation, management, or control of a site through foreclosure or under the terms of a security interest held by the financial institution, under the terms of an extension of credit made by the financial institution, or any successor in interest thereto."[41]

The act also reforms lender liability under general environmental law of Illinois. Through its definition of "owner or operator," the act provides that a financial institution will be an owner or operator only if it "takes possession of the vessel or facility and . . . exercises actual, direct, and continual or recurrent managerial control in the operation of the vessel or facility that causes a release."[42]

Lender-Liability Protections Specific to Brownfields Cleanup Statutes

In addition to Pennsylvania, Ohio, and Illinois, at least twenty-three other states currently have brownfields programs.[43] Many of these programs provide, either explicitly or implicitly,[44] protection for banks lending to brownfields projects. State programs with explicit protections typically include within the state brownfields statute specific protection for lenders.[45] The provisions in brownfields legislation in Michigan, Maine, and Minnesota are given as examples below. The approaches of these three states may be compared with the provisions that reform state environmental laws generally, discussed earlier. State programs with implicit protections for lenders typically provide in the state brownfields statute for a release of liability for the developer of a brownfields site. The presumption is that the lender is not likely to be held liable when the underlying borrower has been released from liability. For lenders, these implicit protections may be as solid as explicit exemptions.[46] The implicit provisions in Indiana's brownfields statute are detailed as an example below. In any given situation, the brownfields statute of a particular state should be analyzed to determine the relevant provisions and the extent to which lenders receive protection.

Explicit Protections

MICHIGAN. In Michigan, only nonresponsible parties may avail themselves of the protections offered by the state's brownfields program.[47] Michigan's brownfields statute provides explicit protections for lenders that seek to protect their security interests during the period before completion of brownfields redevelopment projects.[48] Amendments to the Michigan Superfund statute have also substantially modified provisions of the law regarding liability for the cleanup of environmental contamination.[49] Under Michigan's amended scheme, certain nonresponsible parties who become owners or operators of contaminated property (including lenders pursuant to foreclosure) on or after June 5, 1995 (the effective date of the amendments) will not be responsible parties if they conduct baseline environmental assessments of such contaminated property.[50]

MAINE. Maine passed voluntary cleanup legislation in 1993.[51] Only nonresponsible parties may take advantage of Maine's cleanup program. A party that completes a voluntary cleanup under the Maine plan will not be deemed a responsible party for any additional cleanup work and will not be subject to state orders or other enforcement proceedings regarding the subject of the cleanup.[52] Maine's brownfields statute provides that, in addition to the person undertaking the voluntary cleanup, the protection from liability covers, among others, any

> person providing financing to the person who undertakes and completes the response actions or who acquires or develops the identified property; [and] lender or fiduciary . . . who arranges for the undertaking and completion of response actions.[53]

A party (including a lender) can be held liable, however, for any contamination that it causes, contributes to, or exacerbates.[54]

Unfortunately, the Maine statute provides no guidance on the actions a lender may take to protect its security interest during the period before the completion of the voluntary cleanup project. Accordingly, a lender that considers foreclosing on a brownfields project before completion (and thus before any release from liability is granted) must, in each instance, ask the state about what foreclosure actions would subject it to liability. In this respect, the Maine statute suffers the same deficiency as the federal CERCLA secured-creditor exemption, which can also leave a lender potentially liable for cleanup costs if it forecloses on contaminated property.

MINNESOTA. Minnesota passed the first brownfields legislation in the country in 1988.[55] Initially, only nonresponsible parties were eligible for protection from liability under the Minnesota program. Amended in 1993, Minnesota's Voluntary Cleanup Program now permits responsible persons to also conduct cleanups and gain liability protection for themselves and associated parties who are not otherwise liable for a release, including lenders.[56] Under the Minnesota plan, nonresponsible or responsible parties that complete a voluntary cleanup will not be deemed responsible for any additional cleanup work and will not be subject to state orders or other enforcement proceedings regarding the subject of the cleanup.[57] The benefit of the Minnesota statute stems from the release it provides from liability for future, unanticipated cleanup of contamination. Thus, through remediating contaminated property under the program, even responsible parties are protected from subsequent claims based on retroactive liability for site contamination.

Minnesota's brownfields statute provides protection from liability to, among others, any

> person providing financing for response actions or development at the identified real property after approval of the response action plan, whether the financing is provided to the person undertaking the response actions or other [sic] person who acquires or develops the property.[58]

Unlike in Maine, lenders in Minnesota are granted liability protection as soon as a response action plan is approved, not merely upon the completion of cleanup work. Accordingly, a lender that considers foreclosing on a brownfields project before completion can do so without worry, as long as an approved response action plan is in place, even though it has not yet been carried out. The protection from liability will be lost by any party (including a lender) who aggravates or contributes to a release or threatened release at the subject site.[59]

Implicit Protections: Indiana

Although the Indiana Voluntary Cleanup Program[60] contains no explicit lender protections, several of its provisions implicitly protect lenders. The program provides for certificates of completion and covenants-not-to-sue for completed brownfields cleanups.[61] These protections are given to the party that completes the voluntary remediation work.[62] The program implicitly protects lenders, because when a borrower is assured there will be no future liability for cleanup costs, a lender can be more assured that it will not itself face liability. In short, when a borrower is no longer liable, a lender is less likely to be liable, even if it were to participate in the business affairs of the borrower. Notably, the Indiana program provides no protections until a voluntary cleanup is completed. Accordingly, no implicit protection is afforded lenders who may be forced to protect their security interests during the period before completion of the remediation work. In this regard, the Indiana statute (like the Maine program) suffers the same deficiency as the original federal CERCLA secured-creditor exemption, which can also leave a lender potentially liable for cleanup costs if it forecloses on contaminated property.

Remaining Concerns at the State Level: The Specter of Federal Liability

Designed to reduce the uncertainties associated with brownfields redevelopment under state law, state brownfields programs cannot address the additional uncertainties presented by the possible application of federal laws to these sites. Potential federal liability exists under various federal environmental laws, including, among others, CERCLA, the Resource Conservation and Recovery Act (RCRA), the Clean Air Act, and the Clean Water Act. Both USEPA and Congress have worked on initiatives to calm or eliminate the concerns of lenders in this regard.

Federal Initiatives That Limit Lender Liability

USEPA Initiatives

USEPA has addressed the brownfields problem through its Brownfields Action Agenda.[63] This agenda, announced on January 25, 1995, includes several initiatives designed to aid communities in the redevelopment of abandoned, idle, or underused industrial land. The initiatives include the following:

- Removing 25,000 sites from the CERCLIS Superfund tracking system
- Funding fifty brownfields pilot projects across the country at up to $200,000 each
- Issuing new guidance on prospective-purchaser agreements
- Issuing Superfund and underground storage tank lender-liability guidance
- Delegating certain components of the federal CERCLA program to the states[64]

Although this process has been ongoing for some time, and despite the fact that all 25,000 sites were removed from the CERCLIS within a month of the announcement of the initiative,[65] The Bankers Roundtable survey found that more than 41 percent of lenders were unaware of the existence of the USEPA Brownfields Action Agenda.[66]

Removing Sites from CERCLIS

The 25,000 sites removed from CERCLIS are sites that USEPA has categorized as "no further remedial action planned" (NFRAP). One key concern of lenders is the specter of federal liability associated with properties where there is suspected contamination. Because the mere listing of a site on the Superfund tracking list is enough to deter some lenders,[67] the removal of sites from CERCLIS should help alleviate some of these fears. In many cases, the presence of a site on the tracking list can affect its appraisal value, further reducing the property's attractiveness to a lender.[68] The removal of NFRAP sites from CERCLIS, therefore, should help significantly in removing some of the barriers to brownfields redevelopment.

Prospective-Purchaser Agreements and PPA Guidance

In 1989, USEPA issued its first guidance on prospective-purchaser agreements (PPAs). A PPA generally requires a purchaser of contaminated property to commit to a specific cleanup, as well as make a payment to USEPA in return for a covenant-not-to-sue from USEPA. The original guidance stipulated that such agreements would be entered only when "enforcement action was anticipated and where the new owner provides the government with direct, substantial benefits."[69] Only fourteen such agreements were entered by USEPA between 1989 and 1995.[70]

On May 25, 1995, USEPA issued a new guidance document entitled "Guidance on Agreements with Prospective Purchasers of Contaminated

Property."[71] This guidance establishes new criteria to determine when a PPA may be appropriate. The new criteria are as follows:

- Enforcement action is anticipated by USEPA.
- A direct cleanup benefit or indirect economic benefit will occur.
- The operation of the site will not aggravate or contribute to contamination, or pose health risks to the community or persons likely to be present at the site.
- The prospective purchaser is a financially viable party.[72]

PPAs provide protection from claims by USEPA and related claims by other federal agencies under other environmental laws; they also provide contribution protection related to subsequent third-party claims (such as those from prior landowners or adjoining property owners).[73] PPAs include, as part of the negotiated package, a covenant-not-to-sue from the United States, issued to the prospective purchaser.[74] PPAs do not, however, protect purchasers from conditions that are unknown at the time of the agreement.[75]

From a lender's perspective, a purchaser armed with a PPA is in a better position than other brownfields redevelopers to secure financing. If the site is one that has been thoroughly characterized, the risk of lender liability is significantly diminished.

USEPA Lender-Liability Reform

As part of its Brownfields Action Agenda, USEPA also has attempted to remove lender-liability barriers to brownfields redevelopment. USEPA's limited success in this endeavor was one of the primary motivations behind the lender-liability bills introduced in the 104th Congress.

As mentioned earlier, USEPA first issued a rule regarding lender liability under CERCLA in April 1992, analyzing how the term "participating in management" should be defined, and when a lender would be liable for cleanup costs for property in which it held a security interest. In 1994, the Court of Appeals for the District of Columbia Circuit ruled that USEPA did not have statutory authority to issue the rule.[76] In September 1995, USEPA and the United States Department of Justice (DOJ) jointly issued a new enforcement policy memorandum on lender liability under CERCLA.[77] The new policy provided that both USEPA and DOJ intend to apply as guidance the provisions of the original lender-liability rule.[78] USEPA's position was that the Court of Appeals decision to vacate the original rule as binding regulation does not preclude USEPA from following the rule's provisions as enforcement policy.[79] In support of USEPA's enforcement policy, Congress and President Clinton codified the USEPA's rule by passing the 1996 Lender Liability Law in the fall of 1996.

USEPA had also issued a final lender-liability rule for underground storage tank liability under RCRA.[80] The RCRA lender-liability rule allows a lender that forecloses on an underground storage tank to remain exempt from RCRA liability if it complies with certain minimum requirements. In general, a lender that limits its action at a foreclosed tank to emptying the tank or otherwise ceasing the operations of the tank will not be liable.[81] This exemption also was codified by the 1996 Lender Liability Law.

Finally, USEPA recently announced a policy that will protect a lender that has foreclosed on property contaminated with hazardous substances that have migrated to the property from a neighboring source via an aquifer.[82] Under this new policy, USEPA will not take any enforcement action under CERCLA against the owner of such property, including a lender that has foreclosed.[83]

Unresolved Problems and the Community Reinvestment Act Rule

USEPA's recent effort, though a good start, are not sufficient to satisfy all the concerns of most lenders with respect to brownfields redevelopment. Therefore, to assure lenders they will not be subject to liability at the federal level for brownfields redevelopment projects, most lenders believe that further legislative reform is required.[84]

Before discussing federal legislative efforts regarding lender liability, it is important to note a recent federal rule that provides incentives to lenders to invest in brownfields redevelopment projects. On May 4, 1995, USEPA Administrator Carol Browner introduced a new regulation under the federal Community Reinvestment Act (CRA). CRA requires certain lenders to make capital available in urban areas.[85] Under the new CRA regulation, lenders subject to CRA may claim credit toward their CRA obligation for loans that are made to clean up or redevelop industrial properties in urban areas.[86] It is expected that this incentive will do much to facilitate brownfields redevelopment.[87]

Federal Legislative Reform

On September 30, 1996, President Clinton signed the 1996 Lender Liability Law as part of a large spending package sent to the President from Congress. This law codified the 1992 USEPA lender-liability rule. The 1996 Lender Liability Law limits a lender's liability under CERCLA, as well as under RCRA, to only those lenders who participate in the management of a business that operates on contaminated property. The law defines a lender's management as control of the "decision-making of the Borrower's compliance with environmental laws" or "responsibility for the handling or disposal of hazardous substances."[88] In addition, the law protects lenders who foreclose on collateral property from liability if the lenders divest in the property at the earliest practicable, commercially reasonable time.[89]

Because the 1996 Lender Liability Law was enacted shortly before this book went to press, we are unable to provide a detailed analysis of the law. Nevertheless, although the 1996 Lender Liability Law appears to alleviate certain lenders' concerns, lenders are still subject to potential liability under federal environmental laws other than CERCLA and RCRA, and under various state environmental laws.

During the 104th Congress, several bills were introduced in the House and Senate to reform lender liability under existing environmental laws. Although the 1996 Lender Liability Law addresses some of the concerns that were the object of reform in other bills, other lender-liability proposals remain on the table. Because it is foreseeable that legislators will continue to pursue further lender-liability reform into the 105th Congress, it is useful to review the lender-liability reform measures pending in Congress. These legislative initiatives fall into three categories: general Superfund reform initiatives, lender-liability protection initiatives, and federal banking-reform initiatives. Of these bills, federal banking reform is currently the most comprehensive.

Superfund Reform Initiatives

On September 29, 1995, Senator Smith of New Hampshire introduced a Superfund reform bill that would exclude lenders from the definition of an owner or operator under CERCLA if the lender did not participate in the operational management of the property.[90] Senator Smith's bill provides that the liability of any lender for the release of hazardous substances at a property acquired through foreclosure is limited to

> the excess of the fair market value of a . . . [property] on the date on which the liability of a lender is determined over the fair market value of the . . . [property] on the date that is 180 days before the date on which the response action is initiated, not to exceed the amount that the lender realizes on the sale of the . . . [property] after subtracting acquisition, holding and disposition costs.[91]

The bill also provides a safe harbor for lenders under CERCLA. If a lender meets the criteria of the safe harbor, it will not be deemed to have participated in management as contemplated by CERCLA.[92] Among other things, this safe harbor allows lenders to take a security interest in property, have the ability to control the operation of the business (as long as that ability is not exercised), include in the provisions of an extension of credit terms or conditions relating to environmental compliance, monitor or enforce such terms or conditions, require environmental cleanup, provide financial advice, restructure or renegotiate the agreement, or exercise legal remedies.[93] The bill also contains a provision that precludes protection for any lender that causes or contributes to the release of a hazardous substance on the property.[94]

In the House, on October 20, 1995, Representative Oxley of Ohio introduced a Superfund reform bill that includes provisions protecting lenders from liability.[95] Oxley's bill exempts thirteen specific lender actions from the definition of "participation in management," including the act of holding legal or equitable title acquired through foreclosure.[96] Oxley's bill was approved by the House Commerce Subcommittee on Commerce, Trade, and Hazardous Materials in November 1995, and was expected to be considered by the full committee in March 1996.[97] Representative Regula of Ohio has also introduced a general bill on brownfields redevelopment.[98] Representative Regula's bill seeks to bring more certainty to the site characterization and cleanup processes through USEPA-certified state "partner" Voluntary Cleanup Programs.[99] The bill authorizes states to make final decisions on cleanup and future liabilities for what are considered "low" and "medium" priority contaminated sites, and also eliminates federal intervention.[100] Notably, the bill is devoid of any specific statutory protection for lenders.

Stand-Alone Lender-Liability Bills

Bills have been introduced in both the House and the Senate that target reforming lender liability under existing environmental laws. Senator D'Amato of New York and Representative LaFalce of New York have been the main forces behind similar bills in the past. Both first introduced bills in 1993, and reintroduced them this year.

Similar to Senator Smith's Superfund reform bill, Senator D'Amato's stand-alone lender-liability bill provides that the liability of any lender for the release of hazardous substances at a property (1) acquired through foreclosure, (2) held pursuant to an extension of credit, or (3) subject to financial control or financial oversight is limited "to the *actual benefit* conferred on such institution or lender by a removal, remedial, or other response action undertaken by another party."[101] The bill defines "actual benefit" as the net gain realized by the lender due to the remedial action.[102] The bill also provides a safe harbor for lenders under CERCLA and the Solid Waste Disposal Act. If a lender meets the criteria of the safe harbor, it will not be deemed to have participated in management as contemplated by these acts.[103] The safe harbor allows lenders to take a security interest in property, include terms relating to environmental compliance, monitor or enforce such terms, require environmental cleanup, provide financial advice, restructure or renegotiate the agreement, or exercise legal remedies.[104] Senator D'Amato's bill also contains a provision that precludes protection for any lender that causes or contributes to the release of a hazardous substance on the property.[105]

Representative LaFalce's stand-alone lender-liability bill[106] also amends CERCLA, but takes an approach similar to USEPA's 1992 lender-liability rule, which was codified in the 1996 Lender Liability Law.

In addition to Senator D'Amato's and Representative LaFalce's initiatives, Representatives Upton of Michigan and Tauzin of Louisiana have introduced the Lender and Fiduciary Fairness and Liability Act of 1995, a bill that combines aspects of both the D'Amato and LaFalce approaches. This bill provides a comprehensive definition of "participation in management" as used in CERCLA, which does not include the capacity or ability to influence the operation of a borrower.[107] Like the LaFalce bill, the Upton/Tauzin bill does not provide any explicit guidelines on how long a lender may hold foreclosed property before becoming liable as an owner. This is less of a concern, however, because the bill adopts the D'Amato limitation of liability to actual benefit.[108] Like the D'Amato bill, the Upton/Tauzin bill also precludes use of the exemption by any lender that directly causes or contributes to the release of a hazardous substance on the property.[109]

Federal Banking Reform

The most comprehensive lender-liability reform pending in Congress is the Financial Institution Regulatory Relief Act of 1995, introduced by Representative Leach of Iowa.[110] Title III of the bill directly addresses lender environmental liability.[111] The bill is a general reform that applies to all environmental laws. Regardless of any provisions in any existing federal environmental law, the bill holds a lender liable only if it does not fit the exemption detailed below.[112]

The bill proceeds in the same manner as other lender-liability reforms in providing that a lender is liable only when it actually participates in management of the borrower's activities.[113] The definition of "participating in management" under Leach's bill does not differ significantly from the LaFalce bill or the Upton/Tauzin bill. The most significant language regarding "participating in management" appears in a provision stating that a lender that did not participate in the management of a property before foreclosure will not be liable when it forecloses on the property, sells it, maintains business at the site, winds up operations at the site, or undertakes any response at the site. To retain this protection, the lender must seek to sell the property at the earliest practical and commercially reasonable time on commercially reasonable terms.[114] The liability of any lender liable under federal environmental law is limited to the cost of any response action or corrective action to the extent—and in the amount—that the lender actively and directly contributed to the hazardous-substance release.[115]

Representative Leach's bill seems to most adequately address the concerns of lenders, because it applies to all federal environmental laws and addresses liability concerns occurring before, during, and after foreclosure. It also provides direct guidance about the actions a lender must undertake after foreclosure to divest itself of a property.

Insuring Against Liability

Though many of the federal legislative initiatives look favorable to lenders, none is yet law. Also, there is no guarantee that any pending legislation will pass as currently written or in a form favorable to lenders. New types of environmental liability insurance, however, offer an option to borrowers to limit their potential financial exposure at brownfields sites. By encouraging borrowers to obtain these new types of insurance, lenders can be more confident that there will be other sources beside them available to pay potential cleanup costs.

The need to redevelop brownfields has spurred substantial growth in the environmental liability insurance field.[116] To obtain insurance, parties must often perform comprehensive risk assessments and, for properties where no contamination is found, pay premiums of approximately $10,000 to $50,000 per site covered.[117] Certain other policies (with higher premiums) are available for sites, such as brownfields, with existing contamination. Such policies, referred to as environmental "stop-loss" or "catastrophic environmental loss" policies, provide coverage above a preset cleanup amount.[118] The policyholder accepts a substantial deductible (the "stop-loss amount"), which includes the estimated cleanup costs plus an additional amount (the "cushion") for cost overruns. The insurer bears the risk of cleanup costs in excess of the "stop-loss" amount, up to the policy limit.[119] This type of insurance is an excellent way to protect against "unknown" contaminants at a brownfields site. However, stop-loss insurance carries substantial premiums and deductibles and may be limited in time and have a set termination date that must be considered.[120]

Conclusion

Many developers, especially smaller and midsized groups, cannot underwrite the purchase, investigation, and remediation of brownfields sites. Consequently, outside financing must be secured. Lender liability is, therefore, a critical issue that must be addressed in any brownfields redevelopment program. A variety of brownfields programs are under way throughout the country. The states and federal government are actively encouraging varied approaches to address brownfields sites. Risk averse by nature, lenders have historically avoided projects that could expose them to liability for environmental contamination. By removing lender-liability concerns, state brownfields programs encourage redevelopment projects for properties that might otherwise remain abandoned, idle, or underused. Because state brownfields programs fail to address all the concerns of lenders, USEPA and federal legislative efforts have sought to further reassure lenders that they will not be held responsible for contamination at brownfields sites. In addition, new types of environmental insurance policies can reduce the liability risks of borrowers (and thus lenders) contemplating brownfields redevelopment projects.

In many opportunities for brownfields redevelopment, the absence of lenders to provide financing may well prevent the project from getting off the drawing board. Only by focusing on the concerns of lenders can brownfields programs hope to stimulate brownfields redevelopment, and thereby achieve the laudable goal of returning nonproductive property in our urban centers to productive use.

Notes

1. *See generally* R. Michael Sweeney, *Brownfields Restoration and Voluntary Cleanup Legislation*, 2 The Envtl. Law. (ABA), at 101 (Sept. 1995) (providing general information on brownfields and voluntary cleanup statutes); E. Lynn Grayson & Stephen A.K. Palmer, *The Brownfields Phenomenon: An Analysis of Environmental, Economic and Community Concerns*, 25 Envtl. L. Rep. (Envtl. L. Inst.), at 10,337 (July 1995) (explaining nature of brownfields problem and providing overview of historic, social, economic, and environmental issues that impact brownfields redevelopment); RESOURCES, COMMUNITY, & ECONOMIC DIV., U.S. GEN. ACCOUNTING OFFICE, COMMUNITY DEVELOPMENT: REUSE OF URBAN INDUSTRIAL SITES, GAO/RCED-95-172 (1995) (discussing approaches to brownfields problems and actions of several brownfields public-interest groups) [hereinafter COMMUNITY DEVELOPMENT].

2. 42 U.S.C. §§ 9601 *et seq.* (1996).

3. *See* United States v. Fleet Factors Corp., 901 F.2d 1550 (11th Cir. 1990) (holding lender could be liable for cleanup if lender participated "in the financial management of a facility to a degree indicating a capacity to influence the corporation's treatment of hazardous wastes."), *cert. denied*, 498 U.S. 1046 (1991); *but see In re* Bergsoe Metal Corp., 910 F.2d 668 (9th Cir. 1990) (holding that the fact that a party merely holds title to facility does not make it owner of that facility). Most courts have tended not to follow the *Fleet Factors* decision, but the specter of liability for lenders remains.

4. States with brownfields programs that directly address the concerns of lenders include Ohio, Pennsylvania, and Maine. In total, twenty-five states have now developed Voluntary Cleanup Programs to deal with brownfields sites. *See* GREENFIELDS GROUP, PROTECTING "GREENFIELDS:" THE STATE VOLUNTARY CLEANUP PROGRAM ALTERNATIVE 7 (1995) [hereinafter GREENFIELDS GROUP] (noting also that Voluntary Cleanup Program legislation or regulatory development is currently under consideration in seven additional states); *see also* U.S. OFFICE TECHNOLOGY ASSESSMENT, STATE OF THE STATES ON BROWNFIELDS: PROGRAMS FOR CLEANUP AND REUSE OF CONTAMINATED SITES, OTA-BP-ETI-153, 103d Cong., 1st Sess. (1995) [hereinafter STATE OF THE STATES ON BROWNFIELDS].

5. Public Law No. 104-208 (1996).

6. *See* COMMUNITY DEVELOPMENT, *supra* note 1, at 4; *see also GAO Report Details Federal Agencies' Efforts to Address Brownfields*, SUPERFUND REP., July 26, 1995, at 37–38.

7. The NPL is a list of contaminated sites maintained by USEPA pursuant to CERCLA. One of the important features of CERCLA is the Hazardous Substance Superfund, which is used by USEPA to clean up NPL sites. *See* 40 C.F.R. § 300.425(b)(1). It is to this fund that CERCLA owes its "Superfund" nickname.

8. Alice J. Kryzan, *Brownfields: An Overview*, 15 Envtl. L. Section J. (NYSBA), at 70 (1995).

9. *Id.*

10. Unless a property is the subject of cleanup or corrective action under CERCLA, the Resource Conservation and Recovery Act (RCRA), 42 U.S.C. §§ 6901 *et seq.* (1996), or a state equivalent, there may be limited authority for requiring cleanup unless the site presents an imminent threat to human health or the environment. *See* David Yaussy, *Brownfield Initiatives Sweep Across the Country*, 10 ENVTL. COMPL. & LIT. STRATEGY, NO. 11, at 2 (1995).

11. For example, with limited exceptions, CERCLA can impose strict, joint and several liability on any current owner or operator of contaminated property. *See* 42 U.S.C. § 9607(a)(1).

12. The analysis presented in this chapter, along with many of the sources cited supporting that analysis, was undertaken prior to the passage of the 1996 Lender Liability Law. It is not yet possible to gauge the full effect of that law on brownfields redevelopment.

13. BANKERS ROUNDTABLE, ROUNDTABLE SURVEY: ENVIRONMENTAL LIABILITY OF SECURED PARTIES AND FIDUCIARIES 1 (May 1995) [hereinafter ROUNDTABLE SURVEY]. Membership in The Bankers Roundtable is reserved for the 125 largest United States banking companies. Current membership represents roughly 70 percent of United States banking assets. *Id.*

14. *Id.*

15. *Id.*

16. COMMUNITY DEVELOPMENT, *supra* note 1, at 4.

17. *Id.*

18. *See, e.g.*, 1995 Pa. Laws 3 (exempting from liability all lenders not involved in activity resulting in environmental contamination).

19. 42 U.S.C. § 9601(20)(A) (1996).

20. *See* STATE OF THE STATES ON BROWNFIELDS, *supra* note 4, at 10.

21. *See, e.g.*, WASH. REV. CODE § 70.105D.020 (1994).

22. *See, e.g.*, MONT. REV. STAT. § 75-10-701(8)(a) (1994) ("'Owns or operates' means owning, leasing, operating, managing activities at, or exercising control over the operation of a facility.... (b) The term does not include holding indicia of ownership of a facility primarily to protect a security interest in the facility or other location unless the holder has participated in the management of the facility....").

23. *See, e.g.*, IND. CODE ANN. § 13-7-8.7-8 (West 1994) (requiring actual and direct managerial control by lender before liability attaches). Washington retains the same definition as CERCLA for its secured-creditor exemption, stating that the term "owner or operator" does not include "[a] person who, without participating in the management of a facility, holds indicia of ownership primarily to protect the person's security interest in the facility." WASH. REV. CODE § 70.105D.020(11)(i) (1994). But Washington has amended its statute to define the standard of participation in management required before liability will attach to a lender as well as to provide a safe harbor for foreclosing lenders.

24. *See, e.g.*, N.C. GEN. STAT. § 130A-310.7 (1994).

25. 57 Fed. Reg. 18,344 (1992) [hereinafter USEPA's 1992 Lender Liability Rule]. USEPA's rule was subsequently struck down by the Court of Appeals for the D.C. Circuit, which held that USEPA lacked the statutory authority to promulgate the rule. Kelley v. Environmental Protection Agency, 15 F.3d 1100 (D.C. Cir. 1994), *cert. denied sub nom.* American Bankers Ass'n v. Kelley, 115 S. Ct. 900 (1995).

26. *See, e.g.*, N.H. REV. STAT. ANN. §§ 146-A, 146-C, 147-A, 147-B (1994). A similar bill is apparently pending in Rhode Island. *See* I. Leo Motiuk & Jayne A. Pritchard, *Conducting Due Diligence: 1995, Lender and Fiduciary Environmental Liability*, 886 PLI/Corp 413, 1995, *available in* LEXIS, Nexis Library.

27. 1995 Pa. Laws 11; *see Pennsylvania's Bankers' Association Environmental Liability Legislation Signed by Governor Ridge*, PR Newswire, May 30, 1995, *available in* LEXIS, Nexis Library.

28. 1995 Pa. Laws 3.

29. *Id.*

30. *Id.*

31. OHIO REV. CODE ANN. § 3746.01 *et seq.* (Baldwin 1995).

32. *Id.* § 3746.26(A)(1)(b).

33. *Id.*

34. *Id.*

35. *Id.* § 3746.26(E)(1).

36. *Id.* §§ 3746.26(E)(1)(a)–(b).

37. *Id.* §§ 3746.26(F)(2)(a)–(b). These requirements are identical to the ones that USEPA sought to impose on foreclosing lenders under its lender-liability rule issued in April 1992. *See* USEPA's 1992 Lender Liability Rule, *supra* note 25.

38. 1995 Ill. Laws 89-431.

39. *Id.* tit. 17, § 58.9.

40. *Id.* § 58.9(a)(1).

41. *Id.* § 58.10(d)(10).

42. *Id.* § 22.2(h)(2)(E).

43. GREENFIELDS GROUP, *supra* note 4, at 7. Some of these programs are available only to those parties who were not responsible for existing site contamination. In other words, only parties that did not cause or contribute to contamination at a site may elect to clean up that site under such programs. *But see, e.g.*, VA. CODE ANN. § 10.1-1429.1 *et seq.* (Michie 1993 & Supp. 1995) (providing for immunity to state enforcement action upon completion

of voluntary cleanup for both responsible and nonresponsible parties). USEPA does not currently have a position on whether brownfields programs should apply to nonresponsible and responsible parties alike. Telephone Interview with Linda Garczynski, Director of Outreach and Special Projects Staff, Office of Solid Waste & Emergency Response, USEPA (Oct. 18, 1995) (commenting on how many state voluntary programs include protections for responsible parties to discourage what is termed "warehousing" or nondevelopment of their contaminated property, and that USEPA has, for several such state programs, implicitly approved the involvement of responsible parties by entering memoranda of understanding that give to the states the final authority on all cleanup decisions).

44. These protections often take the form of covenants-not-to-sue and other written assurances of liability protection from a state, which can be recorded in the chain of title and run with the land, in some cases extending the protections to subsequent purchasers of the property. *See generally* Sweeney, *supra* note 1, at 164 (analyzing liability protections under various state voluntary cleanup statutes). By adding value to the lender's underlying collateral, these assurances can eliminate some of the financial concerns of lenders involved with brownfields redevelopment projects. Interview with Mark D. Anderson, Esq., Counsel to The Greenfields Group, in Washington, D.C. (Oct. 17, 1995).

45. Because brownfields programs are being developed at such a rapid pace, reference must be made to the brownfields legislation of a specific state to determine whether it contains explicit protection.

46. *See* STATE OF THE STATES ON BROWNFIELDS, *supra* note 4, at 17 (noting that "[s]ome lenders have voiced approval of certificates of completion and no further action letters [to developers] as easing concerns involving loan decisions").

47. *See* MICH. COMP. LAWS ANN. § 20133(2)(b) (West 1995). Michigan may provide a person who proposes to redevelop or reuse a facility, including a vacant manufacturing or abandoned industrial site, with a covenant-not-to-sue concerning liability under the state's Superfund statute. *Id.* § 20133.

48. *Id.* §§ 20101a, 20101b. Specifically, under Michigan's statute, lenders who engage in policing activities before a foreclosure may (1) require the borrower to undertake response activities at the facility during the term of the security interest, (2) require the borrower to comply with applicable environmental laws, (3) secure or exercise authority to monitor or inspect the facility in which indicia of ownership are maintained, including on-site inspections of the borrower's business or financial condition, during the term of the security interest, and (4) before foreclosure, engage in work-out activities and their equivalents if the lender does not actively participate in management. *Id.* § 20101a(2)(e).

49. 1995 Mich. Pub. Acts 71 (amending 1994 Mich. Pub. Acts 451).

50. The statute provides that a liable party under the act does not include an owner or operator who complies with both of the following:

(i) A baseline environmental assessment is conducted prior to or within 45 days after the earlier of the date of purchase, occupancy, or foreclosure. For purposes of this section, accessing property to conduct a baseline environmental assessment does not constitute occupancy.

(ii) The owner or operator discloses the results of a baseline environmental assessment to the [Michigan Department of Natural Resources] and subsequent purchaser or transferee if the baseline environmental assessment confirms that the property is a facility.

MICH. COMP. LAWS ANN. § 20126(1)(c) (West 1995) (The term "facility" is defined in the statute as "any area, place, or property where a hazardous substance in excess of [state] concentrations . . . has been released, deposited, disposed of, or otherwise comes to be located." *Id.* § 20101(l)). Accordingly, disclosure (not necessarily cleanup) of contamination can be sufficient to avoid liability under the statute.

51. ME. REV. STAT. ANN. tit. 38, § 343-E (West 1995).
52. *Id.*
53. *Id.* §343-E.6.c.
54. *Id.* § 343-E.7.a.
55. STATE OF THE STATES ON BROWNFIELDS, *supra* note 4, at 19.
56. MINN. STAT. ANN. § 115B.175[subdiv. 6A] (West 1992 & Supp. 1994).
57. Minnesota provides written assurances for protection from environmental liability for parties successfully meeting the program's requirements. *Id.* § 115B.175[subdiv. 6]; *see* MINN. POLLUTION CONTROL AGENCY, VOLUNTARY INVESTIGATION AND CLEANUP, GUIDANCE DOCUMENT NO. 4 (Jan. 1994).
58. MINN. STAT. ANN. § 115B.175[subdiv. 6A](c)(2) (West 1992 & Supp. 1994).
59. *Id.* § 115B.175[subdiv. 7](1).
60. IND. CODE ANN. § 13-7-8 *et seq.* (West 1994).
61. *Id.* §§ 13-7-8.9-17(a), 13-7-8.9-18.
62. *Id.*
63. USEPA BROWNFIELDS ACTION AGENDA (1995) [hereinafter BROWNFIELDS ACTION AGENDA].
64. *Id.*
65. USEPA, *EPA Announces Superfund Administrative Reform,* EPA ENVTL. NEWS, Feb. 17, 1995, at 6.
66. ROUNDTABLE SURVEY, *supra* note 13, at 2.
67. *Id.* at 1.
68. *Cf.* John M. Scagnelli, *The Impact of Environmental Contamination on Market Value,* ENVTL. COMPL. & LITIG. STRATEGY, Apr. 1995, at 4–6 (examining how courts calculate stigma damages in cases involving contaminated properties).
69. *See Guidance on Prospective Purchase, State Role Included in Major Reform Announced by EPA,* 25 Env't Rep. (BNA), at 267 (June 2, 1995).
70. Eric Rothenberg, *New EPA Guidance Shields Buyers of Contaminated Sites,* N.Y. L.J., June 12, 1995, at 53.
71. 60 Fed. Reg. 34,792 (1995).
72. *Id.*
73. Rothenberg, *supra* note 70, at 53.
74. A PPA and a covenant-not-to-sue may be assigned or transferred with the permission of USEPA. 60 Fed. Reg. 34,792, 34,797 (1995).
75. *Id.*
76. Kelley v. Environmental Protection Agency, 15 F.3d 1100 (D.C. Cir. 1994), *cert. denied sub nom.* American Bankers Ass'n v. Kelley, 115 S. Ct. 900 (1995).
77. U.S. EPA & U.S. DEP'T JUSTICE, POLICY ON CERCLA ENFORCEMENT AGAINST LENDERS AND GOVERNMENT ENTITIES THAT ACQUIRE PROPERTY INVOLUNTARILY (1995) [hereinafter LENDER LIABILITY GUIDANCE]; *see EPA, Justice Issue Guidance on Lender Liability under CERCLA,* BANKING REP. (BNA), at 617 (Oct. 16, 1995).
78. LENDER LIABILITY GUIDANCE, *supra* note 77, at 1.
79. *Id.* at 3.
80. 60 Fed. Reg. 46,692 (1995); *see EPA Releases Lender Rule under RCRA; Some Issues Clarified, but Others Remain,* BANKING REP. (BNA), at 373–75 (Sept. 11, 1995). USEPA believes that RCRA gives it broader statutory authority than CERCLA to issue such guidance. *Id.*
81. 60 Fed. Reg. 46,692 (1995).

82. 60 Fed. Reg. 34,790 (1995).
83. *Id.*
84. ROUNDTABLE SURVEY, *supra* note 13, at 1.
85. *See* Reed D. Rubinstein, *Shortening the Ten-Foot Pole*, CONN. L. TRIB., May 15, 1995, at 25.
86. *Id.*
87. *Id.*
88. Olaf de Senerpont Domis, *New Law Finally Limits Environmental Liability*, AM. BANKER, Oct. 2, 1996, at 3.
89. *Id.*
90. S. 1285, 104th Cong., 1st Sess. tit. III, § 303 (1995); *see Lender Liability Provisions Included in Banking, Superfund Bills*, BANKING REP. (BNA), at 566–67 (Oct. 9, 1995) [hereinafter *Lender Liability Bills*]. Senator Smith's bill provides that a security-interest holder shall be considered to be participating in management and thus liable under CERCLA only if the security-interest holder has undertaken

> (i) responsibility for the hazardous substance handling or disposal practices of the property; or
> (ii) overall management of the property encompassing day-to-day decision-making over environmental compliance or over an operational function (including functions such as those of a plant manager, operations manager, chief operating officer, or chief executive officer), as opposed to financial and administrative aspects of a property.

S. 1285, 104th Cong., 1st. Sess. tit. III, § 304 (1995). Senator Smith's bill also includes a proposed amendment to the Federal Deposit Insurance Act, which would limit the liability of federal banking agencies such as the Resolution Trust Corporation and the Federal Deposit Insurance Corporation for environmental cleanup. *Id.* It would also extend that exclusion to certain subsequent purchasers of the property from a federal banking agency. *Id.* This bill has reportedly been referred to the Senate Environment and Public Works Committee. *Lender Liability Bills, supra* at 567. If the bill ultimately clears a joint committee, USEPA Administrator Carol Browner is expected to advise President Clinton to veto it due to the controversial general limitations placed on the Superfund program. Elliott P. Laws, Assistant Administrator, USEPA Office of Solid Waste and Emergency Response, Remarks at the Meeting of the ABA's Section on Natural Resources, Energy and Environmental Law on Brownfields (Sept. 27, 1995). (Mr. Laws was one of the key authors of USEPA's Brownfields Action Agenda.)

91. S. 1285, 104th Cong., 1st. Sess. tit. III, § 304 (1995).
92. *Id.*
93. *Id.*
94. *Id.*
95. HR 2500, 104th Cong., 1st Sess. § 302 (1995).
96. *Id.* § 302(a)(2)(L).
97. Jennifer Silverman, *Bliley to Offer Amendments on Liability, Brownfields to CERCLA Reform Bill*, BANKING REP. (BNA), at 165 (Feb. 5, 1996).
98. H.R. 1621, 104th Cong. 1st Sess. (1995).
99. *Id.*
100. *Id.*
101. S. 394, 104th Cong., 1st Sess. tit. II, § 201(a) (1995) (emphasis added).
102. *Id.*
103. *Id.*

104. *Id.*
105. *Id.*
106. H.R. 914, 104th Cong., 1st Sess. § 1(2) (1995).
107. H.R. 200, 104th Cong., 1st Sess. § 2(a)(E)(I) (1995).
108. *Id.* § 2(a)(O).
109. *Id.*
110. H.R. 1858, 104th Cong., 1st Sess. tit. III (1995). The Senate version of this bill, included in a manager's en bloc amendment to the substitute version of S. 650, 104th Cong., 1st Sess. (Economic Growth and Regulatory Paperwork Reduction Act of 1995) at the Senate's Banking Committee markup session on September 27, 1995, limits the liability of lenders under CERCLA to the actual benefit conferred on the lender by a remedial or other corrective action undertaken by another party. *Lender Liability Bills, supra* note 90, at 566. In the Senate bill, "actual benefit" is defined as the net gain realized by the lender due to the corrective action. *Id.*
111. H.R. 1858, 104th Cong., 1st Sess. tit. III, § 301 (1995).
112. *Id.* § 301(a)(1).
113. *Id.*
114. *Id.* § 301(a)(3)(E).
115. *Id.* § 301.
116. *Brownfields, Possible Liability Reforms Spur Growth in Environmental Insurance,* Hazardous Waste News, Feb. 20, 1995, *available in* LEXIS, Nexis Library.
117. *See, e.g.,* ECS UNDERWRITING, INC., ENVIRONMENTAL INSURANCE PRODUCTS AND SERVICES (1994) (describing policies, policy requirements, and fees).
118. *See* Robert J. Gilbert, *Playing It Safe in Today's Real Estate Market with Environmental Insurance,* MASS. LAW. WEEKLY, Dec. 5, 1994, at 1.
119. Mark Vuono & Robert P. Hallenbeck, *Redeveloping Contaminated Properties,* RISK MGMT., Apr. 1995, at 58–59, 62; *see Policy Sets Cap on Cleanup Liability: New Insurance Product Designed to Promote Brownfields Redevelopment,* SUPERFUND REP., July 26, 1995, at 38 [hereinafter *Cap on Cleanup Liability*].
120. *See Cap on Cleanup Liability, supra* note 118, at 38. For example, Eric Underwriters Agency, Inc. offers stop-loss insurance with premiums ranging from $35,000 to $45,000, a limit of $2 million, and a $50,000 deductible.

CHAPTER 9

The Deductibility of Environmental Remediation Costs

BERNARD J. JERLSTROM
ALAN S. DORIS

To Deduct or to Capitalize?—That Is the Question

Understanding the federal income-tax treatment of environmental remediation costs is critical to brownfields development because the dollars involved are staggering. It has been estimated that the cost of remediating known brownfields sites in America could be in excess of $650 billion.[1] Unfortunately, this area of tax law is perplexing, because it has evolved in a piecemeal fashion. The pivotal question that must be addressed concerning cleanup costs and costs attendant to cleanup costs, such as the costs for site investigations, legal fees, and oversight expenses, is whether these costs (1) are currently deductible, or (2) must be capitalized and added to the property's tax basis.

The Statutory Scheme

To be deductible, a remediation expense must qualify under the rules for ordinary and necessary business expenses contained in section 162(a) of the Internal Revenue Code of 1986 (the Code) and the Income Tax Regulations promulgated thereunder (the Regulations). Although Code section 162(a) contains a broad, general rule for the deductibility of "ordinary and necessary" business expenses, this general rule is limited by Code section 263(a), which denies deductions for capital expenditures. Section 263(a)(1) provides that no deduction shall be allowed for permanent improvements or betterments made to increase the value of any property or estate. Because remediation expenses could, theoretically, fall within the purview of either section 162(a) or 263(a), a taxpayer must distinguish between remediation expenses that are properly deductible under section 162(a) and those that are subject to the capitalization rules of section 263(a). (Remediation expenses are not specifically addressed by either section 162(a) or 263(a)).

This distinction is made by analogizing certain remediation expenses to repairs (which are currently deductible), as opposed to permanent improvements or betterments (which must be capitalized and depreciated over their useful life). The Regulations state the following about repairs:

> The cost of incidental repairs which neither materially add to the value of the property nor appreciably prolong its life, but keep it in an ordinarily efficient operating condition, may be deducted as an expense, provided the cost of acquisition or production or the gain or loss basis of the taxpayer's plant, equipment, or other property, as the case may be, is not increased by the amount of such expenditures. Repairs in the nature of replacements, to the extent that they arrest deterioration and appreciably prolong the life of the property, shall either be capitalized and depreciated in accordance with section 167 or charged against the depreciation reserve if such an account is kept.[2]

Accordingly, to qualify as a current deduction, a remediation expense must neither (1) add to the value of property, nor (2) appreciably prolong its life. Although this two-pronged test appears to be straightforward, its application in the area of remediation expenses has spawned considerable controversy. Arguably, all repairs add to the value of property (and perhaps prolong the property's life). However, a large and accepted body of tax law has evolved that stands for the proposition that the cost of a repair is deductible if the predominant result is to restore property to the condition (and value) that existed before the damage (or wear) that necessitated the repair. Yet sometimes, the distinction between a deductible repair and a capital expenditure is not altogether clear. For example, a property owner could deduct, as a repair, the replacement of a portion of a leaking roof to prevent further damage to a building. However, if the property owner replaced the entire roof, such cost would be a capital expenditure.

Although basing its foundation on this rather meager statutory and regulatory scheme, this body of tax law has been largely fashioned by case law and by the pronouncements of the Internal Revenue Service (the Service). Therefore, to understand the current state of the law, one must examine both the case law and the Service pronouncements.

The Case Law

Brief Overview

This section explores six cases that stand for four major principles of taxation. Two of these cases, *Illinois Merchants Trust Co. v. Commissioner*[3] and *Plainfield-Union Water Co. v. Commissioner*,[4] are landmark cases that established the principles for which they are cited. The other cases are illustrations of the principles for which they are cited and were chosen because they have often been cited in the Service's pronouncements.

The *Illinois Merchants Trust Co.* case is the seminal case establishing the principle that repair costs are deductible. In *Illinois Merchants Trust Co.*, the court crafted a distinction between a repair cost and a capital expenditure, and enunciated the principle that the purpose for an expenditure is an important factor in the determination of deductibility. Next, the *Plainfield-Union* case established the principle that in determining whether an expenditure has materially added to the value of property, a comparison must be made between the value of the property after the repair and the value of the property before the onset of the condition that necessitated the repair. The third and fourth cases, *American Bamberg Corp. v. Commissioner*[5] and *Oberman Manufacturing Co. v. Commissioner*,[6] stand for the proposition that neither the permanency nor substantial nature (or cost) of a repair is relevant to the question of deductibility. Finally, *Mountain Fuel Supply Co. v. United States*[7] and *Wolfson Land & Cattle Co. v. Commissioner*[8] illustrate the "plan of rehabilitation" doctrine, which denies deductibility when repair expenditures are part of an overall plan of rehabilitation including capital expenditures.

The Judicial Foundation for the Deductibility of Repair Costs

The case law on the deductibility of repairs goes all the way back to 1926, when the Board of Tax Appeals (the forerunner of the present Tax Court) decided the *Illinois Merchants Trust Co.* case. In this case, a taxpayer incurred costs to shore up a wall and repair a building foundation to prevent the building from collapsing. The Board of Tax Appeals drew a distinction between a repair—characterized as an expenditure for the purpose of keeping property in an orderly and efficient operating condition—and a replacement—an alteration, improvement, or addition that prolonged the life of property, increased its value, or made it adaptable to a different use. The first type of expenditure (a repair) was characterized as a deductible maintenance charge, while the latter types of expenditures were characterized as additions to capital, which should not be applied against current earnings.[9]

In addition to drawing a distinction between repairs and capital expenditures, the court fashioned another major tax principle when it stated that

> [i]n determining whether an expenditure is a capital one or is chargeable against operating income, it is necessary to bear in mind the purpose for which the expenditure was made.[10]

The *Plainfield-Union* Test

The taxpayer in *Plainfield-Union* was a water company that used tar-painted, cast-iron water pipes to supply well water to its customers. In 1950, the taxpayer tied one of its water mains into another water system to provide itself with an additional source of water. The new system used river water, which caused damage to the taxpayer's water pipes. As a result, the taxpayer decided to clean the pipe and reline it with cement. The amount of pipe cleaned and lined was approximately one-half of one percent of the taxpayer's pipes.

The issue, of course, was whether the cost of cleaning and lining the pipe with cement was a deductible repair or a capital expenditure.

The court noted the fact that a repair must *materially* add to the property's value and *appreciably* prolong its life before it can be denied as a deduction. In deciding whether the repair materially added to the property's value, the court stated that any properly performed repair adds value, as compared with the situation existing immediately before the repair. In the court's view, however, such a comparison was not the proper test. Therefore, the court stated that

> [t]he proper test is whether the expenditure materially enhances the value, use, life expectancy, strength, or capacity as compared with the status of the asset prior to the condition necessitating the expenditure.[11]

The court also stated that deductible repairs did not require a relatively sudden, unexpected, or unusual external factor that results in casualty damage.

Whether a Repair Is Permanent or Substantial Is Not Dispositive of the Question of Deductibility

The taxpayer in *American Bamberg* built a large rayon plant not far from a river in Tennessee. In 1940, major cave-ins of the floor took place in the plant's spinning room, causing one machine to disappear completely. The taxpayer tried emergency repairs, but to no avail. It then hired a subsoil engineering firm, which instituted an extensive program calling for drilling through bedrock and injecting a cement grout into the cavities. The expenditures for this program exceeded $900,000.

The Tax Court held that the purpose of the expenditures was not to improve, better, extend, or increase the original plant, nor to prolong its useful life, but rather, to avert a plantwide disaster. Also, the scope of the work did not concern the court, which believed that the work could not have been successful on a smaller scale. Accordingly, the costs were currently deductible.

Oberman Manufacturing involved a taxpayer engaged in the manufacture of men's clothing. The plant it leased in Arkansas developed a leaky roof. The roof was built of steel decking sections covered with a roofing material called perlite and built up with asphalt and gravel. To correct the leaks, the taxpayer had the asphalt, gravel, and perlite roof covering removed (leaving the basic steel decking intact). The company inserted a wood-framed, copper-covered, roof expansion joint that ran the length of the building, and had the roof re-covered with a layer of one-half-inch fiberboard insulation and a twenty-year, four-ply asphalt and gravel top. Once again, the court focused on the work's purpose (to prevent leaks and keep the plant in operating condition). Therefore, as in *American Bamberg*, the court was not concerned by the scope of the work, which included structural change. Hence, the costs for this work were currently deductible.

The Plan-of-Rehabilitation Doctrine

In *Mountain Fuel Supply Co.*, a taxpayer, which was an operator of a natural-gas pipeline system, reconditioned part of its pipeline. The taxpayer operated approximately 5,300 miles of transmission lines, distribution lines, and gathering lines. The pipelines were originally laid in 1929 or 1930 and were reconditioned in 1962, 1963, and 1964. Only about 40 of the 5,300 miles of pipeline were reconditioned. The segments selected for reconditioning were those sections with a history of leaks.

The reconditioning process entailed (1) removing the pipe from the ground, (2) straightening the pipe, and (3) removing the old couplings, expansion joints, valves, and defective welds. The pipe's interior was cleaned and the exterior was recoated and wrapped. The taxpayer treated some of the total reconditioning costs as a currently deductible expense and some of the costs as a "new investment."

The Service challenged the deductions, but the trial court found for the taxpayer because the work was done to repair defects and leaks in the pipe and to maintain the line's reliability. The trial court also found that the work was not part of a general plan of betterment. The Tenth Circuit Court of Appeals reversed, holding that the work constituted a plan of rehabilitation.

Although the pipes that were reconditioned had a history of leaks, the Tenth Circuit found that the reconditioned pipeline sections had reached or passed the period initially assigned as their useful life. According to the court, the record demonstrated that the work was done not to enable the pipeline to continue to operate during its initial estimated life, but to enable the pipeline to begin a new period of expected life. Furthermore, the reconditioned pipe could operate at a greater pressure (almost double) than before reconditioning. The court stated that a general plan of rehabilitation need not encompass all of a taxpayer's properties or even all of one particular category of property.

The next major case discussing the plan-of-rehabilitation doctrine was *Wolfson Land & Cattle Co.* This case involved a cattle ranch that had an irrigation system consisting of ditches and levees that gradually became clogged with sediment. This irrigation system had to be cleaned by periodic draglining. Instead of instituting a maintenance program, the taxpayer elected to operate the system until it was almost dysfunctional, at which point the system was repaired (draglined).

In requiring the costs to be capitalized, the Tax Court analogized the irrigation system to the pipeline system in *Mountain Fuel Supply Co.*, but also stated that the expenditures in the instant case created a separate, amortizable intangible of a useful life much shorter than that of the pipeline system. This finding was extremely important to the court's decision. The Tax Court stated that the expenditures were a part of a systematic plan to dragline the irrigation system, and that a cleaned ditch or reworked levee was more valuable than one in need of repair. Therefore, to the extent that the expenditures had the effect of replacing the previously wasted intangible created by the last draglining of a ditch or levee, there was substantial impact on the value of the system, as well as the creation of a separate item of value.[12] The creation of a separate and distinct asset with a life expectancy in excess of one year invariably results in the capitalization of costs.[13]

Summary of Relevant Principles

- The purpose for which an expenditure is made is important to determining whether it qualifies as a "repair."
- A "repair" keeps property in an orderly, efficient operating condition, and does not materially increase the property's value or appreciably prolong its life beyond its original expected life.
- In determining whether an expenditure adds material value, a comparison must be made between the value of the property after the expenditure and the value of the property before the condition that necessitated the expenditure.
- Neither the permanency of a repair nor its substantiality is dispositive of the question of deductibility.
- An expenditure that is part of a systematic plan of rehabilitation that includes capital expenditures will not constitute a "repair" qualifying for a current deduction.

The Service's Rulings

Since 1992, the Service has issued five technical advice memoranda (TAMs)[14] and one published ruling[15] in this area (hereinafter referred to as rulings). Two of the TAMs were based on situations involving insulation containing asbestos, while the other two involved contaminated land. A review of these rulings demonstrates the Service's current position concerning some aspects of the deductibility question.

The Asbestos Rulings

The initial ruling in this area is TAM 9240004, dated June 29, 1992. It dealt with a taxpayer that had manufacturing equipment which was insulated with asbestos material. In response to state and federal health and safety requirements, and to afford its workers health and safety protection, the taxpayer decided to implement an asbestos abatement program. The taxpayer had two alternatives: (1) it could encapsulate the asbestos, or (2) it could remove the asbestos insulation entirely and replace it with different insulating material. The taxpayer chose the latter alternative. The new replacement insulation was about 10 percent less thermally efficient than the asbestos insulation.

The taxpayer treated the cost of removing the asbestos insulation and replacing it with other material as a currently deductible repair cost. The taxpayer argued that the cost was incidental. Although substantial, the cost was minor in relation to the total repair costs of equipment and to the total value of equipment. The taxpayer relied on the rationale contained in *Plainfield-*

Union and argued that the test as applied to asbestos should be a comparison of the value of the asset following asbestos abatement with its value *before asbestos was known to be a health hazard*. (Apparently the taxpayer felt compelled to use this variation of the rationale set forth in *Plainfield-Union* because the equipment contained the asbestos insulation from inception.)

The Service responded by using an unusual rationale. First, the Service stated that the *Plainfield-Union* test is relevant only in situations where repairs are necessary because the property has progressively deteriorated (an expansive reading of *Plainfield-Union*). Second, the Service stated that the taxpayer decided to remove asbestos because it was a health hazard and because removal was the most cost-effective way of dealing with regulatory guidelines (not because its effectiveness as an insulating material had diminished). The Service then noted that several courts have held that modifications to bring property into compliance with local regulations and requirements increases the value of affected property, and stated that the removal of the asbestos significantly reduced or eliminated the possibility that the taxpayer would be forced to suspend operations because of excessive concentrations of asbestos.

Consequently, the Service reasoned, the property was more valuable because it could continuously operate within the regulatory guidelines. Thus, according to the Service's rationale, the equipment had not intrinsically increased in value, but had increased in value as a result of the aforementioned collateral benefits. Therefore, the costs for this improvement had to be capitalized.

The second asbestos ruling is TAM 9411002, dated November 19, 1993, and involves a factual situation with more facets than the first TAM. This case involved a taxpayer engaged in the sale of rental warehouse space and related services. To secure a bank loan, the taxpayer was required to abate asbestos in its facilities. The taxpayer removed boilers and tanks contaminated by asbestos from a boiler house and cleared the room of asbestos. Thereafter, the boiler house was converted into a garage and office space and was rented out. In addition, the lender required the taxpayer to abate the problem of exposed and damaged asbestos pipe insulation in the warehouse. The pipe was rewrapped and encapsulated.

The Service addressed two issues: (1) the removal of the asbestos from the boiler house, and (2) the encapsulation of the asbestos in the warehouse. As for the boiler house, the Service distinguished the instant situation from the *Plainfield-Union* situation, because the costs incurred to remove the asbestos increased the value of the property as compared with the status of the property in its original asbestos-containing condition. The Service used the same line of reasoning contained in TAM 9240004: the taxpayer's expenditures permanently eliminated the health risks posed by the presence of the asbestos in the boiler room (therefore creating better working conditions), and the expenditures enhanced the property's usefulness and capacity by enabling the taxpayer to provide office and garage space made available by the asbestos hazard's elimination. The Service reasoned that the condition necessitating the expenditure was the property's original condition at acquisition and that the remediation increased the facility's value, use, and capacity. Therefore, expenses for remediating asbestos in the boiler room had to be capitalized.

As for the encapsulation of the asbestos around the pipes in the warehouse, the Service stated that such costs could be currently expensed because they neither appreciably increased the value of the property nor prolonged its useful life. The expenditure merely served to keep the property in good operating condition. The Service based its position on the fact that applying canvas or plastic wrapping over damaged pipe insulation reduced, but did not eliminate, the threat of exposure to airborne asbestos fibers. Accordingly, the encapsulation did not materially enhance the property's value, substantially prolong its life, or adapt such property to a new or different use.

The PCB and Hazardous-Substance Rulings

On December 17, 1992, the Service issued TAM 9315004, the first—and extremely controversial—ruling on polychlorinated biphenyls (PCBs). It involved a taxpayer that used PCB-containing lubricant in some of its equipment. The equipment manufacturer advised the taxpayer that because PCB was a hazardous substance, it would no longer sell lubricant containing PCBs. Accordingly, in 1973, the taxpayer switched to a non-PCB lubricant.

Shortly thereafter, the taxpayer drained and flushed the PCB-contaminated units in numerous earthen pits and trenches on its property. The United States Environmental Protection Agency (USEPA) filed suit and the parties entered an agreement to remediate the contaminated areas. The cleanup program was anticipated to extend for a number of years. The agreement required the taxpayer to perform environmental compliance audits and develop a PCB manual for use by company personnel.

The taxpayer treated all the costs associated with the cleanup activities as a current expense, with the exception of the costs of equipment and associated facilities and groundwater monitoring wells, which costs it capitalized. During the tax years in question (and future years) the taxpayer incurred (or anticipated incurring) costs for the following:

- Soil contamination assessment
- Groundwater contamination assessment
- Remediation, which involved excavating and transporting PCB-contaminated soils and backfilling
- Legal fees to defend against charges by the USEPA and state agencies, and claims by third parties
- Oversight costs for the cleanup operations
- The costs of the environmental audit and the compliance manual
- Research and development expenses for chemical remediation processes that might facilitate the remediation of PCBs

The Service discussed the deductibility issue by separating the costs into two classes: (1) environmental cleanup costs, and (2) legal fees, and oversight and environmental auditing costs. As for the cleanup costs, the Service analogized the situation to the *Wolfson Land & Cattle Co.* case. The Service stated that, like the taxpayer in *Wolfson Land & Cattle Co.* that chose to forego performing annual ditch maintenance, the taxpayer in the ruling failed to institute an ongoing program of waste identification and disposal. The Service believed the taxpayer could have disposed of the PCB-containing residues in a manner that would have obviated the need for "extraordinary clean-up." Further, both the taxpayer in the ruling and the taxpayer in *Wolfson Land & Cattle Co.* made the expenditures as part of a systematic plan involving extensive identification and/or remediation activities. Finally, relying on the reasoning in *Wolfson Land & Cattle Co.*, the Service stated that the property was more valuable after the PCBs had been cleaned up (thus ignoring the rule in *Plainfield-Union* that the comparison must be made to the property before it was contaminated).

The Service sidestepped *Plainfield-Union* by stating that the *Plainfield-Union* court viewed the value test as merely one of the determining factors, and that the decision in that case was based on a full consideration of the entire factual context. The Service also distinguished the facts in the ruling from *Plainfield-Union* by stating that the repairs in *Plainfield-Union* involved a very minor part of that taxpayer's operation and were not part of a general plan.

In deciding that the remediation constituted a general plan of rehabilitation, the Service focused on the fact that the cleanup program was a long-term, systematic program that had many aspects. Therefore, the cleanup operations, taken in their entirety, would result in permanent betterments. The betterments listed by the Service included transforming the land into land that was no longer contaminated, avoiding further government penalties, providing a safe environment for workers and adjoining property owners, and increasing the property's marketability. Thus, the costs for cleanup had to be capitalized.

As for legal fees, the Service held that their deductibility was determined from the lawsuit's character. The Service held that the legal fees were incurred to protect the taxpayer's business by defending it against claims and securing its contractual rights with insurers. Thus, these legal fees were not part of the plan of rehabilitation and, therefore, were currently deductible. However, because the oversight costs were incurred in connection with the cleanup program, the Service held that these costs were part of the overall rehabilitation plan and should be capitalized.

The Service felt that further development by Service field personnel was necessary before a determination could be made about whether the costs associated with the environmental audits and the compliance manual could be deducted. The Service needed more facts to determine whether these items were also part of the plan of rehabilitation.

The second PCB ruling issued by the Service was Revenue Ruling 94-38.[16] In that ruling, a corporation operated a manufacturing plant on property that it purchased in 1970, which was not contaminated when acquired. In 1993, to comply with federal, state, and local environmental laws, the taxpayer decided to remediate soil and groundwater at its property, which had become contaminated by hazardous waste. The soil remediation project began in 1993, with a projected completion date in 1995. The taxpayer also began constructing groundwater treatment facilities, which included wells, pipes, pumps, and other equipment, to extract, treat, and monitor contaminated groundwater.

The Service held, in a ruling with a relatively brief analysis, that the groundwater treatment facilities had to be capitalized. However, the soil remediation expenses and ongoing groundwater treatment expenditures (other than the expenditures to construct the groundwater treatment facilities) could be deducted. The ruling cites *Plainfield-Union*, and states that the proper test in determining whether an expenditure increases the value of property is to compare the value of the asset after the expenditure with the value of the asset before the condition arose that necessitated the expenditure. The ruling also held that costs incurred to evaluate the soil and groundwater contamination were deductible. Finally, the ruling modified Revenue Ruling 88-57[17] to the extent it implied that the value test applied in *Plainfield-Union* cannot be an appropriate test in any case other than one in which there is a sudden and unanticipated damage to an asset.

It is interesting to note that although the remediation project would span a period of at least three years, the Service did not treat it as a plan of rehabilitation (as it did the cleanup project in TAM 9315004). Also, the Service did not focus on the intangible benefits it had cited in TAM 9315004, such as the value derived by the taxpayer because (1) the land would no longer be contaminated, (2) government penalties would be avoided, and (3) a safe working environment would be provided for workers. The holding seems to indicate that the Service has abandoned its position that expenditures compelled by compliance with governmental regulations should be capitalized. Finally, the *Plainfield-Union* test appears to have been firmly reestablished. Although Revenue Ruling 94-38 did not address all the unanswered questions in this area, tax practitioners hailed the ruling as taxpayer friendly and a large step in the right direction.

The next deductibility ruling, TAM 9541005, dated September 27, 1995, involved a subsidiary (in a consolidated group of companies) that owned a piece of property that had originally been used for farming. The subsidiary's predecessor purchased the property, which over two decades became a site for the disposal of industrial waste, such as agricultural chemical wastes and coke oven by-products. The following chronology of events took place:

- Year 3: The subsidiary contributed the land to the county, which planned to use it as a recreational park.
- Year 4: The county discovered the land was contaminated and ceased developing the recreational park.
- Year 5: The county conveyed the land back to the subsidiary for $1.
- Year 6: The state and the USEPA conducted tests on the land, which revealed the presence of hazardous substances.

- Year 7: The land was designated as a Superfund site under the Comprehensive Environmental Response, Compensation, and Liability Act of 1980 (CERCLA).
- Year 8: The subsidiary entered a Consent Order with the USEPA for the purpose of completing a remediation investigation and feasibility study.

The subsidiary incurred the following costs: fees paid to an engineering firm for an environmental study, legal fees paid in connection with the Consent Order and the retention of consulting firms, and fees paid to consulting firms for various activities in connection with the land. The taxpayer treated the costs as environmental remediation costs and, using Revenue Ruling 94-38 as its authority, deducted the costs under Code section 162.

The Service, on the other hand, held that Revenue Ruling 94-38 did not apply to this situation. The Service based its conclusion on a very formalistic application of the restoration principle enunciated in the *Plainfield-Union* case. The Service stated that the restoration principle envisions that a taxpayer acquire property in a clean condition, contaminate the property in the course of its everyday business operations, and incur costs to restore the property to the same condition existing at the time the taxpayer acquired the property. Hence, the Service found that Revenue Ruling 94-38 was inapplicable in this instance because the taxpayer acquired the land in a contaminated condition in Year 5. The ruling also stated that the taxpayer had not provided sufficient facts to determine whether, under the general auspices of Code section 162 (the ordinary and necessary business expense provision), the costs were deductible under other theories relevant to deductibility (for example, that the costs were incurred to avoid litigation).

However, on December 18, 1995, the Service received a request for reconsideration from the District Director with jurisdiction over the taxpayer (who did not dispute the amount or purpose of the costs incurred by the taxpayer). The Service reconsidered the matter and in a subsequent ruling (released by the taxpayer on January 23, 1996), revoked and superseded TAM 9541005. In its subsequent ruling, the Service stated that the theory of Revenue Ruling 94-38, though the facts in that ruling were different because there was no interim break in ownership, applied to the instant situation. This holding was based on the rationale that the same taxpayer that contaminated the land incurred the associated cleanup costs. The ruling stated that the contamination to the land and the taxpayer's liability for remediation were unchanged during the break in ownership. Although a favorable ruling, it probably has limited applicability due to the unique set of facts involved.

Observations

- The *Plainfield-Union* doctrine is applicable only when the property was not contaminated when acquired. The taxpayer that incurs the remediation costs must be the party that has contaminated the property.

- The *Plainfield-Union* doctrine is applied by comparing the property's value after the expenditure to the value of the property before the condition that necessitated the expenditure.
- Not surprisingly, the creation of facilities as part of a remediation project must be capitalized.
- Applying the plan-of-rehabilitation doctrine will result in the disallowance of remediation expenditures when they are part of an overall plan of rehabilitation, which includes capital expenditures.
- Legal and oversight fees incidental to a remediation project should be deductible if the cleanup costs are deductible. Legal fees are also deductible if they are incurred to protect a business.
- An interim break in ownership will not be fatal if the taxpayer that incurred the remediation expenses is the same party that contaminated the property.

Proposed Legislation

Given the Service's nebulous treatment of remediation expense deductibility, President Clinton's 1997 balanced-budget proposal (which was released on March 19, 1996) contains a provision for tax legislation related to environmental remediation costs (the Proposal). The Proposal provides that taxpayers could elect to treat certain environmental remediation expenditures that would otherwise be chargeable to a capital account as deductible in the year paid or incurred. However, the expenditures must be incurred in connection with the abatement or control of hazardous substances at a "qualified contaminated site." The following brief explanation of the Proposal is based upon both the language contained in the Proposal and the Joint Committee on Taxation's description of the Proposal.[18]

"Qualified" sites would be properties that satisfy use, geographic, and contamination requirements. In general, a "qualified contaminated site" would be any property that (1) is held for use in a trade or business, for the production of income, or as inventory, (2) is certified by the appropriate state environmental agency to be located within a targeted area, and (3) contains or potentially contains a hazardous substance.

Targeted areas generally would include (1) empowerment zones and enterprise communities (as designated under current law and to be designated under the Proposal), (2) sites announced before February 1, 1996, as being included in the brownfields pilot project of the USEPA, (3) any population census tract with a poverty rate of not less than 20 percent, and (4) certain industrial and commercial areas that are adjacent to tracts described in (3). Sites that are identified on the National Priorities List under CERCLA would not qualify as targeted areas.

The contamination requirement would be satisfied if hazardous substances are present or potentially present on the property. Hazardous substances would be defined generally by reference to sections 101(14) and 102 of CERCLA, subject to additional limitations applicable to asbestos and similar substances within buildings, certain naturally occurring substances such as radon, and certain other substances released into drinking-water supplies due to deterioration through ordinary use.

Under the Proposal, any deduction for a qualified environmental remediation expenditure that, absent the proposed legislation, would have to be capitalized would be treated as a depreciation deduction and would be subject to the recapture rules of section 1245 of the Code. Accordingly, there would be recapture (as ordinary income) in the event of any sale or other disposition of the property that produces a gain. The Joint Committee on Taxation's explanation states that in general, any expenditure for the acquisition of depreciable property used in connection with the abatement or control of hazardous substances at a qualified contaminated site would not constitute a qualified environmental remediation expenditure. Thus, expenditures for such property would have to be capitalized. The explanation goes on to state, however, that depreciation deductions allowable for property, which would otherwise be allocated to the site under the principles set forth in *Commissioner v. Idaho Power Co.*[19] and Code section 263A, would be treated as environmental remediation expenditures. (*Idaho Power Co.* stands for the proposition that the depreciation costs of equipment used to construct capital improvements must be capitalized as part of the capital improvement.)

The Proposal states that deductible remediation expenses would generally be limited to those paid or incurred in connection with the abatement or control of environmental contaminants. Therefore, expenses incurred for the demolition of existing buildings and other structural components would not qualify for this treatment, except in unusual circumstances when the demolition was required as part of ongoing remediation.

The Proposal would apply to expenditures paid or incurred after the date of enactment, in taxable years ending after such date. It contains a "sunset" provision and would expire on January 1, 2001, if the fiscal dividend for the year 2000 is not at least $20 billion.

Thus, on the deductibility issue, the Proposal would dramatically change the present state of the law, and might all but eliminate controversies between taxpayers and the Service. However, questions may still arise, such as which expenditures are to be treated as depreciation deductions for these purposes. Of course, it is impossible to predict with any certainty whether the Proposal will be enacted into law, when it would be effective, and what its final form would be. If the Proposal does become law there will be a window of opportunity for the period beginning with the effective date of the Proposal until January 1, 2001 (unless the fiscal dividend for the year 2000 is at least $20 billion or there is an extender for any legislation that may be enacted). Practitioners should monitor the Proposal so that they can best advise clients involved in brownfields remediation projects.

State Tax Incentives

Many states offer tax incentives to taxpayers who conduct environmental remediation activities. Although such incentives are not fashioned as tax deductions (they usually take the form of either a tax credit or an abatement of tax), they provide a current benefit and thus merit mention in any discussion of this issue. Practitioners should not overlook these incentives when advising clients about remediation activities. For example, Ohio has passed legislation that provides for tax abatement.[20] A brief description of the Ohio tax-abatement legislation is provided as an example of state legislation of this type.

Pursuant to certain provisions of the Ohio Revised Code, the Ohio Director of Environmental Protection, after issuing a covenant-not-to-sue on a property participating in the Ohio Voluntary Action Program, can determine if remedial activities have commenced or been completed. If such a determination is made, the Director can certify the matter to the Ohio Tax Commissioner and the Director of Development.[21] Upon receipt of the certification, the Tax Commissioner issues an order granting an exemption from real-property taxation for (1) the amount of the increase in the assessed value of the land that is described in the certification, and (2) the increase in the assessed value of improvements, buildings, fixtures, and structures situated on that land at the time the order is issued.[22] The exemption commences on the first day of the tax year in which the order is issued and ends on the last day of the tenth tax year after the issuance of the order.[23]

A transfer of the property (by sale or otherwise) will not affect the exemption, and the exemption continues for the full period stated in the exemption order.[24] If at any time, the Director of Environmental Protection revokes the covenant-not-to-sue, which is the basis for the Tax Commissioner's issuance of an exemption order, the Director notifies the Tax Commissioner and the legislative authority of the municipal corporation and county in which the property is located, and the exemption order is rescinded.[25] Upon revocation of the covenant-not-to-sue, the property's record owner pays the amount of tax that would have been charged against the property had the property not been exempted from taxation, for the period beginning with commencement of the exemption and ending with the date of revocation.[26]

Applicability of the Tax Law to Brownfields

The environmental remediation of brownfields that are to be sold by a current owner can generally be accomplished in one of four ways: (1) the owner can remediate the property and then sell it, (2) the owner can issue an option for the sale of the property and remediate it before sale, (3) the owner can sell the land in its contaminated condition, with the purchaser taking on the task of remediation, or (4) the owner can sell the land in its contaminated condition, but agree to pay the costs of remediation after the sale. Set forth below is a short, general discourse on the applicability of the law, both as it presently stands and under the Proposal, to these four scenarios. Of course, this discourse—as it relates to the Proposal—is conjecture at best, as we have no way of knowing how a tax provision that may evolve from the Proposal would actually be worded. The material set forth below is based on an assumption that brownfields cleanup projects will generally involve only contaminated land;

thus, no consideration is given to the removal of asbestos from machinery or buildings in the discussion that follows.

First Scenario: The Current Owner Remediates the Property

Under the present state of the law in this area, if the current property owner is the party that has contaminated the site, such owner should be able to deduct the actual costs of remediation under the *Plainfield-Union* doctrine. However, the cost of the creation of facilities as part of a remediation project would have to be capitalized. Taxpayers should also remember that the Service might try to invoke the plan-of-rehabilitation doctrine to disallow deductions if the remediation project is extensive.

Under the Proposal, the actual cleanup costs would also be deductible. The Joint Committee on Taxation's explanation states that expenditures for the acquisition of depreciable property used in connection with the abatement or control of hazardous substances at a qualified contamination site would not constitute a qualified environmental remediation expense. Because any legislation would probably contain such a concept, the costs relating to the creation of facilities would probably have to be capitalized.

If the property was located in Ohio, it would be eligible for the Ohio tax-abatement program. Under the provisions of the program, once remediation has been commenced the property would be exempt from property tax for (1) the amount of the increase in the assessed value of the land, and (2) the increase in the assessed value of the improvements, buildings, fixtures, and structures located on the land at the time the order granting the exemption was issued by the Tax Commissioner.

Second Scenario: The Current Owner Grants an Option and Remediates the Property before Its Sale

If the current owner was the party that contaminated the property and is the owner of the property when the remediation occurs, then under current law, the current owner should be able to claim deductions for cleanup costs under the *Plainfield-Union* doctrine. The same caveats would still apply concerning the capitalization of expenditures to create facilities, and concerning the possibility that the Service could invoke the plan-of-rehabilitation doctrine.

Under the Proposal, the result would be the same as that set forth above for the first scenario. Further, if the property was situated in Ohio it would be eligible for the Ohio tax-abatement program, and would be eligible for exemption from property tax for the increase in its value.

Third Scenario: The Current Owner Sells the Property and the New Owner Remediates the Property

It is this third scenario that presents the biggest problem under the current state of the law. Because the Service has taken the position that the *Plainfield-*

Union doctrine is not applicable to property that is contaminated when acquired, a subsequent purchaser would be denied deductions for remediation expenses—obviously a bad result. If the prospective purchaser has agreed to take on the remediation project, it might be better (from a tax standpoint) for the purchaser to contract with the current owner (seller) to undertake the project on behalf of the seller, and to remediate the property before the time of sale.

The result under the Proposal for this scenario is not clear. Neither the Proposal nor the Joint Committee on Taxation's explanation discusses the issue of whether a deduction is available to a party who has not contaminated the property. Based on the reason stated for the Proposal (the need to clean up thousands of neglected or underutilized sites), it would seem that the Proposal is intended as an incentive, and, therefore, that any legislation that might be enacted might allow the deduction to any party (that is, a subsequent purchaser as well as the party who contaminated the land).

As in the first and second scenarios, the property, if situated in Ohio, would be eligible for the Ohio tax-abatement program even if the current owner (the seller) is the party who receives the certification from the Ohio Tax Commissioner, and even if the current owner begins the remediation project before sale. As noted, section 5709.87(D) of the Ohio Revised Code provides that a transfer of the property will not affect the exemption.

Fourth Scenario: The Current Owner Sells the Property and Agrees to Remediate the Property after the Sale

If the current owner was the party that contaminated the property, the current owner should be able to claim deductions for the cleanup costs incurred after the sale. Moreover, if cleanup costs are placed into a qualified settlement fund, such costs might be deductible when the trust fund is established.[27]

Calculation of Actual Tax Savings

The scenarios set forth above do not attempt to determine whether taxes will actually be saved by allowing the present owner to deduct the costs of remediation currently, as opposed to selling the property for a lower price and having the purchaser conduct remediation. For example, if the seller is a United States C corporation, has always paid current taxes at maximum regular corporate income-tax rates (as opposed to alternative minimum tax rates), has no capital loss carry-forwards, and is selling the property at a gain (despite the need for remediation), there is likely no tax advantage to expensing costs of remediation as opposed to having a lower capital gain.

However, if the sale at a lower cost would result in a capital loss, if the taxpayer is an S corporation or has capital loss carry-overs, or if numerous other individual characteristics of the taxpayer should exist, there could be a tax advantage to different treatment. Each company involved in remediation efforts must examine its own unique tax characteristics to determine its own best strategy.

Notes

1. U.S. Gov't Accounting Office, Community Development, Reuse of Urban Industrial Sites, GAO\RCED-95-172, (1995).
2. Treas. Reg. § 1.162-4 (1958).
3. Illinois Merchants Trust Co. v. Commissioner, 4 B.T.A. 103 (1926).
4. Plainfield-Union Water Co. v. Commissioner, 39 T.C. 333 (1962), *nonacq.*, 1964-2 C.B. 8.
5. American Bamberg Corp. v. Commissioner, 10 T.C. 361 (1948), *aff'd*, 177 F.2d 200 (6th Cir. 1949).
6. Oberman Mfg. Co. v. Commissioner, 47 T.C. 471 (1967), *acq.*, 1967-2 C.B. 3.
7. Mountain Fuel Supply Co. v. United States, 449 F.2d 816 (1971), *cert. denied*, 405 U.S. 989 (1972).
8. Wolfson Land & Cattle Co. v. Commissioner, 72 T.C. 1 (1979).
9. *Illinois Merchants Trust Co.*, 4 B.T.A. at 106.
10. *Id.*
11. *Plainfield-Union*, 39 T.C. at 338.
12. *Wolfson Land & Cattle Co.*, 72 T.C. at 17.
13. Note: After the decision in *Indopco, Inc. v. C.I.R.*, 503 U.S. 79 (1992), *aff'g* 918 F.2d 426 (3d Cir. 1990), a separate asset need not be created for costs to be capitalized. If the expenditure will provide a long-term benefit (one that will last beyond the current year), capitalization may be required.
14. TAMs are issued by the Service's National Office upon the receipt of a request from either a District Director or an Appeals Office, and contain guidance regarding the proper interpretation or application of the tax law for technical or procedural questions that have arisen upon the audit of a taxpayer.
15. A letter ruling is a written statement, issued to a taxpayer by the Service's National Office, that interprets and applies the tax law to the taxpayer's specific set of facts. Ruling requests are initiated by taxpayers.
16. Rev. Rul. 94-38, 1994-1 C.B. 35.
17. Rev. Rul. 88-57, 1988-2 C.B. 36.
18. Joint Committee on Taxation's Description of Revenue Proposals Contained in President Clinton's Budget Plan as released on March 27, 1996.
19. Commissioner v. Idaho Power Co., 418 U.S. 1 (1974).
20. Ohio Rev. Code Ann. § 5709.87 (Baldwin 1995).
21. *Id.* § 5709.87(B).
22. *Id.* § 5709.87(C).
23. *Id.*
24. *Id.* § 5709.87(D).
25. *Id.* § 5709.87(E).
26. *Id.*
27. *See* Treas. Reg. § 1.468B-1 (1992).

CHAPTER 10

Building Consensus for the Project

HOWARD C. LANDAU

Land-use discussions, whether regarding brownfields or greenfields, seldom begin with consensus. Money is a powerful motivator in a free society; so any change that is perceived to impact property value—upward or downward—is bound to create controversy. Whether the motive is greed or jealousy, improving a blighted area for the future, or changing its use to a more or less intensive one, the dialogue surrounding the debate can often be heated. To complicate matters further, the media is going to weigh in to do its duty as a third party providing news and information to the public at large.

Given that a contentious scenario is assured, a blueprint can be used to navigate the waters and create both a process and an outcome that will leave a long-term constituency for the new use of a brownfield. An understanding of the roles the various parties typically play will help create the communications process.

The Cast of Characters
The Developer

Let us call the developer "the visionary." It is a better word, and puts him or her in the role of the idea person. Not every idea is a good one, however, so the developer's role must be to educate. Vision without explanation never sells. Many visionaries were considered kooks when they started out, because they did not think about anything or anyone other than their ideas and themselves. Inventors and researchers who work in isolated labs on esoteric experiments may be able to get away with being eccentric, but a real estate developer—especially the developer of a brownfields project—must put forth his or her vision in such a way as to show the logic—indeed the wisdom—of the plan. Without that attitude, the review process is sure to bog down, creating tension and contention, and adding time and cost to the process.

The Allies

The allies are the developer's team—the architect, the planners, the lawyers, and the bank. This team is part of the vision. Its position is clearly understood by all parties—to make the deal happen. That being said, the team has another function: to add credibility and show that the developer's vision is not smoke and mirrors, but doable and in the best interest of the community. The team's role is to aid the education process by demonstrating to all that the concept on the table has been extremely well conceived, that alternatives have been considered, and that the plan is the best possible use for the site. The fact that the site is a brownfield should be positioned as adding to the vision, not subtracting from it. An environmental consultant on the team adds a point of view for the environmental considerations that the regulators and the public are sure to have.

The Skeptics

This group contains everyone else: the government, whose job it is to provide oversight; the neighbors, whose job is it to be nervous; the media, whose job it is to explain the implications; the citizens groups, whose job is it to exert and enhance their political influence; and the Monday-morning quarterbacks, who believe they should be involved because they feel they have something to offer to the discussion. This latter group is the most difficult to predict because it can include surprises, like a person who played in the vacant lot that was there before the use that is there now. Successful development teams uncover these people with unique stories before they unveil their plans, and use them to help demonstrate both their research skills and their sensitivities to the concerns of the community. The skeptics must be sold, and the only way to sell them is to explain the vision in a powerful, yet sensitive, way. That is where good "public relations" comes in.

The Role of Public Relations

The role of public relations in the development of brownfields is to translate, to intermediate between the idea and the execution, to keep the idea as the focal point of the debate, and to not allow the participants to lose sight of it.

Most people think of the public-relations function in terms of media relations. The "PR person" is the one who puts the spin on the story, who writes the glowing press releases, who organizes the fireworks display for the groundbreaking ceremony, and who calls friends at the television station and gets them to come to the public hearing and interview the developer. The media is an important part of public-relations responsibility, but it should be used as a tool in an overall communications program, not as an end in itself.

The developer who believes that he or she can use the media to gain approval for the project is doomed to failure. The media is an indirect means of communication. It cannot tell the story for the developer, and it is bound to turn a complex issue into a boiled-down, mistake-laden litany that unfolds over days or weeks, rather than persuading in a concise, powerful way. The best measure of a successful public-relations campaign is when participants say at the end that they never woke up to a surprise in the newspaper. They heard it from the horse's mouth, and it resonated with them in a positive way.

The Direct Approach

The best way to win friends and influence people is to meet with them directly and show them the power of your ideas. Everyone—especially a person who feels he or she has a stake in the process—wants to be treated specially. It is time consuming to meet with everyone, and it is difficult to draw the line on the page of contacts that have to be reached and say, "enough." If, however, you take the attitude that every important constituency should have the opportunity to hear the idea in the most direct way possible, you will be buying an important insurance policy that will ultimately cut the time and the contentiousness of the approval process.

The Steps

Message Development

Once the core development team is in place, the first step is to create the theme and message for the campaign. A developer should look at the assignment as a political campaign in which he or she lines up allies, creates a powerful, thematic message that captures the vision in "sound bite" form, and then communicates it first to a core group and then to a broader audience.

Think of the proposal as if it were built and ready to occupy. What would it be called? Putting the name on the project from the start will get people thinking about it as a finished reality. If plans are unveiled as a "concept," they will be subject to review and change. If they are presented as a reality, the vision becomes more powerful and harder to tamper with. All the drawings and plans should have the project name on them. A logo or graphic identity is another touch that can add power to a concept. The more ready-to-market the development plan looks, the more real it becomes to the stakeholders.

The Case Statement

Next comes the creation of the case statement. The case statement is the opportunity to explain the project in the developer's own words. This one-page document becomes the piece to give to elected officials, regulators, community groups, and the media. It becomes the rallying point for explaining the benefits of the project to all parties. Once it is created, the other documents become supporting materials to the case; a site plan, an access map, a rendering of the finished project, project specifications, an environmental impact statement, and all the rest.

One of the rules of effective public relations is to have everyone singing from the same hymnal. The case statement is the document that orchestrates the process. If everyone is describing the project in the same way, it can be a very effective way to control the agenda for the debate.

An important component of the case statement is a community impact analysis. This should include facts and figures on the jobs and economic-development impact the project is going to have. This is the real meat of the proposal, and all the stakeholders need to understand that the purpose of development is to take a brownfields site, which is a drain on the resources of the community, and turn it into a paying customer.

Once the message and materials are ready, the next step is to begin the approval process. Other chapters of this book address the specifics of brownfields approvals and regulatory oversight. The remainder of this chapter will deal with building consensus for the development overall.

The First Meetings

Typically, the first stops are at the offices of the mayor and the elected council person from the district where the proposed development is located. These people are going be the most important to success and, from a public-relations point of view, they can help navigate the process with citizens' groups and the media. Understanding their concerns and listening to their advice will get the project off on the right foot. They will also be the first people to whom a reporter will go once the plans become public. Having them say nice things about the project to the reporter will help create a climate of support.

Next Steps

After the first round of one-on-one meetings with local elected officials, the next step usually is to go to a community meeting, often hosted by the council member from the district, to unveil the development plan.

A good strategy is to meet with the leadership of the group first and agree on the agenda and the process for the meeting. Getting their participation (not approval) early is an important signal to send to the community. When people walk into the room to hear the presentation, they should see that the parties know one another and that the mood is not strained. This is the type of environment in which people gain trust and open their minds to new ideas. Other meetings and follow-up reports are then scheduled throughout the process. Developing a constituent mailing list and keeping parties notified of approvals and subsequent meetings is an important administrative step in the ongoing communications process.

The meeting itself has some special considerations. Successful development teams recognize that words alone do not sell vision. When the site is a decaying vacant factory, neighbors have a hard time envisioning a beautiful new shopping center in its place. Before-and-after renderings, actual photographs of the site as it exists (with emphasis on the environmental problems that must be corrected), aerial shots of the site in the context of the community, and proposed site plans and drawings all go a long way to help people share the vision and understand the thoughtfulness of the concept. Providing handouts with the case statement and some of the visual elements of the plan is another way to ensure the participants understand the plan and can share it with their families and friends who are not at the meeting. Viewing each person at the meeting as an emissary for the vision is a public-relations technique that is most successful.

The Media

Every community and every local newspaper, radio, and television station is different in the way it covers real estate development issues. Some put stories on a "Metro" page, some in a "Business" area, and others in a special "Real Estate" section. Especially in the case of brownfields, some media outlets may have a specialist who covers environmental issues.

The decision of who covers the story is not generally the developer's to make. However, having existing relationships with key editors and writers at the local news outlets is always good advice. Should the decision be made to assign the story to a reporter who does not have a background in brownfields issues, it would be advisable to have a discussion with him or her about the differences between developing brownfields and other types of property. The reporter must understand that this is a unique case, with costs, benefits, and requirements different from standard "not in my backyard" rezoning or development matters.

When to involve the media is an important strategic decision that often has to be kept flexible. If, after the first meeting with local elected officials, the developer believes the officials are not going to be receptive to the concept, then the developer may want to go to the editorial board of the local newspaper and ask for a meeting to unveil the plans before any word of the concept leaks. If, on the other hand, the reception is positive, it is always preferable to let more opinion leaders see the plans before they become public knowledge.

In any event, the key to dealing with media people is to respect their roles in the process. Treat them fairly and be accessible to answer their questions. After all, a visionary should always welcome an opportunity to impart his or her vision. "No comment" is not the quote for which the developer of an important brownfields site should be known. It is always a good idea to have one person on the development team designated as the media spokesperson. This policy helps keep the message consistent and makes it easy for the reporters to get comments during the process. Printing the contact's name and telephone number (office and home) on the case statement makes clear where the media should go for information.

During the entire process, the designated spokesperson should keep the media apprised of the status of the plan. As different approvals are garnered, press releases should be put out as a way of keeping the public informed that the plan is going forward. When the public is not in favor of approval, different strategies must be employed to keep the process moving without public disclosure, and oversight, of each and every step.

Building Consensus: An Ongoing Process

If you are the developer in this process, congratulations! You have done a terrific job of creating and communicating the vision of your plan and have received approval. But the job of public relations is not over. It is really just beginning.

Now the process shifts from winning approval to marketing the new use for the former brownfields site. The ground breaking, topping off, and grand opening are all events in the development process when the developer has an opportunity to recognize all the people who helped make this milestone possible. Keeping a good database of all the people who participated in the approval process now proves its worth, by changing into an invitation list to all the public events that will be part of the construction and/or renovation process.

Once the project is completed, the new property owners and tenants will want to be good neighbors in the community that is helping to support their efforts by providing construction workers, permanent employees, shoppers, tenants, and the like. The public-relations process never really ends; and, for a site that was once environmentally contaminated, it will be wonderful to look back some day and see how everyone's efforts resulted in something that is returning tangible value to the community for future generations.

CHAPTER 11

Environmental Insurance in the Brownfields Transaction

WILLIAM McELROY
TODD S. DAVIS

Brief History of Environmental Insurance

Environmental insurance is, quite simply, a contractual vehicle to transfer environmental risks. It is a subdiscipline of the casualty-insurance business, even though some of the applicable concepts are derived from the property-insurance business. Generally, environmental insurance is not treated in the context of other casualty contracts because of the technical risk elements peculiar to environmental losses and the commonly held notion that environmental risk analysis requires highly specialized training.

Environmental insurance, as currently practiced, can be directly tied to the enactment of the Resource Conservation and Recovery Act of 1976 (RCRA).[1] A landmark piece of legislation, RCRA established the statutory basis for two concepts that have become the fundamental foundation of the environmental-insurance business. By far, the most significant impact of RCRA on the insurance business was the statutory codification of the concept now known as "cradle-to-grave" responsibility for "hazardous waste," as defined by statute. This law was an almost entirely new liability concept: that a "generator" should be liable for events and their consequences that are in some way removed from the immediate and/or proximate intervention of the liable party. Secondarily, RCRA imposed a specific regulatory obligation on certain businesses involved in the generation, storage, or management of hazardous waste. To obtain an operating permit for certain facilities regulated under RCRA subtitles C, D, and J, the operator was required to demonstrate "financial assurance" for the ability to take corrective action in response to environmental events and to compensate injured parties. Environmental insurance has become the option of choice for RCRA "financial assurance" compliance for many regulated operators. The products and processes developed to meet this specific need have evolved into other applications, including brownfields projects.

Most insurance disciplines rely heavily on the statistical analysis of historic loss trends. However, the statistical basis of insured environmental-loss data is not sufficient to conduct traditional actuarial rate analysis. To address this lack of fundamental loss data, environmental underwriting has evolved in a different direction. The underwriting of environmental risk has emphasized detailed engineering analysis of each unit of risk rather than analysis and interpretation of historical experience in the aggregate.

The enactment of the Comprehensive Environmental Response, Compensation, and Liability Act of 1980 (CERCLA)[2] substantially expanded concepts of environmental liability and, therefore, the horizons of environmental insurance. Specifically, provisions related to owner/operator, transporter, or arranger liability, under section 107 of CERCLA, have significantly increased the business risks of owning property that may in some way be contaminated with hazardous substances or pollutants.

Applicability of Environmental Insurance to Brownfields Projects

With the emerging sociopolitical drive to reinvest in former industrial areas, those persons intending to build businesses through the development of commercial, industrial, or residential projects on environmentally impacted property will confront unique problems. In addition to bearing all the business risk attendant to a commercial development enterprise, a developer will be consciously exposing its business to environmental risk. A portion of this risk will be expected and can be addressed in the planning process. For example, project financing can be designed to accommodate the expected cost to remediate existing site conditions. Some elements of environmental risk are, on the other hand, fortuitous and totally unforeseeable. In many, if not most, cases of exposure to toxic materials, the manifestations of damages are hidden for years. This observation is particularly true for cases that allege bodily injury to neighbors of, or transients on, the subject property.

The sciences of environmental toxicology and epidemiology lack the tools that would permit early injury recognition. Moreover, it remains extraordinarily difficult to establish a cause-and-effect relationship in many areas of suspected environmentally related disease. As a result, many bodily injury claims are based on statistical conclusions that a specific injury is "likely" to have been caused by a specific environmental condition.

The cost to defend environmentally related bodily injury actions can be very substantial, to the point of undermining the economic viability of a development project. Litigation is often protracted, occupying management time and resources for years. Awards in successful claims can be very large indeed. Although there are defenses, due to evolving standards of liability and jurisdictional variability, a developer must assume that standards of liability to third parties could follow the statutory liability for cleanup imposed by federal and state law. Collectively, these factors create a major barrier to brownfields investment.

Perhaps more problematic is the quantification of the cost and degree of cleanup necessary to assure that the environmental condition of a piece of property is sufficient to protect human health and the environment. Though generally regarded as a matter of technical analysis, this conclusion is influenced as much by sociopolitical judgment as science. Certainly, sociopolitical mores change with time, as does the quality of our scientific understanding of the issues. These changes can influence a project while in the process of development, but also must be considered against an unlimited time horizon. After all, most commercial developments have no real limit on their useful lives.

Insurance can be a vehicle to transfer many of these environmental risks. The relative uncertainty inherent in any particular project can be sufficient to make financing difficult. Laying off this uncertainty to stabilize the economics of a particular project against a large risk pool is a primary function of environmental insurance in the brownfields transaction. In some cases, attracting tenants can be influenced by concerns about the ability of the sponsor, owner, or developer to absorb environmental costs over the long term. Third-party financial assurance in the form of environmental insurance can add tremendous value in the marketing process, as well as providing direct financial security to a developer.

Insurance Underwriting Risks for Environmentally Impacted Properties

The interaction of risk and time creates unique insurance underwriting problems. Commercial insurance contracts are usually issued for a limited term, typically one year. In additional to paralleling the traditional financial planning and evaluation term of companies, a year represents a reasonable time frame to allow for fortuitous events without generating unacceptable administrative cost burdens. Moreover, even companies with poor loss performance can usually rely upon the availability of standard property and casualty insurance at the traditional time of renewal, because of the breadth of capacity in the insurance market. For very complex reasons, the availability of insurance to cover environmental risks from year to year has always been questioned by buyers of this coverage.

Unlike traditional projects needing property and casualty coverage, brownfields projects require a different thought process. Development of the property may need to wait until the completion of an environmental project, or proceed concurrently. It is the actual cleanup and monitoring of environmental conditions at subject properties that often require several years to complete. Though experience gained during the past fifteen years should create little doubt of the future viability of the environmental-insurance market, in some cases it will be difficult to secure funding and governmental support without longer-term commitment from specific insurers. Moreover, a project could be jeopardized if an event creates a minor loss early in its life, reducing the attractiveness of the project to other insurers.

Underwriting risks in brownfields projects can be broken down into three basic categories. For the sake of discussion, the risks can be classified as project risks, technological risks, and regulatory risks. Traditional project risks are those fortuitous risks, inherent in all development projects, that could cause a loss. These project risks include fire, workplace accidents, and design errors or omissions and the like. Technological risks are those risks associated with scientific advancement and limitations, such as the ability to detect substances at increasingly lower levels and the ability to address the presence of contaminants. Finally, regulatory risk describes the risk of changes in laws, regulations, and policies that require reaction or impose liability on an insured party. The underwriter's job is to assess these risks and charge an appropriate premium to cover the true risks posed in each of these risk categories.

The Underwriting Process

Underwriting assessment of brownfields projects must incorporate more functions than most insurance policies, due to the unique risks identified above. It is vitally important that an underwriting company have sufficient quantitative data to evaluate environmental risk in the abstract, as is true of all lines of insurance. The process of laying off project level uncertainty is intellectually sound only if there is a relative degree of certainty in a large population. This evaluation requires significant investment in research and interpretation of quantitative information dealing with environmental-loss experience. Additionally, successful underwriters must be looking to the future and contemplating likely loss scenarios years or decades ahead.

The most important part of the underwriting process for environmental insurance dealing with brownfields projects may be the evaluation of a project's economics. Brownfields projects are, by definition, characterized by a higher degree of risk. Their financial character is often more speculative, with a developer expecting a higher return on investment to compensate for the higher risk.

The risk of "moral hazard"[3] makes it imperative that the carrier have confidence in the business integrity of the applicant. As a result, significant time is committed to review the credentials, experience, and references of a prospective client. This type of review is more closely associated with surety underwriting than traditional insurance underwriting. However, as noted, the nature of environmental risk is not traditionally within the purview of standard lines of insurance.

Quantifying the environmental risk is only part science. Hard data on environmental costs is thin, but available. It requires a great deal of effort merely to acquire credible data. Interpretation of available data and translating data into meaningful risk-rating information requires a nontraditional underwriting skill set. However, the most difficult part of the function is addressing the artistic side of risk quantification. Using historical data to forecast future loss potential is a mixture of technical knowledge, insight, and skepticism.

Ongoing monitoring of the project is likely to be a part of any insurance program supporting a brownfields transaction for the foreseeable future. As a project goes forward, more details related to cost and risk will necessarily emerge. Cooperation in the early recognition of potential problems will also give a sense of confidence to the insurance carrier. Though this cooperation may have value in a current project, it is sure to create value for future projects through the development of goodwill, which is too often lacking in insurance relationships.

Appropriate Coverage and the Insurance Contract

Brownfields transactions are complex commercial real estate ventures, which involve many competing interests. All parties to a deal must reach a contractual agreement for a project to proceed. In many cases, environmental insurance will be a crucial element of that contractual agreement. As a result, arriving at an agreed insurance contract occurs through negotiation. Every insurance program must be customized to suit the needs of the parties. There is no single insurance product that can be identified as *the* environmental-insurance product that facilitates brownfields projects. This is likely to remain the case as long as there are differences in opinion about the nature of real estate contracts, indemnity arrangements, and risk tolerance in these complex endeavors.

With that in mind, a number of coverage parts generally should be considered in any brownfields project. The first coverage part should always be coverage for liability to third parties and defense costs. Second, coverage for required cleanup resulting from newly discovered conditions or the reemergence of previously remediated conditions should certainly be evaluated. Third, coverage to protect against unforeseen escalation in the cost to correct conditions that are a known part of the project (conditions that generated the brownfields label to begin with) should be reviewed.

Products to address these risks can be recognized under a variety of names, including, but not limited to, to following: environmental impairment liability, property transfer liability, remediation stop-loss, and cost cap. As the marketing of these concepts increases in its sophistication, one can expect more colorful labels for what is essentially the same insurance protection. The insurance "products" themselves are not that important, because of the nature of these complex transactions. It is very safe to assume that any "product" that does not respect the technical, legal, or contractual peculiarities of a particular project is not worthy of consideration.[4] Summaries of the primary types of coverage follow.

Environmental-Impairment Liability Coverage

Environmental-impairment liability (EIL) insurance covers third-party bodily injury and property damage claims that result from environmental impairment at a covered location. It also may cover first-party cleanup costs. This coverage includes costs associated with investigation, defense, and remediation related to the cleanup of contamination on the insured's property and any migration of contaminants from the insured's property onto adjacent property. An important exclusion from coverage is costs associated with the presence of contamination identified before policy inception. Therefore, in practice, Phase I and Phase II investigations of the property are conducted by approved environmental consultants to identify the presence of environmental liabilities and to quantify attendant liabilities at a covered site. For example, the environmental site assessment may identify the existence of an underground storage tank (UST) release on a quarter acre of a twenty-acre site. This UST liability would be excluded from EIL coverage. However, if additional USTs were discovered on the site subsequent to this investigation and after the EIL policy was issued, costs associated with the contamination stemming from this previously unidentified UST system, including bodily injury and third-party property damage, would be covered.

Remediation Stop-Loss/Cost-Cap Coverage

Cost-cap coverage assures that cleanup costs above a self-insured retention level (SIR) will be capped, subject to the available policy limit. This policy indemnifies the insured for financial losses that arise when the anticipated cost of a remediation project is exceeded. Based upon a scheduled remediation project defined through an environmental site investigation, this type of coverage addresses unanticipated cost overruns. Typically, this policy provides coverage after the applicable SIR, which equals the expected cost of cleanup plus a buffer. For example, the developer of a project with expected remediation costs of $1,000,000 may carry a $250,000 buffer (SIR) and purchase a $2,000,000 cost-cap policy. If actual cleanup costs for the project are $2,000,000, then when the costs of the project run over the $1,250,000 anticipated costs plus SIR, the cost-cap policy would respond by paying the unanticipated $750,000.

This type of coverage is designed to address the risk and uncertainty associated with the cost of environmental remediation projects. Therefore, it acts essentially as a stop-loss policy.

Asbestos/Lead-Paint Liability Coverage

These policies are designed to protect building owners against claims resulting from releases of asbestos or lead paint at covered locations. For instance, building owners with an operation and maintenance program for asbestos-containing materials can purchase coverage to protect against bodily injury and property damage claims resulting from a release of the asbestos on an occurrence basis.

Nonowned Disposal-Site Coverage

This insurance protects against exposure at specified disposal sites owned by others. Exposure at such sites could lead to liability stemming from Superfund cost-recovery actions. Coverages are very specific and can be limited.

Owner-Controlled Environmental Contractors Insurance

Especially in the context of brownfields redevelopment, property owners and prime contractors demand coverage for liability associated with environmental remediation on specified projects. As a condition precedent to bidding on a project, subcontractors and consultants must provide evidence of blanket annual aggregate coverage for all work performed during the course of a year. In some cases, owners and prime contractors may not be able to obtain adequate protection for their potential exposures, because the policy limit of liability for the contractor's blanket program is shared with other projects and may have terms and conditions unacceptable to the owner or prime contractor. Additionally, many contractors may not carry insurance for environmental loss exposures. Therefore, owner-controlled insurance provides a single insurance program with consistent coverage for all contractors on the environmental remediation project. This product may offer substantial benefits through the provision of standardized forms and relief from administrative burdens associated with coordinating insurance coverage.

Finite-Risk Programs

These programs can provide both financing options and true insurance coverage for prospective and historical environmental liabilities. Programs may be designed to address closure and postclosure liability funding and financial assurance regarding future environmental remediation. For example, funding for future cleanups may be established through payment of equal premiums over a designated time period. Additionally, stop-loss coverage may be blended into this program to cap costs to prescribed levels.

The Cost and Value of Environmental Insurance

The most frequently asked question regarding environmental insurance in brownfields projects is, "How much does it cost?" This is similar to asking, "How much does it cost to purchase a parcel of property?" The only correct answer to these questions is, "It depends." Without a specific frame of reference regarding both the actual project and the insurance needs and desires of the buyer, one cannot reasonably price environmental insurance. For the reasons identified above, prices vary dramatically based on the physical conditions at the subject property, the capabilities of the parties in a project, the details of coverage purchased, the duration of the project, past and future experience, and other factors. It can be irresponsible to even speculate on price in the absence of exposure-specific information, because of the likelihood that business plans will be developed and investment decisions made on the basis of totally irrelevant costing assumptions. Notwithstanding these disclaimers, generally the cost of purchasing the most desired types of coverage has become considerably more affordable over the past five years.

Due to its very nature, the value of environmental insurance is also difficult to quantify. However, the primary benefit undoubtedly is that environmental insurance significantly eliminates uncertainty associated with environmental liabilities. This greater degree of certainty is translated in a number of specific areas, including the following:

1. *Cost:* Environmental insurance provides the ability to establish project costs more accurately by eliminating an area of risk or quantifying costs associated with identified risks.
2. *Facilitating transactions:* In certain cases, the ability to provide environmental insurance may facilitate the closing of a transaction, as the parties to the deal may look to a viable third-party (insurance company) source for funding environmental liabilities rather than relying only on the viability of the parties to the transaction. Further, in certain circumstances, it may be possible to substitute environmental-insurance policies for traditional indemnification provisions and/or environmental escrow agreements. This may provide a significant benefit to the economics of the transaction.
3. *Balance sheet impacts:* New Securities and Exchange Commission reporting regulations requiring public companies to quantify contingent environmental liabilities more aggressively may avoid reporting these issues if environmental insurance is obtained to address such contingencies.

In essence, in return for the payment of a one-time fixed price in the form of an insurance premium, business people have the ability to quantify the costs of environmental risks and to transfer those risks to a third party willing to accept them.

Limitations to Available Insurance

There are some things that environmental insurance just cannot do. As some risks are transferred through insurance, there may be a tendency for brownfields deals of an increasingly speculative character to be pursued. Insurance cannot reduce the basic business risks associated with a real estate transaction. There seems to be the impression in some areas of the market that insurance can make a project viable though it might not otherwise be undertaken. Insurance certainly can provide a vehicle to ameliorate risk and encourage investment, but it cannot turn a bad deal into a good deal.

Because of the nature of actual insurance policies used in support of brownfields transactions and the need for flexibility to customize programs, the environmental-insurance business generally will be restricted to the surplus-lines portion of the insurance market. This result has a number of implications. First, surplus-lines contracts are subject to much less supervision by state insurance regulatory authorities than other insurance contracts. The policies will not have been reviewed by insurance regulators and are not required to meet some standards applicable to the so-called "admitted" insurance contracts. It is vital that a buyer have the assistance of a licensed and competent surplus-lines insurance brokerage professional. In addition, surplus-lines contracts are typically taxed directly, at a higher rate than admitted policies. This adds administrative-cost friction to deals that may already be financially stressed. Last, surplus-lines insurance policies are not generally secured through state insurance guarantee programs, making the long-term solvency of the issuing surplus-lines carrier an issue worthy of consideration in the purchase decision.

Statutory liability limitations can change over time, but this does not override the specific limitations of a contract between an insurer and an insured. Although public policy initiatives are clearly one of the driving forces in the brownfields arena, there is no assurance that this movement is an irreversible trend. Front-end design of a contract that may run for years becomes a key limitation. One should not expect a carefully constructed insurance policy to address some types of future regulatory and liability shifts.

A Comment on the State of the Environmental-Insurance Market

There is a ubiquitous and uniformed impression that only a limited market exists for environmental insurance. This mistaken notion bespeaks a lack of understanding of certain macroeconomic principles and familiarity with the richness of the insurance marketplace. To any single buyer of a product, a small number of providers may seem to offer options that appear insufficient to generate the benefits of price competition. However, in any market, one can characterize the adequacy of supply only in the context of demand. The broad insurance market is a vibrantly competitive place. Indeed, it can be argued that in the mid-1990s the capacity of the insurance market clearly oversupplies the demand for insurance, on the basis of the financial performance of the insurance industry.

The environmental portion of the insurance business is clearly lost in the financial performance of the overall industry. There is little doubt that environmental risk has been a financial disaster for the industry, though the new discipline of environmental insurance has contributed favorable results to the companies that practice it thoughtfully. Though the market is far from mature, at least two dozen companies in the primary insurance market are legitimate providers of specialized environmental insurance. Notwithstanding this marketplace, there are approximately half a dozen insurance companies participating in the discussion of brownfields insurance issues.

As with providers of any product or service, the providers of environmental insurance can be classified based on the value they provide. Unfortunately, as noted, the value of insurance is poorly understood. It may be very difficult for buyers to discriminate among carriers on the basis of quality. Therefore, buyers of environmental insurance in the brownfields redevelopment context may benefit tremendously by seeking appropriate legal and professional insurance advice before they purchase any particular type of coverage.

Notes

1. 42 U.S.C. § 6901 *et seq.* (1988).
2. 42 U.S.C. § 9601 *et seq.* (1988).
3. "Moral hazard" is the risk resulting from uncertainty about the insured's honesty.
4. For the foregoing reasons, much of the insurance dealing with brownfields will be transacted in the surplus-lines insurance market. This allows for customization of individual policies, but does so at the expense of surrendering much insurance-department control and supervision of carriers.

CHAPTER 12

Using Old Insurance Policies as Weapons

DIANE R. ARCHANGELI

Insurance Archaeology
What Is It?

Insurance archaeology is the systematic recovery and analysis of insurance policies as far back as records will allow. Environmental liabilities that arise out of past business operations on a site may be offset by insurance coverage purchased at the time the worst pollution was taking place. This is the case because general-liability policies written from 1945 through 1985 often offer a great possibility for coverage. Once overlooked because they had long since expired, old policies are now recognized as vital assets worth millions of dollars that can make or break a real estate deal. Successful business managers are beginning to understand the benefits of taking advantage of brownfields statutory schemes while at the same time pursuing insurance proceeds to assist in funding site cleanup.

How Is It Done?

The first step is to look for insurance policies in the client's own records. The best evidence of coverage is the actual, executed insurance policy. If this type of primary evidence cannot be found, often an insurance archaeologist will locate "secondary evidence" of coverage, such as insurance certificates; partial policies; letters—with policy numbers on them—that are stored in garages, warehouses, and basements; management reports; corporate records; financial ledger entries; umbrella and excess-policy schedules; records maintained at Lloyd's of London; and correspondence, to or from insurance agents, that refers to coverages for particular time periods. Files maintained by brokers, agents, and reinsurer intermediaries are also places to locate coverage; the files often contain premium registers, declarations pages for coverages ceded to reinsurers, and other documents that evidence the existence of coverage. Corporate correspondence often provides valuable leads about the identity of past employees or brokers who have information about the company's insurance history. In addition, corporate lawyers and officers, as well as former risk managers, will provide a wealth of information. Once the most knowledgeable people are located, the archaeologist will conduct interviews to identify as many leads as possible to piece together the insured's insurance history.

Generally, the burden of proving that insurance coverage exists is on the insured, as the party claiming coverage under the policy.[1] Once the insured can show the existence of a policy and can establish its loss, the contents of a lost or missing policy may be proven indirectly through secondary evidence, such as testimony or other documents tending to support the substance of the policy.[2] Insurance companies often assert they do not have a copy of the policy or know its contents because of document destruction procedures that are part of their normal course of business. However, for reinsurance purposes, many insurance companies keep records of policies they have issued for many decades after the expiration dates of the policies. Requesting that carriers specifically search their underwriting, claims, and reinsurance records should be part of any request for policies and evidence of policies.

Making a Claim under Old Commercial General-Liability Policies

The Legal Basis for Such a Claim

The legal foundation for claims for cleanup costs to remediate a contaminated site is based in contract law. Insurance contracts, like other legal contracts, are construed to give effect to the intent of the parties at the time the parties entered the contract. In most states, disputes that arise concerning the meaning of certain policy provisions, exclusions, and conditions are resolved in accordance with established contract-interpretation principles. Included in these principles is the well-established rule that the policy language should be construed to protect the reasonable expectations of the insured. Any ambiguity in an insurance policy is to be resolved against the insurance carrier.[3] This basic rule of contract construction is often referred to as the doctrine of *contra preferentem*.

Types of Policies under Which You Can Make the Claim

Claims for damages, including costs associated with the cleanup of property contaminated by pollutants, can be made under a number of different types of insurance policies. Though the earliest claims were tendered to carriers for payment under the comprehensive general liability (CGL) policies, there is a growing body of law that has interpreted cases in which policyholders seek coverage under other types of policies, such as automobile and garage-liability policies, environmental-impairment liability policies, first-party property policies, and "personal injury" coverages.

Assignability of Insurance Policies

There is ample case law that examines when one corporation can be held the successor-in-interest to another corporation. A corporation that merges with, or purchases assets of, another corporation may become liable for the liabilities of the corporation with which it merged or from which it purchased assets. This raises two issues: (1) whether the successor corporation has the right to claim coverage under the predecessor corporation's policies, and (2) whether the successor corporation can recover under its own policies for liabilities that arose out of activities of the predecessor.

Most insurance policies contain a term providing that any assignment of interest under the policy does not bind the insurance company until the insurance company provides consent. Even with this "no assignment clause," courts have found that the clause does not prevent a successor from recovering under a predecessor's policies in the event of a statutory merger or consolidation.[4] Courts have held that the "no assignment clause" contained in the predecessor's policy will not bar coverage for liability transferred to a successor if the predecessor's policy would have provided coverage had the transaction not taken place. A major rationale in these cases has been that the insurance coverage should follow the underlying liability, in spite of a "no assignment clause," when there is no increase in risk to the insurer.[5]

The issue of whether a successor corporation can claim coverage under its own policies for damage that arose out of a predecessor's activities is also important to understand when faced with the problem of contaminated property. Generally, an insurance company will not be required to provide coverage under the policies issued to the successor corporation that inherits a predecessor's liabilities, unless the policies issued to the successor are independently triggered by an insurable event.[6]

Coverage under CGL Policies

The CGL form was first written in 1940 and had a very broad scope. The insuring agreement of such policies provided coverage for "bodily injury" and/or "property damage" resulting from an "accident." Most courts that have considered old "accident" policies have found that policies issued from 1940 to 1966 cover pollution damage resulting from gradual causes.[7] In 1966, the National Bureau of Casualty Underwriters changed the coverage on the standardized form from "accident" to "occurrence." This significant change is discussed later in this chapter. In 1970, the Insurance Rating Board filed a pollution-exclusion endorsement that barred coverage for environmental damage unless the injury resulted from sudden and accidental happenings, commonly referred to as the "sudden and accidental" pollution exclusion. In 1973, the Insurance Services Office included the "sudden and accidental" pollution exclusion in its standardized CGL form. In 1985, the insurance industry adopted what is commonly referred to as the "absolute pollution exclusion." Though policies written after 1970 present additional issues for environmental-insurance recovery efforts, they are often issues that, like the many other defenses insurance companies consistently assert, can be overcome or at least compromised to attain the mutual goal of resolving such liabilities.

Coverage under Automobile Policies

Most commercial automobile-liability policies contain an insuring agreement providing as follows:

> The Company will pay on behalf of the Insured all sums the Insured shall become legally obligated to pay as damages because of
>
> a. bodily injury or
> b. property damage
>
> to which this insurance applies, caused by an occurrence and arising out of the ownership, maintenance or use, including loading and unloading of any automobile.

Motor vehicles are often used to transport waste to landfills. Policyholders contend that when a motor vehicle is involved in the discharge of pollutants, the resulting environmental liability may be said to arise out of the ownership, maintenance, or use, including loading and unloading, of a vehicle. Few cases have directly dealt with the applicability of motor-vehicle coverage in the environmental context. However, in at least one case, a court found that allegations that hoses were used to transfer materials between the underground tanks and trucks "arguably fall within the policy provisions of the automobile policies at issue."[8]

Coverage under Garage-Liability Policies

Garage-liability policies generally apply only to injury or damage arising out of the use of, or in connection with, the policyholder's garage operations. Very few courts have interpreted what constitutes "garage operations" in the environmental context. Nevertheless, garage policies are still a potential avenue to pursue in making a claim.

Coverage under Environmental-Impairment Liability Policies

Environmental-Impairment Liability (EIL) policies were written in the 1970s and were intended to cover situations that were not covered under CGL policies, which generally began to include standard pollution exclusions by 1972. There were several different types of EIL policies written in the 1970s, with most providing coverage for gradual, nonsudden contamination. Unlike occurrence-based CGL policies, which provide coverage for injuries that take place during the policy period regardless of when the loss is reported, EIL policies were almost always written on a "claims-made" basis. Claims-made policies provide coverage only when the claim against the insured is asserted against the insured during the effective policy period. The insuring agreement section of many claims-made policies requires that the insured also must report the claim to the insurer during the same policy period. Policies mandating that (1) a claim be asserted against the insured during the policy period and (2) the claim be reported to the insurance carrier during the policy period often result in more restricted coverage than the name of the coverage implies.

Coverage under Personal-Injury Section of the CGL

Typical personal-injury coverage contains an insuring agreement that provides as follows:

> The insurer will pay on behalf of the insured all sums the insured shall become legally obligated to pay as damages because of injury (herein called "personal injury") sustained by any person or organization and arising out of one or more of the following offenses committed in the conduct of the named insured's business:
>
> Wrongful entry or eviction, or other invasion of the right of private occupancy . . . if the offense is committed during the policy period.

Environmental actions are often pursued by private parties or governmental agencies on theories of nuisance, trespass, or other theories related to interference with the use and enjoyment of property. In such cases, courts have recognized that trespass and nuisance will be deemed to have taken place as long as the pollutant is on the property or is interfering with the use or enjoyment of the property, from initial contamination through abatement.[9] In addition, personal-injury coverage generally contains its own specific terms, conditions, and exclusions. Significantly, several courts have held that the CGL pollution exclusion does not apply to personal-injury coverage sections contained in the CGL, which can strengthen a recovery strategy in a state with procarrier law on the pollution exclusion.[10]

Coverage under First-Party Property Insurance

First-party property insurance provides coverage to the insured for damage to property the insured owned or leased at the time the damage took place. The key to recovery is tendering a claim in which there has been direct physical loss to covered property, during the policy period, that is caused by an insured peril. In a "named perils" policy, the specific perils covered are enumerated, while "all risk" policies provide coverage for all perils unless specifically excluded. "All risk" policies ordinarily provide explicitly that the insurer will pay expenses for the removal of debris of covered property caused by a covered cause of loss. As construed by some courts, such debris-removal clauses cover the cost of cleaning up contaminated property.[11]

Accident versus Occurrence Policies

Before 1966, the standard-form CGL policy provided coverage on an "accident" basis. That is, pre-1966 CGL policies provided coverage for personal injury or property damage caused by an "accident," a term that in most cases was not defined. Since 1966, the standard-form CGL policy has provided coverage for all sums the insured becomes legally obligated to pay as damages due to an "occurrence." In the 1966 revision, "occurrence" was generally defined as an accident, including "injurious exposure" to conditions that, during the policy period, results in bodily injury or property damage neither expected nor intended from the standpoint of the insured. Subsequent CGL policies often define "occurrence" to include a continuous or repeated exposure to conditions that results in bodily injury or property damage. The "occurrence" language is generally understood to provide broader coverage than the former "accident" language.[12]

Defenses Insurance Companies Often Use to Avoid Liability under CGLs

Who Qualifies as an Insured?

Determining who qualifies as "the insured" has been the source of much debate and is especially important if the company named as the insured in the old CGL policy is no longer in existence or is bankrupt. The first place to turn is the definitions section of the policy. Many courts have held that a successor corporation may succeed to the benefits of the insurance polices issued to the acquired corporation despite "no assignment" clauses.[13]

Late Notice

CGL policies generally contain a provision that requires the insured to provide written notice to the insurer of an occurrence "as soon as practicable" and to give notice of a claim or suit "immediately." The standards for determining whether notice was given on a timely basis differ from state to state and will depend upon the facts of the particular case. The trend has been to move away from a strict contractual construction of the provision toward a more reasonable rule requiring the insurer to show prejudice by reason of the late notice to bar coverage. States requiring prejudice to the insurer to bar coverage include Alaska, Arizona, California, Connecticut, Delaware, Florida, Hawaii, Idaho, Indiana, Iowa, Kansas, Kentucky, Louisiana, Maine, Maryland, Michigan, Minnesota, Mississippi, Missouri, Montana, Nebraska, New Hampshire, New Jersey, New Mexico, North Carolina, North Dakota, Massachusetts, Oklahoma, Oregon, Pennsylvania, Rhode Island, South Carolina, Texas, Utah, Washington, West Virginia, and Wisconsin.[14]

Cleanup Costs Constitute Equitable Remedies, Not Legal Damages

For several years insurers argued, and many courts agreed, that response costs recoverable under environmental statutes constituted equitable (injunctive) relief and thus were not "legal damages" as required under the insuring agreement of liability policies. However, there has been a swing in favor of policyholders, with the majority of courts now finding that sums an insured is required to pay for remediation, either directly or as reimbursement to a governmental agency, constitute "damages" as required under the insuring agreement of most policies.[15]

No Property Damage

The definition of property damage is also used by insurance companies to argue that coverage is precluded under CGL policies. CGL policies issued after 1973 often contain the following definition of property damage:

> (1) Physical injury to or destruction of tangible property which occurs during the policy period, including the loss of use thereof at any time resulting therefrom, or (2) Loss of use of tangible property which has not been physically injured or destroyed provided such loss of use is caused by an occurrence during the policy period.

Insurers argue, with limited success, that costs incurred by a policyholder to comply with governmental orders to remediate soil and groundwater contamination do not meet the definition of property damage but rather are claims for economic loss. Several courts have held that costs incurred by a policyholder to modify business operations to prevent future emissions of pollutants do not constitute claims for property damage.[16] Even carriers who concede that response costs constitute legal damages because of property damage generally assert that they may not be liable to indemnify the policyholder because the property damage did not take place during the policy period, as required under the definition of property damage. This "trigger" argument has resulted in courts creating four different theories regarding when bodily injury or property damage takes place, so they can determine which policies are triggered.

1. *Exposure theory:* Most often associated with bodily injury cases, including the first asbestos cases, coverage is said to be triggered in all policies in effect during the time exposure to harmful substances took place.[17]
2. *Manifestation theory:* Coverage under the manifestation theory is triggered by the policy in effect at the time the bodily injury or property damage is discovered.[18] The seminal case that first applied the manifestation theory in an asbestos bodily injury case is *Eagle-Picher Industries, Inc. v. Liberty Mutual Insurance Co.*, 682 F.2d 12 (1st Cir. 1982), *cert. denied*, 460 U.S. 1028 (1983). The manifestation theory often ends up being the most restrictive theory in pollution cases because only

one policy is ordinarily triggered, as whereas under other theories the loss will typically trigger several carriers' policies. In addition, the policy triggered is often one that was in effect after 1985, when the so-called "absolute pollution exclusion" was commonly contained in CGL policies.
3. *Injury-in-fact theory:* Under this theory, each policy in effect during the time the bodily injury or property damage is sustained, up until the date of the discovery of such damage or injury, is said to be triggered.[19]
4. *Continuous (or triple trigger) theory:* Under this theory, all policies on the risk, beginning at the time of first exposure through the date of manifestation of the injury or damage, are triggered. This theory was also first set out in an asbestos bodily injury case.[20] A California court extended the continuous trigger to include "all policies in effect from the first exposure until date of death or date of claim, whichever occurs first."[21]

Lack of an Occurrence

CGL policies generally obligate the insurers to indemnify the insured for all sums the insured becomes legally obligated to pay as damages because of bodily injury or property damage caused by an occurrence. Most CGLs define an occurrence as

> an accident, including a continuous or repeated exposure to conditions, which results, during the policy period, in bodily injury or property damage *neither expected nor intended from the standpoint of the insured.*

Insurance coverage disputes involving CGL policies often turn on the issue of whether the injury or damage for which an insured seeks coverage was expected or intended by the insured. In determining whether bodily injury or property damage was unexpected or unintended, and therefore within the definition of an occurrence, several courts have looked to the resulting damage rather than the act that gave rise to the damage. A number of courts have found an occurrence in cases in which the act that gives rise to the damage was intentional, but the resulting damage was unintentional.[22] However, if an insured knew or should have known that there was a substantial probability that certain injurious results would follow from its acts or omissions, coverage may be barred. In recent years, the issue of whether a subjective or objective standard should be applied has been hotly debated, with no clear trend in the courts.

Application of the Pollution Exclusion

Insurers almost always assert the pollution exclusion as a bar to coverage in policies that contain a pollution exclusion. The standard pre-1986 pollution exclusion in most CGL policies provides that coverage does not apply to bodily injury or damage arising out of the discharge, dispersal, release, or escape of pollutants or contaminants unless the discharge is sudden and accidental. Drafting-history documents concerning the meaning of the pollution exclusion have been introduced to support the policyholder view that the exclusion is merely a restatement of the definition of occurrence and bars coverage only for environmental impairment that is expected or intended.[23] A growing number of courts have found that the term "sudden" in the exception to the pollution exclusion includes a temporal element.[24] Another frequently litigated issue is whether the sudden-and-accidental exception to the pollution exclusion refers to the discharge in the initial dumping of waste or the discharge of leaching contaminants from the site.[25] Most courts have indicated that it is the initial release or discharge into or upon land that must be sudden and accidental to avoid the application of the pollution exclusion.[26]

After 1986, CGL policies generally contain what is commonly referred to as an "absolute" pollution exclusion. This exclusion eliminates the sudden-and-accidental exception language contained in the pre-1986 pollution exclusion. The majority of courts have held that this exclusion bars coverage for bodily injury and property damage arising out of pollution claims. However, some courts have held that the so-called "absolute" pollution exclusion is not a bar to coverage in all situations.[27]

Application of the Owned-Property Exclusion

CGL policies generally contain an "owned-property" exclusion, also known as the "care, custody, and control" exclusion. This exclusion precludes coverage for damage to property "owned, occupied by, or rented to" the insured, or to property in the "care, custody, or control of the insured, or to premises alienated" by the insured. The exclusion was intended to clarify that policies do not provide coverage for first-party situations that warrant coverage under the first-party property section of the policy. However, some courts have held that if there is groundwater contamination, the exclusion does not apply because contamination to groundwater goes beyond damage to property that is owned by the insured. Similarly, some courts have also held that the owned-property exclusion does not apply in instances where on-site remediation is being performed to prevent off-site contamination.[28]

Application of the Voluntary-Payment Provision

The typical voluntary-payment provision contained in CGL policies before 1986 provides as follows:

> The policyholder shall not, except at his own cost, voluntarily make any payment, assume any obligation or incur any expense other than for first aid to others at the time of accident.

The primary purpose of the voluntary-payment provision in CGL policies is to protect the insurer from the risks of collusive settlements between the policyholder and an injured third party.[29] Several courts have held that insurers must indemnify policyholders for costs incurred by policyholders who remediate a site to avoid litigation with governmental entities. Some courts have also indicated that the insurer must indemnify the policyholder who voluntarily remediates a site to avoid fines or penalties from a governmental entity. This can be a significant issue, because under the Comprehensive Environmental Response, Compensation, and Liability Act of 1980 (CERCLA),[30] a person who fails, without sufficient cause, to provide removal or remedial action is potentially liable for up to three times the costs incurred by the government, if the government undertakes the cleanup itself. In at least one case, the court held that even though the policyholder materially breached its obligation to obtain the insurers' consent to pay for cleanup costs at a site, that breach did not allow the insurers to avoid coverage without a showing by the insurers that they had been substantially prejudiced.[31]

How Much Coverage Can You Get?

The answer to how much coverage one can tap depends on several issues, including (1) the number of policies triggered, (2) the number of "occurrences," and (3) whether the applicable policies contain aggregate limits for operations coverage. Once multiple policies are held triggered, either because of multiple "occurrences" or continuing property damage, courts must determine how to apply the limits of the multiple policies to the underlying claim.

In addition to an aggregate (or total) limit of liability, most policies have a "per occurrence" limit of liability. In some cases, policyholders have been successful in increasing the number of occurrences, thereby expanding the amount of coverage available. Determining the number of occurrences causing property damage for which the policyholder may be liable can have a significant impact on the amount of insurance available to a policyholder. A majority of courts have looked to the cause of the property damage rather than the number of claims that result from the property damage. Under such an analysis, courts have found only one occurrence when there is one proximate, uninterrupted, and continuing cause that results in all the injuries and damage, even though several discrete items of damage resulted.[32] The focus of the analysis is the cause of the accident rather than its effect. A minority of jurisdictions look to the results of the event in calculating how much coverage will apply.

Stacking

In the area of cumulative-injury, toxic-tort cases (including hazardous waste and pollution claims), stacking policy limits means that if more than one policy is triggered by an occurrence, each policy can be called upon to respond to the claim, up to the full limits of the policy. Stacking allows the limits of every triggered policy to be added together to determine the amount of coverage available for a claim.

Allocation issues that deal with the ultimate financial implications of the environmental claim become important only after resolving the issues of which policies are triggered. Several recent decisions addressing allocation issues in the environmental area have applied methodologies that apportion responsibility to the policyholder for those periods with gaps in coverage. Coverage gaps may be the result of various factual scenarios, including periods of time in which a policyholder was self-insured or coverage was not purchased or no longer exists due to previous exhaustion of limits by claim payments or settlements, insolvencies, missing policies, and commutations. The application of the continuous trigger of coverage resolves insurance companies' concerns about being singled out to pay all the costs associated with a cleanup under the policyholders' preferred joint and several liability theory.[33]

Additional allocation issues are raised by the applicability of excess coverages. In *Owens-Illinois, Inc. v. United Insurance Co.*, a case involving the interpretation of when an excess carrier's policies must pay, the court adopted a continuous trigger in determining that the fair method of allocation includes an analysis of time on the risk and degree of risk assumed.[34] As excess carriers assume a smaller degree of risk than the insurers that issued primary coverage, the implication seems to be that the excess carrier's liability should reflect the lesser risk that it assumed.

In a case interpreting *Owens-Illinois*, the court in *Schering Corp. v. Evanston Insurance Co.* found that any insured that settles with its insurance carriers for less than the full amount of the policy limits cannot obtain recoveries from the nonsettling carriers in excess of their proportionate share.[35] The court also held that when there are different layers of coverage, all available primary coverage must be exhausted before excess coverages can be tapped. The court's rationale for triggering excess coverage is based on an asbestos property-damage case, *United States Gypsum Co. v. Admiral Insurance Co.*, in which the court cited the "other insurance" provision to determine that all triggered and available primary coverage must be exhausted before tapping into coverages provided by excess policies.[36] The "other insurance" provision typically provides that if other valid and collectible insurance with any other insurer is available to the insured, covering a loss also covered by the policy at issue, the insurance afforded by the policy at issue shall be in excess of such other insurance.

Good Approaches in Negotiating Settlements with Insurance Companies

Before negotiating with the carrier or carriers who issued policies that may respond to property damage at a site, all the information related to costs that have been incurred to remediate a site, as well as estimates of what will be spent in future cleanup efforts, must be gathered. All the insurance information and history of the site or sites must then be collected. Then, under various allocation methods, the best demand for each carrier with potential exposure should be calculated. Although there are many different ways to allocate the carrier's exposure, some of the most well-established methods include pro rata share by time on the risk, pro rata by limits, pro rata by time multiplied by limits, and per capita (equal shares). There are endless ways to calculate a carrier's exposure, and knowing where to begin will depend largely upon how many sites are involved and how much money is at stake, as well as other factors. For instance, if there is excess coverage available but the amount of the loss does not reach the excess level, allocation methodologies that take into account excess layers of coverage will be of little benefit.

A meeting with the carrier should be arranged to discuss resolving any coverage issues and to present a settlement demand. It may be helpful not to threaten immediate court action by joining forces with an expert in negotiating claims rather than litigating coverage issues. Negotiating directly with carriers without resorting to litigation can be fruitful if a representative that understands insurance issues and environmental issues goes to the negotiating table. However, if litigation must be initiated to get the carriers to the negotiating table, it should be stayed to reduce litigation costs while negotiating with the carriers. Critical terms and conditions of the settlement should be resolved before discussing settlement amounts. Very often, negotiations will fall apart after a number is reached if the parties find there was no meeting of the minds regarding what that number represents.

There are many types of settlements that carriers routinely enter with policyholders in the environmental context, including total policy "buybacks." This is the broadest type of release and should not be entered without careful consideration. The settlement should be limited to a property-damage release related solely to the site(s) at issue. A settlement agreement that releases the insurance carrier for all bodily injury and property damage that arise out of the site should be avoided, especially without knowing about the possibility of future bodily injury claims related to the site. The scope of any release entered with the carrier must be as narrow as possible.

Structuring the settlement should also be considered. If the amount needed over a certain number of years is known (for example, for the long-term pumping and treating of groundwater at a site, or operation and maintenance costs), then the carrier's purchase of an annuity for future liabilities could be suggested.

Another key to negotiating an environmental claim is to be well informed about all parties to the negotiations. When developing a strategy to settle a claim or seek reimbursement of cleanup costs, one must consider some of the factors that affect the insurance industry as a whole, as well as what may be motivating the particular insurance companies at the negotiations.

Insurers are clearly interested in exchanging uncertainty for certainty when it comes to environmental claims, because they realize that to stay competitive (and/or not get gobbled up by stronger carriers), they must prove to market forces that they have recognized their liabilities and can pay environmental claims. Many insurance companies that issued commercial CGL coverage from the 1940s until the mid-1980s increasingly have been bowing to the growing pressure on the property-casualty industry—from securities regulators, insurance-ratings agencies, accountants, and shareholders—to quantify their environmental liabilities accurately. The property-casualty insurance industry today needs $40 billion to pay off its future environmental liability claims, and current reserving does not come close to this figure, according to a report released by Standard & Poor's.[37]

Though it is important to understand where the property-casualty insurers stand as an industry on environmental issues, it is also important to get as much "intelligence" as possible on the specific carriers at the negotiation table. For instance, if negotiating with Nationwide Mutual Insurance Company and its affiliates, it may be useful to know that Nationwide recently increased its reserves for potential environmental and asbestos-related liabilities by $1.1 billion.[38] This is not to suggest that Nationwide will be paying every environmental claim that is tendered to it, without an exhaustive investigation of each claim, but it is certainly a relevant piece of information to have at the bargaining table.

For the party involved in remediating brownfields, or simply a party with environmental liabilities, old insurance policies can be part of the solution. In many cases, insurance coverage will provide monies to remediate contaminated property. Therefore, the identification of all insurance policies and pursuit of coverage under those policies should be part of any plan to address contaminated property.

Notes

1. 21 JOHN A. APPLEMAN, INSURANCE LAW AND PRACTICE § 12,094 (1980).
2. 19 COUCH, COUCH ON INSURANCE 2D § 79:6 (1983).
3. Liverpool & London & Globe Ins. Co. v. Dearney, 180 U.S. 132, 135–36 (1901). *See, e.g.*, Arkwright-Boston Mfrs. Mut. Ins. Co. v. Wausau Paper Mill Co., 818 F.2d 591 (7th Cir. 1987); Westchester Resco Co. v. New England Reinsurance Corp., 818 F.2d 2 (2d Cir. 1987); Union Ins. Soc'y, Ltd. v. William Gluckin & Co., 353 F.2d 946, 951 (2d Cir. 1965).
4. Imperial Enters., Inc. v. Fireman's Fund Ins. Co., 535 F.2d 287 (5th Cir. 1976); Brunswick Corp. v. St. Paul Fire & Marine Ins. Co., 509 F. Supp. 750 (E.D. Pa. 1981); Chatham Corp. v. Argonaut Ins. Co., 334 N.Y.S.2d 959 (Sup. Ct. Nassau County 1972); *see also* Maryland Casualty Co. v. W.R. Grace & Co., 794 F. Supp. 1206 (S.D.N.Y. 1991).
5. *But see* Quemetco, Inc. v. Pacific Auto Ins. Co., 29 Cal. Rptr. 2d 627 (Ct. App. 1994) (holding that company that succeeded to CERCLA liabilities as "mere continuation" of company that generated hazardous wastes at site could not obtain coverage under predecessor's CGL policies because transfer of insurance benefits would result in increased risk to insurers).
6. Upjohn Co. v. Aetna Casualty & Sur. Co., No. K. 88-124 CA (W.D. Mich. Sept. 9, 1991), 1991 WL 490026; Johnson Controls, Inc. v. Employers Ins. of Wausau, No. 89 CV 016174 (Cir. Ct. Mil. Cty. Wis. Dec. 22, 1993); Idaho v. Bunker Hill Co., 647 F. Supp. 1064 (D. Idaho 1986); Textron, Inc. v. Aetna Casualty & Sur. Co., No. 92-650 (R.I. Sup. Ct. Mar. 11, 1994); Aetna Life & Casualty Co. v. United Pac. Reliance Ins. Co., 580 P.2d 230, 232 (Utah 1978).
7. Coverage under insurance policies sold from 1940 to 1970 includes pollution liability. *See* Morton Int'l, Inc. v. General Accident Ins. Co., 629 A.2d 831 (N.J. 1993), *cert. denied*, 114 S. Ct. 2764, *reh'g denied*, 115 S. Ct. 25 (1994).

Finding coverage: *see, e.g.,* Anchor Casualty Co. v. McCaleb, 178 F.2d 322 (5th Cir. 1949); Aetna Casualty & Sur. Co. v. Martin Bros. Container & Timber Prods. Corp., 256 F. Supp. 145 (D. Or. 1966); Moffat v. Metropolitan Casualty Ins. Co., 238 F. Supp. 165 (M.D. Pa. 1964); City of Myrtle Point v. Pacific Indem. Co., 233 F. Supp. 193 (D. Or. 1963); Employers Ins. Co. v. Rives, 87 So. 2d 653 (Ala. 1955), *cert. denied*, 87 So. 2d 658 (Ala. Ct. App. 1956); Moore v. Fidelity & Casualty Co., 295 P.2d 154 (Cal. Ct. App. 1956); The Travelers v. Humming Bird Coal Co., 371 S.W.2d 35 (Ky. 1963); White v. Smith, 44 S.W.2d 497 (Mo. Ct. App. 1969); City of Kimball v. St. Paul Fire & Marine Ins. Co., 206 N.W.2d 632 (Neb. 1973); Cosmopolitan Mut. Ins. Co. v. Packer's Supermarket Inc., 340 N.Y.S.2d 461 (Sup. Ct. 1972); Lancaster Area Refuse Auth. v. Transamerica Ins. Co., 263 A.2d 368 (Pa. 1970); Taylor v. Imperial Casualty & Indem. Co., 144 N.W.2d 856 (S.D. 1966); Massachusetts Bonding & Ins. Co. v. Orkin Exterminating Co., 416 S.W.2d 396 (Tex. 1967).

Denying coverage: *see, e.g.,* Leggett v. Home Indem. Co., 461 F.2d 257 (10th Cir. 1972); American Casualty Co. v. Minnesota Farm Bureau Serv. Co., 270 F.2d 686 (8th Cir. 1959); Farmers Elevator Mut. Ins. Co. v. Burch, 187 N.E.2d 12 (Ill. App. Ct. 1962); United States Fidelity & Guar. Co. v. Briscoe, 239 P.2d 754 (Okla. 1951); Town of Tieton v. General Ins. Co., 380 P.2d 127 (Wash. 1963); Clark v. London & Lancashire Indem. Co., 124 N.W.2d 29 (Wis. 1963).

The Wisconsin Supreme Court reversed *City of Milwaukee v. Allied Smelting Corp.*, 344 N.W.2d 523 (Wis. Ct. App. 1983), *State v. Mauthe*, 419 N.W.2d (Wis. Ct. App. 1987), and *Wagner v. Milwaukee Mut. Ins. Co.*, 427 N.W.2d 854 (Wis. Ct. App. 1988) in *Just v. Land Reclamation, Ltd.*, 456 N.W.2d 570 (Wis. 1990).

8. USF & G v. Thomas Solvent Co., 683 F. Supp. 1139 (W.D. Mich. 1988).

9. Kirk A. Pasich, *Insurance under Personal Injury Provisions*, 3 ENVTL. CLAIMS J. 449 (1991).

10. *E.g.*, City of Edgerton v. General Casualty Co., 493 N.W.2d 768 (Wis. Ct. App. 1992), *rev'd on other grounds*, 517 N.W.2d 462 (Wis. 1994); Titan Holdings Syndicate, Inc. v. City of Keen, 898 F.2d 265 (1st Cir. 1990); Lincoln Properties, Ltd. v. CIGNA Ins. Co., No. 238274 (Cal. App. Dep't Super. Ct. Mar. 4, 1992) (no absolute pollution exclusion for personal-injury coverage); *but see* American Universal Ins. Co. v. Whitewood Custom Treaters, Inc., 707 F. Supp. 1140 (D.S.D. 1989).

11. Lexington Ins. Co. v. Ryder Sys., Inc., 234 S.E.2d 839 (Ga. Ct. App. 1977); Daniels v. Aetna Life & Casualty Co., No. IP 81-1413-C, 1983 WL 13684 (S.D. Ind. June 29, 1983).

12. United States Fidelity & Guar. Co. v. Morrison Grain Co., 734 F. Supp. 437, 443 (D. Kan. 1990), *aff'd*, 999 F.2d 489 (10th Cir. 1993) (quoting 11 G. COUCH, COUCH ON INSURANCE 2D, § 44:285, at 437 (1982)).

13. Brunswick Corp. v. St. Paul Fire & Marine Ins. Co., 509 F. Supp. 750 (E.D. Pa. 1981); *see also* Paxton & Vierling Steel Co. v. Great American Ins. Co., 497 F. Supp. at 580; Imperial Enters., Inc. v. Fireman's Fund Ins. Co., 535 F.2d 287, 292–93 (5th Cir. 1976); Aetna Life & Casualty v. United Pac. Reliance Ins. Co., 580 P.2d 230 (Utah 1978); Chatham Corp. v. Argonaut Ins. Co., 334 N.Y.S.2d 959 (Sup. Ct. Nassau County 1972).

14. Irene C. Warshauer et al., *Late Notice in New York: Time for a Change*, 5 ENVTL. CLAIMS J. 200 (Winter 1992/93).

15. Dianne Dailey & Margaret M. VanValkenburg, *Environmental Liability Insurance Claims: Considerations in "Third-Party" Insurance*, 1995 Defense Research Institute 39.

16. *See* AIU Ins. Co. v. Superior Court, 799 P.2d 1253 (Cal. 1990); Boeing Co. v. Aetna Casualty & Sur. Co., 784 P.2d 507 (Wash. 1990); Hazen Paper v. United States Fidelity & Guar. Co., 555 N.E.2d 576 (Mass. 1990).

17. *See* Insurance Co. of North Am. v. Forty-eight Insulations, 633 F.2d 1212 (6th Cir. 1980).

18. Safeco Ins. Co. v. Federated Mut. Ins. Co., 915 F.2d 1565 (4th Cir. 1990); Suburban Constr. Co., Inc. v. Hartford Fire Ins. Co., No. 90-379-D (D.N.H. July 23, 1992); Transamerica Ins. Co. of Mich. v. Safeco Ins. Co., 472 N.W.2d 5 (Mich. Ct. App. 1991); West Am. Ins. Co. v. Tufco Flooring E., Inc., 409 S.E.2d 692 (N.C. Ct. App. 1991); Board of Educ. of Cleveland v. R.J. Stickle Int'l, 602 N.E.2d 353 (Ohio Ct. App. 1991).

19. *See, e.g.*, Maryland Casualty Co. v. W.R. Grace & Co., 23 F.3d 617 (2d Cir. 1994); Inland Waters Pollution Control, Inc. v. National Union Fire Ins. Co., 943 F.2d 52 (6th Cir. 1991); Dow Chem. Co. v. Associate Indem. Corp., 724 F. Supp. 474 (E.D. Mich. 1989); Cortland Pump & Equip. Inc. v. Firemen's Ins. Co. of Newark, 604 N.Y.S.2d 633 (App. Div. 1993); Harford County v. Harford Mut. Ins. Co., 610 A. 2d 286 (Md. 1992).

20. *See* Keene Corp. v. Insurance Co. of N. Am., 667 F.2d 1034 (D.C. Cir. 1981), *cert. denied*, 455 U.S. 1007, *reh'g denied*, 456 U.S. 951 (1982).

21. *In re* Asbestos Ins. Coverage Cases, Judicial Coordination Proceeding No. 1072 (Cal. Super. Ct. San Francisco May 29, 1987).

22. *See, e.g.*, Farmer in the Dell Enters. v. Farmers Mut. Ins., 514 A.2d 1097 (Del. 1986); Clemmer v. Hartford Ins. Co., 587 P.2d 1098 (Cal. 1978); McGroarty v. Great Am. Ins. Co., 329 N.E.2d 172 (N.Y. 1975); Seniuk v. United States Fidelity & Guar. Co., 432 N.Y.S.2d 213 (App. Div. 1980); Town of Huntington v. Hartford Ins. Group, 415 N.Y.S.2d 904 (App. Div. 1979).

23. *See* Steven G. Bradbury, *Original Intent, Revisionism and the Meaning of the CGL Policies*, 1 ENVTL. CLAIMS J. 279, 283–84 (1988).

24. *See, e.g.*, Dimmitt Chevrolet Inc. v. Southeastern Fidelity Ins. Corp., 636 So. 2d 700 (Fla. 1993) ("sudden" has a temporal meaning), *reh'g denied*, 636 So. 2d 700 (Fla. 1993); Hybud Equip. Corp. v. Sphere Drake Ins. Co., Ltd., 597 N.E.2d 1096 (Ohio 1992) (same holding); Upjohn Co. v. New Hampshire Ins. Co., 476 N.W.2d 392 (Mich. 1991) (same holding); Lumbermans Mut. Casualty Co. v. Belleville Indus. Inc., 555 N.E. 2d 568 (Mass. 1990) (same holding); Waste Management of Carolinas Inc. v. Peerless Ins. Co., 340 S.E.2d 374 (N.C. 1986) (same holding).

25. *See, e.g.*, Broderick Inv. Co. v. Hartford Accident & Indem. Co., 954 F.2d 601 (10th Cir.), *cert. denied*, 113 S. Ct. 189 (1992); Hussey Plastics Co., Inc. v. Continental Casualty Co., C.A. No. 90-13104-WD (D. Mass. June 17, 1993); United States Fidelity & Guar. Ins. Co. v. Morrison Grain Co., 734 F. Supp. 437 (D. Kan. 1991), *aff'd*, 999 F.2d 489 (10th Cir. 1993); Nat'l Ins. Co. of Omaha v. City of Woodhaven, 476 N.W.2d 374 (Mich. 1991); *but see* Sylvester Bros. Development Co. v. Great Central Insurance Co., 480 N.W.2d 368 (Minn. Ct. App.), *review denied* (Minn. Mar. 26, 1992); South Macomb Disposal Auth. v. Westchester Fire Ins. Co., No. 84-2686-CZ, at 22–23 (Mich. Cir. Ct. Macomb Cty. Jan. 31, 1994), *aff'd*, 1994 WL 687230 (Mich. Ct. App. Nov. 18, 1994).

26. *See, e.g.*, Broderick Inv. Co. v. Hartford Accident & Indem. Co., 954 F.2d 601 (10th Cir.), *cert. denied*, 113 S. Ct. 189 (1992) (initial discharge into ponds); Hussey Plastics Co., Inc. v. Continental Casualty Co., C.A. No. 90-13104-WD (D. Mass. June 17, 1993) (insured's transfer of waste material to site); United States Fidelity & Guar. Ins. Co. v. Morrison Grain Co., 734 F. Supp. 437 (D. Kan. 1991) (abandonment of pesticides and fertilizers in deteriorating building and burying similar materials in drums underground); Nat'l Ins. Co. of Omaha v. City of Woodhaven, 476 N.W.2d 374 (Mich. 1991), *aff'd*, 999 F.2d 489 (10th Cir. 1993); (initial spraying of pesticides); *but see* Sylvester Bros. Development Co. v. Great Central Insurance Co., 480 N.W.2d 368 (Minn. Ct. App.), *review denied* (Minn. Mar. 26, 1992) (question is whether release from landfill, not whether initial dumping, is sudden and accidental); South Macomb Disposal Auth. v. Westchester Fire Ins. Co., No. 84-2686-CZ, at 22–23 (Mich. Cir. Ct., Macomb Cty. Jan. 31, 1994) (for purposes of summary judgment mo-

tion, court will assume that insured "intended for the subject landfills to act as containers," and therefore, filling of landfills did not constitute discharge into environment as defined in pollution exclusion), *aff'd*, 1994 WL 687230 (Mich. Ct. App. Nov. 18, 1994).

27. *See* John A. MacDonald & Eugene R. Anderson, *The Pollution Exclusion: The Industry's Inconsistent Regulatory and Judicial Representations*, FOR THE DEFENSE, May 1995.

28. *See* Intel Corp. v. Hartford Accident & Indem. Co., 952 F.2d 1551, 1565–66 (9th Cir. 1991); Allstate Ins. Co. v. Quinn Constr. Co., 713 F. Supp. 35, 41 (D. Mass. 1989), *settled*, 784 F. Supp. 927 (D. Mass. 1990); Paul Patz v. St. Paul Fire & Marine Ins. Co., 817 F. Supp. 781 (S.D. Wis. 1993), *aff'd*, 15 F.3d 699 (7th Cir. 1994); United States Aviex Co. v. Travelers Ins. Co., 336 N.W.2d 838, 843 (Mich. Ct. App. 1983).

29. Roberts Oil Co. Inc. v. Transamerica Ins. Co., 833 P.2d 222, 229 (N.M. 1992); Kindervater v. Motorists Casualty Ins. Co., 199 A. 606 (N.J. 1938); Augat, Inc. v. Liberty Mut. Ins. Co., 571 N.E.2d 357 (Mass. 1991).

30. 42 U.S.C. § 9601 *et seq.* (1988).

31. Roberts Oil Co. v. Transamerica Ins. Co., 833 P.2d 222 (N.M. 1992).

32. BARRY R. OSTRAGER & THOMAS R. NEWMAN, HANDBOOK ON INSURANCE COVERAGE DISPUTES § 7.04, 163 (1994).

33. Northern States Power v. Fidelity & Casualty Co. of New York, 523 N.W.2d 657 (Minn. 1994) (court found that when damages resulting from soil and groundwater contamination are continuous, damage should be spread evenly over relevant period of time, with allocation on "pro rata by time of the risk basis"); Montrose Chem. Corp. v. Admiral Ins. Co., 42 Cal. Rptr. 2d 324 (Cal. 1995) (applying continuous trigger of coverage, court ruled that allocation would require contribution among all insurers on risk in proportion to their respective policy liability limits or time periods covered under each such policy).

34. Owens-Illinois, Inc. v. United Ins. Co., 650 A.2d 974 (N.J. 1994).

35. Schering Corp. v. Evanston Ins. Co., No. 40,481, No. L-97311-88 (N.J. Super. Ct. Law Div. Union County Jan. 24, 1995), *appeal denied* (N.J. Super. Ct. App. Div. May 1995), motion for leave to appeal (N.J. filed June 1995).

36. United States Gypsum Co. v. Admiral Ins. Co., 643 N.E.2d 1226 (1st Dist. 1994).

37. L. H. Otis, *Insurers Need $40 B to Pay Pollution Claims, S & P Says*, NAT'L UNDERWRITER, Oct. 30, 1995, at 1.

38. Joe Niedzielski, *Nationwide Adds $1.1B to Reserves*, NAT'L UNDERWRITER, Dec. 11, 1995, at 1; *Nationwide to Boost Reserves by $1.1 Billion for Its Environmental and Asbestos Exposures*, BEST WEEK, PROP./CASUALTY EDITION, Dec. 11, 1995, at 1.

CHAPTER 13

Hiring the Right Laboratory

CHRISTINE DiCATO-THAXTON

Partnering with the right laboratory can minimize legal liabilities and overall costs, and maximize project efficiency. By choosing the right laboratory, the parties of interest are given quality data that enables them to make prompt and correct decisions. The most common misconception of environmental professionals is that *all* laboratories produce legally defensible data and follow the same procedures outlined under the regulated methodologies of the United States Environmental Protection Agency (USEPA). This misconception may lead the regulated environmental community directly into contracting with the least expensive laboratory, which can be an expensive error.

Analytical Expenses

Analytical costs can be one of the largest expenses in an environmental investigation. These costs can range from 20 to 40 percent of the total investigative costs. Thousands of dollars are spent, based on the data attained throughout an environmental investigation. The environmental consultant must take multiple samples, in various areas, which are analyzed for multiple lists of contaminants. These environmental consultants must follow a strict quality-control protocol that requires field blanks, trip blanks, background sampling, duplicate sampling, and equipment blanks. The number of samples, quality-control requirements, and analyses requested are a few of the contributing factors in the total analytical expense.

Coordination between the Client, Environmental Consultant, and Laboratory

Coordination between the client, the environmental consultant, and the laboratory is essential for all environmental investigations, especially brownfields projects. This coordination should be viewed as building a valued team to ensure the project is completed correctly, in accordance with applicable regulations. Over the long term, this investment spent up front may be the key to the success of brownfields projects and profit. The process begins with finding the right environmental consultant and laboratory. Once the environmental consultant has been retained, the responsibility for choosing the laboratory can be allocated to the client or the environmental consultant.

After the laboratory has been chosen, the team—consisting of the client, laboratory, and environmental consultant—should meet and devise confidentiality agreements, contracts, and all other necessary paperwork. These activities require open discussions among all parties. By encouraging open discussions, the team's expertise will be maximized and overall costs will be minimized. Moreover, this type of interaction uses each member of the team to his or her maximum potential, reduces any potential for miscommunication, and facilitates information exchange and the education of each team member. Communication, honesty, and open discussions will enable the team to complete the project correctly and minimize legal liabilities.

Selecting the Appropriate Laboratory

Identifying laboratories that meet analytical needs and specialize in the matrix types desired is the first step in selecting a laboratory. Brownfields projects often require both soil and groundwater analyses. If laboratory specializes in waste or wastewater analyses, it may not be the best fit for a brownfields project. No business can be good at everything, and the laboratory business is like every other business. The laboratory that specializes in environmental analyses and routinely runs the methods the project requires should be sought.

Brownfields projects are certification-intensive and have specific requirements set forth by the USEPA. Not all laboratories are aware of specific states' brownfields programs. Being "brownfields certified" requires acceptance of specific legal liabilities. For instance, the Ohio Environmental Protection Agency maintains a list of laboratories certified to run brownfields projects. Each laboratory has a certification number and a list of the analytical methods for which it holds certification. These laboratories have specific responsibilities to archive analytical data for extended periods of time (up to ten years). The laboratory must also maintain an open door to the USEPA for review of brownfields data. Certification takes place only after the laboratory has been audited and completes biannual performance evaluation samples.

Whoever contacts the laboratory should ask for the brownfields project manager or the person responsible for tracking brownfields projects, and then request a copy of the laboratory's certifications, quality-assurance/quality-control policy, and client-services document. Questions should be asked about the brownfields program and whether the laboratory has run analytical data for any other brownfields sites. All this information should be available upon request, with the exception of confidential information concerning other sites, and is critical for choosing the right laboratory. The investigation should also include comparing at least three laboratories' documents and compiling a list of questions to be addressed by the laboratories' quality assurance managers, brownfields managers, and/or technical directors.

Auditing

Auditing a laboratory is, by far, one of the most important exercises in selecting the appropriate laboratory. Auditing should be scheduled with the laboratory to ensure that key personnel are available during the auditing process. Audits are done to ensure data quality, inspect the laboratory physically, and meet with the laboratory's staff. This is the time to compare the quality-assurance/quality-control policy with the actual information attainable in the laboratory. The auditing process will determine (1) whether the laboratory conforms to its published quality-assurance/quality-control policies, and (2) whether the laboratory performs as its personnel claims it does.

The easiest way to review the entire laboratory process is to track a sample from delivery to reporting. This review will provide valuable information about how samples are handled, extracted, batched for quality-control purposes, and analyzed, and how data is evaluated. The audit should include review of quality-control log books, maintenance logs, standard logs, and calibration information. The auditor should determine whether all quality-assurance steps are taken, and should talk with chemists who analyze samples. Chemists should be asked about how much time is spent for data review, and who evaluates the data—the computer or the analyst. (Analytical data should always be personally evaluated, especially when automation is involved. Automation should be used to free the chemist or analyst for the data-review process.) These questions and discussions will provide insight about how the facility views data quality. *The laboratory must follow good laboratory practices.* Does the laboratory do what it says it does?

The quality-assurance/quality-control manual is the most important document to review and understand. This document describes the procedures used by the laboratory to ensure the data generated is technically valid and legally defensible. A quality-assurance/quality-control manual should reflect three distinct measurements of quality control: accuracy, precision, and completeness. This manual should include the following information:

- Philosophy, focus, or specialty
- Services and objectives
- Laboratory organization
- Handling and tracking procedures
- Analytical work flow
- Quality-assurance/quality-control protocol
- Analytical procedures:
  —Log books
  —Instrument maintenance
  —Sample preparation
  —Methods used
  —Instrument calibration procedures or calibration verification
  —Quality control
  —Corrective actions
  —Detection/reporting limits
- Data/Reduction/Reporting

- Performance evaluations
  —External audits
  —Internal audits

Selecting the Appropriate Analytical Tests

Selecting the appropriate analytical methods is a difficult and time-consuming task without open communication between the laboratory and environmental consultant. The environmental consultant should outline the potential analytical plan and then meet with the laboratory representative. The laboratory personnel can review the plan, make sure the test methods are appropriate for the information requested or required, and discuss any areas of concern. The laboratory should then be able to give the environmental consultant analytical options and define the qualitative and quantitative techniques available.

The laboratory's technical staff should have input on the analytical plan, explaining all the tools available to them to produce the type of data the environmental professional needs to evaluate the site. Brownfields sites are unique because they all have suspected or documented contamination. These sites are not restricted to specific analytical test methods nor are they permitted. Therefore, the laboratory must be creative, technically sound, and capable of method modification and development to produce defensible data. For example, all environmental laboratories can (or should be able to) run volatile organic compounds by the SW846 methods. The real question is, can that laboratory produce data that is defensible by deviating from the published method to attain information that is essential to the remediation of the site? Also, because communication with regulatory agencies is necessary, the laboratory staff must understand the applicable regulations and be able to defend their decisions.

The Most Common Analytical Tests

The most common analytical tests required in an environmental investigation are as follows:

- Volatile organic compounds SW846-8240B/8260A
- Semivolatile organic compounds, both the acid fraction and the base neutral fraction SW846-8270A
- Metals SW846-6010A
  —RCRA metals
  —Priority pollutant metals
  —Target analyte list metals (This list of metals seems unnecessary but is often requested. This group includes multiple metals that are naturally occurring in the earth's crust.)
- PCBs SW846-8270 or SW846-8081
- Pesticides and herbicides (These may need to be run if they are potential contaminants for a site.)

If the methods identified above do not cover a compound or compounds of interest, then a performance-based method must be used. Devising performance-based methods requires creativity and technical knowledge. For instance, Ohio law provides for development of performance-based methods; this allows the laboratory to develop or modify an analytical method for specific compounds. The laboratory must meet quality-control criteria outlined by the Ohio Environmental Protection Agency to prove the method is valid and defensible. Performance-based methods will need to be developed as more compounds are regulated, and developing methods to meet specific analytical requirements is time consuming and expensive. Many laboratories simply do not have the technical knowledge or other capabilities for method development.

The SW846 methods are currently being updated by the USEPA and have had multiple revisions and additions. The laboratory needs to know and understand the brownfields laws in the applicable state to be aware of which revision it may need to apply. These revised methods allow for longer compound lists, which have the potential to complicate the analytical plan.

The team must outline the information that needs to be attained on the brownfields site and work from that starting point. A high-quality laboratory should communicate any overlooked contaminants. For example, just because a decision was made to run the SW846-8240B method and that compound list came up below detection levels does not mean the samples are clean; it means only that those specific compounds were not detected at or above the stated reporting limit. In this situation, information exchange is essential and the client must be made aware of any and all potential contaminants. This information is imperative if the environmental consultant is to develop appropriate actions and remediation systems.

In Ohio, brownfields certification requires the Ohio Environmental Protection Agency to have access to all data generated on brownfields sites. If the regulating agency audits the site and collects samples, it may run additional analyses or find contamination in areas not previously considered. This issue highlights the need for communication, cooperation and information exchange. The process of communication and education must continue throughout the entire project. Lack of communication can lead to multiple problems for the entire team.

Insurance, Contracts, and Confidentiality

Insurance is one of the deciding factors in choosing a specific laboratory. The laboratory must have insurance coverage for general and commercial liability, errors and omissions, or professional liability. The errors-and-omissions insurance cannot have a pollution exclusion.

General Liability

A request should be made to be listed as "additionally insured" on the general-liability policy. This practice provides the client some comfort, as coverage then exists if samples are in the possession of the laboratory and the laboratory somehow loses, destroys, or mishandles them.

Professional Liability or Errors and Omissions

Laboratories have difficulty finding professional-liability insurance. Laboratory services are a niche business; but policies do exist. If the laboratory is serious about its business and really wants coverage, it will find a policy. With brownfields sites, it would be in the best interest of the entire team for all parties to maintain professional-liability insurance. Pollution exclusions are common in environmental professional-liability policies. These exclusions are unacceptable in the environmental business. However, many companies are not aware of their pollution exclusions. To avoid problems down the road, the laboratory's policy exclusions should be reviewed.

Contracts

Contracts are usually negotiated once the laboratory has been chosen. Contracts should include payment terms, insurance requirements, confidentiality agreements, indemnification provisions and limitations, procedural information, and dispute-resolution provisions. If a mobile laboratory is being used, the laboratory will need to meet the environmental consultant's insurance requirements.

Data Validation, Certifications, and Prices

Data Validation

The first step in data validation is to request quality-control reports along with the analytical data. The second step is to review the quality-assurance/quality-control policy. If the data is released without meeting the criteria in the quality-assurance/quality-control policy, the data should be questioned. This is the time to see if the laboratory actually follows its own quality-assurance/quality-control procedures. These procedures are to be revised and updated as the laboratory increases technical capability and the chemists update their control limits and run method detection limit studies. If the data eluded the quality-assurance/quality-control policy, an explanation or narrative must be given. This is an easy way to evaluate the laboratory's quality.

Certifications

Does the state have a laboratory certification for the brownfields program? Does the laboratory hold any other brownfields certifications? Certifications are one way of evaluating a laboratory. Many states have certification programs, which vary from state to state. The most common certification held by laboratories is for the analysis of drinking water. A list of certifications held by the laboratory and current copies of the certifications should be requested.

Has the laboratory ever lost any certifications and, if so, why? (Have the certifications been revoked or did the laboratory choose not to participate due to profitability concerns?)

With certifications come performance-evaluation samples. Some laboratories are certified in multiple states for multiple programs. These facilities may be running six, seven, or more performance-evaluation samples annually. Many laboratories participate in the Water Pollution Study and Water Supply Study. Results of those studies are valuable in determining the laboratory's accuracy.

Prices

Prices are a direct reflection of quality and service. A common misconception is that excellent service and high-quality data can be obtained for the lowest price. Analytical services are the same as any other service—*you get what you pay for!* These projects are time consuming and management intensive. Once the time has been taken to evaluate laboratories, price will become less important. The right laboratory will have already saved time and money.

Comfort Levels and Responsiveness

The laboratories to consider are those that focus on high-quality analytical data, service, consistent turnaround times and, most of all, technical support. Many state brownfields programs provide unique opportunities for the right team of players to develop specific site needs and use technical capabilities in the most innovative and creative form. The team can actually do what it does best without limitation. The environmental consultant takes personal responsibility for his or her decisions. The laboratory must realize the legal ramifications of the program and be willing to defend and accept those responsibilities.

Establishing a comfort level is critical. Complications with the analyses of the samples taken at brownfields sites are to be expected. How a laboratory reacts to those complications and how it uses the technology available to help solve those difficult analytical challenges are deciding factors. Responding to problems and encouraging open thought, communication, and creativity are the key elements to consider when hiring the most qualified laboratory.

CHAPTER 14

Rebuilding Communities through Brownfields Redevelopment

SCOTT D. GARSON

Community development corporations (CDCs) are nonprofit organizations established to revitalize neighborhoods by increasing investment and employment opportunities. As an example, MidTown Corridor, Incorporated (MidTown) began in 1982 as a private-sector initiative to revitalize a one-square-mile area located just east of downtown Cleveland, Ohio. Supported by 250 contributing members, MidTown assists in all aspects of neighborhood revitalization, including marketing, real estate and business development, visual quality, security, and employment.

Throughout the 1970s, the area now served by MidTown was devastated by the movement of small, medium, and large businesses from the central city to the edge of the county, or even outside northeastern Ohio. With that movement came a contraction of the job base and the business tax base, which is a critical source of funds for city services.

There are a variety of reasons why many neighborhoods have experienced declines in security, physical structures, and visual quality. The recent recognition of brownfields sites within neighborhoods has exacerbated the problem of redevelopment considerably. To steer neighborhoods away from further decline, CDCs must address the issue of brownfields redevelopment, along with traditional efforts to improve the physical appearance and security of inner-city neighborhoods.

At the present time, urban areas remain largely unprepared to compete with suburban and rural communities in business attraction and retention. Most importantly, economic-development programs to retain and expand job opportunities in the central cities do not go far enough to close the competitive gap. The gap exists because of the higher costs to acquire land, as well as the need for environmental assessment and remediation. The missing ingredient in the redevelopment strategy is good product: land and buildings. If programs are not developed that address the issue of urban brownfields redevelopment, then we, as a country, run the risk of further deterioration of urban centers. The tax base and employment opportunities will continue to diminish and the gap between the "haves" and the "have nots" will increase dramatically.

The Need for a Community Brownfields Strategy

This problem's complexity demands that a number of issues be addressed simultaneously, and indeed, a wide variety of resources have been developed to enhance economic development. Many of these programs are showing results and are at work within MidTown, as well as other neighborhoods. The true potential of urban neighborhoods will not be realized, however, without a comprehensive strategy for the central city that includes the conversion of brownfields sites, including vacant land and buildings, into productive tax-producing properties. The investment in land, demolition, and environmental remediation will bring market forces to urban redevelopment. The subsidy needed for environmental remediation is similar to infrastructure improvements in less urban sections of the state and will enable the development process to move forward on brownfields sites.

A major obstacle to keeping up with the competition from greenfields is the inability to move ahead on development projects in a timely manner. For a variety of reasons, including convoluted ownership issues and environmental constraints, development in the central city typically takes a long time to evolve. Land assembly is a complicated process, involving the combination of vacant land and obsolete commercial and residential properties. Although there is a large amount of vacant land and buildings, the number of easily developable sites is small. The ability of CDCs and the private sector to conduct environmental assessment and cleanup in a timely fashion is a critical element in the effort to assemble land for future development opportunities.

It has become evident in MidTown that the city is often at a cost-competitive disadvantage with suburban greenfields sites. The largest impediment is an assemblage of clean, cost-competitive land. MidTown has done well over the past fourteen years to counteract past problems of safety and visual quality that exist in almost all inner-city neighborhoods. As a result, several companies are willing to expand in the MidTown area and hire local residents in their new facilities. Furthermore, there has been extensive interest from companies outside MidTown and Cleveland in developing new facilities in the area. Unfortunately, opportunities for urban redevelopment are compromised because of the excessive costs of land assembly and cleanup. Until a cost-effective approach toward brownfields cleanup is established, urban areas will struggle to meet the demands of the private marketplace. Land assemblage must take place in a proactive manner, before the development occurs, so that CDCs and governments can respond in a timely fashion to requests for developable sites.

The Need for Infrastructure Subsidies

Recently in the MidTown area, several development projects have illustrated the difficulties of redeveloping brownfields sites. One current project is receiving considerable public assistance because of the increased costs for expanding at its current urban location. A detailed study by the company concluded that the cost to move operations to a *new* facility in the suburbs would be $1.2 million *less* than expanding its current manufacturing facility. Without providing a subsidy, Cleveland would lose fifty current manufacturing jobs, the prospect of twenty-five new jobs, and the income and sales tax generated by the company. In addition, the city and MidTown would be in the difficult position of marketing another vacant, obsolete, and potentially contaminated facility.

Another major project preparing for construction will happen as a result of significant public subsidy. The costs to assemble and remediate a nine-acre site resulted in a total land cost of almost three times a suburban greenfields location of similar size. A recently completed $6 million construction of a new distribution and corporate-headquarters facility occurred with public subsidy as well. In this project, the land costs had to be written down ten times to compete with a greenfields site being considered. The high land costs were a direct result of the clearance of obsolete buildings and environmental remediation.

A current project that MidTown has been working on for over five years has been stymied by the property's environmental condition. Early discussions about the property involved the clearance of all buildings to make land available for development. An environmental assessment revealed that the building contained significant amounts of asbestos. The cost of cleanup before demolition would be over $800,000, resulting in a land cost of over $1.6 million per acre. Current land costs of suburban greenfields sites are in the $20,000 to $150,000 range.

This discovery changed MidTown's strategy on the property. The environmental condition of the existing property can be resolved through proper encapsulation techniques. However, the cost to renovate the building is $90 per square foot. The rents that are needed to support this development are significantly higher than the market can bear, necessitating a significant amount of public subsidy for the project to move forward.

Limited Resources

A variety of circumstances limit a CDC's ability to accomplish the goal of redeveloping brownfields sites. CDCs are nonprofit entities that most often have limited budgets. In a few isolated cases, MidTown has been able to raise dollars through foundation grants to obtain Phase I and Phase II environmental audits on a variety of properties that had been targeted for redevelopment. The organization's leadership quickly learned that to spend these funds without a fairly secure development in mind would be imprudent.

To secure financing for a large multiuse or owner-occupied project, it is critical to secure tenants in advance. Unfortunately, it is difficult to attract tenants before environmental issues are resolved. The cost of environmental cleanup plays a large role in developing the total project costs and thus affects the rental rate or the debt service for an owner-occupied property. This information is critical for a business to make a location decision. Compounding matters is the reluctance of many businesses to pay up-front, predevelopment costs in the form of environmental audits, without a firm commitment to locating at the brownfields site. It takes a firm commitment to redevelopment to rationalize the use of funds for environmental audits.

If a CDC acts as the developer, the same issues can come into play. If an environmental audit uncovers extensive problems, the CDC incurs the burden of cleanup. Even if the persons responsible for the environmental problems are discovered, the CDC's limited resources make it difficult to pursue legal action. Though legal costs may be recovered, there are no guarantees. Therefore, the up-front costs can make it nearly impossible for a CDC to pursue a redevelopment strategy that includes its participation as a developer.

As stated earlier in this chapter, it is important to develop a land-assemblage program in urban areas to respond to market demands in a timely fashion. However, recent court decisions have made it difficult to acquire and assemble sites for future development, due to the carrying costs associated with land banking property. In Ohio, the state's position is that the holding of land by CDCs is not a public purpose and is therefore subject to property taxes. Additional carrying costs include the simple acts of cutting grass, carrying liability insurance, and keeping vacant buildings boarded and free of code violations. With limited budgets, CDCs must have a thorough plan for maintaining property and absorbing carrying costs before acquiring property and pursuing redevelopment opportunities.

State Voluntary Cleanup Programs: Costs and Benefits

Programs like Ohio's Voluntary Real Estate Reuse and Cleanup Program (Voluntary Action Program) attempt to ease the difficulty of redeveloping brownfields sites. Although the Voluntary Action Program is a very positive step toward brownfields redevelopment, several items included in Ohio's legislation fall short of remedying the problem. As stated earlier, the ability and/or desire of CDCs or private individuals to conduct environmental audits on a speculative basis is difficult at best. The incentive to do so is very small when greenfields sites are available at a lower cost. The Ohio legislation may offer loans to conduct these audits. However, to provide significant incentives to rehabilitate brownfields sites, a grant program must be developed to entice local governments, nonprofit organizations, and the private sector to conduct predevelopment environmental audits.

Another Ohio financing tool is tax abatement on properties participating in the Voluntary Action Program. Because of the poor condition of the Cleveland Public School System, this form of financial incentive is not viewed favorably. The political ramifications of using tax abatement today could result in less attractive financial packages for the construction and rehabilitation of property.

State Voluntary Cleanup Programs that provide different sets of standards for different types of property use offer much-needed clarification for brownfields redevelopment to succeed. In one project in MidTown, the developer of a distribution facility was required to test the groundwater 150 feet deep. Though this type of testing may be necessary if a day-care facility with an outdoor play area were being developed, it is much too stringent an exercise for industrial development.

However, as an area such as MidTown realizes new investment, the entire area experiences an upgrade in perception and demand. As demand increases, the potential for higher and better property uses is created. Therefore, one downside to tying cleanup standards to specific uses would be a permanent limitation on the character of development within a community. For instance, if a brownfields site was previously an industrial property (as is the case with almost all brownfields) the future use of the site would be limited to industrial use (unless a "higher level" of cleanup was conducted). If a commercial or residential project were in demand, the cost associated with this upgrade in development would be difficult to justify. Therefore, tying cleanup standards to intended land use may seriously limit certain opportunities for redevelopment of brownfields sites.

The Role of CDCs: Educator, Broker, Facilitator

Although there are limitations on what can be done effectively to reuse existing brownfields sites, there are many things that CDCs can do to stimulate brownfields redevelopment. One of the main roles a CDC can play is that of educator. There is a perception that brownfields sites are unmanageable and present excessive liability concerns. However, if the right steps are taken, many of the brownfields sites that exist can be reused. It is critical that the CDC community help to educate potential developers of brownfields sites on the opportunities for reuse.

One of the best ways for CDCs to educate is through the brokering of information regarding public programs, available sites for redevelopment, and additional resources that can assist in the remediation of environmental problems. CDCs should have lists of certified contractors that are able to conduct environmental audits and cleanups. The goal must be to make brownfields redevelopment as easy as possible, considering all the hurdles that must be overcome to build on these sites. Many people in the private sector are not aware of funding opportunities or where to turn for qualified technical assistance on brownfields issues.

CDCs can also act as facilitators between the private sector and government entities. A CDC's familiarity with government, and its expertise in working through the government process, can be extremely helpful to a business that will be working on real estate development as a one-time experience. CDCs are usually great sources of history regarding activity on specific sites and can provide information to government entities that will be involved in the process. Frequently, the CDC is able to work with the private sector and devise a redevelopment package before involving government officials, saving time and money all around.

CDCs must familiarize themselves with the rules and processes involved in turning brownfields into greenfields. MidTown has had a great deal of success in accessing a variety of public funding sources that have helped attract new private-sector investment and create employment opportunities in the urban core. It has been MidTown's philosophy that the private sector should concentrate on keeping its own businesses profitable. It is the job of the CDC to lead the private sector through the governmental process, to help the private sector accomplish its goal of relocation or expansion. If the business fails due to excessive attention to the government process, we all lose. Therefore, the CDC community must learn the nuances of applicable state Voluntary Cleanup Programs to assist the private sector properly.

Experience at MidTown has been that speculative development is very difficult. All the recent successes in MidTown have involved a business owner acting as developer and occupying the redeveloped property. More resources can be brought to bear in these situations to help overcome the increased costs of development in the urban core. Furthermore, because of the higher costs of development associated with brownfields, the profit margins on leased properties are very slim. The rents that are needed to make a project profitable often price the project outside the market.

One way the development community can be helpful is to participate in turnkey development projects. These projects involve the development of real estate that conforms to the specific needs of an end-user. Once the project is completed, the developer sells the property to the end-user and realizes profit at the point of sale. These are not necessarily speculative deals. CDCs can work with developers in identifying the potential end-user. The end-user is then freed of the burden of redevelopment and the liability associated with acquiring and remediating property. For many in the private sector, this type of development would eliminate many of the concerns associated with relocation or expansion at brownfields sites.

Although there are many concerns about the environmental condition of land and buildings in this country, there is considerable movement toward rectifying the situation. We cannot continue to allow businesses to flee the urban core where most brownfields properties are located. The government, private sector, and CDC community must work together to counteract the destructive flow of investment and employment opportunities out of the urban core. The rise of state Voluntary Cleanup Programs provides one more tool to be used in these efforts.

CHAPTER 15

Community Participation in Brownfields Redevelopment

JOHN C. CHAMBERS
MICHELLE A. MEERTENS

Brownfields redevelopment involves a collaborative process affecting the interests of a variety of stakeholders, including investors, developers, financial institutions, and community members. Though all these parties have significant vested interests in brownfields redevelopment, more attention has traditionally been paid to business interests. This focus is an understandable consequence of the need to encourage more business investment, but the interest of the community in the process of redevelopment is important and should not be overlooked.

The United States Environmental Protection Agency (USEPA) recognized the importance of community participation when it issued its Brownfields Action Agenda (Action Agenda) in January 1995.[1] The Action Agenda outlines USEPA's future plans and strategies to help states and localities carry out brownfields redevelopment. The Action Agenda delineates USEPA's intentions to clarify liability and cleanup issues, create partnerships and promote outreach, and conduct job development and training.[2] The main focus of the Action Agenda, however, is USEPA's Brownfields Economic Redevelopment Initiative (Brownfields Initiative). The Brownfields Initiative is a pilot program under which USEPA gives funding to states and local municipalities to assist them in conducting environmental assessments of selected brownfields sites. This process is forward looking and is designed as a prelude to the eventual cleanup and redevelopment of these brownfields sites.

One of the most important aspects of the Brownfields Initiative is the call for active community involvement. USEPA hopes to use the brownfields pilot program as a way to identify effective working models for meaningful public participation, which can then be implemented around the country.[3] To this end, USEPA makes the adequate planning for, and actual participation of, the community one of the criteria it uses when it selects brownfields grant recipients. Before and after the grant is awarded, USEPA performs community involvement checks by telephone to get updates on the level of community participation at various brownfields sites around the country.[4] USEPA also promotes public participation in the Brownfields Initiative by publicizing activities and providing assistance to local organizing groups so they can hold public dialogues and town meetings. This coordinated effort is significant because it recognizes the necessity of giving individuals a true voice in a process that will affect the future of their communities.[5] According to USEPA officials, "[t]he U.S. EPA is committed to building partnerships with states, cities and community representatives to develop strategies for promoting public participation and community involvement in brownfields decision making."[6]

Although the Brownfields Initiative approaches community participation with renewed vigor, the concept of involving the community in the process of environmental remediation is not a new one. For example, there are provisions for public participation under the Comprehensive Environmental Response, Compensation, and Liability Act of 1980 (CERCLA). Federal law requires that USEPA provide public notice of plans for removal or remediation within a specified number of days and that it set aside an adequate period of time for public comment.[7] In addition, technical-assistance grants are available for local communities to ensure that participation is knowledgeable and meaningful.[8] These provisions, however, have traditionally been underutilized. For instance, since 1988, USEPA has awarded only 151 technical-assistance grants to local communities.[9] Many individuals and community leaders have charged that despite CERCLA's public-participation provisions, community involvement has been minimal.[10] Thus, although the government has put forth the concept of community involvement in environmental restoration projects, it has never been manifested in its fullest potential.

The Brownfields Initiative makes active public participation paramount. The goal of this chapter is to examine community involvement in the Brownfields Initiative. It will discuss the concerns about the Brownfields Initiative, which have been expressed by communities near brownfields sites. It will also evaluate the effectiveness of the mechanisms for community involvement that have been used to date.

The Community Perspective: A Historical Grounding

During the past few decades, urban centers have undergone a huge transformation. Many of the large institutions and manufacturing companies that once employed a great percentage of the surrounding population are no longer in business or have relocated. The removal of these blue-collar jobs left many people out of work. Additionally, because many of them lacked the requisite educational background and training, they were unable to compete for skilled-service positions. The result was a dramatic increase in the level of unemployment. Correspondingly, poverty levels rose. Thus, the end of the industrial era played a major role in creating the conditions that are now a familiar part of the inner-city landscape.

Despite these depressed conditions, many urban residents maintain the hope for positive change to better themselves and their communities. One source of hope for revitalization and change has always come from the collaboration of developers, property owners, and financial backers with "a plan." The "plan" has often taken the form of new housing, retail stores, infrastructure, and even waste-disposal facilities or industrial factories. Too frequently, however, the plan for revitalization and change materializes without creating any benefits for the community. New facilities are built using outside labor, not labor from the community. If the new facilities are retail oriented, the community often patronizes the stores, but the stores infrequently give anything of benefit back to the community. If these newly constructed facilities include housing, they often serve to "improve" the community so much that they start a process of gentrification. This process ultimately pushes out the poor because they can no longer afford to live there. If the new facilities involve the placement of waste-disposal facilities or industrial factories, these operations are often the source of additional environmental hazards.

These historical experiences form the backdrop for the myriad of responses many urban communities have to the Brownfields Initiative. These responses are valid expressions of concern. Though the optimists in these communities see great potential for the Brownfields Initiative to generate positive change, the cynics remain skeptical about whether that potential will ever be realized. Some fear that the project will not only fail to produce any tangible benefits for their communities, but possibly harm them as well. If the Brownfields Initiative is to achieve its goal of revitalizing urban communities with active community involvement and participation, local-community concerns must be taken into consideration.

Community Concerns

Economic Development

One of the major concerns for urban communities is stimulating economic development. There is a widespread feeling that cities must begin to make better use of economic resources and become better able to compete to survive. Many people feel that urban residents can no longer afford to depend on big outside institutions to create low-skilled jobs. The industrial era has ended. Most large institutions and manufacturers have relocated their plants elsewhere, often abroad where labor is less costly. The majority of available work in the marketplace, therefore, is skilled labor. Consequently, many feel that community members must learn marketable skills to compete on an equal footing for these jobs. In addition, the community must learn to create and maintain its own businesses. These businesses should be owned and operated by community members. In this way, urban communities can begin to achieve greater economic self-sufficiency.

Many community leaders view brownfields as viable tools to help achieve economic self-sufficiency for urban communities. Brownfields redevelopment offers the opportunity to bring contracts and jobs into the community. These resources and opportunities can be helpful, provided they are given to the people of the community and not to outsiders. Unfortunately, many communities too frequently have had negative experiences with outside developers who have promised revitalization. Consequently, there is the fear that the Brownfields Initiative will become just another "get rich" tool for wealthy investors and developers. Positive-thinking community leaders, however, want to ensure that this does not occur.

One of the ways communities can reap economic benefits from the Brownfields Initiative is through jobs, skills training, and career development. Much of the work that accompanies a brownfields project is contract driven. The initial work is oriented toward environmental assessment. The later work is oriented toward planning, surveying, and construction. Communities want to ensure they will get first priority at receiving these jobs. In addition, many community leaders would like the brownfields project to provide them with funding so they can organize programs to give community members who currently do not have the requisite skills the proper training to enable them to work. If communities are given the proper education and skills, they can begin to take care of these sites themselves. Providing members of the community with concrete skills and experience is one tangible benefit the Brownfields Initiative can give to communities, which they can use long after the brownfields pilot project is complete.

Gentrification

Another concern many leaders in the community have voiced is the possibility of gentrification. Many communities that were once blighted and depressed have lived through "redevelopment" and "revitalization" that, while serving to better the neighborhood, also served to push out the poor because they could no longer afford to live there. "I see it happening now in West Oakland,"[11] commented Allen Edson, a community leader at the African-American Development Association in Oakland, California. Mr. Edson is an active participant in the area's local brownfields project. He was speaking about a community in the San Francisco area currently undergoing redevelopment. The neighborhood, as he described it, is poor, mostly African-American, with an extremely high unemployment rate. According to Edson, the community is favorably located, very near San Francisco, and metro accessible. "[N]ow we're seeing young Asian families and young white couples beginning to move back into the area . . . and the people there are being pushed out. . . ."[12]

Communities in the urban center are not only aware of the gentrification effects of redevelopment, they are cognizant of the underlying factors that cause it to happen. For instance, Edson comments, "One other thing that I've noticed is that the target cities [of the Brownfields Initiative] are [located] on the prime real estate. Emeryville is on the Bay. Richmond is on the Bay. San Francisco, East Palo Alto, they're all on the Bay. Stockton is on the River. . . . So you have developers and real estate [people] speculating and chomping on the bit. . . . [Our] inner cities are under siege."[13] In addition, he noted, because communities are frequently prevented from playing an active role in the process, they are helpless to affect it.

Though Edson admits that the Brownfields Initiative has been good at giving the community a forum in which to voice concerns, according to him, an active role in the process means more than just a chance to have views aired. The community must be given a real chance to participate on an active level. This chance to participate is something Edson claims he has not seen. According to him, the community is, and has been, at the bottom of the pecking order in the Brownfields Initiative. He says all the meetings to date have been dominated by developers, investors, and lawyers. In addition, he notes that all the money given out by the Brownfields Initiative so far has been given to the states, cities, and municipalities, not to the communities. From his point of view, the community has not yet seen any tangible benefit from the Brownfields Initiative and he fears the worst. What he would like to see is money or resources given to the communities for technical assistance so that they can begin to understand and control the process themselves, instead of being passive participants along for the ride. For community leaders like Edson, the only effective difference between brownfields redevelopment and any other redevelopment thus far is that the community is more aware of what is happening; but it is not necessarily more able to affect it.

Environmental Justice

Another primary concern for individuals who live in communities near brownfields sites is environmental justice. Studies have shown that, historically, a disproportionate amount of waste-disposal facilities and industrial factories have been located in neighborhoods of color. Many urban communities are located near more than one of these facilities. For instance, the community of Bay View/Hunter's Point, California, another brownfields community in the San Francisco area, is the location of not one, but two, Superfund sites in addition to its brownfields pilot site.[14] Calling attention to these perceived injustices and finding ways to rectify them has been the aim of the environmental justice movement. The importance of environmental justice was recognized by President Clinton in his Executive Order 12,898, in which he stated that one of the goals of his administration was that "[n]o segment of the population, regardless of race, color, national origin, or income, as a result of U.S. EPA's policies, programs, and activities, suffer disproportionately from adverse human health or environmental effects, and that all people live in clean, healthy and sustainable communities."[15]

Many people in brownfields communities see the Brownfields Initiative as a way to rectify past environmental injustices. They would like to transform abandoned brownfields into productive greenfields. In the opinion of many community members, redeveloping a brownfields site to facilitate further industrial use does nothing to rectify past environmental injustice. The determination of the future use of a site is, therefore, a very important decision. Differences in circumstances and needs may color each decision. For instance, while one community may need or desire affordable housing, another may need or desire a community center, a recreational park, or office space. Despite minor differences in circumstances, however, most members of the environmental justice movement feel very strongly that brownfields sites should be redeveloped to support positive, environmentally clean uses. Although this goal is a worthy one, it often runs counter to the primary monetary interests of developers and investors, because it increases the level of cleanup required. This additional remediation has the effect of increasing the cost of a project that may have little potential for making a profit. Thus, there is often tension between the concerns of the community in obtaining the most environmentally safe and healthy use of the site and the concerns of developers in obtaining the largest possible profit.

Community Health and Cleanup Standards

Another primary area of concern for individuals living in communities near brownfields sites is community health and cleanup standards. The health of the community is a paramount concern for most local residents. No one wants to live in an area beset by health hazards. As noted above, however, a great number of urban communities exist in neighborhoods that are saturated with brownfields sites. An important concern for these communities is the proper cleanup of these sites and their return to an environmentally safe, non-health-threatening condition. The determination of cleanup standards, therefore, is very significant.

The cleanup standards set for a brownfields site dictate how environmentally safe it has to be before any redevelopment can take place. Most members of the community want the brownfields sites in their neighborhoods to be returned to the cleanest possible condition, regardless of the planned future use. They want the cleanup standards to be set high. Many community members, however, are extremely skeptical of the likelihood of this occurring. They fear that environmental standards will be lowered, not heightened, to decrease the cost of cleanup and to encourage investors to participate in brownfields redevelopment. They fear that their health and the future health of the community will be sacrificed in favor of cutting costs and making a profit.

Unfortunately, this fear cannot be alleviated by the application of a universal environmental standard of cleanup for all brownfields sites. The process of determining the appropriate standards is something that must be accomplished on a case-by-case basis. The final decision is made by whatever governmental municipality has authority over the site. Several factors contribute to the determination of the appropriate environmental cleanup standard. For instance, considerations of cost affect the determination of the appropriate environmental standard for a site. Because the grant funding currently being provided by USEPA is slated for environmental assessment purposes only, the actual cleanup costs must be absorbed by investors, developers, or the communities themselves. Given these considerations, the amount of money available to conduct a cleanup will be limited. This, in turn, naturally affects how thorough a cleanup job can be undertaken. The present state of contamination is another factor that affects the determination of the appropriate environmental standard for a site. The more polluted the site, the greater the cost of cleanup. Last, the intended future use affects the determination of the appropriate environmental standard for a site. Not all future uses require an environmental cleanup to the site's original pristine state. Although many community members would like every brownfields site to be returned to this heightened level of cleanliness, this approach is not practical. Thus, there are several factors to be considered in the determination of the appropriate environmental standard for the cleanup of a brownfields site.

Regardless of these different factors, however, the determination of the appropriate environmental standards should not be made without representation from the community. It is safe to assume that the interests of business will always be represented. The interests of the community, however, will be represented only if community members are given the opportunity to participate up front and are properly equipped to engage in meaningful and knowledgeable interaction. To do this, however, "the community must understand the process itself," says John A. Rosenthall, Director of Environmental Justice at the NAACP National Office.[16]

One of the tasks Rosenthall frequently undertakes is conducting workshops and seminars on brownfields and brownfields-related issues for communities and community leaders around the country. According to him, one of the first things about which communities must be made aware are the ramifications of having a brownfields site in their neighborhood. The community must be informed of the risks posed by brownfields sites. Moreover, it must be informed of federal policies relating to brownfields sites, as well as the funding and technical-assistance resources available to help. Once a community becomes aware of the issues involved with brownfields redevelopment and begins to participate actively in the planning process, there is a greater chance that an acceptable agreement on cleanup standards can be reached.

Current Methods of Participation

There is no single method of public participation universally used in brownfields redevelopment. Each community, therefore, elicits public participation differently. Some of the more common methods of public participation, however, are the pubic dialogue and the working group.

Public Dialogues

The public dialogue is an effective method of eliciting community participation because it gives community members an opportunity—in a structured format—to interact and voice their concerns regarding brownfields redevelopment to USEPA, government officials, and other stakeholders. The National Environmental Justice Advisory Council (NEJAC), a subcommittee of

USEPA, made extensive use of the public-dialogue format in the summer of 1996, in an attempt to encourage and elicit public participation in brownfields redevelopment. NEJAC held five major dialogues in selected cities near brownfields pilot projects across the country. These dialogues were held in Boston, Philadelphia, Detroit, Oakland, and Atlanta. The dialogues were day-long events structured in two tiers. First, citizens were provided with an opportunity to voice their concerns about brownfields redevelopment, as well as their visions and ideas about comprehensive ways to revitalize their communities. Second, representatives from government agencies were asked to address how their respective organizations might assist the community in achieving these visions. Finally, the dialogue provided several structured opportunities for interactive discussion and debate. Many community leaders who attended the dialogues thought they were very successful. The discussions were well publicized beforehand and many members of the community attended. Furthermore, the atmosphere of the dialogues was respectful and the community members who spoke were made to feel that their views, opinions, and concerns were significant and worthy of consideration. It was in the course of the discussions that many of the above-mentioned issues and concerns were raised.

The success and effectiveness of the public-dialogue format in general, and the NEJAC public dialogues specifically, are demonstrated not only by how the dialogues have elicited community comment on several brownfields-related issues, but also by how they have elicited community comment on the very process of community participation itself. For instance, during the NEJAC public dialogues, several suggestions about how to improve community involvement were offered. USEPA has already made use of some of these ideas.[17] Furthermore, in the wake of the dialogues, NEJAC published a report entitled *Environmental Justice, Urban Revitalization, and Brownfields: The Search for Authentic Signs of Hope,* which not only summarizes the proceedings but identifies recommendations for specific action on all the topics and concerns that were raised.[18]

In its report, NEJAC made several recommendations on the issue of public participation. For instance, NEJAC recommended the creation and support of structured mechanisms for community participation, such as public dialogues and community advisory boards at all levels of brownfields redevelopment (that is, national, regional, and local). NEJAC also encouraged the support and promotion of *substantive* public participation. Substantive participation, as defined by NEJAC, includes participating in relevant activities such as reviewing research projects and developing grant proposals. According to NEJAC, such involvement is necessary and much more valuable than merely having access to information or having an opportunity to provide comment. In addition, NEJAC recommended that innovative and nontraditional methods of outreach be used to disseminate educational information to the community. For example, in addition to the normal use of posters, and advertisements in local papers, community newsletters, and electronic mail, NEJAC recommended making use of existing social and cultural networks, such as schools, churches, and civic organizations. NEJAC also recommended holding meetings in more accessible locations, at more convenient times, and perhaps providing day care and transportation. Most importantly, however, NEJAC's report stressed that the community members must be educated enough to not only understand the process, but influence it.

Working Groups

Another method of soliciting public participation is the use of the working group. The working group consists of a small number of community leaders who work in close connection with USEPA and other government officials to represent the community's interests in the remediation and redevelopment process. Although the public dialogue is an effective way to achieve active community participation and discussion in a brainstorming format, because it usually involves a great number of people, it is not the most conducive mechanism for decision making. In contrast, the working group is a much more flexible and efficient tool. Community leaders who participate in working groups represent the community by remaining in contact with its members and relaying their concerns and opinions back to the working group.

The working-group format is used by several brownfields projects around the country. For instance, in Bay View/Hunter's Point, California, a working group of approximately fifteen members comprises community leaders and officials from the USEPA and state and local governments. This group meets once a month to discuss current issues related to local brownfields development. The community members of this group regularly keep the larger community apprised of events and relay any arising needs or concerns back to the group for discussion or evaluation. According to community leaders, this format is an effective one for decision making and policy planning.

Conclusion

All parties acknowledge that the goal of community involvement in brownfields redevelopment is an important and worthy one. The traditional view that community participation is satisfied by a mere opportunity to review and comment on government decisions and policies is defunct. A question remains, however, about the level of community participation that USEPA's Brownfields Initiative can truly achieve. The public dialogues that took place during the summer of 1996 proved that successful community participation is possible. Yet, in the opinion of many (NEJAC included), this level of participation is not enough. The mere opportunity to air opinions and concerns early in the process does not amount to substantive community participation. Substantive community participation can be achieved only when the community is properly educated and given an active role in the actual planning and decision-making process. Although the groundwork for such participation exists in mechanisms such as working groups and advisory committees, their use in many brownfields projects is still formative. Additionally, many community leaders and citizens have varying degrees of optimism about their success. One thing is certain, however. The vision of a community driven and directed urban revitalization will be achieved only with great commitment and perseverance on the part of all stakeholders.

Notes

1. *EPA Administrator Announces Relief Package for Cities and Towns*, EPA ENVTL. NEWS, Jan. 25, 1995, at 1.
2. *The Brownfields Action Agenda*, U.S. EPA web-site document, http://www.epa.gov/swerosps/bf/ascii/action.txt, at screen 1 (accessed July 25, 1996).
3. *Answers to Frequently Asked Questions*, U.S. EPA web-site document, http://www.epa.gov/swerosps/bf/answers.htm#26, at screen 1 (accessed July 25, 1996).
4. Telephone Interview with Katherine Dawes, Environmental Protection Specialist, USEPA Office of Solid Waste & Emergency Response (Aug. 7, 1996).
5. Exec. Order No. 12,898, 3 C.F.R. 859 (1995) ("Those who live with environmental decisions—community residents, State, Tribal, and local governments, environment groups, business—must have every opportunity for public participation in the making of those decisions. An informed and involved community is a necessary and integral part of the process to protect the environment.").
6. *The Brownfields Action Agenda*, U.S. EPA web-site document, http://www.epa.gov/swerosps/bf/ascii/action.txt, at screen 7 (accessed July 25, 1996).
7. 40 C.F.R. §§ 300.155, 300.415, 300.430, 300.435 (1995).
8. *Reform of the Superfund Act of 1995: Hearings on H.R. 2500 Before the House Subcomm. on National Economic Growth, Natural Resources and Regulatory Affairs of the Comm. on Government Reform and Oversight*, 104th Cong. (1966) (statement of John C. Martin, Inspector General).
9. *Id.*
10. *Reform of the Superfund Act of 1995: Hearings on H.R. 2500 Before the House Subcomm. on National Economic Growth, Natural Resources and Regulatory Affairs of the Comm. on Government Reform and Oversight*, 104th Cong. (1996) (statement of Florence T. Robinson, North Baton Rouge Environmental Association, Louisiana Environmental Action Network, Mississippi River Basin Alliance Communities at Risk Network) ("The major stakeholder in the Superfund process has been denied meaningful participation and input in the decision-making process of Superfund. Public participation needs to come earlier, resources need to be provided to the community to effectively participate and communities need to be given access to and decision-making power regarding their sites. The entire process, from beginning to end must include community participation.").

See also Al Knight, *Mining and the Environment "Disaster" at Summitville Was after the EPA Arrived*, DENVER POST, Apr. 28, 1996, at E-01; Jennifer Oulette, *Superfund: A Call for Change, Responsible Care*, 249 CHEMICAL MARKETING REP., SR 10, 1996.

11. Telephone Interview with Allen Edson, African-American Development Association (Aug. 7, 1996).
12. *Id.*
13. *Id.*
14. Telephone Interview with Romel Pascual, Urban Habitat (Aug. 6, 1996).
15. Exec. Order No. 12,898, 3 C.F.R. 859 (1995).

16. Telephone Interview with John A. Rosenthall, Director of Environmental Justice, NAACP (Aug. 13, 1996).

17. Telephone Interview with Katherine Dawes, Environmental Protection Specialist, USEPA Office of Solid Waste & Emergency Response (Aug. 16, 1996). One of the recommendations offered at the public dialogues was to emphasize the adequate planning for, and actual participation of, the community as a criterion for receiving USEPA funds. USEPA incorporated this suggestion in its most recent round of grant awards by revising its grant-application brochure to clarify and explain its concept of community participation and by instituting telephone checks on levels of community involvement before and after the granting of a brownfields funding award.

18. The interim draft version of this document is available at the U.S. EPA web site, http://www.epa.gov/swerosps/bf/ascii/nejacbrn.txt (accessed July 25, 1996).

PART III

Scientific Concepts Used to Address Contaminated Property

~

CHAPTER 16

The Science of Brownfields

DAN B. BROWN

Brownfields sites are not natural phenomena. They occur as the result of historical use at industrial and commercial facilities. Brownfields can trace their roots to the dawn of the Industrial Revolution in the late 1800s.[1] Since that time, cities, in particular, have seen continuous industrial and commercial use of numerous tracts of land. The soil and groundwater at these sites have been repeatedly impacted by multiple contaminants in various concentrations and amounts. To understand the approach to investigating and remediating these sites, one must first understand the basic science of the subsurface environment.

Subsurface Geology and Hydrogeology

The subsurface comprises three major components: soils, bedrock, and groundwater.[2] Although these three components are interrelated, each plays its own role in determining the fate and transport of contamination as it enters the subsurface. Soil is typically present as a result of alluvial or other weathering processes. Soils are also typically associated with near-surface deposits.[3] Bedrock is the result of geologic phenomena that, under pressure and temperature, compressed soils and other materials into a solid mass. Bedrock is often associated with deeper deposits, but can also be present in near-surface scenarios.[4] Groundwater is present in both soil and bedrock in intergranular pore spaces or fractures. Groundwater is present everywhere below the ground's surface, but its depth may vary significantly.[5]

Soils

Soils can be classified into three primary categories and can be defined by their associated permeabilities: sands (highest permeability), silts (marginal permeability), and clays (lowest permeability).[6] Permeability can be described as the rate at which water will move through a permeable medium.[7] Subsurface soils can be compared to a sponge. Regardless of permeability, soils, like a sponge, will absorb contamination. For example, imagine placing contamination of some kind, one drop at a time, onto the top of a sponge. The sponge would continue to absorb the contamination until just one drop would be released from the bottom of the sponge. As contamination is released into soils at a site, this same phenomenon occurs. As would be suspected, soils with higher permeabilities (that is, greater space between the pores of each grain of soil) will tend to transmit contamination through the soil horizon more quickly.

Bedrock

Bedrock is different from soils in that it is consolidated. Bedrock has a very different impact on contamination as it reaches the subsurface. Bedrock does not typically act like a sponge. Instead, contamination seeks fractures and other secondary pore spaces within which to travel. Therefore, when dealing with bedrock and contamination, a primary concern is determining the location and direction of fracturing and other secondary porosity systems so that contamination can be tracked and removed.[8] In addition, because bedrock does not typically exhibit the same absorptive characteristics as soil, it has the capability of transmitting contamination in an expedited fashion, as compared with soil.

Groundwater

Although groundwater is often considered a concern because of its use as a drinking-water source, this discussion focuses primarily on its importance as a mechanism of transport for contamination. Groundwater in the subsurface can be referred to as an aquifer. An aquifer is a geologic formation with sufficient interconnected porosity and permeability to store and transmit significant quantities of water under natural hydraulic gradients.[9] Beneath the land surface, groundwater is present at various depths. The groundwater exists in the subsurface as a result of the hydrologic cycle (Figure 1).

Through the process of evaporation of oceanic and lake water, the subsequent condensation and precipitation of that water on the land surface, and the infiltration of groundwater to the subsurface, aquifers are developed.[10] This groundwater zone (aquifer) may be imagined as a huge reservoir in rocks or soil whose capacity is the total volume of pores or openings that are filled with water.[11] When contamination in a sufficient amount percolates through subsurface soil or bedrock, it will reach the groundwater surface. Once at the groundwater surface, dissolution and the force of gravity can mobilize the contamination at significantly greater rates than in soil or bedrock.

After groundwater infiltrates the subsurface as described above, it collects in the primary and secondary pore spaces of subsurface soils and bedrock, creating saturated conditions. As groundwater accumulates in these various media, it seeks a level that is exactly equal, but opposite, to that of atmospheric pressure.[12]

FIGURE 1. The Hydrologic Cycle[13]

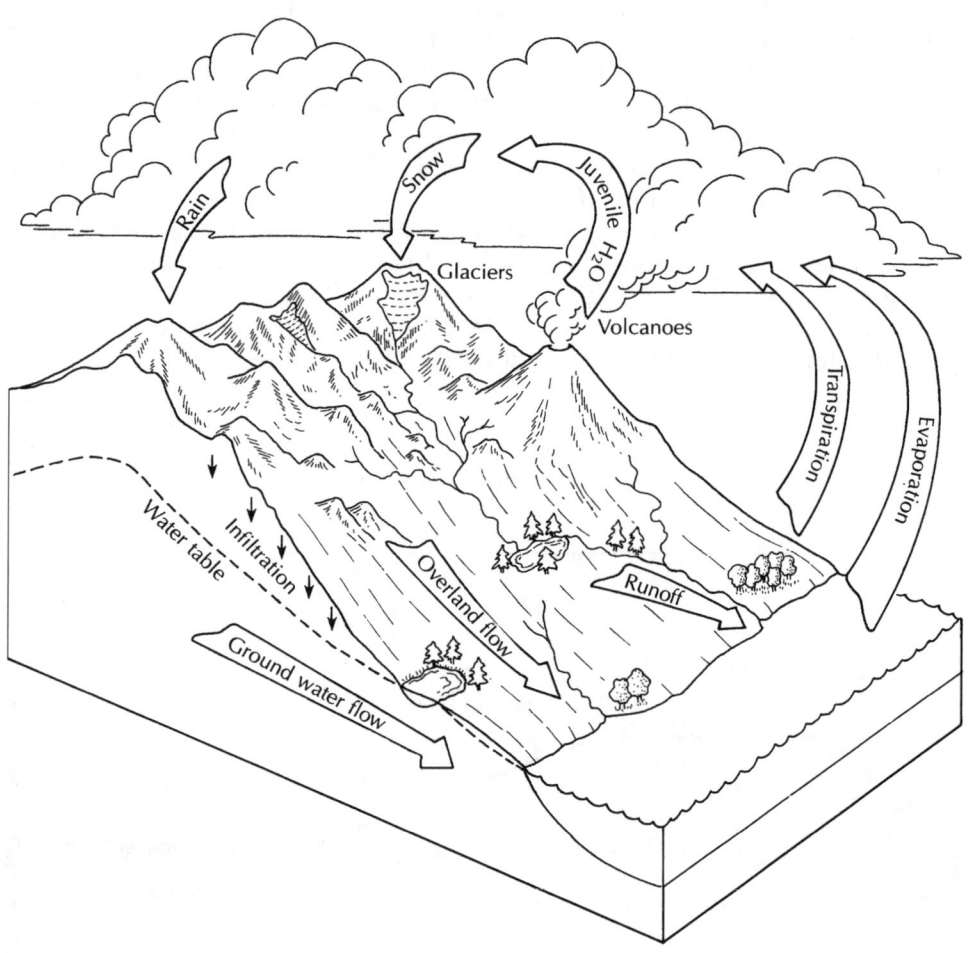

APPLIED HYDROGEOLOGY 2/E by Fetter, C.W., © 1988.
Reprinted by permission of Prentice-Hall, Inc., Upper Saddle River, NJ.

Once stabilized at this level, which varies based on numerous hydrogeologic and physical characteristics, groundwater then flows through soils and bedrock under the force of gravity. Groundwater at higher elevations in the subsurface will tend to flow toward groundwater that occurs at lower elevations.[14] This difference in elevation may often be minor. For example, the difference in elevation across a large industrial site may be only a few feet. However, it is this phenomenon that determines the direction in which groundwater flows and, subsequently, the direction in which contamination on or in the groundwater will move. (In Figure 2, groundwater would flow in the unconfined aquifer from left to right.)[15]

Groundwater moves very slowly compared with surface water. A high groundwater velocity would be one to three feet per day, while a rapid stream may move at one to three feet per second.[16] Another characteristic of groundwater is its residence time. Residence time is the amount of time groundwater will remain in the subsurface from infiltration (for example,

FIGURE 2. Groundwater Flow

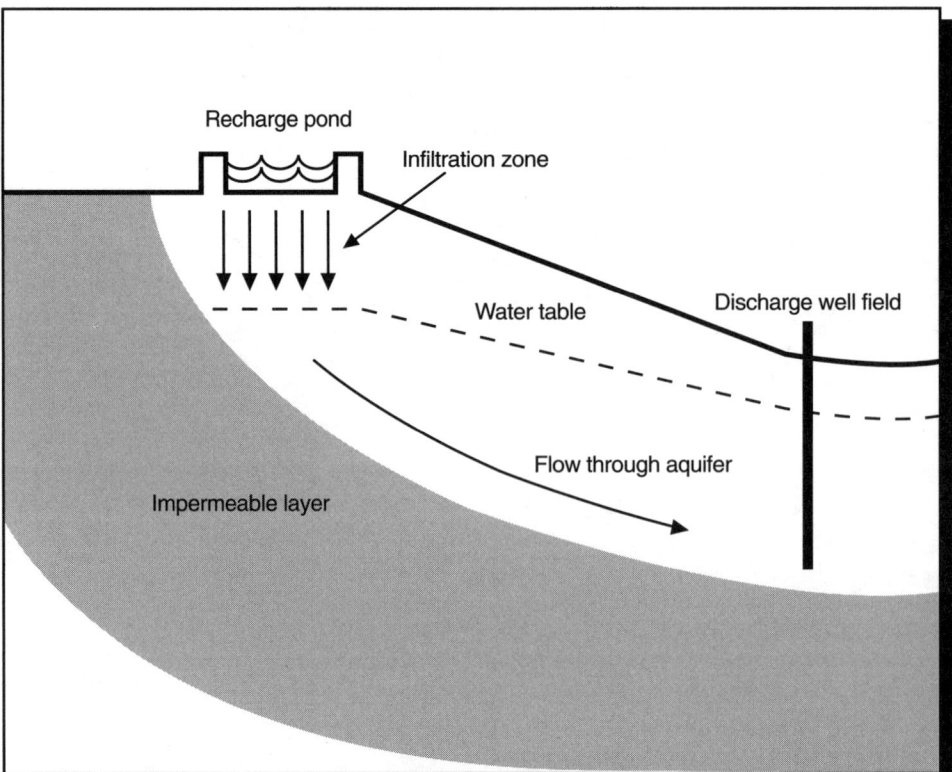

rain) to discharge (into rivers, lakes, or oceans). The average residence time of groundwater throughout the world is approximately 280 years.[17]

Phases of Contamination

In addition to the hydrogeologic forces described above, the fate and transport of contamination is affected by its various states once it reaches the subsurface. Contamination typically occurs in one of four phases: (1) liquid/solid, (2) dissolved, (3) absorbed, or (4) vapor. Often, contaminants such as volatile organic compounds occur in multiple phases at a single site.

Liquid/Solid Phase

Because contaminants are frequently released in their liquid or solid phases, these phases are grouped together. These phases are often highly concentrated and may represent a significant mass of the initial release. Over time, the liquid/solid phase will decrease in its relative percentage as the contaminant breaks down and enters the dissolved, vapor, or absorbed phases.

Obviously, solids are not highly mobile once released. However, liquid contamination released to the environment can be highly mobile. Liquid contamination in the subsurface will migrate, under the forces of gravity, to the groundwater surface. Once reaching the groundwater surface, the contaminant's chemical characteristics will play an important role in the final disposition of the contamination.

Primary among these characteristics is the property of specific gravity. Specific gravity is the ratio of the weight of a given volume of a substance to that of an equal volume of water.[18] Contaminants with a higher specific gravity than groundwater—dense nonaqueous phase liquids—will sink through the aquifer until reaching an impermeable layer (such as clay or bedrock), where they will collect. These contaminants will then migrate under the force of gravity along the impermeable layer. Contaminants with a lower specific gravity than groundwater—light nonaqueous phase liquids—will float and collect on the groundwater surface. These contaminants will then migrate under the influence of the direction of groundwater flow.

Dissolved Phase

Contaminants that reach the groundwater surface may exhibit the ability to dissolve. Based on the chemical's water solubility, chemical activity will dissolve the liquid or solid contaminant, allowing contamination to spread through the groundwater. In a relative sense, the volume or mass of the chemical that enters the dissolved phase is significantly less than the chemical volume or mass in liquid/solid phase. Once in the groundwater, contamination will spread by two primary means: groundwater flow (described above) and chemical transport. Chemical transport is a function of the aquifer characteristics, but can allow contaminants to migrate counter to groundwater flow and at rates exceeding groundwater flow.

Absorbed Phase

Before reaching the groundwater surface, contaminants will typically migrate through near-surface soils. Their migration will be dependent primarily on the effective porosity of the soils. Effective porosity can be defined as the interconnectedness of soil grains.[19] Soils with higher porosity will allow contaminants to move more easily through them (Figure 3). In addition, as contaminants move through soil, chemical and physical action will cause significant amounts of the contaminants to remain within the intergranular spaces of the soil. This phenomenon was previously described in the sponge example. Because of this sponge-like characteristic, absorbed contamination in soil often represents the greatest volume of contamination at a site.[20] (Imagine wringing out the sponge from our example and collecting the

FIGURE 3. Porosity

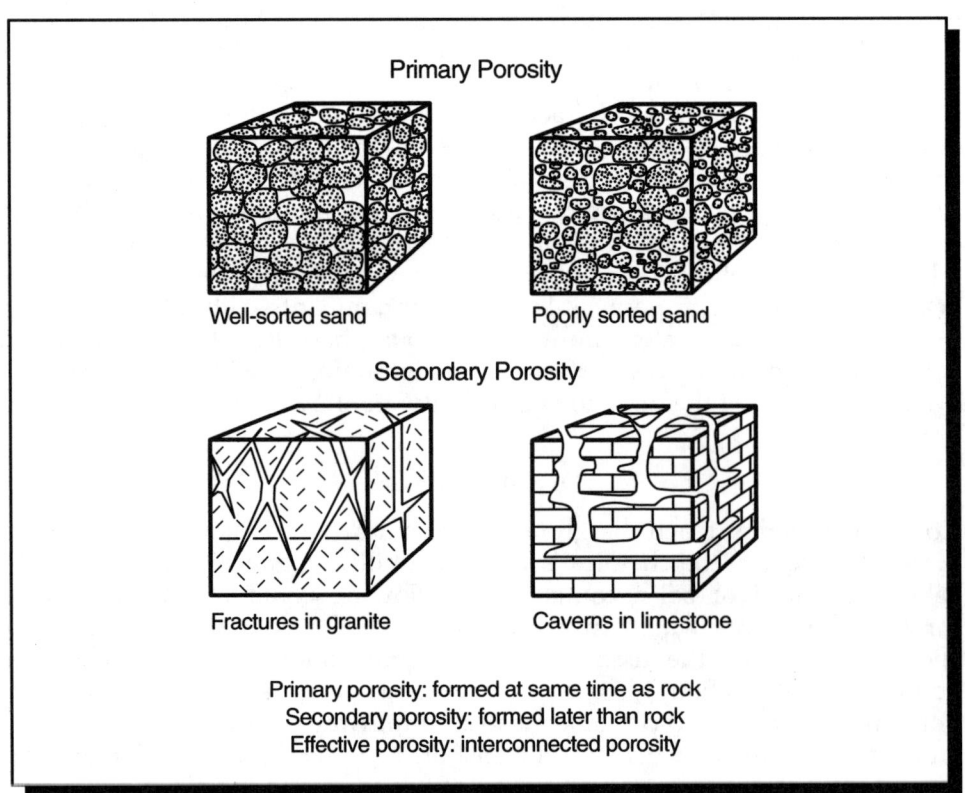

resulting liquid.) Once a contaminant has moved through a soil medium, remobilization of that contaminant is not typically significant, because the contaminant becomes trapped in the pore spaces by surface tension and chemical reactions.

Vapor Phase

The final phase of a contaminant is the vapor phase. The primary characteristic for volatilization of a contaminant is its vapor pressure. Vapor pressure can be defined as the ability of a compound or element to vaporize into the air.[21] Two common issues associated with the vapor phase of a contaminant are the possibility of inhalation by a worker at a facility or the ignition of explosive vapors that have accumulated.

Vapors are the most mobile phase of contamination. Vapors can migrate against gravity, through pore spaces, and into synthetic and natural conduits. Vapors also usually represent the smallest volume of contaminant mass following a release. However, because of their characteristic mobility, vapors must be considered in any investigative decision and remedial process.

Site Investigation Process

Many states have, or are adopting, standardized investigation procedures as part of their Voluntary Cleanup Programs. This section illustrates a model site investigation using the Ohio Voluntary Action Program (VAP) as an example.

Under VAP the site investigation process involves four primary stages.[22] The initial step is a Phase I Property Assessment (Phase I). The purpose of this step is to identify whether contamination may be present on-site. If present, the volunteer would undertake the second step in the process, a Phase II Property Assessment (Phase II). Phase II would identify whether contamination was present on-site or off-site and, if so, its extent both vertically and horizontally. Once the extent of contamination has been defined, remedial alternatives would be considered under the third step. After remediation is complete, the final steps would be to obtain the no-further-action (NFA) letter, from a certified professional, and a covenant-not-to-sue (CNS) from the Ohio Environmental Protection Agency (Ohio EPA).

Phase I Property Assessment

Phase I is the first step in determining (1) whether a property is contaminated with hazardous substances or petroleum, and (2) which potential environmental issues are present at the site. If the results of the Phase I identify that hazardous substances or petroleum have been released at the site, that past and/or current practices indicate the likely release of hazardous substances or petroleum, or that surrounding properties have been found to impact or be likely to impact the subject site, then a Phase II must be conducted.[23]

Most environmental consultants are familiar with the American Society for Testing Materials (ASTM) Standard E 1527-94 for Phase I environmental assessments. In undertaking a VAP-Phase I, the ASTM standard can be viewed as a reasonable starting point. The primary differences between the ASTM standard and the VAP-Phase I are that the VAP-Phase I requires additional research regarding both site history (back to the first commercial development) and surrounding property use. The certified professional must also have visited the site.[24]

Briefly, the Phase I should include the following information:[25]

- A determination of the first use of the site and surrounding property, using reasonably reviewable information
- An identification of present and past owners of the property, back to when the property was first used for business or industry, to determine the property's historical use
- A review of various regulatory agency files to determine if the property was used in a manner consistent with applicable environmental laws
- Interviews with personnel who are familiar with current and historical operations at the facility
- An inspection of the entire property (the certified professional must visit the property), looking for signs of released hazardous substances and petroleum
- A report of findings, which includes a recommendation by the certified professional about whether (1) an NFA letter should be written, or (2) a Phase II should be conducted

A certified professional may issue an NFA letter after a reliable Phase I has been conducted within the past 180 days, he or she visits the site, and he or she believes that both the following are true:[26]

1. All reasonable and available information about the contamination has been considered; and
2. there is no evidence that any hazardous materials or petroleum have contaminated the soil, surface water, sediments, or groundwater under the property.

Phase II Property Assessment

A Phase II is required after a Phase I demonstrates that hazardous substances or petroleum may have been released at a property. The purpose of the Phase II is to collect data that either confirms or refutes that hazardous substances or petroleum, above applicable standards or background conditions, is present at the site. If contamination is confirmed at the site, the Phase II process continues until the nature and extent of the release are defined.[27]

When a comparison to background standards is not used, the volunteer can apply the generic-standards rule[28] or the site-specific, risk-assessment rule.[29] The generic standard is a listing of various select contaminants and their acceptable concentrations in soils. These standards are based on predetermined generic site conditions. During the Phase II process, it is important to identify that the subject site meets the generic conditions described in the standard. If the site does not meet these standards or a site-specific cleanup goal is desired, the volunteer may undertake a site-specific risk assessment. The risk-assessment process uses site-specific information to determine an appropriate site-specific cleanup goal.[30]

The Phase II process can be separated into two distinct steps. Tier 1 includes determining whether hazardous substances or petroleum is present above applicable standards. Tier 2 is used to determine the extent of hazardous substances or petroleum identified during Tier 1.[31]

Tier 1 Investigations

The Tier 1 investigation[32] is used to pinpoint areas of the property, identified in the Phase I, where chemicals of concern are likely to be present. The first step in this process is to develop a field sampling plan. The plan should describe the technical features of the investigation program including, but not limited to, criteria for selecting sampling locations, the analytical program, the quality-assurance/quality-control program, and various investigative methodologies. All laboratory analyses during the Phase II must be conducted by a VAP-certified laboratory.[33] Other important information in the Tier 1 investigation includes classification of groundwater,[34] determination of exposure pathways, current and future land use, and subsurface hydrogeologic characteristics.

Once all sampling is complete and the analytical data has been reviewed, the results may be compared with background or applicable standards. If the concentrations of the chemicals of concern are equal to or less than background levels or applicable standards, then an NFA letter may be issued by the certified professional. If the background conditions or applicable standards have been exceeded for the chemicals of concern, then a Tier 2 investigation is required.

Tier 2 Investigations

A Tier 2 investigation is conducted to determine the extent of chemicals of concern detected above background levels or applicable standards during the Tier 1 investigation.[35] As with a Tier 1 investigation, the first step is to develop an investigation plan. Although much like the Tier 1 plan, this plan differs in its focus. The goal of the Tier 2 plan is to delineate the chemicals of concern, not merely locate them. The information collected in the Tier 2 investigation may be used to conduct a site-specific risk-assessment. Therefore, basic information necessary to provide input to the risk assessment models must also be collected. This information may include potential receptors and exposure pathways, chemical and physical properties of the subsurface media, future land-use information, and groundwater classification and properties.[36]

The results of the Tier 2 analytical data are then compared with background conditions or applicable standards. If desired, a risk assessment may be completed to determine site-specific applicable standards. If the analytical data indicates that chemicals of concern are below applicable standards (which may be determined by a risk assessment) or background conditions, then no further action is needed for those identified areas. If the chemicals of concern are above applicable standards or background conditions, then remedial alternatives must be considered and implemented to obtain a CNS.[37] In addition, the Tier 2 investigation must confirm that an exposure pathway does not exist at the property for the chemicals of concern. If an exposure pathway does not exist and chemicals of concern are below applicable standards, then an NFA letter can be issued by a certified professional.

Determining Remedial Alternatives for the Property

In the event that a Phase II identifies chemicals of concern above background conditions or applicable standards, remedial alternatives must be evaluated. Today, many remedial alternatives exist for environmental cleanups. The historic remedial approaches, such as "muck and truck," "hog and haul," or "yank a tank," have been replaced by bioremediation, air sparging, and nutrient injection augmented by in-situ (in place) bioventing. Although these approaches seem much more complex than the older excavation approach, costs and long-term liability can be decreased with these new technologies. However, the decision-making process for the appropriate remedial technology may not start with technical issues. The basis for making sound remedial decisions for a site is often found in the limitations posed by costs, remedial goals (or end point desired), time constraints, and the regulatory approval processes.

Cost

Environmental remediation of brownfields sites can be costly. (If remediation was not potentially costly, then brownfields might not exist.) A Voluntary Cleanup Program is not a panacea that somehow eliminates the need for costly remediation. These programs do, however, take an aggressive approach to determining an acceptable risk for chemicals of concern at a site. To the extent that those site-specific, risk-based levels exceed background conditions, cost savings may be realized by using current technologies and approaches.

The primary cost-savings mechanism, with today's technology, is the ability to remediate contaminants in-situ (in place). With the growing limitations on landfill space, in-situ remedial alternatives are becoming more and more cost-effective to implement. In-situ methodologies attempt to remediate contaminants in the subsurface without removing them to the surface. The advantages to these methodologies include (1) continued, uninterrupted operation of the facility, (2) elimination of long-term liability associated with landfill disposal, (3) lowered risk of worker exposure during the remedial process, and (4) less exposure to unwanted publicity due to on-site activity.

In-situ remediation is primarily supported by two widely accepted approaches that have developed in recent years: risk-based cleanup criteria and regulatory recognition/utilization of biological activity. Risk-based cleanup levels have allowed remedial technologies to move from removal to management. By determining an appropriate risk-based cleanup goal, the party conducting the cleanup uses the site's natural conditions to help in the remedial effort. Sites that do not pose a risk to public health or the environment do not require remediation. In-situ technologies apply this concept by leaving contamination in place and using the natural site conditions to augment remediation and/or limit migration.

The driving force behind risk-based cleanup levels is the recognition that naturally occurring bacteria and other organisms help to break down contaminants in the subsurface. Aerobic and anaerobic bacteria in the subsurface have been proven to aid in the destruction of various contaminants.[38] The use and augmentation of these bacteria are crucial to the positive results achieved with in-situ technologies. This phenomenon is often referred to as bioremediation. This is a broad term covering many different methodologies, but the common denominator among bioremediation technologies is the use of naturally occurring bacteria to break down contaminants in the subsurface.

Effectiveness

The effectiveness of the chosen remedial approach depends heavily on the determination of accurate on-site conditions and reasonable goals. Before ever contemplating remedial alternatives, the project's end point should be visualized. The following questions should all be discussed and answered before remedial alternatives are determined: What is the cleanup standard? How much contamination is present? Has the source of contamination been eliminated? It is also important to understand that multiple technologies exist for each contamination scenario. Knowing the "hows" and "whys" of the site will be useful in making good choices.

From a technical standpoint, the information needed to design an effective remedial system rests in the proper evaluation of on-site conditions. Cutting corners and leaving open issues during the investigative portion of the project will serve only to create more unknowns when developing a remedial approach. Further, pilot testing of various options should be conducted to determine site-specific reactions to various technologies.

Reviewing the information basis commonly used for remedial decision making might be illustrative. A typical industrial site may cover twenty acres. During the investigation, thirty borings with a total of fifty separate analyses may be conducted by a laboratory. Taking the volume of potentially contaminated soil at a site of twenty acres over a thickness of ten feet results in a volume of over three hundred thousand yards of potentially contaminated soil.[39] During such an investigation, the total volume of soil analyzed by a laboratory is less than one yard, or 0.000001 percent of the potentially contaminated soil volume at the site. With such a small snapshot of actual site conditions, the effectiveness of the remedial approach will be significantly dependent on the data collected. The use of practical, yet comprehensive, investigative techniques and pilot testing of remedial approaches will ensure the best possible results.

Time Considerations

Conducting brownfields investigations and implementing remedial alternatives take time. A Voluntary Cleanup Program should not be viewed as a quick turnaround, first-blush evaluation of a site that results in obtaining an NFA letter. The evaluation of remedial alternatives or risk-based standards can be a lengthy process. People undertaking brownfields projects must be comfortable with this fact. Time frames, however, are also relative. Factors that will serve to shorten the remedial time frame include the use of risk-based standards to prevent the unnecessary cleanup of sites, consolidated permitting, and limited regulatory interaction. Once a determination is made that remedial technologies are needed, the evaluation of appropriate time frames for cleanup will have a significant impact on the outcome and cost of the chosen technology.

Often, shorter time frames for environmental cleanups result in the highest costs. Immediate removal technologies (such as excavation) frequently take the shortest amount of time and typically result in significant costs for removal operations, transportation, and landfilling. In-situ methodologies, such as groundwater extraction and treatment, soil-vapor extraction and treatment, and in-situ bioremediation involve longer time frames.

For typical sites, remedial time frames may range from three to five years. When remedial technologies are considered, the break-even point between long-term operational costs and immediate removal costs must be evaluated. The remedial solution that often results in the lowest cost, but the longest time frame, is bioremediation. Bioremediation processes happen on a microscopic level. Considering the mass of the contaminant released and other physical properties of the site, biological processes may take a considerable time period. Site cleanup times using bioremediation may extend beyond five years.

Often overlooked is the remedial option of risk assessment. A risk assessment is a passive remedial approach. Further, risk assessments may be combined with active remediation to determine the remedial system end point. By using a risk assessment to determine the closure level for the site, cleanup time frames and costs may be reduced, while ensuring protection of public health and the environment.

It may not be necessary to remediate releases at the site to background or applicable standards. The Ohio brownfields program allows for the use of engineering and institutional controls to obtain both an NFA letter and a CNS.[40] In the case of engineering controls, a system may be installed that effectively controls the migration of contamination or eliminates pathways of exposure. For example, a fence could be placed around a particular portion of the site to prevent an exposure pathway, or a groundwater pumping system could be installed that controls groundwater flow at the site, preventing contamination from leaving the site. The volunteer must ensure that the system maintains operation and compliance. However, once control is demonstrated, an NFA letter can be issued.

Institutional controls can also be used for the same purpose. A volunteer could place a deed restriction on the property, limiting its use to industrial purposes. This would effectively eliminate potential exposure pathways associated with residential occupancy. If this form of control were formally established through an attachment to the deed for the site, an NFA letter could be obtained.

Regulatory Considerations

As with any remedial approach, regulatory permitting is required. Under the Ohio program, consolidated permits can be obtained from the Ohio EPA.[41] A volunteer, who would otherwise be required to obtain all permits, plans, licenses, plan approvals, or other approvals from the Ohio EPA, may obtain a consolidated permit for the activities in connection with conducting the voluntary action. Permitting is often a lengthy process and may involve overlapping agencies. To avoid this problem, the Ohio program allows for the consolidation of that process under one agency and eliminates the time-consuming process of resolving multiple permitting requirements.

Established Protocols in the Ohio VAP

The Ohio VAP establishes comprehensive and specific guidelines for conducting brownfields projects. Each of these internal standards was developed with the input of multidisciplinary committees and subcommittees, including the leading brownfields experts in the state. As such, these guidelines will be heavily relied upon to direct all aspects of the brownfields process. These standards have been developed based on the most current technical information available in the environmental field.

The Phase I protocol relies heavily on ASTM Standard E1527-94. The actual requirements vary somewhat from ASTM, but the primary conclusion, that of the determination of any recognizable environmental concerns, remains the same. The Phase II process relies upon the development of a preliminary plan. This plan must conform to general industry standards. The primary issue in this standard is developing a comprehensive, yet practical, quality-assurance/quality-control (QA/QC) plan. The QA/QC plan ensures that proper procedures are followed during the actual sampling program and that the data is accurate.

Because of the individual nature of the concerns at each site, a specific procedure for conducting investigations has not been developed. Rather, the program relies upon the certified professional to develop a plan for the Phase II process and then ensures that proper procedures are followed by requiring adherence to the plan. The risk-assessment guidelines establish criteria for determining acceptable levels of risk. These guidelines are based on mathematical models provided in the risk-assessment rule. The rule also specifies certain input data to the model and other appropriate assumptions.

It is critical that these standards and others be considered when undertaking a brownfields cleanup. Although a certified professional may prepare an NFA letter for the volunteer, the CNS must be issued by the Ohio EPA. Before issuing a CNS, the Ohio EPA's primary function will be to ensure that applicable standards have been met. These standards pertain to the cleanup levels as well as the remedial methodologies used.

It is important to realize that the Ohio EPA is not conducting a technical evaluation or interpretation of the data. The Ohio EPA's task is to evaluate the environmental condition of the site in relation to standards applicable to the property.[42] This is the primary difference between the Ohio VAP and other Ohio regulatory programs. In the VAP, the certified professional conducts all evaluation and interpretation of data. The Ohio EPA established a significant certification and penalty process to ensure the quality of the interpretations made by certified professionals. However, the interpretations are at the discretion of the certified professionals. Ohio EPA's role in review of the NFA letter is limited to determination of whether applicable standards at the property have been achieved. Therefore, the historically time-consuming and negotiation-intensive review process should be eliminated under this program.

Audits

In addition to conducting the standard review process to obtain a CNS, the Ohio EPA will also audit properties participating in the Ohio VAP. Any site obtaining a CNS will be subject to inclusion in an audit pool. A limited number of sites from this pool will be chosen for audits each year. At a minimum, 25 percent of all CNSs issued each calendar year will be audited. These audits will range from a review of the certified professional's files to on-site inspections. As indicated in the review process, the goal of the audits will be to ensure that applicable rules were followed during the voluntary cleanup process.[43]

The Context of the Cleanup Standards

Although many people view the voluntary nature of the Ohio VAP as its primary innovation in the environmental field, it is the use of risk-based standards that has been the true innovation. Though risk-based cleanup standards have been in use, to some extent, for many years, never has a statewide environmental program so heavily relied on their validity. The idea of risk is important to this program, because it moves us away from cleanups based on comparison to background levels and into an era when actual risk to human health and the environment is the basis for environmental cleanups.

To understand how risk-based standards work, it is necessary to first understand the relative context of the concentrations being evaluated. The terms "part per billion" (ppb) and "part per million" (ppm) are often quoted when establishing concentrations of contamination at a site. These standards identify allowable contaminant levels in various concentrations. Although these standards are intended to establish guidelines that are protective of public health and the environment, the definition of "protective" must first be determined. To better understand the protective nature of the brownfields standards, the following analogies place the concepts of ppb and ppm in context.

media	ppm	ppb
time	1 minute in 2 years	1 second in 32 years
money	1 cent in $10,000	1 cent in $10,000,000
golf	1 bogey in 3,500 rounds	1 bogey in 3.5 million rounds
water	1 cup in a pool	1 drop in a pool

Based on previous regulatory thinking, one drop of contamination in a pool is grounds to remediate the entire pool, because the one drop of contamination exceeds background levels. "Protective" in this sense is to eliminate anything above background levels. The brownfields approach focuses on the risk to human health as a result of that one drop in the pool. If the one drop of contamination does not pose a significant risk, then there may not be a need to remediate the pool. This approach is the true innovation being adopted not only in Ohio, but throughout the nation as Voluntary Cleanup Programs rapidly develop.

Conclusion

To understand brownfields cleanups, it is important to first understand the science behind remediation strategy. The subsurface is both complex and diverse. As a result, subsurface contamination may react in complex and diverse manners. Before undertaking a brownfields cleanup, one must first understand the phenomena of contaminant transport, hydrogeology, and the phases of contamination. To gain this understanding, certain investigative methods should be used. Most brownfields programs use the Phase I property assessment to initiate this process. If the potential for contamination is identified, a Phase II property assessment is conducted to define the extent and magnitude of the contamination. Once this is known, remedial alternatives can be considered.

In terms of brownfields cleanups, the most prominent remedial alternative is risk assessment. The use of risk assessments in the brownfields process, to determine whether a contaminant must be removed from a site, represents the primary innovation presented by these new programs. Risk assessments help us understand the relative significance of various contaminants, and therefore help us conduct cost-effective, efficient brownfields cleanups.

Notes

1. This is not to say that brownfields are solely the result of industry, but rather to point out that brownfields are typically located in long-standing, multiple, industrial-use settings.
2. DRISCOLL, GROUNDWATER AND WELLS, CH.3 (2D ED. 1986).
3. *Id.*
4. *Id.*
5. *Id.*
6. *Id.*
7. National Water Well Ass'n, Review of Well Hydraulics 20 (1989) (seminar course notes).
8. DRISCOLL, *supra* note 2, ch. 3.
9. The Princeton Course, Groundwater Pollution and Hydrology (1990) (seminar course notes).
10. *Id.*
11. DRISCOLL, *supra* note 2, ch. 5.
12. The Princeton Course, *supra* note 9.
13. C. W. FETTER, JR., APPLIED HYDROGEOLOGY 6 (1980).
14. DRISCOLL, *supra* note 2, ch. 5.
15. *Id.* ch. 22.
16. The Princeton Course, *supra* note 9.
17. *Id.*
18. DRISCOLL, *supra* note 2.
19. National Water Well Ass'n, Review of Well Hydraulics (1989).
20. Hinchee et al., Enhanced Bioreclamation; Soil Venting and Ground-Water Extraction: A Cost Effectiveness and Feasibility Comparison (Nov. 1987) (National Water Well Association/API Conference on Petroleum Hydrocarbons).
21. AMERICAN HERITAGE DICTIONARY 1416 (8th ed. 1982).
22. Summary of Proposed Rule for Applicability and Eligibility, Ohio EPA (proposed Dec. 26, 1995).
23. Summary of Proposed Rule for Phase I Property Assessment, Ohio EPA (proposed Dec. 26, 1995).
24. Proposed draft of Phase I Property Assessment, at this writing, the Phase I rule remains in draft form; all references are based on the Ohio EPA December 26, 1995 rule summary listed above.)
25. *Id.*
26. *Id.*
27. Summary of Proposed Rule for Phase II Property Assessment, Ohio EPA (proposed Feb. 14, 1996).
28. Proposed draft, OHIO ADMIN. CODE § 3745-300-08, Ohio EPA (proposed Dec. 26, 1995). (At this writing, the Generic Standards Rule remains in draft form. All references are based on the Summary of Proposed Rule for Generic Numeric Standards).

29. Proposed draft, OHIO ADMIN. CODE § 3745-300-09, Ohio EPA (proposed Dec. 4, 1995). (At this writing, the Risk Assessment Rule remains in draft form.)

30. A further discussion of site-specific risk assessment appears in Chapter 17.

31. Summary of Proposed Rule for Phase II Property Assessments, Ohio EPA (proposed Feb. 14, 1996).

32. *Id.*

33. A further discussion of certified laboratories appears in Chapter 13.

34. Proposed draft, OHIO ADMIN. CODE § 3745-300-10 (proposed Dec. 4, 1995). (At this writing, the Groundwater Classification Rule remains in draft form.)

35. Summary of Proposed Rule for Phase II Property Assessment, Ohio EPA (proposed Feb. 14, 1996).

36. Proposed draft OHIO ADMIN. CODE § 3745-300-09, Ohio EPA (proposed Dec. 4, 1995).

37. The process for obtaining a CNS is further described in Chapter 39 (the Ohio program).

38. Hinchee et al., *supra* note 20.

39. The calculation producing the result obtained in the example is as follows: 1 acre = 43,560 square feet; (43,560 square feet × 20 acres × 10 feet = 8,712,000 cubic feet; 8,712,000 cubic feet ÷ 27 (1 yard = 27 cubic feet) = 322,666 yards.

40. OHIO ADMIN. CODE § 3745-300-01(19) (1995).

41. OHIO REV. CODE ANN. § 3746.05 (Baldwin 1995).

42. *Id.*

43. OHIO ADMIN. CODE § 3745-300-14 (1995).

CHAPTER 17

The Role of Risk Assessment in Redeveloping Brownfields Sites

MICHAEL L. GARGAS
THOMAS F. LONG

The decade before the establishment of federal and state environmental regulatory programs is memorable for the number and type of environmental issues confronting the nation. Images of dead lakes, burning rivers, belching smokestacks, untreated wastewater, smog alerts, species extinction, and the like were striking and highly visible representations of the degree of environmental degradation with which the nation was struggling. Tackling and eliminating many of these problems represented an immense commitment of time, money, and other resources and was marked by notable early successes in the improvement of environmental quality. For the most part, however, these were engineering problems and the early successes enjoyed were the result of improved technology that markedly reduced, but did not entirely eliminate, pollution.

As the highly visible, acute effects of pollution were eliminated, the Law of Diminishing Marginal Returns began to exert its influence and further improvements were accompanied by ever-increasing costs in return for uncertain benefits. Among these issues were the questions first raised by Rachel Carson in *Silent Spring* about the consequence, to the public health and welfare, of long-term exposure to low levels of chemical and physical agents in the environment.[1] These concerns were subsequently and repeatedly brought to the public consciousness by names like Love Canal, Times Beach, Agent Orange, Minamata, Seveso, and highly emotional images of leaking barrels and frightened, angry, and ill people. These became the rallying points for the public, advocacy groups, regulators, and scientists convinced that an epidemic of environmentally induced diseases (typically cancer) was imminent, if not already underway.

The governmental response to these concerns was manifested in the form of new and revised regulatory agendas, such as the Comprehensive Environmental Response, Compensation, and Liability Act of 1980 (CERCLA, or Superfund), the Toxic Substances Control Act (TSCA), the Resource Conservation and Recovery Act (RCRA), and others. What was lacking was an understanding of what exposure to trace levels in the environment actually meant. Although a relatively clear understanding of the effects of short-term exposure to high levels of chemical and physical agents existed due to extensive experiments with laboratory animals and studies of accidental or occupational human exposures, neither the scientific nor regulatory communities had the knowledge or tools needed to assess the impact of low-level, long-term exposures to substances in the water, air, soil, or food supply. It was from this need that the discipline known as "risk assessment" was born and has evolved.

What Is Risk?

Most simply stated, risk is uncertainty. An unloaded gun held to the temple and fired presents no uncertainty. Neither, however, does a fully loaded gun in the same circumstance. The outcome is certain, yet different, in both cases. The only time risk enters this example is if the gun is partially loaded. Russian Roulette typifies risk, because the outcome is uncertain. In this sense, all human endeavors carry an element of risk, although typically the term "risk" is associated with a negative outcome (Table 1). With enough information, the probability of the desired (or undesired) outcome can be calculated. This is the basis of actuarial science, structural-failure analysis, or odds-making, in which a sufficiently complete understanding of the system and a large historical database combine to result in probability estimates for the various possible outcomes. Thus, a state can design lotteries that make more money than they lose. Without an understanding of the system and sufficient information regarding possible outcomes, the probability (or risk) of being wrong increases. For instance, the early history of the insurance industry is replete with instances of financial collapse as a result of little experience with customers' life expectancies. Even today, new enterprises involving unproven

TABLE 1. Annual Average Individual Risk of Death from Various Causes

Activity	Risk
All causes	1 in 100
Heart disease	6 in 100
Cancer	1 in 1,000
Auto accident	2 in 10,000
Suicide	2 in 10,000
Murder	1 in 10,000
Drowning	4 in 100,000
Accidental shooting	1 in 100,000
Electrocution	5 in 1,000,000
Tornado	7 in 10,000,000
Lightning	5 in 10,000,000
Bee sting	1 in 10,000,000
Botulism poisoning	5 in 1,000,000,000
Meteor strike	1 in 150,000,000,000,000

technologies attract venture capital by promising higher than normal profits as compensation for the uncertainty of success (risk of failure). For an interesting historical perspective on risk, risk assessment, and risk-management practices, the reader is directed to the excellent review by Covello and Mumpower.[2]

In the area of environmental hazards, risk is a function of two different and complementary concepts: toxicity and exposure. Toxicity, as first noted by Paracelsus in the sixteenth century, is the inherent property of a compound to produce adverse effects, if the dose (or concentration) is sufficient. Dose here is taken to mean the frequency and duration of contact with a compound at a given concentration (that is, exposure). Exposure is simply the conditions under which the material is contacted. Thus, though all substances are toxic, it is only when exposure is possible that a hazard may exist. In short, the dose makes the poison. Risk is the probability that, under given conditions of exposure to an agent, an adverse outcome will result. This concept can be expressed simply as:

RISK = TOXICITY × EXPOSURE

It should be evident that the degree of risk is influenced by both toxicity and exposure. A change in either parameter, or both, results in an equivalent alteration in the risk. A compound of relatively low toxicity but for which exposure is large may, therefore, pose a higher risk to an individual or population than a highly toxic compound to which there is little or no exposure. It is the goal of risk assessment to integrate knowledge regarding the toxicity of an agent or agents with estimates of the exposure in the short and long term to identify if a risk exists and its relative degree. An equally important, if usually unfulfilled, goal is to provide an estimate of the uncertainty surrounding the risk estimate, which should presumably become smaller as knowledge and time increase. For more detailed discussion on the principles and practice of toxicology, the reader can refer to any of the many reviews and references on the subject, including that by Eaton and Klaassen.[3]

The Rise of Environmental Risk Assessment

Although most of the concepts necessary to estimate and assess risk have been known and used for centuries in various areas of human endeavor,[4] the application of a risk-assessment methodology to environmental health issues is only slightly more than ten years old. The need for a consistent, rational risk-assessment policy to assess low-level exposures in the early 1980s resulted in the landmark report by the National Academy of Sciences (NAS), *Risk Assessment in the Federal Government: Managing the Process*, in which risk assessment was defined and the format for all subsequent risk assessments was developed.[5] In its report, NAS defined risk assessment as:

> The characterization of the potential adverse health threats of human exposure to environmental hazards. Risk assessments include several elements: description of the potential adverse health effects based on an

evaluation of results of epidemiologic, clinical, toxicologic, and environmental research; extrapolation from those results to predict the type and estimate the extent of health effects in humans under given conditions of exposure; judgments as to the number and characteristics of persons exposed at various intensities and durations; and summary judgments on the existence and overall magnitude of the public health problem. Risk assessment also includes characterization of the uncertainties inherent in the process of inferring risk.[6]

Significantly, NAS also made clear the distinction between risk assessment and risk management by offering the following definition:

> The Committee uses the term *risk management* to describe the process of evaluating alternative regulatory actions and selecting among them. Risk management, which is carried out by regulatory agencies under various legislative mandates, is an agency decision-making process that entails consideration of political, social, economic, and engineering information with the risk-related information to develop, analyze, and compare regulatory options and to select the appropriate regulatory response to a potential chronic health hazard. The selection process necessarily requires the use of value judgments on such issues as the acceptability of risk and the reasonableness of the costs of control.[7]

Unfortunately, because of the many uncertainties in the risk-assessment process then and now, value judgments with the potential to warp the process and the final outcome are made throughout the process, in terms of the assumptions used in the place of knowledge about both toxicity and exposure.

The Risk Assessment Process

Risk assessments always follow the basic format first identified in the 1983 NAS report and recently expanded on in a National Research Council Report, *Science and Judgment in Risk Assessment*.[8] This includes four major divisions: hazard identification, dose-response evaluation, exposure assessment, and risk characterization (Figure 1).

Hazard Identification

Hazard identification is the first and most easily identified step in the process. It is a qualitative description of the toxicity of the agent as influenced by its physical and chemical properties and its fate in the environment. In essence, the question to be answered is, "What adverse effects can the agent cause, and what is the likelihood that they might occur in exposed humans?" The strength of causation from all available sources should be reviewed in this step, although typically the most available and useful information is derived from animal studies. All information, negative and positive, should be used to make a qualitative judgment about the weight of evidence supporting the type of hazard(s) posed by the agent in question.

FIGURE 1. Conceptual Approach to Risk Assessment

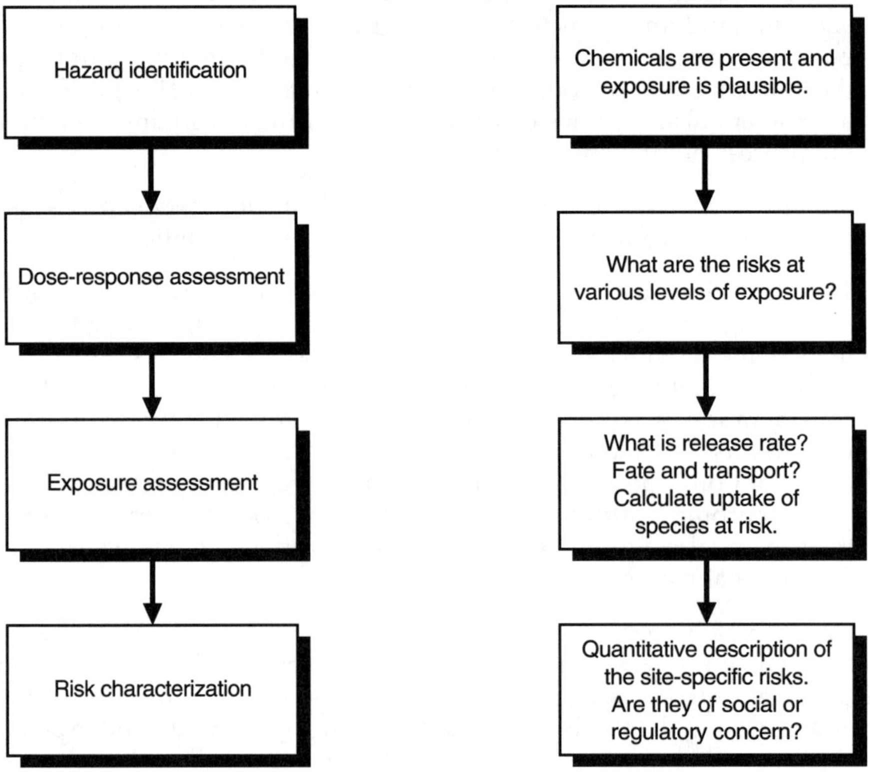

```
┌─────────────────────────┐         ┌─────────────────────────┐
│  Hazard identification  │         │ Chemicals are present and│
│                         │         │  exposure is plausible. │
└───────────┬─────────────┘         └───────────┬─────────────┘
            ▼                                   ▼
┌─────────────────────────┐         ┌─────────────────────────┐
│ Dose-response assessment│         │   What are the risks at │
│                         │         │ various levels of exposure?│
└───────────┬─────────────┘         └───────────┬─────────────┘
            ▼                                   ▼
┌─────────────────────────┐         ┌─────────────────────────┐
│                         │         │  What is release rate?  │
│  Exposure assessment    │         │  Fate and transport?    │
│                         │         │  Calculate uptake of    │
│                         │         │  species at risk.       │
└───────────┬─────────────┘         └───────────┬─────────────┘
            ▼                                   ▼
┌─────────────────────────┐         ┌─────────────────────────┐
│                         │         │ Quantitative description of│
│  Risk characterization  │         │  the site-specific risks.│
│                         │         │  Are they of social or  │
│                         │         │  regulatory concern?    │
└─────────────────────────┘         └─────────────────────────┘
```

The risk-assessment process consists of four interrelated steps: hazard identification, dose-response evaluation, exposure assessment, and risk characterization. The product of a risk assessment is provided to risk managers for decisions on actions required to reduce the risks.

Dose-Response Assessment

Dose-response assessment is the process of characterizing the quantitative relationship between the dose (or intake) of an agent and the resultant biological response. The fundamental concept underlying toxicology and risk assessment is that as the dose increases so does the response, and as the dose decreases, the response likewise drops (Figure 2). This relationship holds for all biological responses, whether it be an acute (or immediate) response like alcohol intoxication or a chronic (or delayed) response like liver cirrhosis or cancer. The absence of a dose-response relationship is a signal that the observed effect is unrelated to the agent under scrutiny. The evaluation should consider such factors as intensity and duration of exposure and other important variables, such as age, sex, lifestyle, and modifying factors.[9] Risk

FIGURE 2. The Dose-Response Relationship

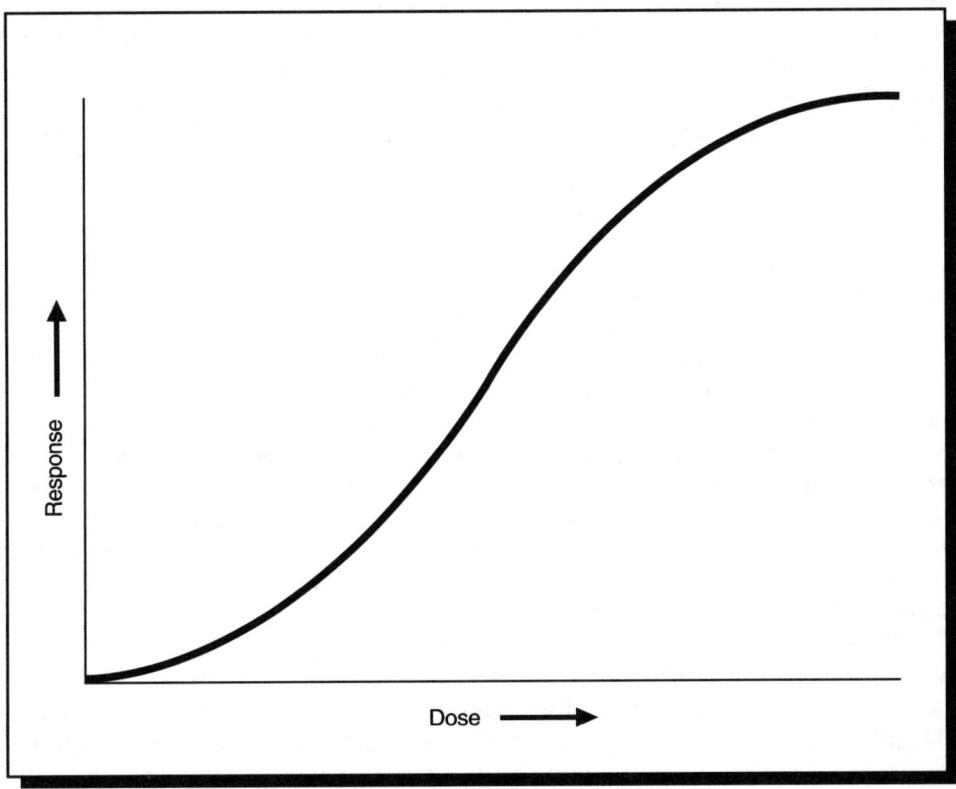

The fundamental concept in toxicology is the dose-response relationship, which implies that—for any end point measured—as the dose or exposure increases, so does the level of response of the organism.

assessments typically require extrapolation both between species (that is, animal to human), and from the high doses used in such tests to the low doses typically encountered in the environment. Both of these steps are fraught with uncertainty.

Exposure Assessment

Exposure assessment is the process of measuring or estimating the quantitative intake of an agent as a result of contact with it in various environmental pathways (that is, air, water, food, or soil). Because of the number of possibilities for exposure that might exist for individual members in a given population, this is typically performed as a series of hypothetical exposure scenarios involving conservative assumptions about various environmental pathways (such as groundwater, soil, or fish, for example), populations at risk (such as children, adults, the elderly, or workers), and durations of exposure (that is, lifetime or some fraction thereof). Again, because environmental contamination, populations, and behaviors may vary markedly over time and space, assumptions regarding exposure potential usually introduce high levels of uncertainty in the resultant risk assessments. The desire to be protective of the public health and err on the side of caution makes the use of such assumptions understandable and acceptable until better information is available.

Risk Characterization

Risk characterization is the final step in the risk-assessment process and entails combining the results of the previous steps to provide a quantitative estimate of the risks under the various conditions of exposure postulated. This is done both by comparing the results with environmental standards or guidelines defined in regulations (for example, maximum contaminant levels, or MCLs, for drinking water), or with levels of intake or risk deemed acceptable by regulatory bodies (for example, a one in a million, or 1×10^{-6} excess, cancer risk). This is also the step in which the uncertainties that exist in the risk-assessment should be identified and their effects on the outcome of the risk assessment clearly stated. In practice, this is rarely done well, if at all. More detailed reviews of the risk assessment process are available in articles by Dennis J. Paustenbach[10] and Frederick Johannsen.[11]

The need for such a tool is a function of the general smallness or uncertainty of the postulated risk, as risks would be readily apparent and easily measured if they were large, immediate in their consequences, and easily identifiable. The underlying notion for using risk assessment as a decision-making tool is that by performing it in a consistent manner, using the same techniques and assumptions, the degree of risk posed by various activities will be accurately identified as related to one another. In other words, although the true risk will rarely be known by such techniques, risk assessment can function to screen various risks and prioritize them for future action. Another advantage of standardizing techniques and assumptions is that risk assessments become "transparent." Anyone with a basic grounding in the terms and methodologies can understand and reproduce the logic and steps that go into the risk assessment under this theory, thus making the resulting decisions accessible and understandable to all stakeholders. Despite this reproducibility, risk assessments involving the exact same data or situations often markedly differ in their outcomes when conducted by different organizations or even different assessors in the same organization. This is due primarily to the conservative nature of the process and the "softness" of the assumptions used.

Risk Assessment and the Creation of Brownfields

By the mid-1980s, the risk-assessment culture was thoroughly entrenched at the United States Environmental Protection Agency (USEPA) and other federal agencies, and was spreading through state regulatory agencies. Armed with a consistent methodology and a variety of well-funded, risk-based mandates, the government was prepared to launch an assault on chronic environmental hazards and reduce or eliminate them. It was acknowledged by all involved that the process involved significant uncertainties, but it was the best and most logically defensible available approach to the problem. The uncertainties were addressed by conservative assumptions intended to protect the most sensitive segments of the population. Proponents thought that if an error was to occur, it was better to err on the side of caution. Proponents also assumed that as time went on, research would improve the quality of the risk assessments by replacing the assumptions with real information and thus reducing the uncertainties in the estimates. In practice, however, neither reduction of risks nor improvement in the quality of risk assessments occurred very quickly.

The developers of risk assessment failed to account for two problems in their vision. First, although the practitioners of risk assessment thoroughly understood the mechanics of the process, the underlying philosophy was lost. They knew how to do a risk assessment and get an answer, but did not know what the answer meant. The process became the goal rather than the means to an end. Second, the desire to standardize the process led to codification of risk-assessment methodologies and assumptions by regulatory agencies.[12] Inclusion of specific methodologies, assumptions, and acceptable risk levels in environmental laws and regulations accomplished this standardization, but also created a rigid system that was unable or unwilling to respond to an expanding understanding of environmental risks by including the best information available. Imposing such strictures on a young and evolving discipline is the intellectual equivalent of legislating sorcery as the cause of disease and ignoring the subsequent identification of microbes as causative agents. Such an approach had a chilling effect on the development and incorporation of new information and the improvement of the risk-assessment process. This resulted in the repeated, consistent prediction of worrisome risks from various environmental insults usually as a consequence of the highly conservative, and often worst-case, assumptions that were used in the assessments.[13]

The impact of this was the flood of "risks du jour" faithfully reported by the media over the last several years, contributing to the fixation that "everything causes cancer" in the public mind. Risk-management decisions based on such risk assessments often led to remedial alternatives that were incredibly costly in terms of both time and money, as typified by the massive soil excavation and disposal projects of the 1980s. Aside from failing numerous "reality checks," the standard risk assessments began to conflict with the risk assessments produced by nongovernmental organizations that used better information and newer—and arguably improved—assumptions and methodologies.[14] This further bogged down an already slow process by increasing the adversarial, often litigious nature of negotiations to apportion responsibilities and costs. If war has been defined as long periods of boredom interspersed with short periods of intense excitement, then the environmental cleanup programs had become war. The image of the various cleanup programs at the beginning of the 1990s was one of much fanfare and little action. Hundreds of millions of dollars had been spent in return for very little in the way of environmental restoration and the amelioration of attendant risks. Due to the limited integration and utilization of innovative approaches to cleanups, the costs of remediation are usually not balanced with the benefits to society and scarce resources are inefficiently allocated.

An unintended and unforeseen consequence of the environmental cleanup programs of the 1980s was the creation and spread of brownfields. As the costs of cleanups and liability issues mounted, the number of idle industrial properties around the country also increased. Property owners faced with huge investigatory and remediation costs on marginal sites simply shut down some industries and, in extreme cases, declared bankruptcy and abandoned the properties. Developers and lenders were dissuaded from reusing such properties by the fear of becoming parties to the seemingly endless liability issues that resulted in few properties being declared "clean." A vicious cycle began, in which the creation of a brownfields site resulted in loss of present and future jobs and a decline in the tax base (Figure 3). This in turn has led to a loss of services and amenities, and devaluation of properties followed by further declines in the tax base. Additionally, the industries that might have redeveloped brownfields sites were instead driven to seek properties that had little in the way of environmental liabilities associated with them. These so-called "greenfields" are converted to industrial use and plant the seed for future brownfields. The spread of brownfields in new areas or

FIGURE 3. The Vicious Circle

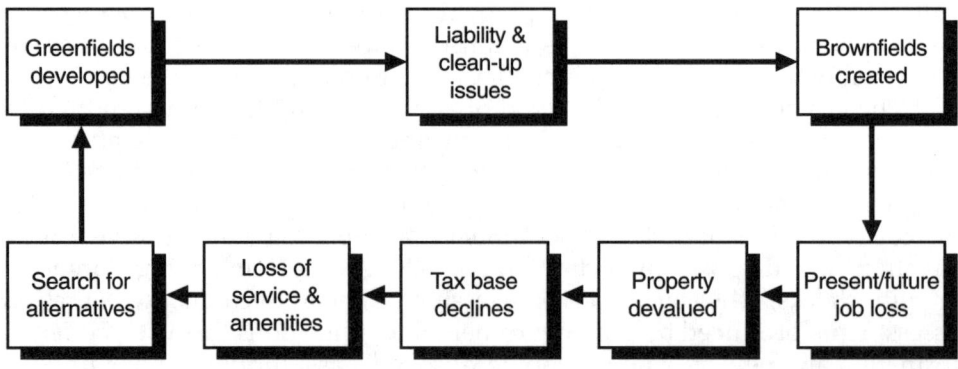

The concern about liability and costs associated with environmental cleanups has created a vicious circle in which brownfields are created both as the result of abandoning old industrial properties and developing previously pristine areas for industrial use to avoid past problems.

communities is thus ensured, with attendant loss of quality of life in the affected areas.

Risk Assessment and the Elimination of Brownfields

The same tools that created the brownfields problem are necessary to solve it. Unless one wishes to remediate contaminants to background or nondetectable levels, risk assessment remains the only feasible approach. What is needed, however, is the judicious application of the best scientific techniques in combination with site-specific and use-specific information, and with the joint goals of remediation and redevelopment clearly in the minds of all stakeholders. This is the strategy taken in developing risk-based corrective action (RBCA) for brownfields sites or any other environmental action. RBCAs retain the goal of protecting human health and the environment, but do so by integrating exposure and risk-assessment techniques at the beginning of the traditional remedial-action process. Such an approach results in a practical, streamlined, consistent, and technically defensible process, while ensuring the selection of the most appropriate and cost-effective remedies and allocation of resources.

The issue of leaking underground storage tanks presents an excellent example of the problem and a possible solution to the issue of brownfields. In the late 1980s, concern about underground storage tanks (USTs) and their impact on groundwater and soil elevated USTs to the status of a full-fledged program within the USEPA and state agencies. Approximately one billion dollars was added to the Superfund reauthorization to begin to address the issue. Unfortunately, the scope of the problem has threatened to outstrip the available resources.

In 1994, there were more than 1.2 million USTs known to be in use nationwide.[15] (The number of abandoned or unreported tanks is unknown.) Less than a third of that number have been replaced or upgraded in advance of a statutorily mandated 1998 deadline. Over 250,000 leaking USTs have been reported thus far, with another 1,000 releases reported weekly. Far less than half of those releases have been corrected to date. UST remediation can run as high as $100,000 for soil contamination and $1,000,000 for groundwater contamination. Because most of these sites involve small businesses with limited resources and little or no environmental insurance coverage, bankruptcy and abandonment has not been an uncommon outcome. State assurance funds were created to make resources available to small businesses to complete cleanups, but the huge number of cases threatened the solvency of these funds as well. The huge caseload (estimated as high as 500 projects per manager), the use of restrictive cleanup standards, and the difficulty achieving these targets slowed the process to a crawl. Real estate transactions collapsed as a result of liability fears and time delays. Properties with USTs were left undeveloped and abandoned, and the tax base decreased proportionately.

In California, it was projected that the state UST cleanup fund would have a $1.5 billion shortfall by 2005, with a significant portion of work remaining. This realization triggered a review and analysis of the UST process and policies in California. This study, the *California Leaking Underground Fuel Tank (LUFT) Historical Case Analysis*, was conducted by the Lawrence Livermore National Laboratory in response to legislative mandate and reviewed 1,500 LUFT sites.[16] The findings were quite revealing. Only 0.4 percent of public water supplies in the state were impacted by benzene (used as a marker of fuel contamination) and only 0.5 percent of LUFT sites impact any drinking-water well. Less than 0.0005 percent of the total groundwater basin storage capacity in California was found to be at risk from LUFTs. This was accounted for, in part, by the fact that 95 percent of all plumes from LUFTs were less than 250 feet in length and were prevented from extending further by both the natural assimilative capacity of many subsurface soils and the relatively small quantity of source material typically lost to the environment. It was found that in most cases the current drinking-water well construction codes were protective, and that passive bioremediation could readily reduce contamination to acceptable levels. A review of the LUFT program revealed that decisions regarding soil remediation were inconsistent, and groundwater decisions, though consistent, were based on generic standards (no detection of contaminants above the drinking-water standards), regardless of the specific circumstances or settings. Common remedial techniques were found to be generally ineffective at meeting the standards set, with the result that typical restoration costs translated to approximately $650,000 per acre-foot of resource as compared with the marginal costs of about $900 per acre-foot for producing drinking water.

This realization triggered a reevaluation of the LUFT effort in California. The State Water Resources Control Board issued new guidance, which included aggressively closing cases that involved only low-risk soil contamination, and replaced active remediation with monitoring in low-risk groundwater cases. Passive bioremediation was used whenever possible once the source was removed, and the LUFT framework was modified to allow risk-based cleanup goals at higher than the drinking-water standards. These same issues led to the development of the American Society for Testing and Materials (ASTM) *Guide to Risk-Based Corrective Action at Petroleum Release Sites* (E1739-95), which provides a classification for sites, a tiered approach to remedial action, and alternate compliance points within a consistent framework.[17] All of this is based upon the integrated use of risk and exposure concepts throughout the process, resulting in more effective resource utilization.

A similar initiative (E50.04) is being developed under ASTM guidance, extending the RBCA concept to USEPA's Voluntary Cleanup Program. A review of 1,300 records of decision by the National Environmental Policy Institute showed that if the Superfund remedial alternative process was based on an RBCA framework that took into account reasonable future land use to assess risk, and balanced risk reduction with aesthetic restoration, then more cleanups would be completed faster, with an overall cost reduction of 60 percent, and greater protection of the public health would result.[18] The ASTM effort is being developed by a working group made up of representatives from industry, consulting, and federal and state regulatory agencies. This standard will specifically address issues of site assessment, ecological and human risk-assessment methodologies, important chemical-property considerations (fate and transport, as well as toxicity), closure and release letters, risk communication and public participation, future land use and institutional controls, modeling and monitoring, and guidance on how to develop an RBCA program, with appropriate examples and case studies. The first draft of a standard is due to be completed during 1997, and the final document will be based on stakeholder response and comment.

It is obvious then that there is recognition of the need to reach agreement on a cost-effective remediation that results in a "safe" environment and the return of land to productive use. To make effective use of an RBCA approach in the current regulatory atmosphere, the approach must be negotiated in the beginning of the process and several issues should be addressed:

1. Understand the concerns of the regulators and the public.
   - What is the history of the situation?
   - What agencies are involved?
   - What policies/programs/regulations are involved?
   - What is the public/political/advocacy interest in the situation?
   - What is the regulated party's history with the above?
2. Identify the scientific issues.
   - What is the quality of the environmental database?
   - What is the health/ecological outcome of concern?
   - What group or receptors are at risk?
   - What is the scientific rationale for this concern?
3. Assemble the pertinent information.
   - What is known about the situation?
   - How well does the existing information support the concerns?
   - What are the number and characteristics of the receptors at risk?
   - What data are missing and how important are they to the outcome?
4. Identify important parameters that are not "cast in stone."
   - Does the potential for a hazard truly exist?
   - Was the sampling properly done and interpreted?
   - Was environmental fate and behavior considered?
   - Are the exposure parameters appropriate and site-specific or use-specific?
   - Is the dose-response evaluation appropriate?
   - Have the risks been properly characterized?
5. Develop a reasonable and plausible RBCA.
   - Does the proposal and approach satisfy the agency objective?
   - Is the product consistent with previous regulatory actions?
   - Does the proposal contain several "negotiable" items?
   - Will the solution satisfy all parties, at least in part?

FIGURE 4. Overall Remediation Costs Decrease Using RBCA

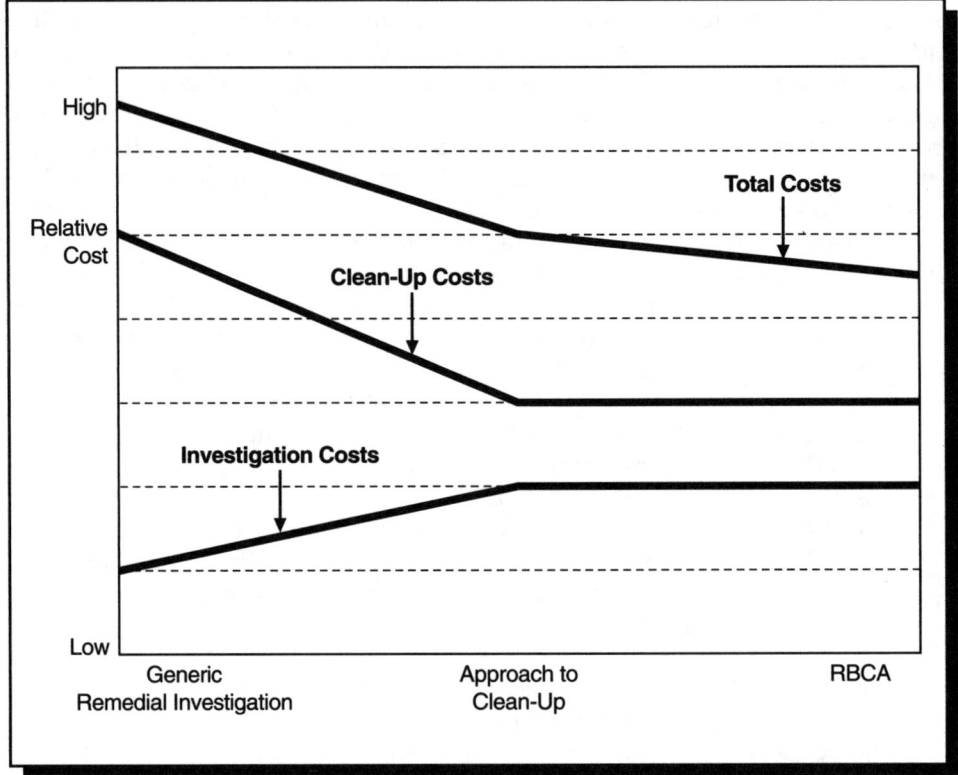

Although the up-front costs of an RBCA are initially higher than that of a remedial investigation, the overall costs of cleanup are far lower when this course of action is followed.

Because of the greater initial effort required to develop the information and approach for an effective RBCA program in brownfields initiatives, such efforts always entail a greater initial expense (Figure 4). However, because an RBCA program will often result in a no-action or reduced-action alternative, this greater initial cost is more than offset by the cost savings in time and money required to gain a release from future regulatory action (Table 2).

Pitfalls and Promises in Risk Assessment for Brownfields

The risk-assessment process available today contains both pitfalls and promises that will influence its success or failure in the brownfields initiatives on both the federal and state levels.[19] RBCA programs should be carefully scrutinized to avoid the pitfalls and take full advantage of the promises.

TABLE 2. Differences in Costs between Standard Remedial Actions and Risk-Based Corrective Actions

Site	Original Remedial Cost Estimate	Remedial Costs after RBCA
Pesticide formulator	$40,000,000	$0
Solvent recovery	$2,000,000	$0
Wood treatment	$1,000,000	$0
Fuel oil storage	$600,000	$100,000
Refinery	$2,000,000	$500,000

Hazard Identification

Hazard identification involves determining the contaminants of concern at a given site. Because industrial or disposal sites often contain a large number of agents, it is a practical necessity to limit the evaluation to those substances that will contribute most markedly to the risk. This is most often performed by combining measures of toxicity with that of environmental occurrence under the notion that those compounds with the highest toxicity and concentration will present the greatest potential hazard. Aside from toxicity, issues of detection frequency, location, levels, persistence, and mobility are important determinants in the selection of the chemicals of potential interest at a given site. Because evaluating site hazards generally results in the need to conduct further evaluation of the identified contaminants and furnishes the concentration terms used in subsequent dose estimates, it is a critical step that influences the outcome of the remainder of the risk-assessment process. However, a number of pitfalls exist at this step.

First, the sampling strategies for remedial investigations are not typically designed or performed with risk assessment in mind, but are intended solely to determine the nature and extent of contamination at a site. This often results in a type of sampling bias, due to intentionally collecting data from areas of high (usually visible) contamination and arguably limited accessibility. These problems may be compounded by inappropriate statistical treatment of the data collected (that is, use of arithmetic instead of geometric means). Using such data in the hazard identification and subsequent steps may result in compounds being inappropriately included in—or excluded from—the assessment, and inappropriate concentration terms developed for use in the exposure-assessment step.

Second, failure to take into account the background levels of the potential chemicals of interest may again skew the selection process and dose estimates. Most often, the comparison with background levels is useful for determining whether the concentrations of inorganic compounds on the site are elevated;[20] however, in industrialized areas, an argument can also be made for comparing the occurrence of synthetic organic compounds on-site and off-site. If, for instance, the same contaminant is found upgradient of the site in groundwater or in upwind soils at similar concentrations, an argument can be made that it would be inappropriate to include it as a site-related chemical of interest in the assessment.

Third, lack of attention to assumptions regarding the form or fate of compounds under consideration may influence the outcome of the hazard identification. For instance, chromium in soils is most often analyzed as total chromium, but assessed as the carcinogenic hexavalent chromium. This is in spite of the fact that, even when hexavalent chromium might be expected to occur, much of the chromium is in the less toxic, noncarcinogenic trivalent state.[21] Similarly, barium is often identified as a chemical of interest because of the potential toxicity of barium in solution (based on the toxicity of barium chloride). Such an assessment fails to take into account that barium generally combines with sulfur in the environment to form the insoluble salt, barium sulfate, which the body does not absorb well.[22] The form and size of lead particles influence their uptake by the body as well.[23] If exposure to contaminants on the soil surface is a concern, use of data from soil cores of varying depth may be inappropriate if the majority of the chemical load is contained within the subsurface portion, as might be the case with any number of volatile or semivolatile chemicals.[24]

Finally, the toxicity criteria on which hazard-identification decisions are made are usually drawn from lists developed by USEPA or other regulatory agencies. It is always conceivable that an incomplete evaluation of the data was initially performed or that better data has been published since the development of those criteria. The tendency to neglect the weight-of-evidence approach in evaluating numerous data sets, or to treat carcinogens of differing potencies as equally hazardous, are examples of potentially overestimating the hazards posed by compounds at a site. Though a complete review of a large list of compounds is a lengthy and costly process, many times it is apparent that one or two compounds are the driving force in the hazard identification, and it can be of benefit both in this step as well as the dose-response evaluation to review and update the data for these compounds to take advantage of all relevant information.

Hazard identification's promise in a RBCA lies in correcting past oversights and identifying the potential hazards that exist at a site in relation to the potential for exposure and the anticipated future land use. The risk assessment thus becomes a primary focus of the site assessment and remedial investigation, rather than an afterthought that forces the risk assessor to make use of data that is not completely appropriate for this use and skews the results toward identifying problems that may not exist. Toward this end, the risk assessor should develop a site-specific conceptual risk model early in the process, and it should be included in the planning stages of the remedial investigation to ensure that useful and appropriate sampling is performed. A well-designed, statistically valid, and risk-based sampling plan is required to ensure that the representative results are obtained.[25] Such samples should be collected in a random manner and in sufficient numbers to allow statistical analyses to be conducted. The number needed to ensure statistical validity is dictated by the variability between results. The greater the variance, the larger the number of samples needed. Because most environmental data are log-normally distributed, the geometric mean and geometric standard deviation are the most useful in evaluating potential hazards. For the results to be valid and useful, both in selecting chemicals of concern and other steps in the risk assessment, the sampling must be both random and representative.

Consideration of fugacity (changes over time) and chemical and physical properties, along with professional judgment involving site-specific or use-specific exposure scenarios, should be used to identify the key media for sampling. Chemical speciation may also be important, not only because it may substantially influence the chemical's behavior, but also as a means for "fingerprinting" the contamination. This relatively new technique is useful for metals and other persistent chemicals and can identify sources of contamination and the relative contribution of different sources to the environmental residues detected.[26]

Finally, one should review the strength and weight of evidence linking the chemicals of interest to the end points of concern, and the information should be used to determine whether the chemical(s) should be retained in the risk assessment after the hazard identification step.

Exposure Assessment

Exposure assessment is the step that quantifies the intake of the chemicals of concern to receptors in the various exposure pathways developed for the site. A pathway is considered complete if there is (1) a release source, (2) a transport mechanism, (3) a point of contact, and (4) a receptor. If any one of the four components is missing, then exposure by that pathway is incomplete and need not be considered further in the risk assessment. If groundwater is contaminated, but no wells exist or are likely to exist downgradient, then this pathway should not play a role in the risk assessment.

Within each completed pathway, numerous exposure scenarios may be constructed. For instance, groundwater may be used domestically for drinking, cooking, and washing over extended periods by adults and children, or commercially for industrial purposes by adults over limited periods, or as a source of water for irrigation or gardening intermittently by various individuals. Each of these groups has variable exposures, due to different rates of intake, age, length of exposure, and physiological, behavioral, and social factors. The differing contributions from each completed pathway in each exposure scenario are determined and summed to assess the overall exposure dose for that receptor. For instance, the child at home may be exposed to contaminants in the drinking water and exterior soils, but none in workplace air, whereas his or her father may be exposed to contaminants in water at home and work, air in the workplace, and fish harvested from the local river, but not from any other source. Though variations abound depending on the specific circumstances and needs,[27] the typical calculation for this step takes the following form, with the variables defined as indicated:

$$Dose = \frac{EC \times IR \times AF \times EF \times ED}{BW \times AT \times 365 \frac{days}{year}}$$

Dose = exposure intake, expressed in milligrams per kilogram per day
EC = the concentration of the chemical of interest found in the media evaluated, expressed in milligrams per kilogram (solids), milligrams per liter (liquids), milligrams per cubic meter (air), or milligrams per square centimeter (skin contact)
IR = intake or contact rate, expressed in kilograms per day (solids), liters per day (liquids), cubic meters per day (air), or surface area exposed (square centimeters per day)
AF = absorption factor, expressed as percent absorbed (unitless) or mass absorbed per unit area per unit time (milligrams per square centimeter per minute)
EF = exposure frequency in days per year
ED = exposure duration expressed in years
BW = body weight, expressed in kilograms
AT = averaging time; for carcinogens, it is set to equal 70 years; for noncarcinogens, it is set equal to the exposure duration

Development of a quantitative exposure estimate requires the risk assessor to select the values for each of the variables for each completed pathway in each scenario.[28]

The choice of these variables and manipulation of the calculation thus becomes critical to the accurate expression of exposure and, ultimately, risk. It is in this area that tremendous pitfalls can befall the unwary,[29] but also where the careful understanding of the tools and techniques available to assess exposure provide great promise of providing better exposure estimates.

Typically, most exposure estimates have suffered from the "worst-case scenario" logic. As previously stated, because of the uncertainties surrounding risk assessments, a conscious decision was often made to err strongly on the side of caution, to assure that sensitive individuals or populations were protected. To accomplish this goal, extreme values (usually the upper 95 percent statistical bound on an estimate) were often employed in the exposure equations. Because they were extreme values with a low probability of being true individually, the combination of many such variables created an exposure estimate that defied credulity, but was, by definition, certainly protective.[30]

The worst-case scenario is typified by the vision of a naked child growing to naked adulthood, never moving from the contaminant source over its lifespan, all the while ingesting contaminated soil, water, and foodstuffs, breathing contaminated air, and absorbing contaminants through the skin.

The single most important driver in risk assessments of the past ten years has been the estimated amount of soil ingested by children (and to a lesser extent, adults). In 1984, a qualitative guess was made for Times Beach that a child might ingest as much as 10,000 milligrams of soil per day.[31] The use of this number in this and other risk assessments quickly identified the soil and the soil ingestion pathway as the largest contributor to risk at most cleanup sites,[32] and was responsible for the "dig and haul" mentality that marked many remedial decisions of the period. Subsequent research has indicated that childhood soil ingestion is actually much smaller, on the order of 4 to 40 milligrams per day.[33] Regulators have responded by reducing their reliance on the extreme values, typically relying on values between 100 and 200 milligrams per day.[34] Failure to embrace the best available estimates, however, still results in contaminated soils driving most risk assessments. This amounts to a risk-management decision embedded in the risk assessment, contravening the recommendations of NAS.[35]

The pitfalls in exposure assessment revolve around reliance on extreme values or scenarios that have resulted in the creation of roadblocks to addressing brownfields. As the result of sampling biased toward finding the highest contamination, the value(s) used to estimate the concentration(s) encountered in the environment may be markedly higher than the average concentration(s). For example, a statement that concentrations as high as 150 ppm were measured on a 1,000-acre waste site could result in the use of that value as the concentration term for the exposure estimate. Closer examination may reveal that this value was found in only 1 out of 200 samples from a 20-acre portion of the site. The geometric mean value of the compound in this area was 2 ppm. This value, together with the reduced area of concern, provide quite a different perspective on the potential risk posed to receptors.[36]

Failure to consider the fate of the compound in the environment is another typical pitfall.[37] Organic compounds tend to degrade over time and may disappear from exposed surfaces relatively quickly. Inorganic compounds typically change over time as well, and these changes may influence how well they can be absorbed into the body. For example, the assumption is typically made that 100 percent of a contaminant is absorbed following ingestion of soil. It has been convincingly demonstrated that the toxicity of various compounds is lower in soil as a result of binding to soil particles and reduced availability for absorption.[38] Simple experiments have likewise demonstrated that compounds do not dissolve well in simulated gastric fluids,[39] which indicates they will not be available to be absorbed into the body (reduced bioavailability). Similar experiments have also been performed to assess dermal absorption.[40]

Finally, the exposure scenarios and pathways selected typically are not critically evaluated. This can be crucial for the success or failure of a brownfields initiative, if the sites are assessed as though their future use was intended to be residential or farming instead of industrial or commercial (Table 3).

TABLE 3. Differences in Human/Ecological Exposure Parameters for Plausible Future-Use Scenarios

Parameters	Residential	Industrial
Human:		
Duration	70 years	40 years
Frequency	24 hours/day, 365 days/year	8 hours/day, 250 days/year
Soil ingestion	200 milligrams/day	20 milligrams per day
Inhalation	20 cubic meters/day	10 cubic meters/day
Foods	Eats local meat/fish/crops	No consumption
Groundwater ingestion	2 liters per day	No consumption
Ecological:		
Crops	Gardens/farming	No agriculture
Runoff	Significant	Minimal
Grazing animals	Uptake/human consumption	No grazing
Fish	Uptake/fishing	No fish/fishing

The exposure estimates can be improved markedly by using the best science and logic to assess the risk. The value of a risk-based sampling plan was previously discussed as having advantages in the hazard-identification step.[41] It should be clear that such an approach plays a critical role in the exposure-assessment step as well. The exposure-assessment step can also be improved by (1) relying on better (or site-specific or use-specific) estimates of intake, exposure frequency and duration, or absorption, (2) avoiding repeated or inappropriate use of worst-case assumptions, (3) developing site-specific or use-specific exposure scenarios that are based on reasonable future land use, and (4) incorporating information of environmental fate or bioavailability into the assessment. In some instances, it can be beneficial to conduct biological monitoring to confirm the results of the exposure estimates.[42] If the body burdens of exposed individuals are not elevated above expected background, it may not be necessary to carry the assessment further.[43]

One of the most powerful tools introduced to risk assessment in recent years is the application of Monte Carlo analysis to environmental issues. This approach is particularly useful for exposure assessment.[44] It should be evident that all the variables in the exposure calculation exist as ranges, rather than point values. No population is exposed to a single concentration; drinks, ingests, or breathes at a single rate; is exposed for the same length of time; or weighs the same amount.[45] The range of values for a given variable is known as its distribution, and it has different shapes according to where the individual values occur (such as normal, log-normal, or uniform). Monte Carlo analysis allows the distributions of the variables involved in the exposure assessment to be combined into a "distribution of distributions." Such an approach provides more information than the simple point estimate, because it uses all the data available. As such, not only are average and worst-case scenario results available to the risk assessor and manager, but their relation to each other and the entire range of exposure is clarified (Figure 5). This approach is gaining acceptance in regulatory circles[46] and has been suggested as

FIGURE 5. Traditional Point Estimate Exposure Analysis Compared to Monte Carlo Exposure Analysis

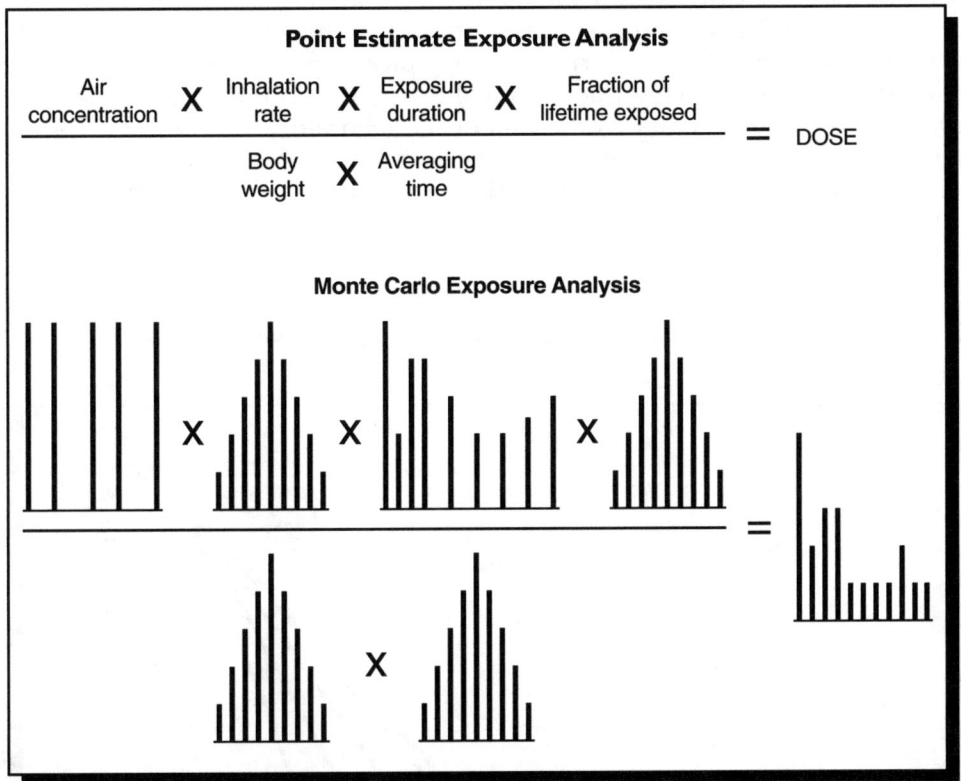

Standard exposure assessments use point estimates of the different variables in the exposure equation to develop a single-dose number. Use of a Monte Carlo approach to assessing exposure allows all available information about the ranges of variables previously expressed as point values to be combined and expressed as a distribution of possible exposures, with an estimate of their probability of occurrence.

an acceptable methodology for addressing brownfields in certain state Voluntary Cleanup Programs.[47]

Dose-Response Evaluation

Dose-response evaluation is the development of the quantitative relation between the exposure to a compound and the biological response elicited. Most risk assessments do not actually conduct dose-response evaluations for their chemicals of concern, but rely instead on information and toxicity values derived from regulatory agencies. The USEPA's Integrated Risk Information System (IRIS) and Health Effects Assessment Summary Tables are only two such examples of sources of this information.[48]

Because of the presumed differences in mechanisms of action, compounds are divided into carcinogens and noncarcinogens. Any exposure to a carcinogen is assumed to carry a finite risk of causing cancer because, in theory, a single molecule of the agent interacting with a single molecule of DNA could cause sufficient damage to result in the production of a tumor. Toxicity values for carcinogens are derived most typically from the application of a mathematical model to high-dose animal tests. This dose-response data is extrapolated from the high-dose results of the laboratory test to the low-dose range that is typical of human exposures in the environment (Figure 6). A variety of such models exists and all fit the observed data well, but diverge

FIGURE 6. High-Dose to Low-Dose Extrapolation for Carcinogens

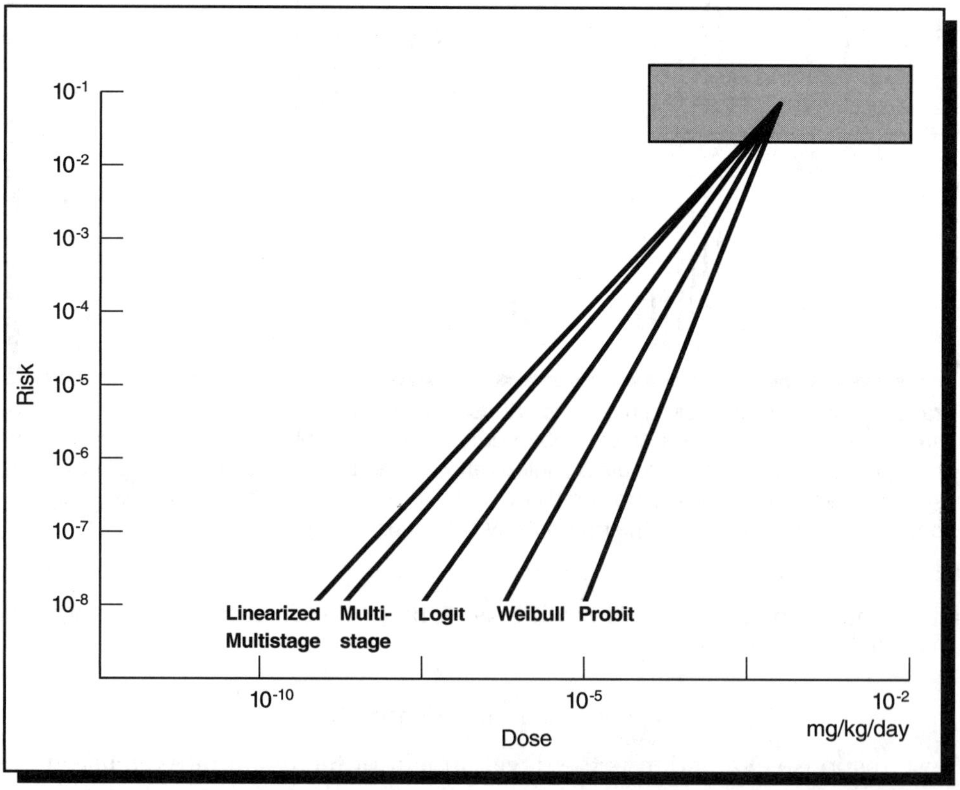

Cancer risk assessment relies on the use of mathematical models to extrapolate from the observed experimental range (gray area) to the unobservable environmental exposure range. All models agree in the high-dose range, but diverge widely in the low-dose range. USEPA currently relies on a linearized multistage model, one of the most conservative of the dose-extrapolation models.

widely in the low-dose range. The difference in the predictions from the most to the least conservative models spans seven or more orders of magnitude.[49]

USEPA currently employs a linearized version of the multistage model, which is among the most conservative of the models available. Two types of related values can be derived from such modeling: unit cancer risks (UCRs) and potency factors (PFs). A UCR is the added cancer risk resulting from lifetime exposure to a unit amount of the carcinogen (for example, 1 microgram per cubic meter of air, or 1 microgram per liter of water). If the UCR is expressed as 1E-4 or 1×10^{-4} at 1 microgram per liter of water, this is interpreted as meaning that an additional 1 cancer out of 10,000 exposed individuals may result for every microgram of the contaminant per liter of water consumed over a lifetime. PFs, on the other hand, are derived from the slope of the modeled dose-response curve and are used in combination with the exposure estimates to calculate added risk at that dose level. For example, the PF for polychlorinated biphenyls is 7.7 (milligrams per kilogram per day)$^{-1}$. This value is multiplied by the estimated dose (in milligrams per kilogram per day) to provide a risk estimate (for example, 1×10^{-4}). The potency or slope factor is actually the upper 95 percent statistical bound on the best estimate of the model. This is selected to provide an additional measure of safety. The difference between the best estimate of the model and the upper-bound estimate may be an additional six or more orders of magnitude.[50] The larger the slope factor, the more potent the carcinogen. A PF is needed to develop a UCR number, and the UCR number can provide the PF by back calculation.

Noncarcinogens differ from carcinogens in the assumption of the existence of a threshold for noncarcinogens. A threshold is defined as the dose below which no effect is evident, and thus no harm occurs. For noncarcinogens, the highest dose for which no effect is seen is termed the "no observed effect level" (NOEL), while the lowest dose eliciting an effect is referred to as a "lowest observed effect level" (LOEL). The NOEL value is preferred in deriving the desired toxicity criteria, but the LOEL is often used in its absence. The toxicity values, called the "reference dose" (RfD) or "reference concentration" (RfC), are derived by dividing the NOEL or LOEL by a number of uncertainty factors (Figure 7). The size of the uncertainty factor used is inversely related to the degree of comfort that the risk assessor feels the data has for its intended purpose or target. The approach is illustrated by the following formula:

$$RfD = \frac{NOEL}{Uncertainty\ Factor(s)}$$

The uncertainty factors, in multiples of 10, are selected based on the quality of the data in relation to the intended use: a chronic intake value protective of human health. Thus, for a NOEL derived from a human study of long duration, the uncertainty factor used would be only 10 to account for interindividual variation, because such a study is, in all other ways, directly relevant to the population and end point of interest. If, on the other hand, the NOEL was derived from a long-term animal study, the uncertainty factor

FIGURE 7. Reference Dose (RfD) Development for Noncarcinogens

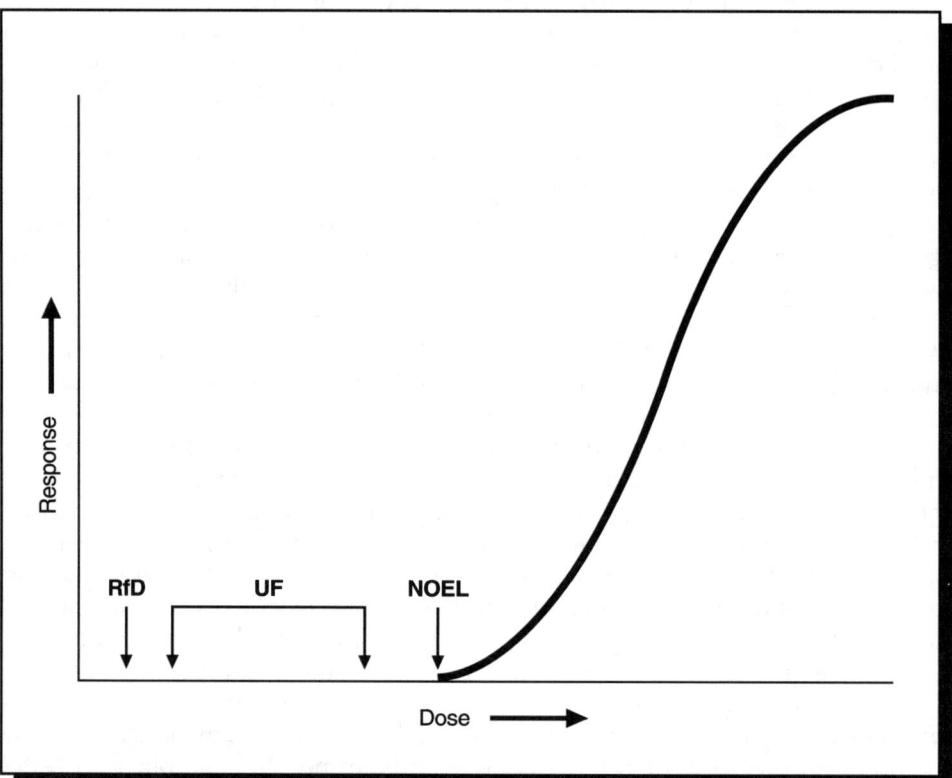

Among the most common approaches to the assessment of noncarcinogens is to determine a NOEL from an animal or human study and divide this value by an uncertainty factor (UF), ranging from 10 to 1000 or more (depending on the level of confidence one has in the data), to derive an RfD, the dose to which one can be exposed daily without risk of adverse effects.

would likely be 100, to account for both interspecies variability (10) and interindividual differences (10). Other uncertainty factors may be used as needed to account for the uncertainty introduced by using LOELs instead of NOELs, or for the results of short-term tests as opposed to long-term results. Uncertainty factors of 100 are most common, although 1,000 may be used in some cases. Data for which uncertainty factors of 10,000 or higher are proposed are probably inappropriate for use in risk assessment.

As with the other steps in risk assessment, numerous pitfalls that will influence the outcome of a risk assessment exist in the area of dose-response evaluation. The mathematical extrapolation models used are typically chosen based on their degree of conservatism rather than their biological relevance.

The failure to consider risk estimates from other equally plausible models places potentially inappropriate emphasis on a result certain to be conservative. This can be easily seen when risk-assessment predictions are compared with the results of epidemiological studies.[51] The results of risk assessments employing the multistage model always predict cancer rates higher than that observed in exposed human populations. This results not only from the use of the most conservative models, but also from the typical presentation of risk as the statistical upper bound instead of the best estimate followed by the upper and lower bounds on the value.[52] Additional problems may be encountered by the failure to adjust appropriately for biological variability and scale-up between species. Finally, many of the various toxicological criteria have been in place for a considerable period and are based on data that have been supplanted by more recent or higher-quality information. The continued use of these values in the face of better information may result in the final estimates either overpredicting or underpredicting risk.

In this step, the review of the important toxicity criteria is essential, as an alternate value may result in a substantially different outcome to the risk assessment. For instance, recent reviews of IRIS values for a variety of compounds have revealed that the RfDs and PFs were based on inappropriate data or assumptions, with the result that the values were withdrawn and replaced with values based on the more up-to-date information.[53] Because the risk assessment is usually driven by only a few compounds, the investment in review of these drivers may pay substantial dividends. Additional benefits may be derived by examining and presenting the predictions of different low-dose models, and using a weight-of-evidence approach to select the most appropriate model for the compound(s) of concern, as not all compounds are best described by the same model.

In the area of dose-response evaluation, one of the most important breakthroughs in recent years has been the development of physiologically based, pharmacokinetic (PBPK) models that allow better evaluation of human risks from animal studies.[54] Conventional risk assessments use linear extrapolation of the administered (external dose) in combination with an interspecies factor, based on body weight or surface area, to arrive at the toxicity values and, ultimately, risk. In PBPK models, the absorbed (internal) dose is calculated through the integration of information on the administered dose, the important physiological compartments and processes of the animal, and the biochemical parameters of the chemical(s) of concern. Predicted internal doses in the animal can be correlated to the toxicological end point of concern (for example, cancer) to yield hypotheses about mechanisms of action. Once the model is validated, the animal parameters are scaled up to human levels and the model is rerun to determine if the same pathways are followed and if the internal doses are similar in people. If the same end products are not produced or the internal doses are lower than seen in animal studies, the risk to humans may not exist or be substantially reduced.

When a PBPK model was developed for methylene chloride, it was found to predict a cancer risk 100 times less than that predicted by the conventional USEPA approach.[55] This finding resulted in USEPA's withdrawal of the methylene chloride potency factor and its replacement with a value partially reflecting the PBPK model results. Validated PBPK models have now been developed for a variety of compounds, including volatile organic solvents, metals, pesticides, and other pollutants of potential concern in brownfields. In general, the results of PBPK modeling lower the estimated risks associated with exposure to compounds of concern. The results of such modeling are increasingly accepted by regulators.

Risk Characterization

Risk characterization is the final step in the risk-assessment process, and involves integration with the other steps. Exposure estimates are combined with the appropriate toxicity values to estimate the added risk quantitatively. For carcinogens, this takes the following form:

$$\text{Added Lifetime Risk} = \text{Exposure (mg/kg/day)} \times \text{PF (mg/kg/day)}^{-1}$$

The units cancel out and a unitless value representing the upper bound on the added risk from lifetime exposure remains. For noncarcinogens, a hazard quotient (HQ) is calculated by dividing the exposure estimate by the RfD or RfC:

$$HQ = \frac{\text{Exposure (mg/kg/day)}}{\text{RfD (mg/kg/day)}}$$

In either case, exposure to multiple agents is addressed additively. Cancer risks or HQ scores for each chemical in a pathway are added together, and all pathways in a given scenario are then added together to get a total scenario risk. These values are then compared with an acceptable risk level for cancer. This value is most often cited as one in one million (1×10^{-6} or 1E-6) added cancer risk, which is viewed as a *de minimis* risk because it is approximately the level of voluntary and involuntary risks commonly accepted by the public (for example, deaths due to lightning strikes or bee stings; Table 1). For noncarcinogens, a hazard is not considered to exist if the sum of the HQ, known as the hazard index or HI, remains below 1.0. Acceptable levels of exposure and remediation goals can also be determined by back calculation. For example, the calculation of a remedial goal for a carcinogen in soil would be as follows:

$$\text{Soil Remediation Goal} = \frac{AR \times BW \times AT \times 365 \frac{\text{days}}{\text{year}}}{PF \times IR \times AF \times ED \times EF}$$

AR = acceptable cancer risk (for example, 1×10^{-5})
PF = potency factor $(\text{mg/kg/day})^{-1}$

In a similar way, an acceptable level for a noncarcinogen in soil can be calculated using the reference dose as follows:

$$\text{Soil Remediation Goal} = \frac{RfD \times BW \times AT \times 365 \frac{days}{year}}{HQ \times IR \times AF \times ED \times EF}$$

RfD = reference dose in mg/kg/day, or the reference concentration (RfC) in mg/m^3/day
HQ = hazard quotient (equal to 1.0)

Remedial goals can be set for all environmental media and exposure scenarios using site-specific information and variables. Careful consideration of the potential exposure and plausible future uses can result in substantial savings of time and money (Table 4).

Most often, the pitfalls in risk characterization result from failure to interpret the meaning of the estimates properly. The cancer-risk estimates resulting from conservative dose-response modeling and worst-case exposure assumptions are not predictions of future happenings, but rather, represent a plausible upper-bound estimate of that risk. For instance, the proper interpretation of a risk assessment that provides an estimate of 4×10^{-5} is not that there will be an additional forty cancer cases per million people exposed over their lifetimes, but rather, that there will be no more than forty additional cancers over a lifetime, probably less, and maybe zero. For noncarcinogens, an excursion above an HI score of 1.0 is not a signal of imminent hazard. HI scores below 1.0 present no risk since the measured or estimated environmental levels of the compounds being assessed cannot, by definition, exceed their respective acceptable levels (i.e., RfDs or RfCs), either individually or collectively. However, an HI score that is greater than 1.0 does not automatically imply that an unacceptable risk is present either.

While such an outcome indicates that the level of at least one compound exceeds an acceptable guideline, it is important to recall that such guidelines are derived from a dose that had little or no adverse effect on test animals

TABLE 4. Risk-Based Preliminary Remediation Goals for Soils (Residential versus Industrial)

Compound	Residential (mg/kg)	Industrial (mg/kg)	Reduction in Effort/Cost
Arsenic	0.45	3.3	8 fold
Chromium VI	312	1530	5 fold
Aroclor 1260	0.034	0.75	20 fold
Benzene	2.96	18.4	9 fold
Tetrachloroethene	189	518	3 fold

(i.e., a LOEL or NOEL) divided by a large uncertainty (or safety) factor (typically 100 or greater). Therefore, the RfDs or RfCs include in their make-up a substantial margin of safety, and while an excursion may reduce this margin of safety somewhat, it is rare that any such exposure will even approach a level at which an adverse response is likely to occur (Figure 7). Thus, the proper interpretation of HI scores is not that, at 1.0 and above, an adverse response is certain, but rather, at 1.0 and below, no adverse effects will occur.

In this light, an HI score above 1.0 may call for additional evaluation of the assumptions and data that lead to this result, but it does not necessarily call for immediate action on the part of regulators or responsible parties. This type of information is critical to an understanding of the risk assessment but is rarely included. Also left out of most risk assessments is a discussion of the uncertainties that surround the risk estimates and their influence on the outcome. The implication that a one-in-a-million cancer risk is a significant public-health problem is inappropriate, because even if accurate, it represents a less than 0.0001 percent increase in the background rates of cancer.[56]

The use of additivity to account for exposures to multiple agents may also inflate risk estimates. The only compounds for which risk estimates could conceivably be added together would be those that share target organs or have mechanisms of action in common. In fact, given the low levels of compounds generally encountered in the environment, sufficient damage may not occur to make such interactions of any consequence in most cases.

Risk characterization can be the most powerful step in the entire process because it can be used to explain the outcome of the assessment and function as an educational tool for the regulators and the public. Failure to take full advantage of its potential is a serious oversight. Use of Monte Carlo analysis in combining the distribution of values from both exposure assessment and dose-response evaluation not only illuminates the range of risks from the worst-case point estimates (that populate most risk assessments) to the most likely risk faced by a population based on the average values (Figure 8), but identifies the factor(s) that introduce the greatest uncertainty into the assessment.[57] Such techniques can also be employed to determine how effective or protective regulatory decisions or standards have been.[58] Improvement in these parameters through additional research may further reduce the uncertainty involved and improve the risk assessment, with substantial savings in terms of time and money.

More recently there has been a recognition that the levels of acceptable risk ought to be tied to the intended future use of the resource. Thus, higher than one-in-a-million added risks have been allowed in a variety of settings, including state brownfields cleanup programs.[59] This can be aided by the recognition that background levels of exposure from other sources are often much higher than the contribution from the site.

A number of questions should be asked when reviewing the results of a risk assessment. A "no" answer to any of them raises questions about the relevance and value of the assessment and ought to signal a reevaluation of the assumptions and approaches. These questions include the following:

FIGURE 8. Monte Carlo Analysis of Cancer Risk Associated with Household Exposure to Tetrachloroethylene in Tapwater

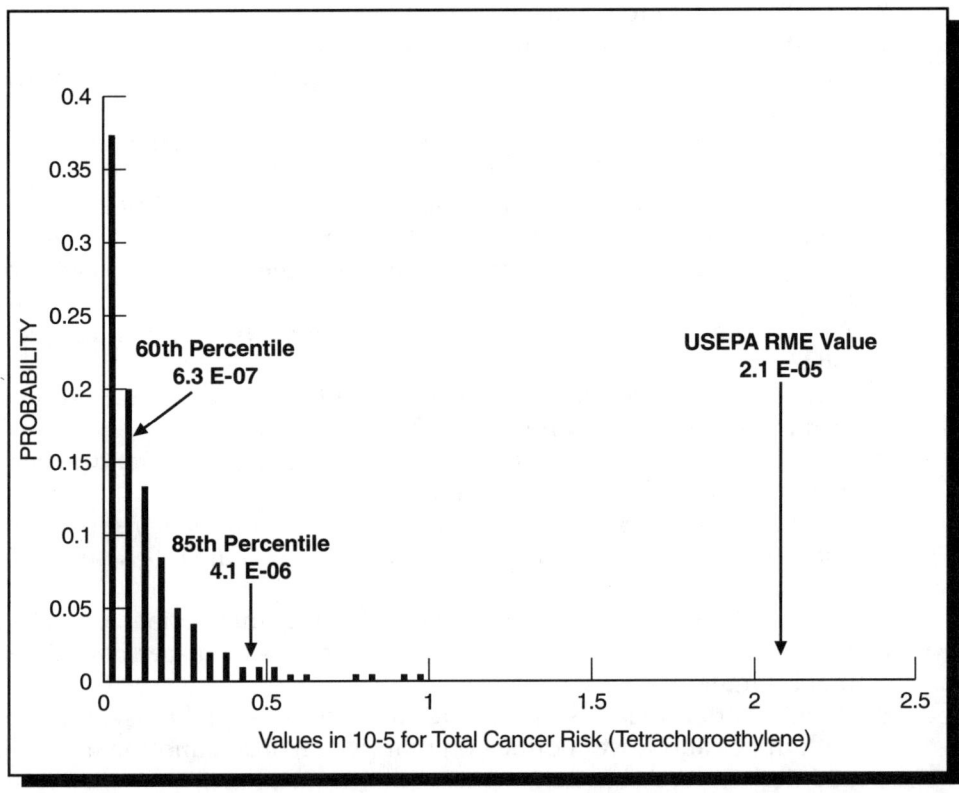

Combining a Monte Carlo analysis of both exposure and toxicity allows the likely distribution of risk to be completely displayed. This includes the risk that results from the use of USEPA-type of default point or reasonable maximum exposure values (RME), which typically uses the upper 95% statistical bound on an estimate. In this example, the USEPA RME risk is 2.1×10^{-5} which implies that we are 95% certain the risk is no higher than 2 in 100,000 excess cancers over a lifetime. Since this value exceeds the 1 in 1,000,000 nominal acceptable cancer risk, regulatory action is possible. The additional information from fully displaying the risk range and the probability associated with each risk allows the risk manager (or the public) to see that the most likely risk is most probably well below this level (6×10^{-7} or less).

1. Were all the adverse outcomes of equal severity and consequence?
2. Did the dose-response model take into account the time the response occurred?
3. Is the stated probability of response inflated by ignoring competing routes and risks?
4. Does the dose-response data contain enough information to allow the model to reflect the expected curvature of the line(s)?

5. Has the dose level been expressed on a biologically relevant scale?
6. Are the experimental data at the high doses relevant to low-dose behavior?
7. Are the animal models relevant to humans?
8. Are the experimental results consistent across species?
9. Have relevant biological differences among species been taken into account?
10. Have the differences in experimental and environmental routes of exposure been taken into account?
11. Are exposure patterns and durations in the experimental studies similar to those of the human population at risk?
12. Are the results from short-term tests consistent with the long-term results?
13. Are results of experimental data consistent with epidemiological data?
14. Have the exposures been completely identified in relation to routes, durations, dose levels, and pattern?
15. Has statistical variability in all the sampling, exposure, and experimental data been taken into account?
16. Have the assumptions, policy decisions, value judgments, and uncertainties been clearly stated and their influence on the risk assessment evaluated?
17. Are risks characterized in understandable and appropriate terms?
18. Are the stated risks the actual model estimates or upper bounds on that estimate?
19. If the uncertainty is described in terms of bounds on the risk, then are both upper and lower bounds on that risk recorded and the meaning and derivation discussed?
20. Is the risk assessment based on the most current and relevant data and protocols?
21. Are the risks based on site-specific and use-specific exposure scenarios that take into account plausible future land use?

Brownfields Case Studies

An appreciation of the value of a well-done risk assessment in achieving rapid, cost-effective closure on brownfields sites can be gained by examining case studies in which risk assessment played a critical role in causing, and then resolving, the problems.

Case Study #1

A small parcel of land previously housing a fuel oil distributor and transfer station is needed for expansion of an adjacent manufacturing facility. The expansion must be completed within two years. A preliminary investigation of the site identified soil and groundwater contamination by total petroleum hydrocarbons (TPH), polyaromatic hydrocarbons (PAHs), and chlorinated and aromatic solvents. The initial study was conducted with the intent of locating sources and identifying the contaminants; thus, likely areas of contamination were preferentially sampled. The high TPH content of some samples interfered with the analyses, thus raising the detection limits for PAHs in soil.

Use of the default approach of assigning one-half the detection limit to nondetects and assuming all PAHs were equivalent in potency to the carcinogenic benzo(a)pyrene resulted in high risk from exposure to surface soils. Additionally, exposure to the low-level groundwater contamination by vinyl chloride, benzene, and trichloroethylene was assumed and also contributed to the risk. Use of an exposure scenario assuming redevelopment as residential property, complete with private wells and seventy-year occupancy, resulted in a risk in excess of 1×10^{-4}. Generic groundwater protection standards (that is, drinking water standards) suggested a remedial alternative involving extensive soil removal and incineration, with an estimated remedial cost of nearly $30 million and a completion time of thirty-six months or more. The sale and redevelopment of this land is blocked until the remediation and liability issues are addressed to the agency's satisfaction.

The first step is always to meet with the regulatory agency to negotiate the use of a risk-based corrective action. In this case, it was agreed to (1) institute a random and representative sampling plan for surface soils, subsurface soils, and groundwater to supplement the existing data, (2) develop a method to provide better detection limits for the PAHs or a mathematical relationship between TPH content and PAH levels, (3) use exposure assumptions in keeping with the anticipated future use of the land, (4) use the results of improved fate and transport models, (5) incorporate Monte Carlo analysis into the exposure assessment, (6) use more realistic absorption factors for dermal and gastrointestinal availability, and (7) use a toxic equivalency factor (TEF) approach for the PAH mixtures.[60]

The revised sampling plan determined that soil contamination was confined to a relatively small portion of the property near bulk loading and storage areas. Groundwater contamination was traced to leaking storage tanks, which were then removed. No groundwater use existed in the area, and the plume was small and confined by heavy clays. Regression analysis between TPH content and PAH levels in soil indicated that the actual PAH concentration was approximately 100 times below the detection limit that resulted from the analytical interference by TPH. The alternate exposure scenario for the intended property use was an adult worker exposed to soil, dust, and vapors for no more than thirty years, and it incorporated environmental half-lives, absorption values derived from the recent literature, and a Monte Carlo analysis of the resulting exposure distribution. No residential scenario, or scenario involving use of groundwater, was employed. The use of the TEF approach for PAHs reduced the dose estimate by another factor of ten.

The resultant risk characterization for this site under the revised parameters found the overall risk to be no more than 7×10^{-7}. The fact that no significant risk was posed by the compounds of interest under conditions of plausible future use resulted in the agency choosing a limited-action alternative involving limited soil removal, monitoring of groundwater, and savings of well over $25 million and eighteen months.

Case Study #2

A former steel-mill site of 560 acres is being readied for sale to the municipality for redevelopment as a industrial park. The estimated value of the land is between $5 and $10 million. It borders a residential area and a major waterway. Slag was used as fill and covers the site to depths between four and twelve feet. A site assessment for property transfer purposes identified PAHs and heavy metals, including lead and beryllium, as contaminants of concern. A residential future-use scenario identified dermal contact with contaminated soil (beryllium and PAHs), inhalation of contaminated dusts (lead, beryllium, and PAHs), and ingestion of soil (lead, beryllium, and PAHs) as the dominant contributors to risk. The overall cancer risk was in excess of 1×10^{-4} based on the most conservative scenario. Use of the USEPA's Integrated Exposure Uptake Biokinetic Model (IEUBK) indicates that lead in soil above 400 ppm may pose a problem for children on-site.[61] Lenders involved in the redevelopment refused to proceed due to uncertainty regarding the extent and cost of remedial action required.

The responsible party agreed to enter the state Voluntary Cleanup Program and negotiated a risk-based corrective action to evaluate the appropriate course of action. The approach used settled on the following points: (1) a risk-based sampling plan to gather sufficient samples to provide a statistically valid profile of contaminant distribution in horizontal and vertical planes of the soil column, (2) measurement of the contaminant content of wind-borne dust, particularly the respirable fraction, (3) benchtop studies of the extractability of the metals in soil, sweat, and gastric fluid to determine their availability for absorption,[62] (4) adoption of more realistic future-use exposure scenarios, (5) use of a TEF approach for assessing the risk from multiple carcinogenic PAHs,[63] and (6) use of an alternate level of acceptable risk (1×10^{-5} versus the default of 1×10^{-6}).

The sampling found that although the metal contamination was well-distributed throughout the site, the highest levels were in the subsurface (greater than six inches below grade). Surface soil levels were 30 percent lower on average than the subsurface material, probably due to the past use of "clean" fill to cover the slag. PAH content was similar throughout the soil column.

Previously, the state assumed that the contaminant content of the wind-borne dust could be estimated by multiplying the contaminant level in the soil by the total suspended particulate level measured at the nearest ambient-air monitoring station. Actual measurement found that the metal and PAH content of airborne dusts to be only slightly above background. This difference in findings was attributed to the particle size of the surficial material, the lower contaminant level in the surface as opposed to the total soil column, and the fact that surficial materials contribute 50 percent or less to the content of airborne dusts. Extraction studies of the metals in simulated sweat and gastric fluid revealed that only 0.0034 percent of the beryllium could be extracted in sweat, and only 7 percent of the beryllium and 49 percent of the lead could be extracted under simulated gastric-extraction procedures, as opposed to the agency assumption that both metals were 100 percent available. This information resulted in a reduced absorbed dose for both compounds in the exposure scenarios.

The residential future-use scenario was abandoned in favor of a more realistic industrial/commercial use scenario, with adults exposed twelve hours a day, two hundred and fifty days per year, for a maximum of thirty years. This scenario assumed limited surface-soil contact and exposure to dusts. Because the IEUBK model addressed only childhood lead exposure, an alternate PBPK model for lead[64] was used to predict soil lead levels that would not result in elevation of blood lead in adults. The TEF assessment of the PAH content of soils and dusts resulted in lowering the risks by a factor of 10 or more.

The overall cancer risks from PAHs and beryllium using the new data and exposure scenarios were found to be below the alternate *de minimis* cancer risk level of 1×10^{-5}, and the new data and alternate model for lead indicated that a soil level of 1000 ppm would be protective. Only three small areas on the property were determined to be in excess of 1000 ppm lead in soil. Following a minor removal action and addition of deed restrictions limiting the future use of the land, the state concurred that no further action was needed and issued a closure letter. Had soil removal down to the 400 ppm lead level taken place as originally suggested, the preferred remedial action would have cost more than $55 million and taken over three years to complete. The cost of the RBCA was approximately $500,000, mostly as a result of the sampling, extraction studies, and laboratory analyses required to affect the improvement in the exposure estimates.

Conclusion

It is clear that over the past several years, regulatory agendas and private-sector fears created an atmosphere in which brownfields were inevitable. The use of selective and biased sampling, conservative default assumptions regarding exposure in mechanical risk assessments, and generic cleanup standards or guidelines, identified many sites as potential health or environmental risks. The length and cost of the remedial process proved a substantial impediment to the redevelopment of industrial property, due to investor and lender fears of being held financially responsible for the sins of the past, with their projects held hostage to an uncertain regulatory future. In such an environment, the smart money moves on and, to quote Nietzsche, "the wasteland grows."

As the number of brownfields expanded, the problem was reflected in a decline in the tax base of many areas already suffering from shifting economic priorities, further accelerating the urban decline so characteristic of the past ten years. The need to spur economic recovery by reinvestment in, and redevelopment of, former industrial properties has required a reevaluation of priorities and identified the need to strike a balance between environmental and economic health. One of the simplest methods to accomplish this goal is to reexamine and refine the methods by which the risks were identified in the past.

The techniques of risk assessment, first developed in the early 1980s, were understandably conservative, created as they were in the face of much uncertainty and an immediate need for an assessment tool. However, many practitioners forgot or never learned that risk assessment was intended to be a dynamic, technology-forcing process in which default-assumptions and methodologies would be supplanted by new knowledge and understanding of exposure and response to trace environmental contaminants. Because most of such advancements, as expected, resulted in lowering the calculated-risk estimates, attempts to incorporate these changes were met with regulatory resistance, either because they conflicted with codified protocols or because they were viewed as being less protective of the public health and environment. In fact, the risks had not changed, only the information used in the risk assessment.[65]

The Brownfields Initiatives now being developed by both the USEPA[66] and state regulatory agencies[67] seek to link the goals of remediation with those of redevelopment. Successful programs must, therefore, function to reduce the costs and time needed to effect cleanups while still maintaining adequate levels of protection for health and the environment. Risk assessment in the form of RBCA is the only tool available to accomplish these ends.

Such assessments must employ the best available science and most current techniques to reduce the uncertainty around the risk estimates. These techniques include (1) a risk-based sampling plan, to understand the spatial distribution of contaminants and the potential for contact, (2) analysis of background levels of compounds of concern, (3) incorporation of environmental half-lives and other physico-chemical parameters that influence contaminant persistence and behavior, (4) use of exposure parameters and site-specific and use-specific scenarios that reflect plausible future land use, including the incorporation of Monte Carlo techniques to illustrate both the range and likelihood of the various risk estimates, (5) incorporation of pharmacokinetic parameters, like extractability and bioavailabilty, to understand the relationship between the residues in the environment and the internal dose, (6) use of the most up-to-date and pertinent toxicity information and extrapolation techniques to assess the significance of any exposures, and (7) use of alternative target risk or cleanup levels for different exposure scenarios and an appreciation for the need to identify clearly the uncertainties involved, their likely influence on the outcome of the assessment, and the most important sources of the uncertainty. This approach, in turn, suggests the information and experiments needed to improve on the estimates. Beyond improvements in the estimates, however, must come a commitment on the part of all stakeholders to adopt a cooperative attitude, allowing the goal of timely and cost-effective remediation and redevelopment to take precedence over the process. This requires a willingness to work together and consider innovative approaches to RBCA as a legitimate and useful tool in solving the brownfields problem.

Notes

1. RACHEL CARSON, SILENT SPRING (1962).
2. Vincent T. Covello & Jeryl Mumpower, *Risk Analysis and Management: A Historical Perspective*, 5 RISK ANALYSIS 103–120 (1985).
3. David L. Eaton & Curtis D. Klaassen, *Principles of Toxicology, in* TOXICOLOGY: THE BASIC SCIENCE OF POISONS 13 (1996).
4. Covello & Mumpower, *supra* note 2.
5. NAT'L ACADEMY SCIENCES, RISK ASSESSMENT IN THE FEDERAL GOVERNMENT: MANAGING THE PROCESS (1983).
6. *Id.* at 18.
7. *Id.* at 18.
8. NAT'L RESEARCH COUNCIL, SCIENCE AND JUDGMENT IN RISK ASSESSMENT (1994).
9. Eaton & Klaassen, *supra* note 3, at 26.
10. Dennis J. Paustenbach, *A Survey of Environmental Risk Assessment, in* THE RISK ASSESSMENT OF ENVIRONMENTAL AND HUMAN HEALTH HAZARDS 27 (1989).
11. Frederick R. Johannsen, *Risk Assessment of Carcinogenic and Non-carcinogenic Substances*, 20 CRIT. REV. TOXICOL. 341–47 (1990).
12. 51 Fed. Reg. 33,992–34,003 (1986); 51 Fed. Reg. 34,042–34,054 (1986); 51 Fed. Reg. 34,028–34,041 (1986); 51 Fed. Reg. 34,004–34,027 (1986).
13. U.S. EPA, RISK ASSESSMENT GUIDANCE FOR SUPERFUND, HUMAN HEALTH AND EVALUATION MANUAL (PART A), INTERIM FINAL, OFFICE OF EMERGENCY & REMEDIAL RESPONSE, 540/1-89/002 (1989).
14. Dennis J. Paustenbach, *Health Risk Assessments: Opportunities and Pitfalls*, 14 COLUM. J. ENVTL. L. 379–410 (1989).
15. Ross L.M. MacDonald, An Overview of Risk-Based Corrective Action for Petroleum Release and Waste Site, Address at New Techniques in Risk Assessment—State of the Art Methods and Future Directions (Feb. 14–16, 1996).
16. LAWRENCE LIVERMORE NAT'L LABORATORY, CALIFORNIA LEAKING UNDERGROUND FUEL TANKS (LUFT) HISTORICAL CASE ANALYSIS (1995).
17. AMERICAN SOCIETY FOR TESTING & MATERIALS, GUIDE TO RISK-BASED CORRECTIVE ACTION AT PETROLEUM RELEASE SITES (E1739-95) (1995).
18. STEVEN J. MILLOY, SCIENCE-BASED RISK ASSESSMENT (1995).
19. Paustenbach, *supra* note 14.
20. Hansford T. Shacklette & Josephine G. Boerngen, Element Concentration in Soils and Other Surficial Materials of the Conterminous United States (1984) (U.S. Geol. Survey Prof. Paper No. 1270).
21. Patrick J. Sheehan et al., *Assessment of Human Health Risks Posed by Exposure to Chromium Contaminated Soils*, 32 J. TOXICOL. ENVTL. HEALTH 161–201 (1991).
22. AGENCY TOXIC SUBSTANCES & DISEASE REGISTRY, TOXICOLOGICAL PROFILE FOR BARIUM U.S. DEP'T HEALTH & HUMAN SERV., PUB. HEALTH SERV. (1991).
23. D. Barltrop & F. Meeks, *Absorption of Different Lead Compounds*, 51 POSTGRADUATE MED. J. 805–809 (1975); D. Barltrop & F. Meeks, *Effects of Particle Size on Lead Absorption from the Gut*, 34 ARCH. ENVTL. HEALTH 280–85 (1979).

24. AGENCY TOXIC SUBSTANCES & DISEASE REGISTRY, PUBLIC HEALTH ASSESSMENT GUIDANCE MANUAL, U.S. DEP'T HEALTH & HUMAN SERV., PUB. HEALTH SERV. (1992).

25. AMERICAN CHEM. SOC'Y, PRINCIPLES OF ENVIRONMENTAL SAMPLING (Lawrence H. Keith ed., 1988).

26. Richard J. Wenning & Gerald A. Erickson, *Interpretation and Analysis of Complex Environmental Data Using Chemometric Methods*, 13 TRENDS ANAL. CHEM. 446–457 (1994); Lyle G. Bruce & Gene W. Schmidt, *Hydrocarbon Fingerprinting for Application Forensic Geology: Review with Case Studies*, 78 AAPG BULL. 1692–1710 (1992).

27. 57 FED. REG. 22,888-22,938 (1992).

28. Id.; EXPOSURE ASSESSMENT GROUP, OFFICE HEALTH & ENVT'L ASSESSMENT, U.S. EPA, EXPOSURE FACTORS HANDBOOK EPA/600/T-95/002A (1995).

29. Paustenbach, *supra* note 14.

30. L. Daniel Maxim, *Problems Associated with the Use of Conservative Assumptions and Exposure and Risk Analysis*, in THE RISK ASSESSMENT OF ENVIRONMENTAL HUMAN HEALTH HAZARDS 526 (Dennis J. Paustenbach ed., 1989); David E. Burmaster & Robert H. Harris, *The Magnitude of Compounding Conservatisms in Superfund Risk Assessments*, 13 RISK ANALYSIS 131–34 (1993).

31. Renate Kimbrough et al., *Health Implications of 2, 3, 7, 8-TCDD Contamination of Residential Soil*, 14 J. TOXICOL. ENVTL. HEALTH 47–93 (1984).

32. John K. Hawley, *Assessment of Health Risk from Exposure to Contaminated Soil*, 5 RISK ANALYSIS 289–302 (1985).

33. Edward J. Calabrese et al., *How Much Soil Do Young Children Ingest: Epidemiological Study*, 10 REG. TOXICOL. PHARMICOL. 123–37 (1989).

34. USEPA, *supra* note 28.

35. NAT'L ACADEMY SCIENCES, *supra* note 5.

36. Paustenbach, *supra* note 14.

37. AMERICAN CHEM. SOC'Y, FATE OF CHEMICALS IN THE ENVIRONMENT (Robert L. Swan & Alan Eschenroeder eds., 1983).

38. Martin Alexander, *How Toxic Are Chemicals in Soil?*, 29 ENVTL. SCI. TECH. 2713–17 (1995).

39. Michael V. Ruby et al., *Lead Bioavailability under Simulated Gastric Conditions*, 26 ENVTL. SCI. TECH. 1242–48 (1992).

40. Brent L. Finley & Susan D. Horowitz, *Using Human Sweat to Extract Chromium from Chromite-Processing Residue: Applications to Setting Health-Based Clean-Up Standards*, 40 J. TOXICOL. ENVTL. HEALTH 585–99 (1993).

41. AMERICAN CHEM. SOC'Y, *supra* note 25.

42. Richard A. Anderson et al., *Designing a Biological Monitoring Program to Assess Community Exposure to Chromium: Conclusions of an Expert Panel*, 40 J. TOXICOL. ENVTL. HEALTH 555–83 (1993).

43. Edward L.J. Baker et al., *Metabolic Consequences of Exposure to Polychlorinated Biphenyls in Sewage Sludge*, 112 AM. J. EPIDEMIOL. 553–60 (1980).

44. Roy L. Smith, *Use of Monte Carlo Simulation for Human Exposure at a Superfund Site*, RISK ANALYSIS, Vol. 14 433–40 (1994).

45. Brent L. Finley et al., *Evaluating the Adequacy of Maximum Contaminant Levels as Health Protective Clean-Up Goals: An Analysis Based on Monte Carlo Techniques*, 18 REG. TOXICOL. PHARMICOL. 443–55 (1993).

46. U.S. EPA *supra* note 28.

47. OHIO EPA GENERIC NUMERIC STANDARDS/RISK ASSESSMENT RULES, OHIO VOLUNTARY ACTION PROGRAM (1996).

48. U.S. EPA Integrated Risk Information System (1995); Office Emergency & Remedial Response, U.S. EPA Health Effects Assessment Summary Tables, Annual FY 95 (1995).

49. Robert L. Sielken, *Cancer Dose-Response Extrapolations*, 21 Envtl. Sci. Tech. 1033–39 (1987).

50. *Id.*

51. Robert James, Using Human Data to Estimate Cancer Risk, Address at New Techniques in Risk Assessment—State of the Art Methods and Future Directions (Feb. 14–16, 1996).

52. Sielken, *supra* note 47.

53. Brent L. Finley, et al., *An Alternative to the USEPA's Proposed Inhalation Reference Concentrations for Hexavalent and Trivalent Chromium*, Vol. 16 Reg. Toxicol. Pharmicol. 161–76 (1992).

54. Nat'l Academy Sciences, *Pharmacokinetics in Risk Assessment, in* 8 Drinking Water and Health, (1987).

55. Melvin E. Andersen et al., *Physiologically Based Pharmacokinetics and the Risk Assessment Process for Methylene Chloride*, 87 Toxicol. Appl. Pharmcol. 185–207 (1987).

56. Kathryn E. Kelley & Nanette C. Cardon, The Myth of Ten to the Minus Six and the Definition of Acceptable Risk (or "In Pursuit of Superfunds Holy Grail"), Address at the 84th Annual Meeting of the Air and Waste Management Association (June 1991).

57. William J. Cronin et al., *A Trichloroethylene Risk Assessment Using a Monte Carlo Analysis of Parameter Uncertainty in Conjunction with Physiologically-Based Pharmacokinetic Modeling*, 15 Risk Analysis 555–66 (1995).

58. Finley & Horowitz, *supra* note 40.

59. Ohio EPA, *supra* note 47.

60. Office Research & Dev., USEPA, Provisional Guidance for Quantitative Risk Assessment for Polycyclic Aromatic Hydrocarbons, EPA/600/R-93/089 (1993).

61. Office Emergency & Remedial Response, U.S. EPA, Guidance Manual for the Integrated Exposures Uptake Biokinetic Model for Lead in Children (version 0.99D), EPA/540/P-93/081 (1994).

62. Finley & Horowitz, *supra* note 40; Ruby et al., *supra* note 39.

63. USEPA, *supra* note 60.

64. Jaroslav Polak et al., *Evaluating Lead Bioavailability Data by Means of a Physiologically Based Lead Kinetic Model*, 29 Fund. Appl. Toxicol. 63–70 (1996).

65. Dennis J. Paustenbach, *The Practice of Health Risk Assessment in the United States (1975-1995): How the U.S. and Other Countries Can Benefit from that Experience*, 1 Human Ecol. Risk Assess. 29–79 (1995).

66. Office Emergency & Remedial Response, USEPA, Superfund Accelerated Cleanup Model, EPA/540/P-93/081 (1994).

67. Ohio EPA, *supra* note 47.

CHAPTER 18

The Risk-Based Corrective-Action Process

JAMES R. ROCCO
LESLEY HAY WILSON

Corrective action has traditionally been a sequence of steps taken at a site,[1] where hazardous substances or petroleum have been released, to restore the site to a prerelease condition. The corrective-action steps include site assessment, risk assessment, design and installation of remedial-action equipment, operation and maintenance of remedial-action equipment, and progress monitoring. Under many current regulatory programs, each step in the process must be completed and approved by the regulatory agency before the next step can begin. This has resulted in each step taking on a life of its own, with objectives existing for the sake of that step, disconnected from the overall corrective-action objectives. Under these circumstances, site assessments alone have taken years to complete, and remedial action—to levels acceptable to the regulatory agencies—has taken even more time. Unfortunately, the complexities and costs of the traditional corrective-action process have left many properties[2] abandoned or in perpetual corrective action.

Recently, however, there has been growing recognition that the traditional corrective action process must be modified to encourage faster and less costly corrective actions. The most promising solution that has emerged is a more cost-effective, site-driven approach that incorporates and integrates site-specific, risk-based decision making into the corrective-action process. This chapter will discuss the evolution of the traditional corrective-action process to a site-driven process.

The Traditional Corrective-Action Process

To put the site-driven paradigm in context, it is important to understand the evolution of the traditional corrective-action process. By way of illustration, discussing the corrective-action process as it relates to the underground storage tank (UST) regulatory program is helpful.[3] Corrective action at UST sites had been considered relatively simple and not as complex or challenging as that at larger industrial properties. However, experience since the promulgation of the federal UST regulations has proven UST corrective action to be as complex and challenging as other corrective-action activities (and maybe more so because of the large number of properties regulated by this program). This illustration is not intended to be a criticism of the program or the regulatory approaches that have been taken. In fact, the UST corrective-action program is probably one of the most successful federal environmental regulatory programs. The significance and magnitude of the UST corrective-action program has brought corrective action practitioners to the approaches and issues that are being addressed today in Voluntary Cleanup Programs in many states.

The Federal UST Program

In December 1988, the United States Environmental Protection Agency (USEPA) regulations governing USTs became effective. These regulations were far-reaching, impacting over 1,200,000 USTs and hundreds of thousands of businesses of all sizes.[4] Even more significant is that these regulations were not applied only to operating UST systems, but also to sites having abandoned or former USTs. Before the promulgation of the UST regulations, USTs were considered to be an asset and in many cases were transferred with a property upon a sale. Today, USTs are often considered a liability and the cause of many delays in the sale and redevelopment of property.

A significant part of the regulations is the implementation of corrective action for releases at UST sites. The regulations require that where a release from a UST system has been confirmed, corrective action must be conducted to "provide adequate protection of human health, safety, and the environment" and the corrective actions taken must "consider: (1) the physical and chemical characteristics of the regulated substance, including toxicity, persistence, and potential for migration; (2) the hydrogeologic characteristics of the facility and the surrounding land; (3) the proximity, quality, and current and future uses of groundwater; (4) an exposure assessment; and (5) the proximity, quality, and current and future uses of surface waters."[5]

The federal UST regulations were designed to be implemented on a state level, resulting in a wide variety of approaches to the program implementation. However, commonalties exist among the state regulations. Corrective-action regulations typically have been very prescriptive, with significant (and in some cases frequent) regulatory-agency oversight (Figure 1). Generally, once a release is confirmed, a site assessment is conducted to define the full extent of the release. Under the traditional corrective-action process, "full extent" usually means installing groundwater monitoring wells and collecting soil and groundwater samples around the area(s) where the release occurred and then stepping out until a "zero line"[6] can be drawn around the entire site. The assessment is submitted to the state regulatory agency for review and approval, and in many cases several iterations are required before final approval is received.

Activities such as site assessment have become processes unto themselves with their own starting and ending points, disconnected from the overall corrective-action objectives. Once there is approval of the site assessment,

FIGURE 1. Traditional Stairway to Corrective Action

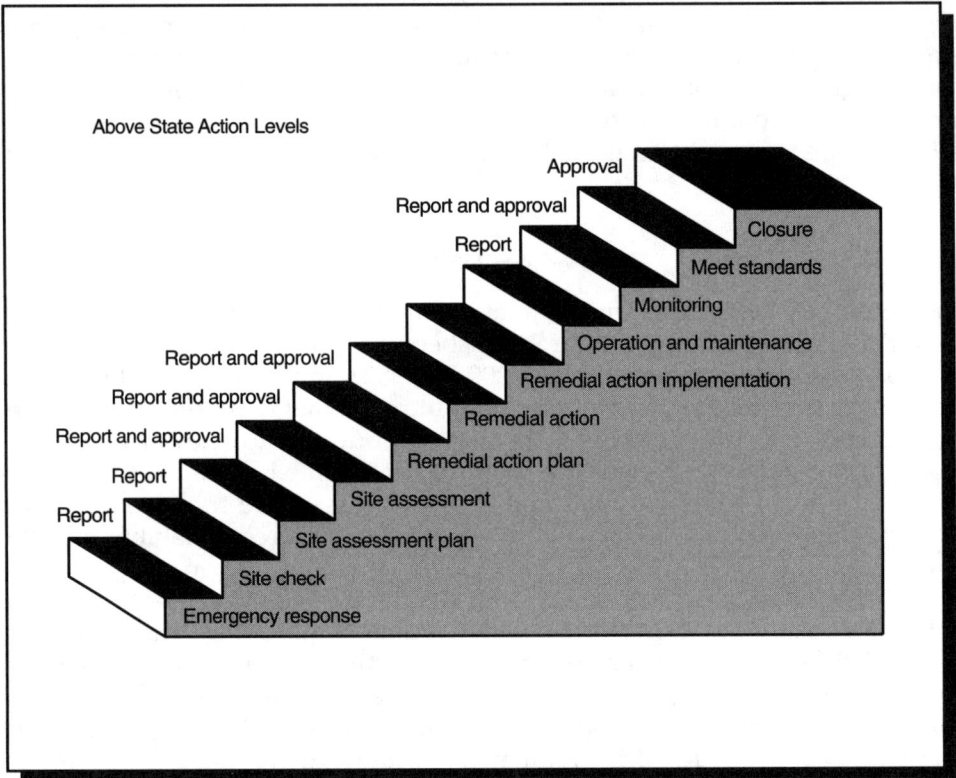

a remedial action plan is developed and submitted to the state regulatory agency for review and approval. Again, final approval in many cases requires several iterations. Finally, after the remedial action plan is approved, the remedial action system is installed and operated until the state target levels[7] are met. Often, however, additional site assessment needs are defined during the development of the remedial action plan, further delaying the installation and start-up of the system.

Because of the uncertainties surrounding the corrective action performance standards (that is, corrective action target levels) for protection of human health and the environment, the state programs are typically designed to ensure protection under the most sensitive conditions (worst-case scenario) by applying a very conservative approach to all sites. As a result, corrective action target levels are generally set at very low concentrations, such as analytical detection limits or background levels. The term "contamination" has become synonymous with any detection of chemicals in the soil or groundwater. In many cases, technological limitations are not considered during the remedial-action decision, and remedial action continues until there is a demonstration that there are no technologies that could achieve the required target levels. Although the ending point appears to be well defined, the ability to achieve that point is not assured.

In addition to the technical regulations, USEPA also promulgated regulations requiring owners/operators of UST systems to demonstrate the ability to be financially responsible for corrective-action activities.[8] Unfortunately, commercial insurance was not generally available and the many small UST owners/operators at first were unable to make the financial responsibility demonstrations. To meet the UST financial responsibility requirements and to pay for the corrective action activities, state UST funds were created. As money became available through the UST funds, hundreds of thousands of dollars (and millions of dollars nationwide, with cost estimates to complete corrective action in the billions of dollars) were spent on small sites to achieve the state corrective-action target levels. To a large extent, the creation of the UST funds minimized the economic impact of extensive corrective action and allowed cost considerations to become secondary to completing the corrective action as quickly as possible.

Another compounding factor is that state "no further action" (NFA) letters have become essential to complete a property transfer for a former or current UST location. The former or current existence of a UST at a property stigmatizes the property as "contaminated," often times regardless of the actual condition of the UST or the property. Buyers and lenders are making significant demands on property owners concerning corrective action. Property transfers for these sites are dependent on the completion of corrective-action activities. Sellers want to complete corrective action quickly so they can complete sales.

Problems with the Federal Approach

Given the development of the UST corrective-action program, the problems have become significant: a large number of sites require corrective-action under current state regulation, state regulatory agency and owner/operator resources (both financial and manpower) are limited, state UST funds are overextended and insolvent, property transfers are stalled, corrective-action costs are substantial (averaging $50,000 to $500,000 for one site), and the progress toward completion of corrective action is slow. These problems have prompted many people to suggest that significant change is necessary.

The primary driving force for change in the state UST corrective-action programs is the insolvency or impending insolvency of many of the state UST funds, resulting from the high costs of corrective action and the pressure from property owners, buyers, and lenders for NFA decisions from the states. However, the fundamental issue is not how to avoid corrective action, but rather, how to meet the objectives of the federal UST corrective-action regulations within the limitations of available funding and manpower.

The overriding solution is in modifying the corrective-action process to allow resources to be directed to their highest and best use through the effective and efficient use of technology and science. These modifications to the

UST corrective-action process will apply not only to corrective action required under the federal UST regulations, but also to any corrective-action program—whether required by regulation or initiated voluntarily by a property owner.

Lessons Learned in the Development of the Corrective-Action Process

The solution can be found in the lessons learned over the past seven years of UST corrective action. What has been learned? First, not all sites require the same level of effort or pose the same level of risk. Therefore, the same level of effort or technology should not (and cannot) be applied to all sites. Second, there are limitations to technology. The effectiveness and efficiency of technologies vary from site to site. Technologies change—they are improved and new technologies are developed. Yet, there is no one technology that is the ultimate corrective-action panacea. Third, the objectives of the program must be understood and the uncertainties and conservatism that are built into the regulatory programs must be defined. Uncertainties and conservatism can be overcome once they have been identified. Excessive conservatism is a deterrent to voluntary and cost-effective corrective action. Fourth, more information will not always make the answer more obvious. Decisions must be made by qualified individuals, based on reasonably available or obtainable information. Finally, the cost of corrective action should be judged on the full life cycle of the corrective action, not on individual steps or activities. Increases in the cost of some activities may ultimately reduce the overall cost of the project. As a simple example, collecting soil-characteristic data or conducting simple aquifer tests during the site assessment may allow a better review of remedial options and minimize the need for more extensive pilot studies or duplicate data collection.

Defining Corrective Action

To many, corrective action is synonymous with remedial action. Therefore, before going further, it is important to define corrective action. Corrective action is a process that integrates a series of activities (including site assessment, remedial action, risk assessment, exposure assessment, and monitoring), typically accomplished over a period of time in stages or with the possibility of repetition. Those activities evaluate a release of hazardous substances or petroleum, determine levels of chemicals of concern[9] that are protective of human health and the environment, and, when appropriate, remove, reduce, or control concentrations of chemicals of concern in the environment (soil and/or groundwater) to a level that is protective of human health and the environment. Remedial action defines a group of activities implemented as part of a corrective action to remove, reduce, or control the concentrations of chemicals of concern at a site, when concentrations of chemical(s) of concern are above levels that are protective of human health and the environment.

FIGURE 2. Corrective-Action Puzzle

As an analogy, corrective action can be viewed as a puzzle composed of a number of pieces (Figure 2). Each puzzle creates a unique picture. Each piece of each puzzle will fit in only one location and must be the right size and shape. However, the puzzle is incomplete without all the pieces and all the pieces of the puzzle are necessary to hold the other pieces in place. As with a puzzle, each corrective action presents a different picture. The activities conducted during the corrective action, like pieces of the puzzle, come in all sizes and shapes and will be different for each corrective action. As an example, data collection will be focal to the corrective action; however, the amount and extent of data collection will be dependent on the site.

Corrective action is an iterative process—a series of loops that are interconnected. Information or data collection is the core or common thread for the corrective-action process. The quantity and quality of data collected will be dependent on the objectives of the corrective-action activity, the site conditions, and the information necessary to perform the evaluation (for example, fate and transport analyses, or pilot studies). Data collection is an activity that continues throughout the corrective-action process.

FIGURE 3. The Corrective Action Process

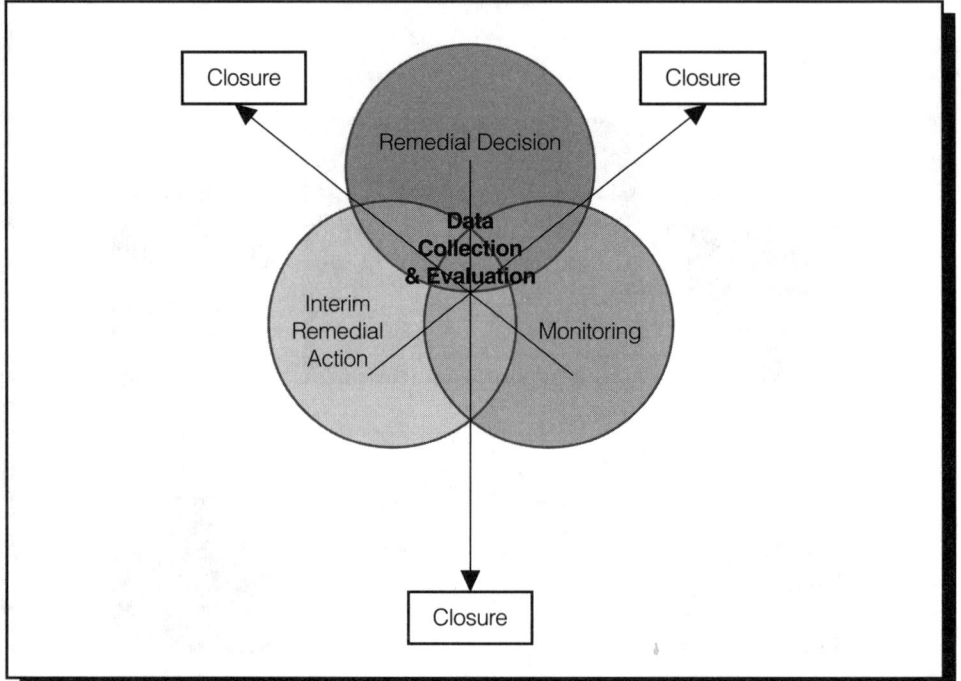

Initial data collection may be qualitative, identifying the conditions surrounding a site. Quantitative data collection will typically begin with the initial site assessment and continue through remedial action, monitoring, and closure. Full delineation of a chemical of concern should be based on the target levels and physical constraints associated with the site, not the identification of every molecule of a chemical that may exist. Decisions on further data collection, remedial action, or closure should be made based on the information collected, when the data justifies the decision.

Interim remedial actions should be taken as the need is identified. Closure can be achieved through a combination of remedial actions, interim remedial actions, or monitoring. Monitoring should be used to validate decisions and assumptions made during the corrective-action process.

Corrective action must also be a collaboration or team effort among the property owner, the regulatory agency, other stakeholders, and the contractors/consultants. The collaboration should identify the mutual objectives for the corrective action, identify the roles each individual is to play in the process, obtain the facts necessary to define the objectives, and establish trust among the members of the team. An effective collaboration will keep the appropriate individuals involved and minimize the second guessing and the desire for extensive regulatory oversight.

The Move toward Risk-Based Corrective Action

To address the needed changes and to develop a standard framework for integrating the traditional components of corrective action with risk and exposure assessment practices, a multifunctional and multidisciplinary group of American Society for Testing and Materials (ASTM) subcommittee members representing diverse interests developed ASTM E-1739, "Guide to Risk-Based Corrective Action for Petroleum Release Sites,"[10] commonly referred to as RBCA (pronounced "rebecca"). RBCA is a streamlined and consistent approach in which exposure and risk-assessment practices are integrated with traditional components of the corrective-action process. The objective of RBCA is to ensure that appropriate and cost-effective remedies that are technically defensible and protective of human health and the environment are selected. The RBCA approach is intended to be flexible and to provide state regulators with a framework that can be modified to meet the needs of individual states. In addition, the RBCA approach intends that the state regulatory agency define the process in a manner that allows the responsible party to conduct corrective action with minimal regulatory oversight.

Tiered Approach to Corrective Action

Initial Site Assessment and Site Classification

The RBCA process is a tiered approach that begins with an initial site assessment, with initial data-collection efforts focused on determining the potential risk posed by the presence and migration of chemical(s) of concern. An important objective of the initial site assessment is to identify and evaluate exposure pathways[11] to determine whether a complete pathway exists. This evaluation includes identifying current and reasonably potential future land use, groundwater use, current and reasonably potential future receptors, potential source area(s),[12] potential complete exposure pathways,[13] potential chemicals of concern, concentrations of chemicals of concern in the environmental media,[14] and site media characteristics.

As information is collected during the initial site assessment, the site is classified based on the immediate potential for harm to human health and the environment. The site classification relates specific site conditions to appropriate actions that should be taken to mitigate the potential for harm from the identified chemicals of concern. The actions taken as a result of the site classification are referred to as interim response actions. The value of the site classification is the identification of any potential for harm or concerns posed by chemicals of concern early in the site-assessment process, and the quick implementation of appropriate interim response actions.

Tier 1

When complete exposure pathways have been identified, the concentrations of chemicals of concern collected during the initial site assessment are compared with risk-based screening levels (RBSLs) for the complete pathways. The RBSLs are conservative, generic concentrations of chemicals of concern that are developed for each potential exposure pathway based on exposure routes,[15] generic site characteristics, and generic exposure factors.[16] The list of RBSLs is referred to as a Tier 1 Look-Up Table.

The RBSLs can also include concentrations of chemicals of concern, such as maximum contaminant levels (MCLs) developed under the Safe Drinking Water Act[17] or other non-site-specific concentrations. The RBCA standard envisions that a state agency would develop a table of RBSL concentrations that would apply across the state. Many states currently have defined generic or statewide standards as their corrective-action goals. However, these standards are typically limited to a few exposure pathways, are based on very conservative assumptions, and are often based on the analytical method detection limits or background concentrations. If the state does not develop RBSL concentrations, the user could develop its own RBSL Look-Up Table. Ultimately, the user is responsible for determining that the RBSLs used are applicable for the conditions of the specific site.

The comparison of concentrations of chemicals of concern collected during the initial site assessment with the Tier 1 RBSL Look-Up Table is referred to as a Tier 1 analysis. The Tier 1 analysis typically assumes that the source area(s), point(s) of exposure,[18] and the point(s) of compliance[19] are all in the same location. In addition, the highest site concentration of the chemical(s) of concern for the comparison with the RBSLs are used for this comparison. This results in a very conservative evaluation.

The Tier 1 analysis is also limited to the comparison of the site concentrations of the chemical(s) of concern with the RBSLs for the exposure pathways that have been determined to be complete. If the site concentrations of the chemical(s) of concern are below the RBSLs for a complete exposure pathway, then no further action is necessary for that pathway. If the site concentrations of the chemical(s) of concern are above the RBSLs for a complete exposure pathway, then further evaluation is warranted.

Tier 2

Options for further evaluation include (1) conducting a Tier 2 analysis to determine site-specific target levels (SSTLs) for the exposure pathways identified for further evaluation, (2) conducting interim remedial actions to remove or reduce concentrations of the chemical(s) of concern at the source area(s) or to eliminate an exposure pathway, (3) conducting remedial actions using the RBSLs as the corrective-action target level, or (4) a combination of these options. The option(s) selected will be dependent on the appropriateness of the assumptions used to derive the RBSLs and the potential for further site-specific evaluation to develop SSTLs substantially different from the RBSLs. If no further action is determined for all complete exposure pathways, then no further action is required for the site.

Tier 2 and Tier 3 analyses are conducted for specific exposure pathways identified for further evaluation in a previous tier using site-specific characteristics. The SSTLs developed under the Tier 2 or Tier 3 analysis are concentrations of the chemical(s) of concern calculated by using site-specific information and evaluation of fate and transport of the chemical(s) of concern for the identified complete exposure pathways.

Under the Tier 2 analysis, SSTLs can be developed in several ways. First, SSTLs can be developed by replacing the default, generic assumptions used to develop the RBSLs with site-specific parameters. Second, SSTLs can be developed by applying the RBSLs at the actual point(s) of exposure and back-calculating,[20] the corresponding concentrations of the chemical(s) of concern at the source area(s) by accounting for natural attenuation.[21] Finally, a statistical representation of the source area(s) concentration(s) of the chemical(s) of concern can be compared with SSTLs or RBSLs, as applicable.

As with Tier 1, if the site concentrations of chemicals of concern are below the SSTLs for a complete exposure pathway, then no further action is necessary for that pathway. If the site concentrations of chemical(s) of concern are above the SSTLs for a complete exposure pathway, then further evaluation is warranted. Options for further evaluation include (1) conducting a Tier 3 analysis to determine SSTLs for the exposure pathways identified for further evaluation, (2) conducting interim remedial actions to remove or reduce concentrations of chemicals of concern at the source areas or to eliminate an exposure pathway, (3) conducting remedial actions using the SSTLs as the corrective-action target levels, or (4) a combination of these options. The option(s) selected will be dependent on the appropriateness of the assumptions used to derive the SSTLs and the potential for further site-specific evaluation in Tier 3 to develop SSTLs substantially different from the SSTLs developed under Tier 2. If no further action is determined for all complete exposure pathways, then no further action is required for the site.

Tier 3

If a Tier 3 analysis is determined to be appropriate, a more complex and sophisticated evaluation of chemical fate and transport using site-specific numerical models, Monte Carlo analysis, or other sophisticated analytical tools is conducted to determine the SSTLs. This is typically a costly evaluation, requiring sophisticated resources. As with Tier 2, if the site concentrations of chemical(s) of concern are below the SSTLs for a complete exposure pathway, then no further action is necessary for that pathway. However, if the site concentrations of chemical(s) of concern are above the SSTLs for a complete exposure pathway, then the options are limited to (1) conducting interim remedial actions to remove or reduce concentrations of chemical(s) of concern at the source area(s) or to eliminate some or all of the exposure pathways and "address the most significant concerns"[22] to enable a reevaluation of the site classification and the Tier 3 analysis, (2) conducting remedial actions using the Tier 3 SSTLs as the corrective-action target levels, or (3) a combination of these options. No additional refinement of the assumptions or methods can be used to change the SSTLs. If no further action is determined for all complete exposure pathways, then no further action is required for the site.

Remedial Action

If remedial action is appropriate, then the method chosen should be based on its ability to achieve the RBSLs or SSTLs, as determined in the applicable Tier analysis, for a complete exposure pathway when the site concentrations of chemical(s) of concern are above the RBSLs or SSTLs. Remedial options include source removal, design and installation of cleanup equipment, natural attenuation processes, engineering controls,[23] and institutional controls.[24]

Monitoring Program

Finally, when no further action is determined, a monitoring program should be implemented to verify the assumptions and fate and transport conclusions used in the Tier 2 and Tier 3 analyses, and to ensure that the RBSLs or SSTLs, as appropriate, are met. When remedial action is appropriate, a monitoring program should be implemented to evaluate the progress of the remedial action. The monitoring program for remedial action should also include operation and maintenance of the remedial action equipment and engineering controls.

A Case Study of the RBCA Approach

The following case study of an RBCA evaluation is presented for illustration purposes. The case study is necessarily simplified for presentation here and is not intended to provide an understanding of the RBSL or SSTL calculations. Information contained in the ASTM E-1739-95 standard and its appendices is used for this evaluation.

Facts

A former service-station property is to be redeveloped as a fast-food restaurant. The property owner has agreed with the state agency to use the RBCA process to evaluate the need for, and scope of, remedial action in conjunction with the property redevelopment. The USTs, piping, and dispenser islands were removed a number of years earlier. Because gasoline was stored in the USTs at the former service-station property, benzene, toluene, ethylbenzene, and xylenes have been identified as the chemicals of concern likely to be present in the soil or groundwater if a release had occurred. For purposes of this case study, benzene will be the only chemical of concern discussed.

The property is located in a mixed commercial and residential area in which public water service is available, but hook-up to the public water is not mandated by the municipality (see Figure 4). Groundwater flows in the direction of the restaurant across Sunset Drive. Monitoring wells have been placed in the source area (MW1) and in Sunset Drive (MW2). A hypothetical drinking-water supply well has been identified across Sunset Drive in the residential area.

Initial Site Assessment and Site Classification

An initial site assessment has been conducted and concentrations of benzene were identified. Soil samples were taken during the installation of the moni-

FIGURE 4. Case Study Site Plan

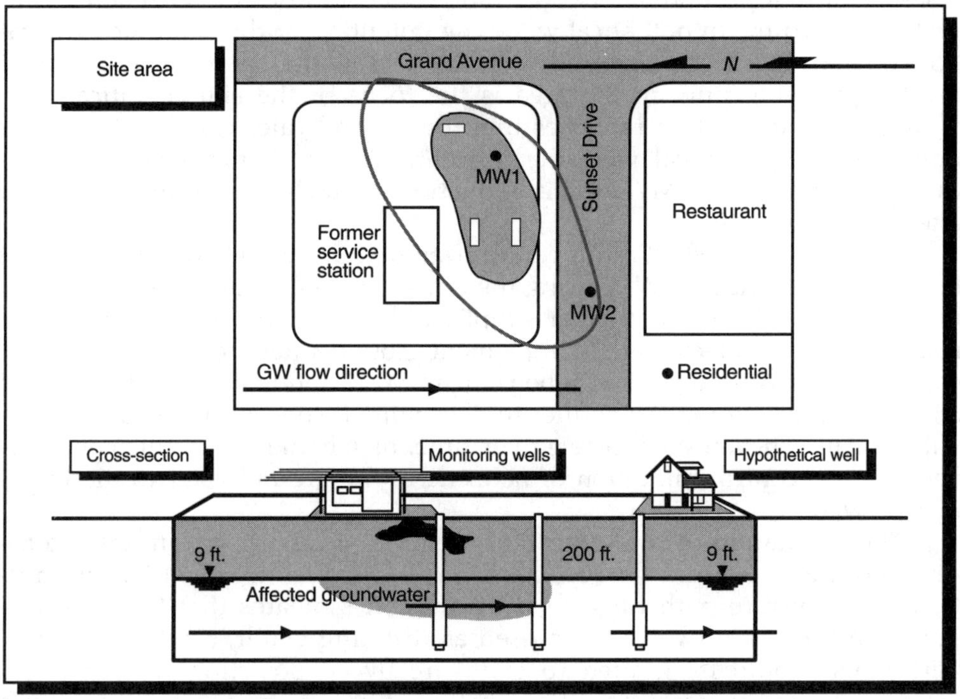

toring wells. Groundwater samples were taken after the monitoring wells were installed. The following additional information about the property was collected during the initial site assessment:

Groundwater Concentration Data for Benzene

Monitoring well (MW) 1	5000 parts per billion[25] (ppb)
Monitoring well (MW) 2	10 ppb

Soil Concentration Data for Benzene

Soil sample (SS) 1	1 part per million (ppm)
Soil sample (SS) 2	0.5 ppm

Site Data

Hydraulic conductivity (K)	60 feet per day
Groundwater gradient (I)	0.005 feet per feet
Groundwater velocity (V)	310 feet per year
Distance to point of exposure	200 feet
Depth to groundwater	9 feet
Soil type	Fine sands

No free-phase gasoline has been identified on the groundwater surface. However, it is apparent from the information collected during the initial site assessment that groundwater containing concentrations of benzene has migrated toward the hypothetical well. The potential for the continued migration of benzene in the groundwater to a well that may be used for a drinking-water supply in the area is the focus of the site classification.[26] Though the information indicates that the groundwater flows in the direction of the hypothetical well, the concentrations of benzene drop significantly from MW1 to MW2. This reduction is likely the result of natural attenuation processes.

Based on this reduction in concentration and the low concentrations of benzene measured in MW2 during the initial site assessment, there is no immediate threat to drinking-water supplies. Therefore, the appropriate initial response actions selected are to implement groundwater monitoring to verify continued low concentrations of benzene in MW2 and to evaluate the natural attenuation capacity of the aquifer to determine if concentrations of benzene will remain constant or decrease over time, or whether hydraulic control is necessary to control migration of benzene in the groundwater to the hypothetical well.

The evaluation of the information collected during the initial site assessment indicates that the source area is in the vicinity of MW1. In addition, for exposure pathways, the information indicates that (1) there is a potential for groundwater to be used as drinking water, (2) direct contact with soils should be limited to exposure by construction workers, and (3) inhalation of vapors should be evaluated based on the commercial use of the property (the proposed fast-food restaurant). These assumptions are important in determining the appropriate RBSLs to use in the Tier 1 analysis. Based on this evaluation, the potentially complete exposure pathways to be evaluated are (1) potential use of groundwater for drinking water from the hypothetical well at the residence in the near vicinity (groundwater ingestion), (2) exposure of construction workers to benzene in soils during the proposed development (direct contact), (3) vapor migration from the source area (both from groundwater and from soil) to indoor air at the proposed fast-food restaurant (inhalation), and (4) soils leaching benzene to groundwater, providing an ongoing source to the groundwater ingestion pathway identified in (1) above (soil-to-groundwater leaching). All other exposure pathways were determined to be incomplete or not applicable to the site.

Tier 1

Once the potentially complete exposure pathways have been identified, and using the information developed during the initial site assessment, a comparison of groundwater and soil benzene concentrations with the state RBSL Look-Up Table[27] is made for each exposure pathway as follows:

Exposure Pathways	State Benzene RBSL	Site Benzene
Groundwater		
Ingestion	5 ppb	5000 ppb
Inhalation	70 ppb	5000 ppb
Soil		
Inhalation	0.011 ppm	1 ppm
Soil-to-groundwater leaching	0.017 ppm	1 ppm
Direct contact	10 ppm	1 ppm

Based on the Tier 1 RBSL Look-Up Table comparison, the site concentrations for benzene are above the RBSL for the groundwater ingestion pathway, the soil and groundwater inhalation pathways, and the soil-to-groundwater leaching pathway. These pathways will be further evaluated in a Tier 2 analysis. The site concentrations for benzene are below the RBSL for the direct contact pathway. Therefore, the direct contact pathway with soils during construction for the property redevelopment does not pose an unacceptable risk. No further action is necessary for this pathway and it is not carried through to the Tier 2 analysis.

Tier 2

Tier 2 analysis for the exposure pathways with concentrations of benzene above the RBSL is appropriate in this case. It is likely that further evaluation of the natural attenuation process in the soil and groundwater will result in a SSTL substantially higher than the RBSL, possibly eliminating some or all of the remaining exposure pathways from further consideration. Once Tier 2 analysis has been determined to be appropriate, the next step in the process is to conduct further site assessment to collect additional site data to calculate the appropriate SSTL for soil and groundwater for the identified exposure pathways.

The additional site assessment selected for this property will focus on (1) determining whether the existing site conditions will result in concentrations of benzene vapors present in the area of the proposed building and (2) collecting information necessary to evaluate natural attenuation at the site. A soil-vapor survey[28] is conducted in the area surrounding the former service-station building and in the area of the proposed fast-food restaurant building to determine if concentrations of benzene in the soil vapor are present in these areas. No detectable levels of benzene are recorded during this survey and the vapor migration pathways for soil and groundwater are therefore empirically documented to be incomplete.

To evaluate natural attenuation, groundwater samples are also collected from the site monitoring wells and analyzed for dissolved oxygen. The reported results indicate that dissolved oxygen levels outside the area containing the concentrations of the chemicals of concern are higher than the dissolved oxygen levels within the area of highest concentration. This is an indication that aerobic biodegradation is occurring in the groundwater. Soil samples are also collected for the site and analyzed for their fraction of organic carbon content (f_{oc}). This value is important in determining the magnitude of the natural tendency of the soils to retain organic materials within the soil, rather than allow them to migrate to the groundwater.

As stated earlier, the Tier 1 evaluation assumes that the source area(s) and point(s) of exposure are in the same location or, in other words, that the groundwater in the source area at the site would be used for drinking water, and the concentration of benzene in the groundwater throughout the site must be below the RBSL for groundwater ingestion. For the Tier 2 analysis, the point of exposure is assumed to be at the location of the hypothetical well and the RBSL will be applied at this point. It is important to note that the location of the point of exposure could vary based on the regulatory requirements. Ideally, the location of the point(s) of exposure should be at an actual, or reasonably potential, point where exposure to the chemical(s) of concern will likely occur. However, if circumstances warranted an alternate location, the point of exposure for the groundwater ingestion pathway could be placed at the former service-station property line, the residence property line, or a selected distance from the source area. Regardless of the location of the point(s) of exposure, the process and calculations are the same.

The SSTL for benzene in the source area that will result in concentrations of benzene below the RBSL at the point of exposure can be calculated. This calculation is accomplished through a fate and transport evaluation (in this case study, the Domenico solution,[29]) using site-specific measured or estimated values (for example, hydraulic conductivity, effective porosity, f_{oc}, an estimate of the size of the source area or the distance to the point of exposure) to predict the allowable site concentrations of benzene in the groundwater at the source area based on the allowable concentration of benzene (RBSL) at the point of exposure (the hypothetical well).

The SSTL for benzene in soil at the source area that will result in concentrations of benzene in the soil-to-groundwater leachate in the source area below the SSTL can also be calculated. Leachate from soil to groundwater is calculated by substituting site-specific data into the equations used for the RBSL calculations to predict the allowable concentrations of benzene in the soil at the source area based on the allowable concentrations of benzene in the groundwater at the source area (SSTL for groundwater calculated above). The calculated SSTL values are as follows:

Exposure Pathways	Tier 2 SSTL for Benzene	Site Benzene
	Groundwater	
Ingestion	840 ppb	5000 ppb
	Soil	
Soil-to-groundwater leaching	6.4 ppm	1 ppm

The resulting SSTL for benzene calculated for both exposure pathways during the Tier 2 analysis is considerably higher than the RBSL. However, the site concentration of benzene in the groundwater in the source area is above the Tier 2 SSTL for the groundwater ingestion pathway. The concentration of benzene in the soil in the source area is below the SSTL for the soil-to-groundwater leaching pathway. Because the soil-to-groundwater leaching pathway does not pose an unacceptable risk, no further action is necessary for this pathway.

However, for the groundwater ingestion pathway, based on the significant difference between the site concentration of benzene and the Tier 2 SSTL, it is unlikely that further analysis will either be cost-effective or result in a substantially higher SSTL. Therefore, remedial action is proposed to reduce groundwater concentrations in the source area to be protective of the hypothetical drinking-water well 200 feet down gradient of the property. The SSTL calculated for the drinking-water ingestion pathway during the Tier 2 analysis will be the target level to be achieved in the groundwater at the source area by the remedial action.

Remedial Action and Monitoring Program

Because the concentration of benzene in the soil-to-groundwater leachate is below the SSTL for that pathway, interim remedial action (for example, removal of soil with concentrations of benzene in the source area) is not appropriate. Alternatives to address reduction of the concentration of benzene in the groundwater are reviewed. These alternatives include a groundwater pump and treat system or an air sparging system. Each alternative is evaluated based on (1) its cost-effectiveness, (2) ability to achieve the SSTL in a timely manner and be protective of the hypothetical drinking-water well, and (3) its implementation feasibility, considering the property redevelopment plans. Installation of the remedial-action equipment is planned to be coordinated with the redevelopment of the property to minimize interference with the use of the property.

A monitoring program is initiated to develop a database to demonstrate that the site-specific natural attenuation rates are consistent with the rates predicted during the Tier 2 analysis, and to monitor the performance and effectiveness of the remedial-action system. In addition, an operation and maintenance program is implemented to ensure optimal performance of the remedial-action system.

The results of the Tier 1 and Tier 2 analyses along with the plans for remedial action and the monitoring program, are discussed and agreed upon with the regulatory agency before implementation. For the site redevelopment, a soil management plan is developed and implemented to address any soils removed during the construction of the fast-food restaurant that may require special handling. The excavated soils will be screened with an organic-vapor analyzer. Based on the screening results, soils will be segregated so that soils containing levels of benzene are removed from the site. In addition, the construction workers are to be properly trained to meet Occupational Safety and Health Administration requirements and use personal protective equipment during the construction phase.

Conclusion

In summary, as the implementation of RBCA, brownfields Voluntary Cleanup Programs, and Superfund reforms move forward, three principles for effective corrective action should be followed:

1. Challenge the traditional processes, technologies, and science. There is always a better way.
2. Create an environment that allows collaboration, leveraging individual involvement to obtain effective and efficient corrective action.
3. Look to the full life cycle of corrective action for cost-effective decisions.

Notes

1. A "site" is the area defined by the extent of migration of chemical(s) of concern.
2. "Property" is an area of land defined by the legal parcel boundaries.
3. Technical Standards and Corrective Action Requirements for Owners and Operators of Underground Storage Tanks, 40 C.F.R. § 280.
4. USEPA, OFFICE UNDERGROUND STORAGE TANKS (1994).
5. Corrective Action for UST Systems Containing Petroleum, 40 C.F.R. § 280 subpt. F (1994).
6. The "zero line" is the point where the released substance can no longer be detected in the soil or the groundwater. This has typically been based on the lowest detection limit achievable by the analytical method used.
7. "Target levels" refer to the concentration of a chemical that must be achieved by a corrective action. Other terms with similar meanings include action levels, cleanup levels, groundwater standards, or soil standards.
8. Financial Responsibility, 40 C.F.R. § 280 subpt. I.
9. "Chemical of concern" is a specific constituent of the hazardous substance or petroleum product released for evaluation during the site assessment and risk-evaluation process.
10. ASTM E-1739 was preceded by ASTM Emergency Standard ES-38, "Guide to Risk-Based Corrective Action at Petroleum Sites." The development of these guides is under the control of ASTM Subcommittee E-50.01.
11. An "exposure pathway" is the path a chemical takes from the area of the release to a point where an individual or population will come in contact with that chemical.
12. The "source area" is the location of the highest concentration of the chemical(s) of concern in the soil or groundwater or free-phase hazardous substance or petroleum.
13. A complete exposure pathway requires a source and mechanism for a chemical release into the environment, a transport medium (for example, soil or groundwater), a point of contact with a receptor, and an uptake route (for example, ingestion).
14. Common environmental media include soil, groundwater, and air.
15. Common exposure routes include ingestion, inhalation, and contact with the skin.

16. Common exposure factors include duration of exposure and frequency of exposure.

17. 42 U.S.C. § 300f (1992).

18. The "point of exposure" is the point at which an individual or population will come in contact with a chemical of concern resulting from a release at a site.

19. The "point of compliance" is a location selected between the source area(s) and the point(s) of exposure used to monitor conditions at a site to verify assumptions and measure progress.

20. A "back-calculation" is the calculation of the acceptable concentration of the chemical of concern at the source area(s) that will result in a concentration of the chemical of concern at the point(s) of exposure that is protective of human health and the environment.

21. "Natural attenuation" is the reduction in the concentration of a chemical of concern in the media with distance and time, due to processes such as dilution, dispersion, sorption, chemical degradation, and biological degradation.

22. ASTM E-1739-95, § 6.7.1.2.

23. "Engineering controls" are controls such as slurry walls, capping, or hydraulic control to eliminate or reduce the potential for exposure to a chemical of concern.

24. "Institutional controls" are the restrictions on use or access to a site to eliminate or reduce the potential for exposure to a chemical(s) of concern.

25. Parts per billion = micrograms per liter.

26. The example site classification system presented in the ASTM E-1739 document describes sample scenarios and corresponding response actions. The case study presented in this chapter uses the ASTM example description to determine the initial response actions that are to be taken at the location.

27. The state RBSL for benzene for the identified exposure pathways in the case study is taken from the example RBSL Look-Up Table presented in Appendix X-2 of the ASTM E-1739-95 standard. The benzene ingestion RBSL is the maximum contaminant level for benzene.

28. The soil-vapor survey is conducted using an organic-vapor analyzer that detects a large number of volatile organic compounds. If the levels of organic vapors measured by the analyzer are consistent with the background conditions in the soils, then the concentration of benzene must not be above detectable levels.

29. The Domenico solution calculates the reduction of concentrations of chemicals of concern in groundwater over a distance, taking into account retardation and other natural attenuation processes. The Domenico solution incorporates biodegradation of chemicals of concern in the aquifer. P.A. Domenico, *An Analytical Model for Multi-Dimensional Transport of a Decaying Contaminant Species*, 91 J. HYDROL. 49–58 (1987).

CHAPTER 19

Risk Assessment—A Physician's Introduction

RICHARD L. CRISTEA, M.D.

Many new state Voluntary Cleanup Programs rely on risk assessment to evaluate and set cleanup goals for brownfields sites. But, from a physician's perspective, what is a risk assessment? Risk assessments characterize the potentially deleterious effects of human exposure to environmental hazards.[1] Further, risk assessments use statistics as the basis for determining such risks.[2] This chapter will briefly define the different aspects of medical risk in simple terms, providing a foundation for understanding the medical basis of risk assessments.

The Jargon of Health Risk Assessment

Risk-Assessment Terms

Health risk assessment is the process through which toxicology data collected from animal studies and human epidemiology are combined with information about the degree of exposure to a toxic substance to predict quantitatively the likelihood that a particular adverse response will be seen in a specific human population. Epidemiology is the study of the relationships between the various factors that determine the frequency and distribution of diseases in human and other animal populations.[3] Toxicity is an adverse effect related to an exposure of a chemical substance. The goal of a risk assessment is to project the likelihood of an adverse effect on humans, wildlife, or ecological systems from a specific level of exposure to a chemical or toxic agent.

Several statistical terms should be defined before proceeding further with a more detailed discussion of health risk assessment. The first is probability. Many phenomena are random and are not necessarily related, in the case of risk, to an exposure. Specifically, probability is an event, described as any specific collection of the possible outcomes of a random phenomenon. Explained in a different way, the probability of an event is, if in a long sequence of repetitions, the relative frequency of the event approaching a fixed number. This number is the probability of the event. Probability can also be defined as long-term relative frequency.

Another principle of risk assessment that requires explanation is the term "confounding." The effects of two variables on a dependent variable are said to be confounded when the effect of each one cannot be distinguished from the other. As a result, each of these variables are independent or extraneous factors. This concept can be further explained as an association or outcome. A clinical event, or the medical reaction of an individual to an exposure to a toxin, can be due to confounding, as well as causation. Therefore, because different combinations of toxins affect people differently and have various clinical expressions in different groups of people, caution should be used when drawing conclusions about the potential effects of a toxic exposure event.

One must also understand the difference between sensitivity and specificity. Sensitivity is the proportion of individuals with a positive test result for the disease that the test is intended to reveal. This can be expressed in a proportion as true positives divided by false negatives. Studies with high sensitivity give a true positive result. On the other hand, specificity is the proportion of individuals with negative test results for the disease that the test is intended to reveal. The proportion in this instance is true negatives divided by false positives.

Human Genetics

To fully understand health risk assessment, one must understand the risk of exposure in "real" terms. Throughout our lives and on a daily basis, we are exposed to an inordinate amount of chemical compounds. These include exposures through food products, ultraviolet light and radiation, and lifestyle selection. There is an interaction between these chemical exposures and our genetic code. The most important component of our genetic code, DNA, is found in the nuclei of animal and vegetable cells. DNA is the repository of all human hereditary characteristics.

DNA is a fragile substance that can be damaged in many ways, including exposure to toxic substances. Medical research has demonstrated that specific genes have areas that are sensitive to damage by environmental toxins. Many compounds found in the environment react with DNA at specific susceptible regions of the DNA strand, damaging gene expression. When our genetic code is disrupted or damaged, our health is affected. This impact may range from changes in our skin (such as skin cancer due to ultraviolet radiation) to changes in the development of our blood cells (specifically, leukemia).

The genetic code plays a distinct role in human response to toxic exposures in the environment. For example, some patients have a genetic predisposition to Parkinson's disease. When confronted with certain environmental exposures, their genetic code "decompensates" and allows the clinical expression of this disease. Another example of the environment causing a disease is multiple sclerosis. Patients with multiple sclerosis are often traced to regions of temperate climates. The environment changes an individual's immune system so that it attacks itself. However, humans have an innate defense mechanism to many kinds of exposures, whether it be environmental or as a metabolic by-product of normal bodily function. This defense mechanism is called "reserve capacity." The difficulty with this concept as it relates to risk assessment is that individuals have different reserve capacities. As a human ages, his or her defense mechanisms and reserve capacity also diminish. Reserve capacity can be expressed on an individual basis as how susceptible that person is to the clinical effects of a toxic exposure. An identical exposure may lead to differing clinical effects in different individuals.

Chemical Transmission

Another way that exposure to toxic compounds may affect a human is at the level of chemical transmission. Chemical transmission is one of the main mechanisms of biological communication within a human being. These chemicals are usually identified as neurotransmitters or hormones. A neurotransmitter is any specific chemical agent released by a presynaptic cell, upon excitation, that crosses the synapse to stimulate or inhibit the postsynaptic cell.[4] A hormone is defined as a chemical substance formed in one organ or part of the body and carried in the blood to another organ or part.[5] These biological messengers are active from the time of conception through death and can be damaged by exposure to toxic compounds. The region or address of where these chemicals are transmitted may also be at risk for damage due to toxic exposure. Other subcellular areas that may be attacked by exposure to toxic substances are intracellular organelles and proteins. Intracellular organelles, such as mitochondria, are responsible for the metabolism of a cell. Proteins control specific functions within a cell, including production of enzymes for cell function.

Epidemiology and Study Design

In the future, as the development of science progresses, more detailed information will be available to determine actual human risk to exposure to toxic substances. Human-health risk assessments were developed for government agencies to use in quantifying the possible health effects of contamination and toxic exposure. Initial studies were completed in animals, followed by retrospective studies in humans. A retrospective study uses a group of individuals with specific health disorders and looks for common links in the development of these disorders. Currently, prospective studies are being evaluated as a tool for risk-assessment analysis. A prospective study is a study that is designed in advance of its initiation (generally following a group of individuals with a predetermined exposure to a potentially toxic substance). In a prospective study, variables can be controlled. Prospective studies tend to be more accurate and specific than retrospective studies.

Risk assessments use information provided by epidemiological studies. Epidemiological studies provide important information that is used in determining the amount of risk to humans from environmental exposure to toxic substances. Exposures may range from solid substances (asbestos) to gases (volatile organic compounds). Unfortunately, effects from exposure to toxins, whether they be liquid, solid, or gaseous, may take years to develop. This introduces an inherent flaw in the epidemiological outcome data when exposure is a controlling factor in the study. It may take many years for an exposure to cause a clinical response.

From a statistical standpoint, epidemiological studies should use a long exposure interval. This long time frame helps remove confounding bias by limiting variables and decreasing the risk of a chance event. Therefore, if an epidemiological study shows an excess of clinical abnormalities, it generally represents a long duration of environmental exposure and/or multiple exposures.

Another problem in epidemiological studies is the overlapping effects of multiple toxins or exposures. Additive effects may be produced. An example would be cigarette smoking combined with asbestos exposure, causing a specific type of thoracic cancer, mesothelioma. It should also be stressed that over prolonged periods of time, dose exposure may fluctuate, again making it very difficult to interpret epidemiological data.

There can be specific clinical effects related to toxic exposure. However, the specificity of a clinical response may or may not be related to a toxic exposure. For example, in the field of neurotoxicology, neurotoxicity is caused by more than 150 different chemicals. Therefore, the nervous system is very sensitive to toxic exposure; however, distinct clinical findings are not necessarily specific to individual toxins. An example includes seizures, which clinically may look identical. The seizures may be caused by exposure to many different chemicals. Environmentally related illnesses cannot be determined using quantitative environmental risk assessments alone. Diseases of the human body are usually multifactorial in nature, making it difficult to ascertain the percentage of the disease caused by environmental factors.

Conclusion

As the previous discussion supports, caution should be exercised when epidemiological studies are used as part of a regulatory mechanism such as risk assessment. Therefore, risk assessments should involve more than the use of existing epidemiological studies in the management of toxic exposure. To the greatest extent possible, large populations of individuals with specific and limited toxic exposures should be examined when used as part of the human-health risk-assessment analysis performed in connection with a contaminated site. These specific and potential exposures should be narrowly tailored to the site-specific conditions.

The definition of risk is the probability that an adverse event will occur as a result of a specific stimulus. There are risks associated with the simplest of human tasks, such as sleeping. Many heart attacks occur in the early morning hours after awakening. Should we stay asleep? Strokes also occur with increased frequency when patients are sleeping. Should we stay awake?

Environmental or health risk assessments evaluate the level of health risk to which we are subject as a result of exposure to a contaminated site. The risk assessment transforms a public-policy determination that we can all live with some health risk into a scientific evaluation. It is important that these evaluations accurately measure the true risks presented and not phantom concerns. As a result, caution and careful use of the scientific underpinnings for these studies should always be part of any risk assessment that is used as part of any brownfields development.

Notes

1. One traditional, formal definition of risk assessment is "the use of the factual base to define the health effects of exposure of individuals or populations to hazardous materials and situations." NAT'L ACADEMY SCIENCES, COMM'N LIFE SERVS., RISK ASSESSMENT IN THE FEDERAL GOVERNMENT: MANAGING THE PROCESS 3 (1983).

2. An example of an environmental risk would be the likelihood of a clinical disorder related to exposure of an individual to a toxin.

3. STEDMAN'S MEDICAL DICTIONARY 522 (25th ed. 1990).

4. *Id.* at 1050.

5. *Id.* at 724.

CHAPTER 20

Remediation Strategies for Brownfields Redevelopment

EDWARD J. CICHON

Remediation of environmentally impaired sites has undergone significant changes during the 1990s. Historically, the cleanup of impacted soils, sediments, and groundwater had been the primary objective of remediation programs, with little or no consideration to the end-use of the affected property. In contrast, the current remediation paradigm, which has been strengthened by brownfields initiatives, has shifted the emphasis of remedial programs so that the intended end-use of the affected property becomes the ultimate objective, and the cleanup of impacted media is simply one element in achieving this new, broader goal.

Though brownfields programs have begun to focus on the redevelopment of environmentally impaired properties, full exploitation of the regulatory and economic incentives associated with these programs will require more practical and cost-effective remediation efforts. Furthermore, the luxury of having the results of comprehensive remedial site investigations before embarking on the cleanup of affected media will be the exception as the brownfields programs mature. The demands exerted by the commercial aspects of brownfields property redevelopment will force remediation methods and costs to be delineated earlier in the remedial process and with a relative paucity of site data. These same commercial factors will also require remediation strategies that do not delay or interrupt the property's income and cash flow.

A summary of the remediation issues associated with brownfields redevelopment is provided in the following list:

1. Cleanup criteria are dependent on intended property use.
2. Commercial factors prohibit inordinately long site investigations.
3. Redevelopment financing must have "hard dollar" costs for cleanup.
4. Cleanup of soil and groundwater must be synergistic with property build-out activities.
5. Involvement of the community is paramount.

General Remedial Strategy

Redevelopment of brownfields properties requires a different remedial strategy and approach than has been employed in most federal, state, and private programs. When remediation of a site was the ultimate objective, a sequential, multiphase remedial program, consisting of the following steps, was standard practice:

- Preliminary site assessment
- Site investigation/risk assessment
  —Phase 1
  —Phase 2
- Feasibility study
- Remedial design/engineering
- Remediation of affected media

This fragmented approach to site remediation generally resulted in the site investigation and feasibility study phases becoming protracted at the expense of delaying remedial actions. A corollary of this phenomenon was that a disproportionately high amount of funds was consumed by studies, as opposed to being spent on cleanups.

In contrast, the commercial pressure to realize income and cash flow as soon as possible after embarking on brownfields property redevelopment requires a more efficient remedial approach. The accelerated remediation process (ARP) outlines such an alternative (Figure 1).

Activity point A-1 is an intense evaluation of existing site data, including, but not limited to, future property use, site history, analytical data, and applicable cleanup criteria. There are no additional field data gathering exercises associated with this activity.

The result of A-1 is to satisfy decision points D-1 and D-2, concerning the need for remediation or monitoring, respectively. If remediation is required, activity point A-2 is activated and the following actions performed:

- Delineation of two to three remedial alternatives that will achieve site objectives
- Development of "hard-dollar" costs for implementing each alternative without further site investigations
- Analysis of commercial risk and cost sensitivities associated with each alternative
- Determination of the best approach for implementing the remediation and managing risk

The preferred remediation approach is to avoid additional site investigations and thereby accelerate the overall remedial schedule. Uncertainties associated with the extent of impacted soil can be addressed through an observational approach during remediation, while analogous groundwater variables may be accommodated by minimal additional site investigations in conjunction with remedial system flexibility. In the ARP, this observational approach is reflected by A-2, when the remedy is implemented, by A-4, when

FIGURE 1. Accelerated Remediation Process (ARP)

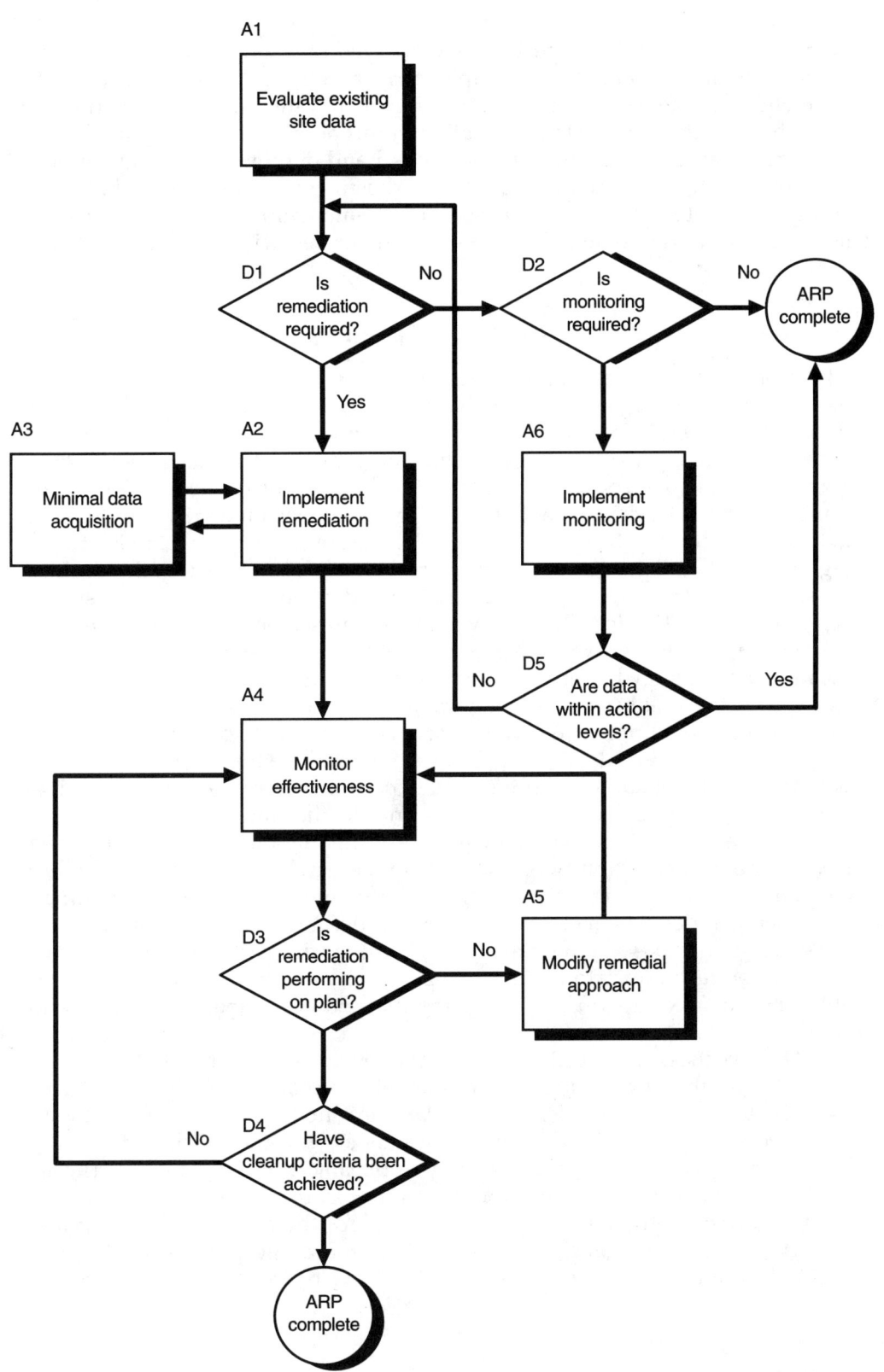

field observations and monitoring are employed to assess the effectiveness of the remedy in relation to the cleanup criteria, and by A-5, when real-time adjustments are made to the remedial approach or system to address variances from the expected performance. Implicit in this accelerated remediation plan is the fact that remedial solutions developed and deployed must have inherent flexibility to accommodate operating adjustments in the field. In both soil and groundwater situations, the associated remediation costs may be established earlier in the remedial process and integrated with the overall site redevelopment costs.

Risk-Based Cleanups

At the onset of the remediation industry in the United States, the use of risk-based cleanup strategies was overshadowed by the emotional public outcry regarding the perceived risk posed by hazardous-waste sites across the country. Once industry experts began quantifying the expenditures required to address the various contaminated sites on the basis of perceived risk, it became apparent that the concept of inherent-versus-perceived risk would have to be integrated into the overall remedial program in order to establish funding priorities, and to achieve a more reasonable cost/benefit relationship. This increasing reliance on inherent risk-based remedial strategies is synergistic with the ARP described above, and with the conditions necessary for realizing maximum brownfields site redevelopment potential. A summary of the remedial strategy hierarchy is provided in Figure 2.

Under the risk-based remedial strategy or cleanup approach, the data from site investigations or assessments is evaluated in the context of intended end-uses of the affected property. Delineation of the expected use of brownfields properties enables environmental experts to determine the risks resulting from existing contaminants using scenarios that more accurately depict future site activities. Future commercial or industrial operations, as opposed to residential or community operations, are generally able to tolerate higher contaminant levels while concomitantly maintaining the same risk to human health and the environment. In addition to defining a property's intended end-use, the risk-based remedial strategy also encompasses the presumptive remediation techniques to assure that the mechanics of achieving an acceptable level of cleanup do not in themselves pose an unacceptable risk to nearby receptors.

The specific commercial or industrial operations at a brownfields site are factored into the human and environmental receptor risk analysis to derive the acceptable cleanup levels. The resultant cleanup levels are compared with any generic state or regional criteria that may exist. If the risk-based cleanup levels are significantly higher than any applicable state or regional criteria, the incremental costs to achieve the latter criteria are contrasted to the probability and cost of gaining regulatory approval for the risk-based cleanup levels. Environmental agencies are becoming more receptive to risk-based cleanup strategies, due in part to property redevelopment pressures.

FIGURE 2. Remedial Solution Hierarchy

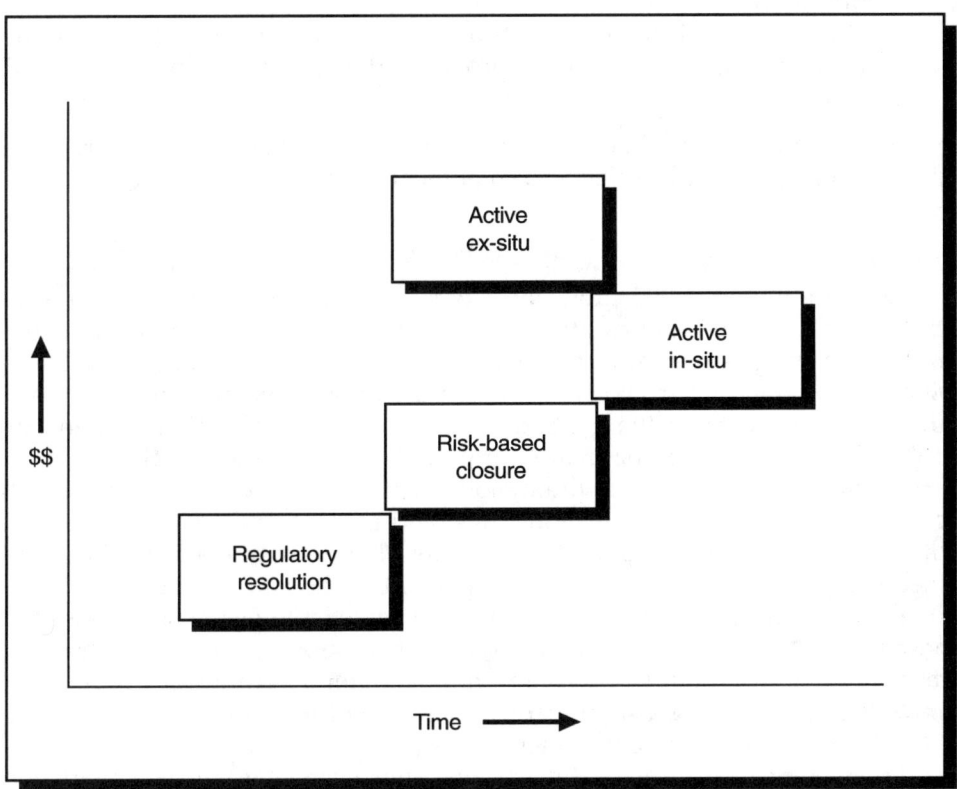

Soil Remediation

The demand for a higher degree of cost certainty earlier in the remedial process, coupled with inherent site-redevelopment logistics, changes soil remediation strategies for brownfields sites. Strategies and technologies that may have been cost-effective under the old sequential remediation paradigm are rendered less than optimal in the brownfields arena. The important criteria for evaluating the relative merits and costs of brownfields soil remediation strategies are as follows:

- *Flexibility:* The resultant technique must have the ability to adapt to a wider range of site characteristics, such as contaminant types and concentrations, due to less site investigation data.
- *Speed:* In response to commercial pressure for expeditious cleanup, the selected soil remediation strategy must achieve the cleanup goals in the shortest time practical.

- *Compatibility:* The location of most brownfields sites dictates that the soil remediation strategy should not be disruptive to the surrounding community.
- *Cost:* Successful financing of brownfields site redevelopment is predicated on soil remediation approaches that can limit and control soil cleanup costs.

Incorporating these criteria into the development of a comprehensive remediation strategy produces a select group of techniques for soil cleanup.

Soil Washing

Sites with more than 5,000 cubic yards of soil impacted with heavy metals or semi-volatile organics are candidates for a remediation strategy encompassing soil washing—an ex-situ, water-based process for mechanically separating clean soil fractions from contaminated fractions. ("Ex-situ" means the soil must be removed from the ground to conduct remedial activities, and "in-situ" means the soil may be treated in place, without removal.) Soil washing is predicated on the fact that the majority of chemical contaminants in soils actually reside in the fine silts and clays, and in the humic materials of soils with appreciable organic content. These silts, clays, and humic materials combined usually represent 5 percent to 20 percent of the total soil volume. Traditional soil remediation techniques have addressed the entire soil mass, and consequently have resulted in excessive remediation costs. In contrast, soil washing separates the silts, clays, and humic materials from clean cobble and sands, thereby isolating a significantly reduced volume and mass of contaminated material for ultimate treatment or disposal.

For instance, one successful soil-washing system is an ex-situ, water-based process for mechanically separating clean, oversized soil fractions from highly contaminated silts, clays, and humic materials. In addition, this system can isolate xenophobic materials, such as lead shot, plastic, and ammunition casings, all of which can impede soil remediation. One of the systems is shown in Figure 3. Soil excavated from the site undergoes scalping to remove large, clean debris, followed by primary separation to remove clean, oversized material. The contaminated undersized material from primary separation may be subjected to secondary separation when humic materials are removed and silts and clays are separated from clean sands. The concentrated fines may be treated on-site or transported off-site for treatment or disposal. This soil-washing system typically reduces the volume of soil requiring treatment by 75 percent to 95 percent, resulting in lower remediation costs. In addition, the reduced volume of soil requiring ultimate treatment or disposal translates to fewer hazardous-waste vehicle trips through communities surrounding brownfields sites.

Stabilization/Fixation

Cost-effective attenuation of the risk posed by site contaminants may be achieved using either in-situ or ex-situ chemical fixation. In this approach, the

FIGURE 3. Hydro-Sep 2500 System Flow Diagram

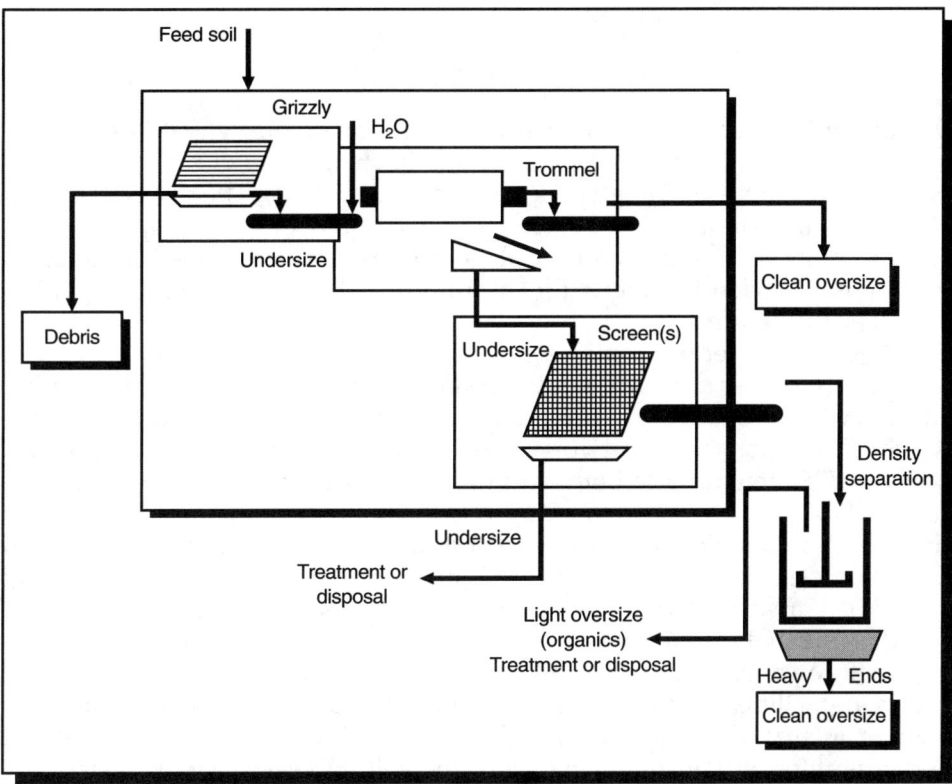

affected soil or sediment at the site is mixed with reagents that chemically bind the target contaminants, thereby rendering them less available to uptake by potential human or environmental receptors. Environmentally distressed sites with heavy-metal contamination have been most conducive to remediation using chemical fixation. Recently, fixation technologies have been developed for treating soil and sediments with relatively high organic content.

When used as an on-site remediation technique, chemical fixation eliminates or significantly reduces the amount of contaminated material transported through surrounding neighborhoods. In the ex-situ approach, the affected soil or sediment is excavated, processed to remove clean oversized material, and fed to a portable pug mill in which the contaminated fractions are mixed with the fixation reagents. The resulting mixture is allowed to cure before being reused at the site as nonstructural fill. Use of enhanced chemical-fixation technologies allows the treated material to pass Toxicity Characteristic Leachate Procedure criteria, in addition to being able to be used as structural backfill on the site.

In the in-situ option for chemical fixation, the affected soil or sediment is treated in place, using a mixing auger through which the chemical reagents are introduced into the subsurface. In-situ fixation is typically more expensive than ex-situ fixation, but has the advantage of addressing sites with physical constraints.

Vapor Extraction

One of the more common remedial challenges facing brownfields sites is the cost-effective attenuation of volatile organic chemicals (VOCs) in soil from leaking underground storage tanks and solvent releases. The control of BTEX, TCE, PCE and similar VOCs is generally required before occupancy of redeveloped industrial or commercial properties can occur. In-situ vapor extraction is a proven technique for meeting this challenge of removing VOCs from unsaturated soils.

There are several variations of in-situ vapor extraction, with the most common involving the placement of a subsurface network of collection pipes through which vapors are extracted by vacuum to an above-ground treatment system. The subsurface network may consist of vertical wells, horizontal wells, or a combination of both. At brownfields sites with existing buildings, use of horizontal extraction systems may facilitate the removal of VOCs from beneath foundations. At sites where buildings have been razed, a soil vapor extraction system may be engineered and installed to allow for redevelopment to occur with little or no impedance by the remediation system.

It is unusual for the VOC levels in the soils to reach the no-further-action level before completion of site redevelopment. However, the above-ground treatment facilities for the extracted vapors are relatively small and can be located in an area of the site so as not to interfere with commercial activities. Periodic monitoring and maintenance of the soil vapor extraction system also can be conducted in an unobtrusive manner.

Groundwater Remediation

Brownfields sites with contaminated groundwater pose unique challenges to the overall site redevelopment process. Before examining remedial strategies for groundwater, it is imperative to identify and then remove or contain the source of contaminants. Remediation methods such as soil washing can remove concentrated pockets of soil contamination, while chemical fixation can transform sources of heavy-metal contamination to prevent further leaching to the groundwater.

Many sites experiencing groundwater contamination by organic chemicals have "pools" of free product where the respective organic chemicals have been released and transported through the subsurface in such a manner as to form repositories of nearly pure chemicals, which in turn serve as sources for groundwater and soil contamination as the groundwater contacts these pools. This discussion of groundwater remediation strategies for brownfields sites is predicated on free-product control or removal having been implemented. Remedial options for free-product situations are examined in the next section.

Because groundwater typically flows through the subsurface of a given site, it represents a dynamic remedial challenge, in contrast to the relatively static conditions surrounding soil remediation. In addition, the groundwater remediation strategies for brownfields sites must accommodate many of the factors that surfaced for soil remediation:

- *Flexibility:* The resultant technique must have the ability to adapt to a wider range of site characteristics, such as hydrogeological conditions, contaminant types, and contaminant concentrations, due to fewer site investigation data points.
- *Speed:* In response to commercial pressure for expeditious cleanup, the selected groundwater remediation strategy must achieve the cleanup goals in the shortest time practical.
- *Compatibility:* The resulting groundwater remediation technique may need to encompass commingling of contaminant plumes from adjacent properties and/or prevent the migration of contaminated groundwater from the subject site to adjacent parcels. In situations where contaminated groundwater has migrated beyond the property boundaries, and must undergo remediation, the remedial strategy must acknowledge the potential impact of the remedial system on any contiguous property.
- *Cost:* Successful financing of brownfields site redevelopment is predicated on groundwater remediation approaches that can be accurately quantified.

Incorporating these criteria into the development of a comprehensive remediation strategy produces a select group of techniques for groundwater cleanup.

In-Well Stripping

This groundwater remediation technique is a derivation of classical pump-and-treat systems, using air-stripping for VOC contamination. In this approach, the need to pump groundwater to the surface is eliminated. Extraction wells are typically installed in the groundwater contaminant plume, and air is delivered to the bottom of the wells. As the air bubbles rise in the well casings, the VOCs are stripped from the groundwater. A deflector in the well casings allows the VOC-laden air to flow to the surface where it is treated, while the clean groundwater is redirected back to the aquifer.

In addition to having lower capital and operating costs when compared with conventional pump-and-treat systems, in-well stripping has less impact on groundwater levels, thereby making it more acceptable in areas where there is a strong relationship between groundwater levels and surface waters. A further advantage for brownfields sites is that in-well stripping systems can be used to contain a contaminant plume or to clean up a "hot spot." Finally, in-well groundwater remediation systems generally effect cleanup at a greater rate when compared with pump-and-treat systems, thereby providing better synergy with site redevelopment and property transfer.

In-Situ Biodegradation

During the past several years, considerable effort has been directed toward developing more efficient bioremediation technologies for groundwater contaminated with organic chemicals. The challenges, including those related to brownfields site projects, have been (1) to effect degradation within a relatively short period of time, and (2) to prevent the buildup of undesirable byproducts, such as vinyl chloride, when addressing chlorinated organic contaminants. Success has been achieved with the discovery and application of in-situ anaerobic bioremediation for groundwater contaminated with halogenated organic chemicals.

At a Texas site with initial PCE concentrations of 1,500 parts per billion (ppb), enhanced in-situ anaerobic bioremediation was able to produce groundwater with nondetectable levels of PCE, TCE, DCE, and vinyl chloride within two years. This method contrasts with conventional pump-and-treat techniques, which would have taken approximately ten years, or conventional bioremediation techniques, which would have produced unacceptable levels of vinyl chloride. Therefore, the enhanced in-situ anaerobic bioremediation technology can be an integral component of brownfields cleanup strategies, given its ability to effect treatment of contaminants within a shorter time than conventional techniques.

Nonaqueous Phase Liquid Removal

Brownfields sites with a history of leaking underground storage tanks or high-volume solvent releases, and the appropriate geological conditions, have a higher probability for the presence of nonaqueous phase liquids (NAPLs) or free product than sites without these situations. As mentioned in the preceding section, NAPL pools of nearly pure organic contaminants, such as TCE or fuel oils, represent a significant reservoir of contaminants and, if left unchecked, may result in escalated cleanup costs. Therefore, it is usually cost-effective to remove, or at a minimum contain, NAPLs before embarking on soil or groundwater remediation. The presence of NAPLs can be ascertained through a combination of site investigation data and site history.

The remedial strategies for NAPLs at sites targeted for redevelopment need to be cost-effective and expeditious. The NAPL strategy for a given site may encompass one or more of the following techniques:

- Dual-phase vacuum extraction
- Surfactant enhanced flushing
- Physical containment

These techniques are the most likely to be employed at brownfields sites with NAPL removal issues.

Conclusion

Environmentally distressed brownfields sites present new challenges for remediation of soil, sediment, and groundwater. Commercial objectives for the redevelopment of these properties dictate that the overall remedial process be streamlined. Protracted, linear remedial programs involving conventional remedial investigations, feasibility studies, engineering, and remedial actions are not conducive to cost-effective rejuvenation of brownfields sites. Accelerated remedial action programs, which rely on risk-based cleanup criteria and embrace existing presumptive remedies, are consistent with the goals of brownfields redevelopment. If common sense prevails, the development and implementation of remediation strategies for brownfields sites should allow more funds to go to cleanup and restoration of properties than to endless, and sometimes nonproductive, studies.

PART IV

State Voluntary Cleanup Programs

~

CHAPTER 21

Arizona

JOEL L. HERZ

Arizona does not have a single, comprehensive, voluntary environmental cleanup program akin to those enacted in other states. Nevertheless, a voluntary environmental cleanup can still be accomplished in Arizona pursuant to the provisions of several unrelated statutes. There are three major sources of authority for voluntary cleanups in Arizona:

1. Arizona's Water Quality Assurance Revolving Fund (WQARF) Voluntary Program, codified at Arizona Revised Statutes (A.R.S.) sections 49-281 *et seq.*, particularly A.R.S. section 49-285(B), and administratively at Arizona's administrative code (A.A.C.) sections 18-7-100 *et seq.*, which allows for voluntary cleanup of environmentally impacted real estate.
2. The Underground Storage Tank (UST) Voluntary Program, codified at A.R.S. sections 49-1001 *et seq.*, and administratively at A.A.C. sections 18-12-101 *et seq.*, which creates the opportunity for a voluntary UST cleanup.
3. A.R.S. section 49-152, Arizona's statute authorizing interim soil-remediation standards, provides that an owner who has voluntarily elected to remediate a property for nonresidential uses must record with the county recorder, in the county where the property is located, a voluntary environmental mitigation use restriction (VEMUR), limiting the area necessary to protect public health and the environment to nonresidential uses. It has effectively created a VEMUR Voluntary Program.[1] Emergency interim soil-remediation standards, established pursuant to A.R.S. section 49-152, were adopted by the Arizona Department of Environmental Quality (ADEQ) on December 15, 1995, and codified administratively at A.A.C. section 18-7-2. They set forth specific remediation standards and criteria for VEMUR and other cleanups. The interim soil-remediation standards, however, do not set forth a separate voluntary program. Indeed, these standards apply to the other voluntary programs that are available—the WQARF Volun-

tary Program—and the UST Voluntary Program, in addition to other nonvoluntary environmental programs in Arizona.[2]

Also available are prospective-purchaser agreements, in which Arizona authorities enter prepurchase agreements to guarantee liability protection if proper cleanups are undertaken.[3] Entities entering such agreements can avoid future liability under relevant Arizona environmental statutes.[4]

A more comprehensive or more formal stand-alone Voluntary Cleanup Program to coordinate and streamline these separate programs may be expected in Arizona in the near future.

Summary of Major Provisions of the Arizona Programs

WQARF Voluntary Program

The WQARF Voluntary Program permits any person who undertakes a remedial action to request that ADEQ approve the remedial action before, during, or after the remedial action.[5] Anyone intending to seek recovery of remedial-action costs from a responsible party under the WQARF statutes should seek ADEQ preapproval,[6] although this is not mandatory.

A party undertaking a preapproved remedial action must file a remedial-action plan (RAP) with ADEQ.[7] Within ninety days, ADEQ will review the RAP for completeness.[8] Thereafter, ADEQ will inform the person undertaking the voluntary remedial action that the plan is approved, approved with modification, or rejected, and will issue a letter of determination concerning whether the plan is approved or what is required to receive approval.[9] The person requesting ADEQ's approval of a remedial action is required to pay the ADEQ's reasonable costs in reviewing the RAP.[10] ADEQ may waive some or all of its costs in appropriate circumstances.[11] If the RAP is approved by ADEQ, the remedial action will be deemed to be in substantial compliance with the rules establishing when funds may be used from the WQARF.[12] In addition, any amounts paid to ADEQ for review of the remedial action are recoverable in a cost-recovery action against third parties.[13]

An additional benefit available under the WQARF Voluntary Program is that if the remedial action is approved by ADEQ and the remedial action is done entirely on site, then Arizona permitting requirements are waived.[14]

Finally, under the WQARF Voluntary Program, ADEQ has the authority (rarely exercised) to reimburse a party for the costs of specific remedial actions that are undertaken in a cleanup.[15]

UST Voluntary Program

The Arizona UST Voluntary Program allows a person who comes into possession or control of a property where a UST is located to remove voluntarily, or otherwise close, the tank in a safe and secure manner that prevents releases of regulated substances.[16] The person undertaking such a voluntary action cannot be the "owner" of the tanks, as defined by Arizona law. As soon as practicable, and if the owner has failed to do so, the volunteer must notify ADEQ of (1) the tank's size, location, and use[17], and (2) any release or suspected release from the tank.[18]

A person who conducts a closure or removal action under this program will be eligible to receive reimbursement of costs under the UST Grant Account, up to a limit of $100,000.[19] In addition, a person who conducts corrective action under this program will be eligible to receive reimbursement of costs under the UST Assurance Account, up to $1 million.[20] A further benefit is that one undertaking corrective action under the UST Voluntary Program need not pay the statutorily required deductible amounts for reimbursement eligibility, which can be as high as $25,000.

VEMUR Voluntary Program

The VEMUR Voluntary Program (promulgated under Arizona's emergency interim soil-remediation standards), though not technically a "program," may be the most useful tool for a brownfields redeveloper. The interim soil-remediation standards permit remediation to a lesser standard of cleanup, as long as the redeveloper agrees to certain constraints on the property—the most important of which is that the property cannot be used for residential purposes of any kind.[21] To qualify for the VEMUR Voluntary Program, any contamination left after remediation must remain on the property at either of the following:

1. predetermined-risk based remediation standards for other than residential exposure assumptions; or
2. concentrations resulting in a hazard index greater than one, or a risk of carcinogenic health effects greater than the range of risk levels set forth in the Code of Federal Regulations (40 C.F.R. § 300.430 (e)(2)(1)(A)(2)).[22]

These standards can be achieved in three ways: (1) by remediating to background concentration levels, (2) by remediating to interim soil-remediation standards containing nonresidential, health-based guidance levels (HBGLs), or (3) by remediating to site-specific standards. The rules permit remediation of nonresidential property according to non-residential HBGLs if a VEMUR is recorded in the county in which the property is located, the VEMUR is filed with ADEQ, and any purchaser of the property is notified of the VEMUR before closing.[23] Similarly, those remediating to site-specific remediation levels may remediate to a carcinogenic risk level more protective than one in ten thousand (1×10^{-4}) but less protective than residential protections (1×10^{-6}), as long as that person records and files a VEMUR as set forth above and notifies any purchaser of the VEMUR before closing.[24] One can also remediate to background concentration levels; however, that is not usually a cost-effective cleanup method and is not fully discussed here.[25]

As a technical matter, a "voluntary" cleanup can take place under the interim soil-remediation standards only when ADEQ has no regulatory authority over the person who is undertaking the cleanup.[26] That will encompass two situations.[27] The first occurs when the contamination at a site does not reach a level that violates Arizona law.[28] The second occurs when the individual who is undertaking the remedial action has no obligation to do so.[29] These remediators, the only ones who are truly volunteers, will participate under the program when they want a "Close-Out Document," formerly known as a Letter of Determination.[30] The Close-Out Document certifies that the remediation was completed in compliance with the interim soil-remediation standards,[31] and is often used to facilitate the sale of a property after it has been remediated. In a certain sense, a Close-Out Document is akin to a "no further action" letter issued in other states.

In practice, a "voluntary" cleanup can take place under the interim soil-remediation standards when ADEQ has regulatory authority over the person who is undertaking the cleanup. Even if the level of contamination at a site violates Arizona law and the individual who is undertaking the remedial action has an obligation to do so, one can still remediate to the lesser nonresidential standards and record a VEMUR.[32] These remediators, who are not "volunteers," can still participate in the program when they want a Close-Out Document or if they want to remediate before ADEQ or the United States Environmental Protection Agency (USEPA) begins an enforcement action. Indeed, even those under an enforcement order can still be eligible to remediate under the lesser nonresidential standards by recording a VEMUR.

Memorandum of Agreement

Currently, Arizona does not have a Memorandum of Agreement with USEPA for any of its voluntary programs, but there is negotiation concerning one. Therefore, a person participating in an Arizona voluntary program could be subject to federal liability even after receiving a Close-Out Document. Participants in Arizona programs may seek "comfort letters" from USEPA or enter prospective-purchaser agreements with the federal government to minimize the threat of federal liability.[33]

Liability Protection

WQARF Voluntary Program

In 1996, Arizona substantially amended the state Superfund program, the WQARF. The 1996 amendments to WQARF were prompted (1) by concerns of the lending and fiduciary communities, who viewed with increasing alarm a series of judicial interpretations of the Comprehensive Environmental Response, Compensation and Liability Act (CERCLA) holding secured creditors and fiduciaries personally liable for cleanup costs under certain circumstances, and (2) by logistical difficulties with prospective-purchaser agreements.

Previously, ADEQ had been hampered in its ability to resolve the liability of prospective-purchasers because of a lack of statutory authority to make prospective purchaser agreements assignable to subsequent titleholders. The amended statute expressly provides such authority. As with prior practice under WQARF and CERCLA, the amended statute authorizes ADEQ to confer covenants-not-to-sue under WQARF and CERCLA and contribution protection (immunity from further private-party civil litigation) to prospective purchasers when the agreement "will provide a substantial public benefit."[34] That benefit can include an agreement to perform or fund remedial measures, or to provide for productive reuse of brownfields property. Under the amendments, joint liability is removed from WQARF from July 20, 1996, until July 31, 1997. On August 1, 1997, joint liability is restored.

The amendments also added a definition of "orphan shares," namely "the equitable shares of the cost for remedial actions that are uncollectible because those costs are allocated to persons who are unable to pay for those shares."[35] In addition, WQARF funds are authorized for payment of remedial-action costs if the responsible party is insolvent, is not subject to claims due to bankruptcy, or is otherwise protected from judgment.[36] These provisions are intended to prevent potentially responsible parties (PRPs) from paying the uncollectible shares of other PRPs.

The amendments also clarify how a purchaser of previously contaminated property would "associate himself with the release," thereby losing an exemption from liability.[37] "Associated with the release" is defined as "having actual knowledge of the release and taking action or failing to take action that the person is authorized to take and that increases the volume or toxicity of the hazardous substance that has been released."[38] Additional amendments provide other safe harbors to fiduciaries and those holding security interests.[39] This provision should provide some comfort to potential buyers of contaminated property.

UST Voluntary Program

Under the Arizona UST regulations, a person who undertakes voluntary closure or corrective action will not incur the liability of an owner.[40]

VEMUR Voluntary Program

The interim soil-remediation standards do not provide any specific liability protections for voluntary remediators. However, a remediator can enter a consent decree with ADEQ and receive a covenant-not-to-sue from ADEQ upon completion of a cleanup using a VEMUR.[41]

Cleanup Standards

WQARF Voluntary Program

A cleanup under the WQARF Voluntary Program must meet, when applicable, surface-water quality standards adopted by ADEQ at A.A.C. sections

18-11-204 and 18-11-205, the groundwater quality standards at A.A.C. section 9-21-403, the drinking water aquifer water quality standards at A.R.S. section 49-223(A), the hazardous-waste corrective-action rules at A.A.C. section 18-8-264 (for those facilities requiring a hazardous-waste permit), the corrective-action requirements pertaining to releases from USTs, and the interim soil-remediation standards set forth in A.A.C. section 18-7-201 *et seq.*[42] In addition, the cleanup must be appropriate under the circumstances, taking into account the threat to the environment and public health.[43]

UST Voluntary Program

A closure under the UST Voluntary Program must meet the standards set forth in A.A.C. section 18-12-101 *et seq.*, which are largely designed to protect the environment from future releases. Cleanup actions will need to meet ADEQ standards[44] and those set forth in the Code of Federal Regulations. (40 C.F.R. §§ 280.60–280.67 (1995).)

Interim Soil-Remediation Standards

Arizona's interim soil-remediation standards are just that—interim rules. The interim rules are based primarily on exposure through ingestion. Exposure from inhalation, saturation, explosivity, corrosivity, and ignitability may be considered in the permanent rule. In addition, the interim rules do not account for a chemical's ability to migrate through subsurface soils, resulting in a threat to groundwater and therefore making the property ineligible for a VEMUR. It is expected that the permanent rule will include remediation standards protective of groundwater quality.

The current rules allow alternative approaches for determining appropriate soil-remediation standards and set forth specific requirements for each approach. A person may use a predetermined standard, or may use site-specific data to determine the appropriate standard. The predetermined standard is an HBGL developed by the Arizona Department of Health Services. Alternatively, a person may use a site-specific risk analysis of the contaminant's fate and transport effects on human health and the environment or may remediate to background concentration levels.

HBGL Cleanups

An HBGL represents a predetermined, acceptable contaminant concentration in soils. This concentration is based on the toxicological characteristics of each specific substance. A list of the chemicals and acceptable concentration levels is set forth in Appendix A to A.A.C. section 18-7-2. The residential HBGLs are designed to protect against toxic doses of systemic toxicants and risk of developing cancer from a carcinogen. Both residential and nonresidential HBGLs limit to one in one million (1×10^{-6}) the excess cancer risk level for carcinogenic compounds. The variation between residential and nonresidential

HBGLs reflects the difference between the two uses of soil ingestion rates, exposure frequency, and exposure duration. Nonresidential HBGLs are calculated by multiplying chemicals with residential HBGLs based upon a carcinogenic risk by 4.2, and by multiplying residential HBGLs based upon a noncancer health effect by 3.5.

Site-Specific Standards

In cases when the remediator chooses not to use HBGLs, when there is no HBGL, or when through the terms of an enforcement action, soil remediation standards must be determined through the use of a site-specific risk assessment, the remediator must undertake a risk assessment. The risk assessment must take into consideration the present concentration of a contaminant, its toxicity, the present and future receptors, and the available pathways between the contaminant and the receptor. Present and potential impacts on human health are to be evaluated by comparing the projected contaminant concentration with established standards and by analyzing the impact on public health and the environment.

The risk-assessment methodology may be either "deterministic" or "probabilistic," or any alternative methodology, as long as it is commonly accepted in the scientific community. The deterministic risk assessment assumes distinct contaminant levels and distinct exposure doses. This provides distinct estimates of cancer risk and noncancer hazard for the exposed population. In contrast, the probabilistic risk assessment assumes a statistical distribution of contaminant levels and exposure doses and provides a statistical distribution of risk.

Using the same data, a deterministic model often represents a worst-case scenario, while a probabilistic model predicts the most likely outcome. Because of this difference, the interim cleanup standards require that remediation standards based on a probabilistic risk assessment be no less protective than the ninety-fifth percentile upper-bound estimate. For a nonresidential cleanup in which a VEMUR will be filed, that number requires a person to remediate the soil to a contaminant concentration with an acceptable risk range between one in ten thousand and one in a million, with a noncancer hazard index of no greater than one. ADEQ can further reduce these standards based on site-specific conditions or facts.

Certified Professionals and Laboratories

A.R.S. section 36-495 *et seq.*, and the rules promulgated thereunder require all environmental laboratory analyses of samples to be performed by a laboratory licensed by the Arizona Department of Health Services. The UST regulations require that the individual who performs, directly supervises, or manages a UST corrective action must have appropriate experience in other UST corrective actions and appropriate educational experience or licenses as set forth in A.A.C. section 18-12-601 *et seq.*

The Mechanics of Participating in the Arizona Voluntary Programs

WQARF Voluntary Program

The WQARF Voluntary Program allows any person who undertakes a remedial action to request that the ADEQ approve the remedial action before, during, or after the remedial action.[44] It should, however, be noted that anyone intending to seek recovery of remedial-action costs from a responsible party under A.R.S. section 49-285 should seek ADEQ preapproval.[45] The party undertaking a preapproved remedial action must file a RAP with the ADEQ.[46]

The RAP must set forth (1) the name, title, address, and telephone number of the person submitting the RAP, (2) the location and a legal description of the site, (3) a description of the nature of the release or threatened release, (4) a description of the routes and types of exposures, (5) a description of the purpose and schedule of the remedial action to be performed, including a work plan, (6) a notarized statement concerning whether the person intends to seek cost recovery against a third party and, if so, the name and basis of such liability, (7) a description of how the remedial-action goals will be met and how the remedial-action criteria were considered, and (8) an explanation concerning how the plan is expeditious.[47] In addition, the RAP, to the extent applicable, must also contain (1) a preliminary site assessment, (2) remedial investigation results, (3) a risk assessment, (4) a health-effects study, (5) a feasibility study, (6) a description of the cleanup methods and how they comply with the WQARF Voluntary Program requirements, and (7) a description of operation and maintenance needs.[48] These additional requirements can be performed in phases, with the permission of ADEQ.[49]

Within ninety days, ADEQ will review the plan for completeness.[50] Thereafter, ADEQ will inform the person undertaking the voluntary remedial action that the plan is approved, approved with modification, or rejected, and will issue a letter of determination concerning whether the plan is approved or what is required to receive approval.[51] The person requesting ADEQ's approval of a remedial action is required to pay ADEQ's reasonable costs in reviewing the RAP.[52]

UST Voluntary Program

The UST Voluntary Program allows a person who comes into possession or control of a property where a UST is located to remove voluntarily, or otherwise close, the tank in a safe and secure manner that prevents releases of regulated substances. As soon as practicable, and if the owner has failed to do so, the volunteer must notify ADEQ of (1) the tank's size, location, and use,[53] and (2) any release or suspected release from the tank.[54]

ADEQ has yet to enact formal rules under the UST Voluntary Program for proceeding with a voluntary tank closure or cleanup. However, the remediation rules for soil cleanups under the VEMUR Voluntary Program apply to UST Voluntary Program cleanups[55] and WQARF Voluntary Program cleanups.

VEMUR Voluntary Program

Under the VEMUR Program, in addition to any other program requirements under the UST Voluntary Program or the WQARF Voluntary Program, a person conducting a remediation project based on predetermined remediation standards or background concentrations must provide to ADEQ, in writing, an initial notice of the intent to remediate the property in question.[56] The initial notice must set forth the nature of the remediation project, including the site location and site characteristics, and a statement that the property is not and will not be used for residential purposes.[57] The initial notice must also contain the rationale for the selection of remediation levels and the technologies to be used at the site.[58] If a site-specific risk assessment will be used, ADEQ approval of the risk assessment methodology must be obtained.[59]

Within forty-five days, ADEQ must respond to the request for approval of the risk-assessment methodology by approving, denying, or requesting more information.[60]

After the remediation activities at the site are completed, a person conducting a VEMUR remediation project must submit to ADEQ a final report of the postremediation site conditions.[61] The report must contain (1) a description of the actual activities, techniques, and technologies used to remediate the site, including the use of engineering controls, (2) a statement that the project applied the relevant cleanup standards and that such standards have been followed, (3) soil sample results that are representative of the entire site, (4) proof of compliance with local zoning requirements, (5) a statement from the owner or the owner's representative that the report does not contain any false or inaccurate statements, (6) a certification from the owner, stating that the owner is aware of the VEMUR notice, filing, and purchaser disclosure requirements, and (7) a copy of the VEMUR signed by ADEQ and recorded with the county recorder's office.[62]

Within sixty days after receiving the final report, ADEQ must issue a Close-Out Document, deny the final report, or request additional information necessary to issue a final close-out document.[63] Even if a close-out document is issued, ADEQ can still request that additional work be done at a site if future evidence would alter its determination.

Cost Recovery

WQARF Voluntary Program

Any person who is a defendant in an enforcement proceeding brought under the WQARF statute may join in the action any other person who is, or may be, a responsible party. Following adjudication of liability and recovery of remedial-action costs in an action brought under this article, any person held liable for such costs may bring a separate action to require any other person who is, or may be, a responsible party to contribute to the payment of such costs.[64]

If any person in an action brought under the WQARF statute establishes by a preponderance of the evidence that responsibility for release or threatened release is divisible, the person is liable only for his or her portion of the release.[65]

The WQARF statute does not affect or modify the obligations or liability of any person, by reason of subrogation or otherwise, under any other provision of state or federal law, including common law, for damages, injury, or loss resulting from a release of any hazardous substance or for remedial-action costs, except that any person who receives compensation for remedial-action costs under WQARF is precluded from recovering compensation for the same remedial-action costs pursuant to any other federal or state law. Any person who receives compensation for remedial-action costs pursuant to any other federal or state law is precluded from receiving compensation for the same remedial-action costs provided in the WQARF statute.[66]

UST Voluntary Program

Under the UST Voluntary Program, a person who undertakes voluntary closure or corrective action has the right to seek contribution for the reasonable costs of the corrective action.[67] Allocation factors are the duration and percentage of ownership during the release, the amount and nature of the substance released, the degree of care exercised by each person, the ability to divide responsibility for any release, and any other factors under the circumstances.[68]

Conclusion

Opportunities are ripe for brownfields redevelopment in Arizona. Although Arizona does not have a single stand-alone statute for brownfields redevelopment, when all the formal and informal programs are placed together, ample authority allows one to approach a brownfields property confidently.

Arizona VEMUR Program

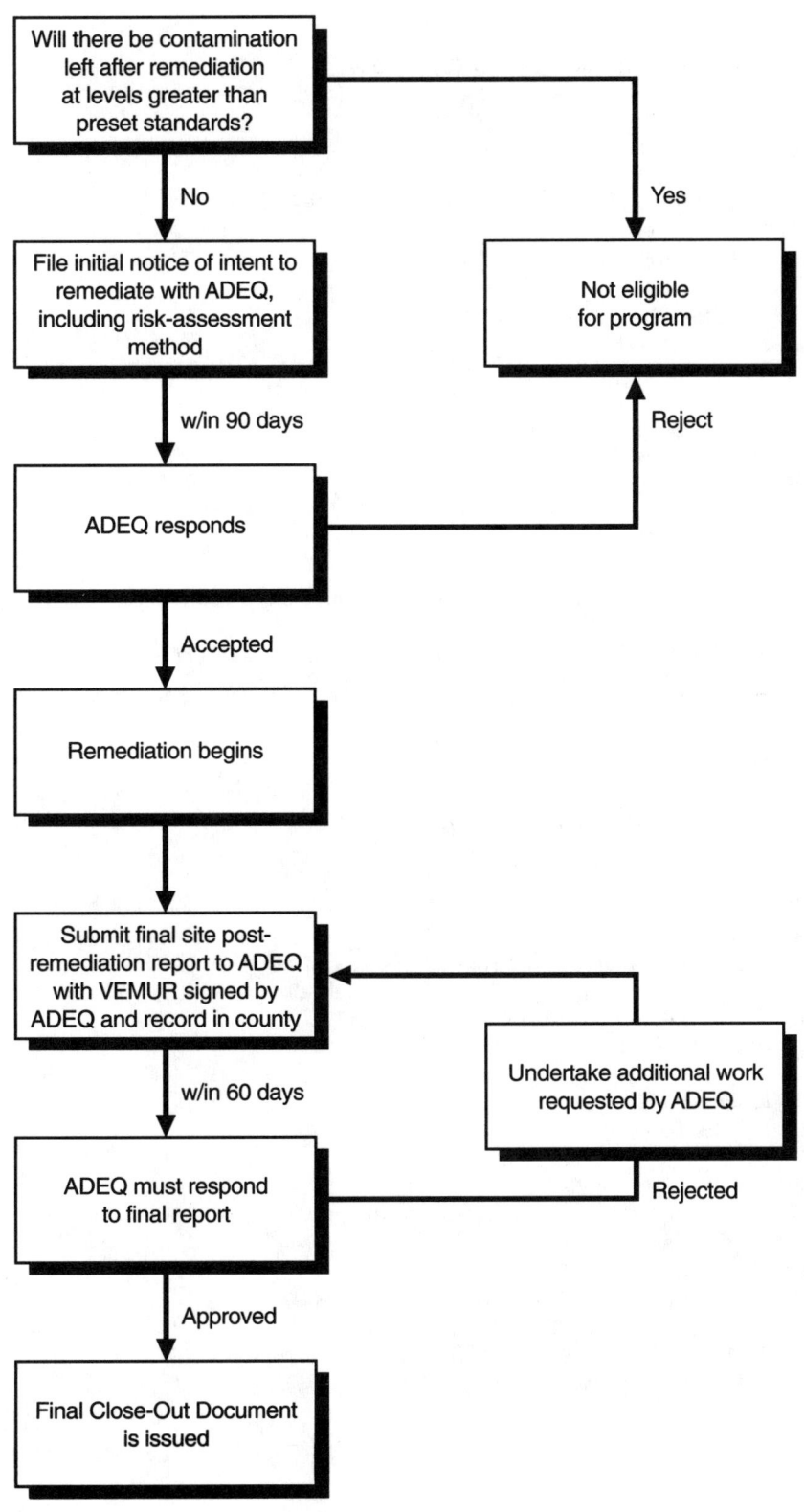

Notes

1. ARIZ. REV. STAT. ANN. § 49-152(B) (1995).
2. ARIZ. COMP. ADMIN. R. & REGS. § 18-7-202 (1995).
3. ARIZ. REV. STAT. ANN. § 49-285.01 (1995).
4. *Id.*
5. *Id.* § 49-285(B).
6. ARIZ. COMP. ADMIN. R. & REGS. § 18-7-107 (1995).
7. *Id.* § 18-7-108.
8. *Id.* § 18-7-108(C).
9. *Id.* § 18-7-108(D)–(F).
10. ARIZ. REV. STAT. ANN. § 49-285(B) (1995).
11. *Id.*
12. *Id.*
13. *Id.*
14. *Id.* § 49-290(A).
15. *Id.* § 49-296(B).
16. *Id.* § 49-1016(C)(3).
17. *Id.* § 49-1016(C)(1).
18. *Id.* § 49-1016(C)(2).
19. *Id.* § 49-1071.
20. *See id.* § 49-1054 (showing amounts reimbursable and deductible levels depending on when costs incurred).
21. *Id.* § 49-152(B); ARIZ. COMP. ADMIN. R. & REGS. § 18-7-201 *et seq.* (1995).
22. *See* ARIZ. REV. STAT. ANN. § 49-152(B) (1995).
23. ARIZ. COMP. ADMIN. R. & REGS. § 18-7-205 (1995).
24. *Id.* § 18-7-206.
25. *Id.* § 18-7-204.
26. *See* ADEQ, EXECUTIVE SUMMARY OF THE CONCISE EXPLANATORY STATEMENT FOR THE EMERGENCY INTERIM SOIL REMEDIATION STANDARDS, at 1 (1995).
27. *Id.*
28. *Id.*
29. *Id.*
30. *Id.*
31. ARIZ. COMP. ADMIN. R. & REGS. § 18-7-201(4) (1995).
32. *Id.*
33. *See* 60 Fed. Reg. 34,792 (1995).
34. ARIZ. REV. STAT. ANN. § 49-285.01 (1995).
35. *Id.* § 49-281(4).
36. *Id.* § 49-282(D).
37. *Id.* § 49-283(B)(3).
38. *Id.*
39. *Id.* § 49-283(H)—(M).
40. *Id.* § 49-1016(D). *See also id.* § 49-1001.01 (providing exemptions to security holders and fiduciaries as being owners of USTs).
41. *Id.* § 49-285.01.
42. ARIZ. COMP. ADMIN. R. & REGS. § 18-7-109(D)(2) (1995).
43. *Id.* § 18-7-109(D)(1), 18-7-109(D)(3).
44. ARIZ. REV. STAT. ANN. § 49-285(B) (1995).
45. ARIZ. COMP. ADMIN. R. & REGS. § 18-7-107 (1995).

46. *Id.* § 18-7-108.
47. *Id.* § 18-7-107(A).
48. *Id.* § 18-7-107(B); ARIZ. REV. STAT. ANN. § 49-285.01 (1995).
49. ARIZ. COMP. ADMIN. R. & REGS. § 18-7-107(B) (1995).
50. *Id.* § 18-7-107(B) (1995).
51. *Id.* § 18-7-108(D)–(F).
52. ARIZ. REV. STAT. ANN. § 49-285(B) (1995).
53. *Id.* § 49-1016(C)(1).
54. *Id.* § 49-1016(C)(2).
55. *See* ADEQ EXECUTIVE SUMMARY OF THE CONCISE EXPLANATORY STATEMENT FOR THE EMERGENCY INTERIM SOIL REMEDIATION STANDARDS, at 59 (1995).
56. ARIZ. COMP. ADMIN. R. & REGS. § 18-7-208(B) (1995).
57. *Id.* § 18-7-208(B)(1).
58. *Id.* § 18-7-208(B)(2)–(3).
59. *Id.* § 18-7-208(C).
60. *Id.* § 18-7-208(D).
61. *Id.* § 18-7-208(E).
62. *Id.* § 18-7-208(E)–(F).
63. *Id.* § 18-7-208(G).
64. ARIZ. REV. STAT. ANN. § 49-285(C) (1995).
65. *Id.* § 49-285(D).
66. *Id.* § 49-285(E).
67. *Id.* § 49-1019(A).
68. *Id.* §§ 49-1017(D), 49-1019(C).

CHAPTER 22

Arkansas

ALLAN GATES

Hazardous-substance remedial actions in Arkansas are generally governed by the Remedial Action Trust Fund Act (RATFA).[1] RATFA is the Arkansas analogue to the federal Superfund statute. In 1995, the Arkansas agency that administers RATFA, the Arkansas Department of Pollution Control & Ecology (ADPC&E), secured passage of a brownfields redevelopment amendment to RATFA.[2] The amendment allows a prospective purchaser of an abandoned industrial site to limit its remedial liability to measures necessitated by the actual land use planned by the purchaser (the Program or the Brownfields Program). ADPC&E has received several contacts expressing interest in the Program, but as of June 1, 1996, no one had actually undertaken a redevelopment project under the Arkansas brownfields amendment. The low level of interest in the Arkansas brownfields amendment is probably due in large part to the provision that requires a previously nonliable prospective purchaser to accept open-ended remedial liability for a contaminated site without receiving any meaningful incentive in exchange for accepting such a burden. Despite the disappointing initial response, ADPC&E views the brownfields amendment as a good first step in promoting redevelopment of contaminated industrial sites, and ADPC&E is continuing to explore other ways to promote and facilitate brownfields redevelopment in Arkansas.

Participation in the Arkansas Brownfields Program

Eligibility for the Program

The Arkansas Brownfields Program is available only to a "prospective purchaser of an abandoned industrial site."[3] The statute defines the phrase "prospective purchaser" as a

> person who expresses a willingness to acquire an abandoned industrial site and is not responsible for any preexisting pollution at or contamination on the site."[4]

Two aspects of this definition merit attention. First, the definition limits eligibility for participation in the Program to individuals who have no liability under the statute for the site in question. Second, participation in the Arkansas Brownfields Program is limited to purchasers. Long-term lessees and other nonpurchasers are not eligible to participate under the amendment.

The Arkansas brownfields amendment defines the phrase "abandoned industrial site" as

> a site on which one had an industrial activity and for which no responsible person can reasonably be pursued for a remedial response to clean up the site.[5]

The statute defines "industrial activity" so broadly that almost any commercial property under consideration for brownfields redevelopment should qualify as an "industrial site."[6] The requirement that a site be "abandoned" is more problematic. Under RATFA, the current owner is ordinarily one of the parties liable for remedial action.[7] Many owners of idle, contaminated industrial sites have at least some capacity to respond to the contamination on their property. Moreover, if a prospective purchaser pays the owner anything for the property that is to be redeveloped, the sales proceeds would seem to be a logical target for the ADPC&E or some other claimant to pursue. Under these circumstances it is unclear what proof would be necessary to satisfy the requirement that a site have "no responsible person [that] can reasonably be pursued for a remedial response to clean up the site."[8] ADPC&E is considering the possibility of seeking an amendment to the statute that would relax or eliminate the requirement that a site be "abandoned" to participate in the Program.

The Arkansas brownfields amendment imposes one additional eligibility requirement. Under the amendment, a brownfields redeveloper must "reuse or redevelop the property for industrial activities to create employment expansion."[9] As noted previously, the definition of "industrial activity" is quite broad; but it appears that at least some redevelopment projects, such as residential or commercial retail development, may be ineligible. It is also unclear what the phrase "employment expansion" may add to this eligibility requirement. For example, if a local manufacturer wishes to move from its existing location to a brownfields site in the same community without adding any new jobs, how would that manufacturer satisfy the "employment expansion" criterion?

Memorandum of Agreement

The state of Arkansas currently does not have a Memorandum of Agreement with the United States Environmental Protection Agency (USEPA) covering its RATFA program or the brownfields amendment. As a consequence, anyone participating in a RATFA remedial action, whether under the brownfields amendment or otherwise, is still subject to at least a theoretical possibility of a federal enforcement action or a cost-recovery claim by USEPA or private parties. USEPA Region VI currently takes the position that it will not initiate federal response actions at sites in the region where remediation under state law is already under way, unless conditions at the site present an imminent and substantial endangerment.[10] USEPA Region VI has not yet issued any "comfort letters" or prospective-purchaser agreements for sites in Arkansas, but according to the regional staff, it would entertain a request to do so under appropriate circumstances.[11]

Liability Protection

Lenders, Trustees, and Fiduciaries

The Arkansas brownfields amendment extends no special liability protection to lenders, trustees, or fiduciaries, but the basic provisions of RATFA do include a secured-lender exemption that is analogous to the secured-lender exemption in the federal Comprehensive Environmental Response, Compensation, and Liability Act (CERCLA) (42 U.S.C. § 9601).[12] There is no statutory language or case law in Arkansas explicitly dealing with general trustee or fiduciary liability, but it is likely that ADPC&E and Arkansas courts would tend to follow federal-court decisions under CERCLA on this point.[13] It should be noted that RATFA does include a specific exemption for grantors of trusts, and officers, directors, and shareholders of corporations, that have the sole purpose of conducting remedial action at a hazardous-substance site under RATFA.[14]

Brownfields Redevelopers

The Arkansas brownfields amendment provides redevelopers no significant protection against liability under state law. However, the amendment does provide that a redeveloper participating in the Brownfields Program

> shall not be responsible for paying any fines or penalties levied against any person responsible for contamination on the abandoned industrial site prior to the consent administrative order with the Department.[15]

This protection against fines and penalties is largely meaningless because a person cannot participate in the Brownfields Program if he or she has any liability for the preexisting contamination at the site.[16]

ADPC&E's original concept for the Arkansas brownfields amendment was to have a prospective purchaser perform a comprehensive site assessment and then negotiate a consent administrative order specifying the remedial action to be taken. Under this concept, the prospective purchaser would have been able to choose whether it wished to proceed with the brownfields project after it knew all the remedial actions ADPC&E would require. Unfortunately, the Arkansas brownfields amendment, as ultimately adopted, fails to limit the prospective purchaser's liability to the remedial measures in the consent order. Instead, the amendment explicitly states that the consent order

> shall not relieve the purchaser of any liability under law for preexisting problems not identified in the comprehensive site assessment.[17]

Because a redeveloper must purchase the abandoned industrial site to be eligible to participate under the Arkansas brownfields amendment, the redeveloper would almost always be liable as a current owner for any additional remedial action required to address conditions discovered after completion of the site assessment.

The redeveloper's exposure to liability for newly discovered conditions is enhanced by the statutory definition of the phrase "site assessment," which provides that the

> assessment shall, *at a minimum*, identify the location and extent of contamination, the quantity or level of contamination, the type of contamination, the probable source of contamination and the risk or threat associated with the contamination.[18]

Because a site assessment must, "at a minimum," identify the location, extent, quantity, type, and probable source of all contamination, and all risks and threats associated with the contamination, it seems obvious that ADPC&E could argue that a site assessment and remedial action cannot satisfy the minimum statutory requirements unless they identify and address all contamination and risks at the site.

Against this background, it seems fair to conclude that the Arkansas brownfields amendment requires a redeveloper, who by definition has no previous liability for the site, to sign a consent administrative order that makes it responsible for cleaning up all known contamination at the site, and leaves the redeveloper, as a current owner/operator, exposed to joint and several liability for any subsequently discovered contamination, even if the contamination is attributable entirely to conduct of others and occurred before the brownfields redevelopment took place.

Subsequent Purchasers of Brownfields

The Arkansas brownfields amendment also fails to provide any meaningful protection to subsequent purchasers. The amendment provides that a redeveloper may transfer its rights and liabilities under the consent administrative order to a subsequent purchaser upon written notice to ADPC&E, as long as the subsequent purchaser did not cause or contribute to any release or threatened release of hazardous substances on the site.[19] The amendment also provides that any subsequent owner must be given a copy of the consent administrative order, which will describe all applicable land-use restrictions and prohibits subsequent owners from engaging in any use of the property inconsistent with the land use specified by the order.[20]

Financial Incentives

There are no financial incentives currently associated with the Arkansas Brownfields Program. The only incentive in the Program—assurance that investigation and cleanup requirements will be based on the redeveloper's actual land use—has thus far proven inadequate to prompt any firm proposals for redevelopment. ADPC&E is continuing to explore the possibility of adding financial or other incentives to make the program more attractive.

Cleanup Standards

Before the 1995 brownfields amendment, Arkansas had no explicit statutory or regulatory RATFA cleanup standards. Instead, the numerical cleanup triggers and the categorical nature of remedial actions were determined on a case-by-case basis, usually in reliance on formal or informal risk assessment, or an applicable regulatory standard, such as water quality standards in affected surface waters. The 1995 brownfields amendment established, for the first time, three explicit statutory cleanup requirements.

First, the amendment requires a redeveloper to

> remediate, remove, and properly dispose of or manage, consistent with applicable requirements, any containerized wastes existing on the site at the time of purchase, including drummed waste, lagoons, and impoundments and wastes in aboveground and underground tanks.[21]

The brownfields amendment makes clear that any disposal or management of waste on the site must comply with all applicable regulatory requirements.[22]

Second, the amendment requires a redeveloper to

> take all necessary steps to prevent migration of hazardous substances beyond the property boundary.[23]

Third, the amendment requires a redeveloper to "remedy any releases of hazardous substances."[24] Although the term "release" is defined in both RATFA and CERCLA, the 1995 amendment includes a special definition of the term "release"[25] for purposes of this particular brownfields cleanup requirement:

> For purposes of subsection (g) [the requirement to "remedy any releases"], releases of hazardous substances are those conditions which pose either:
> (1) (A) An unacceptable risk, either acute or chronic, to the health of employees or any other person likely to be exposed to the release at the site, based upon the intended [land] use described ... in the consent administrative order[; or]
> (2) An unacceptable risk to degrade either groundwaters or surface waters or any risk to degrade extraordinary resource waters of the State of Arkansas.[26]

In selecting the remedial action to meet the three statutory cleanup requirements, the brownfields amendment directs ADPC&E to include five specific factors in its consideration:

(A) The intended and allowable use of the abandoned industrial site;
(B) The ability of the contaminants to move in a form and manner which would result in exposure to humans and the surrounding environment at levels considered to be a significant health risk . . .;
(C) Consideration of the potential environmental risks of proposed alternative remedial action and its technical feasibility, reliability, and cost effectiveness;
(D) When an imminent and substantial endangerment is posed; and
(E) Whether institutional or engineering controls eliminate or partially eliminate the imminent and substantial endangerment or otherwise contain or prevent migration.[27]

The brownfields amendment also provides that remedial action under subsection(g)—the requirement to remedy all releases—is not required if the contamination will not pose an "unreasonable risk" in light of the restrictions on use proposed by the prospective purchaser:

(3) Remedial actions pursuant to subsection (g) of this section are not required to provide for the removal or remediation of the conditions or contaminants causing a release or threatened release on the abandoned industrial site if:
(A) Contaminants pose no unacceptable risk as described in subdivisions (h)(1)(A) and (2) of this section, or the remedial actions proposed in the assessment and intended uses of the industrial site will eliminate unacceptable risks as described in subdivisions (h)(1)(A) and (2) of this section; or
(B) Activities required to allow the intended reuse or redevelopment of the industrial site are done in a manner which will protect public health and the environment as described in subdivisions (h)(1) and (2) of this section.[28]

The statutory cleanup standards described above are confusing and potentially contradictory. For example, subsection (f) of the brownfields amendment requires a prospective purchaser to *"take all necessary steps to prevent migration* of hazardous substances beyond the property boundary" (emphasis added), but subsections (h) and (j) appear to limit the remedial obligation to only the releases that pose "unacceptable risks" to human health, "unacceptable risks" of degradation of ordinary surface water or groundwater, or "any risk" of degradation of extraordinary resource waters. It is difficult to predict how these seemingly contradictory cleanup standards might be applied in selecting a remedy in a brownfields consent administrative order. At a minimum, however, it seems clear that a remedy selected by ADPC&E under the brownfields amendment will be tied to the intended use of the property. ADPC&E currently has no numerical cleanup standards for specific contaminants, or any specific risk goals for selecting site remedies. Site-specific risk assessment has been accepted by ADPC&E as the basis for remedy selection in other RATFA cleanups, and ADPC&E staff confirm that approach will be accepted in a brownfields project as well. ADPC&E does not specify any particular procedures or methodology for performing a risk assessment. ADPC&E has accepted institutional and engineering controls in selecting remedies in ordinary RATFA remedial actions; therefore, one would expect such approaches to be available in appropriate circumstances in a brownfields project.

The Mechanics of Participating in the Arkansas Brownfields Program

Any prospective purchaser who is interested in participating in the Arkansas brownfields program must first submit a proposal to ADPC&E for conducting a comprehensive site assessment to "establish the baseline of existing contamination on the site."[29] ADPC&E will review and comment on the site-assessment proposal. If the prospective purchaser elects to proceed after receiving ADPC&E's comments, it must perform the comprehensive site assessment consistent with ADPC&E's comments.[30] Upon completion of the comprehensive site assessment, ADPC&E will review the results to determine "whether the assessment adequately identifies the environmental risks posed by the abandoned industrial site."[31] Although the brownfields amendment does not address what should happen if ADPC&E concludes that a site assessment does not adequately identify the environmental risks posed by the site, it seems probable that additional assessment would be required before a redeveloper could proceed further. Once ADPC&E approves the site assessment, the next step is for the prospective purchaser and ADPC&E to negotiate a consent administrative order that "shall establish the intended use of the property;" and "shall establish [the redeveloper's] cleanup liabilities and obligations for the abandoned industrial site."[32]

If the redeveloper wishes to proceed on the basis of the consent administrative order it has negotiated with ADPC&E, the redeveloper must sign the order, publish it in a local newspaper, and formally record it in the county where the site is located.[33] The redeveloper becomes obligated to undertake remedial action when it acquires legal title to the site.[34] If the redeveloper wishes to convey ownership of the site to another, the transferee must be given a copy of the consent administrative order, and any use of the property by the transferee or any other subsequent owner must be consistent with the consent administrative order.[35] The redeveloper, upon written notice to ADPC&E, may also transfer its "rights and cleanup liabilities" under the consent administrative order to any person who did not cause or contribute to any release of hazardous substances on the site.[36]

Covenant-Not-to-Sue

The Arkansas brownfields amendment contains no express provisions regarding covenants-not-to-sue. As no one has utilized the Arkansas Brownfields Program, one can only speculate what a brownfields consent administrative order might provide. It is unclear whether ADPC&E could agree to a broad covenant-not-to-sue or any other meaningful limitation on the redeveloper's liability, because the statute expressly provides that the consent administrative order shall not relieve the redeveloper of any liability under law for preexisting problems not identified in the comprehensive site assessment.[37]

Deed Restrictions

The brownfields amendment requires a participant to record the consent administrative order in the county where the site is located, and to record deed restrictions that limit use of the property to the land uses upon which the remedy is based.[38] The amendment also requires that subsequent owners of the property receive a copy of the consent administrative order, and the statute prohibits subsequent owners from using the site in any manner inconsistent with the land use described in the order.[39]

The brownfields amendment expressly contemplates that the consent order may be amended. Therefore, it appears that the land-use restrictions could be subject to change by way of amending the deed and consent order. Obviously, changing the land-use restrictions upon which a remedy was based might also require change in the remedial action required at the site. As noted previously, the brownfields amendment authorizes the transfer of a redeveloper's rights and liabilities upon written notice to ADPC&E. Because the rights under such an order may be negligible, and the liabilities could be substantial, a subsequent purchaser should carefully examine the statute and the order before accepting such a transfer.

Disclosure and Discovery

There is no protection under the Arkansas Brownfields Program against disclosure or discovery of facts relating to the contamination or the cleanup involved. Indeed, public notice of brownfields cleanup projects must be given, and the consent administrative order, which specifies the terms of the cleanup, must be recorded publicly.[40] Although Arkansas has an audit privilege statute, it is unlikely that information regarding a brownfields project would qualify for the audit privilege.[41] In fact, Arkansas has a very strong Freedom of Information Act (FOIA), and it is virtually certain that ADPC&E would consider most or all facts related to a brownfields project to be available to any member of the public upon request under the state FOIA.[42]

Fees, Approvals, and Certifications

There are no special fees required under the Arkansas Brownfields Program, but under ADPC&E Regulation No. 23, ADPC&E charges a fee of approximately $60 per hour for review of corrective-action documents and technical proposals. ADPC&E indicates that it will charge this fee for the review of documents submitted as part of the Program. Fees for review of such documents are currently limited to a $15,000 per year maximum for any one unit or group of units that will be remediated, or a single corrective-action management unit.[43]

The only certifications or approvals specified by the brownfields amendment are (1) ADPC&E's approval of the site assessment, and (2) entry of the consent administrative order specifying the land use and directing the redeveloper to undertake remediation. Arkansas law does require use of certified laboratories for analytical results.[44] A registered professional engineer must approve any plans and specifications involving the practice of engineering that are filed for public record,[45] and a registered engineer or registered professional geologist must approve subsurface studies or plans involving the practice of geology that are filed for public record.[46] Beyond these general requirements, there are no special limitations or prescriptions on the qualifications of the professionals used in the Arkansas Brownfields Program.

Remedy Failure or Ineffectiveness

The brownfields amendment is silent about what happens if a remedy selected by ADPC&E in a brownfields consent administrative order fails or is ineffective. The first four subparagraphs of the brownfields amendment seem to suggest that the consent administrative order will specify a remedy, and that the redeveloper's only obligation is to install and operate that remedy.[47] Subsequent provisions of the amendment, however, suggest that the redeveloper is responsible for satisfying three statutory cleanup requirements, regardless of whether the remedy selected by ADPC&E in the consent administrative order proves to be ineffective.[48] This confusion is compounded by the fact that there is no express mechanism in the brownfields amendment for ADPC&E to approve a completed remedial action or even for determining when a redeveloper's remedial efforts are deemed complete. These problems might be at least partially resolved by including appropriate provisions in the consent administrative order.

Cost Recovery

The brownfields amendment includes a provision that explicitly authorizes a redeveloper to seek contribution from liable parties for the costs of the cleanup.[49] Contribution actions are subject to a three-year statute of limitations, which runs from the date the order regarding the remedial action is entered.[50] Such actions may be commenced as soon as the person proposing to bring the action has begun undertaking remedial action pursuant to a RATFA order or settlement.[51] In any contribution action, a party challenging the adequacy of the remedy must show that the remedy is arbitrary and capricious, based upon the administrative record.[52] Liability in a contribution action is to be apportioned by the court, "using such equitable factors as the court determines are appropriate."[53]

There are no reported cases addressing the character of liability under RATFA, but one recent decision explicitly states that RATFA should be read in appropriate circumstances to have a meaning similar to analogous provisions in CERCLA.[54] Under this decision, it seems likely that an Arkansas court hearing an RATFA claim would consider the federal case law under CERCLA on strict, joint, and several liability to be at least relevant, and perhaps directly applicable. It should be noted, however, that RATFA also contains a provision that explicitly departs from the CERCLA model of joint and several liability and declares, at least for purposes of government cost-recovery actions, that the state should bear the cost of the orphan share, that viable private parties should pay only costs attributable to their own actions, and that costs should be allocated, at least presumptively, in proportion to the volume of wastes contributed by each party to the site.[55]

It is unclear how a court would apportion costs between an otherwise nonliable brownfields redeveloper and parties, like generators and prior owner/operators, who are liable under RATFA. Presumably the redeveloper would argue for apportioning all costs to the parties who caused the contamination; the liable parties presumably would argue that a redeveloper would enjoy an inappropriate windfall if it acquired property at a discount due to the presence of contamination and then recovered all the costs it expended cleaning up the property.

Conclusion

The Arkansas brownfields amendment stands as clear evidence of ADPC&E's interest in developing a successful voluntary cleanup program. Unfortunately, the brownfields amendment has not attracted any interest, and can be considered only as a good first step on the path toward facilitating significant voluntary cleanup activity. ADPC&E appears to recognize the need for further revisions to the relevant statutes and regulations, and it has demonstrated a substantial commitment to further action necessary to adopt the appropriate changes.

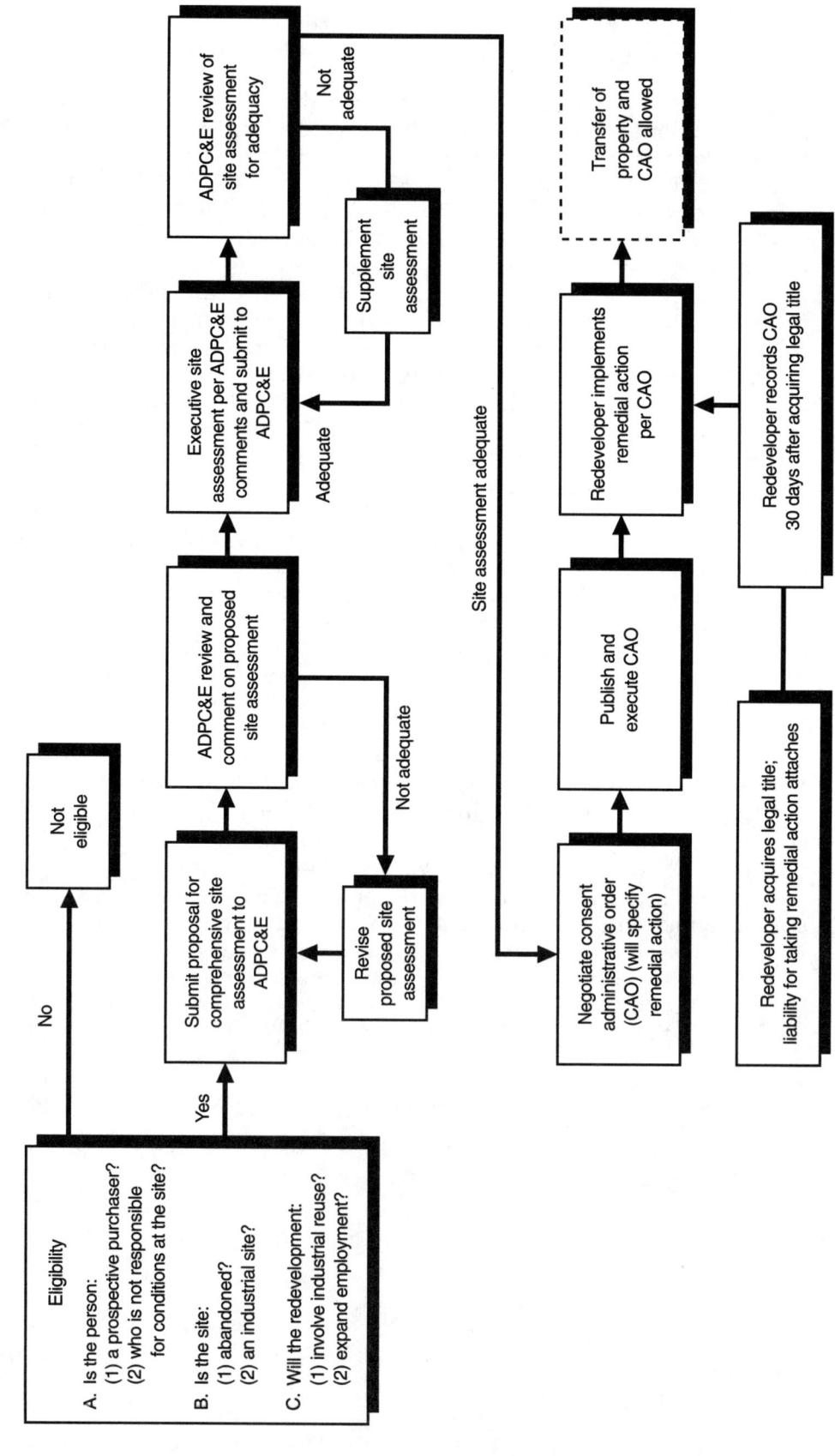

Notes

1. Ark. Code Ann. § 8-7-501, *et seq.* (Michie 1993 & Supp. 1995).
2. 1995 Ark. Acts 125, codified at Ark. Code Ann. § 8-7-523 (Michie Supp. 1995).
3. Ark. Code Ann. § 8-7-523(a)(1) (Michie Supp. 1995).
4. *Id.* § 8-7-503(16).
5. *Id.* § 8-7-503(13).
6. The brownfields amendment defines "industrial activity" as follows: "`Industrial activity' means commercial, manufacturing, or any other activity done to further either the development, manufacturing, or distribution of goods and services, including, but not limited to, research and development, warehousing, shipping, transport, remanufacturing, repair, and maintenance of commercial machinery and equipment." *Id.* § 8-7-503(14).
7. *See id.* § 8-7-512(a)(1).
8. *Id.* § 8-7-503(13).
9. *Id.* § 8-7-523(a)(3).
10. Interview with Stan Hitt, Brownfields Coordinator, USEPA Region VI (Apr. 25, 1996).
11. *Id.*
12. Ark. Code Ann. § 8-7-508(a)(2) (Michie 1993).
13. *See infra* note 53 and accompanying text.
14. Ark. Code Ann. § 8-7-522 (Michie 1993).
15. *Id.* § 8-7-523(l).
16. *Id.* § 8-7-523(a)(2) (providing that brownfields amendment is available only a person who "Did not, by act or omission, cause or contribute to any release or threatened release of a hazardous substance on or from the identified abandoned industrial site or is not otherwise considered to be a responsible party pursuant to [RATFA.]"). Similar language appears in the definition of "prospective purchaser." *See id.* § 8-7-503(16).
17. *Id.* § 8-7-523(m).
18. *Id.* § 8-7-503 (17)(B) (emphasis added).
19. *Id.* § 8-7-523(p).
20. *Id.* § 8-7-523(q).
21. *Id.* § 8-7-523(e)(1).
22. *Id.* § 8-7-523(e)(2).
23. *Id.* § 8-7-523(f).
24. *Id.* § 8-7-523(g).
25. *See id.* § 8-7-503(7) (general definition of "release" in RATFA); *see also* 42 U.S.C. § 9601(22) (1991) (definition of "release" under CERCLA).
26. Ark. Code Ann. §§ 8-7-523(h)(1)–(2) (Michie Supp. 1995).
27. *Id.* § 8-7-523(j)(2).
28. *Id.* § 8-7-523(j)(3).
29. *Id.* § 8-7-523(b)(1).
30. *Id.* § 8-7-523(b)(2).
31. *Id.* § 8-7-523(c).
32. *Id.* § 8-7-523(d).
33. *Id.* §§ 8-7-523(d)(3), 8-7-523(r).
34. *Id.* §§ 8-7-523(e)–(g).
35. *Id.* § 8-7-523(q).
36. *Id.* § 8-7-523(p).
37. *Id.* § 8-7-523(m).
38. *Id.* § 8-7-523(o).

39. *Id.* § 8-7-523(q).

40. *Id.* §§ 8-7-523(d)(3), 8-7-523(o), 8-7-523(r).

41. The Arkansas audit privilege statute provides that communications are not eligible for the audit privilege if they must be made available to ADPC&E under any statute or order, or if the information was obtained from a source independent of an environmental audit. *Id.* § 8-1-305.

42. Under the Arkansas audit privilege statute, information that is privileged is still subject to disclosure under the state FOIA. The availability for public disclosure does not alter the fact that the information may not be used in any proceeding. *See id.* § 8-1-312(b).

43. *See* Ark. Reg. Docket No. 014.09.96–001 (commonly referenced as ADPC&E Regulation No. 23, § 6(t)).

44. ARK. CODE ANN. § 8-2-206(a)(1) (Michie 1993) (laboratory certification required for data submitted by consulting laboratories).

45. *Id.* § 17-32-310(b).

46. *Id.* § 17-30-303(a)(2).

47. *See id.* §§ 8-7-523(a)–(d).

48. *See id.* §§ 8-7-523(e)–(g).

49. *Id.* § 8-7-520(a).

50. *Id.* § 8-7-520(g).

51. *Id.* § 8-7-520(a).

52. *Id.* § 8-7-520(h). The precise meaning of this subparagraph is difficult to determine because the original act and the code both contain what appears to be a typographical error that directs the court to disallow only those costs that are inconsistent with the terms of an administrative or judicial order or settlement found to be arbitrary or capricious. Stated differently, this would seem to allow recovery of only those costs that are consistent with an arbitrary and capricious remedy.

53. *Id.* § 8-7-520(d).

54. Gurley v. Mathis, 856 S.W.2d 616 (Ark. 1993).

55. ARK. CODE ANN. § 8-7-513 (Michie 1993). This section opens with an introductory phrase that appears to limit its application to cost-recovery actions by the state: "Any party found liable for any costs or expenditures *recoverable under §§ 8-7-512, 8-7-514, 8-7-515, and 8-7-517. . . .*" *Id.* § 8-7-513(a)(1) (emphasis added). The enumerated sections all deal with claims brought by the government, not by private parties.

CHAPTER 23

California

JANE B. KROESCHE

Brownfields in California are varied and complex, and include virtually every imaginable past uses of properties, from now-closed (or closing) military installations, to aerospace and "high-tech" manufacturing sites, to locations of historical industrial uses (such as railyards and gas manufacturing plants). In many instances, these industrial sites are—or would be—considered prime real estate but for the real or perceived contamination located on, under, or about them. As a result, the climate is ripe in California for brownfields redevelopment initiatives.

A lingering recession from the late 1980s, coupled with stringent (and in many cases procedurally or economically crippling) environmental cleanup standards, in large measure spurred the drive, both public and private, within California to invigorate the voluntary and cooperative redevelopment of contaminated properties. In the early 1990s, the legislature and various agencies within the state began devising means to revitalize contaminated urban properties, well before the term "brownfields" became common coinage. Amid a groundswell of support, governmental agencies, private entities, and citizens groups alike have mobilized to coordinate cleanup efforts and facilitate property redevelopment and reuse, while at the same time ensuring the protection of human health and safety and the environment. These efforts have resulted in numerous legislative and administrative reforms, as well as private initiatives, within the state to encourage the redevelopment of brownfields.

As a result of these several efforts, brownfields initiatives in California (as with other regulatory schemes) are not neatly packaged in a single statute

---

Ms. Kroesche would like to thank Karen L. Peterson, an associate, and Gregory N. Mandel, a summer associate, at Skadden, Arps, Slate, Meagher & Flom in San Francisco, for their assistance in the preparation of this chapter. Special thanks goes to Megan Cambridge, Chief of the Expedited Remedial Action Program, California Environmental Protection Agency, Department of Toxic Substances Control, for her input and interest in the preparation of this chapter.

or regulatory program. The various programs, however, provide significant options and opportunities in California for increased flexibility and innovation in addressing the rehabilitation of contaminated property. As of the fall of 1996, the brownfields redevelopment landscape in California remains in tremendous flux. Anyone seeking to take full advantage of brownfields programs in California is cautioned to check with the appropriate regulatory agencies and industry groups for the current status of existing programs and new innovations in this area.

This chapter will focus on the primary brownfields redevelopment vehicles currently being implemented or developed in California. Those programs include the Department of Toxic Substances Control's Voluntary Cleanup Program and Expedited Remedial Action Program. Other existing or pending reforms also will be discussed, including recent revisions to the State Water Resources Control Board's policies regarding contaminated aquifers and pending changes regarding underground storage tank cleanups, as well as prospective-purchaser protections, permit streamlining, and lender-liability protections. It should be noted that in many cases, more than one program—or a combination of programs—may be appropriate for addressing brownfields redevelopment in California.

Voluntary Cleanup Program (VCP)

The California Environmental Protection Agency, Department of Toxic Substances Control (DTSC) has been at the forefront in promoting brownfields redevelopment in California. DTSC's primary vehicle for this purpose is the Voluntary Cleanup Program (VCP). Formerly known as the "Walk-In Program," the VCP was created administratively in 1993 by DTSC, under the authority of the Carpenter-Presley-Tanner Hazardous Substance Account Act (the HSAA, also known as the state Superfund law).[1] Section 25355(b) of the California Health and Safety Code allows DTSC to initiate removal or remedial actions unless those actions are undertaken in a proper and timely manner by a "responsible party."[2] Section 25355.5(a)(1)(C) permits DTSC to enter an enforceable agreement with a potentially responsible party, which requires the party to take necessary corrective action either to remove the threat of a release or to determine the nature and extent of the release, prepare a remedial action plan, and complete the necessary removal or remedial actions.

The VCP was created to allow project proponents to investigate and/or clean up contaminated properties that are not high-priority sites under the HSAA (or other state or federal laws), with regulatory oversight and, ultimately, state sign-off by DTSC.[3] Under the VCP, a project proponent, which need not be a responsible party, by agreeing to pay DTSC's oversight costs, may enter an agreement with DTSC to remediate or remove the threat of a release, characterize a site, prepare a remedial action plan or removal action work plan, and complete any necessary response actions.[4] Under the VCP, then, the project proponent undertakes site remediation and the DTSC provides review, oversight, and approval of site cleanup.

The major components of the VCP include the following:

- DTSC oversees cleanup activities at sites not included in DTSC's Annual Workplan (AWP),[5] provided DTSC is paid its oversight costs.
- A project proponent, though subject to the same cleanup standards designed to protect human health and the environment as set forth in the state Superfund law, may conduct the remediation in a manner and according to a schedule established largely by the project proponent.[6]
- Site-specific risk analyses and land-use restrictions may be used by the project proponent to tailor the remediation to the planned permanent land use.[7]
- Future liability risks are minimized by the issuance of a "no further-action" letter or "remedial-action certification."[8]

Eligibility for Participation in the VCP

If a project proponent is interested in conducting an investigation or cleanup under the VCP, it must submit an application to DTSC.[9] The application must identify the project proponent and the site to be investigated, and include details about site conditions, proposed land use, and potential community concerns. No fee is required to apply for the VCP; however, at the outset, the project proponent must agree to pay DTSC's estimated costs.[10] As of mid-1996, over two hundred projects had entered the program; approximately half of those have completed remediations. In many cases, the process from application to agency sign-off has been less than one year.

For the most part, only those sites that are ineligible for the program will not be admitted into the program. Sites that are not eligible to participate in the program include those sites that are listed on the DTSC's AWP (that is, state Superfund sites), sites on the National Priority List (federal Superfund sites), federal facilities,[11] and sites that fall outside DTSC's jurisdiction (such as sites that contain only petroleum products or waste) or that are under the oversight of another state or local regulatory program.[12]

The Voluntary Cleanup Agreement

Negotiating the Voluntary Cleanup Agreement

After DTSC accepts a project proponent's application, the project proponent must meet with DTSC's Regional Site Mitigation staff to negotiate the scope and terms of the voluntary cleanup agreement.[13] The agreement can cover any combination of services, including site investigation and characterization, sampling plans and work plans, health and safety plans, a risk assessment, a preliminary endangerment assessment (PEA),[14] remedial technology selection and implementation strategy, remediation goals, public participation, and, ultimately, site cleanup certification.[15] The agreement must specify the scope of work to be undertaken, a project schedule, and the DTSC oversight services and payment structure.[16] By entering a voluntary cleanup agreement, the project proponent does not admit to legal liability for the site.[17]

Eligible sites are divided into two categories, based on the estimated total cost of the cleanup (not including DTSC's oversight costs). For removal actions with an estimated cost of under $1 million, the project proponent must prepare a removal action work plan (RAW).[18] For larger cleanups, the project proponent must prepare a remedial action plan (RAP), which, among other things, must include a statement of reasons for the remedial action selected, an evaluation of alternative actions, and a nonbinding preliminary allocation of responsibility among all identifiable, potentially responsible parties.[19] Although both the RAP and the RAW must provide for agency and community notification and input,[20] the procedural and scoping requirements for a RAP are more extensive, including a mandatory thirty-day public comment period.[21] Both the RAW and the RAP are subject to DTSC approval.[22]

Site Activities under the Voluntary Cleanup Agreement

Once the project proponent and DTSC enter the voluntary cleanup agreement, on-site activities may begin, subject to the oversight of a DTSC project manager. DTSC will not provide any services under the VCP or the agreement until (1) the voluntary cleanup agreement has been executed by both the project proponent and DTSC, (2) DTSC has received advance funding for its oversight costs from the project proponent, and (3) the project proponent has committed to paying all project costs (including any additional DTSC oversight costs).[23] Any party to the agreement may terminate the voluntary cleanup agreement for any reason by giving thirty days' written notice to the other parties.[24]

The VCP requires the collection of all data necessary to identify the nature and extent of contamination. Generally, this is done through the preparation of a PEA,[25] which consists of an initial assessment and a risk-based analysis and is used to determine if the property warrants further work.[26] The VCP specifically contemplates site-specific, risk-based remedies and land-use restrictions as components of the remedial activity at the site.[27]

Cleanup Standards

As noted above, the required cleanup standards under a voluntary cleanup agreement may be tied to the planned use of the property.[28] If the site investigation demonstrates that it is not cost effective, technologically feasible, or necessary to the planned land use to remediate the site fully, then alternative cleanup levels may be used.[29] The applicable cleanup levels and the planned uses of the property will then be enforced by land-use controls through deed restrictions or by operation and maintenance requirements.[30]

In any event, all remedial actions undertaken pursuant to the VCP must comply with the HSAA, which, in turn, incorporates the requirements of the National Oil and Hazardous Substances Pollution Contingency Plan (National Contingency Plan or NCP).[31] As a result, all procedures, standards, and documentation regarding public participation, risk assessment, remedial investigation/feasibility studies, and land-use review must comply with the HSAA and the NCP, and all procedures and standards for deed restrictions and operation and maintenance programs must be consistent with the HSAA.[32]

Site Certification

Upon satisfactory completion of a PEA or equivalent report in which DTSC determines that certification is unnecessary or inappropriate, DTSC will issue the project proponent a no-further-action (NFA) letter.[33] Upon satisfactory completion of a remedial project (whether pursuant to an RAP or an RAW), DTSC will issue the project proponent a "Remedial Action Certification," which certifies the cleanup as complete and satisfactory to DTSC.[34] Both the NFA letter and the remedial action certification may be issued once all source removal at the site is completed, even if there is ongoing groundwater monitoring or other operations and maintenance activities anticipated under the voluntary cleanup agreement.[35]

The NFA letter or Remedial Action Certification will specify acceptable or unacceptable land uses, operation and maintenance requirements, and deed restrictions. Neither the NFA letter nor the Remedial Action Certification, however, constitutes a release or covenant-not-to-sue.[36]

Enforcement Actions under the VCP

An action for judicial review of the RAP or the RAW must be brought as a traditional mandamus action pursuant to section 1085 of the California Code of Civil Procedure.[37] The reviewing court will uphold the agency's determination regarding the RAP or the RAW if the agency's determination was based on "substantial evidence."[38]

DTSC retains its authority to issue emergency abatement orders or institute removal actions at the site, or even to list the site on the state Superfund list, if, at any time, it determines that the site poses or may pose an imminent or substantial endangerment to human health and safety or the environment.[39] In addition, DTSC retains its authority to take any other enforcement action authorized under the HSAA and other applicable laws, including the right to seek cost recovery, monetary penalties, and punitive damages.

Because the cleanup under the VCP is required to be consistent with the NCP, the project proponent may seek contribution from, or cost recovery against, other potentially liable parties that are not signatories to the voluntary cleanup agreement under the federal Comprehensive Environmental Response, Compensation, and Liability Act (CERCLA or Superfund).[40]

Expedited Remedial Action Program

The California Expedited Remedial Action Reform Act of 1994 (ERAP)[41] was enacted by the California legislature "to establish a pilot program to determine if expedited procedures for carrying out response actions at response action sites are appropriate and protective of human health and the environment."[42] ERAP is specifically designed to promote efficient and expedited contaminated-site remediation, and to encourage alternative approaches to cleanup outside the federal and state Superfund programs.[43] In contrast to the VCP, ERAP is entirely based in statute and, in fact, deviates markedly from other, more traditional approaches to site investigation and remediation.[44] ERAP also coordinates with, and in part integrates, other statutory schemes.[45]

Under ERAP, subject to certain limitations, up to thirty sites may participate in the program.[46] Once in the program, project proponents have significant latitude in designing a remediation program for their site that is tied to the planned permanent use of the property. Major components of ERAP include:

- Project proponents are not bound by the NCP or the HSAA, and may select a cleanup remedy consistent with compatible and permanent land use.
- With certain limitations for "hot spots" and unanticipated "interim endangerment," project proponents may design their own cleanup remedy, consistent with equitable factors, to protect human health and the environment.
- Potentially liable parties and equitable allocation of liability for the cleanup are determined early in the process, including possible *de minimis* settlement opportunities.
- Agency review of the response action is streamlined and consolidated, with resulting "super" certification for the completed cleanup.
- Participating responsible persons receive a covenant-not-to-sue from DTSC and qualified immunity from future liability.
- Orphan-share funding is available for up to ten sites under the program, and up to five state Superfund sites may be eligible to transfer into the program.
- Specific statutory and regulatory guidelines establish procedures that provide certainty and statutory presumptions of compliance.
- Dispute resolution mechanisms are available to program participants and the public.

Selection for the ERAP Program

To participate in ERAP, one or more responsible persons must submit a Request for Site Selection to DTSC. The application must include a Notice of Intent, stating that the project participant will comply with all ERAP requirements, enter an enforceable agreement with DTSC, and pay all response costs allocated to that party by DTSC in connection with the cleanup.[47] The application also must include a completed PEA (or equivalent report) that discusses the scope and nature of the release, documents the need for a remedial action, and discusses the affected community.[48]

The completed application is then submitted by DTSC, together with a recommendation by DTSC, to the Site Designation Committee (SDC) and to the affected local governmental agency.[49] The SDC may accept the application if it finds that (1) DTSC will be the "Administering Agency,"[50] (2) the findings in the PEA establish that remedial action is warranted at the site, and (3) the site is eligible under ERAP.[51]

Ineligible sites include federal Superfund sites, federal facilities, sites at which there is a "known condition of interim endangerment existing at the site at the time it is selected,"[52] and, with limited exception, state Superfund sites or other sites at which there is a judicial or administrative enforcement action pending.[53] In addition, before it can approve a site for cleanup under ERAP, the SDC must find that there are sufficient funds available to cover any orphan shares[54] at the site, unless at least one responsible party agrees (and financially guarantees) to pay for any unfunded orphan shares.[55]

The ERAP Agreement

Once the SDC approves an ERAP application, DTSC assumes jurisdiction over the cleanup. Under ERAP, DTSC, as the Administering Agency, assumes responsibility for administering all laws, regulations, and policies of all state and local agencies for the cleanup, and is authorized to enforce the cleanup and, ultimately, issue a binding certification that the cleanup is complete.[56]

Upon approval, DTSC is required to conduct a search of, and notify, all appropriate state and local agencies with jurisdiction over the site, notify all known potentially responsible parties for the cleanup at the site, and conduct a public hearing regarding the planned use of the site.[57] Within ninety days of the SDC's approval, DTSC is required to hold a Site Conference with the identified potentially responsible persons to discuss the nature of the cleanup and the procedures for carrying out the response action at the site.[58]

Within ninety days of the Site Conference,[59] DTSC may enter, with one or more of the responsible persons (RPs), an enforceable agreement that (1) requires the RPs to take all necessary response actions required by ERAP, (2) requires the RPs to pay all the state's response costs related to the site, and (3) contains a covenant-not-to-sue between all the parties to the agreement.[60] The enforceability of the covenant-not-to-sue shall be expressly conditioned on the successful completion of the response action and any other obligations set forth under the agreement.[61]

The RAP

After the ERAP agreement is reached, a RAP must be prepared.[62] The RAP and resulting response action are governed by the planned use of the site, which shall be determined by the local governmental agency.[63] The remedy selected under the RAP must (1) be protective of human health and the environment, (2) provide long-term reliability at a reasonable cost, (3) reasonably protect the waters of the state, and (4) remedy the site in a manner that will allow it to be permanently used for its planned use without significant risk to human health or potential for environmental damage.[64] Provided the foregoing standards are met, with limited exception, the RPs may use various self-selected remedial actions and controls for accomplishing the cleanup.[65] DTSC will designate and supervise the response for any discrete areas (or "hot spots") within the site that are present in high concentrations or are highly mobile and cannot be reasonably contained.[66]

The statutory requirements for the preparation and content of the RAP are detailed and specific. The RAP must include the selected response action alternative(s) (including a statement of remedial alternatives considered and rejected), a site-specific assessment, a description of the site characteristics, a cost analysis, a feasibility analysis, and an "appropriateness" analysis of the selected remedy.[67] The RAP also must contain a detailed risk analysis, a study of long-term effects, and a summary of legal compliance.[68] Finally, the RAP must provide a detailed description of all proposed remedial action measures, and establish an implementation schedule and any needed long-term operation and maintenance plans.[69] In setting forth the response action, the RAP may include land-use controls to limit or restrict site use consistent with the permanent planned use of the property.[70] Any land-use controls must be recorded to run with the land.[71] Any violation of the land-use controls by a person who knew or reasonably should have known about the restrictions may result in a civil penalty of up to $25,000 per day.[72]

Public Participation

Public participation is a key element of ERAP. ERAP mandates that DTSC ensure that the public has the opportunity to participate in any response actions under ERAP, including RAP approval and apportionment of liability for the site.[73] To this end, before approving the RAP for the site, DTSC must issue public notification of the RAP and any allocations thereunder, provide a thirty-day period for comment, and conduct at least one public meeting regarding the proposed response action.[74]

Approval of the RAP

Following the close of the public comment period, DTSC has sixty days to either approve the RAP or issue a notice of deficiency.[75] In a notable departure from other regulatory approval programs, under ERAP, DTSC may not reject the RAP outright. Rather, the RPs may modify the RAP in accordance with the notice of deficiency and resubmit the RAP to DTSC.[76] DTSC then has another sixty days to approve the RAP or issue another notice of deficiency. Alternatively, the RPs may, within thirty days, reject the deficiencies identified by DTSC and appeal to the Secretary of the California Environmental Protection Agency (Cal/EPA).[77] Failure by Cal/EPA to act on the appeal within thirty days entitles any RP to seek judicial review in a traditional mandamus action.[78]

Once DTSC has approved the RAP, it becomes final and effective sixty days after notice of the approval is issued, unless a petition for review is filed before that time.[79]

Apportioning Liability and Orphan Shares

At the same time DTSC considers the RAP, it must apportion liability for the site.[80] DTSC's ability to apportion liability for site cleanup under ERAP takes advantage of emerging principles in environmental cleanup standards. Specifically, DTSC is mandated to take into consideration "equitable factors and fairness principles" (the "Gore Principles") in allocating liability for the site.[81] This allocation includes weighing factors "appropriate" to the circumstances, such as the amount of hazardous waste attributable to each RP, the toxicity of the hazardous substances at the site, the degree of involvement and care of each RP for the site, and the degree of cooperation by each RP with regulatory agencies.[82] Unless the RPs agree among themselves to an alternate allocation of liability, DTSC's allocation of liability is determinative and subject only to limited review.[83]

At the time it reviews the RAP, DTSC also must determine whether there are orphan shares that will require funding from the expedited site-remediation trust fund (the Trust Fund).[84] Orphan shares are those site cleanup costs attributable to RPs that are insolvent or cannot be identified or located.[85] Up to, but no more than, ten sites participating in ERAP may include orphan-share funding.[86] RPs, however, may agree to assume responsibility for all orphan shares, backed with sufficient financial assurance, to have the site included under ERAP.[87]

DTSC may enter a *de minimis* settlement agreement with RPs if it determines that such an agreement is in the best interests of the public.[88] DTSC may not enter a *de minimis* settlement agreement if the total amount of *de minimis* liability exceeds 10 percent of the projected response action costs.[89] Factors that would warrant a *de minimis* agreement include (1) minimal contribution (presumed to be less than 1 percent) of toxic or hazardous substances to a site by an RP in comparison with the volumetric total, and (2) minimal contribution by an RP to the release at the site.[90] RPs that enter a settlement agreement with DTSC will receive contribution protection.[91]

Interim Endangerment

After the RAP is approved, the RPs may begin site remediation. If, at any time during this process, an "interim endangerment" occurs, immediate action must be taken by either the RPs or DTSC to contain or eliminate the endangerment.[92] An interim endangerment means "conditions at a site which pose a significant risk either of harm to human health or of serious environmental damage unless immediate response action is initiated."[93] Response actions pursuant to the RAP may not continue until the interim-endangerment condition is abated. The RPs are allowed the opportunity to initiate response actions appropriate to control the interim endangerment condition; otherwise, DTSC may take any actions necessary to control or eliminate the condition.[94] Failure to take appropriate action by any RP ordered to abate an interim endangerment is subject to a civil penalty of up to $25,000 per day for each day of noncompliance.[95]

Arbitration

The statute provides for mandatory arbitration of certain disputes that arise in the ERAP process. The arbitration process is governed through Cal/EPA's Office of Environmental Health Hazard Assessment (OEHHA), by a panel of prequalified arbitrators who sit on OEHHA's Hazardous Substance Cleanup Arbitration Panel.[96] The disputes submitted for arbitration include those concerning the RAP, DTSC's proposed apportionment of liability, any proposed *de minimis* settlements, DTSC's approval or denial of a change in land use under an existing RAP, or DTSC's approval or denial of a certificate of completion.[97] The statute specifies the standards of review, as well as the specific statutory provisions, that apply to the arbitration proceedings.[98] All arbitration proceedings include public hearings and must follow established quasi-judicial procedures.[99] Judicial review of decisions by the arbitration panel must be brought in an action for administrative mandamus.[100]

Certificate of Completion

Once all response actions set forth in the RAP are completed, the RP must file a request with DTSC for a certificate of completion.[101] The RP may file the request, and DTSC may issue the certification, even if long-term operation and maintenance is required or other long-term remedial activities (such as groundwater monitoring) remain to be done.[102] DTSC may issue the certification once it determines that the RAP has been completed, the site now is suitable for its planned permanent use, and there is no significant risk to human health or of future environmental damage.[103] DTSC is required to issue its determination within ninety days after the request is filed.[104] If DTSC denies the request, an appeal may be submitted to the arbitration panel.[105] DTSC must provide public notice of its decision on the request for certification; its determination will become final thirty days thereafter unless appealed.[106]

An RP may choose at any time not to continue the site cleanup under ERAP. However, before abandoning the program, the RP must take any necessary actions to stabilize the site and protect the public health and the environment.[107] If the RP does not stabilize the site, then it will be removed from the program and the RP will not be entitled to the statutory protections against further liability for the site. In addition, DTSC may pursue any appropriate enforcement action to clean up the site, including placing the site on the state Superfund list.[108]

Any RP who has complied with its ERAP agreement and the cleanup at the site may bring a contribution action against any RP that has not so complied. The party seeking contribution may also seek treble damages for the recalcitrant party's failure to comply under ERAP.[109] One-half the damages recovered must be paid to the Trust Fund.[110]

Protections against Future Liability

As noted above, the ERAP agreement contains a covenant-not-to-sue.[111] This covenant becomes effective upon completion of the RAP and receipt of the certificate of completion from DTSC.[112] Further, pursuant to the Unified Agency Review Law,[113] the certificate of completion issued under ERAP by DTSC constitutes "super" certification by all state and local agencies,[114] thereby entitling the RP to qualified immunity and protections from future liability set forth under the Unified Agency Review Law.[115]

It should be noted that subsequent purchasers, because they are not "responsible persons" as defined under ERAP, are not subject to liability thereunder, except to the extent that they are subject to the land-use controls established for the site.[116] Purchasers of sites at which a certificate of completion has been issued will be entitled to the same protections under the Unified Agency Review Law as the RPs under ERAP.[117]

Additional Measures to Facilitate Brownfields Redevelopment

Although the VCP and ERAP are specific programs designed to promote brownfields redevelopment in California, several other measures have been, or are being, undertaken throughout the state to facilitate streamlined review and rehabilitation of contaminated or suspect properties, or provide regulatory assurances to purchasers, developers, lenders, and others engaging in real property transactions or management.[118] Each such measure can have significant benefits to those seeking to rehabilitate brownfields in California. These measures are especially important in the context of real property development, where the value of time is money, and the value of regulatory assurances can be priceless. Several of the measures currently under way in California are discussed briefly below.

The Unified Agency Review Law

California is noteworthy for its numerous environmental agencies with overlapping jurisdictions for permitting and cleanup matters. For many years, a common complaint among the regulated community has been the inability to obtain complete assurance when conducting an environmental cleanup that all agencies with jurisdiction over the matter had been involved or would provide the necessary cleanup approvals. In response to this concern, in 1993 the California legislature passed the Unified Agency Review Law.[119]

Under the Unified Agency Review Law, an RP that commits to conduct a site investigation and cleanup may submit a request to the SDC for the designation of one agency to supervise and approve the site cleanup when complete.[120] The RP may request a specific agency to serve as the Administering Agency for this purpose.[121] All other agencies with an interest in the cleanup would become "support agencies," to which the Administering Agency, rather than the RP, would be responsible.[122]

Within forty-five days after the request is submitted, the SDC must either approve the RP's request or appoint an alternative agency to serve as the Administering Agency.[123] The SDC may deny the request only if (1) no single agency has the expertise to oversee the response action, (2) the designation would reverse prior agency orders or response actions, (3) the designation would interfere with federal laws or regulations, or (4) the RP and the proposed Administering Agency are part of the same political subdivision.[124] The SDC's determination is final and is not subject to administrative or judicial review.[125]

Once the SDC issues its approval, the designated Administering Agency assumes "sole jurisdiction over all activities that may be required to carry out a site investigation and remedial action necessary to respond to the hazardous materials released at the site."[126] After the cleanup is complete and the Administering Agency determines that a permanent remedy has been implemented at the site, the Administering Agency is authorized to issue a "super" certification for the cleanup to the RP, which is binding on all state and local agencies and provides qualified immunity to the RP.[127] This "super" certification may not be overturned except under certain specified circumstances that would warrant reopener.[128]

The Private Site Manager Program

In 1995, the California legislature passed a law that would streamline contaminated property cleanups even further. The Private Site Management Program Act[129] provides that certain site cleanups may be supervised by an independent "class II environmental assessor" (or private site manager).[130] The private site manager must meet specific requirements to hold this credential.[131] A private site manager may conduct site investigations and submit recommendations based on such investigations to DTSC.[132]

If the private site manager determines that no remedial action is required at the site, and if DTSC does not issue a timely notice of disagreement with the recommendation, DTSC shall be bound by the private site manager's recommendation, subject only to limited exceptions.[133] If the private site manager recommends that a response action be undertaken at the site, the project proponent may submit a request to DTSC that the response action be managed by a private site management team.[134] The private site management team must prepare an RAW or RAP, as appropriate, as well as any remedial designs, for approval by DTSC before undertaking any remediation at the site.[135]

Once the remediation is complete, the private site manager must file with DTSC a request for a certificate of completion.[136] The request for certification must include a detailed summary of the response action taken and confirmation that the site has been remediated, together with other information required by DTSC.[137] DTSC has thirty days to approve the request or issue a notice of deficiency to the project proponent.[138] If DTSC approves the request, it must issue a certificate of completion and any requirements for ongoing monitoring or operations and maintenance.[139] DTSC also must provide public notice of the certification.[140]

Public Agency Cleanup Measures That Benefit Brownfields Redevelopment

In certain cases, public agencies are provided statutory incentives to clean up and redevelop contaminated sites, including special funding, qualified immunity, or special authority for the cleanup. Certain benefits of these agency authorizations can be passed on or can provide additional incentives to owners or developers of contaminated sites. Current relevant programs are outlined below.

The California Community Redevelopment Law

In 1990, the California legislature amended the California Community Redevelopment Law to provide protections to redevelopment agencies conducting hazardous-substance release cleanups.[141] Pursuant to the amendments, California redevelopment agencies that undertake hazardous-substance response actions before January 1, 1999, are granted qualified immunity from liability under state or local law, provided the agency follows specific procedures and conducts the response action pursuant to an RAP approved by DTSC or the regional water quality control board.[142] The immunity derived as a result of the specific response action will extend to, among others, bona fide developers, their successors, and persons providing financing for redevelopment.[143]

The Mello-Roos Community Facilities Act

The Mello-Roos Community Facilities Act was amended in 1990 to create the first form of tax-exempt funding for hazardous-substances cleanups.[144] The amendments facilitate contaminated-site redevelopment by authorizing community facilities districts, subject to voter approval, to levy special taxes and issue tax-exempt bonds to operate a revolving fund to finance response actions for the cleanup of hazardous-substance releases.[145]

Local Health Agency Cleanups

In 1995, the California legislature passed a law that, in effect, recognized a practice that already was being undertaken by local agencies. This new law, dubbed the Local Health Agency Cleanup Law,[146] allows a local health agency to enter an agreement for noncomplex remedial actions with an RP.[147] Under this arrangement, the local agency will supervise the remediation and issue a certification that the cleanup has been accomplished when the remedial goals set forth in the cleanup agreement have been met.[148]

Lender-Liability Protections

On September 18, 1996, the California legislature enacted a bill to protect lenders and fiduciaries from liability for environmental contamination caused by, or occurring at the property of, their borrowers.[149] As of January 1, 1997, the new law provides protections to lenders in a manner similar to federal lender-liability protections,[150] exempting from liability for environmental contamination lenders and fiduciaries that do not participate in the management of their borrowers' property or otherwise fall under any of the other enumerated exceptions of the law. It is expected that these new statutory protections will encourage lenders to provide financing for, among other things, brownfields redevelopment.

The CalSites Validation Program

CalSites is an information database that is used by DTSC to track activities at sites that had come to DTSC's attention as possible hazardous-substance release sites.[151] DTSC created the CalSites database in 1991 by combining lists from previously existing agency databases. At its height, the CalSites database included approximately 26,500 sites.[152] In many cases, actual releases from many of the sites listed on CalSites have never been confirmed. Notwithstanding this fact, CalSites is viewed generally as an indicator of environmental problems and potential enforcement risk. Lenders and others conducting environmental assessments of properties, as a result, have tended to "red-line" properties simply because they were on CalSites, adversely affecting their marketability.[153]

In 1993, DTSC undertook to review the CalSites listings and remove from the list those sites at which no contamination had been confirmed or that otherwise were not under DTSC's jurisdiction.[154] This effort was spurred by DTSC's recognition that brownfields, in effect, were being created simply by the listing alone.[155] The result of this effort, which concluded in March 1996, was the deletion of approximately 23,500 properties from the CalSites database.[156] More than 2,000 other listings have been referred to other agencies or accorded designations on CalSites that indicate they are not of concern to DTSC.[157] DTSC is actively working to address the issues at the remaining sites.[158]

Prospective-Purchaser Agreements and Similar Measures

On June 25, 1996, DTSC issued its long-awaited prospective-purchaser policy, which establishes the criteria under which DTSC will enter prospective-purchaser agreements.[159] This policy is intended to provide new owners and occupants of brownfields properties with certainty regarding site cleanup remedies and responsibilities, as well as other assurances against future liabilities at the site. DTSC's prospective-purchaser policy establishes eligibility criteria and includes a prospective-purchaser agreement application and a model agreement.[160] This policy was based on two prospective-purchaser agreements previously entered by DTSC and criteria outlined in an earlier "informal" policy statement by the agency.[161]

Elements contained in the new policy are (1) broad eligibility criteria for applicants (including those under DTSC's VCP), (2) inclusion of more sites than only those at which response actions have been or are being conducted, and (3) a requirement that there be significant public benefits at stake (such as job creation, tax incentives, and community investment).[162] A prospective-purchaser agreement may include a covenant-not-to-sue, provided the purchaser fulfills its remedial and other obligations under the agreement, as well as certain state and local contribution protections.[163]

Short of actually entering prospective-purchaser agreements, DTSC has taken other administrative measures to encourage the redevelopment of contaminated properties. Among those measures is the issuance of "clean parcel" letters that would allow noncontaminated portions of properties to be developed and put into productive use while other portions of the site continue to be remediated.[164] Originally created to encourage the redevelopment of closing military bases in California, the program has been extended to other sites as well.[165] Other measures include the issuance by DTSC in 1990 of its policy not to pursue landowners of properties that overlie contaminated aquifers that are not sources of the contamination.[166]

Reforms to California Water-Quality Cleanup Standards

Like DTSC, the State Water Resources Control Board (SWRCB) and several Regional Water Quality Control Boards (RWQCBs) in California have taken a leading role in reevaluating regulatory cleanup standards and pursuing reform measures that are protective of human health and the environment, and yet technologically feasible in light of new and ongoing studies and increased scientific and empirical data. Although many of the measures undertaken by the SWRCB may not have been instigated by brownfields initiatives, they nonetheless find application in addressing the myriad concerns facing would-be brownfields developers in California. Two of the most significant issues now being addressed by the SWRCB and RWQCBs in California that have implications for brownfields are: the SWRCB's groundwater containment-zone policy and reforms to underground storage tank cleanup standards.

The SWRCB's Containment-Zone Policy

The SWRCB, together with the nine RWQCBs, is charged with protecting water quality within California.[167] Under this mandate, in 1992 the SWRCB promulgated resolution Number 92-49.[168] Resolution Number 92-49 sets forth the

SWRCB's policy regarding the cleanup and abatement of discharges of wastes into waters of the state.

Beginning in 1994, the SWRCB proposed amendments to resolution Number 92-49, largely in response to actions by some of the RWQCBs to amend their basin plans, to allow for "groundwater nonattainment zones." These nonattainment zones were determined to be areas of contaminated groundwater that pose no significant risk to public health or the environment, but that could not feasibly and economically be cleaned up to existing, stringent water-quality standards.[169] The proposed basin plan amendments had been submitted to the SWRCB for approval; however, in lieu of approving the several plans, the SWRCB instead adopted a uniform policy governing such nonattainment areas.

Following several revisions to its proposed amendments, the SWRCB issued its final "containment zone" policy in October 1996.[170] This policy, which is expected to be effective in early 1997, significantly amends the SWRCB's discharge requirements to allow for the creation of groundwater "containment zones" where applicable water-quality objectives cannot reasonably be achieved due to technological or economic infeasibility.[171] Essentially, the proposed amendments will allow risk-based assessments to be considered in connection with groundwater remediation efforts, in place of traditional, standard-based cleanup goals.

The impact of these amendments on groundwater cleanup in California in general, and brownfields redevelopment in particular, is significant. The designation of a containment-zone will allow dischargers to limit their ongoing costs and obligations for remediation of groundwater contamination where the contamination cannot reasonably be cleaned up. Further, the discharger will be able to pass on certain protections created by the containment-zone designation, as long as the conditions for that designation continue to be met, to later owners of the overlying land.

CONTAINMENT-ZONE REQUIREMENTS. A discharger or local agency may apply to the appropriate RWQCB for the designation of a containment zone.[172] The RWQCB may designate a containment zone if (1) the zone is limited in vertical and lateral extent, (2) it is as protective as reasonably possible of human health and safety and the environment, and (3) it is not expected to result in violation of water-quality objectives outside the containment zone.[173] Specific procedural requirements must be met, including interagency and public review requirements.[174]

A containment zone may be no larger than necessary under the circumstances, and the designation must include a detailed evaluation of the contamination and the risks to water quality, human health, and the environment, and include a detailed cleanup and management plan.[175] The designation may not allow for significant adverse effects to human health or the environment; nor may a containment zone be designated in a critical water recharge area.[176] Further, a containment zone may not be designated if it would be inconsistent with a local groundwater management plan or other provisions of law or court order.[177]

In addition, the person or entity requesting the designation must commit to do the following: (1) complete all remedial measures that are feasible to prevent water-quality degradation, (2) implement an ongoing management plan, subject to the RWQCB's approval, to assess, clean up, abate, and monitor the containment zone, and (3) provide appropriate and reasonable mitigation measures to offset any significant adverse impacts resulting from the residual pollutants in the containment zone.[178] The containment-zone designation may include management controls, including land-use restrictions, engineering controls, or overlying landowner agreements, to ensure that the limitations imposed by the designation run with the land.[179]

A containment-zone designation will be made through the issuance of a cleanup and abatement order by the RWQCB.[180] The order must include proposed time schedules imposed on the discharger, together with prescribed civil penalties to be imposed if the discharger does not comply with the order in accordance with the schedule.[181] With limited exceptions, the discharger will be responsible for the RWQCB's oversight costs.[182] Overall implementation of the containment-zone program is under the oversight of the "Containment Zone Review Committee," comprised of representatives of the SWRCB and the RWQCBs.[183]

CONTAINMENT-ZONE LIABILITY PROTECTIONS. Once an area is designated a containment zone, no further cleanup is necessary and the discharger will be relieved of further costs associated with remediating the contamination for which the containment-zone designation was sought, unless a discharger fails to comply with the requirements of the cleanup and abatement order, or if monitoring indicates that the contamination cannot be, or is not being, contained in the containment zone (in which case the containment-zone designation will be subject to revocation by the RWQCB and the discharger— including its successors—may be subject to further enforcement action under applicable law).[184] In the event that water-quality objectives are later met, the RWQCB will rescind the containment-zone designation and issue a closure letter following public notice.[185]

Reforms to Underground Storage Tank Cleanup Requirements

The Barry Keene Underground Storage Tank Cleanup Trust Fund Act of 1989[186] requires cleanup of unauthorized releases from underground storage tanks (USTs). Under the UST cleanup law, an RWQCB or local agency may require owners, operators, or other persons responsible for a leaking UST to take corrective action in response to an unauthorized release from the UST.[187] Alternatively, the RWQCB or local agency may undertake corrective action and seek cost recovery against the RPs.[188] The standards applicable to corrective actions under the UST cleanup law are those deemed most protective of the quality of the waters of the state.[189] Depending on the particular circumstances, those standards derive from applicable drinking-water standards,[190] the SWRCB's "nondegradation" policy,[191] and the SWRCB's Leaking Underground Fuel Tank Manual.[192] In most cases, the cleanup standards in effect have required cleanup to background levels.

In 1994, the California legislature enacted Senate Bill 1764,[193] which required the SWRCB to convene an advisory committee (the Advisory Committee) to conduct a comprehensive review of the cleanup standards applicable to leaking USTs pursuant to the UST cleanup law.

The statute requires that the Advisory Committee recommend to the SWRCB changes believed necessary to ensure that applicable cleanup standards are both technologically feasible and necessary to ensure the protection of human health and safety and the environment.[194] The statute also requires that by March 1, 1997, the SWRCB promulgate new regulations, taking into account the Advisory Committee's recommendations, for corrective actions undertaken pursuant to the UST cleanup law.

THE LAWRENCE LIVERMORE REPORT. At the same time the Advisory Committee was convened, a working group headed by Lawrence Livermore National Laboratory (LLNL) undertook a review of leaking underground fuel tank (LUFT) cleanups in California. The working group had been commissioned by the SWRCB in July 1994, before the creation of the Advisory Committee. LLNL's charge was to evaluate California's LUFT program, including the regulatory framework and cleanup process applied to LUFTs in California. LLNL issued its report in the Fall of 1995.[195]

The LLNL report focuses on low-risk soils and low-risk groundwater cases[196] that have been impacted by leaking USTs. The report acknowledges that its findings should not apply to higher-risk sites that may require immediate response actions and remediation to protect human health and the environment.[197] The LLNL report concludes that low-risk soils and groundwater impact cases do not warrant the extensive remediation efforts that have occurred to date, in light of its findings that natural attenuation has been determined to perform as efficient and effective, and more cost-effective, a job as active remediation.[198]

As a result, the LLNL report recommends that for low-risk sites, passive bioremediation should be used as a remediation alternative whenever possible, that SWRCB policies should be modified to allow for risk-based analyses to control in determining appropriate LUFT cleanups, and that passive bioremediation consistent with industry and state standards be the preferred remedial measure in such cases.[199]

On December 8, 1995, in response to the LLNL report, the SWRCB issued an interim guidance to all RWQCBs to proceed aggressively to close low-risk-soil-only cases and to replace active remediation with monitoring for low-risk groundwater cases.[200] The SWRCB proposal directs RWQCBs and local oversight programs to prioritize the review and closure of low-impact cases that require little or no action to be closed (for instance, when the tank has been pulled, there is no free product, and only soil has been impacted), and allow passive bioremediation and monitoring for low-risk groundwater-affected cases.[201]

Pursuant to the SWRCB's directive, the RWQCBs have begun to implement procedures for closing low-risk LUFT sites in their regions.[202] For example, on January 5, 1996, the RWQCB for the San Francisco Bay region issued "Supplemental Instructions," which implement the SWRCB's directive and establish additional criteria to be applied in low-risk cases.[203] This RWQCB's instructions make clear that the guidance applies only to petroleum leaks and low-risk cases, and that the burden is on the RP for the LUFT to establish its eligibility for the alternate cleanup standards under the LLNL report.[204] Outside of these cases, the existing cleanup standards apply.[205]

THE ADVISORY COMMITTEE REPORT. Following review of the LLNL report, and largely in response to it, the Advisory Committee issued its report on May 31, 1996.[206] The Advisory Committee made sixteen recommendations to the SWRCB, including recommendations regarding (1) the use of risk-based corrective actions,[207] (2) passive bioremediation, (3) drinking-water source protection, beneficial-use designation, and water-quality objectives, (4) economic considerations, (5) contamination from fuel additives, (6) independent review and oversight, and (7) standard setting.[208]

Essentially, the Advisory Committee report agreed with the LLNL report that risk-based corrective actions are preferable, because they will allow limited resources to be used on the highest-priority sites. However, it took issue with the LLNL report's methodology, particularly concerning certain fuel contaminants, and its potential for use beyond true low-risk cleanup sites.[209] In particular, the Advisory Committee report recommended that additional analyses be performed in connection with contamination from MTBE and other fuel oxygenates.[210]

In response to the Advisory Committee's recommendations concerning MTBE, the SWRCB has commissioned a new LLNL panel to conduct an extensive scientific study of MTBE and to provide recommendations regarding cleanup of contamination from MTBE.[211] That study is expected to be completed in late 1997 or early 1998.

THE SWRCB'S UST POLICY. As required by Senate Bill 1764, the SWRCB is now undertaking to review and revise its UST corrective action program. A draft policy, which would formally adopt risk-based standards for UST corrective actions, is expected to be issued for public review by early 1997.[212]

NEW UST LEGISLATIVE MANDATES. On September 18, 1996, new legislation implementing additional reforms to California's UST cleanup program was enacted.[213] This new law affirmatively recognizes the propriety of risk-based cleanups, requires the SWRCB to adopt a uniform closure letter for UST cleanups, establishes a preapproval process and time limitations for determining costs eligible for state UST fund reimbursement, provides affirmative mechanisms for RPs to obtain closure of UST cleanups, and creates a $10 million fund for the cleanup of commingled plumes.[214] The law takes effect January 1, 1997, pending promulgation of implementing regulations and policies by the SWRCB.

IMPLICATIONS FOR BROWNFIELDS REDEVELOPMENT IN CALIFORNIA. Before the issuance of the Advisory Committee report and the LLNL report, all groundwater cleanups in California were required to meet stringent water-quality standards. The recent and pending changes in regulatory cleanup standards represents a significant change under California law, with resultant savings in time, costs, and regulatory approval processes.

Conclusion

The legislature and governmental agencies in California, together with private entities and individuals, have seized on several fronts the initiative to embrace, explore, and exploit the possibilities that brownfields redevelopment can have for the rehabilitation of contaminated properties in the state. As a result of the wide-ranging and diverse programs that have been developing in California, there are numerous opportunities for governmental agencies and the private sector alike to turn such properties once again into desirable, marketable, and profitable centers of activity and commerce. Brownfields programs in California are at a still-nascent stage; in spite of this, or perhaps because of it, the prospect for the success and continued growth of such programs is great. No doubt, the success of many such programs over time—coupled with the large number of properties in California where such programs can undergo experimentation—will set new standards for dealing with and refurbishing contaminated properties in the future.

The California Voluntary Cleanup Program

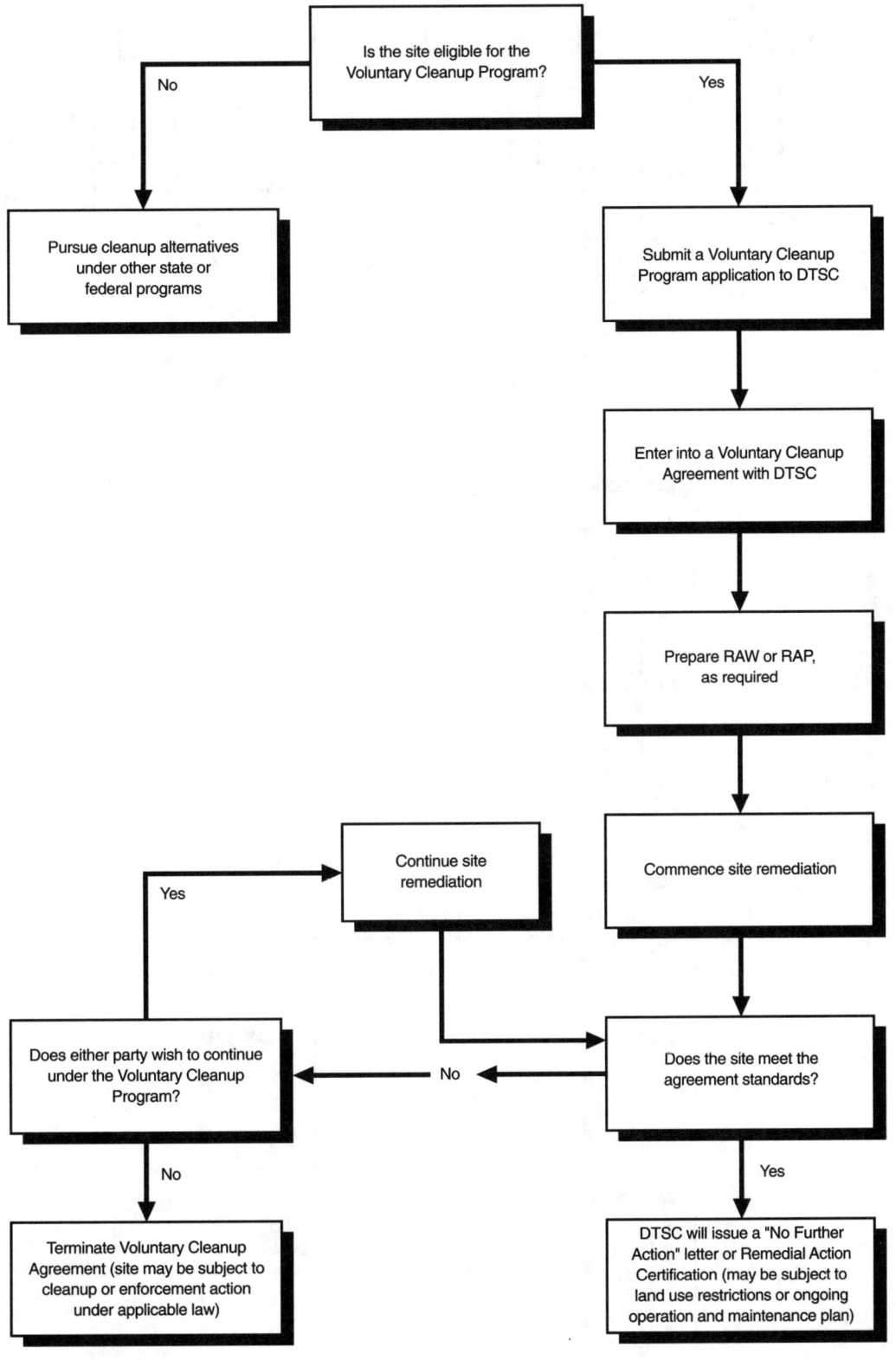

The California Expedited Remedial Action Program

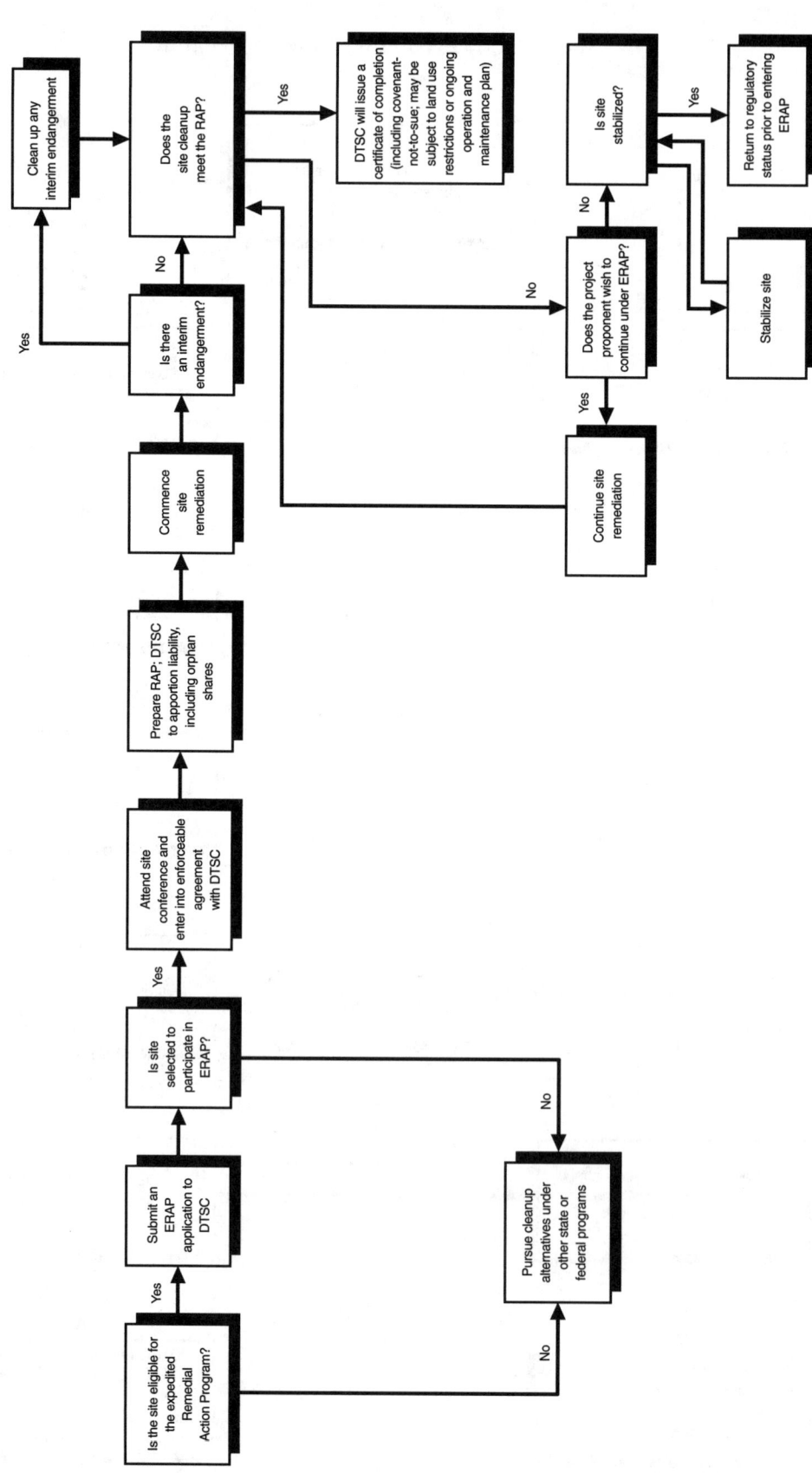

Notes

1. *See* CAL. HEALTH & SAFETY CODE, ch. 6.8, § 25300 *et seq.* (West 1992 & Supp. 1997).

2. Subject to certain stated exceptions, the term "responsible party" is the same as "potentially responsible party" under the federal Superfund law. *See id.* § 25323.5 (West Supp. 1997); *cf.* 42 U.S.C.A. § 9607(a) (1995).

3. *See* CAL. DEP'T TOXIC SUBSTANCES CONTROL, NO. EO-95-006-PP, POLICY AND PROCEDURE FOR MANAGING VOLUNTARY SITE MITIGATION PROJECTS, THE VOLUNTARY CLEANUP PROGRAM 1 (1995) [hereinafter DTSC VCP Policy]. DTSC's VCP Policy supersedes its previously issued policy outlining the "Walk-In Program." *See* CAL. DEP'T TOXIC SUBSTANCES CONTROL, OPP NO. 87-9, POLICY ON MITIGATING UNSCHEDULED WASTE SITES AND MANAGING WALK-IN BUSINESS (1988).

4. *See* DTSC VCP Policy, *supra* note 3, at 1–2.

5. The AWP lists those sites for each fiscal year that are undergoing mitigation activities by DTSC or by responsible parties with DTSC oversight. AWP sites are also known as state Superfund sites. *See id.* at 1.

6. BARBARA COLER & STEVE KOYASAKO, REDEVELOPMENT AND REVITALIZATION OF BROWNFIELDS, DEPARTMENT OF TOXIC SUBSTANCES CONTROL INITIATIVES 2 (Oct. 1995).

7. *Id.* Land-use controls may be warranted when the cleanup is tied to the planned use of the property. Land-use restrictions must be recorded with the deed to run with the land and bind future heirs, successors, assignees, and others. *See* CAL. HEALTH & SAFETY CODE § 25355.5(a)(1)(C) (West 1992); *see also id.* § 25222.1 (West 1992); CAL. DEP'T TOXIC SUBSTANCES CONTROL, OFFICIAL POLICY AND PROCEDURE ON DEVELOPMENT AND IMPLEMENTATION OF LAND USE COVENANTS (1990). A new California law, effective January 1, 1996, also allows a landowner to impose its own legal "restrictions" on the use of the land in order to protect present or future human health, safety or the environment due to the presence of hazardous substances on the land. *See* CAL. CIV. CODE § 1471 (West Supp. 1997). If recorded, the restrictions will be binding on successive landowners.

8. COLER & KOYASAKO, *supra* note 6, at 2. Though the VCP itself does not include covenants-not-to-sue or a complete release of liability, the DTSC may consider entering into such assurances on a case-by-case basis. See the section of this chapter entitled "Prospective-Purchaser Agreements and Similar Measures."

9. DTSC has prepared application forms for the VCP.

10. DTSC VCP Policy, *supra* note 3, at 3.

11. Generally, all federal facilities are deemed excluded from the VCP program. However, the official DTSC policy statement excludes only "military" facilities. *Compare* DTSC VCP Policy, *supra* note 3, at 3 (military facilities excluded from program) *with* COLER & KOYASAKO, *supra* note 6, at 2 (federal facilities excluded from program).

12. DTSC VCP Policy, *supra* note 3, at 2. Sites that are under another agency's oversight may enter the VCP for specific services only with that agency's consent.

13. *Id.* at 4.

14. *See infra* note 25 and accompanying text.

15. DTSC VCP Policy, *supra* note 3, at 2.

16. *Id.* at 3.

17. *Id.*

18. *See* CAL. HEALTH & SAFETY CODE §§ 25356.1(h), 25323.1 (West Supp. 1997).

19. Certain remedial actions with an estimated cost of less than $2 million may be relieved of some of the applicable RAP requirements. *See id.* § 25356.1(h)(3) (West Supp. 1997).

20. *See id.* § 25358.7 (West 1992); CAL. DEP'T TOXIC SUBSTANCES CONTROL, No. EO-94-002-PP, PUBLIC PARTICIPATION POLICY AND PROCEDURES MANUAL (1994).

21. CAL. HEALTH & SAFETY CODE § 25356.1(e) (West Supp. 1997).

22. DTSC must comply with the California Environmental Quality Act (CEQA) to the extent that activities required under the VCP agreement are "projects" as defined by CEQA. CAL. PUB. RES. CODE § 21000 *et seq.* (West 1996 & Supp. 1997). A discussion of the applicability of CEQA to brownfields projects in general is beyond the scope of this chapter. However, if project remediation or development activities fall under CEQA, the project proponent must prepare an initial study, and DTSC will determine if a negative declaration or environmental-impact report should be prepared.

23. DTSC VCP Policy, *supra* note 3, at 5.

24. *Id.* at 3. Any monies previously paid by the project proponent but not yet incurred will be refunded.

25. CAL. HEALTH & SAFETY CODE § 25319.5 (West 1992); *see* CAL. DEP'T TOXIC SUBSTANCES CONTROL, PRELIMINARY ENDANGERMENT ASSESSMENT GUIDANCE MANUAL (1994).

26. COLER & KOYASAKO, *supra* note 6, at 2.

27. *Id.; see supra* note 7 and accompanying text.

28. DTSC VCP Policy, *supra* note 3, at 3; *see* CAL. HEALTH & SAFETY CODE §§ 25355.5(a)(1)(C), 25356.1(d) (West 1992 & Supp. 1997).

29. DTSC VCP Policy, *supra* note 3, at 3.

30. *See* CAL. HEALTH & SAFETY CODE § 25355.5(a)(1)(C) (West 1992); COLER & KOYASAKO, *supra* note 6, at 2; *supra* note 7 and accompanying text.

31. *See* 40 C.F.R. § 300.61 *et seq.* (1996); CAL. HEALTH & SAFETY CODE § 25356.1(d) (West Supp. 1997); *see also* DTSC VCP Policy, *supra* note 3, at 3.

32. *See* DTSC VCP Policy, *supra* note 3, at 3.

33. *Id.*

34. Neither the NFA letter nor the remedial action certification are binding on other agencies, including the United States Environmental Protection Agency (USEPA). DTSC and USEPA currently are in the process of negotiating state Memoranda of Agreement (SMOAs) for both the VCP and the expedited remedial action program (ERAP) that would provide assurance that, subject to certain reopeners, a sign-off by DTSC would also carry the endorsement of USEPA. *See* BARBARA COLER, REDEVELOPMENT AND REVITALIZATION OF BROWNFIELDS, DEPARTMENT OF TOXIC SUBSTANCES CONTROL INITIATIVES (March 1996); U.S. ENVTL. PROTECTION AGENCY, REGION 9 BROWNFIELD ACTION AGENDA (1996).

35. DTSC VCP Policy, *supra* note 3, at 2; *cf.* CAL. DEP'T. TOXIC SUBSTANCES CONTROL, OPP. NO. 86-22, PROCEDURES FOR CERTIFYING COMPLETION OF HAZARDOUS WASTE SITE REMEDIAL ACTION (1987) (establishes standards for issuance of DTSC cleanup certifications).

36. Prospective-purchaser agreements have been used in isolated instances and may be used in the future to provide this assurance. DTSC recently developed a formal prospective-purchaser policy for this purpose. See the section of this chapter entitled "Prospective Purchaser Agreements and Similar Measures."

37. CAL. HEALTH & SAFETY CODE § 25356.1(g) (West 1992); *see also id.* § 25356.8 (West 1992).

38. *Id.* § 25356.1(g)(2) (West 1992). A full discussion of traditional mandamus and other forms of judicial review under California law is beyond the scope of this chapter.

39. *See* DTSC VCP Policy, *supra* note 3, at 4.

40. 42 U.S.C.A. § 9601 *et seq.* (1995 & Supp. 1996); *see* COLER & KOYASAKO, *supra* note 6, at 2.

41. CAL. HEALTH & SAFETY CODE, ch. 6.85, § 25396 *et seq.* (West Supp. 1997).

42. *Id.* § 25396.5(a) (West Supp. 1997).

43. *See* ALLEN K. WOLFENDEN & MEGAN CAMBRIDGE, VOLUNTARY PROGRAM PROMOTES EQUITABLE AND EXPEDITED REMEDIATION OF CONTAMINATED PROPERTIES (undated memorandum).

44. In 1995, DTSC issued emergency regulations to implement ERAP. Those emergency regulations are set forth in the California Code of Regulations, title 22, sections 67401–67401.12. As of this writing, permanent regulations are pending final promulgation. In addition, DTSC has prepared policies and procedures, as well as applicable forms and a model agreement, for ERAP.

45. *See* Unified Agency Review of Hazardous Materials Release Sites, ch. 1184, § 1, 1993 Cal. Stat. 5452 (codified at CAL. HEALTH & SAFETY CODE, ch. 6.65, § 25260 *et seq.*) (West Supp. 1997) [hereinafter Unified Agency Review Law]; see the section of this chapter entitled "The Unified Agency Review Law."

46. CAL. HEALTH & SAFETY CODE § 25396.5(b) (West Supp. 1997). As of the fall of 1996, ten sites had been designated under the ERAP program.

47. *Id.* § 25396.6(c)(3) (West Supp. 1997).

48. *Id.* § 25396.6(b) (West Supp. 1997); *see* CAL. DEP'T TOXIC SUBSTANCES CONTROL, EXPEDITED REMEDIAL ACTION PROGRAM: SITE SELECTION POLICY AND PROCEDURE (undated); Wolfenden & Cambridge, *supra* note 43.

49. *See* CAL. HEALTH & SAFETY CODE §§ 25261, 25396(x), 25396.5(b) (West Supp. 1997); *see* Wolfenden & Cambridge, *supra* note 43. The site designation committee is established pursuant to the Unified Agency Review Law. See the section of this chapter entitled "The Unified Agency Review Law."

50. *See* CAL. HEALTH & SAFETY CODE §§ 25236, 25396.6 (West 1992 & Supp. 1997); see also the discussion of the Unified Agency Review Law in the section of this chapter bearing that title.

51. CAL. HEALTH & SAFETY CODE § 25396.6 (West Supp. 1997).

52. *Id.* § 25396.6(c)(4); *see id.* § 25396(k) (West 1992).

53. ERAP provides that up to five state Superfund sites may be included by the SDC in the program, as long as the remedy under ERAP will not delay cleanup of the site and otherwise is in the public interest. *Id.* § 25396.6(c)(5)(B) (West Supp. 1997).

54. *See* the discussion in the section of this chapter entitled "Apportioning Liability and Orphan Shares."

55. CAL. HEALTH & SAFETY CODE § 25396.6(c)(2) (West Supp. 1997).

56. *See id.* § 25398(a) (West Supp. 1997). DTSC does not yet have authorization to commit the USEPA under ERAP. A SMOA that would provide this authorization is pending between the USEPA and DTSC as of this writing. *See supra* note 34.

57. CAL. HEALTH & SAFETY CODE § 25398(b), 25398(c), 25398(d) (West Supp. 1997).

58. *Id.* § 25398.2(a) (West Supp. 1997).

59. The ninety-day period may be extended by mutual agreement between DTSC and the other parties to the agreement. *Id.* § 25398.2(b)(2) (West Supp. 1997).

60. *Id.* § 25398.2(b) (West Supp. 1997).

61. *Id.* § 25398.2(c) (West Supp. 1997). Neither ERAP nor the covenant-not-to-sue will protect any responsible party against natural-resource damages claims brought pursuant to the federal Superfund law. *Id.* § 25398.2(b)(1)(C) (West Supp. 1997); *see* 42 U.S.C.A. § 9607(f) (1995).

62. CAL. HEALTH & SAFETY CODE §§ 25398.2(e), 25398.6 (West Supp. 1997).

63. *See id.* § 25398(d) (West Supp. 1997). Specific procedures apply if disputes arise concerning the local governmental agency's determination.

64. *Id.* § 25398.6(a) (West Supp. 1997).

65. *Id.* § 25398.6(b)–(c) (West Supp. 1997).

66. *Id.* § 25398.6(d) (West Supp. 1997).

67. *Id.* § 25398.6(e), 25398.6(h) (West Supp. 1997).

68. *Id.* § 25398.6(f) (West Supp. 1997).

69. *Id.* § 25398.6(g) (West Supp. 1997).

70. *Id.* § 25398.7(a) (West Supp. 1997). Once land-use controls are in place, specific procedures, including public notice and comment and, as necessary, provisions for additional response actions at the site, must be followed to modify any such land-use controls. *Id.* § 25398.7(c) (West Supp. 1997).
71. *Id.* § 25398.7(c)(3) (West Supp. 1997).
72. *Id.* § 25398.7(b) (West Supp. 1997).
73. *E.g., id.* §§ 25397.2, 25398.6(i)(4), 25398.8(a)(1) (West Supp. 1997).
74. *Id.* § 25398.6(i) (West Supp. 1997).
75. *Id.* § 25398.6(l) (West Supp. 1997).
76. *Id.*
77. *Id.*
78. *Id.*
79. *Id.* §§ 25398.6(m), 25398.10(b) (West Supp. 1997).
80. *Id.* § 25398.8 (West Supp. 1997).
81. *Id.* § 25398.8(b) (West Supp. 1997).
82. *Id.* § 25398.8(c) (West Supp. 1997).
83. *Id.* §§ 25398.8(e), 25398.8(f) (West Supp. 1997); *see also id.* § 25399.
84. *Id.* § 25398.8(a)(2) (West Supp. 1997); *see id.* § 25399.1 (West Supp. 1997). The Trust Fund is expected to be funded at $5 million annually. *See* Wolfenden & Cambridge, *supra* note 43.
85. CAL. HEALTH & SAFETY CODE § 25396(m) (West Supp. 1997).
86. *Id.* § 25396.5(b) (West Supp. 1997).
87. *Id.* § 25396.6(c)(2) (West Supp. 1997).
88. *Id.* § 25398.9 (West Supp. 1997).
89. *Id.* § 25398.9(b) (West Supp. 1997).
90. *Id.* § 25398.9(a) (West Supp. 1997).
91. *Id.* § 25398.9(d) (West Supp. 1997).
92. *Id.* § 25398.3 (West Supp. 1997).
93. *Id.* § 25396(k) (West Supp. 1997).
94. *Id.* § 25398.3 (West Supp. 1997).
95. *Id.* § 25398.3(d) (West Supp. 1997).
96. *Id.* § 25398.10(a) (West Supp. 1997); *see also id.* §§ 25356.2, 25398.10(c)(4) (West Supp. 1997).
97. *Id.* §25398.10(a) (West Supp. 1997).
98. *Id.* § 25398.10(b)–(c) (West Supp. 1997).
99. *Id.* § 25398.10(d) (West Supp. 1997).
100. *Id.* § 25398.10(e) (West Supp. 1997); *see* CAL. CIV. PROC. CODE § 1094.5 (West Supp. 1997). DTSC has indicated that short of arbitration, it intends to pursue informal means of resolving disagreements with RPs, project proponents, and others under ERAP whenever possible. *See, e.g.,* Cal. Dep't Toxic Substances Control, Draft Policy Statement— Expedited Remedial Action Program: Site Conference and Agreement Policy and Procedure 5 (June 16, 1995) [hereinafter Draft ERAP Policy Statement].
101. CAL. HEALTH & SAFETY CODE § 25398.15(a) (West Supp. 1997).
102. *Id.*
103. *Id.* § 25398.15(b) (West Supp. 1997).
104. *Id.* Failure by DTSC to act may be appealed first to the Secretary of Cal/EPA; then—if the Secretary fails to act within thirty days—to the courts, in an action for traditional mandamus. *Id.; see* CAL. CIV. PROC. § 1085 (West 1980).
105. CAL. HEALTH & SAFETY CODE §25398.15(b) (West Supp. 1997); *see id.* § 25398.10 (West Supp. 1997); see also the previous discussion in the section entitled "Arbitration" and accompanying notes.

106. CAL. HEALTH & SAFETY CODE § 25398.15(c) (West Supp. 1997).
107. *See* Draft ERAP Policy Statement, *supra* note 100, at 5.
108. *Id.*
109. CAL. HEALTH & SAFETY CODE § 25398.17 (West Supp. 1997).
110. *Id.* § 25398.17(b) (West Supp. 1997).
111. See the previous discussion in the section entitled "The ERAP Agreement."
112. *See* CAL. HEALTH & SAFETY CODE § 25398.2(c) (West Supp. 1997).
113. See the section of this chapter entitled "The Unified Agency Review Law."
114. As noted previously (*see supra* note 34), DTSC and the USEPA currently are negotiating an SMOA that would include the USEPA's consent and commitment to a certification issued under ERAP.
115. CAL. HEALTH & SAFETY CODE § 25398.15(d) (West Supp. 1997); see the discussion in the section entitled "The Unified Agency Review Law."
116. *See* CAL. HEALTH & SAFETY CODE § 25398.7(b) (West Supp. 1997). DTSC's emergency regulations specifically include bona fide purchasers in the program and, hence, allow them the same assurances (covenants-not-to-sue) and protections (qualified immunity) as RPs. *See* CAL. CODE REGS. tit. 22, § 67401.12 (1996). DTSC's pending permanent regulations do not include a bona fide purchaser provision; however, DTSC has indicated its management policy to extend the benefits of ERAP to bona fide purchasers. *See* Wolfenden & Cambridge, *supra* note 43.
117. See the sections of this chapter entitled "The Unified Agency Review Law" and "Prospective-Purchaser Agreements and Similar Measures."
118. In that regard, in 1995, the "California Senate passed a resolution, titled the California Land Use Accord," which provides guidance intended to serve as a model for managing the remediation of contaminated property and returning it to productive use. The guidance incorporates principles of joint private and public decision making and communication as its cornerstones to facilitate property redevelopment. S. Res. 29, 1995-96 Reg. Sess. (Cal. 1995).
119. Unified Agency Review Law, ch. 1184, 1993 Cal. Stat. 5451 (codified at CAL. HEALTH & SAFETY CODE, ch. 6.65, § 25260 *et seq.*) (West Supp. 1997). In addition, the Environmental Protection Permit Reform Act of 1993, permits Cal/EPA to coordinate the permitting process when multiple permits are required for a single facility. CAL. PUB. RES. CODE § 71000 *et seq.* (West 1996 & Supp. 1997).
120. CAL. HEALTH & SAFETY CODE § 25262 (West Supp. 1997).
121. *Id.* § 25262(b) (West Supp. 1997). Certain response actions require mandate oversight by a particular agency. *See id.* § 25262(c)(2) (West Supp. 1997).
122. *Id.* § 25264(a)(3) (West Supp. 1997); CAL. DEP'T TOXIC SUBSTANCES CONTROL, THE SITE DESIGNATION PROCESS UNDER THE UNIFIED AGENCY REVIEW OF HAZARDOUS MATERIALS RELEASE SITES, FACT SHEET & IMPLEMENTATION GUIDE 1–2 (1995).
123. CAL. HEALTH & SAFETY CODE § 25262 (West Supp. 1997).
124. *Id.* §§ 25262(a)(1)–(3) (West Supp. 1997); *see* Hazardous Substance Cleanup: Redevelopment Agencies, ch. 623, 1996 Cal. Stat. 2843 (codified in part at CAL. HEALTH & SAFETY CODE § 25262(a)(4)) (West Supp. 1997).
125. CAL. HEALTH & SAFETY CODE § 25262(d) (West Supp. 1997).
126. *Id.* § 25264(a) (West Supp. 1997).
127. *Id.* §§ 25264(b)–(c); *see also* COLER & KOYASAKO, *supra* note 6, at 5.
128. CAL. HEALTH & SAFETY CODE §§ 25264(c)(1)–(6) (West Supp. 1997).
129. Private Site Management, ch. 820, § 1, 1995 Cal. Stat. 4876 (codified at CAL. HEALTH & SAFETY CODE §§ 25395.1 *et seq.*) (West Supp. 1997). The private site manager program is expected to take effect in 1997, pending promulgation of required regulations by DTSC and OEHHA.

130. CAL. HEALTH & SAFETY CODE § 25395.1(a) (West Supp. 1997).
131. *See id.* § 25570.3(c) (West Supp. 1997). Regulations further delineating the requirements for a qualified private site manager under the act are pending. *See* COLER, *supra* note 34, at 1.
132. CAL. HEALTH & SAFETY CODE § 25395.2 (West Supp. 1997).
133. *Id.*
134. *Id.* § 25395.3 (West Supp. 1997). Notably, a private site management team may be authorized to coordinate response actions at sites that are under ERAP. *See id.* § 25395.5(a) (West Supp. 1997).
135. *Id.* §§ 25395.5, 25395.6 (West Supp. 1997); *see* Memorandum from Paul D. Blais, Deputy Director, Cal. Dep't Toxic Substances Control (Feb. 9, 1996) (entitled "The Private Site Management Program". Regulations establishing performance standards and other criteria to implement this program are pending. *See* COLER, *supra* note 34, at 1.
136. CAL. HEALTH & SAFETY CODE § 25395.8(a) (West Supp. 1997).
137. *Id.* § 25395.8 (West Supp. 1997).
138. *Id.* § 25395.8(c) (West Supp. 1997).
139. *Id.* § 25395.8(d) (West Supp. 1997).
140. *Id.*
141. *See* Hazardous Substance Release Cleanup, ch. 113, 1990 Cal. Stat. 4043 (codified at CAL. HEALTH & SAFETY CODE § 33459 *et seq.*) (West Supp. 1997).
142. *See* CAL. HEALTH & SAFETY CODE §§ 33459.1, 33459.2, 33459.3 (West Supp. 1997).
143. *Id.* §§ 33459.3(a), 33459.3(d)(2)–(4) (West Supp. 1997). Persons responsible for the contamination, however, may not derive immunity from the government's actions. *Id.* § 33459.3(e) (West Supp. 1997). Also see the section of this chapter entitled "Lender-Liability Protections."
144. Community Facilities Districts: Hazardous Substances Cleanup, ch. 175, § 1, 1990 Cal. Stat. 1019 (codified at CAL. GOV'T CODE § 53311 *et seq.*) (West Supp. 1997).
145. *See* CAL. GOV'T CODE §§ 53313(f), 53314.6 (West 1983 & Supp. 1997). Persons responsible for the contamination will remain liable for the costs incurred for the response action. *Id.* § 53314.7(a).
146. The Local Health Agency Cleanup Law was reorganized and recodified in 1996. *See* CAL. HEALTH & SAFETY CODE § 101480 *et seq.* (West Supp. 1997).
147. *Id.* §§ 101480(b)–(c) (West Supp. 1997). "Noncomplex" in this context refers to sites that do not involve multiagency or multimedia issues.
148. *Id.* § 101480(e) (West Supp. 1997).
149. Hazardous Materials Liability of Lenders and Fiduciaries, ch. 612, 1996 Cal. Stat. 2797 (codified at CAL. HEALTH & SAFETY CODE, ch. 6.96, § 25548 *et seq.*) (West Supp. 1997).
150. *See* Policy on CERCLA Enforcement against Lenders and Government Entities That Acquire Property Involuntarily, 60 Fed. Reg. 63,517 (1995); *see also* Assets Conservation, Lender Liability and Deposit Insurance Protection Act of 1996, P.L. No. 104-208 (Sept. 30, 1996).
151. In this respect, CalSites is similar to the federal Comprehensive Environmental Response, Compensation, and Liability Information System (CERCLIS) list. *See* 40 C.F.R §§ 300.5, 300.420(b) (1996); Cal. Dep't Toxic Substances Control, Fact Sheet, CalSites Validation Program (March 1995) [hereinafter CalSites Fact Sheet].
152. CalSites Fact Sheet, *supra* note 151.
153. COLER & KOYASAKO, *supra* note 6, at 5.
154. *Id.*
155. *Id.*
156. Unlike a similar endeavor undertaken by USEPA concerning the CERCLIS list, a site deletion from CalSites constitutes confirmation that there is no evidence of release

from that site, and that DTSC anticipates no investigation or cleanup. *See* CalSites Fact Sheet, *supra* note 151.

157. *Id.*

158. *Id.*

159. CAL. DEP'T TOXIC SUBSTANCES CONTROL, NO. EO-96-005-PP, PROSPECTIVE PURCHASER POLICY (1996) [hereinafter PROSPECTIVE PURCHASER POLICY]. Similar agreements have been developed and entered by certain of the regional water quality control boards in California that provide assurances to purchasers when contaminated waters of the state (groundwater in particular) are involved.

160. *Id.*

161. *See* Letter from Sarah Morrison, Staff Counsel, DTSC, to Pamela Nehring, Atchison, Topeka & Santa Fe Railway Co. (Dec. 12, 1994).

162. *See* PROSPECTIVE PURCHASER POLICY, *supra* note 159.

163. *Id.*

164. CAL. DEP'T TOXIC SUBSTANCES CONTROL, NO. 92-4, APPROVAL OF A PARTIAL SITE CLEANUP (1992); *see* COLER, *supra* note 34, at 2.

165. *See* COLER, *supra* note 34, at 2.

166. CAL. DEP'T TOXIC SUBSTANCES CONTROL, NO. 90-11, RESPONSIBLE PARTIES—OWNERSHIP OF PROPERTY OVER CONTAMINATED GROUND WATER, (1990); *see* COLER & KOYASAKO, *supra* note 6, at 5–6. State water pollution control agencies have provided similar assurances, at least insofar as committing not to hold primarily liable those overlying landowners that have not contributed to the contamination. As of this writing, DTSC also is considering regulatory reforms that would reduce cleanup requirements for soils contaminated with "naturally occurring substances" and modified cleanup standards for non-RCRA wastes undergoing remediation. No date for the issuance of these policies has been established as of this writing.

167. Among other things, the SWRCB establishes statewide policies to protect water quality. *See* CAL. WATER CODE § 13140 *et seq.* (West 1992 & Supp. 1997). Implementation and enforcement of state water quality is the responsibility of the RWQCBs. *See id.* § 13200 *et seq.* (West 1992 & Supp. 1997).

168. *See* State Water Resources Control Bd., Resolution No. 92-49, Policies and Procedures for Investigation and Cleanup and Abatement of Discharges under Water Code Section 13304 (June 18, 1992, as amended Apr. 12, 1994).

169. *Id.*

170. *See* State Water Resources Control Bd., Amended Resolution No. 92-49, Policies and Procedures for Investigation and Cleanup and Abatement of Discharges Under Water Code Section 13304 [hereinafter Amendments]. The first amendments initially proposed by the SWRCB in 1994 were substantially revised and reproposed three times, following extensive input from the regulated industry, water suppliers, and citizens groups, as well as the RWQCBs. The Amendments were followed by a memorandum issued by the SWRCB's Office of Chief Counsel, which addressed specific points raised by the RWQCBs. *See* Memorandum from William R. Attwater, Chief Counsel, State Water Resources Control Bd., to Walt Pettit, Executive Director, State Water Resources Control Bd., Regional Water Board Discretion Under the Containment Zone Policy (Oct. 3, 1996).

171. Technological feasibility means those technologies that have been demonstrated to be effective under given hydrogeologic conditions. Economic feasibility is determined by balancing the incremental costs of further cleanup against the benefits derived therefrom; the economic infeasibility analysis does not include the financial capability of the discharger, except to the extent it may affect the determination of reasonable compliance schedules. Amendments, *supra* note 170, ¶ III.H.1.

172. *Id.* ¶ III.H.2.a.; *see id.* Appendix to Section III.H., "Application for a Containment Zone Designation." In certain cases, when the contamination is the result of leaking under-

ground petroleum tanks, local agencies authorized under the state Underground Storage Tank Law (see section of this chapter entitled "Reforms to Underground Storage Tank Cleanup Requirements") are required to forward their files and recommendations for cleanup to the RWQCB charged with carrying out the requirements of the containment-zone policy. Amendments, *supra* note 170, ¶ III.H.7. Certain cleanup actions undertaken pursuant to CERCLA and RCRA will substitute for the requirements set forth in the containment-zone policy. *Id.*

173. Amendments, *supra* note 170, ¶ III.H.3.
174. *Id.* ¶¶ III.H.8–9.
175. *Id.* ¶ III.H.2.
176. *Id.* ¶¶ III.H.3.c–d.
177. *Id.* ¶ III.H.3.d; *see* CAL. WATER CODE § 10750 *et seq.* (West 1992 & Supp. 1997).
178. Amendments, *supra* note 170, ¶¶ III.H.2.b–d.
179. *Id.* ¶ III.H.2.d.
180. *Id.* ¶ III.H.5.
181. *Id.*
182. *Id.* ¶ III.H.13.
183. *Id.* ¶ III.H.11.
184. *Id.* ¶ III.H.4.
185. *Id.* ¶ III.H.12.
186. *See* CAL. HEALTH & SAFETY CODE, ch. 6.75, § 25299.10 *et seq.* (West 1992 & Supp. 1997).
187. *Id.* § 25299.37(c) (West Supp. 1997).
188. *Id.* § 25299.37(f) (West Supp. 1997).
189. *See id.* § 25299.37(b) (West Supp. 1992).
190. *See* State Water Resources Control Bd., Resolution No. 88-63, Sources of Drinking Water (1988).
191. State Water Resources Control Bd., Resolution No. 68-16, Statement of Policy with Respect to Maintaining High Quality of Waters in California (1968).
192. *See* LEAKING UNDERGROUND FUEL TANK TASK FORCE, STATE WATER RESOURCES CONTROL BD., LEAKING UNDERGROUND FUEL TANK FIELD MANUAL: GUIDELINES FOR SITE ASSESSMENT, CLEANUP, AND UNDERGROUND STORAGE TANK CLOSURE (1989).
193. Hazardous Substances: Petroleum Underground Storage Tanks, ch. 1191, § 3, 1994 Cal. Stat. 5966 (codified at CAL. HEALTH & SAFETY CODE § 25299.38) (West Supp. 1997).
194. CAL. HEALTH & SAFETY CODE § 25299.38(a) (West Supp. 1997).
195. *See* Lawrence Livermore National Laboratory, *Recommendations to Improve the Cleanup Process for California's Leaking Underground Fuel Tanks (LUFTs)* (Oct. 16, 1995), and accompanying *Letter to State Water Resources Control Board* (Nov. 2, 1995) [hereinafter *LLNL Report*].
196. Low-risk soils cases are those occurring when only on-site soils only have been impacted by the LUFT. Low-risk groundwater-impact cases are those involving shallow groundwater with a maximum depth of 50 feet and no drinking-water wells within 250 feet. *See* Letter from Walt Pettit, Executive Director, State Water Resources Bd., to All Regional Water Board Chairpersons, All Regional Water Board Executive Officers, All LOP Agency Directors (Dec. 8, 1995).
197. *Id.* at 1.
198. *Id.* at EX-2, 15–17.
199. *Id.* at EX-3–4, 19–20.
200. *See id.* at 1.

201. *Id.*

202. *See, e.g.,* Letter from Benjamin D. Kor, Executive Officer, North Coast Region, Regional Water Resources Control Bd., to Bob Vasquez, Southland Corporation, et al. (Dec. 20, 1995).

203. San Francisco Bay Regional Water Resources Control Bd., Supplemental Instructions to State Water Board, December 8, 1995, Interim Guidance on Required Cleanup at Low Risk Fuel Sites (Jan. 5, 1996) (as referenced in Letter from Loretta K. Barsamian, Executive Officer, Regional Water Resources Control Bd., San Francisco Bay Region, to San Francisco Bay Area Agencies Overseeing Underground Storage Tank Cleanup (Jan. 5, 1996)).

204. *Id.* at 1.

205. The San Francisco Bay RWQCB notes that in appropriate cases, land-use restrictions may be appropriate to ensure public health and safety in such cases. *Id.* at 3.

206. State Water Resources Control Bd., Senate Bill 1764 Advisory Committee, Recommendations Report regarding California's Underground Storage Tank Program (May 31, 1996) [hereinafter Advisory Committee Report]. The report consists of the Advisory Committee's recommendations, as well as a "Minority Opinion" (in three parts) and a "Minority Report."

207. See the American Society for Testing and Materials standard for risk-based corrective action, ASTM E-1739-95.

208. *Id.* at 3–23.

209. *Id.* at 24–28.

210. *Id.* at 15–17, 25. The Minority Opinion urged, in addition, that the SWRCB reconsider its anti-degradation policy (*see supra* note 191) and that state funding program criteria include requirements for the use of risk-based corrective-action standards. Advisory Committee Report, *supra* note 206, at MO-1-5. The Minority Report primarily took issue with the Advisory Committee Report's failure to adopt affirmatively the LLNL report's low-risk site closure recommendations in accordance with risk-based standards. *Id.* at MR-1-7.

211. *See* CAL. ENVTL. PROTECTION AGENCY, 5 REPORT NO. 7, 5–6 (1996).

212. *See* Draft State Water Resource Control Bd. Resolution No. 1021b (Oct. 29, 1996). *See also* Memorandum from Loretta K. Barsamian, Executive Officer, San Francisco Bay Regional Water Resource Control Bd., to Harry Schueller, Chief, Div. of Clean Water Programs, State Water Resource Control Bd. (Nov. 15, 1996) (entitled "Initial Review of SWRCB-CWP's October 29th Draft `Policy for Investigation and Cleanup of Petroleum Discharges to Soil and Groundwater'").

213. Thompson-Richter Underground Storage Tank Reform Act, ch. 611, 1996 Cal. Stat. 21780 (codified at CAL. HEALTH & SAFETY CODE § 25289 *et seq.*) (West Supp. 1997).

214. *Id.*

CHAPTER 24

Colorado

DAVID F. GOOSSEN

Colorado's Voluntary Cleanup and Redevelopment Act[1] (Voluntary Cleanup Act, or Act) was enacted in 1994 to protect human health and the environment and to foster the transfer, redevelopment, and reuse of previously contaminated facilities and sites.[2] In connection with this purpose, the Act establishes a program (Voluntary Cleanup Program, or Program), administered by the Hazardous Materials and Waste Management Division of the Colorado Department of Public Health and Environment (CDPHE), for the approval of voluntary cleanup plans and the issuance of no-action determinations.[3] The Program permits and encourages voluntary cleanups of contaminated properties by providing a framework for determining site-specific cleanup responsibilities for such properties. It facilitates the transfer and redevelopment of contaminated properties by using a streamlined review and approval process.

Summary of Major Provisions

Colorado's Voluntary Cleanup Program is intended to eliminate the impediments to the transfer, redevelopment, and reuse of contaminated properties and encourage voluntary cleanups by providing property owners with a mechanism for determining site-specific cleanup responsibilities in advance of a sale or redevelopment plan.[4] The Program has two major components: (1) the approval of voluntary cleanup plans, and (2) the issuance of no-action determinations.

Major provisions related to the approval of voluntary cleanup plans include the following:

- Providing for practical and reasonable cleanups by taking the following into consideration when selecting among remedial alternatives:
  - present or proposed uses of the site
  - mobility of the contaminants and the resulting harm to human health and the environment

- potential risks associated with proposed cleanup alternatives in light of the economic and technical feasibility of such alternatives
- Minimizing the cost of applying for approval of voluntary cleanup plans by, for example, accepting risk-based cleanup levels that have been developed elsewhere rather than requiring site-specific risk assessments

A major provision related to the issuance of no-action determinations is:

- Providing no-action determinations if contamination originates from an off-site source, and the person or entity responsible for the source is, or will be, taking appropriate action

Participation in the Program

Participation in the Colorado Program is voluntary.[5] The Act specifies that no person, financial institution, or other entity financing a commercial real estate transaction shall require a purchaser to participate in the Program, and that no entity of Colorado state government regulating any person, financial institution, or other entity shall require evidence of participation in the Program to be a component of standard real estate loan documentation.[6]

Eligibility For the Program

The owner of a contaminated site is the only party that may initiate participation in the Program.[7] Unless specifically excluded, any person who owns real property that has been contaminated with hazardous substances or petroleum products may submit an application for the approval of a voluntary cleanup plan.[8] Unless excluded, any property owner may file a written petition with CDPHE to request a no-action determination for his or her property.[9] Eligibility for participation in the Program is not limited to certain sites or geographic areas.

Sites Ineligible for the Program

The Program is designed to permit and encourage the cleanup and redevelopment of contaminated properties; it is not intended to replace other regulatory programs.[10] As a result, not all properties are eligible to participate in the Program. The Act states that its provisions do not apply to the following:

1. property that is listed or proposed for listing on the National Priorities List of Superfund sites established under the federal act;
2. property that is the subject of corrective-action orders or agreements issued pursuant to part 3 of article 15 of the Act or the federal Resource Conservation and Recovery Act (RCRA) of 1976, as amended;
3. property that is subject to an order issued by, or an agreement with, the Water Quality Control Division, pursuant to part 6 of article 8 of the Act;

4. a facility that has, or should have, a permit or interim status pursuant to part 3 of article 15 of the Act for the treatment, storage, or disposal of hazardous waste; or
5. property that is subject to the provisions of part 5 of article 20 of title 8 of the Colorado Revised Statutes, or of article 18 of the Act.[11]

CDPHE has made clear that properties without a permit or interim status under federal or state hazardous-waste laws, but at which hazardous waste, as defined in these laws and their implementing regulations, was treated, stored, or disposed of any time after 1980, are considered to have required a permit or interim status and are ineligible.[12] Disposal, CDPHE notes, is defined as "any discharge, injection, dumping, spilling, leaking, or placing any solid wastes or hazardous wastes into or on any land or water so that such solid waste or hazardous waste or any constituent thereof may enter the environment."[13] In response to concerns from potential applicants that participation in the Program may put them at risk of enforcement, CDPHE has explained that if a site qualifies for the Program, by definition it will not fall within the authority of other environmental programs.[14]

Memorandum of Agreement

CDPHE has entered a Memorandum of Agreement (MOA) with the United States Environmental Protection Agency (USEPA).[15] The purpose of the MOA is to define the roles and responsibilities of CDPHE and USEPA concerning activities conducted under the authority of the Voluntary Cleanup Act.[16] Under the MOA, CDPHE will implement the Act to allow owners of contaminated properties to propose cleanup action voluntarily, or petition for no-further-action determinations for eligible sites, while including in the Voluntary Cleanup Program the specific elements described in Attachment A to the MOA.[17] USEPA agrees that once an application to clean up a site in accordance with the Program has been received by CDPHE, USEPA will not plan—and does not anticipate taking—any federal action under the Comprehensive Environmental Response, Compensation, and Liability Act (CERCLA)[18] at such a site unless (1) the site is a federal National Priorities List type (NPL Caliber) site or the site poses an imminent and substantial endangerment to public health, welfare, or the environment and exceptional circumstances warrant USEPA action, (2) CDPHE's approval of the cleanup becomes void, or (3) the applicant fails to complete or comply materially with the cleanup plan as approved by CDPHE.[19] USEPA also reserves the right to take action under CERCLA at a site in the event a cleanup plan or no-action determination is deemed "approved" as a result of CDPHE failing to approve or deny an application before the expiration of the forty-five-day time limit in the Act.[20] USEPA agrees to provide technical support to CDPHE to develop and expand the use of the Program.[21]

The specific Program elements described in Attachment A include site screening and communication requirements, public-participation requirements, and verification procedures.[22] Attachment A sets forth a two-step, site-screening procedure for CDPHE. First, sites are screened for eligibility in accordance with the Act. If the site qualifies for participation in the Program, a second screening is performed to determine existing actions proposed by USEPA and USEPA's level of potential interest in the site. The purpose of this screening is to avoid duplication of effort between CDPHE and USEPA. The second screening involves a determination of whether the site is on the CERCLIS list (USEPA's list of properties that may need investigation under CERCLA) or might be considered an NPL Caliber site. NPL Caliber sites generally are those where a significant human exposure to hazardous substances has been documented or where sensitive environments have become contaminated.[23] If the site is on CERCLIS, CDPHE will request that USEPA suspend activities at the site to allow the cleanup to proceed under the Program.[24] If CDPHE determines a site to be an NPL Caliber site, it will notify the applicant. CDPHE and the applicant will then jointly decide whether to inform USEPA of this determination and request USEPA's review of, and concurrence with, the cleanup plan and application.[25] If CDPHE and the applicant jointly decide to seek USEPA's review and approval, USEPA will provide its comments as quickly as possible.[26] If CDPHE and the applicant decide not to solicit USEPA's approval, and CDPHE approves the application without USEPA's review and concurrence, the applicant may implement the cleanup plan but USEPA's forbearance to plan or undertake federal action under CERCLA is void.[27]

The Act has no requirements for public participation or review of applications. However, to obtain USEPA's forbearance to plan or undertake any action under CERCLA as contained in section III, paragraph 2 of the MOA, the applicant must—within thirty days of approval of its Voluntary Cleanup Program application—provide adequate public notice of its cleanup plan.[28] This should include publication of the availability of the cleanup plan in a local newspaper, or posting of any public-notice plan required by permit or zoning ordinance procedures.[29] For large sites, CDPHE may request, in accordance with the MOA, that the applicant hold a public meeting to explain its cleanup plan.[30]

Similarly, under the Act, verification of the completion of a cleanup under the Program is left to the applicant. The Act does not provide for inspections by CDPHE. To obtain USEPA's forbearance to plan or undertake any action under CERCLA as contained in section III, paragraph 2 of the MOA, however, the applicant must submit a written petition requesting a no-further-action determination on the subject property following the completion of the cleanup plan.[31] This petition must include a completion report that describes how the applicant has complied with the initial or modified cleanup plan as approved by CDPHE.[32] CDPHE will review the report to assure compliance with the approved cleanup plan and may conduct an inspection of the subject property to obtain readily available information concerning the property's current condition.[33]

Protections Granted by the Colorado Program

If CDPHE approves a voluntary cleanup plan, it will provide written notification of such approval to the applicant.[34] This written approval must include the following statement:

> Based upon the information provided by [the applicant], it is the opinion of the Colorado Department of Public Health and Environment that upon completion of the voluntary cleanup plan no further action is required to assure that this property, when used for the purposes identified in the voluntary cleanup plan, is protective of existing and proposed uses and does not pose an unacceptable risk to human health or the environment at the site.[35]

The Act requires that similar language be included in the written notification of a no-action determination.[36]

Unlike the voluntary cleanup laws of some other states, there are no provisions in Colorado's Voluntary Cleanup Act stating that lenders, trustees, fiduciaries, or any other participants in the Program will be granted liability protection following the implementation of a voluntary cleanup plan or approval of a no-action petition. Moreover, the Act states that persons participating in the program are not absolved from obligations under other laws or regulations, "including any requirement to obtain permits or approvals for work performed under a voluntary cleanup plan."[37]

The Act provides that failure to comply materially with the voluntary cleanup plan approved by CDPHE[38] or submission of materially misleading information by the applicant in the context of the voluntary cleanup application[39] or a no-action petition will render the approval void.[40] It further provides that the approval of a voluntary cleanup plan or a no-action petition by CDPHE applies only to conditions on the property and state standards that exist at the time of submission of the application or petition.[41]

Financial Incentives

The Colorado Program is intended to operate in such a way as to eliminate impediments to the sale or redevelopment of previously contaminated property and to encourage and facilitate prompt cleanup activities but still minimize administrative processes and costs.[42] As a result, the Act does not contain specific financial incentives for the cleanup of contaminated property. It does not provide for tax abatement or low-interest loans for the cleanup of contaminated properties. Instead, the Act includes clear mechanisms to determine cleanup responsibilities and specific deadlines within which CDPHE must either approve or disapprove a voluntary cleanup plan or request for a no-action determination.[43] Under the Act, voluntary cleanup plans or petitions for no-further-action determinations must be approved within forty-five days, unless otherwise agreed.[44] In this way, the Act encourages private redevelopment at no direct cost to the public.

Cleanup Standards

The Act requires that remediation alternatives be based on "the actual risk to human health and the environment currently posed by contaminants on the real property."[45] This determination is made considering the following factors:

1. the present or proposed uses of the site;
2. the ability of the contaminants to move in a form and manner that would result in exposure to humans and the surrounding environment at levels exceeding applicable standards, representing an unacceptable risk to human health or the environment;
3. the potential risks associated with proposed cleanup alternatives, and the economic and technical feasibility and reliability of such alternatives.[46]

CDPHE is required to approve a voluntary cleanup plan when, considering these factors, it concludes that the plan will either "attain a degree of cleanup or control of hazardous substances or petroleum products, or both, that complies with all applicable state requirements, regulations, criteria or standards" or, for constituents for which there are not applicable standards, "reduce concentrations such that the property does not present an unacceptable risk to human health or the environment based upon the property owner's current use and any future uses proposed by the property owner."[47] To allow it to make this determination, CDPHE requires that an applicant include in its application a risk assessment, performed in accordance with standard USEPA policy, or a calculation of appropriate cleanup levels using CDPHE's "Interim Final Policy and Guidance On Risk Assessment For Corrective Action at RCRA Facilities," dated November 16, 1993.[48] CDPHE has stated that it will evaluate this analysis based on an acceptable excess cancer risk of 1×10^{-6} or a hazard index of less than one.[49]

The Mechanics of Participating in the Program

Property owners are the only persons who may initiate an application for approval of a voluntary cleanup plan or a petition for a no-action determination.[50] Before deciding to either submit a voluntary cleanup plan for approval or request a no-action determination, a property owner should consider whether his or her property is eligible. Sites that fall under existing regulatory programs are ineligible.[51] The Act specifically excludes Underground Storage Tank sites; sites that have or should have an RCRA permit for treatment, storage, or disposal of hazardous waste or are under an RCRA corrective-action order; sites that are listed or proposed for listing on the National Priorities

List; and sites that are subject to an order from the Water Quality Control Division under the Clean Water Act.[52]

If a property owner determines that his or her site is eligible for the Program, the landowner may submit an application for approval of (1) a voluntary cleanup plan pursuant to section 25-16-304, when remediation may be necessary to protect human health and the environment in light of the current or proposed use of the property,[53] or (2) a no-action petition pursuant to section 25-16-307, when remediation is complete or not necessary to protect human health and the environment in light of the current or proposed use of the property.[54] CDPHE is granted access to the subject property during the period of review of no-action petitions and applications for approval of voluntary cleanup plans.[55]

Applications for approval of voluntary cleanup plans and petitions for no-further-action determinations must be accompanied by a filing fee. Under the Act, CDPHE is directed to determine the amount of these filing fees.[56] CDPHE is also directed to establish hourly rates for review charges performed in connection with such applications and petitions.[57] Under this authority, CDPHE has determined that the filing fee should be $2,000,[58] and has established hourly rates of $75 an hour.[59] Review charges are billed against the $2,000 filing fee, with the surplus returned to the applicant within thirty days after CDPHE's approval or denial.[60] Twenty-seven of forty-three applications received as of March 7, 1996, received a rebate on the initial $2,000 fee.[61]

Voluntary Cleanup Plans

The Act states that a voluntary cleanup plan must include

1. an environmental assessment that describes the contamination, if any, on the property, and the risk the contamination currently poses to public health and the environment;
2. a proposal, if needed, to remediate any contamination or condition unacceptable to human health or the environment, considering the present—and any differing proposed—uses of the property, and a timetable for implementing the proposal and for monitoring the site after the proposed measures are completed; and
3. a description of applicable state standards establishing acceptable concentrations of constituents in soils, surface water, or groundwater and, for constituents present at the site for which such state standards do not exist, a description of proposed cleanup levels and any current risk to human health or the environment based upon the current or proposed use of the site.[62]

Application requirements are made more specific in CDPHE's "Voluntary Cleanup and Redevelopment Act Application."

Once it has received an application for approval of a voluntary cleanup plan, CDPHE is required to approve or deny the application within forty-five days.[63] Unless granted an extension, the applicant is required to implement the plan within one year, and complete the plan within two years of approval.[64] Verification that the cleanup plan has been completed must be submitted by a qualified environmental professional within forty-five days of completion.[65] Property owners desiring to implement a voluntary cleanup plan after the applicable time limits must submit a written petition for reapplication, with a written certification of a qualified environmental professional that the conditions on the subject real property are substantially similar to those that existed at the time of the original approval.[66] Reapplications are subject to limited review by CDPHE and must be completed within thirty days.[67]

If the voluntary cleanup plan is not approved, CDPHE is required promptly to provide the property owner with a written statement of the reasons for the denial.[68] If the denial is based on a failure to submit the information required by section 25-16-304, CDPHE is required to notify the applicant of the specific information omitted.[69]

No-Action Petitions

The Act provides that a no-action petition must be granted when

1. the environmental assessment described in section 25-16-308, performed by a qualified environmental professional, indicates the existence of contamination that does not exceed applicable promulgated state standards, or contamination that does not pose an unacceptable risk to human health and the environment; or
2. CDPHE finds that contamination or a release or threatened release of a hazardous substance or petroleum product originated from a source on adjacent or nearby property, if a person or entity responsible for such a source of contamination is, or will be, taking necessary action to address the contamination.[70]

Review of no-action petitions is limited to a review of materials submitted by the applicant and documents readily available to CDPHE.[71] Except in exceptional circumstances, the review must be completed in forty-five days or the petition is deemed approved.[72] If the no-action petition is not approved, CDPHE must state why,[73] and if the denial is based on the applicant's failure to submit information, CDPHE must state what information is missing.[74]

Environmental Assessments

Environmental assessments, required for either a no-action petition or an application for approval of a voluntary cleanup plan, must be prepared by a qualified environmental professional.[75] The assessment is required to include the following:

1. the legal description of the site and a map identifying the location and size of the property;

2. the physical characteristics of the site and areas contiguous to the site, including the location of any surface-water bodies and groundwater aquifers;
3. the location of any wells located on the site or on areas within a one-half mile radius of the site, and a description of the use of those wells;
4. the current and proposed use of on-site groundwater;
5. the operational history of the site and the current use of areas contiguous to the site;
6. the present and proposed uses of the site;
7. information concerning the nature and extent of any contamination and releases of hazardous substances or petroleum products that have occurred at the site, including any impacts on areas contiguous to the site;
8. any sampling results or other data that characterizes the soil groundwater or surface water on the site; and
9. a description of the human and environmental exposure to contamination at the site based upon the property owner's current use and any future use proposed by the property owner.[76]

Risk Assessments

Risk assessments should be performed in accordance with standard USEPA policy, or appropriate cleanup levels may be calculated using CDPHE's "Interim Final Policy and Guidance On Risk Assessment For Corrective Action at RCRA Facilities," dated November 16, 1993.[77] CDPHE has stated that it will evaluate this analysis based on an acceptable excess cancer risk of 1×10^{-6} or a hazard index of less than one.[78]

Enforceability of Voluntary Cleanup Plans

Voluntary cleanup plans are not enforceable against a property owner.[79] Nonetheless, if CDPHE can demonstrate that a property owner who initiated a voluntary cleanup under an approved plan has failed to implement that plan fully and properly, CDPHE may require further action if the action is authorized by other laws or regulations.[80]

The Act does not include civil or criminal penalties. In fact, the Act provides that information supplied by a property owner to support a voluntary cleanup plan or no-action petition does not provide CDPHE with an independent basis to seek penalties from the property owner pursuant to state environmental statutes or regulations.[81] Moreover, the Act states that

> [i]f, pursuant to state environmental statutes or regulations, [CDPHE] initiates an enforcement action against the property owner subsequent to a submission of a voluntary cleanup plan or no action petition regarding that contamination addressed in the plan or petition, the voluntary disclosure of the information in the plan or petition shall be considered by

the enforcing authority to reduce or eliminate any penalties assessed to the property owner.[82]

Conclusion

Colorado's Voluntary Cleanup and Redevelopment Act is a positive development for owners of contaminated real property and the public in general. Its nonbureaucratic design, which is intended to foster a cooperative relationship between applicants and CDPHE, encourages cleanups that benefit property owners as well as those who may be exposed to contaminants from a subject property. Although it does not contain specific financial incentives, the Program's philosophy of approving practical and reasonable cleanups is beginning to show results. CDPHE has approved more than twenty-five voluntary cleanup plans, and applications are on the rise. Although somewhat limited in scope, as compared with programs in other states, the Act has been successful in accomplishing its objectives.

Colorado's Voluntary Cleanup Program

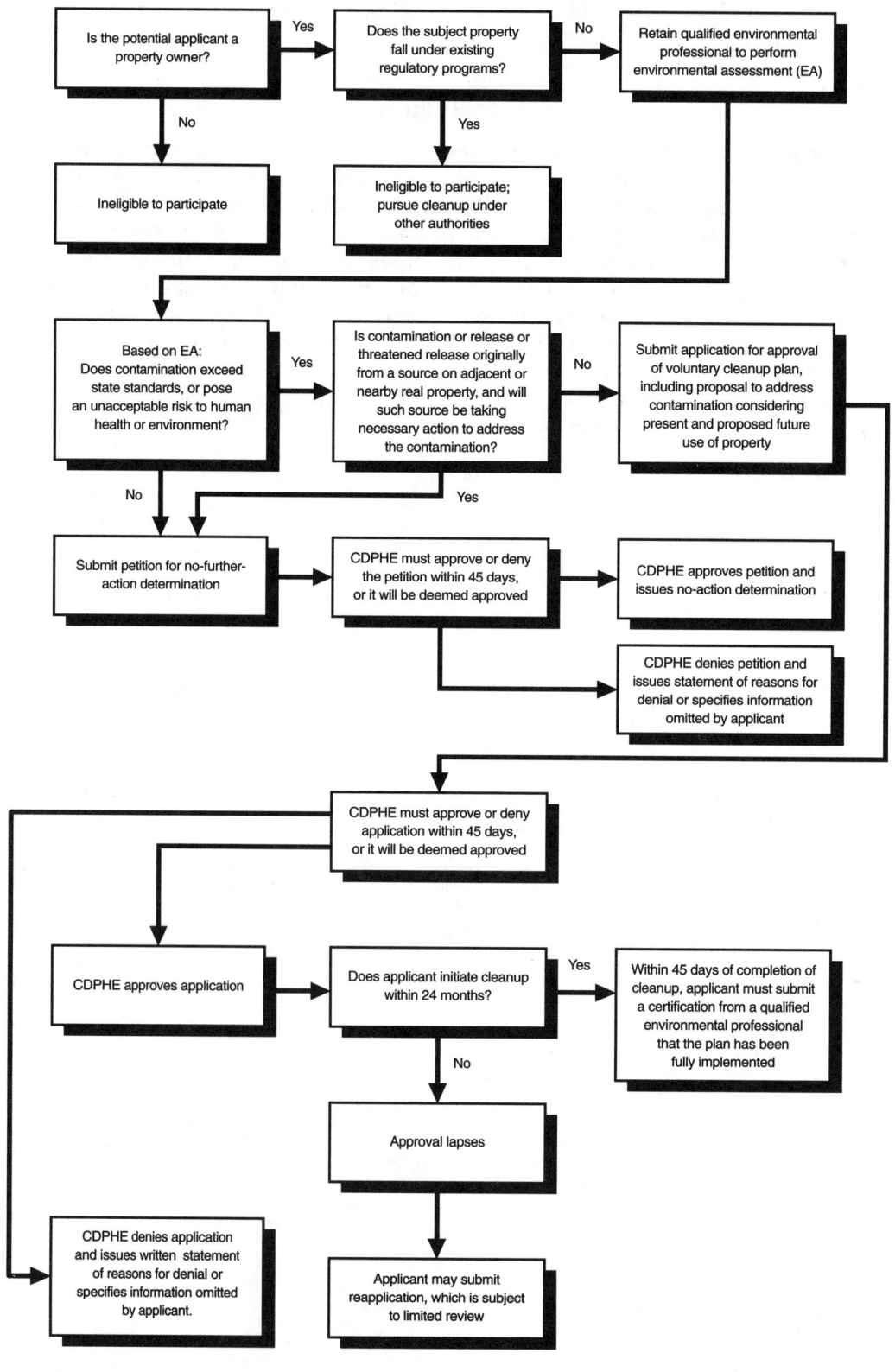

Notes

1. H.B. 94-1299 (codified at COLO. REV. STAT. § 25-16-301 *et seq.*) (1995).
2. *Id.* § 25-16-302(1).
3. *Id.* § 25-16-303.
4. *Id.* § 25-16-302(1).
5. *Id.* § 25-16-303(3)(a).
6. *Id.* § 25-16-303(2).
7. *Id.*
8. *Id.* § 25-16-304(1).
9. *Id.* § 25-16-307(1).
10. *Id.* § 25-16-302(1).
11. *Id.* § 25-16-303(3)(a).
12. COLORADO DEP'T PUB. HEALTH & ENV'T, HAZARDOUS MATERIALS & WASTE MANAGEMENT DIV., VOLUNTARY CLEAN-UP & CERTIFICATION. (1995).
13. *Id.*
14. *Id.*
15. Memorandum of Agreement between the Colorado Department of Health and Environment and the United States Environmental Protection Agency, Region VIII (1996) [hereinafter Memorandum of Agreement].
16. *Id.* § II.
17. *Id.* § III, at 1.
18. 42 U.S.C. § 9601 *et. seq.* (1995).
19. Memorandum of Agreement, *supra* note 15, § III, at 2.
20. *Id.* § III, at 4.
21. *Id.* § III, at 6.
22. *Id.*, Attachment A.
23. *Id.*
24. *Id.*
25. *Id.*
26. *Id.*
27. *Id.*
28. *Id.*
29. *Id.*
30. *Id.*
31. *Id.*
32. *Id.*
33. *Id.*
34. COLO. REV. STAT. § 25-16-306(2) (1994).
35. *Id.*
36. *Id.* § 25-16-307(2)(b).
37. *Id.* § 25-16-309(1).
38. *Id.* § 25-16-306(3)(a).
39. *Id.* § 25-16-306(3)(b).
40. *Id.* § 25-16-307(2)(c).
41. *Id.* §§ 25-16-306(1)(d), 25-16-307(2)(c).
42. *Id.* § 25-16-302(1).
43. *Id.* §§ 25-16-306(1)(a), 25-16-307(1).
44. *Id.*
45. *Id.* § 25-16-305(1).

46. *Id.*
47. *Id.*
48. Colorado Dep't Pub. Health & Env't, Hazardous Materials & Waste Management Div., Voluntary Cleanup and Development Act Application 6 (Draft July 11, 1994).
49. *Id.*
50. COLO. REV. STAT. § 25-16-303(3)(a) (1994).
51. *Id.* § 25-16-303(3)(b).
52. *Id.*
53. *Id.* § 25-16-303(1)(a).
54. *Id.* § 25-16-303(1)(b).
55. *Id.* § 25-16-303(5).
56. *Id.* § 25-16-303(4)(a).
57. *Id.* § 25-16-303(4)(b).
58. Under the Act, an applicant cannot be charged more than $2,000.
59. Daniel L. Sheppers, P.E., Colorado Department of Health and Environment, Hazardous Materials and Waste Management Division, Address at "Brownfields Redevelopment: Cleaning Up the Urban Environment, How to Get the Deals Done" (March 7, 1996).
60. COLO. REV. STAT. § 25-16-303(4)(b) (1994). All moneys collected pursuant to subsection 4 of the Act are transmitted to the state treasurer, who credits such moneys to the Hazardous Substance Response Fund. *Id.* § 25-16-303(c). They are subject to general appropriation only to defray the cost of the program. *Id.*
61. Sheppers, *supra* note 59.
62. COLO. REV. STAT. § 25-16-304(2) (1994).
63. *Id.* § 25-16-306(1).
64. *Id.* § 25-16-306(4)(a).
65. *Id.* § 25-16-306(5).
66. *Id.* § 25-16-303(4)(b).
67. *Id.* § 25-16-306(4)(c).
68. *Id.* § 25-16-306(1)(c).
69. *Id.*
70. *Id.* § 25-16-307(2)(a).
71. *Id.* § 25-16-307(1).
72. *Id.*
73. *Id.* § 25-16-307(4).
74. *Id.*
75. *Id.* § 25-16-308(1).
76. *Id.* § 25-16-308(2).
77. Colorado Dep't Pub. Health & Env't, *supra* note 48.
78. *Id.*
79. COLO. REV. STAT. § 25-16-310(1) (1994).
80. *Id.*
81. *Id.* § 25-16-310(2).
82. *Id.*

CHAPTER 25

Connecticut

MARGARET MURPHY

Since 1992, Connecticut has been active in developing statutes and regulations that encourage the development of brownfields. Although the state has not formulated one, streamlined Voluntary Cleanup Program, it has three separate initiatives that foster brownfields development: the Urban Sites Remedial Action Program,[1] the 1995 amendments to Connecticut's Transfer Act,[2] and the 1995 Voluntary Site Remediation Programs.[3] These brownfields initiatives are further enhanced by the newly adopted cleanup standards of the Connecticut Department of Environmental Protection (CTDEP).[4]

Summary of Connecticut's Brownfields Initiatives

Connecticut's first brownfields initiative, the Urban Sites Remedial Action Program (Urban Sites Program), began as a pilot project in 1992 and has grown into a full-fledged brownfields redevelopment program.[5] The Urban Sites Program provides (1) for Urban Sites Program staff to work with willing parties who own contaminated land to expedite the review and approval of environmental cleanups, (2) for the acquisition by the Connecticut Department of Economic Development (CTDED) of polluted properties, the assumption of liability, and the authority to lease or sell the property to a developer after remediation has been completed by the state, and (3) for the "use of state bond funds to undertake investigation and remediation of sites where a responsible party is either unwilling or unable to undertake the work."[6] Sites eligible for the use of bond funds for cleanup must be located in distressed communities or be in an "enterprise zone," and must have economic development potential as determined by the CTDED.[7] This bond fund provides the state with a source of funding to clean up certain contaminated properties and subsequently recover its costs from responsible parties. Although this results in the cleanup of contaminated properties, the program does not provide incentives specific to developers and property owners to clean up brownfields and, therefore, will not be further addressed in this chapter.

Connecticut's Public Act 95-183 amends Connecticut's Hazardous Waste Establishment Transfer Act (Transfer Act) to encourage redevelopment. In general, the Transfer Act requires any party who transfers the operations of an establishment where hazardous waste may have been released to file a form declaring the environmental status of the site.[8] Under the Transfer Act, a Form I is filed if there has not been any release of hazardous waste on the premises being transferred; a Form II is filed if there was such a release on the premises, which has been cleaned up with CTDEP's approval; a Form III is filed if there has been a release that has not yet been cleaned up, but will be cleaned up by either the transferor or the transferee; and a Form IV is filed if the establishment was contaminated by the release of hazardous waste and all remedial actions have been taken except for postremediation monitoring.[9] The 1995 amendments to the Transfer Act under Public Act 95-183 allow the investigation and remediation of a transferred establishment to be conducted by a private, state-licensed environmental professional[10] and authorize the CTDEP to enter a covenant-not-to-sue with certain parties who have filed a Form III or Form IV.[11]

Connecticut's Public Acts 95-183 and 95-190 (establishing the "95-183 Program" and the "95-190 Program," respectively) provide covenants-not-to-sue to certain volunteers who clean up property using a licensed environmental professional.[12] These statutes (collectively, the Voluntary Site Remediation Programs) establish programs similar to each other and overlap considerably because the programs were set up in separate legislative bills enacted in the same year. Nonetheless, there are important differences between the two programs, as more fully discussed below.

Although the Urban Sites Program began in 1992, the majority of Connecticut's brownfields initiatives are newer. For example, the CTDEP has issued only a few covenants-not-to-sue and the CTDED has never taken possession of contaminated property pursuant to the Urban Sites Program.[13] The CTDEP has, however, worked with many private parties and already-owned government sites to remediate contaminated land for development purposes under the Urban Sites Program. CTDEP has not yet developed a policy for dealing with Connecticut's Public Acts 95-183 and 95-190. As a consequence, it is difficult to predict how to interpret, in practice, the overlapping provisions and sometimes vague references in Public Acts 95-183 and 95-190.

The remainder of this chapter will address the various aspects of the three types of Connecticut brownfields initiatives separately within each topic heading below and, when appropriate, how they have developed in practice.

Eligibility for Connecticut's Brownfields Programs

This section discusses the eligibility of potential participants in Connecticut's various brownfields initiatives. Eligibility for a covenant-not-to-sue at the completion of a program is discussed in the section of this chapter entitled "Liability Protection for Brownfields Volunteers."

Urban Sites Program

Voluntary Cleanups

A party who (1) is occupying property that may be contaminated, has a developer who is interested in property that may be contaminated, or is currently working with CTDEP to evaluate a site for reuse, and (2) has the resources available to begin an environmental investigation or cleanup, is eligible to apply to the CTDED for review of the property and a ranking of the property against other properties in the state according to economic benefit and need.[14]

State Purchase and Remediation

The CTDED may acquire and remediate a site if the following conditions exist:

1. "the commissioner of economic development finds that the state owns the site or otherwise has or obtains the power to approve the type of development which first occurs on the site after remediation";[15]
2. "the commissioner of environmental protection is unable to determine the responsible party for the pollution or the cleanup of the site, or the responsible party is not in timely compliance with orders issued by the commissioner to provide remedial action, or the commissioner has not issued a final decision on an order to a responsible party to provide remedial action because of (i) a request for a hearing on an order, or (ii) an order issued is subject to an appeal pending before a court";[16] and
3. the property is either (i) polluted, undeveloped, and zoned commercial or industrial or (ii) polluted, developed, and zoned commercial or industrial and is abandoned or underused, and "the commissioner finds that the remediation of such property will assist with the retention or expansion of an existing manufacturing or economic base business or businesses operating on such property."[17]

Although the statute does not directly prohibit responsible parties from participating in the Urban Sites Program, it is unlikely that a responsible party (a term that includes the owner of contaminated property[18]) who is unknown to the CTDED, not in compliance with remedial orders, or currently opposing a final decision on an order for remedial action would be accepted by the CTDED to enter this program.

Transfer Act

A party who is transferring contaminated property and must file a Form III or Form IV under the Transfer Act may benefit from the 1995 amendments to the Transfer Act in two ways. First, a party may receive a covenant-not-to-sue from CTDEP after the remediation of the contaminated property. Eligibility to obtain a covenant-not-to-sue will be discussed below in the section entitled "Liability Protection for Brownfields Volunteers." Second, a party may receive CTDEP approval to use a licensed environmental professional to expedite the cleanup process. In determining whether a party should be eligible to use a licensed environmental professional, the CTDEP considers the following factors:[19]

- The potential uncontrolled loss, seepage, or filtration of hazardous waste on the parcel
- The degree of environmental investigation at the parcel
- The proximity of the parcel to significant natural resources
- The character of the land uses surrounding the parcel
- The complexity of the environmental condition of the parcel
- Any other factor the commissioner deems relevant

Voluntary Site Remediation Programs

A party may be eligible to obtain CTDEP approval of a cleanup under the 95-183 Program if the subject property is listed "on the inventory of hazardous waste disposal sites maintained pursuant to section 22a-133c" of the Connecticut statutes.[20] A site may be placed on the inventory of hazardous-waste disposal sites if the CTDEP determines that the site is a threat to the environment or the public health.[21] The CTDEP is generally made aware of such sites through reporting under the Transfer Act, reports made under federal laws such as the Resource Conservation and Recovery Act[22] (RCRA) and the Comprehensive Environmental Response, Compensation, and Liability Act[23] (CERCLA), and information gathered pursuant to Connecticut's underground storage tank program.[24]

A party may be eligible to obtain approval from the CTDEP of a final remedial-action report under the 95-190 Program if the party uses a licensed environmental professional to conduct the remediation, the property to be remediated is in an area classified for groundwater as "GB" or "GC" by the CTDEP, and the property is not the subject of a state order or judgment regarding the contamination to be remedied.[25] A "GB" area is one classified by the CTDEP as an urban area in which the groundwater is most likely already contaminated, and a "GC" area is one classified by the CTDEP as an area that is permitted for pollution discharge into the groundwater.[26]

Memorandum of Agreement

Currently, Connecticut does not have a Memorandum of Agreement with the United States Environmental Protection Agency (USEPA) for participation in its various brownfields programs. Therefore, a person participating in Connecticut's Voluntary Site Remediation Programs could theoretically be subject to federal liability, even after obtaining a covenant-not-to-sue from the state. As a result, currently there is no statutory or other basis to preclude private parties from bringing a federal cause of action against a Connecticut brownfields program participant. USEPA, however, is currently developing a formal policy under which it will not take action related to a site that is participating in a state brownfields program.[27] In addition, to minimize the threat of federal liability, parties may seek "comfort letters" from USEPA's Region I office.[28]

Liability Protection for Brownfields Volunteers

Voluntary Site Remediation Programs

The 95-183 Program and the 95-190 Program both give CTDEP the authority to provide covenants-not-to-sue to certain parties who remediate brownfields. Under Public Act 95-183, the CTDEP may enter into a covenant-not-to-sue with a party if

1. Such person did not establish or create a facility or condition which reasonably can be expected to create a source of pollution to the waters of the State for purposes of section 22a-432 of the general statutes, and prior to the date of such person's acquisition of the parcel, did not maintain any such facility or condition for purposes of section 22a-432, and such person is not responsible pursuant to any other provision of the general statutes for any pollution or source of pollution on the parcel;
2. Such person is not affiliated with any person responsible for such pollution or source of pollution through any direct or indirect familial relationship or any contractual, corporate or financial relationship other than that by which such person's interest in such property is to be conveyed or financed;
3. The parcel is or will be remediated in accordance with standards adopted by the commissioner pursuant to section 22a-133k of the general statutes;
4. Such person will redevelop the parcel for productive use; and
5. The covenant-not-to-sue is in the public interest.[29]

A covenant-not-to-sue given under the 95-183 Program may release only those claims "related to pollution or contamination on or emanating from the subject parcel, which contamination resulted from a discharge, spillage, uncontrolled loss, seepage or filtration on such parcel prior to the date the recipient of such covenant acquired the subject parcel."[30]

Under Public Act 95-190, the CTDEP may enter into a covenant-not-to-sue with the owner or lessor of a property for which the CTDEP has approved a final remediation report prepared by a licensed environmental professional who has determined, in his or her sole discretion, that continued monitoring of the property is not required.[31] The covenant issued under the 95-190 Program may exempt the property owner, lessor, or security-interest holder from liability for any spill on or at the property before the date of the covenant.[32] In addition, unlike the covenant-not-to-sue under Public Act 95-183, Public Act 95-190 does not prohibit the CTDEP from entering a covenant-not-to-sue with a party who is responsible for pollution on the subject property or on other property in the state. A covenant-not-to-sue under Public Act 95-190, however, is not available to a party who wishes to clean up property that is under a state order or judgment to remediate contamination.

Neither of the covenants-not-to-sue issued under Public Act 95-183 or Public Act 95-190 provide for any "reopeners" or other mechanisms for rescinding a covenant-not-to-sue based on fraud or newly discovered historical contamination.[33]

Urban Sites Program

Cleanup Volunteers

A party who is voluntarily cleaning up property under the Urban Sites Program is eligible to receive, at the CTDEP's discretion, a covenant-not-to-sue from CTDEP, pursuant to Public Act 95-183.[34]

State Purchase and Remediation

A party who has entered an agreement to develop property that the CTDED will purchase, remediate, and lease to a developer under the Urban Sites Program may also receive liability protection. Under this program, the CTDED assumes liability as owner of the property during the cleanup and leasing period, freeing the developer from ownership liability during that time. When the developer purchases the remediated property, the developer is eligible to receive, at the CTDEP's discretion, a covenant-not-to-sue from CTDEP, pursuant to Public Act 95-183 as discussed above.[35] A developer's covenant-not-to-sue under this program, however, may be restricted due to the statutory requirement that the developer "assume responsibility for any direct costs in excess of $15 million."[36] This potential restriction on a release from liability obtained by the developer depends in part on the interpretation of "direct costs" and whether such costs include liability for historical contamination.

Transfer Act

A covenant-not-to-sue may also be issued pursuant to Public Act 95-183 to an eligible party who is developing a site for which a Form III or Form IV has been filed.

Financial Incentives

95-190 Program

Parties who wish to clean up property under the 95-190 Program may receive loans from the CTDED for (1) the cost of having a licensed environmental professional conduct a Phase II environmental site assessment or Phase III investigation, (2) the cost of any demolition undertaken to prepare the property for development, and (3) expenses related to administration of the 95-190 Program.[37] Loan applications under this program are considered based on a variety of criteria, including anticipated commercial value of the property, potential tax revenue to the relevant municipality, environmental or public health risk posed by the spill, potential community or economic development benefit to the relevant municipality, environmental or public health risk posed by the spill, potential community or economic development benefit to the relevant municipality and potential restoration of an abandoned property.[38]

Parties who borrow funds pursuant to the 95-190 Program must repay the loan, on terms set forth by the CTDED, upon the sale or lease of the property or upon the approval by the CTDEP of a final remedial-action report completed by a licensed environmental professional.[39] The interest rate on a loan under the 95-190 Program cannot exceed "the interest cost to the state on such loans."[40] As security for loans under this program, the CTDED will place a lien on the subject property to cover the amount of money borrowed from the CTDED.[41]

Cleanup Standards

Public Act 95-190 amended Connecticut's statutes to allow the CTDEP to consider the use of the remediated property in determining cleanup criteria. In January 1996, the CTDEP adopted use-based cleanup standards.[42] For example, the standards for soil remediation generally require the use of residential direct-exposure criteria for each pollutant. If, however, access to the property is limited to individuals working at or temporarily visiting the property and an environmental land-use restriction prohibiting residential use is in effect for the property, a party may use industrial/commercial direct-exposure criteria for all substances, except for polychlorinated biphenyls.[43] Connecticut's cleanup standards also allow the CTDEP to grant certain variances from the cleanup standards, including the use of engineering controls, such as placing an impermeable cap over contaminated soil. For instance, variance for capping may be granted if

- The CTDEP authorized the disposal of solid waste or polluted soil at the subject area.
- The soil at such area is polluted with a substance for which remediation is not technically practicable.
- The CTDEP has determined that removal of the substance would create an unacceptable risk to human health.
- The CTDEP has determined that a proposal by the owner of the parcel to use a cap is acceptable because the cost of remediating the soil is significantly greater than the cost of installing a cap, and that the significantly greater cost outweighs the risk to the environment and human health if the cap fails to prevent mobilization of the substance or human exposure to the substance.[44]

A party who wishes to request a variance for use of an engineering control must submit to the CTDEP a written request detailing that party's engineering plans, maintenance plans, assurances of safety, and confirmation that an environmental land-use restriction is or will be in effect over the subject area.[45]

The Mechanics of Participating in Connecticut's Brownfields Initiatives

Urban Sites Program

Voluntary Cleanups

A volunteer under the Urban Sites Program must first file an Informational Form #DED-CRID-T1P-002 with the CTDED.[46] The volunteer is asked to provide a brief history of his or her property, a copy of an existing site plan, evidence of local and community support of cleanup, and the proposed site plan, including the type of development that is anticipated on the property.[47] The CTDED will identify properties that may be significant to the state's economy and will then prioritize the volunteer's property, based in part upon the environmental conditions and economic information available.[48] According to the CTDED, the "primary benefit of this program is that it is designed to be a mechanism to expedite the review of environmental remedial action plans from a willing owner or developer [whereas] [i]n the past, the reviewing processes could continue for a number of years if not given priority status."[49]

State Purchase and Remediation

Before the CTDED will purchase polluted property for remediation, the party who wishes to develop the property must enter an agreement with the CTDED. As of the spring of 1996, the CTDED had not yet implemented this program, and therefore had never entered an agreement under which it would purchase contaminated property to lease and eventually sell to a developer. Although the CTDED has guidelines from the Urban Sites Program statute and the regulations, it is unclear what would be expected of a developer under an agreement. However, according to the regulations, the agreement must contain the following provisions:

- A promise to conduct a manufacturing or economic base business at the site
- Payment of all necessary and appropriate legal costs incurred by the CTDED related to the project
- Payment of local property taxes for the property, costs incurred by the CTDED in administering the program, and costs incurred by CTDEP in assessing and remediating the property
- Long-term commitment by the applicant to use the property for certain specified economic-development purposes and any other conditions the CTDED finds necessary
- Obligation of the applicant to assume responsibility for any direct cost in excess of $15 million

- Agreement that the applicant will assume full title to the property upon full repayment of all appropriate costs to the state
- Agreement that the state will not be liable for any contamination that occurs after the date of completion of the site assessment conducted for the property[50]

Although the regulations make clear that certain CTDED and CTDEP costs will be paid, it is not clear who must pay those costs. CTDED and CTDEP costs could be paid in whole or in part by the Urban Site Remediation Fund, which was established under the Urban Site Program.[51] CTDED and CTDEP costs could also be paid in whole or in part by the developer as the costs are incurred, or could be paid over time as lease payments. Even if the developer pays all the CTDED and CTDEP costs, the CTDED could agree to transfer the property, after remediation is complete and costs have been paid, for a nominal price that may make the investment worthwhile. It is also unclear what constitutes the "direct costs in excess of $15 million" for which the developer must retain responsibility—the costs of cleanup, any future liability, or both.

If and when a developer and CTDED come to terms with these regulations and formulate an agreement pursuant to this program, the CTDEP may provide the developer with a covenant-not-to-sue.[52]

Transfer Act

A party involved in the transfer of contaminated property who must file a Form III or Form IV with the CTDEP must submit with his or her form an initial filing fee of $2,000, plus additional fees of up to $21,000 if the CTDEP does not allow a licensed environmental professional to verify the cleanup.[53] The CTDEP will, within fifteen days of receipt of a Form III or Form IV, notify the party whether the form is complete or incomplete.[54] The CTDEP will then notify the party within forty-five days of receiving a complete form whether remediation must be completed by the CTDEP or may be completed by a licensed environmental professional.[55] If the CTDEP approves the use of a licensed environmental professional, the party must, within thirty days of such approval, submit a schedule for investigating and remediating the property.[56] The party also must

1. publish notice of the remediation in a newspaper having a substantial circulation in the area affected by the establishment;
2. notify the director of health of the municipality where the parcel is located of the remediation;
3. either (A) erect and maintain for at least thirty days, in a legible condition, a sign not less than six feet by four feet on the parcel, which must be clearly visible from the public highway, and must include the words "Environmental Clean-up in Progress at This Site. For Further Information Contact: . . ." or (B) mail notice of the remediation to each owner of record of property that abuts the parcel, at the address for

such property on the last-completed grant list of the municipality where the parcel is located.[57]

After the CTDEP approves in writing the remediation of the property transferred pursuant to the Form III or Form IV, the commissioner may provide the party with a covenant-not-to-sue under the 95-183 Program.[58]

Voluntary Site Remediation Programs

A party who wishes to remediate contaminated property pursuant to the 95-183 Program may, at any time, submit to the CTDEP an environmental condition assessment form (a form describing an environmental condition of the property) with an initial fee of $2,000.[59] The CTDEP will then notify the party, within thirty days, whether the property requires remediation and, if so, whether such remediation can be performed and verified as complete by a licensed environmental professional or must be reviewed and approved by the CTDEP.[60] If and when the CTDEP or a licensed environmental professional has approved any necessary remediation of the property, the CTDEP may provide the party with a covenant-not-to-sue.[61] In addition, once remediation has been approved pursuant to the 95-183 Program and the volunteer wants to transfer the remediated property, an approval made under the 95-183 Program may be used as the basis for submitting a Form II pursuant to the Transfer Act, provided there has been no subsequent environmental contamination of the property.[62]

A volunteer who wishes to remediate contaminated property pursuant to the 95-190 Program may have a licensed environmental professional conduct a Phase II assessment and/or a Phase III assessment and prepare a Phase III remedial-action plan for any eligible property.[63] Before beginning any remedial work, the volunteer must submit the Phase III remedial-action plan to the CTDEP and must provide notice of such remediation by

1. publishing notice in a newspaper having a substantial circulation in the town where the property is located; and
2. either (A) erecting and maintaining for a least thirty days, in a legible condition, a sign not less than six feet by four feet on the property, which must be clearly visible from the public highway, and must include the words "Environmental Clean-up in Progress at This Site. For Further Information Contact: ... " or (B) mail notice of the remedial action to each owner of record of property that abuts such property, at the address on the last-completed grant list of the relevant town.[64]

After a licensed environmental professional has completed remediation of the property, he or she must submit a final remedial-action report to the CTDEP, which is to be deemed approved if the CTDEP does not respond within sixty days requiring an audit of the cleanup.[65] If an audit is required, it must be completed within six months of the determination that an audit is required.[66] When the final remedial-action report has been completed, the volunteer is eligible for a covenant-not-to-sue from the CTDEP under Public Act 95-190.[67] To obtain the covenant-not-to-sue, the volunteer must pay a fee of three percent of the value of the remediated property as appraised after the remediation.[68]

Liability Protection for Lenders

Since 1985, the Connecticut statutes have provided some protection to mortgagees who hold security interests in contaminated property. Specifically, the law provides that

> a mortgagee who acquires title to real estate by virtue of a foreclosure or tender of a deed in lieu of foreclosure, shall not be liable for any assessment, fine or other costs imposed by the state for any spill upon such real estate beyond the value of such real estate, provided such spill occurred prior to the date of acquisition of title to such real estate by such mortgagee.[69]

This mortgagee liability protection is good for all mortgagees, not only those holding interests in properties undergoing remediation. In addition, lending institutions that take a security interest in property that is part of the 95-190 Program may receive liability protection from Connecticut under the property owner's or lessor's covenant-not-to-sue.[70]

Public Act 95-183, which provides covenants-not-to-sue to participants involved in the Urban Sites Program, the Transfer Act, and the 95-183 Program, does not specifically mention lender protection pursuant to a covenant-not-to-sue. Assuming, however, that the covenant-not-to-sue issued to the remediating party under Public Act 95-183 would run with the land, a lender who takes possession of such land would benefit from that covenant.

Recording

Pursuant to Connecticut's regulations for cleanup standards, the CTDEP, in most instances, requires parties who do not remediate property using residential cleanup criteria to place an environmental land-use restriction (Land-Use Restriction) on the subject property.[71] A party placing a Land-Use Restriction on property must apply to the CTDEP for approval of the Land-Use Restriction. An application must include

1. a draft declaration of the Land-Use Restriction in the form set forth in Appendix 1 or 2 to section 22a-133q-1 of the Connecticut regulations;
2. a Class A-2 survey of the parcel, or portion thereof, that is the subject of the proposed Land-Use Restriction;
3. a proposed "Decision Document," which sets forth the type and location of pollutants, the provisions of the Land-Use Restriction, and a description of the reason for the Land-Use Restriction; and
4. a certified copy of the notice required to appear in a newspaper of general circulation.[72]

An applicant for a Land-Use Restriction must also obtain subordination agreements from each person who holds an interest the property.[73] Any party who has obtained approval to place a Land-Use Restriction on his or her property must record the restriction on the land records in the municipality in which the property is located and must send, by certified mail, a copy of the Land-Use Restriction to the chief administrative office of the town where the property is located, to the chair of the municipal zoning commission, to the local director of health, and to any other person who submitted comments on the Land-Use Restriction.[74]

Admission of Liability

The Connecticut brownfields initiatives do not address the issue of protection from disclosure of voluntary environmental audits. Though many states have enacted bills that create privileges for voluntary environmental audits, the Connecticut legislature has yet to introduce a bill providing for such a privilege.[75]

Under common law, a party working through its lawyer on a voluntary cleanup may argue that an environmental audit is subject to protection from disclosure under the attorney-client privilege. This privilege may be waived, however, if the contents of the audit report are disclosed to third parties. Given the close interaction a party working under a Connecticut brownfields initiative may have with the CTDEP concerning the details of the environmental condition of the subject property, it is likely that any privilege asserted for the protection of an environmental audit would be waived as having been disclosed to a third party.

Civil and Criminal Penalties

Connecticut may impose a civil penalty of up to $100,000 on a party who gives false information on any document required by the program, or who fails to comply with the Transfer Act or the Voluntary Site Remediation Programs.[76]

Cost Recovery

Private or public redevelopers of brownfields sites in Connecticut may pursue a cost-recovery action against the party responsible for causing the pollution or contamination. The Connecticut statutes contain a general, private cost-recovery provision which states that

> [a]ny person, firm, corporation or municipality which contains or removes or otherwise mitigates the effects of oil or petroleum or chemical liquids or solid, liquid or gaseous products or hazardous wastes resulting from any discharge, spillage, uncontrolled loss, seepage or filtration of such substance or material or waste shall be entitled to reimbursement

from any person, firm or corporation for the reasonable costs expended for such [remediation if such pollution] resulted from the negligence or other actions of such person, firm or corporation. When such pollution or contamination or emergency results from the joint negligence or other actions of two or more persons, firms or corporations, each shall be liable to the others for a pro rata share of the costs of containing, and removing or otherwise mitigating the effects of the same and for all damage caused thereby.[77]

Conclusion

Connecticut's brownfields programs, although relatively untested, provide several incentives to, and opportunities for, those who wish to invest in brownfields. The Urban Sites Program, the 95-183 Program, and the 95-190 Program offer covenants-not-to-sue at the completion of voluntary site remediation. In addition, the 95-190 Program offers loans to volunteers for the cost of site assessments, demolition, and expenses related to the administration of the program; the Urban Site Program provides the state with funding for extra staff to expedite review of brownfields projects, the use of bonds to investigate and remediate contaminated sites, and the power to purchase, redevelop, and lease brownfields sites. Parties interested in Connecticut's brownfields programs should contact the Permitting, Enforcement, and Remediation Division of the CTDEP at (860) 424-3705.

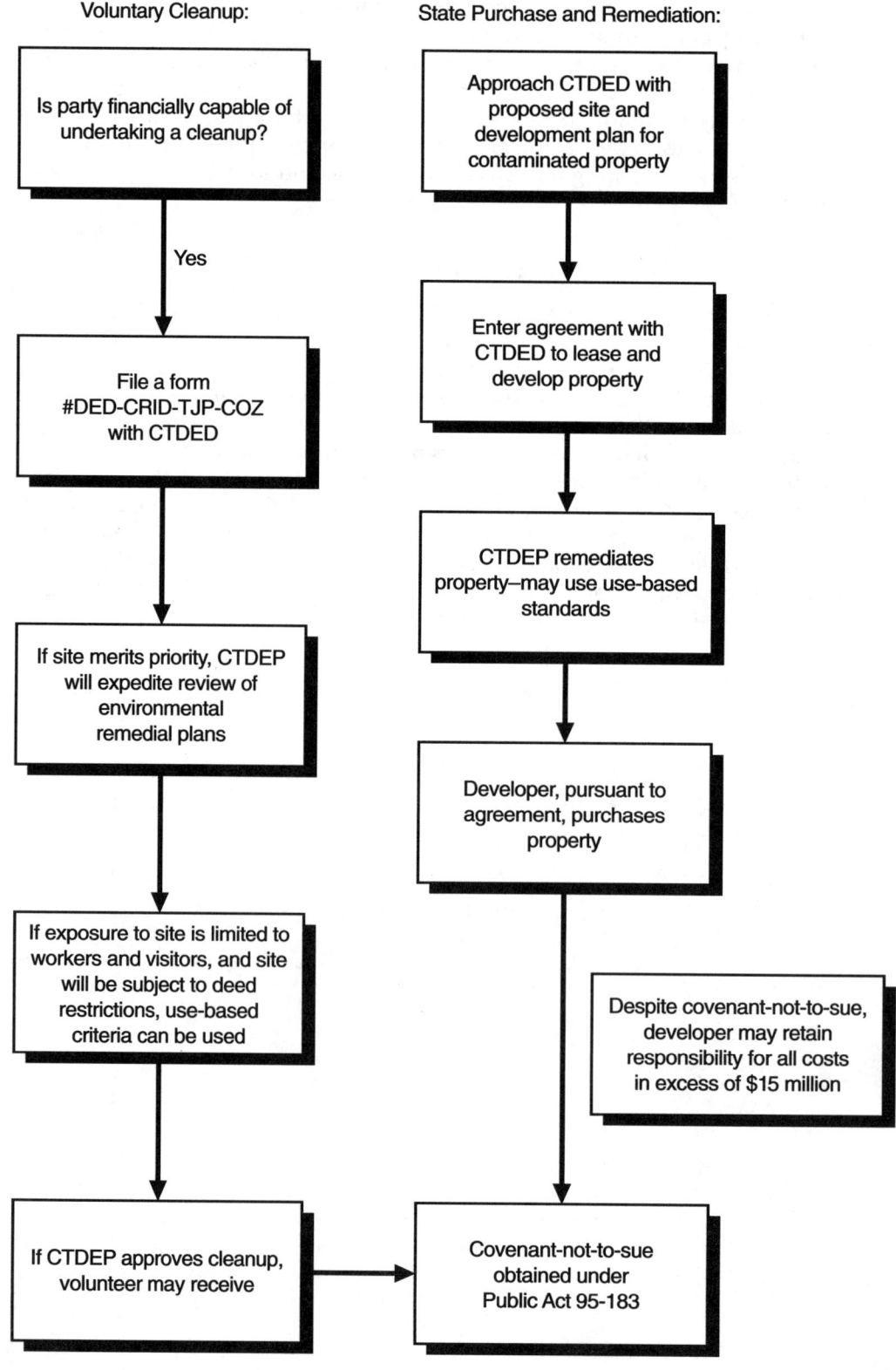

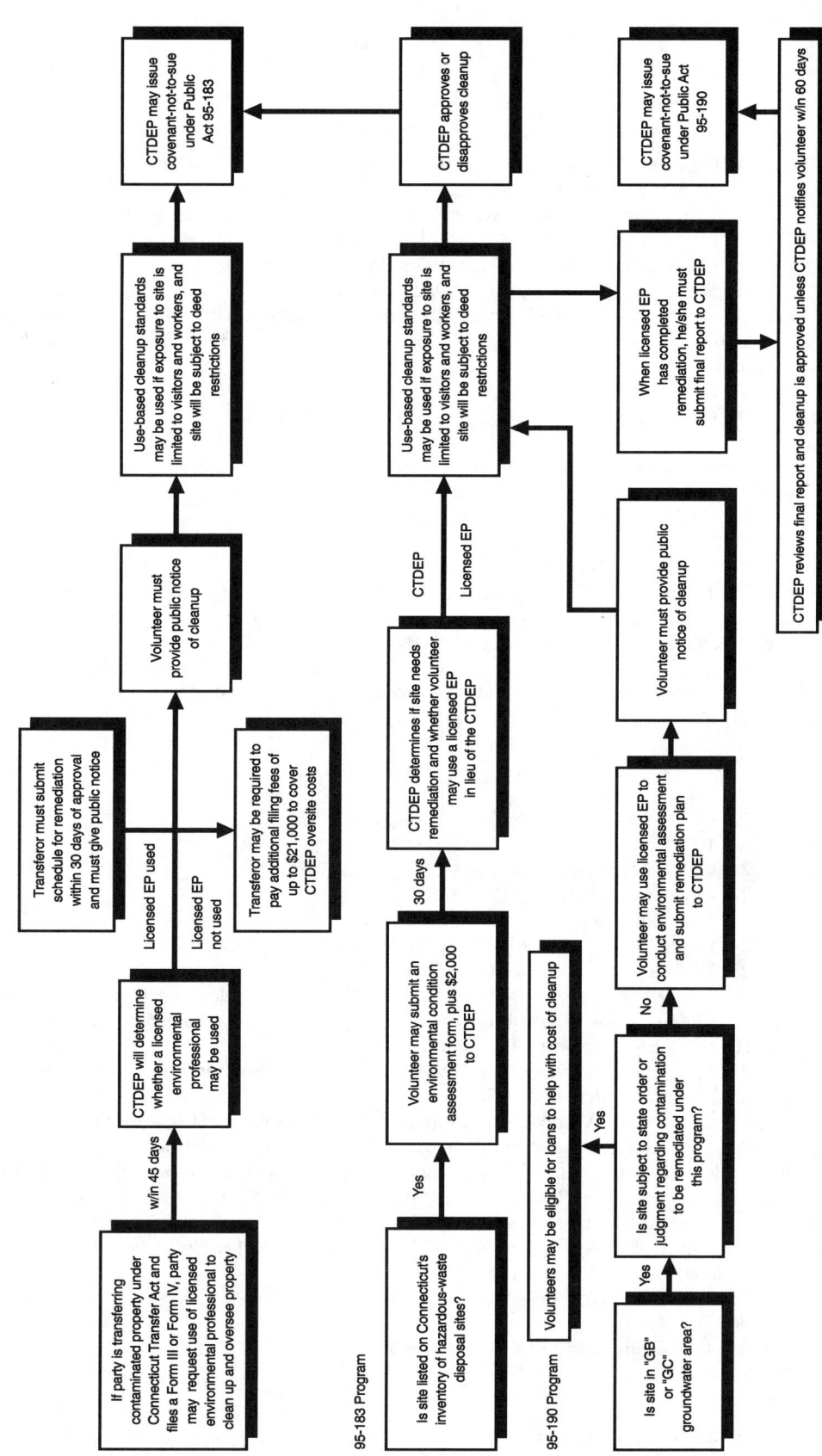

Notes

1. CONN. GEN. STAT. ANN. § 22a-133m (West 1995).
2. 1995 Conn. Acts 183 § 5 (Reg. Sess.).
3. 1995 Conn. Acts 183 § 3 (Reg. Sess.); 1995 Conn. Acts 190 §§ 1–6 (Reg. Sess.).
4. CONN. AGENCIES REGS. § 22a-133k-1-3 (1996).
5. CONN. GEN. STAT. ANN. § 22a-133m (West 1995).
6. *See* Betsey Wingfield, *Two Public Acts Target Remedial Goals: A Discussion of Connecticut's Urban Sites Remedial Action & Voluntary Remediation Programs*, CONN. ENV., Feb. 1996, at 6–7. (Betsey Wingfield is Director of Connecticut's Urban Sites Remedial Action Program).
7. *Id.*
8. CONN. GEN. STAT. ANN. § 22a-134 *et seq.* (West 1995).
9. *Id.*; *see also* Terry J. Tondro, *Reclaiming Brownfields to Save Greenfields: Shifting the Environmental Risks of Acquiring and Reusing Contaminated Land*, 27 CONN. L. REV. 789, 793 (1995).
10. 1995 Conn. Acts 183 § 5 (Reg. Sess.).
11. *Id.* at 9 (Reg. Sess.).
12. 1995 Conn. Acts 190 §§ 1–6 (Reg. Sess.).
13. Telephone Interview with Richard Hathaway, Staff Member, Permitting Enforcement and Remediation Division of the Bureau of Water Management of the Connecticut Department of Environmental Protection (Mar. 8, 1996).
14. *See* DEP'T ECONOMIC DEV., URBAN SITES REMEDIAL ACTION PROGRAM, ECONOMIC DEV. INITIATIVE 2, THE DEPARTMENT OF ECONOMIC DEVELOPMENT WOULD LIKE TO HELP (1995).
15. CONN. GEN. STAT. ANN. § 22a-133m (West 1995).
16. CONN. GEN. STAT. ANN. § 22a-133m(b) (West 1995).
17. CONN. AGENCIES REGS. § 22a-133m-2 (1996).
18. *Id.*
19. CONN. GEN. STAT. ANN. § 22a-134a (West 1995).
20. 1995 Conn. Acts 183 § 3 (Reg. Sess.).
21. CONN. GEN. STAT. ANN. § 22a-133(c) (West 1995).
22. 42 U.S.C. § 6901 *et seq.* (1995).
23. *Id.* § 9601 *et seq.*
24. CONN. GEN. STAT. ANN. § 22a-133(c) (West 1995).
25. 1995 Conn. Acts 190 § 2 (Reg. Sess.).
26. Telephone Interview with Richard Hathaway, Staff Member, Permitting Enforcement and Remediation Division of the Bureau of Water Management of the Connecticut Department of Environmental Protection (Mar. 8, 1996).
27. Telephone Interview with David Evans, Office of Emergency and Remedial Response Division, United States Environmental Protection Agency (Mar. 18, 1996). (The working title of USEPA's policy is "Guidance for State Voluntary Cleanup Programs.")
28. The USEPA is currently developing a policy under which it would advocate the distribution of "comfort letters" to brownfields volunteers (*see* Vicki Allen, *EPA to Ease City Waste Site Cleanups*, REUTERS N. AM. WIRE (Jan. 25, 1995), while some regional offices, like Region V, already make it a policy to issue "comfort letters" to brownfields volunteers (*see* David Fink, *Born Again: New Approaches to Redevelopment of Urban Industrial Sites*, BUS. DATELINE DETROITER, Mar. 1995, at 24).
29. 1995 Conn. Acts 183 § 9 (Reg. Sess.)
30. *Id.*
31. 1995 Conn. Acts 190 § 3 (Reg. Sess.).

32. *Id.*
33. 1995 Conn. Acts 183 (Reg. Sess.); 1995 Conn. Acts 190 (Reg. Sess.).
34. 1995 Conn. Acts 183 § 9 (Reg. Sess.).
35. *Id.*
36. Conn. Gen. Stat. Ann. § 22a-133m(e) (West 1995).
37. 1995 Conn. Acts 190 § 6(b) (Reg. Sess.).
38. *Id.* § 6(e).
39. *Id.* § 6(c).
40. *Id.*
41. *Id.* § 6(d).
42. Conn. Agencies Regs. § 22a-133k-1-3 (1996).
43. *Id.* § 22a-133k-2(b).
44. *Id.* § 2(f).
45. *Id.*
46. *See* Dep't Economic Dev., *supra* note 14.
47. *Id.*
48. *Id.*
49. *Id.*
50. Conn. Agencies Regs. § 22a-133m-3 (1996).
51. Conn. Gen. Stat. Ann. § 22a-133m(f) (West 1995).
52. 1995 Conn. Acts 183 § 9 (Reg. Sess.).
53. Conn. Gen. Stat. Ann. § 22a-134e (West 1995).
54. *Id.* § 22a-134a(f).
55. *Id.*
56. *Id.* § 22a-134a(h).
57. *Id.* § 22a-134a(j).
58. 1995 Conn. Acts 183 § 9 (Reg. Sess.).
59. *Id.* § 3.
60. *Id.*
61. *Id.* § 9.
62. *Id.* § 3(d).
63. 1995 Conn. Acts 190 § 2 (Reg. Sess.).
64. *Id.* § 2(b).
65. *Id.* § 2(c).
66. *Id.*
67. *Id.* § 3.
68. *Id.*
69. Conn. Gen. Stat. Ann. § 22a-452b (West 1995).
70. 1996 Conn. Acts 190 § 3 (Reg. Sess.).
71. Conn. Gen. Stat. Ann. § 22a-133o (West 1995).
72. Conn. Agencies Regs. § 22a-133q-1 (1996).
73. *Id.* § 22a-133h.
74. *Id.* § 22a-133i–j.
75. *See Environmental Audits: State Privilege Legislation Multiplies in 1995; Predictions Differ About 1996,* Daily Env't Rep. (BNA) (Aug. 29, 1995).
76. 1995 Conn. Acts 190 § 15 (Reg. Sess.).
77. Conn. Gen. Stat. Ann. § 22a-452(a) (West 1995).

CHAPTER 26

Delaware

R. JUDSON SCAGGS, JR.

The remediation of contaminated sites in Delaware is governed primarily by the Hazardous Substances Cleanup Act (HSCA).[1] HSCA was adopted in 1990 and is administered by the Department of Natural Resources and Environmental Control (DNREC), which has promulgated regulations governing the identification, investigation, and remediation of contaminated sites.[2] HSCA was patterned on the federal Comprehensive Environmental Response, Compensation, and Liability Act of 1980 (CERCLA).[3] HSCA contains the same basic liability scheme as CERCLA, including joint and several liability of responsible parties,[4] with changes that make HSCA liability broader than CERCLA in certain respects.[5]

Summary of Major Provisions

Delaware has taken two major steps to provide incentives for the voluntary cleanup of contaminated sites. First, in July 1995, Delaware enacted "brownfields" legislation. It provides corporate tax incentives for the development of vacant, contaminated property and grants protection from liability to persons cleaning up contaminated sites or purchasing remediated sites. Second, DNREC is operating a Voluntary Cleanup Program (VCP) under HSCA, which provides persons owning or acquiring contaminated property the opportunity to clean it up at their own pace without coercive orders from DNREC. The brownfields legislation provides substantive incentives for redevelopment of brownfields sites and the VCP provides a procedural mechanism to simplify and expedite cleanups.

The section in this chapter on tax incentives for brownfields redevelopment in Delaware was provided by John S. McDaniel, who is a partner in the firm of Morris, Nichols, Arsht & Tunnell in Wilmington, Delaware.

Brownfields Legislation

The Delaware brownfields legislation has two major elements: tax incentives and liability protection.

Tax Incentives

Delaware's Blue Collar Jobs Credit[6] and Targeted Areas Credit[7] provide corporate income-tax credits and a phase-in of applicable gross-receipts taxes. The credits are available to persons engaged in certain specified businesses who make significant investments in qualified facilities at which new qualified employees are employed. The conditions under which both credits may be claimed are more liberal, and the amount of each credit is higher, if the qualified facility is located on a "brownfield."

A brownfield is defined as

> a vacant or unoccupied site with respect to any portion of which the taxpayer has reasonable cause to believe may, as a result of any prior commercial or industrial activity by any person, have been environmentally contaminated by the release or threatened release of a hazardous substance as defined under 7 Del.C. c. 91 in a manner that would interfere with the taxpayer's intended use of such site; provided, however, that such term shall not include any site or facilities with respect to any portion of which enforcement action has been initiated against any person pursuant to Chapter 63, Chapter 74 or Chapter 91 of Title 7; 42 U.S.C. § 6901 et seq.; or 42 U.S.C. § 9606 or § 9607.[8]

Generally, the basic Blue Collar Jobs Credit and a ten-year, gross-receipts tax phase-in is available to corporations that invest at least $200,000 in a qualified facility employing five or more qualified employees in certain enumerated types of businesses.[9] The basic corporate income-tax credit is $400 per additional qualified employee added in the year in which the qualified facility is placed in service, and $400 per $100,000 of qualified investment.[10] Normally, a qualified facility is one used only in certain enumerated types of businesses, and qualified investment is limited to new or expanded facilities consisting of land, buildings, machinery, and equipment.[11]

If the qualified facility is located in a brownfield, however, the credit is liberalized. First, the basic credit is $650 per qualified employee and per $100,000 of qualified investment.[12] Second, any type of business is a qualified activity for purposes of the credit.[13] Last, qualified expenditures are not limited to expenditures for land, buildings, and equipment. Rather, all amounts expended for environmental investigation and remediation of a brownfield are treated as qualified investment.[14]

Under the Targeted Areas Credit, the amount of the credit is increased to $650 per qualified employee and per $100,000 of qualified investment.[15] The increased credit applies if the qualified facility is located in a targeted area.[16] Targeted areas are real properties owned by the state, local governments, or certain charitable organizations, or located in foreign trade zones or in economically depressed areas.[17] If the qualified facility is on a brownfield located in a targeted area, the credit is $900 per qualified employee and per $100,000 of qualified investment.[18] Also, if the qualified facility is located in a targeted area, the gross-receipts tax phase-in is extended from ten to fifteen years, with a complete gross-receipts tax holiday for the first sixty months.[19]

Liability Protection

Before the brownfields legislation, a person acquiring a contaminated site became an owner of the site, and thereby became jointly and severally liable for all costs of cleaning up that site.[20] Now, any person who owns, operates, or otherwise controls activities at a facility, after DNREC issues a certification of completion of remedy, will not—solely by virtue of that later ownership, operation, or control—be liable for the contamination addressed in the certification or for any future release attributable to conditions existing before the issuance of the certification.[21] DNREC is obligated to issue a certification of completion when a remedy is completed under DNREC supervision.[22]

The brownfields legislation also provided DNREC with more flexibility in issuing certificates of completion. DNREC is authorized to issue a certification when the remedy is operational, although not complete in all respects.[23] DNREC can, in the certificate of completion, require additional activity at the site, such as operation, maintenance, and compliance monitoring.[24] For example, DNREC may issue a certificate of completion at the time when a pump-and-treat system for groundwater contamination is installed and condition the continued effectiveness of the certification upon operation of the system until the contamination is brought below specified levels.

The liability protection provisions in the brownfields legislation are designed to enhance the transferability of facilities that have been cleaned up under HSCA. A person who owns or operates a facility after DNREC has issued a certificate of completion will not be liable for any past releases that are addressed in the remedy, or any future releases attributable to conditions existing before the remedy.[25] A subsequent owner will not receive liability protection for past releases that are unknown at the time of the cleanup and, therefore, not addressed in the remedy. A subsequent owner will become liable if it disposes of hazardous substances on the site or interferes with the remedy.[26]

As of November 1996, DNREC was negotiating a memorandum of agreement with the United States Environmental Protection Agency (USEPA) concerning cleanup of sites under Delaware law. At least until an agreement is finalized, parties will be exposed, theoretically, to federal liability in cases in which they have obtained liability protection under Delaware law. As a practical matter, it is highly unlikely that USEPA will take enforcement actions at sites that are undergoing, or have completed, cleanups supervised and approved by DNREC.

Liability protection is also available to any person who, in connection with the sale, lease, or acquisition of a facility, enters an agreement with DNREC to perform a remedy at a facility.[27] To obtain the protection, the purchaser must actually complete the remedy and receive a certificate of completion.[28] Like a person who acquires a property after the completion of a remedy, a prospective purchaser who enters an agreement to conduct a remedy is protected from liability for past releases that are addressed in the remedy and future releases that are attributable to conditions existing before the remedy.[29]

The brownfields legislation also made other important, although less dramatic, changes to HSCA. DNREC can document an agreement to perform a remedy in the form of an administrative order, memorandum agreement, consent decree, or any other appropriate form.[30] Previously, HSCA required DNREC to document all remedies in the form of consent decrees filed with the Superior Court. The brownfields legislation expressly authorizes DNREC to resolve a person's liability under HSCA through the use of a settlement agreement pursuant to CERCLA.[31]

The brownfields legislation also broadened the liability protection provided to lenders. Previously, protection for lenders was limited to commercial lending institutions. HSCA now exempts from liability any person who acquires ownership or control of a property through a security interest.[32] HSCA also protects fiduciaries who have legal title to, or manage, any property for purposes of administering an estate or trust.[33]

VCP

The VCP is designed to provide owners of contaminated properties and persons interested in acquiring contaminated properties with an efficient, flexible process to complete an approved cleanup.

Site Eligibility Requirements

Because the VCP is designed to address contaminated properties that may be transferred or redeveloped and pose no immediate threat to human health and the environment, a site must meet certain eligibility requirements. A site is not eligible for the VCP if

1. it has contamination detected in the soil or groundwater at a risk value greater than a cancer risk of 10^{-4} or a hazard index of 10 or greater;
2. a public or domestic water supply well on the site is contaminated at or above a maximum contaminant level, an excess cancer risk of 1×10^{-5}, or a hazard index of 1;
3. it has either groundwater or soil contamination located near a domestic or public well, which has the potential to contaminate the well;
4. the contamination from the site impacts surface water that is used as a drinking-water source and maximum contaminant levels are exceeded;
5. the contamination from the site impacts surface water that has an exceedance of at least one order of magnitude of the Delaware surface water quality standards;

6. it is subject to a corrective action under the Resource Conservation and Recovery Act; or
7. any other reason exists, as determined by DNREC, that precludes involvement in the VCP.[34]

If a party is interested in cleaning up a site under the VCP, it must submit a VCP application to DNREC.[35] The application is designed to assist the applicant and DNREC in determining whether the site requires remedial activities, as well as whether it is eligible for the VCP.[36]

Overview of the Process

The basic cleanup process employed under HSCA is based upon the CERCLA process. After a contaminated facility is identified, the steps in the cleanup process include remedial investigation/feasibility study, remedial decision, remedial design, and remedial action.[37] A cleanup conducted under the VCP technically must contain the same steps required during an involuntary cleanup. DNREC, however, has the authority to combine several steps in the cleanup process into fewer steps, or even a single step, which can greatly streamline the process.[38]

The VCP Agreement

After DNREC accepts a party into the VCP, the party has two options concerning how to proceed. The party may elect immediately to enter a VCP agreement with DNREC.[39] The VCP agreement is the operative document that sets forth the rights and obligations of DNREC and the VCP participant. DNREC uses a standard form VCP agreement. After signing the VCP agreement, the party will submit the name of its environmental consultant and laboratory for DNREC approval.[40] After DNREC approves the consultant and laboratory, the party will submit a work plan addressing the investigation and/or remediation of the site for DNREC approval.[41]

Alternatively, a party may first enter a letter agreement, whereby the party agrees to pay DNREC's initial oversight costs. The party then obtains approval from DNREC for its consultant, laboratory, and work plan, and then enters the VCP agreement.[42] This alternative provides a party with the ability to find out whether DNREC will approve its consultant, laboratory, and work plan before committing to a VCP agreement.

The VCP agreement obligates the voluntary party to perform the investigation and/or remediation activities described in the agreement. All work must be conducted in compliance with all applicable federal, state, and local laws, including the HSCA regulations. The VCP agreement provides several advantages over unilateral cleanup orders and consent decrees. The VCP agreement

1. contains no provisions for stipulated penalties for noncompliance;[43]
2. does not require public notice;[44]
3. allows the voluntary party to withdraw at any time;[45] and

4. typically will allow the party to proceed with the work on a more relaxed pace than is required during involuntary cleanups.

Consultant Qualification

In June 1994, DNREC issued a policy on the minimum qualification requirements for consultants performing work under HSCA.[46] DNREC has established separate minimum qualifications for consultants performing (1) facility evaluations, (2) remedial investigations/feasibility studies, and (3) remedial design/remedial action oversight.[47] In general, a consulting firm must establish that licensed professionals, such as geologists and engineers, will perform or supervise the work when required.[48] The consulting firm must demonstrate that its personnel have been trained and are familiar with quality assurance and quality-control protocols, site control, health and safety requirements, hazard monitoring, Occupational Safety and Health Administration requirements, and analytical service requirements.[49] Additionally, the firm must demonstrate that it possesses adequate equipment to perform the investigation and/or remediation.[50]

Site Investigation and Screening Levels

Upon entry into the VCP, a participant is required to provide DNREC with all studies, reports, and data available about the site.[51] DNREC may decide that due to the existing data, no—or limited—additional investigation is needed.[52]

DNREC has developed screening levels for soil and groundwater contaminants.[53] The screening levels are based upon USEPA's proposed soil-screening guidance (document 9355.4-1, PB95-963530, EPA 540/R-94/101), risk-based concentration table (dated March 7, 1995 and published by USEPA Region III), and the maximum contaminant levels promulgated under the Safe Drinking Water Act.[54] DNREC has established four types of screening levels: residential surface soil, industrial surface soil, groundwater, and subsurface soil.[55]

Properties with contaminant levels that are below screening levels in both the soil and groundwater may receive a no-further-action letter.[56] In deciding whether to issue a no-further-action letter, DNREC will review sampling data in relationship to the physical characteristics of the site, including the site's proximity to public water supply sources, surface water bodies, residential areas, sensitive habitats, and agricultural areas.[57] A no-further-action letter is not the equivalent of a certificate of completion, which a party receives after completing a cleanup. A no-further-action letter does not provide the liability protection afforded by a certificate of completion. Instead, a no-further-action letter evidences DNREC's intent not to pursue further investigation or remediation at the site.[58] If contaminants on the site exceed screening levels, the party must proceed with the VCP process, including further investigation and remediation.

DNREC recommends that owners report any exceedance of screening levels that are discovered during site investigations or evaluations.[59] A mandatory reporting obligation is triggered when a party discovers the release of a "reportable quantity" of a pollutant.[60] Delaware does not have an environmental audit privilege, although DNREC currently is working on an environmental audit policy.

Selection of Remedy and Remediation

DNREC has several mechanisms that it can use to expedite the remediation process in the VCP. DNREC can allow parties to clean up properties to screening levels in lieu of performing a health-based risk assessment in accordance with HSCA regulations.[61] DNREC is authorized to employ remedies that isolate contaminants from humans and the environment, as opposed to removing the contaminants or treating them. Engineering controls to seal in contaminants, such as an asphalt cap, are acceptable.[62] Institutional controls, such as deed restrictions prohibiting any change in land use without DNREC approval, are also acceptable, provided they are not used as a substitute for remedial actions that otherwise would be technically practicable.[63]

HSCA regulations specifically allow DNREC to authorize interim response activities at contaminated sites.[64] DNREC may authorize an interim response action without the documentation and public participation that is applicable to a final remedy decision.[65] If an interim response activity resolves all the contamination problems at the site, DNREC later can fulfill public notice requirements and approve the interim response activity as the final remedy. For example, DNREC can authorize the removal of contaminated soil as an interim action and, after soil sampling to confirm that all contaminants have been removed (or brought below screening levels), DNREC can approve the soil removal as the final remedy.

Cost Recovery and Financial Assistance

Cost Recovery

Any person who expends money performing any remedial action under HSCA, including the VCP, may bring a cost-recovery action against any responsible party who has not entered a settlement agreement with DNREC.[66] The party seeking reimbursement is entitled to recover all costs that are consistent with HSCA, or all funds expended to reimburse the state for its expenses.[67] Responsible parties include (1) any person who owned or operated the facility at any time, (2) any person who owned a hazardous substance and arranged for disposal or treatment of it at the facility, (3) any person who arranged with a transporter for the transportation, treatment, or disposal of a hazardous substance at the facility, (4) any person who generated, disposed of, or treated a hazardous substance at the facility, (5) any person who accepted a hazardous substance for transportation to the facility when the facility was selected by the transporter, and (6) any person who is responsible in any other manner for a release of a hazardous substance.[68]

Neither HSCA nor the regulations promulgated under it provide guidance concerning how cleanup costs should be allocated among responsible persons. Because HSCA is based upon CERCLA, one would reasonably expect Delaware courts to look to federal case law for guidance on allocation issues.

Although no Delaware cases specifically address the issue, one would expect the elements of a prima facie case under HSCA to follow closely the elements required for a prima facie case under CERCLA. Consequently, a prima facie case under HSCA probably would consist of a showing that (1) the defendant falls within one of the six categories of responsible parties, (2) there has been a release or threatened release of a hazardous substance from a facility, and (3) the plaintiff has expended money in performing a remedial action consistent with HSCA in response to the release.[69] The statutory defenses to liability under HSCA are essentially the same as those found in CERCLA.[70]

HSCA does not contain a specific statute of limitations for cost-recovery actions. The Delaware statute of limitations of three years for actions "based on a statute" would probably apply to cost-recovery cases under HSCA.[71]

Contractual allocation of HSCA liability is not addressed directly in Delaware statutes, regulations, or case law. Because HSCA is based upon CERCLA[72] and Delaware case law recognizes the enforceability of indemnification agreements,[73] Delaware courts probably will enforce a contract allocating HSCA liability between private parties. Such a contract probably could not affect the state's rights against a responsible party.

Financial Assistance

The Delaware Economic Development Office is empowered to make matching grants to parties for conducting environmental investigations of vacant, unoccupied, or underutilized sites that have been environmentally contaminated in a manner that would interfere with the intended use of the site.[74] The maximum amount of assistance is $25,000 or 50 percent of total costs, whichever is less.[75] To qualify for a matching grant, a party must demonstrate that its proposed environmental investigation will have the potential to serve a public purpose by expanding employment, diversifying business, or increasing the tax base in the state.[76] An environmental investigation that is considered by the authority as normal and ordinary in the course of a real estate transfer transaction is not eligible for a matching grant.[77] The investigation of any site that has been the subject of an enforcement action by DNREC is not eligible for a matching grant.[78]

Conclusion

Brownfields cleanup and development in Delaware has now created useful cleanup tools, liability protections, and development incentives in the recently passed brownfields legislation and the DNREC's VCP. With these aids, Delaware, like other states that are taking progressive approaches to brownfields redevelopment, has established a structure that encourages and supports private investment and interest in brownfields projects.

Delaware Voluntary Cleanup Program

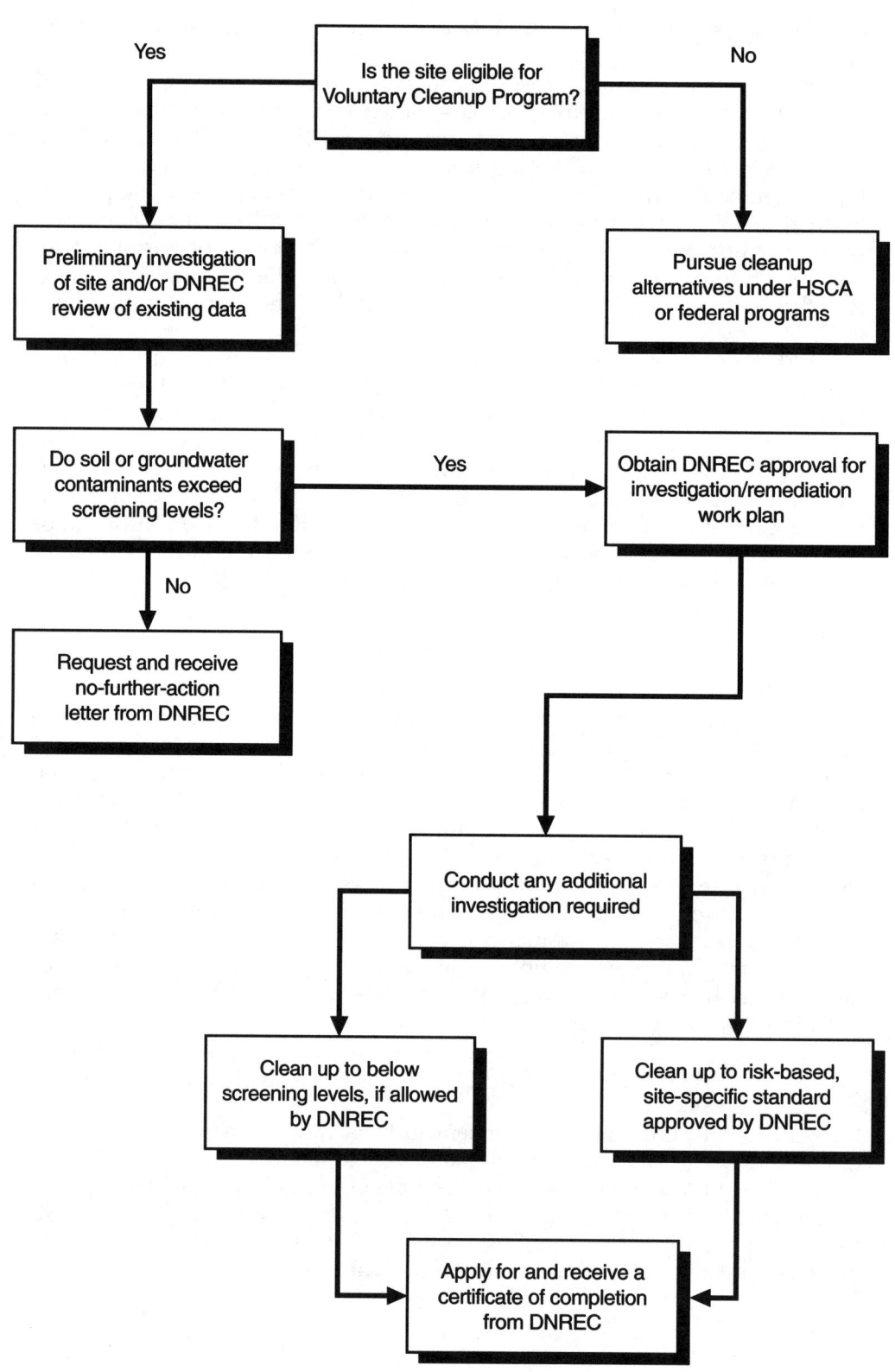

Notes

1. DEL. CODE ANN. tit. 7, ch. 91 (1991)
2. Del. Dep't Natural Resources & Envtl. Control, Delaware Regulations Governing Hazardous Substance Cleanup (rev. Sept. 1996) [hereinafter HSC Regs.].
3. 42 U.S.C. §§ 9601–75 (1995).
4. DEL. CODE ANN. tit. 7 § 9105(b) (1991).
5. Under HSCA, any person who owned or operated a facility "at any time" is liable, as opposed to CERCLA, under which liability from ownership or operation is limited to current owners and operators and those on hand when disposal of the hazardous substance took place. *Compare* 42 U.S.C. §§ 9607(a)(1)–(2) *with* DEL. CODE ANN. tit. 7, § 9105(a)(1) (1991). HSCA provides a "catch-all" category of liability for "any person who is responsible in any other manner for the release or imminent threat of release." DEL. CODE ANN. tit. 7, § 9105(a)(6) (1991). HSCA defines "hazardous substance" in the same manner as CERCLA, but does not contain an exclusion for petroleum. *Compare* 42 U.S.C. § 9601(45) (1995) *with* HSC Regs., *supra* note 2, § 2.1.
6. DEL. CODE ANN. tit. 30, §§ 2010–14 (Supp. 1996).
7. *Id.* §§ 2020–24.
8. *Id.* § 2010(16).
9. *Id.* §§ 2011(a)–2012.
10. *Id.* §§ 2011(b)(1)–(2).
11. *Id.* §§ 2010(1)–(5).
12. *Id.* § 2011(l)(iii).
13. *Id.* § 2011(l)(ii).
14. *Id.* § 2011(l)(i).
15. *Id.* § 1021(a).
16. *Id.*
17. *Id.* § 2020(1).
18. *Id.* § 2021(d).
19. *Id.* § 2022(a).
20. DEL. CODE ANN. tit. 7, § 9105(a)–(b) (1991).
21. *Id.* § 9105(e).
22. *Id.* § 9108.
23. *See id.* § 9108(a).
24. *Id.*
25. *See id.* § 9105(e).
26. *Id.*
27. *Id.* § 9105(f).
28. *Id.*
29. *Id.* § 9105(e).
30. *Id.* § 9107(b).
31. *Id.* This provision is consistent with, and should compliment, the efforts by USEPA to liberalize the use of prospective-purchaser agreements at the federal level. *See* 60 Fed. Reg. 34,792 (1995).
32. DEL. CODE ANN. tit. 7, § 9105(c)(3) (Supp. 1996).
33. *Id.*
34. Del. Dep't Natural Resources & Envtl. Control, Delaware Voluntary Cleanup Program Guidance, at 4 (rev. Feb. 1995) [hereinafter VCP Guidance]. The VCP has not been codified either statutorily or in regulations. The program is currently governed by guidance and policy documents issued by DNREC.
35. *Id.* at 5.
36. *Id.*
37. *See* HSC Regs., *supra* note 2, § 1.2.

38. *Id.* § 8.1(5).
39. VCP Guidance, *supra* note 34, at 5–7.
40. *Id.* at 6.
41. *Id.* at 7.
42. *Id.* at 6–7.
43. *Id.* at 6.
44. *Id.*
45. *Id.*
46. Del. Dep't Natural Resources & Envtl. Control, Policy on Minimum Qualification Requirements for Consultants/Contractors Performing Work under the Hazardous Substance Cleanup Act (eff. June 17, 1994; rev. June 6, 1996).
47. *Id.*
48. *Id.*
49. *Id.*
50. *Id.*
51. VCP Guidance, *supra* note 34, at 6.
52. *See* HSC Regs., *supra* note 2, §§ 5.1(2), 8.3(2).
53. Del. Dep't Natural Resources & Envtl. Control, Interim Guidance on Screening Levels for Hazardous Substances Discovered during Site Assessments under the Delaware Hazardous Substance Cleanup Act (Oct. 1995, rev. Mar. 1996) [hereinafter Interim Guidance on Screening Levels].
54. *Id.*
55. *Id.*
56. VCP Guidance, *supra* note 34, at 7.
57. Interim Guidance on Screening Levels, *supra* note 53, at 1–2.
58. *Id.*
59. *Id.* at 1.
60. DEL. CODE ANN. tit. 8, § 6028 (1991); Delaware Regulations on Reporting of a Discharge of a Pollutant or an Air Contaminant § 2.1 (1995).
61. VCP Guidance, *supra* note 34, at 8.
62. *See* HSC Regs., *supra* note 2, §§ 8.5–8.7.
63. *See id.* § 8.10.
64. *Id.* § 8.2.
65. *Id.*
66. DEL. CODE ANN. tit. 8, § 9105(d) (1991).
67. *Id.*
68. *Id.* § 9105(a).
69. *See id.* §§ 9105(a), 9105(d).
70. *Compare* DEL. CODE ANN. tit. 8, § 9105(c) (1991 & Supp. 1996) *with* 42 U.S.C. 9607(b) (1995).
71. DEL. CODE ANN. tit. 10, § 8106 (1975).
72. Under CERCLA, parties are not able contractually to "transfer" liability to the government from one party to another, but they can allocate liability between themselves. *See, e.g.,* American Int'l. Inc. v. International Forgiving Equip., 982 F.2d 989, 994 (6th Cir. 1993).
73. *See, e.g.,* Aetna Casualty & Surety Co. v. Security Ins. Co. of Hartford, 267 A.2d 582, 585 (Del. 1970).
74. DEL. CODE ANN. tit. 7, § 9113(c)(6) (Supp. 1996); Del. Economic Dev. Office, Regulations on Brownfield Assistance (1995) [hereinafter Brownfield Assistance Regs.].
75. Brownfield Assistance Regs., *supra* note 74, ¶ I.B.
76. *Id.* ¶ A.1.
77. *Id.* ¶ II.B.2.
78. *Id.* ¶ II.B.3.

CHAPTER 27

Illinois

DAVID J. ENGEL

As a result of legislation enacted in late 1995,[1] the Illinois Voluntary Cleanup Program is in transition. The new legislation, effective December 15, 1995, added a comprehensive voluntary remediation program to the Illinois Environmental Protection Act (the Act), which will eventually replace the more informal program that began in 1989 (the 1989 Program).[2] The legislation also requires the Illinois Environmental Protection Agency (the Agency) to propose regulations to the Illinois Pollution Control Board (the Board) by September 15, 1996, to implement numerous legislative requirements. The Board in turn must promulgate regulations by June 16, 1997.[3] Because there are no regulations or judicial opinions interpreting the 1995 legislation, many implementation issues and ambiguities have yet to be resolved.

Contributing to the uncertainty is the fact that the 1995 legislation did not completely eliminate the 1989 Program. Sites enrolled in the 1989 Program that did not complete remediation as of December 15, 1995, can remain in the 1989 Program or opt into the new program.[4] Because two Voluntary Cleanup Programs are currently in effect, this chapter will address both programs, but place primary emphasis on the recent statutory changes.

Site Remediation Program Created by the 1995 Legislation

Summary of Major Provisions

The major provisions of the voluntary program created by the 1995 legislation include the following:

- Requirement that the Board develop generic soil and groundwater cleanup standards by June 1997 (The legislation also mandates the development of regulations containing a risk-assessment methodology that will be used to derive property-specific numeric cleanup objectives.)

- Liability protection to volunteers in the form of a no-further-remediation letter (This letter will release the volunteer and related entities from liability for further environmental investigation or remedial activities.)
- Board review of Agency decisions
- Use of non-Agency professional engineers to review and approve, with appropriate Agency oversight, site investigations and remedial activities.

Participation in the Program

Eligibility for the Program

The 1995 Site Remediation Program (the Program, or Voluntary Cleanup Program) is contained in Title XVII[5] of the Act and contains requirements for conducting voluntary site investigations and remedial activities at sites where there is a release, threatened release, or suspected release of hazardous substances, pesticides, or petroleum.[6] To participate in the Program, a person simply must intend to perform investigative or remedial activities. Portions of sites, or only some of the contaminants present at a site, can be enrolled in the Program.[7] The legislation also permits entry into the Program at any point in the site investigation or remediation process, including upon completion of remedial activities, provided that all activities were conducted in accordance with Program requirements.[8]

The new provisions of the Act characterize Program participants as "Remediation Applicants." Remediation Applicants can be site owners, operators, or other persons who have the owner's or operator's consent to enter a site in the Program.[9] The Program is administered by the Agency; however, certain functions (described below) can be performed by non-Agency professional engineers selected by Remediation Applicants and overseen by the Agency.

Sites Ineligible for the Program

Not all sites can participate in the Program. Specifically excluded are the following properties:

- Sites listed on the federal National Priorities List under the federal Comprehensive Environmental Response, Compensation, and Liability Act (CERCLA)[10]
- Permitted treatment, storage, or disposal facilities, or facilities subject to closure requirements under federal or Illinois solid or hazardous-waste laws
- Sites subject to federal or Illinois underground storage tank (UST) laws
- Sites undergoing investigation or remedial activities pursuant to a federal court or United States Environmental Protection Agency (USEPA) order.[11]

Liability Protection under Illinois Law

Proportionate-Share Liability

The new 1995 provisions of the Act changed the liability scheme previously available to the state to recover response costs or order investigative or remedial activities. Section 58.9 of the Act, effective July 1, 1996 (the proportionate-share liability legislation), precludes the state from seeking remedial activities or response costs from anyone other than a person whose act or omission caused the release of hazardous substances or petroleum.[12] The new legislation also repealed joint and several liability under the Illinois Superfund program, and altered other liability provisions of the Act and the Illinois Groundwater Protection Act.[13]

The proportionate-share liability legislation requires the Board to adopt implementing regulations by January 1, 1998.[14] The rules must contain criteria for determining apportioned responsibility and procedures permitting liable parties to petition the Board for a determination of their respective liabilities at a specific site.[15]

Lenders

Since 1991, the Act has included liability protection for lenders. Financial institutions that acquire ownership, operation, management, or control of a facility through foreclosure or under the terms of a security interest or an extension of credit are not considered "owners" or "operators" for purposes of Illinois Superfund liability.[16] The exception to this rule is if the financial institution takes possession of the facility and "exercises actual, direct, and continual or recurrent managerial control" of operations that "causes a release or substantial threat of release of a hazardous substance or pesticide."[17] However, the proportionate-share legislation may have largely rendered the lender-liability provision meaningless, with the exception of costs incurred by the state before July 1, 1996.[18]

Liability Protection for Program Participants and Related Entities

The 1995 Voluntary Cleanup Program legislation offers Remediation Applicants, who complete an Agency-approved cleanup or an investigation demonstrating that no remedial activities are warranted, the ability to obtain a no-further-remediation letter (NFRL) from the Agency.[19] The NFRL constitutes a release from further responsibilities under the Act and is "prima facie evidence that the subject site does not constitute a threat to human health and the environment" and requires no further remediation under the Act.[20] The NFRL applies to the following:

- The site owner and operator
- Any parent corporation or subsidiary of the site owner
- Any co-owner or any other entity sharing a legal relationship with the site owner

- Any holder of a beneficial interest of a land trust or inter vivos trust, whether revocable or irrevocable, involving the site
- Any mortgagee or trustee of a deed of trust of the site owner or any assignee, transferee, or successor-in-interest
- Any successor-in-interest of the site owner
- Any transferee of the site owner
- Any heir or devisee of the site owner
- Any financial institution that acquired ownership, operation, management, or control of the site (1) through foreclosure or under the terms of a security interest held by the financial institution, or (2) under the terms of an extension of credit made by the financial institution
- In the case of a fiduciary, the estate, trust estate, or other interest in property held in a fiduciary capacity, and a trustee, executor, administrator, guardian, receiver, conservator, or other person who holds the site in a fiduciary capacity or a transferee of such party.[21]

As discussed below, if the NFRL contains land-use restrictions or requires engineering controls, the Agency can void the NFRL if the site is used in a manner inconsistent with the restrictions or the controls are not maintained.[22]

Program Financial Incentives

Currently, there are no financial incentives to induce participation in the Program.

Cleanup Objectives

Perhaps the most significant feature of the 1995 legislation is the requirement that cleanup objectives must be based on consideration of present and future site uses.[23] Until the Board adopts regulations, the legislation directs the use of cleanup objectives that were previously adopted by the Board pursuant to the Illinois UST program.[24] Both engineering controls (such as caps, slurry walls, and clean fill) and institutional controls (such as restrictive covenants and negative easements) are acceptable approaches to meeting applicable remediation objectives.

Interim Objectives

Unfortunately, the UST regulations contain soil cleanup objectives for only the relatively few substances that are likely to be released from a petroleum UST. As such, soil objectives exist for only seven metals (arsenic, barium, cadmium, chromium, lead, mercury, and selenium) and six organics (benzene, toluene, ethyl benzene, xylenes, naphthalene, and benzo(a)pyrene).[25] The groundwater cleanup objectives contained in the UST regulations are based on the Board's most stringent groundwater standards or, for constituents for which no standard exists, on the procedures used to calculate those standards.[26]

On November 14, 1994, the Agency published guidance containing soil cleanup objectives for constituents not included in the UST regulations.[27] These objectives must be used until additional objectives are added to the UST regulations. As with the soil objectives contained in the UST regulations, the soil objective for each constituent varies depending on the distance of the compliance point (either the property boundary or two-hundred feet, whichever is less) from the point of release.

Final Objectives

Cleanup objectives derived under the new legislation will be based on background contaminant concentrations or objectives that are developed by using a three-tiered process, modeled in part after the risk-based corrective action (RBCA) procedures of the American Society for Testing and Materials.[28] The purpose of this approach is to ensure that cleanup objectives protect human health and the environment, but also reflect a realistic assessment of current and expected land use. Persons deciding whether to perform remediation or proposing cleanup objectives to the Agency will be able to select which tier or tier combination to use. The legislation specifies that groundwater consumption, human inhalation, and human ingestion are the minimum exposure pathways that must be addressed in each tier. For carcinogens, soil and groundwater objectives for residential use cannot exceed a target risk of one in one million,[29] while the maximum acceptable target cancer risk permitted for industrial and commercial sites is one in ten thousand. However, in no case can a groundwater objective be more stringent than the applicable groundwater quality standard contained in Board regulations.[30]

Background concentrations are defined as concentrations that are "consistently present" in the vicinity of the site as a result of natural and human activities, but not solely the result of a release at the site.[31] In general, cleanup objectives cannot be established at levels more stringent than site background concentrations.[32] The only exception is when the Agency determines that a constituent's background concentration poses an "acute threat to human health or the environment" after consideration of the postremedial land use.[33] In this circumstance, the Remediation Applicant must develop appropriate risk-based cleanup objectives using the three-tiered process.

Similar to RBCA, Tier I will contain soil and groundwater cleanup objectives for residential, commercial, and industrial uses.[34] The Agency's proposed regulations contain Tier I groundwater cleanup objectives based on the Board's stringent groundwater standards. Groundwater cleanup objectives for contaminants for which no Board groundwater standard exists will likely be calculated using the health advisory procedures contained in subpart F of the Board's groundwater regulations.[35] Tier I objectives can be used without any site-specific information.

Tier II will have more variables, including the equations used to derive Tier I objectives, as well as the input parameters used in the equations. This will allow Remediation Applicants to modify certain input parameters to reflect site conditions.[36] For example, the regulations will likely permit the use of site-specific soil saturation limits, groundwater dilution factors, and soil-air and soil-water partition coefficients in place of the values used to calculate Tier I objectives. It is also likely that the regulations will contain default values, which will be input parameters that cannot be modified (such as the amount of drinking water consumed each day).

The legislation also authorizes cleanup objectives to be based on risk assessments, which is the Tier III process.[37] The Board has been directed to adopt a risk-assessment methodology for residential and other land-use categories. The Board's promulgation of a risk-assessment methodology may not preclude Remediation Applicants from using other "nationally recognized" methodologies. Tier III will likely provide Remediation Applicants the most flexibility to modify default parameters used in Tier I equations, eliminate one or more pathways of concern (for example, soil-to-groundwater migration if the site is underlain by extensive clay deposits), request "common sense" objectives because further remediation is not practical, and accommodate situations that are substantially different from those assumed in Tiers I and II.

The legislation also alters the Agency's past practice of generally prohibiting, without site-specific regulatory relief from the Board, groundwater cleanup objectives that are less stringent than Board groundwater regulations. It specifies that the Agency can approve less stringent groundwater cleanup objectives if the Remediation Applicant demonstrates that implementation of the cleanup plan has, to the extent practical, (1) minimized the degree to which the groundwater standard was exceeded, and (2) returned the groundwater to beneficial use. Also, the Remediation Applicant must demonstrate that any threat to human health or the environment has been minimized.[38]

The Agency will likely continue to require groundwater management zones (GMZs) at sites where groundwater exceeds applicable cleanup objectives during the course of remediation. A GMZ is a three-dimensional region containing groundwater being managed to mitigate impairment and shield the Remediation Applicant from liability for violating the Board's groundwater regulations while the site is undergoing remediation. The criteria for establishing a GMZ are contained in those regulations.[39]

The Mechanics of Program Participation

Investigations, Reports, and Plans

Once a decision is made to enter a site in the Program, all investigations, remedial activities, plans, and reports must be conducted and prepared in accordance with legislative and regulatory requirements. All these activities must be undertaken and prepared under the supervision of a licensed professional engineer.[40] Moreover, all documents must include a certification from the engineer that, "to the best of his or her knowledge and belief, the work described in the plan or report has been completed in accordance with generally accepted engineering practices, and the information presented is accurate and complete."[41]

The first step in the process is a site investigation that must assess the nature, concentration, direction, and rate of movement of site contaminants, as well as the physical features that may affect contaminant transport.[42] The results of this investigation will be contained in a "Site Investigation Report." Next, the Remediation Applicant must prepare a remediation objectives report.[43] If objectives are proposed using Tier II or Tier III methodologies or background concentrations, the report must document how these objectives or concentrations were derived. Although not addressed in the legislation, if Tier I objectives are proposed, the remediation objectives report will likely contain only the applicable Tier I cleanup objectives.

If Agency-approved cleanup objectives are less than concentrations existing at the site, the Remediation Applicant must prepare a remedial action plan. That plan must "describe the selected remedy and evaluate its ability and effectiveness to achieve the remediation objectives approved for the site."[44] The plan will also contain any applicable engineering and institutional controls proposed for the site to meet cleanup requirements.[45]

The final step in the process is the preparation of a remedial action completion report. That report must demonstrate that the remedial action was completed in accordance with the remedial action plan, and demonstrate that site cleanup objectives were attained.[46] If contaminating constituents are present at levels at or below cleanup objectives, the remedial action completion report must only document this information.

The legislation allows a Remediation Applicant to submit these reports and plans to the Agency concurrently or following the completion of each activity.[47] A Remediation Applicant must evaluate the risks and benefits of submitting this information concurrently, including the possibility that the Agency could assert that an investigation reviewed by the Agency after completion of remedial activities was inadequate. Regardless of the sequence of document submission, the Agency's review must proceed in accordance with section 58.7 of the Act, entitled "Review and Approvals."

Review and Approvals

All plans and reports must be reviewed for compliance with legislative and regulatory requirements, either by an Agency professional engineer or a "Review and Evaluation Licensed Professional Engineer" (RELPE) retained by the Remediation Applicant. The RELPE will act "on behalf of and under the direction of the Agency."[48] The Remediation Applicant must notify the Agency before retaining the RELPE, "discuss" with the Agency the terms of the contract between the Remediation Applicant and the RELPE, and pay for the RELPE's services. Moreover, the RELPE cannot be an employee of the Remediation Applicant, the site owner or operator, or any other person with whom the Remediation Applicant has contracted to perform services at the site.

The benefits of using a RELPE are questionable. Only the Agency has the authority to approve, disapprove, or approve with conditions a Remediation Applicant's plans and reports.[49] Although an early draft of the Voluntary Cleanup Program legislation included a provision that would have allowed non-Agency professional engineers to approve all documents, this concept was not included in the final statute because of Agency concerns about the lack of appropriate oversight. In fact, during the initial phase of Program implementation, the use of a RELPE could increase the time required to complete a voluntary cleanup, simply because there will be another entity inserted into the process.

All plans and reports submitted to the Agency or a RELPE must be reviewed and the decision communicated to the Remediation Applicant within sixty days, unless the Remediation Applicant authorizes additional time.[50] If the Agency or RELPE does not act, the Remediation Applicant may seek Board review within thirty-five days of the decision date. Under these circumstances, the Remediation Applicant would likely request an order finding that the report or plan met Program requirements. Because of the time that will be consumed during a Board proceeding, the benefits of an appeal are also questionable.

If the Remediation Applicant establishes that the investigation, any remediation, and all reports and plans conform to Program requirements, the Agency must issue an NFRL.[51] If the Agency fails to issue the NFRL within thirty days after its approval of the remedial action completion report, the NFRL will be issued by operation of law.[52]

Public Participation

The legislation requires the Agency to develop guidance to assist Remediation Applicants in implementing a community relations plan for sites enrolled in the Program.[53] The Remediation Applicant may also engage the services of the Agency's experienced Community Relations Office to assist it, but this is not a condition of Program participation.

Appeals of Agency Decisions

The Agency must give the Remediation Applicant written notice of its decisions. If the Agency disapproves a plan or report, or conditions its approval, the Agency must identify the section(s) of the Act or Board regulations that may be violated if the plan or report was approved and explain why any conditions are required.[54] The Agency must also identify any information that it believes should have been provided to it. The Remediation Applicant can challenge the Agency's determination by filing an appeal to the Board within thirty-five days.[55]

Recording the NFRL

A Remediation Applicant must record the NFRL with the appropriate county recorder's office for the NFRL to be effective.[56] The Remediation Applicant must also provide the Agency with a certified copy of the NFRL after it is recorded.[57]

Activities at the site must conform to any restrictions contained in the NFRL. If further remedial action or subsequent investigation demonstrates that any restriction is no longer appropriate, a new NFRL must be obtained from the Agency and recorded.[58] Therefore, a site owner or operator will be required to receive the Agency's concurrence that any contamination remaining on the site that resulted in a land-use restriction or engineering control either no longer remains or is presently below concentrations that led to the restriction or control.

Voidance of an NFRL

An NFRL must contain a description of any engineering control or institutional restriction, and a prohibition against the use of the site in a manner inconsistent with those requirements (absent additional remediation or further investigation demonstrating those limitations are no longer required).[59] The Agency may void an NFRL if activities at a subject site are inconsistent with any restriction contained in the NFRL or if controls are not maintained. Moreover, the NFRL is voidable if the Remediation Applicant failed to pay all appropriate fees (discussed below) or if contaminants not identified during the investigation or remedial activities are subsequently discovered and pose a threat to human health or the environment.[60]

To void an NFRL, the Agency must notify the site owner and the Remediation Applicant. Unless the Agency's determination is appealed to the Board within thirty-five days of receipt of the notice, the NFRL will be void.[61] If an appeal is not filed or if the Board affirms the Agency's decision, the Agency must include a notification of this determination in the site's chain of title.[62]

Fees for Participation

Remediation Applicants will incur two fees as a result of Program participation, and all such monies are deposited in the state's Hazardous-Waste Fund. First, Remediation Applicants must pay an advance partial payment to the Agency in an amount determined by the Agency, which will be $5,000 or one half the total estimated Agency oversight costs, whichever is less. Additional fees must be paid at the conclusion of the Agency's review to reimburse the Agency fully for its oversight services.[63]

Second, within thirty days of receiving an NFRL (or after issuance of an NFRL by operation of law), the Remediation Applicant must pay a no-further-remediation assessment. The assessment will be the lesser of $2,500 or the amount of the fee charged by the Agency for its oversight services.[64] This second fee will provide monies to assist the Agency in responding to contamination of other sites caused by entities that are not financially viable or unidentifiable. The state included this additional fee in the 1995 legislation because of the repeal of joint and several liability under the Illinois Superfund program.

Admission of Liability

A Remediation Applicant can withdraw from the Program at any time and for any reason.[65] The only requirement will be to fully reimburse the Agency for its oversight services. Nothing in the Act precludes the Agency or third parties from using information that Remediation Applicants submit to the Agency for any purpose.

Illinois enacted environmental audit privilege legislation in January 1995, which creates a qualified statutory privilege against disclosure of environmental audit reports generated as a result of an audit of a facility or process.[66] However, the privilege likely will not be extended to voluntary cleanup reports and plans submitted to the Agency. Those documents will not be confidential, which is a prerequisite to invoking the environmental audit privilege.

Memorandum of Agreement

The Agency and USEPA Region V executed a Memorandum of Agreement (MOA) on April 6, 1995, concerning sites completing cleanups in accordance with the 1989 Program. Although the MOA was executed before the 1995 legislative changes, those changes do not appear to affect the terms or conditions of the MOA. To date, there has been no attempt to rescind the MOA or restrict its applicability to investigations and cleanups completed only in accordance with 1989 Program requirements.

The principles of the MOA apply to (a) sites for which the Agency, after reviewing the results of an investigation, has determined that no remediation is necessary, or (b) sites at which Agency-approved remediation has been completed. For both classes, the MOA specifies that the site "will not be expected to require further response actions" and that Region V will not take action under CERCLA unless "the site poses an imminent threat or emergency situation."

Consequently, it is unlikely that contamination that remains upon completion of Program requirements will be the subject of USEPA interest. If a site owner or other entity desires assurances beyond those contained in the MOA, it could seek a "comfort letter" from Region V or enter a prospective-purchaser agreement with the federal government.[67] However, neither the MOA nor these other protections would preclude private parties from bringing a CERCLA claim against a Remediation Applicant or subsequent site owners as a result of remaining contamination.

Cost Recovery

Although CERCLA provides a clear private right of action for an entity that engages in voluntary investigation and remedial activities, Illinois statutory law is much murkier. Three possible causes of action exist to recover the cost of these activities: (1) the volunteer could attempt a cost-recovery action under section 22.2 of the Act, which is the state Superfund liability provision (Illinois Superfund Statute)[68]; (2) the volunteer could bring a contribution action; or (3) the volunteer could bring a citizen suit under provisions of the Act other than the Illinois Superfund Statute, such as sections 12(a), 12(d), and 21(a). Unfortunately, Illinois law remains unsettled concerning whether any of these avenues are available in a form that would benefit volunteers.

Like CERCLA, the Illinois Superfund Statute establishes four classes of liable parties, and the statute defines these classes with language that is virtually identical to the language of CERCLA.[69] Also, as under CERCLA, the defined parties are liable subject only to specifically enumerated defenses. However, the Illinois Superfund Statute provides only that these parties are liable "for all costs of removal or remedial action incurred by the State of Illinois or any unit of local government."[70] The text of this section suggests that, unlike under CERCLA, private cost-recovery actions may not be available under the Illinois Superfund Statute. However, section 31(b) of the Act provides a general private right of action "against any person allegedly violating this Act or any rule or regulation thereunder. . . ."[71] Though section 31(b) may create a private right of action to recover any costs that the state could recover under section 22.2, a defendant in an action brought by a Program volunteer would likely argue that liability under the State Superfund Statute does not constitute a "violation" of the Act. Unfortunately, no Illinois cases have considered this issue.

If the volunteer is the site owner, it could also attempt a contribution action. Illinois allows a "tortfeasor" to seek contribution from another tortfeasor for "liability in tort" under the Illinois Joint Tortfeasor Contribution Act (Contribution Act),[72] and the Illinois Supreme Court has held that "tort" liability for this purpose includes liability under the Act.[73] As a result, a right of action in contribution exists for a potentially responsible party (PRP) site owner against another PRP, at least assuming that the site owner did not intentionally contaminate the property.[74] However, two problems may exist for site owners who conduct voluntary cleanups and attempt to recover costs via the Contribution Act. First, the Contribution Act may permit contribution only for claims arising out of occurrences on or after March 1, 1978.[75] As a result, volunteers may not be able to recover costs resulting from disposal or releases occurring before that date. Second, the Illinois courts have held that a Contribution Act cause of action exists when the plaintiff has been sued and paid more than its fair share of liability.[76] Because in most circumstances a volunteer has not been sued and, therefore, would not have paid more than its fair share of liability, a contribution action may not be available.

Finally, in addition to cost recovery under the Illinois Superfund Statute or contribution, a site owner might attempt to recover costs through a general citizen suit under the Act. Apart from the Illinois Superfund Statute, the previous owner or operator of a site enrolled in the Program may have violated other portions of the Act, particularly section 12(a) (causing water pollution), section 12(d) (depositing contaminants on land that create a water pollution hazard), or section 21(a) (open dumping of waste). As noted above, section

31(b) allows any person to bring a citizen suit before the Board for violating any provision of the Act. Although the Board ordinarily issues cease-and-desist orders and imposes penalties in response to violations, it has held that it has the power to award cleanup costs.[77] However, the Board has also held that cleanup costs are not automatically available as a remedy, but are available only if appropriate after the Board considers several statutory factors, including the character and degree of injury a defendant has caused, and the social and economic value of the defendant's activity.[78]

As a result, cost recovery under the Act's citizen suit and other provisions remains uncertain. Volunteers should carefully evaluate causes of action available under federal law and not assume that Illinois law will afford them the ability to recover any investigation or remedial costs.

The 1989 Program

The Illinois General Assembly added section 22.2(m) to the Act in 1989, to authorize the Agency to oversee private-party environmental investigations and remedial activities at sites where hazardous substances or pesticides may be present.[79] The Board was drafting regulations to implement the 1989 Program, but this effort was preempted by the 1995 legislation.

Owners or operators (or other entities who have received the permission of the owner or operator) may request Agency oversight of these activities. Although not specified in the legislation, the Agency as a matter of practice excludes from the 1989 Program contamination that is subject to response activities under any other state or federal environmental program.

The legislation authorizes the Agency to (1) require participants to submit work plans for actions at the site, (2) visit the site, (3) receive a commitment from the applicant to implement Agency-approved work plans, (4) collect Agency oversight expenses, and (5) charge an advance partial payment for these expenses, which is the lesser of $5,000 or one half the expected cost of the Agency's services.[80]

The Agency's practice is to require persons seeking enrollment in the 1989 Program to submit to the Agency a completed application form that is available from the Agency's Bureau of Land. The application requires general information about the site, its location, and the contamination that will be the subject of Agency review. Upon receipt of a complete application, the Agency will make an eligibility determination, request the applicant to complete a review and evaluation services agreement, and make a partial payment of anticipated Agency oversight expenses. The Agency's receipt of the completed agreement and the fee will secure enrollment in the program. The Agency will typically issue an enrollment letter acknowledging site acceptance.

In implementing the 1989 Program, the Agency has often required many of the documents (and other information) that are prescribed by the 1995 legislation. Thus, participants typically submit to the Agency investigation and remedial-action work plans and completion reports. These documents are usually submitted before their implementation, to ensure that the Agency is comfortable with the planned activity.

Before the 1995 legislation, nothing in the Act addressed cleanup objectives. The Agency's practice has been to set groundwater cleanup objectives based on the Board's groundwater regulations. When a contaminant appears at a site that is not listed in the groundwater standards, other health-based standards are used. Moreover, the Agency has often used these standards as soil cleanup objectives unless, using risk-assessment methodologies, the participant presents information showing that less stringent levels would be protective of human health and the environment. The Agency's Office of Chemical Safety will review a risk assessment to ensure that it was done in accordance with acceptable practices. In recent years, the Agency has also looked to the Board's UST cleanup objectives as guidance for reviewing cleanup objectives for sites enrolled in the 1989 Program.

The Cleanup Objectives Review and Evaluation Group, which consists of senior managers in the Agency's Bureau of Land, reviews the participant's proposed cleanup objectives, taking into account site-specific information. Engineering and institutional controls are acceptable. After the Agency has completed its review of a site, it will require the program participant to reimburse the Agency fully for its oversight services.

Before the regulations implementing the 1995 legislation are effective, it may be beneficial to communicate with the Agency frequently, particularly for sites that have not opted into the 1995 Program. If 1989 Program participants want to preserve their ability to opt into the Program created by the 1995 legislation, they should prepare reports and plans in accordance with the requirements prescribed in the 1995 legislation, to the extent possible, to minimize the chance of needing to redo any activities.

The Act authorizes the Agency to release any person completing a voluntary cleanup under the 1989 Program from further responsibility for "preventive or corrective action under [the] Act."[81] Nothing in the Act before the 1995 legislation provided that the release applied to anyone other than the participant, or that the release "ran with the land." This inadequacy was the primary factor leading to the inclusion of detailed language in the 1995 legislation on this issue.

Any participant can opt out of the 1989 Program at any time. If this occurs, the Agency will invoice the former participant for full payment of its oversight services.[82] The Agency would retain its statutory authority to require the former participant to respond to any contamination identified but not addressed.

Conclusion

Illinois has been in the forefront of the efforts to develop programs to assist the owners of contaminated properties with developing cost-effective and protective remediation strategies. The 1995 legislative changes are another example of these efforts. The new legislation, combined with the MOA between USEPA and the Agency should provide further incentives to redeveloping contaminated sites.

Obtaining a No-Further-Remediation Letter under the Illinois Site Remediation Program

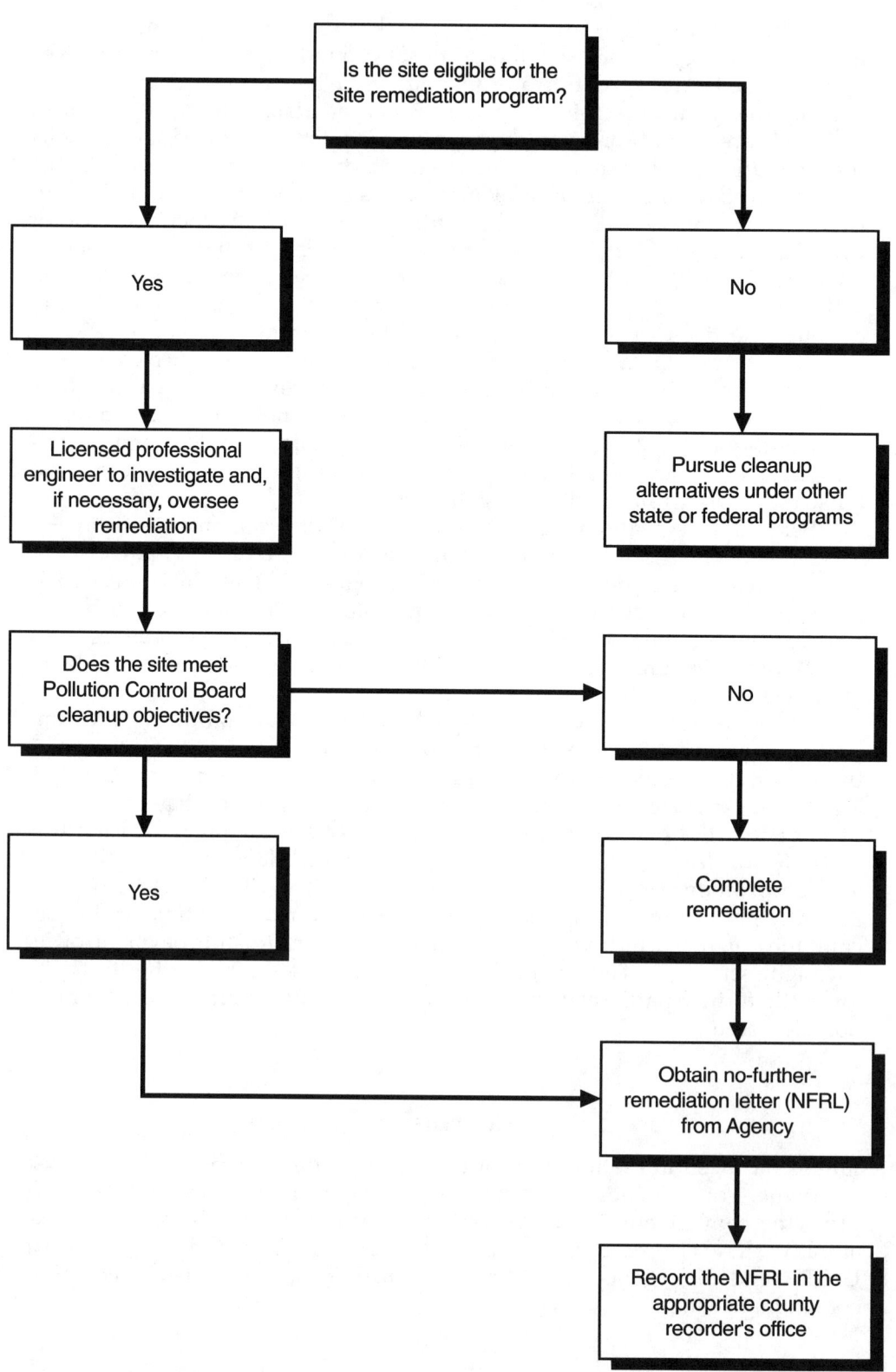

Notes

1. 415 ILCS 5/58.1–58.12 (West 1995).
2. *Id.* 5/22.2(m).
3. The legislation establishes a ten-member committee to perform a number of functions, including assisting the Agency and the Board in developing these regulations. *Id.* 5/58.11. On September 16, 1996, the Agency proposed two regulations to implement the Site Remediation Program. Proposed Part 742 would contain the process used to set soil and groundwater cleanup objectives. Part 740 would contain the regulations applicable to all other aspects of the program.
4. *Id.* 5/58.1(b).
5. *Id.* 5/58–58.12.
6. *Id.* 5/58.1(a)(1). Nothing in the 1995 legislation or any other section of the Act prohibits investigation and remedial activities outside of Program requirements and Agency review. *See id.* 5/58.6(g).
7. *Id.* 5/58.10(b)(10).
8. *Id.* 5/58.6(f).
9. *Id.* 5/58.2 (see definition of "Remediation Applicant").
10. 40 C.F.R. § 300, Appendix B (1996); 42 U.S.C. § 9601 (1994).
11. 415 ILCS 5/58.1(a)(2) (West 1995). The legislation specifies that to the extent permitted by federal law, sites excluded from Program participation may utilize Program provisions, including the procedures for establishing risk-based remediation objectives. The legislation does not preclude from Program participation sites undergoing investigation or remedial activities pursuant to a *state* court or Agency order. (The Agency's policy in implementing the 1989 Program was to exclude those sites from participation.)
12. *Id.* 58.9(a).
13. *Id.* 55/1–9.
14. The Agency and the ten-member site advisory committee established by the Voluntary Cleanup Program legislation will assist the Board in developing proportionate-share regulations. *Id.* 5/58.11(b)(2). It is likely that this deadline will be extended by one year.
15. *Id.* 5/58.9(d).
16. *Id.* 5/22.2(h)(2)(E).
17. *Id.*
18. The proportionate-share liability section states that it does not apply to any cost-recovery action brought by the state to recover costs incurred before July 1, 1996. *Id.* 5/58.9(f).
19. *Id.* 5/58.7(d)(4). The legislation further specifies that if the Agency fails to issue an NFRL within thirty days after approving documentation of completion of remedial activities, the NFRL will be issued by operation of law. *Id.* 5/58.10(b).
20. *Id.* 5/58.10(a).
21. *Id.* 5/58.10(d).
22. *Id.* 5/58.10(e)(2).
23. *Id.* 5/58.
24. *Id.* 5/58.5(f). The Board's UST regulations are codified at Illinois Administrative Code, title 35, section 732. The cleanup objectives are contained in Tables A and D of section 732.
25. ILL. ADMIN. CODE tit. 35, § 732, Appendix B (1994).
26. *Id.*
27. ILL. ADMIN. CODE tit. 35, § 732, Appendix B Soil Remediation Methodology Objectives (Draft Nov. 19, 1994).

28. 415 ILCS 5/58.5 (West 1995).
29. *Id.* 5/58.5(d). In the event background concentrations exceed residential cleanup objectives, remediation will not be required, but the property cannot be used for residential purposes. *Id.* 5/58.5(b)(2).
30. ILL. ADMIN. CODE tit. 35, § 620 (1994).
31. 415 ILCS 5/58.2 (West 1995) (definition of "area background").
32. *Id.* 5/58.5(b)(1). The regulations that will be adopted by the Board will contain procedures for determining background contaminant levels. *Id.* 5/58.5(c)(3). The Agency's Bureau of Land published guidance in August 1994 entitled *A Summary of Selected Background Conditions for Inorganics in Soil*. This guidance will also assist Remediation Applicants in determining whether constituents exceed background concentrations, particularly before the Board develops regulations.
33. 415 ILCS 5/58.5(b)(3) (West 1995).
34. *Id.* 5/58.5(d)(1).
35. ILL. ADMIN. CODE tit. 35, § 620, subpart F (1994).
36. 415 ILCS 5/58.5(d)(2) (West 1995).
37. *Id.* 5/58.5(d)(3).
38. *Id.* 5/58.5(d)(4).
39. ILL. ADMIN. CODE tit. 35, § 620.250 (1994).
40. 415 ILCS 5/58.6(a) (West 1995).
41. *Id.* 5/58.7(f).
42. *Id.* 5/58.6(b).
43. *Id.* 5/58.6(c).
44. *Id.* 5/58.6(d).
45. The legislation also contains a permit waiver provision. Remedial activities undertaken pursuant to Program requirements are exempt from state permit requirements unless a permit or permit revision is required by federal law. *Id.* 5/58.4.
46. *Id.* 5/58.6(e).
47. *Id.* 5/58.6(f).
48. *Id.* 5/58.7(c).
49. *Id.* 5/58.7(d)(3).
50. *Id.* 5/58.7(d)(5).
51. *Id.* 5/58.7(d)(4).
52. *Id.* 5/58.10(b).
53. *Id.* 5/58.7(h).
54. *Id.* 5/58.7(d)(4).
55. *See id.* 5/40 (section of Act governing Board appeals).
56. *Id.* 5/58.8(b).
57. *Id.*
58. *Id.* 5/58.8(d).
59. *Id.* 5/58.10(b).
60. *Id.* 5/58.10(e).
61. *Id.* 5/58.10(f).
62. *Id.* 5/58.10(f)(2).
63. *Id.* 5/58.7(b)(1)(D)–(E).
64. *Id.* 5/58.10(g).
65. *Id.* 5/58.7(b)(3).
66. *Id.* 5/52.2. Unlike several other states, nothing in Illinois's audit legislation or any other provision of the Act provides immunity for disclosure of violations of the Act or Board regulations to the state authorities.

67. *See* 60 Fed. Reg. 34,792 (1995).
68. 415 ILCS 5/22.2 (West 1995).
69. *See id.* 5/22.2(f); 42 U.S.C. § 9607(a) (1994).
70. 415 ILCS 5/22.2(f) (West 1995).
71. *Id.* 5/31(b).
72. 740 ILCS 100/2 (West 1995).
73. People v. Brockman, 574 N.E.2d 626, 634 (Ill. 1991).
74. *Id.* at 635 (contribution not available to intentional tortfeasors).
75. 740 ILCS 100/1 (West 1995); *Brockman*, 574 N.E.2d at 633.
76. *See, e.g.,* Hahn v. Norfolk & Western Rwy. Co., 608 N.E.2d 683, 686 (Ill. App. Ct. 1993); Highland v. Bracken, 560 N.E.2d 406, 408 (Ill. App. Ct. 1990).
77. Lake County Forest Preserve Dist. v. Ostro, IPCB No. 92-80, 1994 Ill. ENV LEXIS 484 (Ill. Pollution Control Bd. Mar. 31, 1994).
78. *Id.*; 415 ILCS 5/33(c) (West 1995).
79. 415 ILCS 5/22.2(m) (West 1995).
80. *Id.* 5/22.2(m)(1)(A)–(E).
81. *Id.* 5/4(y).
82. *Id.* 5/22.2(m)(4).

CHAPTER 28

Indiana

ANNE SLAUGHTER ANDREW

In 1992, Indiana adopted the Voluntary Remediation of Hazardous Substances and Petroleum Act[1] (Voluntary Remediation Act). The Voluntary Remediation Act, which went into effect in July 1993, provides a program through which a person, including a prospective purchaser, voluntarily may complete a state-approved remediation of a site on which an actual or threatened release of a hazardous substance or of petroleum has occurred.[2]

Summary of Major Provisions

Indiana has implemented the Voluntary Remediation Act through the Voluntary Remediation Program (VRP), which is administered by the Indiana Department of Environmental Management (IDEM). The following provides a quick look at the VRP:

- *Eligibility:* Few limits on access to the VRP exist.
- *Timing*: Cleanup under the VRP is estimated to take from four to twenty-four months, depending on the site.
- *Costs:* Fees and oversight costs apply.
- *Cleanup standards:* IDEM has cleanup guidance that allows for use-based and risk-based cleanup standards.
- *Public participation:* The VRP allows for public notice and input into the cleanup process.
- *Finality:* A successful cleanup will result in a covenant-not-to-sue relating to the area and contaminants addressed at the site.
- *Cost recovery:* The Voluntary Remediation Act does not include a statutory cost-recovery provision.

Participation in the Voluntary Remediation Program

Within thirty days after IDEM receives an application and the $1,000 application fee for the VRP, IDEM will determine if the applicant is eligible to participate in the VRP.[3] IDEM, at its discretion, may reject an application only for one of the following reasons:[4]

1. A state or federal enforcement action that concerns remediation of the hazardous substance or petroleum described in the application is pending.
2. A federal grant requires an enforcement action at the site.
3. The condition of the hazardous substance or petroleum described in the application constitutes an imminent and substantial threat to human health or the environment.
4. The application is not complete.

If IDEM rejects an application on the grounds that it is not complete, the applicant is provided an opportunity to complete the application. On the other hand, when an application is complete, but IDEM rejects it on other grounds, the applicant has the right to appeal the rejection.[5]

Memorandum of Agreement

During the December 1995 brownfields redevelopment conference in Indianapolis, the Commissioner of IDEM signed a Memorandum of Agreement with the United States Environmental Protection Agency (USEPA) pertaining to Indiana's brownfields and its VRP. In essence, the agreement provides for the following:

> When a site in Indiana has been investigated or remediated in accordance with the practices and procedures of the VRP, and IDEM has issued a Certificate of Completion for the site, Region V will not plan or anticipate any federal action under the Comprehensive Environmental Response, Compensation, and Liability Act of 1980 (CERCLA/Superfund) unless, in exceptional circumstances, the site poses an imminent and substantial threat to human health or the environment. In all cases, the Region V decision will be based strictly on the information available at the time of the IDEM determination. The foregoing principle does not apply to sites listed on the National Priorities List or sites currently subject to orders or enforcement actions under Superfund law.

To encourage the financing, transfer, and redevelopment of industrial and commercial property, USEPA Region V committed to continue working with IDEM to remove concerns related to USEPA's Superfund activity. In addition, Region V agreed to continue technical assistance and financial support to local and state governmental agencies, to facilitate brownfields redevelopment in Indiana.

Lender and Fiduciary Liability Protection

Under Indiana's Hazardous Substances Response Trust Fund Act,[6] Indiana has provided lender-liability protection. Specifically, a secured or unsecured creditor or a fiduciary is not liable under the environmental management laws regarding the release or threatened release of hazardous substances, unless the creditor or fiduciary exercised "actual and direct managerial control over the use, generation, treatment, storage, or disposal of hazardous substances at the facility."[7] Further, Indiana provides that any liability imposed on a fiduciary by the environmental management laws regarding the release or threatened release of a hazardous substance at a facility held by the fiduciary in its fiduciary capacity may be satisfied only from that estate or trust assets held by the fiduciary.[8]

Financial Incentives

Indiana has not yet established financial-incentive programs specifically targeted for brownfields redevelopment. However, Indiana has legislation that allows local units of government to offer financial incentives for redevelopment projects, including brownfields redevelopment. The most typical of these programs are tax-increment financing (TIF) and the economic-development increase tax (EDIT). Generally, TIF permits municipal corporations to use increased tax revenues stimulated by redevelopment to pay for capital improvements, environmental cleanups, or any other "redevelopment purpose" that is needed to induce the occurrence of economic development of the area. For bonds issued to pay for a redevelopment project, many municipal corporations have chosen EDIT as a means to service the debt on such bonds. EDIT is a tax imposed on the adjusted gross income of county taxpayers. This tax is collected by the state and distributed to the counties for use by the cities and towns for their economic-development projects.

Cleanup Standards

The Voluntary Remediation Act does not set forth cleanup standards, but requires IDEM to establish cleanup guidelines.[9] IDEM has developed these guidelines through the efforts of a technical-standards subcommittee. These guidelines have not been adopted by the Solid Waste Management Board and, thus, are not enforceable as law.

The criteria in the technical guidance for cleanups include both performance and quantitative standards.[10] The performance standard requires remediation to result in the prevention or minimization of future releases of hazardous substances and petroleum. This standard is to be accomplished by the removal and/or control of the release, to the extent practical and technologically feasible.

The quantitative standard provides three tiers of cleanup criteria. Stated generally, the first tier would be cleanup to background levels, the second tier would be cleanup to generic risk-based levels, and the third tier would be cleanup to site-specific, risk-based levels. The applicant proposes which tier it intends to use for establishing cleanup criteria. However, to ensure that ecological impacts are considered in the selection of cleanup criteria, a baseline ecological assessment must be submitted along with the proposed cleanup criteria. The levels for each tier are established as follows:

- The Tier I levels will use the procedures similar to those for determining background levels for cleanups regulated under the Resource Conservation and Recovery Act (RCRA).[11] Tier I for synthetic organic chemicals will be the practical quantitation limit. For other contaminants, the Tier I cleanup level will be equal to the background mean value plus three times the standard deviation.
- The Tier II levels are calculated by using standard equations for risk assessment, as established by USEPA for risk assessments under RCRA or CERCLA. The Tier II levels will likely be less stringent than Tier I and will be more appropriate for industrial sites than for sites intended for residential development.
- The Tier III levels will be established on a site-specific basis, using a risk-assessment methodology that is established in the applicant's work plan. Tier III levels may be required for a unique type of contaminant, an environmentally sensitive site, or other unique, site-specific conditions.

The Mechanics of the Voluntary Remediation Program

The Voluntary Remediation Act involves the following steps:

1. The applicant submits the application, fee ($1,000), and environmental assessment.[12]
2. IDEM reviews these documents and determines eligibility, within thirty days after receipt of the application.[13]
3. The applicant submits a voluntary remediation work plan, which must include a description of the work needed to determine the nature and extent of the actual or threatened release, a proposed statement of the work needed to accomplish the remediation in accordance with guidelines established by IDEM, and plans for various aspects of the remediation, including a proposed schedule for implementation.[14]
4. Before IDEM evaluates a proposed work plan, IDEM and the applicant must enter a Voluntary Remediation Agreement (VRA) that sets forth the terms and conditions of the evaluation and the implementation of the work plan.[15]
5. Once the VRA has been signed, IDEM or its contractor will review the work plan and recommend whether the IDEM Commissioner should approve it. Before the IDEM Commissioner makes a decision, there must be opportunity for public comment and possibly a public hearing on the work plan. The comment period must continue for at least thirty days after publication of a notice.[16]
6. If the IDEM Commissioner approves the work plan, the applicant has sixty days to notify IDEM that it intends to proceed with implemen-

tation. IDEM or its contractor will then oversee and review the implementation of the work plan and make regular reports to the Commissioner.[17]

7. Once the work is successfully completed, IDEM issues the certificate of completion[18] and the governor issues the covenant-not-to-sue.[19]

Public Involvement

The Voluntary Remediation Act requires public input before the IDEM Commissioner approves or rejects a proposed work plan.[20] The commissioner is required to notify local government units in a county affected by the proposed plan, assure that a copy of the proposed plan is placed in at least one public library in a county affected by the plan, and publish a notice requesting comments concerning the plan. A comment period of at least thirty days must follow publication of the notice. During the comment period, interested persons may submit written comments on the plan to the Commissioner, and request a public hearing concerning the plan. The Commissioner "may" hold a public hearing in the geographical area affected by the plan if the Commissioner receives at least one written request. Although the Commissioner is required to consider all written comments and public testimony, the act does not require that IDEM prepare any written response to comments upon approving a plan.

Liability Protection

The key feature of Indiana's Voluntary Remediation Act is that on successful completion of a voluntary cleanup, the state of Indiana provides those undertaking such cleanups with a covenant-not-to-sue.[21]

The covenant-not-to-sue, issued by the governor, bars all public or private claims under Title 13 of the Indiana Code in connection with the release or threatened release that was the subject of the work plan.[22] The covenant-not-to-sue, however, does not protect against claims by the federal government based on federal law.[23] The covenant-not-to-sue does not protect against future liability for a condition existing when the work plan was approved and implemented, but not known to IDEM when the certificate was issued.[24]

The covenant-not-to-sue protects anyone receiving the certificate of completion, anyone receiving the certificate through a legal transfer of the certificate, and anyone acquiring the property to which the certificate applies.[25]

A person who implements or completes an approved, voluntary remediation work plan may not be held liable for claims for contribution concerning matters addressed in the work plan or in a certificate of completion.[26]

No Admission of Liability

In the VRA, the applicant and IDEM agree that the applicant's actions under the VRP are not an admission of liability regarding the contamination addressed by the work plan.[27] Further, during the implementation of the work plan, "a person may not bring an action, including an administrative action, against a person implementing [a] Voluntary work plan for any cause of action arising under [Title 13 of the Indiana Code] and relating to the release or threatened release of a hazardous substance or petroleum that is the subject of [a] work plan."[28]

Costs and Cost Recovery

There are substantial costs associated with the VRP over and above the actual costs of the cleanup. In addition to the $1,000 application fee, a participant in the program must agree to pay all IDEM's reasonable costs incurred in the creation of the VRA and in the review and oversight of the work plan.[29] To date, IDEM's costs appear to range from about 10 to 25 percent of the applicant's actual costs of cleanup.

As the Voluntary Remediation Act offers no mechanism for cost recovery, the only statutory vehicle for cost recovery is that provided under CERCLA.[30] Thus, to be eligible for cost recovery, any cleanup performed under Indiana's Voluntary Remediation Act must be consistent with the CERCLA requirements for cost recovery.[31] Although it appears likely that the applicant could propose cleanup standards under the Voluntary Remediation Act that would be consistent with the CERCLA requirements, it may be more difficult to satisfy the procedural aspects of a CERCLA cleanup. Certainly, substantial compliance with those procedures would necessitate many more steps than are required under Indiana's VRP, including more comprehensive public involvement.

Conclusion

As of January 1996, approximately fifty-five sites had entered the VRP; only three had completed the approved cleanups and only one had received a covenant-not-to-sue. To date, sites involved in the process generally have been satisfied with the process and the results. Though a cleanup under the Voluntary Remediation Act may not be the right solution in all cases, there is quiet optimism that Indiana has devised and implemented a "Voluntary" Cleanup Program that works.

Indiana VRP

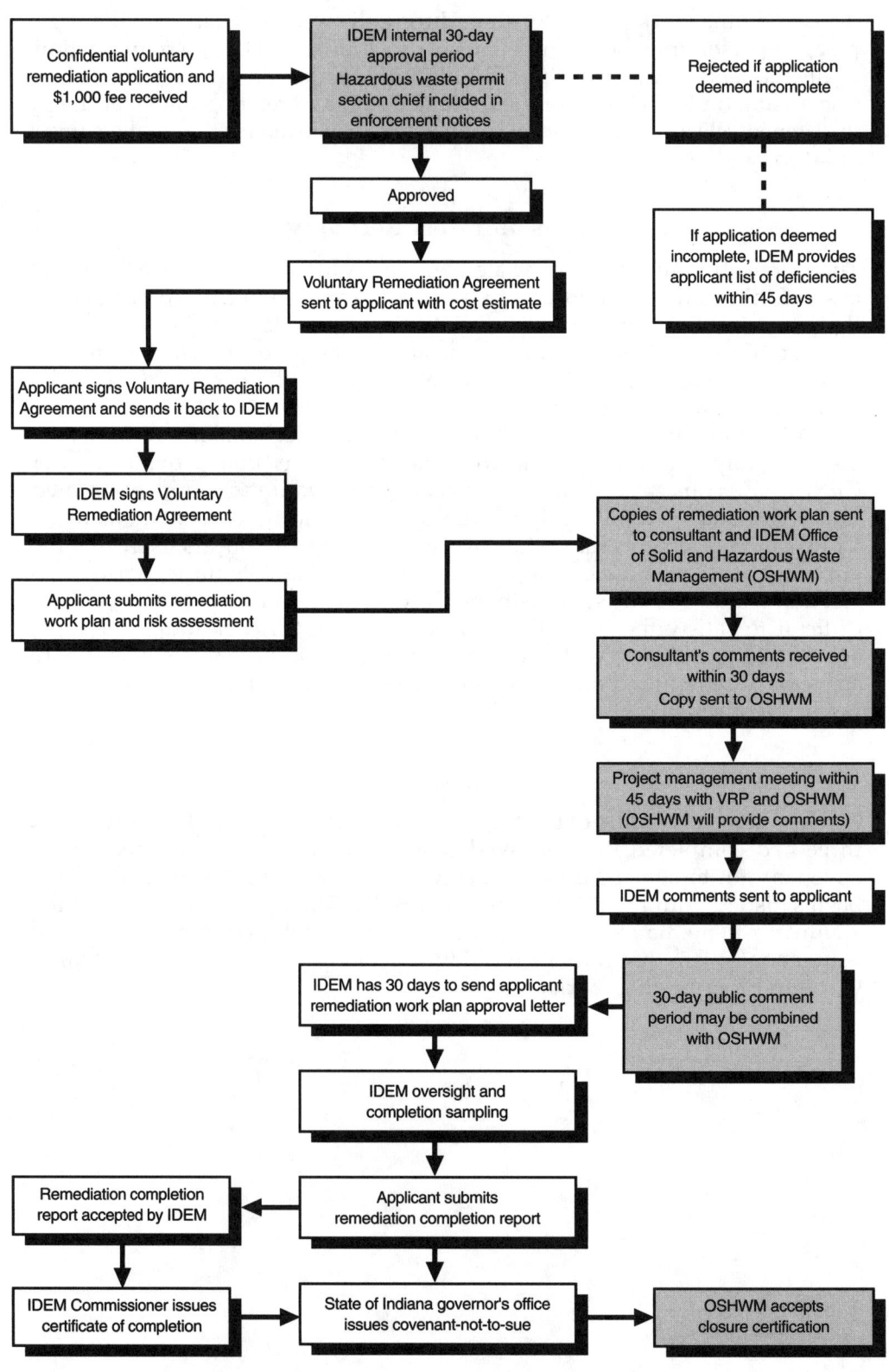

Notes

1. The Voluntary Remediation Act has been amended several times: Public Law 87-1992, section 3; amended in 1993—Public Law 160-1993, section 2; amended in 1994—Public Law 1-1994, section 73, Public Law 84-1994, section 3; and amended in 1995—Public Law 170-1995, section 1. This entire act was recodified at section 13-25-5 of the Indiana Code, effective July 1, 1996. Previously, it was codified at section 13-7-8.9 of the Indiana Code.
2. *See generally* IND. CODE § 13-25-5-2 (1996).
3. *Id.* § 13-25-5-4.
4. *Id.* § 13-25-5-5(a).
5. *Id.* § 13-25-5-6.
6. *Id.* § 13-25-4 (1996).
7. *Id.* § 13-25-4-8(c).
8. *Id.* § 13-25-4-8(d).
9. *Id.* § 13-25-5-23 (1996); *also see* the definition for "remediation" set forth in the former codified version at section 13-7-8.9-5 of the Indiana Code.
10. INDIANA DEP'T OF ENVT'L MANAGEMENT, OFFICE OF ENVT'L RESPONSE, RESOURCE GUIDE TO VOLUNTARY REMEDIATION PROGRAM (1996).
11. 42 U.S.C. § 6901 *et seq.* (1995)
12. IND. CODE § 13-25-5-2 (1996).
13. *Id.* § 13-25-5-4.
14. *Id.* § 13-25-5-7.
15. *Id.* § 13-25-5-8.
16. *Id.* §§ 13-25-5-9 through 13-25-5-13.
17. *Id.* §§ 13-25-5-14, 13-25-5-15.
18. *Id.* § 13-25-5-16.
19. *Id.* § 13-25-5-18.
20. *Id.* § 13-25-5-11.
21. *Id.* §§ 13-25-5-18, 13-25-5-20.
22. *Id.* § 13-25-5-18.
23. *Id.*
24. *Id.*
25. *Id.*
26. *Id.* § 13-25-5-20.
27. *See* INDIANA DEP'T OF ENVT'L MANAGEMENT, *supra* note 10.
28. IND. CODE 13-25-5-18(e) (1996).
29. *Id.* § 13-25-5-8.
30. 42 U.S.C. § 9607(a)(4)(B) (1995).
30. *Id.* The National Contingency Plan provides, in relevant part, as follows:

For the purpose of cost recovery under section 107(a)(4)(B) of CERCLA: (i) a private party response action will be considered "consistent with the NCP" if the action, when evaluated as a whole, is in substantial compliance with the applicable requirements in [40 CFR § 300.700(c)(5) and (6)] and results in a CERCLA-quality cleanup.

40 C.F.R. § 300.700(c)(3) (1996).

CHAPTER 29

Kentucky

HENRY L. STEPHENS, JR.

Although initially lagging behind other states in promoting the use of brownfields through appropriate legislation, the 1996 Kentucky General Assembly's passage of Senate Bill 219 (codified at sections 224.01-450–465 of the Kentucky Revised Statutes) marks the beginning of Kentucky's efforts to promote redevelopment of environmentally contaminated properties. As a consequence, the legislation is a potentially powerful tool in securing the beneficial reuse of urban industrial properties.

Summary of Major Provisions

The major provisions of sections 224.01-450–465 of the Kentucky Revised Statutes include the following:

1. Public entities may apply to the Kentucky Natural Resources and Environmental Protection Cabinet (NREPC) for a "no-further-remediation" (NFR) letter upon the submission of, *inter alia*, a proposed plan to remediate environmental contamination and apply appropriate use restrictions applicable to the property after remediation.
2. The NFR letter signifies that the public entity has been released from further responsibilities under the remediation plan, as well as any further responsibilities for remediation of the site under Kentucky law.
3. The NFR letter serves as prima facie evidence that the site does not constitute a threat to human health or the environment if the site is used in accordance with the terms of the NFR letter.
4. The NFR letter may be used by virtually any assignee, transferee, or successor of the public entity in whose favor the letter was issued.

Background Surrounding the Passage of Kentucky's Brownfields Legislation

In addition to any remedial obligations a responsible party may have in connection with contaminated property pursuant to the Comprehensive Environmental Response, Compensation, and Liability Act of 1980 (CERCLA),[1] Kentucky state law imposes remedial obligations, principally contained in section 224.01-400 of the Kentucky Revised Statutes. Section 224.01-400(18) contains the grist of Kentucky's environmental remedial obligations by providing as follows:

> [A]ny person possessing or controlling a hazardous substance, pollutant or contaminant which is released to the environment, or any person who caused a release to the environment of a hazardous substance, pollutant or contaminant shall characterize the extent of the release as necessary to determine the effect of the release on the environment, and shall take actions necessary to correct the effect of the release on the environment. Any person required to take action under this subsection shall have the following options:
> (a) demonstrating that no action is necessary to protect human health, safety and the environment;
> (b) managing the release in a manner that controls and minimizes the harmful effects of the release and protects human health, safety and the environment;
> (c) restoring the environment through the removal of the hazardous substance, pollutant or contaminant; or
> (d) any combination of paragraphs (a) through (c) of this subsection.[2]

Passed in the 1992 session of the Kentucky General Assembly, section 224.01-400(18) is intended to make clear that persons required to take action as a result of a release of a hazardous substance, pollutant, or contaminant may do so through (1) demonstrating that no action is necessary, or (2) proposing a risk-based approach to environmental management of the site, rather than restoring the environment to historical "background levels" that NREPC had insisted upon before 1992. Accordingly, section 224.01-400(21) confirms the existence of the risk-based remediation option by stating as follows:

> [A] person required to take action under subsection (18) of this section who does not restore the environment through removal of the hazardous substance, pollutant, or contaminant in accordance with subsection (18)(c) of this section shall demonstrate to the cabinet that the remedy is protective of human health, safety, and the environment, by considering the following factors:
> (a) The characteristics of the substance, pollutant, or contaminant, including its toxicity, persistence, environmental fate and transport dynamics, bioaccumulation, biomagnification, and potential for synergistic interaction and with specific reference to the environment into which the substance, pollutant, or contaminant has been released;
> (b) The hydrogeologic characteristics of the facility and the surrounding area;
> (c) The proximity, quality, and current and future uses of surface water and groundwater;

(d) The potential effects of residual contamination of potentially impacted surface water and groundwater;
(e) The chronic and acute health effects and environmental consequences to terrestrial and aquatic life of exposure to the hazardous substance, pollutant, or contaminant through direct and indirect pathways;
(f) An exposure assessment; and
(g) All other available information.[3]

NREPC promulgated, in minute specificity, detailed regulations describing the obligations of parties seeking approval for risk-based remediation alternatives,[4] notwithstanding the absence of any authority in section 224.01-400 enabling it to do so. Immediately following the promulgation of these proposed regulations, industry groups challenged NREPC's authority to promulgate such regulations, as well as the specific content of the regulations.[5] In addition, industry groups filed several legislative proposals immediately after the 1996 Kentucky General Assembly convened in January 1996. These legislative proposals sought to restrain NREPC's authority to impose tedious site characterization requirements upon those seeking risk-based remediation options, such as the mandated preparation of an ecological (as opposed to human health) risk assessment. As a result of negotiations between industry groups and NREPC, in the spring of 1996, NREPC withdrew its proposed regulations implementing section 224.01-400 in exchange for withdrawal of pending legislation by industry groups. As a consequence, section 224.01-400 remains as it was before the 1996 Kentucky General Assembly session, leaving it to NREPC officials to detail the requirements of risk-based remediation on a site-by-site basis. Nevertheless, NREPC continues to insist upon ecological risk assessments, even on urban brownfields sites, the effect of which is to impose more rigid remediation requirements than the ultimate use of such sites would dictate.

Rationale for Kentucky's Brownfields Legislation

It was against this historical backdrop that the 1996 session of the Kentucky General Assembly considered and passed Senate Bill 219, embodied in sections 224.01-450–465 of the Kentucky Revised Statutes. Championed by the Louisville Chamber of Commerce, this legislation—according to many—would enable public entities to play a more direct role in securing remediation of sites within their borders.[6] Further, the new legislation conveys the subtle message that public entities may be more successful in securing reasonable risk-based remediation obligations than would private industry. Before passage of the new legislation, NREPC did not issue NFR letters and, of course, the legislation limits the issuance of such letters to "public entities."

Implementation of Kentucky's Brownfields Legislation

Kentucky's brownfields legislation, contained in sections 224.01-450–465 of the Kentucky Revised Statutes provides that only "public entities" may apply to NREPC for NFR letters.[7] "Public entities" are defined as the Commonwealth of Kentucky, a county, a city, an urban county government (such as Lexington-Fayette County), a chartered county government (or any of the agencies or departments of the foregoing entities), or a nonprofit corporation organized pursuant to the provisions of section 58.180 of the Kentucky Revised Statutes expressly for the purpose of securing financing for public projects.[8] The legislation provides that NFR letters may be issued only for properties, owned by public entities, upon which a release of a hazardous substance, pollutant, or contaminant has occurred.[9]

As title to relatively few contaminated properties currently resides in public entities throughout the Commonwealth of Kentucky, implicit in the implementation of sections 224.01-450–465 is the notion that private entities holding title to contaminated properties may wish convey such properties to a public entity pursuant to contractual provisions, whereby the public entity, upon receiving an NFR letter, would reconvey the property back to the grantor. In such a case, sections 224.01-450-465 provide that the NFR letter issued to the public entity is akin to a covenant running with the land, which may benefit any subsequent mortgagee, trustee, assignee, transferee, or successor-in-interest of a deed conveying such property from the public entity.[10]

Accordingly, persons wishing to take advantage of Kentucky's brownfields legislation prospectively should contact the economic-development authorities in the area of the Commonwealth that is the desired location for the new facility, to determine whether any economic-development agencies currently hold title to any attractive properties.[11] In the event that no suitable locations are presently owned by public authorities, the current owner and the prospective purchaser of an ideal brownfields site should contact local public authorities, with a view toward having a public entity take title to the property for purposes of obtaining an NFR letter and subsequently convey title back to either the current or prospective owner. As either the prospective purchaser or the current owner will, in all likelihood, need to prepare the appropriate documentation for the public entity to obtain the NFR letter, a discussion of the documents the public entity must file follows.

NFR Letters

Obtaining an NFR Letter

To obtain an NFR letter pursuant to the provisions of section 224.01-460 of the Kentucky Revised Statutes, the public entity (with the likely assistance of either the current or prospective purchaser) must submit to NREPC an application for an NFR letter. The application must contain the following information: (1) a legal description of the property, (2) a copy of the deed to the property, (3) an environmental site assessment, sufficient to characterize the extent of any contamination of the site, (4) a proposed plan to remediate the environmental contamination on the site, and (5) the proposed use of the property intended by the public entity after obtaining the NFR letter.[12] Although section 224.01-400(21) does not allow for consideration of the proposed use of the property in determining appropriate remediation obligations, implicit in the brownfields legislation at section 224.01-460(1)(e) is the notion that properties with use restrictions in place following remediation may be required to undertake less onerous remediation than properties available for maximum human exposure, such as residential properties.

NREPC will put applications for NFR letters out for public comment and thereafter either deny the application or enter negotiations with the public entity to modify the application in a way that will make it acceptable to NREPC.[13] If the public entity and NREPC successfully negotiate an agreement on the proposed remediation plan, NREPC approves the amended application.[14] When the remediation plan is satisfactorily completed, NREPC issues the NFR letter authorized by sections 224.01-450–465.[15]

Effect and Contents of NFR Letters

NREPC's issuance of an NFR letter signifies that the public entity has been released from further responsibilities under the remediation plan, as well as from any further responsibilities under section 224.01-400 to undertake any other remediation of the site. The issuance of an NFR letter is also to be considered prima facie evidence that the site does not constitute a threat to human health and the environment and does not require additional remediation under section 224.01-400, if the site is used in accordance with the terms of the letter.[16]

The NFR letter itself, pursuant to section 224.01-465(2), must contain (1) an acknowledgment that the requirements of the remediation plan were satisfied or are being satisfied,[17] (2) a description of the location of the property by a reference to a legal description, (3) any additional monitoring requirements or land-use limitations imposed as a result of the remediation efforts, (4) language signifying that the letter constitutes a release from further responsibilities under section 224.01-400, and stating that the letter will be prima facie evidence that the site does not constitute a threat to human health or the environment if the site is used in accordance with the terms of the NFR letter, (5) a prohibition against the use of the property by the public entity in a manner inconsistent with any land-use limitation imposed as a result of the remediation efforts, and (6) provisions requiring that any deed conveying the property to a third party contain deed restrictions limiting the use of the land in accordance with the remediation plan.[18] Finally, the NFR letter must contain a description of any preventative, engineering, and institutional controls required in the remediation plan, and a caveat providing that failure to manage and maintain the controls in full compliance with the terms of the remediation plan may result in voidance of the NFR letter.[19] Once issued, the public entity must record the NFR letter with the county clerk in the county in which the property is located.[20]

As a safety valve, section 224.01-465(4) allows NREPC to void an NFR letter if the remediation plan is not carried out or if NREPC determines that any facts upon which the NFR letter was based were unknown at the time of its issuance or were known and either not disclosed or falsely misrepresented.[21]

Entities That May Secure the Benefits of NFR Letters

Kentucky's brownfields legislation expressly contemplates that the NFR letter is akin to a covenant running with the land, and therefore applies to the property for which it is issued in favor of the following persons: (1) the public entity to which the NFR letter was issued, (2) any mortgagee or trustee, or their assignee, transferee, or successor-in-interest of a deed of trust of the public-entity property, (3) any successor-in-interest of the public entity, (4) any transferee of the public entity, whether the transfer was by sale, charitable gift, bequest, bankruptcy proceeding, petition, settlement, or adjudication of any civil action, and (5) any financial institution, or its successor-in-interest, which, after the date the NFR letter was issued, acquired the ownership, operation, management, or control of the property through foreclosure, or under the terms of a security interest held by the financial institution, or under the terms of an extension of credit made by the financial institution.[22]

Conclusion

Kentucky's brownfields legislation, contained in sections 224.01-450–465 of the Kentucky Revised Statutes is very new. As a consequence, the manner in which it is to be implemented by public entities and NREPC is open to speculation. Whether NREPC will require less remediation on sites owned by public entities, the subsequent use of which is restricted by deed, remains to be seen. Further, whether the legislation will have any real effect outside the urban areas of Louisville-Jefferson County, Lexington, and northern Kentucky is unclear. Nevertheless, the enactment of brownfields legislation is a watershed event in the history of Kentucky environmental law, because for the first time, public entities and their transferees have some security from the threat of future remediation responsibilities mandated by section 224.01-400. However, persons obtaining NFR letters should remain cognizant that an NFR letter from Kentucky NREPC in no way eliminates or minimizes any federal remediation responsibilities that might be applicable to the site under the Federal Resource Conservation and Recovery Act[23] or CERCLA.[24]

Kentucky Brownfields Program

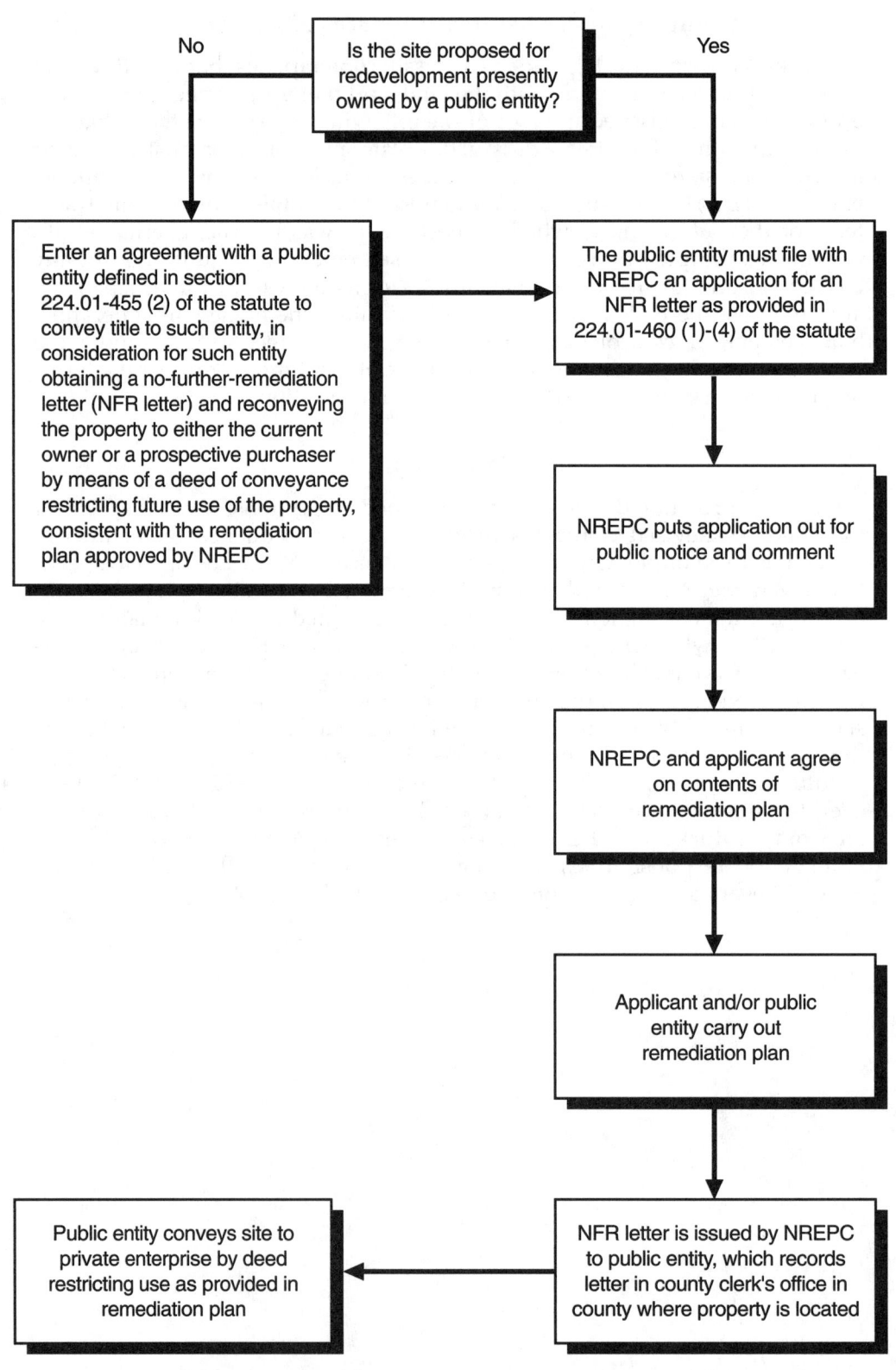

Notes

1. 42 U.S.C. § 9601, *et seq.* (1996).
2. KY. REV. STAT. ANN. § 224.01-400(18) (Michie 1996 Supp.).
3. *Id.* § 224.01-400 (21).
4. *See* 22 KY. ADMIN. REG. § 13 (1995) (to be codified at 401 KY. ADMIN. REGS. 100:050 *et seq.*, proposed July 1995.
5. *See* Associated Indus. of Kentucky, Inc. v. NREPC, Civil Action No. 95-CI-1488 (Franklin Cir. Ct., Oct. 12, 1995).
6. *See* KENTUCKY CHAMBER COMMERCE, Business Bulletin No. 8, at 1 (1996).
7. *See* KY. REV. STAT. ANN. § 224.01-460(1) (Michie 1996 Supp.).
8. *Id.* § 224.01-455(2).
9. *Id.* § 224.01-455(1).
10. *See id.* § 224.01-465(3)(b).
11. Persons interested in obtaining the list of economic-development agencies in the Commonwealth of Kentucky should contact the Economic Development Cabinet, at 502/564-7670.
12. *See* KY. REV. STAT. ANN. §§ 224.01-460(1)(a)–(e) (Michie 1996 Supp.).
13. *See* KY. REV. STAT. ANN. § 224.01-460(2).
14. *Id.*
15. *But see infra* note 18 and accompanying text.
16. *See* KY. REV. STAT. ANN. § 224.01-465(1) (Michie 1996 Supp.).
17. This provision seems to imply that NFR letters may be issued before the actual completion of all remediation, perhaps when a long-term remedy is approved. *See id.* § 224.01-465(2)(a).
18. *See id.* § 224.01-465(2)(a)–(e).
19. *Id.* § 224.01-465(2)(f).
20. *Id.* § 224.01-465(5).
21. *Id.* § 224.01-465(4).
22. *See id.* §§ 224.01-465(3)(a)–(e).
23. 42 U.S.C. § 6972 *et seq.* (1996).
24. 42 U.S.C. § 9601 *et seq.* (1996).

CHAPTER 30

Maine

JAMES T. KILBRETH
JULIET T. BROWNE

In place since June 1993, Maine's voluntary response action program (Voluntary Response Program, or Program)[1] establishes a process whereby a person associated with contaminated property may undertake investigation and cleanup activities at the site without the risk of incurring liability under the Maine environmental laws. Under the Maine Program, an owner, developer, purchaser, or the like who implements a state-approved cleanup plan is exempt from liability under the state's environmental laws. Eligible participants are protected from state enforcement and cost-recovery actions, as well as third-party contribution claims. This protection from liability creates an incentive for developers, purchasers, or other would be-investors in property with known or suspected contamination to develop or reuse properties that might otherwise be of no economic or community value.

Summary of Major Provisions

The highlights of Maine's Program include the following:

- Creation of a unit within the Department of Environmental Protection (DEP) to assist in the development and implementation of appropriate response actions for contaminated properties
- Liability protection, to persons who participate in the Program, in the form of a "no-action" assurance agreement with DEP (These agreements generally provide that DEP will not take action, under the state environmental statutes, against persons who voluntarily clean up a property pursuant to a state-approved plan.)
- Contribution protection to persons who participate in the Program
- Extension of liability protection to other parties associated with the property, including lenders and developers, as long as they are not otherwise responsible for the contamination

- Liability protection for partial cleanups, as long as the cleanup protects human health and the environment and satisfies other statutory requirements

Mechanics of Participating in the Maine Program

The State Agency's Role

The Voluntary Response Program is implemented by DEP. When it enacted the Program, the Maine legislature did not specify detailed procedural requirements or dictate substantive criteria for participating in, and obtaining the benefits of, the Program. For example, the statute does not set any limits on the types of sites eligible for cleanup, nor does it articulate cleanup standards.[2] Instead, the legislature vested DEP with substantial discretion to carry out the operational aspects of the Program.[3] As of June 1996, DEP had not enacted any regulations under the Program.

At the same time it enacted the Program, the legislature provided for the creation of a technical services unit within DEP to provide assistance and oversight in the development and implementation of appropriate response actions under the Program.[4] Personnel within this unit review and comment on proposed cleanup plans and provide additional technical assistance upon request. Generally, DEP becomes involved in a project at an early stage, in an effort to guide the process toward successful completion.

Any person involved in a real estate transaction may seek assistance from DEP in determining whether the subject property has been the site of a discharge, release, or threatened release.[5] Persons who participate in the Program or otherwise seek technical assistance must pay an initial nonrefundable fee of $500.[6] Additionally, DEP may charge an hourly fee, not to exceed $50 per hour per person, for any review or oversight assistance provided in excess of those services covered by the initial $500 fee.[7] In general, most persons participating in the Program do not pay costs in excess of the initial $500 fee. However, if a proposed cleanup involves complex technical issues, the participant may incur additional oversight costs.

Site Eligibility Limitations

There are no specific statutory or regulatory provisions limiting the types of sites eligible for cleanup under the Program.[8] Thus, any contaminated site, including one contaminated with petroleum or petroleum by-products, is potentially eligible for cleanup under the Program.[9]

However, there may be limits on the extent to which sites that are regulated under federal or other state programs may participate in, or obtain the full benefits of, the Voluntary Response Program. For example, generators of hazardous wastes, as well as facilities that treat, store, or dispose of hazardous waste, may be regulated under subtitle C of the federal Resource Conservation and Recovery Act program (RCRA).[10] This program comprises a myriad of regulations that govern the handling, storage, transportation, and disposal of hazardous substances, and includes provisions governing the liability for cleanup of contamination resulting from regulated activities.[11] Additionally, sites with certain types of underground storage tanks may fall within the jurisdiction of subtitle I of RCRA. This RCRA program regulates the construction, operation, and closure of certain underground storage tanks containing either petroleum or hazardous substances, and includes provisions addressing the liability of owners and operators for releases from such tanks and their responsibility for appropriate corrective actions.[12]

Maine is currently authorized to administer both the subtitle C and subtitle I RCRA programs.[13] However, its continued authorization to do so is dependent upon it using standards that are at least as stringent as the federal requirements.[14] Thus, DEP generally cannot provide a release or take other action under the Voluntary Response Program that would be inconsistent with applicable provisions of the federal RCRA program, or that would undermine the state's authority to administer that program. Of particular relevance in this context are those RCRA provisions dictating corrective actions at regulated facilities and the persons potentially responsible for such measures. For example, in the event of a release, the owner or operator of an RCRA-regulated facility may be required to take corrective action that satisfies certain regulatory standards, regardless of when the hazardous substances at issue were placed on the property.[15] Additionally, as discussed below, under Maine law a secured lender not otherwise responsible for contamination at a site is generally exempt from liability under the state's environmental statutes.[16] However, there may be circumstances, particularly following foreclosure, under which a lender may be considered an "owner" under RCRA, and therefore potentially liable for cleanup or other costs.[17]

The determination of whether a particular RCRA or other regulated site is eligible for participation in the Voluntary Response Program is made by DEP on a case-by-case basis.[18] If DEP determines that the site is eligible for participation in the Program, it nonetheless might tailor the no-action assurance agreement to satisfy the relevant requirements of RCRA or other potentially applicable regulations.

Submission and Approval of a Cleanup Plan

Any person who elects to participate in the Voluntary Response Program must submit for DEP review and approval a proposed plan for addressing contamination at the site.[19] A person may submit, and DEP may approve, either a complete or partial cleanup plan. A complete cleanup is one that removes or remedies all known contamination at the site.[20] A partial cleanup addresses some, but not all, contamination on the property.[21] As discussed below, the liability protection provided in connection with a partial cleanup may be more limited than that provided with a complete cleanup.

A cleanup plan submitted for approval under the Program must include an investigative report prepared by a duly qualified professional, which analyzes the nature and extent of contamination at the site and identifies the methods of investigation used, the analytical results obtained, and the professional's evaluation of the information.[22] The statute does not define a "duly qualified professional."[23] However, the person must possess the requisite skills to analyze the particular type of contamination at issue. Typically, DEP requires certified geologists or engineers to be involved in the preparation of cleanup plans.

The statute also does not specify a procedure or impose time limits for reviewing and approving proposed cleanup plans.[24] Typically, DEP reviews the proposed plan and provides substantive comments and feedback to the participant before the plan is finalized. When the proposed plan is complete, it is submitted to DEP for final approval. If a plan is not approved, then a participant is not entitled to the liability protection afforded under the Program.[25] Moreover, if contamination at the site poses a threat to human health or the environment, DEP could identify responsible parties and institute necessary response actions.[26]

The statute does not include any public-notice requirements or provisions for public participation in the review and approval of voluntary cleanup plans.

Issuance of No-Action Assurance Letter

Upon approval of a voluntary cleanup plan, DEP issues a no-action assurance letter, in which it agrees not to take action under Maine environmental statutes against persons implementing the plan, as long as the plan is implemented pursuant to its terms and with the exercise of due care.[27] DEP may condition its issuance of no-action assurance on any reasonable terms, including a requirement that the requestor allow DEP access to the property and permission to undertake activities there, such as installing borings and wells.[28] Additionally, if institutional controls are in place at the site, DEP typically conditions its no-action assurance on the continued maintenance of such controls. Any conditions imposed by DEP must apply to, and be binding on, successors and assigns of the owner.[29] Thus, DEP generally requires that the terms of the no-action assurance letter be recorded in the county where the property is located.[30]

A partial cleanup plan must satisfy the following minimum requirements: (1) the cleanup must be consistent with any proposed reuse or development of the site and must protect public health and the environment, (2) the cleanup must not exacerbate or substantially increase the cost of remediating contamination that is not the subject of the partial cleanup, and (3) the owner of the property must agree to cooperate with DEP and others to avoid any action that would interfere with the cleanup.[31] Additionally, liability protection may be limited to the matters addressed in the partial cleanup plan.[32]

Persons participating in the Program must notify DEP when the cleanup is completed. Upon demonstration that the response plan has been fully implemented in accordance with its terms, DEP will issue a certificate of completion, acknowledging that the participant has satisfactorily completed the cleanup.[33]

Cleanup Standards

There are no cleanup standards specifically governing voluntary cleanups.[34] Instead, DEP establishes cleanup levels and approves remedial actions on a case-by-case basis, through a process that includes both a risk assessment and risk management component, and which will protect human health and the environment.[35]

The state-issued guidance on risk assessments at hazardous substance sites (Guidance Manual)[36] provides some insight on factors DEP might consider in setting cleanup levels and determining appropriate response actions under the Voluntary Response Program. The Guidance Manual provides that the risk-assessment component of determining appropriate response actions consists of "evaluating the toxic properties of chemical(s) and the conditions of human exposure to it to determine the likelihood that exposed individuals will be adversely affected."[37] Thus, the Guidance Manual includes specific input on hazard identification, exposure assessment, dose-response assessment, and risk characterization. Generally, the state guidance is intended to be consistent with federal risk-assessment guidance under the Comprehensive Environmental Response, Compensation, and Liability Act of 1980 (CERCLA), although there are some differences.[38]

DEP does not always require preparation of a formal risk assessment in connection with a voluntary cleanup. If data on the risks posed by contamination at a site is readily available, such as when pertinent information is obtained in connection with another site, DEP may set cleanup levels or approve response actions without the use of a formal risk assessment. If, however, a formal risk assessment is required, it must be conducted in accordance with state-issued guidance on risk assessments.[39]

DEP's determination of target cleanup levels and approval of proposed response actions not only takes into account the risks posed by particular contaminants at a site, but also reflects a policy judgment about the most appropriate means of controlling or eliminating those risks judged to be significant.[40] Thus, in approving response actions, DEP may consider, for example, any of the following: the environmental impact of the site, the cost and technical feasibility of remedial options, public acceptance of alternatives, other regulatory standards, and the intended future use of the site.[41] Moreover, when appropriate, DEP may approve the use of institutional controls, such as deed restrictions, fencing, a prohibition on the installation and use of wells, or a prohibition or limitation on surface excavation, as a means of reducing the risks posed by contamination at a site.[42] When institutional controls are used, the no-action assurance is conditioned on the continued maintenance of such controls and the recording of the agreement.[43]

The DEP regulations governing the investigation and remediation of leaks, spills, or other discharges from underground oil facilities provide additional guidance on factors DEP may consider in setting cleanup standards under the Voluntary Response Program.[44] These regulations specify reporting, investigation, and corrective-action requirements. They require, *inter alia*, the immediate removal of free product, remediation of all saturated soils as well as soils contaminated above an action level established by DEP on a case-by-case basis, and protection of groundwater.[45]

Liability Protection under the Program

Persons Participating in the Program

A person who undertakes and completes a cleanup in accordance with an approved voluntary response action plan is protected from liability under the Maine environmental statutes. Specifically, a person complying with the dictates of the Program "may not be deemed a responsible party," and is not subject to state orders or other enforcement proceedings under state environmental laws concerning the subject property.[46] Protection from state action is memorialized in a written no-action assurance letter, in which the state agrees not to take action under the state environmental laws against persons carrying out an approved cleanup plan, as long as the plan is implemented pursuant to its terms and with the exercise of due care. The no-action assurance is issued at the time the plan is approved.[47] Once the plan is satisfactorily completed, DEP issues a certificate of completion that acknowledges that the cleanup plan has been satisfactorily completed.[48]

Persons implementing approved response action plans are also protected from liability in actions initiated by third parties. Generally, liability for cleanup and other related costs under Maine's environmental laws is premised on a finding that the person is a "responsible party."[49] Because participating persons may not be deemed responsible parties, they are generally protected from liability in actions initiated by third parties as well as the state. Moreover, Maine's Voluntary Response Program specifically provides that persons who are implementing or have implemented an approved cleanup plan are not liable for claims for contribution regarding the site.[50] Thus, persons complying with the dictates of the Program are afforded liability protection from state enforcement or cost-recovery actions, as well as third-party claims for contribution.

Extension of Liability Protection to Third Persons

Liability protection extends to other persons associated with the property, as long as those persons are not otherwise responsible for the contamination. Protected parties include, for example, persons financing the cleanup, acquisition, or development of a property,[51] lenders or fiduciaries who arrange for the cleanup,[52] persons who seek to acquire or develop property and arrange for the cleanup,[53] and successors or assigns of protected persons.[54] Additionally, a person who causes, contributes to, or exacerbates a discharge or release while implementing an approved cleanup plan is not liable under the state environmental laws, provided that the plan was being implemented with due care and the release or discharge is removed or remediated to DEP's satisfaction.[55]

Responsible Parties

An owner or operator who is a responsible party is afforded limited liability protection under the Program. If that person undertakes and completes a plan that fully removes or otherwise remedies all known contamination at the property, he or she is protected from further cleanup requirements, but may be subject to penalties, fines, or natural-resource damages.[56] Persons responsible for contamination at a site that is the subject of a partial cleanup plan that does not require removal or remediation of all known contamination are not eligible for liability protection under the Program.[57] Similarly, persons who cause, contribute to, or exacerbate contamination that was not remedied pursuant to an approved cleanup plan are not eligible for liability protection.[58]

Finally, persons who obtain approval of a voluntary response action plan by means of fraud or intentional misrepresentation are not afforded any liability protection.[59]

Liability Protection for Lenders and Fiduciaries

At the same time it implemented the voluntary response action program, the Maine legislature added a fiduciary exemption to the strict-liability provisions of Maine's environmental laws, and amended the general lender-liability provisions to make them more consistent with the 1992 United States Environmental Protection Agency (USEPA) lender-liability regulations.[60] Thus, fiduciaries—broadly defined to include a full range of financial and personal representatives—are exempt from liability under state environmental laws unless the fiduciary causes or contributes to the contamination, or fails to take appropriate steps after becoming involved with a contaminated site.[61] These steps include notifying DEP of contamination at the site, allowing DEP reasonable access to the property for the purpose of conducting response actions, and taking reasonable steps to control access and prevent imminent threats to public health and the environment.[62] The fiduciary exemption does not extend to the assets of the trust or estate, or to the grantor.[63]

Similarly, a secured lender who does not cause, contribute to, or exacerbate contamination at the site and otherwise complies with the dictates of the secured-lender exemption is afforded protection from liability under Maine's environmental statutes.[64] The exemption requires that the lender not "[participate] in management" of the site before acquiring ownership, "[act] diligently to sell or otherwise divest" the property after acquiring ownership, notify DEP of contamination at the site, allow DEP access to the site for purposes of undertaking response actions, and take reasonable steps to control access and prevent imminent threats to public health and the environment.[65] The law presumes that a secured lender is acting diligently to sell or otherwise divest during the first eighteen months after taking possession, and allows up to forty-two additional months for disposal of the property.[66] Also, unlike its federal counterpart, the Maine secured-lender exemption applies to petroleum-product contamination as well as hazardous-substance contamination.[67]

Liability Protection in Connection with Federal Claims

By its terms, the no-action assurance provided under the Program protects a person from liability only under the Maine environmental statutes. It does not provide protection from claims based on federal law.

However, persons concerned about potential federal liability could seek a "comfort letter" from USEPA Region I. Such letters clarify for a site owner, lending institution, or other interested party that the property in question is not currently considered to be an appropriate candidate for listing on the National Priorities List (NPL), and therefore no further steps will be taken to list the site.[68] Such a letter is "not a release from possible CERCLA liability, but [is] an attempt to clearly and realistically describe [USEPA's] future intentions with regard to the property in question."[69]

The state plays a central role in determining whether a comfort letter is warranted, and Region I will issue such a letter only upon the written request of the state.[70] Upon assurances by the state that the site is not an appropriate candidate for listing on the NPL and that the site either does not pose a substantial threat to human health or the environment, or is being addressed under the Maine waste-site cleanup program, USEPA will generally issue the requested comfort letter. However, USEPA retains the right to take action if circumstances change, such as when conditions deteriorate or subsequent environmental contamination is discovered, or when considerations arise that make a recommendation for listing appropriate. Moreover, a comfort letter would not preclude a third party from bringing an action related to contamination at the site.

Maine is currently discussing with Region I the possibility of obtaining a Memorandum of Agreement that would formalize USEPA's general policy not to take action at a site that is cleaned up under the Program, thereby eliminating the need for site-specific comfort letters.

Even in the absence of a Memorandum of Agreement or a site-specific comfort letter, persons who have implemented an approved cleanup plan may derive some comfort from the fact that USEPA is voluntarily adhering to proposed appropriations language that permits it to list a site on the NPL only with the consent of the governor of the state in which the property is located.[71] Thus, as a practical matter, absent extraordinary circumstances, USEPA is unlikely to take action at a site cleaned up under Maine's Voluntary Response Program.[72]

Miscellaneous Liability Concerns

Under Maine law, there is no specific statutory protection against the admissibility or discovery of material provided to DEP in connection with a voluntary cleanup. Typically, information submitted to state agencies is available to the public.[73] However, there is a statutory mechanism for protecting certain information submitted to DEP from public disclosure. For example, persons may designate as "confidential" hazardous-waste information submitted to DEP pursuant to the waste-management provisions of the state environmental laws.[74] Upon a showing that the designated information is not otherwise publicly available and its disclosure would impair the submitter's competitive position, DEP may protect the information from public disclosure.[75]

Financial Incentives under the Program

Maine has not enacted any specific financial provisions to induce persons to participate in the Voluntary Response Program. However, the Program itself creates a mechanism for eliminating certain liabilities that might otherwise extend to persons interested in purchasing, developing, or otherwise reusing contaminated property. As a result, implementation of an approved cleanup plan may eliminate the stigma attached to contaminated properties and may even create a "value added" market for real estate that has received a certificate of completion from DEP.

Recovery of Cleanup Costs

There are no specific provisions under Maine law governing the right of a participant in the Program to recover the costs of conducting a voluntary cleanup. Therefore, any such recovery would be dictated by broader principles governing cost-recovery actions. Generally, a person who incurs cleanup costs or other damages from environmental contamination may seek redress either through a statutory action, when one is available, or based on common-law principles.

For example, Maine's Hazardous Waste and Hazardous Matter Control Law[76] provides that persons who dispose or treat hazardous waste in circumstances that endanger the health, safety, or welfare of another are statutorily liable for all resulting damages.[77] It is not necessary to prove negligence in such an action.[78] Thus, when the facts support it, a person undertaking a voluntary cleanup could seek recovery of cleanup costs under this provision. Recoverable damages in such a suit are limited to damages to real estate or personal property, or loss of income as a result of the disposal or treatment of hazardous wastes.[79] For the cleanup costs to be recoverable, they would have to fit within this statutory definition. Additionally, a liable person's damages may be reduced if the disposal or treatment was the result of an act of war or God.[80]

Under Maine's Uncontrolled Hazardous Substance Sites Law,[81] responsible persons are jointly and severally liable for all costs incurred by the state for the abatement, cleanup, or mitigation of the threats or hazards posed or potentially posed by an uncontrolled site.[82] An uncontrolled site is defined as an area or location at "which hazardous substances are ... located if it is concluded by [DEP] that the site poses a threat or hazard to the health, safety or welfare of any persons or to the natural environment and that action [is necessary] to abate, clean up or mitigate that threat or hazard."[83] Although the statute does not expressly provide for a private cost-recovery action, persons liable to the state may seek contribution from other responsible parties.[84] Responsible parties include the owner or operator of the uncontrolled site, prior owners or operators of the site since the time hazardous substances arrived there, and certain persons involved in the transport and handling of hazardous substances that arrived at the site.[85] However, a responsible party is not subject to liability if he or she can establish, by a preponderance of the evidence, that the threats or hazards for which he or she would otherwise be responsible were caused solely by (1) an act of God, (2) an act of war, (3) the act or omission of certain third parties, or (4) any combination of the foregoing.[86]

To the extent a participant can demonstrate that cleanup costs incurred under the Voluntary Response Program come within the purview of the Uncontrolled Hazardous Substance Sites Law, he or she may be able to bring an equitable claim for contribution against other responsible parties.

There also may be circumstances under which a person incurring oil-related cleanup costs could seek recovery pursuant to Maine law governing oil discharges. For example, the Maine Coastal and Inland Surface Oil Cleanup Fund, established pursuant to the Oil Discharge Prevention Act[87], is established for use by DEP to cover cleanup costs for, and to pay damages awarded to persons damaged by, oil spills.[88] Persons suffering property damage or actual economic damages as a result of a prohibited discharge may bring a claim in accordance with specified procedures under the act, and in appropriate circumstances may recover either from the fund or from the person responsible for the prohibited discharge.[89] A person might also be able to seek recovery of oil-related cleanup costs pursuant to similar provisions under the Underground Oil Storage Act.[90]

Finally, a participant also might consider seeking recovery of cleanup costs pursuant to a tort-based theory of liability such as negligence, nuisance, or trespass.[91]

The Voluntary Response Program does not include a cost-recovery provision, and therefore participants will have to resort to other statutory or common-law principles to recover their cleanup costs. However, because there is minimal case law on environmental cost-recovery actions under Maine law, it is unclear how courts will view these efforts. Moreover, given the uncertainty of recovering voluntary cleanup costs, a would-be participant in the Program might want to consider instituting one of the above actions in lieu of undertaking a voluntary cleanup, at least when substantial cleanup costs are expected.

Conclusion

Maine's Voluntary Response Program provides a mechanism for persons associated with contaminated property to implement appropriate corrective measures. Action under the Program is dictated by the Maine DEP, which must approve a cleanup plan before a participant is entitled to the liability protection afforded under the Program. Moreover, DEP is vested with substantial discretion, not only for determining what constitutes an appropriate response action, but also for assessing whether there are considerations that effectively render a particular site ineligible for participation in the Program. To date, DEP and persons associated with contaminated property have worked cooperatively to implement response actions that reflect sound environmental and social policy.

Maine's Voluntary Response Action Program

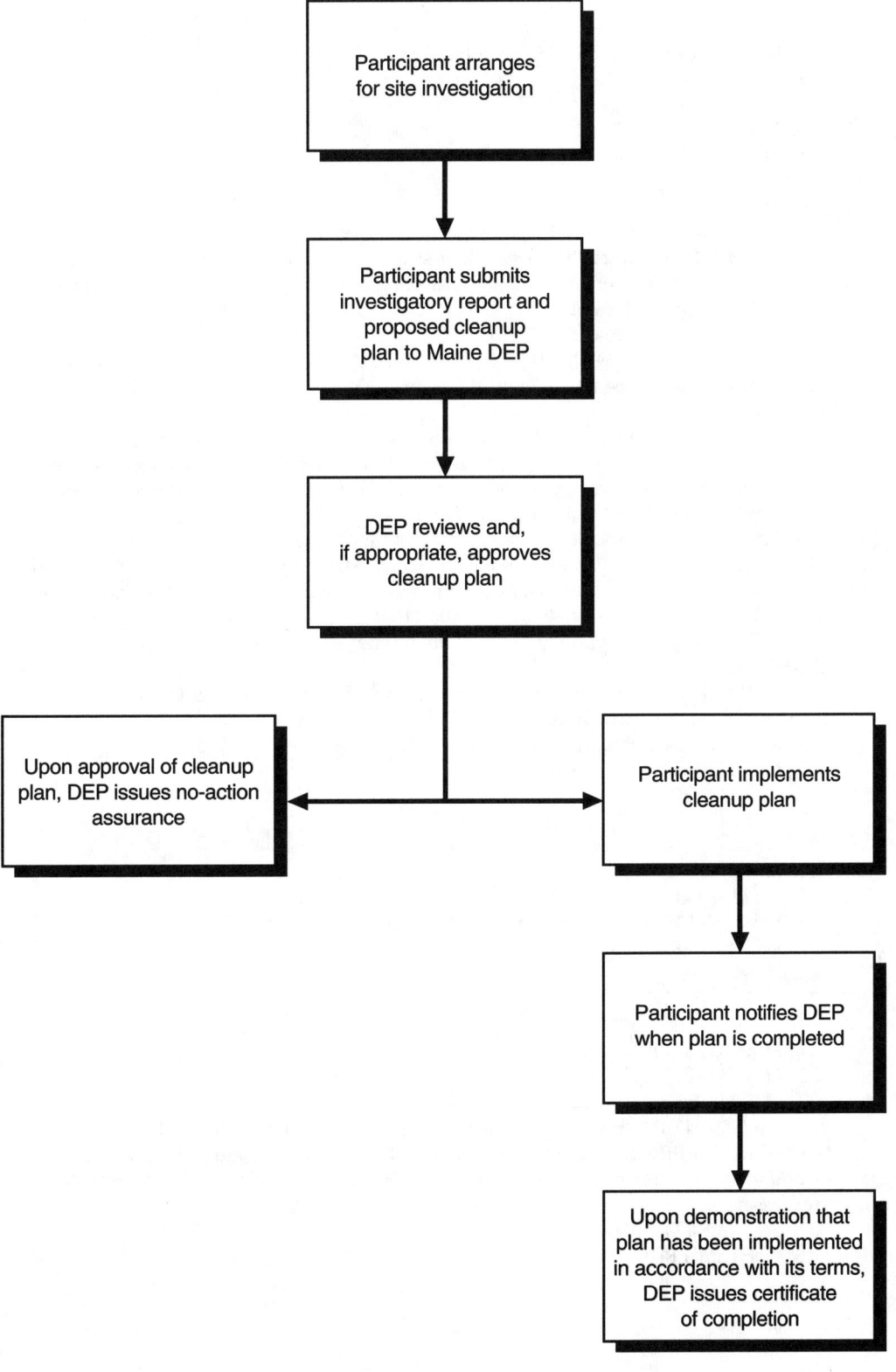

Notes

1. ME. REV. STAT. ANN. tit. 38, § 343-E (West 1995).
2. *Id.*
3. *Id.*
4. *Id.* § 342-15.
5. *Id.*
6. *Id.*
7. *Id.*
8. *See* ME. REV. STAT. ANN. tit. 38, § 343-E (West 1995).
9. Id. § 343-E(1) (referencing "hazardous substance, hazardous waste, hazardous matter, special waste, pollutant or contaminant, including petroleum products or by-products").
10. 42 U.S.C. §§ 6922, 6924 (1995); 40 C.F.R. §§ 262, 264, 270 (1996).
11. 40 C.F.R. §§ 262, 264, 270 (1996).
12. 42 U.S.C. § 6991 (1995); 40 C.F.R. § 280 (1996).
13. 53 Fed. Reg. 16,264 (1988) (subtitle C approval); 57 Fed. Reg. 24,759 (1992) (subtitle I approval).
14. 42 U.S.C. §§ 6926(b), 6929, 6991(c) (1995); 40 C.F.R. §§ 271, 281 (1996).
15. *See, e.g.*, 42 U.S.C. §§ 6924(u)–(v) (1995); 40 C.F.R. subpt. F (1996) (releases from Solid Waste Management Units, closure and postclosure, surface impoundments, and corrective action for Solid Waste Management Units); 40 C.F.R. § 265 (1996).
16. ME. REV. STAT. ANN. tit. 38, §§ 342-B(2), 342-B(4) (West 1995).
17. *Compare* United States v. Fleet Factors Corp., 901 F.2d 1550 (11th Cir. 1990) (lender liability under the Comprehensive Environmental Response, Compensation, and Liability Act [CERCLA]), *cert. denied*, 498 U.S. 1406 (1991); *see also* 60 Fed. Reg. 63,517 (1995) (lender-liability enforcement policy of the United States Environmental Protection Agency [USEPA] CERCLA).
18. DEP has issued no formal guidance on the type of RCRA-regulated sites it considers inappropriate for participation in the Program.
19. ME. REV. STAT. ANN. tit. 38, §§ 343-E(1)–(2), 343-E(4) (West 1995).
20. *Id.* § 343-E(1).
21. *Id.* § 343-E(2).
22. *Id.* § 343-E(4).
23. *See id.* § 343-E.
24. *Id.*
25. *See id.* § 343-E(5), 343-E(9).
26. *See id.* §§ 1361 *et seq.* (Uncontrolled Hazardous Sites Law).
27. *Id.* § 343-E(9).
28. *Id.* §§ 343-E(3), 343-E(9).
29. *Id.* § 343-E(3)(C).
30. *Id.*
31. *Id.* §§ 343-E(2)(A)–(C).
32. *Id.* § 343-E(9).
33. *Id.* § 343-E(5).
34. *See id.* § 343-E. However, DEP is in the process of establishing soil cleanup standards for thirty to fifty common contaminants. These standards are reported to be risk based, and to reflect three exposure scenarios: residential, worker (commercial, industrial), and trespasser.
35. *See* MAINE DEP'T ENVTL. PROTECTION & DEP'T HUMAN SERVS., GUIDANCE MANUAL FOR HUMAN HEALTH RISK ASSESSMENTS AT HAZARDOUS SUBSTANCE SITES ix-xii (1994); *see also*

2 Code Me. R. § 661.12(A)(i) (1995) (cleanup levels for releases from underground oil tanks determined on case-by-case basis, taking into account "the potential for human exposure and for adverse effects on public safety, health and welfare as well as the environment").

36. Maine Dep't Envtl. Protection & Dep't Human Servs., Guidance Manual for Human Health Risk Assessments at Hazardous Substance Sites (1994).

37. *Id*. at ix.

38. *Id*. at x–xi.

39. *Id*. at xi ("when a risk assessment, or any portion of one, is warranted at a given site, DEP and DHS will not consider it unless it is conducted in the manner specified in this document").

40. *See id*. at ix.

41. *See id*. at ix, 1-1, exh. 1.1, 8-2.

42. *See id*. at x.

43. *See* Me. Rev. Stat. Ann. tit. 38, § 343-E(3)(C) (West 1995).

44. 2 Code Me. R. § 691.12 (1995).

45. *See id*.

46. Me. Rev. Stat. Ann. tit. 38, § 343-E(1) (West 1995). (liability protection for persons implementing a partial cleanup plan may be limited to the portion of the property where contamination is removed or otherwise addressed). *See id*. § 343-E(9).

47. *Id*. § 343-E(5).

48. *Id*.

49. *See, e.g., id*. §§ 568, 570, 1367.

50. *Id*. § 343-E(5).

51. *Id*. § 343-E(6)(B).

52. *Id*. § 343-E(6)(C).

53. *Id*. § 343-E(6)(D).

54. *Id*. § 343-E(6)(E).

55. *Id*. § 343-E(6)(F).

56. *Id*. § 343-E(6)(A).

57. *Id*. § 343-E(7)(B).

58. *Id*. § 343-E(7)(A).

59. *Id*. § 343-E(7)(C).

60. USEPA's regulations were subsequently vacated in *Kelley v. United States EPA*, 15 F.3d 1100 (D.C. Cir.), *reh'g denied*, 25 F.3d 1088 (1994), *cert denied sub nom* American Bankers Ass'n v. Kelley, 115 S. Ct. 900 (1995). However, USEPA recently articulated an enforcement policy that adopts as guidance the 1992 lender-liability rule vacated in *Kelley*. *See* 60 Fed. Reg. 63,517 (1995).

61. Me. Rev. Stat. Ann. tit. 38, §§ 342-B(2)–(3) (West 1995).

62. *Id*. § 342-B(3)(B).

63. *Id*. §§ 342-B, 1362(1)(D).

64. *Id*. §§ 342-B(2), 342-B(4).

65. *Id*. §§ 342-B(4)(B)–(C).

66. *Id*. § 342-B(4)(C)(4).

67. *Id*. § 342-B(2).

68. U.S. EPA, New England Comfort Letters Minimum Requirements, Final Guidance (Feb. 20, 1996).

69. *Id*.

70. *Id*.

71. *See, e.g.*, EPA Memorandum from Stephen D. Luftig, Director Office of Emergency and Remedial Response to Regional Offices regarding Updating the National Priorities List—Proposal 20 and Upcoming Final Rule (Feb. 29, 1996).

72. *See Id.; see also* 60 Fed. Reg. 34,792, 34,794 (1995) (in the context of establishing criteria for entering prospective-purchase agreements, USEPA has stated that action at sites undergoing cleanup through a state program is "extremely unlikely").

73. *See* ME. REV. STAT. ANN. tit. 1, §§ 401 *et seq.* (West 1995) (general public-records law); *see also* ME. REV. STAT. ANN. tit. 38, § 1310-B (West 1995) (generally, information obtained by DEP under specified environmental provisions constitutes public record).

74. ME. REV. STAT. ANN. tit. 38, § 1310-B(2) (West 1995). Maine's waste-management provisions are codified at ME. REV. STAT. ANN. tit. 38, §§ 1301–19-X (West 1995). These provisions address, *inter alia*, the licensing and operation of solid waste and disposal facilities and the handling (including the treatment, storage, and disposal) and transportation of hazardous waste. The Voluntary Response Program is not codified under the waste-management chapter. *See id.*

75. ME. REV. STAT. ANN. tit. 38, § 1310-B(2) (West 1995).

76. *Id.* §§ 1317–19-X, 1401–04, 1601–08, 1651–54.

77. *Id.* § 1319-U(5).

78. *Id.*

79. *Id.*

80. *Id.*

81. *Id.* §§ 1361–71. This is Maine's counterpart to the federal Superfund statute.

82. *Id.* § 1367.

83. *Id.* § 1362(3).

84. *See* ME. REV. STAT. ANN. tit. 14, § 156 (West 1995) (comparative-fault statute); Otis Elevator Co. v. Cunningham & Sons, 454 A.2d 335 (Me. 1983) (recognizing equitable right of contribution between joint tortfeasors).

85. ME. REV. STAT. ANN. tit. 38, § 1362(2) (West 1995).

86. *Id.* § 1367.

87. *Id.* §§ 541–60.

88. *See id.* § 551.

89. *Id.* § 551(2)–(5).

90. *See id.* § 568-A (fund coverage requirements and eligibility for recovery under fund).

91. *See* Murray v. Bath Iron Works, Corp., 867 F. Supp. 33, 47–49 (D. Me. 1994) (discussing claims for environmental contamination brought under Maine law); *see also* Hanlin Group, Inc. v. International Minerals Chem. Corp., 759 F. Supp. 925, 932–37 (D. Me. 1990).

CHAPTER 31

Maryland

ANTHONY M. CAREY
JUDITH A. ARMOLD

In 1996, the Maryland legislature passed separate bills in the House of Delegates (HB 5) and the Senate (SB 205), both establishing comprehensive "Brownfields Voluntary Cleanup and Revitalization Programs." The bills were similar in a number of respects, but contained substantial differences. Unfortunately, a conference committee at the end of the session was unable to reconcile the two bills, and the legislation failed to pass. It is likely that legislation will be reintroduced in 1997.[1] This chapter summarizes the Maryland House and Senate bills so that individuals intending to redevelop brownfields in Maryland can anticipate the likely provisions of legislation enacted in the future.

Summary of Major Provisions

The House and Senate bills were similar in that each provided for the following:

1. A voluntary cleanup program or programs administered by the Maryland Department of the Environment (MDE), and a program of financial incentives under the state Department of Business and Economic Development (DBED)
2. Alternative cleanup standards, including natural background, statewide numeric limits, site-specific risk-based limits, or existing state or federal cleanup standards
3. Up to 125 brownfields sites to be designated for financial incentives, including low-interest loans, grants, and property-tax credits
4. Participants in the program to receive a release of liability from cost-recovery actions under state law
5. Safe-harbor provisions for lenders, to encourage participation in brownfields projects by reducing the liability risk under state law

The two bills differed in a number of significant respects:

1. The House bill provided for a single cleanup program and a separate financial-incentives program (labeled the "Brownfields Revitalization Program"); the Senate bill provided for two separate cleanup programs (labeled the "Brownfields Revitalization Program" and the "Voluntary Remediation Program"), only the first of which involved financial incentives.
2. The House bill provided for a release from liability after a cleanup plan had been approved, but before it was implemented; the Senate bill required that the cleanup be completed and approved before the release was granted.
3. Under the single cleanup program in the House bill, a responsible person could participate as long as the person could show that he or she had not contaminated the site through acts that would constitute a criminal violation. In contrast, the Senate bill provided for a Brownfields Revitalization Program, not available to responsible persons, and a separate Voluntary Remediation Program, in which a responsible person could participate as long as the person had not caused a release or threatened release by negligently, knowingly, or willfully violating any requirement or prohibition under the state's hazardous-materials and hazardous-substance laws.
4. Under the House bill, a person undertaking a voluntary cleanup could abandon the project without penalty at any time, except that a responsible person could abandon the cleanup only within thirty days after plan approval. Under the Senate bill, once a party had agreed to clean up a property, it could not back out.
5. Both bills confined voluntary cleanups to sites contaminated with hazardous wastes only. Under the House bill, however, the cleanup of petroleum-contaminated properties would qualify for financial incentives.
6. The House bill mandated property-tax credits for sites cleaned up under the program. The Senate bill provided for discretionary credits, available only for sites cleaned up under the Brownfields Revitalization Program.
7. The Senate bill provided that sites contaminated after October 1, 1996, would not be eligible. The House bill contained no such limitation.

The Cleanup Programs

Eligible Property

House Bill: "Eligible property" is defined as property that is contaminated or suspected of being contaminated by a release, discharge, or threatened release of a "controlled hazardous substance" (CHS). Under Maryland law,[2] CHSs include all substances defined as "hazardous wastes" under the federal Resource Conservation and Recovery Act, plus low-level nuclear wastes and state-listed hazardous wastes such as polychlorinated biphenyls.

The following sites would not be eligible for a voluntary cleanup:

- Petroleum-contaminated sites
- Sites listed on the National Priorities List (NPL) under section 105 of the federal Comprehensive Environmental Response, Compensation, and Liability Act (CERCLA)
- Properties subject to an unresolved enforcement action under the state's hazardous-waste laws
- Properties listed on the Comprehensive Environmental Response, Compensation, and Liability Information System (CERCLIS) maintained by the United States Environmental Protection Agency (USEPA) (though MDE, in its discretion, could find that a property was eligible if it was listed on CERCLIS but designated "no further remedial action planned")
- A regulated unit under a Maryland CHS permit (except that MDE could find that a portion of the property was not part of the regulated unit and was therefore eligible)
- Sites under investigation for a violation of the state's hazardous-waste laws (except to the extent that MDE, in its discretion, decided that the property was eligible)

Senate Bill: "Eligible site" is defined as a site at which there is a release or threatened release of a CHS. The definition does not include a site "suspected of being contaminated." The Senate definition would exclude petroleum-contaminated sites, properties on the NPL, and those subject to hazardous-waste enforcement actions by MDE. Operating facilities subject to MDE regulation could not participate in the Brownfields Revitalization Program, but MDE could allow parts of such facilities to participate in the Voluntary Remediation Program, as long as those parts were not subject to water-discharge permits or were not regulated units under any CHS permit. The Senate bill would also exclude any site contaminated by a release or threatened release caused by the disposal of hazardous substances after October 1, 1996.

Eligible Persons

House Bill: Any person, whether an owner, prospective purchaser, or responsible party, could apply to undertake a voluntary cleanup project, if the person could show that the contamination of the property did not result from any violation of CHS laws or regulations that would subject the person to criminal penalties.

Senate Bill: An "eligible person" is defined as a current owner, lender, developer, or prospective purchaser of an eligible site, except that a "responsible person" would not be eligible for the Brownfields Revitalization Program at all, and such a person could participate in the Voluntary Remediation Program only if the person did not cause a release or threatened release by negligently, knowingly, or willfully violating any requirement or prohibition of the state's hazardous-waste laws.

Cleanup Standards

House Bill: Response action plans are to meet one or more of four cleanup standards: numeric risk-based standards, standards based on a site-specific risk assessment, naturally occurring background levels, or existing and applicable Maryland cleanup standards.

MDE is directed to adopt by regulation numeric risk-based criteria for residential and industrial uses. These criteria are to be reviewed every two years and revised as necessary. There is no time limit for promulgation of numeric standards, but the bill provides that failure of MDE to adopt them shall not prevent its implementing the voluntary cleanup project on an individual basis. For cleanup plans based on site-specific risk assessments, MDE is directed to consider protection of the public health and the environment, cost effectiveness, technical practicability, and the proposed use of the property. The background-level standard requires cleanup to the level of a substance naturally occurring at the site, before any man-made spill or release.

In addition to the four standards, MDE is also given authority to approve the use of "presumptive and generic remedies" to achieve cleanups.

MDE may condition the granting of an assurance letter, a no-further-action (NFA) letter, or a no-action (NA) letter on an applicant's agreement to limit future use of a cleaned site. When future use of the property is limited to commercial or industrial purposes, the plan must be recorded in the land records of the jurisdiction where the property is located. The House bill does not address engineering controls (such as caps), but there would appear to be no limitation on MDE's discretion to approve the use of such controls.

Senate Bill: Remedial action is to be based on one or more of six cleanup standards, as "appropriate and relevant": federal or state drinking-water limits, federal soil and water quality criteria, site-specific risk evaluations, uniform numeric cleanup standards to be determined by MDE, background levels, and any other federal or state standard. When approving cleanup standards, MDE is to consider whether the property is located in an industrial area used for industrial purposes, a residential area used for industrial purposes, or a residential area used for residential or other purposes that require unlimited public access. In addition, MDE must ensure that the cleanup standards will protect the public health and welfare and the environment.

The Mechanics of Participation in the Program

House Bill: An applicant can apply for an assurance letter, an NFA letter, or an NA letter. An assurance letter confirms that work plans, environmental site evaluations, or other technical documents are adequate, that contamination at a site was caused by a release from property other than the eligible property, that a person other than the applicant caused the contamination, or that the property is not contaminated at all. An NFA letter is provided after approval of a proposed response action plan (RAP), whereas an NA letter is issued only after completion of an RAP, and only if MDE finds that the requirements of the RAP have been met and cleanup standards have been achieved. Both the NFA and NA letters include a liability release.

To obtain any of the three letters, an applicant must submit an application to MDE, demonstrate the eligibility of the property and the applicant, describe in summary form the proposed cleanup project (including proposed cleanup criteria), and pay an application fee. The fee is $1,000 for an assurance letter or an NA letter limited to a declaration that the property is not contaminated; the fee is $5,000 for an NFA letter or a regular NA letter. MDE must notify the applicant in writing when the application is complete, and within thirty days thereafter MDE must notify the applicant about whether the application meets the requirements of the program. After MDE approves an application, the applicant must propose an RAP, which includes a report on the environmental conditions at the site and any site evaluations, a selection of the cleanup standards, and a work plan, including provisions for long-term monitoring, operation, and maintenance, if necessary.

After the applicant submits a complete, proposed RAP to MDE, the applicant must publish the plan once a week for two consecutive weeks in a newspaper of general circulation in the area in which the eligible property is located. MDE then receives comments from the public for thirty days and, to the maximum extent practicable, notifies the applicant within sixty days whether the plan is approved, is rejected, or must be modified to receive MDE approval.

After an RAP has been approved and work begins, the applicant is required to submit a cost affidavit every six months, certifying the amount paid for engineering, scientific, or technical work. Within thirty days after submission of an affidavit, the applicant must pay an oversight fee, equal to 10 percent of the certified six-month costs, which goes into a Voluntary Cleanup Fund under MDE's jurisdiction. The oversight fee is not required to be calculated on costs incurred for construction, equipment, materials, or laboratory analyses.

Senate Bill: Under the Senate bill, an NFA letter is issued by MDE after satisfactory completion of an RAP. To participate in the program and receive such a letter, a person submits an application to MDE on the proper form, showing eligibility and including an environmental assessment concerning the nature and extent of known contamination. An application fee of $10,000 must be paid. Within thirty days after receipt of a complete application, MDE must notify the applicant in writing whether the application has been accepted.

After an applicant receives notice of acceptance, it must submit a proposed RAP that describes the methods and results of a site investigation, describes the removal or remedial action to be performed (including long-term monitoring and operation and maintenance measures), selects a cleanup standard or standards, and demonstrates that the completed RAP will protect the public health and welfare and the environment.

Within thirty days after receipt of a proposed RAP, MDE must require the applicant to publish a newspaper notice describing the RAP and allowing thirty days for public comment. The applicant must also mail notice of the RAP to owners of adjacent property. MDE may, but is not required to, hold a public hearing on the proposed RAP, and must hold a public informational meeting if at least five residents from separate households in communities adjacent to the site file a written request for such a meeting. To the maximum extent practicable, within sixty days after completion of the public-participation process, MDE is to notify the applicant of its decision regarding the RAP.

Liability Protection

House Bill: A party submitting an approved RAP and receiving an NFA letter, or a party completing an RAP and receiving an NA letter, will receive a release of liability from actions by the state or third parties under state law for any cleanup costs of the contamination at the eligible property addressed in the RAP. The liability release is ineffective if it is obtained through fraud or misrepresentation, and it does not preclude MDE from taking action or affect MDE's authority under the following circumstances:

- MDE may still prevent or abate an imminent or substantial threat to public health or the environment at the property.
- MDE may still take action against a person who violates limitations, contained in the NFA or NA letter, on the use of the property.
- MDE can act when there is new or previously undiscovered contamination at a site.
- An NA letter does not prevent MDE from requiring a responsible person to take further action if completion of the RAP does not result in achievement of approved cleanup standards.
- An NA letter does not prevent MDE from taking action against a responsible party if the long-term operation and maintenance required by an RAP have not been performed.

Moreover, the House bill expressly provides that neither the liability release provisions nor anything else in the bill prevent an applicant who is not a responsible person from seeking cost recovery against a responsible person or affect a tort action for personal injury.

Senate Bill: MDE must issue an NFA letter after a party has completed an RAP to the satisfaction of MDE. The effect of the NFA letter is generally to prohibit MDE from bringing an "enforcement action" under the CHS subtitle of the Maryland Annotated Code.[3] However, MDE may bring such an action (1) if there is an imminent or substantial threat to the public health or the environment, (2) if the NFA letter was obtained through fraud or material misrepresentation, (3) if new or previously undiscovered contamination is found, (4) if the site fails to meet the applicable cleanup criteria set forth in the RAP, (5) if the recipient of the NFA letter does not comply with limitations on the use of the property, or (6) if the long-term operation and maintenance program set forth in the RAP has not been performed.

Lenders, Trustees, and Fiduciaries

Existing Law: Under current Maryland law,[4] holders of mortgages or deeds of trust on contaminated property, holders of security interests in property located on contaminated sites, and fiduciaries who hold legal title to contaminated sites or property on contaminated sites in connection with the administration of estates or trusts are all exempt from the definition of "responsible person" under the state's hazardous materials and hazardous-substance laws, provided they did not participate in the day-to-day management of the site or property and did not directly cause the discharge of hazardous substances on the site. Similarly, holders of mortgages or deeds of trust who acquire title through foreclosure or deeds in lieu of foreclosure are exempt from responsible-person status, again assuming they were not involved in day-to-day management and were not directly responsible for the release that caused the contamination.

House Bill: The House bill provides an additional safe harbor for a bank in connection with a voluntary cleanup. A bank may, after notice to MDE, complete an RAP or take action to contain, stabilize, or remove hazardous substances to protect or secure eligible property, or property located on eligible property, without being considered a "responsible person" under state law. However, the bank is still a responsible person in connection with contamination that the bank itself causes on the eligible property. MDE is required to adopt regulations establishing liability limits consistent with those set forth in the bill.

Senate Bill: The Senate bill provides a similar exemption from the definition of "responsible person" for the holder of a mortgage or deed of trust who acquires title to a site subject to an RAP and complies with its requirements, and for a lender who extends credit for the performance of any removal or remedial action or who takes action to protect or preserve a security interest at a site by stabilizing, containing, removing, or preventing the release of a hazardous substance in a manner that does not cause or contribute to ongoing releases of a hazardous substance at the site.

Agreements and Withdrawal of an Application or an RAP

House Bill: Within thirty days after notification of approval of an RAP, an applicant who is a responsible person must either sign an agreement with MDE regarding completion of the RAP or notify MDE in writing that the application or the RAP has been withdrawn. A responsible person will not receive the liability protection of an NFA or NA letter if it does not execute and complete an agreement. An executed agreement is enforceable by MDE through a civil action. An applicant who is not a responsible party is not required to sign an agreement.

An applicant who is not a responsible party may withdraw an application or an RAP at any time by notifying MDE in writing. The applicant has no further obligation, except for forfeiture of any application or oversight fees that have been paid. As a general rule, any assurance letter, NFA letter, or NA letter issued to such an applicant becomes void.

Senate Bill: The Senate bill provides that when an RAP is approved, MDE must enter a voluntary remediation agreement with the person who intends to implement the RAP. There is no provision for withdrawal from such an agreement and, once executed, the agreement would be enforceable by MDE through a civil action.

The Financial-Incentive Programs

Both the House and Senate bills provide for a financial-incentive program under the jurisdiction of DBED. Under both bills, DBED, after consulting with MDE, is to publish a list of brownfields sites that are eligible for participation in the program. Financial incentives are offered to parties that undertake and complete cleanups. In selecting eligible sites, DBED is to use both geographic and economic-development criteria, including the feasibility of redevelopment, the level of benefit to the state and public provided by redevelopment, the potential of a redeveloped site creating or retaining jobs, and the absence of an identifiable and financially solvent responsible party. The final list under each bill is to be updated annually and contain no more than 125 sites.

House Bill: Under the House bill, the financial-incentive program is itself referred to as the "Brownfields Revitalization Program." Incentives, including low-interest loans and grants, are available for "brownfields sites," defined to include sites eligible under the MDE cleanup program, as well as sites where there is a release or threatened release of oil. Properties owned or operated by a responsible person that caused the on-site contamination are not brownfields sites. Only those brownfields sites included on DBED's list are eligible for incentives.

In addition, the House bill mandates a five-year property-tax credit, equal to 50 percent of the tax attributable to any increased assessment placed on a property after completion of a voluntary cleanup. (This includes any increased assessment due to improvements built on the site within the five-year period.) In a designated enterprise zone, the state or locality may extend the property-tax credit for an additional five-year period. The property-tax credit is apparently available to any "brownfields site," regardless of whether it is included on DBED's list.

Senate Bill: The Senate bill provides for a separate MDE-administered cleanup program known as the "Brownfields Revitalization Program." Although in most respects this program and the Voluntary Remediation Program, also administered by MDE, would be identical, only those applicants approved for participation in the Brownfields Revitalization Program would be eligible for financial incentives and property-tax credits. Eligibility would require listing on the DBED list.

In contrast to the House bill, the Senate bill does not mandate a property-tax credit for a site cleaned up under the Brownfields Revitalization Program, but leaves it to the governing body of the county or municipal corporation where the property is located to grant such a credit and to determine the amount and duration of any credit.

Other Provisions of the House and Senate Bills

Confidentiality and Admission of Liability

Neither bill establishes a liability privilege or immunity in connection with the conduct of a property audit. The Maryland legislature considered, but did not pass, several audit privilege bills during the 1996 session. Those bills would have provided that the report of a voluntary environmental investigation submitted to MDE would be inadmissible and not subject to discovery in any civil or administrative proceeding, and that violations of environmental law disclosed in the report would not be subject to civil or administrative actions by MDE, if the party conducting the investigation followed certain guidelines. Audit privilege legislation is expected to be introduced again during the 1997 session.

Enforcement and Cost Recovery

The cleanup programs established in the House and Senate bills, including the requirements of any RAP, would be enforceable through an array of legal mechanisms available to MDE under sections 7-256 through 7-268 of the Environment Article of the Maryland Annotated Code.[5] These include rights of entry, administrative search warrants, corrective orders, civil penalties, injunctions, and criminal penalties.

Existing Maryland law also provides for cost recovery in connection with MDE cleanup of contaminated sites. Under sections 7-218 through 7-223 of the Environment Article, MDE may expend funds from the state Hazardous Substance Control Fund for the removal, restoration, or remediation of sites contaminated by hazardous substances and seek cost recovery from responsible persons.[6] The term "responsible person" has substantially the same meaning under state law as it has under CERCLA, except that safe-harbor provisions have been carved out for lenders, fiduciaries, and holders of mortgages, deeds of trusts, and security interests. MDE is directed to seek recovery from responsible persons on an equitable apportionment basis, when there is a reasonable basis for such apportionment.

Under the House and Senate bills, MDE could use funds from application fees, overhead charges, and cost-recovery actions—all of which would go into a Voluntary Cleanup Fund—to complete abandoned cleanup projects, abate immediate or substantial threats to public health or the environment, or address new or undiscovered contamination at eligible sites. MDE could then seek recovery of its costs, including legal expenses and interest, from responsible persons.

In contrast to CERCLA, Maryland law does not provide for a private right of action against a responsible person for recovery of the costs of investigating and cleaning up contaminated property.

Memorandum of Agreement with USEPA

Maryland has not entered a Memorandum of Agreement (MOA) with USEPA regarding a Voluntary Cleanup Program. It is expected that the state will negotiate such an MOA with USEPA Region III after a Voluntary Cleanup Program has been enacted by the legislature.[7]

Conclusion

Each house of the Maryland General Assembly passed a brownfields bill during the 1996 legislative session, but the two bills could not be reconciled before adjournment and therefore no legislation was enacted. However, the differing provisions of the two bills suggest those likely to be included in bills introduced during the 1997 legislative session and in any legislation that is ultimately enacted.

Notes

1. At the time this chapter was written, the content and fate of 1997 legislation could not be predicted. However, before the final publication date, the General Assembly of Maryland unanimously passed, and the Governor signed, emergency legislation establishing a voluntary cleanup program and a financial incentives program. The emergency legislation, embodied in virtually identical Chapters 1 and 2 of the Laws of Maryland 1997 (Senate Bill 340, House Bill 409), took effect February 25, 1997. It differs in a number of respects from the 1996 bills described in this chapter. A copy of the legislation can be obtained from the authors (at the firm of Venable, Baetjer and Howard, L.L.P., 410/244-7620) or from the Maryland Department of Legislative Reference (410/841-3810 or 301/858-3810).
2. MD. CODE ANN. ENVIR. § 7-201 (1996).
3. *Id.* §§ 7-256–68.
4. *Id.* § 7-201(x).
5. *Id.*
6. *Id.* §§ 7-218–23.
7. Before the final publication of this book, MDE and USEPA Region III had entered into an MOA.

CHAPTER 32

Massachusetts

NED ABELSON
WILLIAM M. SEUCH
MAURA McCAFFERY

The Commonwealth of Massachusetts has been working to encourage the cleanup and redevelopment of brownfields sites since the early 1990s. This chapter provides an overview of the twin cornerstones of the Massachusetts brownfields program, a "privatized" regulatory scheme that streamlines the cleanup process and a covenant-not-to-sue initiative that is currently designed to encourage "innocent" prospective purchasers and tenants to become interested in, and involved with, previously contaminated sites.

Summary of Major Provisions

In 1993, the regulations that govern the cleanup of oil and hazardous-materials releases in Massachusetts were substantially revised to streamline and privatize the environmental cleanup process. This regulatory program, known as the Massachusetts Contingency Plan (MCP), provides the regulated community with a detailed blueprint for cleaning up releases of oil and/or hazardous materials *without* significant government involvement. Instead of overseeing every phase of a cleanup, the Massachusetts Department of Environmental Protection (DEP) now relies on private environmental consultants, known as "Licensed Site Professionals," to determine whether the appropriate cleanup standards have been satisfied.

In addition to streamlining the cleanup process, DEP is also now willing to encourage brownfields redevelopment by providing prospective owners and tenants of contaminated properties with certain protections from liability under chapter 21E of the Massachusetts General Laws, the Massachusetts Oil and Hazardous Material Release Prevention and Response Act (Chapter 21E). Under the "Clean Sites Initiative," a prospective owner or tenant of a contaminated property may obtain a covenant-not-to-sue from the Commonwealth in return for a promise to ensure that all known contamination will be cleaned up in accordance with applicable law.

Understanding the Massachusetts Superfund Statute
The Liability Scheme

The Massachusetts Superfund statute, Chapter 21E, originally enacted in 1983 and extensively amended in 1992, is patterned closely after the federal Comprehensive Environmental Response, Compensation, and Liability Act of 1980 (CERCLA). Chapter 21E imposes liability on several classes of "potentially responsible parties" (PRPs), including (1) owners and operators of sites where there has been a release of oil or a hazardous material, (2) past owners or operators of sites at which a hazardous material was released, (3) any person who arranged for the transportation, disposal, storage, or treatment of a hazardous material, (4) any person who transported any hazardous material to a site from or at which there has been a release of that material, and (5) any person who "otherwise caused or is legally responsible for" a release of oil or hazardous material.[1] It is also important to note that Chapter 21E provides causes of action for both private parties and the Commonwealth.[2]

Liability under Chapter 21E is joint, several, and without regard to fault.[3] However, liability for releases of oil falls upon only current owners and operators and those who have "otherwise caused" such releases or threats of releases.[4] In other words, former owners and former operators cannot be held liable for a release of oil unless it can be proven that they "otherwise caused or are legally responsible for" the release.[5] In contrast, liability for releases of hazardous materials is notably broader, as it falls on both current and former owners and operators.[6]

Recoverable Damages

It is important to understand that liability under Chapter 21E can be much broader than liability under CERCLA because, as previously noted, Chapter 21E expressly permits the recovery of damages associated with petroleum contamination. In addition, Chapter 21E establishes rights of action for both "response costs" (costs incurred in assessing and cleaning up releases of oil or hazardous materials), and "property damage."[7]

Although courts in Massachusetts are just now beginning to explore the parameters of the "property damage" remedy available under section 5 of Chapter 21E, it appears that damages for the diminution in a property's market value may be awarded when a property cannot be fully remediated. Courts have also held that the measure of recovery for "property damage" under Chapter 21E is identical to the measure of recovery at common law for damages to real or personal property. Accordingly, damages for lost use or lost rental value also appear to be recoverable under Chapter 21E.[8]

Attorney Fees

Chapter 21E also provides for the recovery of attorney fees. Section 15 of Chapter 21E provides that courts "may award costs, including reasonable attorney and expert witness fees, to any party . . . who advances the purposes of [Chapter 21E]."[9] In addition, Chapter 21E provides for the award of attorney fees against parties who take unreasonable or bad-faith negotiating positions in connection with the prelitigation dispute-resolution process that is mandated by section 4A of the statute.[10]

Statute of Limitations

Actions for property damage under section 5 of Chapter 21E are subject to a three-year statute of limitations, which runs from the date on which the person seeking recovery first suffers the damage or the person seeking recovery discovers or reasonably should have discovered that the plaintiff is a liable person, whichever is later.[11]

In contrast, the statute of limitations for response cost actions under section 4 of Chapter 21E is much more lenient, as a suit must be commenced within three years after (1) the date on which the plaintiff first discovers or reasonably should have discovered that the defendant is a liable person, (2) the date on which the plaintiff first learns of a material violation of an agreement reached pursuant to the dispute-resolution procedures that are set forth in section 4A of Chapter 21E, (3) the date on which the plaintiff incurs all response costs at a site (that is, completes all response actions), or (4) the date on which the person seeking recovery sends notice to a defendant pursuant to section 4A of Chapter 21E, whichever is latest.[12]

The statute of limitations for actions brought by the Attorney General for the Commonwealth of Massachusetts requires that all actions be commenced within (1) five years from the date on which the Commonwealth incurs all response costs at a site, or (2) five years from the date on which the Commonwealth discovers that the person against whom the action is being brought is a person liable for a release or threatened release for which the Commonwealth has incurred response costs, whichever is later.[13]

Contractual Allocation of Liability

Private agreements can be used to allocate environmental liabilities under Chapter 21E between and among private parties. However, private agreements cannot serve to protect a party from liability to the Commonwealth. More specifically, section 5(f) of Chapter 21E provides as follows:

> No indemnification, hold harmless, or similar agreement or conveyance shall be effective to transfer from the owner or operator of any vessel or site or from any person who may be liable for a release or a threat of release of hazardous material under this section, to any other person the liability imposed under this section. Nothing in this paragraph shall bar any agreement to insure, hold harmless, or indemnify a party to such agreement for any liability under this section.[14]

Apportionment of Response Costs

Under Chapter 21E, an owner of contaminated property who did not own or operate the site at the time of the release in question, and did not cause or contribute to such release, can be held liable to the Commonwealth but not to any other party that contributed to the subject release or owned the property at issue at the time of the release. However, Chapter 21E does expressly contemplate the equitable apportionment of response costs between and among responsible parties. Section 4 of Chapter 21E provides as follows:

> If two or more persons are liable . . . for such release or threat of release, each shall be liable to the others for their equitable share of the costs of such response action.[15]

"Safe Harbors" for Lenders and Fiduciaries

Under Chapter 21E, fiduciaries and lenders are provided with certain protections from owner and operator liability. More specifically, the definitions section of Chapter 21E provides that "fiduciaries" and "secured lenders" cannot be held liable as "owners" or "operators" if all of a number of very specific statutory requirements are met.[16] Note, however, that liability under the other three PRP categories is still possible, as is liability under common law.

The requirements that Chapter 21E sets forth for a secured lender not to be considered an owner or an operator before taking title to a property by foreclosure or other similar means are as follows:

1. Neither the secured lender nor its employees or agents can cause or contribute to a release of hazardous materials or oil, or cause a release or threat of release to become worse than it would otherwise have been.
2. The secured lender, before acquiring ownership or possession of the relevant site, must not participate in the management of the site. The statute indicates that "participation in the management of the site" includes "substantially divesting from the borrower or any other person possession or control over those aspects of the operations involving the management of oil or hazardous materials. . . ."[17]

After acquiring ownership or possession of a site and obtaining knowledge of a release or a threat of release of oil or hazardous material, a secured lender must satisfy all the following conditions to not be deemed an owner or an operator:

1. The secured lender must notify DEP immediately upon obtaining knowledge of a release or a threat of release for which notification is required pursuant to the requirements of Chapter 21E or the MCP.
2. The secured lender must provide reasonable access to the site to employees, agents, and contractors of DEP, and to other persons intending to conduct "necessary response actions."
3. The secured lender must undertake reasonable steps (a) to prevent the exposure of persons to oil or hazardous materials by fencing or otherwise preventing access to the site and (b) to contain the further release or threat of release of oil or hazardous materials from a structure or container (presumably at the subject site).

4. If the secured lender elects voluntarily to undertake a response action or a portion of a response action at the site, the secured lender must conduct those activities in compliance with the requirements of Chapter 21E and the MCP.
5. The secured lender must act diligently to sell or otherwise divest itself of ownership or possession of the site.[18]

Understanding the Massachusetts Contingency Plan

The regulations under the MCP govern the assessment and remediation of contaminated sites in Massachusetts.[19] The MCP is designed to encourage the cleanup and redevelopment of brownfields sites by (1) providing private parties with clear rules about reporting obligations, (2) providing the regulated community with flexibility in determining the appropriate pace of most cleanups, and (3) allowing responsible parties to take the planned future uses of a site into account when determining whether a cleanup is necessary.

DEP has often described the MCP as a highway system with fast and slow lanes and a variety of entrance and exit points. In other words, the MCP establishes minimum and maximum speeds for moving through the waste-site cleanup process (that is, performance standards) as well as a variety of "off-ramps" that give PRPs and other parties flexibility in determining when and how they will meet the applicable cleanup standards to exit the system.

The Role of Licensed Site Professionals

At the center of the MCP process is a new class of private environmental professionals, known as "Licensed Site Professionals" (LSPs), who are hired by private parties and charged with ensuring that cleanup efforts at individual sites comply with the requirements of the MCP. To become an LSP, an environmental professional must pass an examination and meet specific standards for technical competence, decision-making experience, and ethical practice.[20]

Instead of relying on its own limited resources to oversee and approve every phase of the assessment and cleanup process, DEP has shifted a tremendous amount of decision-making responsibility to the LSPs. In most cases, the MCP allows private parties to avoid DEP oversight by retaining an LSP to determine whether a cleanup is necessary and, if so, how that cleanup should proceed. Simply put, the use of an LSP allows private parties to avoid the sometimes time-consuming and often frustrating experience of waiting for DEP to focus on a particular site.

Although the MCP does not require that every step in the site assessment/cleanup process be conducted under the direction of an LSP, it does require LSP opinions (or certifications) at several key junctures as sites move through the MCP process. At each of these key points (and at several others), an LSP needs to prepare a special DEP form, sign it, and affix to it his or her LSP certification stamp, indicating that the actions described therein were conducted under the LSP's professional oversight, in a manner consistent with both good professional practice and the requirements of the MCP. For most sites regulated under the MCP, an LSP is required for the following actions:

- Disposal of contaminated soils or debris
- Developing plans for assessment and cleanup
- Preparing Response Action Outcome Statements (as discussed below)[21]

Audit Procedures

Buyers and sellers of real estate in Massachusetts can now look to LSPs for comfort regarding questions about "how clean is clean enough" because the opinions issued by an LSP are assumed to be correct unless and until a DEP audit proves otherwise. Generally, DEP has the ability to audit an LSP's work at a particular site for a period of five years, and DEP is required to conduct random audits of response actions at 20 percent of the sites in the waste-site cleanup program on an annual basis.[22] In addition, DEP conducts specifically targeted audits when compliance problems are anticipated at particular sites or particular kinds of sites.[23] As of January 1996, approximately 92 percent of all randomly selected sites and 81 percent of all specifically targeted sites passed muster without further site work being required.[24]

Notification Requirements

The MCP includes a list of "reportable concentrations"—specific thresholds for determining whether contaminant levels in soil and/or groundwater could pose a risk to human health or the environment and therefore should be reported to DEP. If a release is discovered at a site and the applicable reportable concentration is not exceeded, then no notice to DEP is required and the site does not need to go through the MCP process. Conversely, an appropriate notice must be submitted to DEP if the published reportable concentration for a specific hazardous constituent is exceeded at a site.[25]

Under the MCP, there are three types of reportable releases: releases that require notice to DEP within two hours, seventy-two hours, or 120 days, respectively. These time frames are tied to when a "responsible person" obtains knowledge that a release has occurred. The two-hour notice requirement applies to the most serious releases. Examples include a release resulting in the presence of oil and/or a hazardous material in a private drinking-water well at concentrations greater than the applicable reportable concentration, and a release in any quantity or concentration that results in an "imminent hazard."[26] An example of a release requiring seventy-two-hour notification is one that results in the presence of subsurface, nonaqueous-phase liquid of oil and/or a hazardous material.[27] Generally, releases requiring a 120-day notice include those resulting in the presence of hazardous materials and/or oil in soil or groundwater in an amount equal to or greater than the applicable reportable concentration.[28]

As noted above, a release is "reportable" under the MCP only if the measured concentration of the relevant contaminant is equal to or greater than the applicable reportable concentration. These reportable concentrations are listed in the Massachusetts Oil and Hazardous Material List, which is part of the MCP.[29] The Massachusetts Oil and Hazardous Materials List separates reportable concentrations into two soil categories (RCS-1 and RCS-2) and two groundwater categories (RCGW-1 and RCGW-2). Briefly, the RCS-1 and RCGW-1 categories apply to more sensitive environmental areas, based on the current use and other characteristics of the site that are specified in the regulations. Categories RCS-2 and RCGW-2 apply to sites that are not in the more sensitive categories, serving somewhat as default provisions. The regulations contain specific provisions regarding the RC categories that apply to a particular site for notification purposes.[30]

Limited Removal Actions

The MCP contains provisions concerning limited removal actions (LRAs), which are preliminary, remedial response actions that can be completed before notice to DEP is required.[31] LRAs are designed to allow private parties to remediate minor releases quickly, without ever reporting the contamination to DEP or entering the MCP system.

LRAs include the excavation and off-site recycling, reuse, treatment, and/or disposal of not more than one hundred cubic yards of soil contaminated by oil and not more than twenty cubic yards of soil contaminated by hazardous materials.[32] As these parameters demonstrate, LRAs are restricted in nature, reflecting the continuing tension between DEP's objective of enabling private parties to deal with sites quickly and its concern about allowing complete, unsupervised freedom for those sites.

Preliminary Response Actions

If notice to DEP is required and performing an LRA does not eliminate the need to give that notice, preliminary response actions are the next steps that must be taken under the MCP. Once these preliminary measures have been completed, it is possible to exit the MCP process if adequate results have been achieved.[33]

Preliminary response actions may include the following activities:

1. *Initial Site Investigation Activities:* Initial site investigation activities are just what their name implies. The objective of these investigations is to obtain preliminary information about a site, which may be used to support a response action outcome statement (described more fully below), or as the basis for a more detailed investigation.[34]
2. *Immediate Response Actions:* Immediate response actions are those measures taken to address imminent hazards, which involve serious threats to human health, safety, public welfare, and/or the environ-

ment. An LSP must be used, and DEP approval of some kind is generally required to perform these actions, unless the delay in obtaining that approval would substantially increase the problem.[35]

3. *Release Abatement Measures:* Release abatement measures are intended to allow the voluntary implementation of certain accelerated remedial actions, but they are limited in scope and complexity. The idea is that if a fairly simple remediation effort can solve the problem, then there is no need for the site to go through the complete MCP process. The MCP requires that an LSP be retained when performing a release abatement measure and that written notification of the relevant release be given to DEP before the measure is performed. DEP has twenty-one days to respond to the required notification. If no response is issued within that time frame, DEP's approval is deemed to have been given.[36]

Tier Classification and the Numerical Ranking System

If the preliminary response actions outlined above do not result in the site exiting the MCP process, then private parties are required to classify the site using a detailed numerical ranking system, which is based on the toxicity of detected contaminants and the characteristics of the disposal site at issue. Generally, each site that has not exited the system within one year of the initial notification to DEP of a release must "tier classify" and pay the applicable tier classification fee. Here, the MCP has been intentionally designed to encourage private parties to undertake response actions as soon as possible to avoid the fees and costs associated with tier classification.

A site is classified as either a Tier I or Tier II site. Tier I sites are those with more significant contamination problems. A permit from DEP is required to perform more in-depth response actions at Tier I sites, but no such permit is required at Tier II sites. In addition, Tier I sites are subject to higher annual compliance fees and an increased level of DEP oversight.[37]

Comprehensive Response Actions and Risk Characterizations

Once a site has been classified using the tier classification system, it is possible to proceed with comprehensive response actions, which may include relatively significant remedial efforts and/or more detailed site assessment studies or risk characterizations. Risk characterizations are used to determine whether detected contaminants pose a threat to human health or the environment and whether further comprehensive response actions are required. Comprehensive response actions are used primarily to reduce or eliminate unacceptable levels of risk associated with contaminants that are present at a particular site.[38]

The MCP recognizes two basic approaches to risk characterization. The first is a chemical-specific approach, which compares site concentrations for specific substances in soil and groundwater with published regulatory cleanup standards.[39] DEP has expended considerable effort in developing these chemical-specific numerical standards so that private parties and their consultants can determine whether significant risks are present simply by comparing a published number with the contaminant concentrations present at a particular site. Note, however, that DEP has intentionally made these standards somewhat conservative because of the "cookbook" manner in which they may be used at most sites. The second risk-characterization approach is a much more detailed cumulative-risk approach, which involves comparing site-specific information with cumulative cancer risk and other health-risk information.[40]

The MCP sets forth three methods to be used in characterizing environmental risks at a site. Method 1 relies on the numerical standards described above. To use Method 1, there must be a promulgated Method 1 standard for each contaminant of concern at the site.[41] If no such standard has been promulgated, then the party conducting the risk characterization may develop a site-specific standard using Method 2.[42] Methods 1 and 2 may be used only if contaminants are present in soil and/or groundwater.[43]

If contaminants are present in other environmental media, then a Method 3 risk assessment must be used.[44] Method 3 may also be used when a private party would like to vary the generic Method 1 cleanup standards based on site-specific information. Frequently, the use of a site-specific approach results in less conservative cleanup standards. In other words, a Method 3 risk characterization may be used to demonstrate that the presence of a particular contaminant at levels that may exceed the published Method 1 standards will not pose a threat to human health or the environment at a particular site.[45]

As is the case with reportable concentrations, the Method 1 soil and groundwater categories reflect the sensitivity of an area from an environmental perspective. Groundwater category GW-1 applies to certain more sensitive areas, which are specified in the MCP. Category GW-2 applies to groundwater located within thirty feet of an occupied building in an area where the average depth to groundwater is fifteen feet or less, the idea being that this type of contamination could be a potential source of vapors to indoor air. Category GW-3 is a default category, which must be used if the other two groundwater categories do not apply. Similarly, soil is separated into three categories, S-1, S-2, and S-3. Category S-1 soils are associated with the highest potential for exposure, while category S-3 soils have the lowest potential for exposure.[46]

Response Action Outcomes

A response action outcome (RAO) is the regulatory endpoint that must be achieved before a site is deemed "clean enough" to exit the MCP system. The MCP provides a number of places throughout the process where an RAO may be achieved.[47]

Several types of RAOs are set forth in the regulations. Class A RAOs apply to sites where a level of "no significant risk" has been achieved through the implementation of response actions.[48] Class B RAOs apply to sites where a level of "no significant risk" has been demonstrated, but response actions (other than site assessments) were not necessary.[49] Class C RAOs apply to disposal sites where a level of "no significant risk" has not been achieved, but all substantial hazards have been eliminated. Unlike the permanent solutions that are documented through Class A and B RAOs, a Class C RAO is a temporary solution, which must be reevaluated every five years to verify whether conditions have changed or whether new technologies have been developed that could allow for the implementation of a permanent solution.[50]

Activity and Use Limitations

An activity and use limitation (AUL) is a recorded environmental deed restriction that can be filed in connection with both Class A and Class B RAOs to notify parties that contamination exists at a property.[51] The implementation of an AUL generally allows the owner of a contaminated site to apply a less stringent cleanup standard in exchange for agreeing to limit the future use of the site and the associated potential for human contract with, and/or exposure to, existing contamination. In other words, if only certain specified and safe uses are permitted (and, conversely, other uses are prohibited), less demanding cleanup standards will apply.[52]

In fact, the MCP mandates that an AUL must be implemented whenever an RAO is based on the restriction or limitation of future activities and/or uses at a site, or on the elimination of an exposure pathway (that is, a means by which human or environmental receptors could come into contact with oil or hazardous materials).[53] In addition, an AUL must be implemented if a Method 1 risk characterization is performed and S-1 soil standards (that is, residential standards) have not been satisfied.[54]

The implementation of an AUL is an attractive option at many sites because significant cost savings can be realized when less stringent cleanup standards are applied. However, these title restrictions have not yet gained universal acceptance in the real estate community. An AUL may still have a negative effect on the perceived value of a property, so property owners must carefully consider, on a case-by-case basis, whether the implementation of an AUL makes sense.

The Massachusetts Covenant-Not-to-Sue Program

Working together, several state agencies in Massachusetts have implemented the Clean Sites Initiative. This is a covenant-not-to-sue program and is a pilot redevelopment program for economic target areas in Massachusetts. The Executive Office of Economic Affairs, the Executive Office of Environmental Affairs, DEP, and the Office of the Attorney General, working with a variety of private-sector representatives, have developed the application materials for this program.[55]

The purpose of the Clean Sites Initiative is to encourage prospective buyers and tenants of contaminated properties, particularly those located in areas targeted by the Commonwealth for economic development, to clean up and redevelop those properties. In exchange, the program provides protection from future liability to the Commonwealth under Chapter 21E once an appropriate cleanup has been performed.

Specifically, a potential buyer or tenant must agree that all known contamination will be assessed and remediated as required by Chapter 21E and the MCP. Once these steps have been taken, the Commonwealth agrees not to sue the new owner or tenant if more contamination is found at a later date. The covenant does not, however, bar claims brought by third parties other than the Commonwealth, and it does not relieve the new owner or tenant from responsibility for new releases that occur after the initial cleanup has been completed and the covenant-not-to-sue has taken effect. Passive releases, such as the leaching of contaminants from drums previously buried at the site by others, are generally not considered "new releases" under the program, and the covenant protects against them.[56]

Eligible Projects and Applicants

For a project to be eligible for a covenant-not-to-sue, it must be located within an economic target area, as designated by the Massachusetts Economic Assistance Coordinating Council. Thirty areas have been designated to date, and a list of them is available from the Massachusetts Office of Business Development and from DEP.

A project that is not located within a designated economic target area may still be eligible if the Secretary of Economic Affairs determines that the project presents an "exceptional economic development opportunity." Generally, it appears that any project that will create at least one bona fide full-time position will be deemed eligible by the Secretary. However, the relevant project must involve the reuse or redevelopment of a contaminated property for commercial or industrial activities. Residential projects are not included in the scope of the pilot program at this time.

The applicant must be a prospective owner or tenant of the relevant property, and cannot be a PRP (that is, any party with potential liability under Chapter 21E) in connection with the cleanup of the site. In addition, the applicant must certify that it is not currently subject to any pending administrative or judicial enforcement actions concerning compliance with environmental laws and regulations (or the applicant must at least show that it has established an approved compliance schedule). The applicant must also certify that it is willing and able to ensure that the site will be addressed pursuant to the requirements of Chapter 21E and the MCP.

The application forms currently available are comparatively brief and fairly straightforward. Moreover, the submission of an application does not trigger a public comment period or any other disclosure requirements. Applications should be submitted to the Massachusetts Office of Business Development, where they are first reviewed. The forms are then reviewed by DEP and the Executive Office of Environmental Affairs. The agencies' stated goal is to process applications within thirty calendar days of receipt by the Office of Business Development and to allow applicants to learn whether an application has been approved before they make a contractual commitment to buy or lease a particular property.

The Covenant-Not-to-Sue

Once an application has been approved, the Office of the Attorney General will issue a covenant-not-to-sue, as part of a form agreement. In addition, DEP will issue a certificate of completion promptly upon receipt of a RAO statement under the MCP, which indicates that a permanent solution has been achieved at the subject property. However, it is important to note that the certificate of completion issued only confirms that DEP has received the RAO statement for the site. It does not imply DEP's approval of the adequacy of cleanup actions taken at the site, and it does not block DEP's ability to audit the site.[57]

The covenant-not-to-sue covers only those releases that are fully described in the RAO statement. As noted above, it does not cover any subsequent new releases of oil or hazardous materials. The covenant is void if any false statements or certifications are contained in the application or if the applicant fails to perform any obligations contained in the RAO statement.[58]

Memorandum of Understanding

Currently, Massachusetts does not have a Memorandum of Understanding with the United States Environmental Protection Agency (USEPA). Therefore, persons or entities cleaning up a site in accordance with the requirements of the MCP theoretically could be subject to federal liability under CERCLA, even after an appropriate RAO has been achieved. However, USEPA's current enforcement position does not contemplate taking enforcement actions for sites that are in compliance with the MCP. In addition, PRPs may seek "comfort letters" from USEPA's Region I office or, in some circumstances, enter prospective-purchaser agreements with the federal government to minimize the threat of liability under CERCLA.

Economic Incentives for Brownfields Redevelopment

Businesses that are expanding, relocating, or building new facilities and creating permanent jobs within any of the seventy-three areas in Massachusetts that have been designated in economic opportunity areas (EOAs) may be able to take advantage of a number of economic incentives, including a 5 percent state investment tax credit, a 10 percent abandoned building tax reduction, and priority status for state capital funding. In addition, municipalities in

Massachusetts are now authorized to offer two types of local real estate tax incentives—either a "special tax assessment" or "tax increment financing."[59]

The "special tax assessment" is a five-year program covering both the existing and new value of real estate in an EOA. In year one, the tax is 0 percent of the existing and new assessed value of the real estate. In year two, up to 25 percent of the assessed value is taxed. In year three, up to 50 percent of the assessed value is taxed, and so on. This program is specifically designed to allow property owners to offset cleanup and other development costs associated with urban/industrial sites.

The Massachusetts tax increment financing (TIF) program allows municipalities to provide targeted tax incentives in order to stimulate development. Under the TIF program, a municipality can take steps to reduce a property owner's tax liabilities significantly during the years when remediation costs will be incurred, by providing a tax exemption based on a percentage of the value added through new construction.

Potential Changes to the Massachusetts Brownfields Program

Readers should note that a number of proposals to expand the Massachusetts brownfields program were being discussed as this chapter was being prepared for publication. Accordingly, interested parties should inquire about the status of these proposals at the time a particular project is being considered. The following proposals are being seriously considered in Massachusetts:

1. *Changes in the Clean Sites Initiative:* A number of proposals to modify and expand the Clean Sites Initiative have been circulated. The most important proposals focus on expanding the Commonwealth's covenant-not-to-sue program to include more sites, including sites that will be redeveloped for residential use, and more private parties, possibly including current owners and operators. In addition, the extent of the liability protection provided by the program may also be expanded.
2. *The Industrial Site Recycling Fund:* This is a proposed $15 million fund that would provide loans, credit enhancements, and grants for site assessments or remedial actions at sites located in economic target areas. The maximum amount of assistance for a single project would be $500,000. The project proponent would be required to match some or all of the loan or grant. Under the current proposal, this fund would be administered by the Land Bank, at the Massachusetts Development Finance Agency. The Executive Office of Administration and Finance would oversee the Land Bank's implementation of the fund.
3. *Capital Access/Loan Guarantee Program:* This program would be modeled on the Small Business Capital Access Program (CAP). Under this program, a borrower would pay a specified amount of points into a secondary loan loss pool held by a lender. This amount would be matched by the lender and by the state. Because environmental loans

are more risky than small-business loans, the amount paid into the pool is expected to be greater than that for the CAP (generally 1.5 percent to 3.5 percent greater). The program would support loans only for site assessment and cleanup activities.

Conclusion

Several economic and public-policy objectives have led to increased private and public sector interest in redeveloping brownfields sites in Massachusetts. As a result of this interest, a number of state programs are now available. In particular, the MCP has become a national model in the effort to privatize the waste-site cleanup process. In fact, the Council of State Governments has selected the MCP as one of its 1995 Innovations Award winners.

Because of these programs and the increased level of experience of the players involved, the tools available for the redevelopment of brownfields properties in Massachusetts are increasing in number and effectiveness. As a result, many sites in Massachusetts that were previously written off may now represent real opportunities.

Obtaining a Covenant-Not-to-Sue in Massachusetts

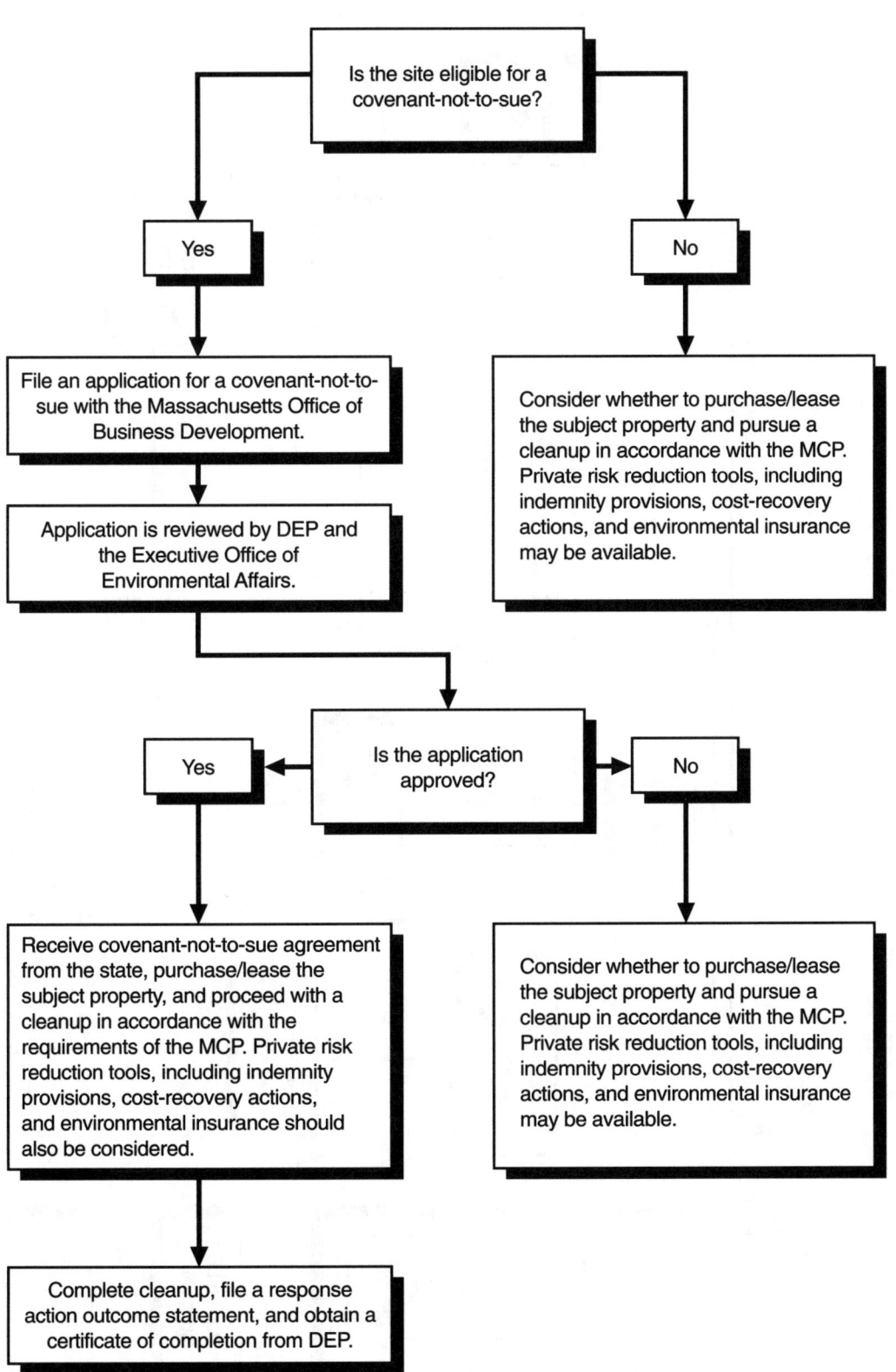

MCP Flowchart

Reporting Trigger | **Front End** — Up to 1 year | **Comprehensive Response Actions** — Up to 5 years

Type of notice:
- 2 hour
- 72 hour
- 120 day

Was an LRA successful? → Yes → No reporting required

→ IRA, RAM and/or risk characterization ← DEP site discovery

IRA, RAM and/or risk characterization → RAO

IRA, RAM and/or risk characterization → Score and tier classification

Score and tier classification → Tier I → List → Permit → DEP Decision → Assess/remediate
- A → RAO
- B → RAO
- C → RAO

List → Publicly funded sites → RAO / Cost recovery

Score and tier classification → Tier II → Assess/remediate → RAO

RAO – Response action outcome
LRA – Limited removal action
IRA – Immediate response action
RAM – Release abatement measure

Notes

1. MASS. GEN. L. ch. 21E, § 5(a) (1994).
2. *Id.*
3. *Id.*
4. *Id.*
5. *See, e.g.,* Marenghi v. Mobil Oil Corp., 624 N.E.2d 561, 563 (Mass. 1993) (proof of causation, not site ownership, is needed to impose liability under Chapter 21E, section 5(a)(5)); Griffith v. New England Tel. & Tel. Co., 610 N.E.2d 944, (Mass. 1993) (fact that site contaminated by oil brought onto site by defendant not sufficient to prove causation). *Accord* Wellesley Hill Realty Trust v. Mobil Oil Corp., 747 F. Supp. 93, 96 (D. Mass. 1990) (ruling plaintiff's § 5(A)(5) allegations sufficient to survive motion to dismiss, but noting that "something more than mere ownership is required to make a past owner of an oil contaminated property liable.").
6. MASS. GEN. L. ch. 21E, §§ (5)(a)(1)–(5) (1994).
7. *Id.* §§ 4, and 5.
8. *See* Guaranty-First Trust Co. v. Textron, Inc., 622 N.E.2d 597, (Mass. 1993) (appropriate measure of recovery for damage to property owner's real or personal property under Chapter 21E, § 5(a)(iii), must be determined by reference to common law; party may recover either (a) permanent diminution in market value or (b) expense of repairs, if less than diminished market value, plus intervening loss of rental value for period reasonably needed to repair injury).
9. MASS. GEN. L. ch. 21E, § 15 (1994). *See also* Sanitoy, Inc. v. Ilco Unican Corp., 602 N.E.2d 193, (Mass. 1992).
10. MASS. GEN. L. ch. 21E, § 4A(f) (1994).
11. *Id.* § 11A(4).
12. *Id.* § 11A(2).
13. *Id.* §§ 11A (1), (3).
14. *Id.* § 5(f). *See also* Griffith v. New England Tel. & Tel. Co., 585 N.E.2d 751 (Mass. App. Ct. 1992), *modified on reh'g,* 1992, Mass. App. LEXIS 430 (1992), *review denied,* 592 N.E.2d 751 (Mass. 1992), *review granted,* 598 N.E.2d 1133 (Mass. 1992), *rev'd, remanded,* 610 N.E.2d 944 (Mass. 1993). Hays v. Mobil Oil Corp., 736 F. Supp. 387, 393 (D. Mass. 1990) ("Indemnification clauses are still permitted to allocate the burdens of risks and costs among otherwise liable parties.").
15. MASS. GEN. L. ch. 21E, § 4 (1994). It is also important to note that "contribution protection" is sometimes available under Chapter 21E. *See* Nestor v. Haley & Aldrich, Inc., No. 90-3618 (Super. Ct., Middlesex County, Feb. 3, 1992) ("A defendant who settles in good faith under c. 21E 'buys its peace' from the plaintiff as well as relieving itself from liability to co-defendants.").
16. MASS. GEN. L. ch. 21E, § 2 (1994).
17. *Id.*
18. *Id.*
19. The Massachusetts Contingency Plan, MASS. REGS. CODE tit. 310, § 40.0000 *et seq.* (1995).
20. MASS. GEN. L. ch. 21A, §§ 19–19J (1994); MASS. REGS. CODE tit. 309, § 4.00 (1995).
21. *See, e.g.,* MASS. REGS. CODE tit. 310, §§ 40.0318(8), 40.0411(2), 40.0441(5), 40.0484(2), 40.0510(2), 40.1056 (1995).
22. *See id.* §§ 40.1101(2), 40.1110(2).
23. *See id.* §§ 40.1101–40.1190.

24. Mass. Dep't Envtl. Protection, Two Years Later: How the New 21E Program is Measuring Up (1996).
25. *See* Mass. Regs. Code tit. 310, §§ 40.0300–40.0371 (1995).
26. *Id.* §§ 40.0311–40.0312.
27. *Id.* §§ 40.0313–40.0314.
28. *Id.* § 40.0315.
29. *Id.* § 40.1600.
30. *Id.* §§ 40.0360–40.0362.
31. *Id.* § 40.0318.
32. *Id.*
33. *Id.* § 40.0406.
34. *Id.* § 40.0405(1).
35. *See id.* §§ 40.0405(2), 40.0411–40.0429.
36. *See id.* §§ 40.0405(3), 40.0440–40.0448.
37. *See id.* §§ 40.0510–40.0590.
38. *See id.* §§ 40.0810–40.0996.
39. *Id.* § 40.0972.
40. *See id.* §§ 40.0990–40.0996.
41. *Id.* § 40.0972.
42. *See id.* §§ 40.0971, 40.0981.
43. *See id.* §§ 40.0972–40.0988.
44. *Id.* § 40.0991.
45. *See id.* §§ 40.0990–40.0996.
46. *See id.* §§ 40.0930–40.0933.
47. *See id.* § 40.1003.
48. *Id.* § 40.1030.
49. *Id.* § 40.1045.
50. *See id.* §§ 40.1030–40.1050.
51. *Id.* §§ 40.1036(2), 40.1046(2).
52. *See id.* §§ 40.1070–40.1099.
53. *Id.*
54. *Id.*
55. It is important to note that the eligibility requirements for the Commonwealth's pilot covenant-not-to-sue program, as well as the scope of the available covenant-not-to-sue, have not been established by statute or regulation. Instead, the relevant requirements and liability protections are set forth in a four-part application package that includes a model covenant. "Clean Sites Initiative Application and Agreement" packages can be obtained from the following address: Massachusetts Clean Sites Initiative, Massachusetts Office of Business Development, One Ashburton Place, Boston, Massachusetts 02108.
56. See DEP's "Application and Agreement" package for the pilot covenant-not-to-sue program (Exhibit B of the Application and Agreement).
57. *Id.* at Exhibit C.
58. Mass. Dep't Envtl. Protection, Clean Sites Initiative Establishes Covenants to Ensure Cleanups, Prevent Future State Lawsuits at Hazardous Waste Sites (1995).
59. Information on all available economic incentives can be obtained from the Massachusetts Office of Business Development, One Ashburton Place, Room 2101, Boston, Massachusetts 02108.

CHAPTER 33

Michigan

GRANT R. TRIGGER

In July 1996, the Michigan legislature enacted new brownfields financial incentives,[1] significantly improving brownfields redevelopment opportunities in Michigan. These incentives, in combination with substantive liability reform and cost reductions implemented in 1995,[2] provide Michigan with the essential ingredients to revolutionize brownfields redevelopment.

The 1995 amendments to the Michigan Environmental Response Act (MERA) implemented critical liability reform and needed relief from unnecessarily costly cleanups. The new financial incentives provide the final component of a system designed to create real opportunities for brownfields success. Not only can buyers acquire, without liability, property that is known to be contaminated, but Michigan communities can now offer tax increment financing specifically for brownfields projects, tax credits against the Michigan single business tax[3] for sites located within a brownfields zone, and loans and grants for site investigation and development activities.

These new incentives should improve the already significant success Michigan has experienced in the first year of the program. In a report issued by the Michigan Department of Environmental Quality (MDEQ) in July 1996,[4] it was reported that twenty of thirty-three Michigan municipalities surveyed had seen an increase in actual development of brownfields properties, resulting in over $221,000,000 in private investment and the creation of more than 2,300 jobs.[5] The most important aspects of the new amendments in encouraging this new development were liability protection for new buyers and reduced cost of cleanup due to more reasonable risk-based standards. In approximately one year, over 425 baseline environmental assessments (BEAs) have been submitted to MDEQ, which is more than ten times the number of covenants-not-to-sue issued in the preceding four years.[6] The new financial incentives are expected to further enhance brownfields successes in Michigan.

Brownfields Financial Incentives

Administrative Organization

The recently enacted "brownfields redevelopment package" created a state board to oversee development of contaminated sites, authorized municipalities to create their own local boards, established various funds to finance redevelopment and pollution prevention projects, and authorized tax credits for certain projects. A brownfields redevelopment board has been established within MDEQ, consisting of the MDEQ director, the director of the state Department of Management and Budget, and the chief executive officer of the Michigan Jobs Commission, or their designees.[7]

In addition, the new public acts create a revitalization revolving loan program[8] within MDEQ and a revitalization revolving loan fund.[9] Local governments meeting certain requirements[10] will be able to borrow from the fund to pay for cleanup and redevelopment of contaminated sites.

Brownfields Redevelopment Authorities

Under the Brownfield Redevelopment Financing Act,[11] municipalities may establish brownfields redevelopment authorities[12] to facilitate local brownfields development by, among other things, (1) paying or reimbursing private and public parties for cleanup activities, (2) leasing, purchasing, and conveying property, (3) accepting grants and donations of property, labor, or other "things of value" from public and private sources, (4) investing the authority's money, (5) acquiring property insurance, (6) borrowing money, and (7) engaging in lending and mortgaging activities associated with property it acquires.[13] The authorities may also establish and administer local site remediation revolving funds to finance redevelopment activities.[14]

Each authority will be overseen by a board, which will be chosen by each municipality's governing body. Each board will be authorized to establish a brownfields zone and implement a brownfields plan to facilitate cleanup and redevelopment in the zone.[15]

Tax Increment Financing

Municipalities are also allowed to capture tax increment revenue on taxes levied after December 31, 1996, and apply that revenue to the redevelopment of contaminated property under certain circumstances. A brownfields redevelopment authority, by resolution of its board, may issue and sell tax increment bonds to finance the implementation of a brownfields plan. To spend revenues captured from taxes levied for school operating purposes, the cleanup activities planned to be financed from such captured taxes must receive MDEQ approval.[16]

Brownfields Plan

The brownfields redevelopment authority has the power to implement a brownfields plan that identifies the eligible properties[17] within the brownfields zone that will be subject to tax increment financing and cleanup activities. The brownfields plan must be approved by the governing body of the municipality and contain the following: (1) a description of the costs of the brownfields plan (intended to be paid with tax increment revenues), (2) an estimate of the captured taxable value and tax increment revenues for each year of the brownfields plan from each eligible parcel, (3) the method of financing the brownfields plan, (4) the maximum amount of note and bond indebtedness to be incurred, (5) the duration of the brownfields plan, (6) an estimate of the impact of tax increment financing on the revenues of all taxing jurisdictions in which eligible property is located, (7) a legal description of each parcel covered by the brownfields plan, (8) an estimate of the number of persons residing on each parcel and the number of individuals to be displaced, (9) a strategy for compliance with the Michigan Relocation for Displaced Persons Act,[18] (10) a priority list for the relocation of persons to be displaced, (11) a description of proposed use of the local site remediation revolving fund, and (12) other pertinent information required by the brownfields redevelopment authority.[19]

Generally, tax increment revenues under a brownfields plan can be used only for costs attributable to the cleanup of eligible parcels, including the cost of repaying obligations issued by a brownfields authority to pay for the costs of cleanup.[20] If a brownfields plan has been approved after notice and an opportunity to comment, a taxing jurisdiction whose taxes are being captured cannot opt out of the brownfields plan.[21]

Local Site Remediation Revolving Fund

A brownfields redevelopment authority may establish a local site remediation revolving fund, from which funds may be used to pay for "eligible activities,"[22] including preparation of BEAs, due-care activities, and additional response activities[23] within a brownfields zone. The money for the local site remediation revolving fund comes principally from certain excess tax increment revenue on properties within the brownfields zone.[24]

Single Business Tax Credit

The Michigan single business tax[25] has been amended to extend a tax credit to nonliable owners and operators of contaminated facilities who conduct cleanup or redevelopment activities on their property. These redevelopment activities may include demolition, construction, restoration, alteration, renovation, or improvement of buildings on eligible property[26] and the addition of machinery, equipment, and fixtures to eligible property pursuant to an approved brownfields plan. The credit is equal to 10% of the cost of eligible investment paid or accrued, up to a maximum cumulative credit of one million dollars.[27]

Liability Reform

Real estate and corporate transactions changed dramatically on June 2, 1995,[28] when Michigan Governor John Engler signed into law substantial and significant amendments to MERA.[29] These amendments (MERA Amendments) implemented a major overhaul of the liability structure applicable to environmentally contaminated property. No longer under Michigan law is the mere ownership of property sufficient to impose liability on those who had nothing to do with any activity causing a release of contamination. The MERA Amendments limit liability, except in certain instances, to those parties that are responsible for an activity causing a release of contamination.

The MERA Amendments revised the liability structure, modified the basis for the cleanup standards, and substantially altered transactional issues relating to contaminated property. Another significant change is the establishment of an affirmative obligation to remediate contamination in certain circumstances.[30] The former MERA imposed such an obligation only if a cleanup order were issued by MDEQ or a court. Other affirmative obligations to prevent exacerbation of existing contamination, to exercise due care to mitigate unacceptable exposures, and to take precautions against the reasonably foreseeable acts or omissions of a third party were also added to MERA.[31] The following discussion presents a general summary of the final legislation as enacted.

Change in Liability Structure

New Definition of "Facility"

Under the MERA Amendments, liability for remediation is imposed, with certain exceptions, only on those who own or operate a "facility" and are/were responsible for an activity causing the contamination at the facility.[32] The definition of "facility" limits liability to those properties that are, in fact, facilities with contamination present in excess of residential criteria. Even if some contamination is present, if that contamination does not exceed the residential criteria, then the site is, by definition, not a facility and therefore none of the related obligations are applicable.[33]

Elimination of "Status" Liability

Strict status liability was essentially eliminated by removing the provisions of the statute that imposed strict liability on those who merely own or operate property.[34] The current owner or operator of a facility or the owner or operator of a facility at the time of disposal of a hazardous substance is liable only if the owner or operator is responsible for an activity causing a release or threat of release.[35] A person who becomes an owner or operator of a facility after June 5, 1995, is liable for the contamination at the facility unless it complies with the BEA procedures,[36] discussed in detail below. Preparation of a BEA in accordance with the statute provides the new owner or operator with an exemption from liability unless the owner or operator is responsible for an activity causing the contamination.[37] Compliance with these requirements, however, does not exempt one from liability for subsequent releases or threats of releases resulting from an activity for which one is responsible.[38] In addition, the MERA Amendments eliminated liability for "interim owners and operators"—persons who formerly owned or operated the property, and who did not own or operate the property at a time when hazardous substances were released at the facility.[39]

Changes in "Arranger" Liability

The liability of persons who arranged for the disposal of a hazardous substance at a facility owned by another is largely unchanged, except that, as with owners and operators, arranger liability is limited to contamination at a location falling within the new definition of "facility."[40] New exemptions from arranger liability are provided for persons who arrange for the sale or transport of a secondary material for use in making a new product[41] and for persons who arrange the lawful transport or disposal of products or containers commonly used in residential households and used in that person's residential household.[42]

The MERA Amendments also created special liability provisions for lenders, fiduciaries, state and local units of government, and other special categories that are addressed below. Specific provisions exempting residential property and lessees who use the property for a retail, office, or commercial purpose are also included in the MERA Amendments.[43]

Burden of Proof

MDEQ bears the burden of establishing a prima facie case that a person is liable under MERA. That person then bears the burden of showing, by a preponderance of the evidence, that he or she is not liable.[44] Of course, the burden of proof is affected by the fact that MDEQ must prove that there is a facility and that the person to be held liable is either an owner or operator who is responsible for an activity resulting in a release at the facility.[45] Alternatively, anyone who becomes an owner or operator after June 5, 1995, and who does not comply with the BEA procedures, would also be liable if MDEQ shows merely that the person failed to prepare a BEA.

Special Liability Protection for Lenders and Others

Definition of "Lender"

The MERA Amendments supplemented and added a number of special liability protections for lenders, fiduciaries, and other categories of potential owners or operators of a facility. The term "lender" replaces the former term "commercial lending institution" and now specifically includes other types of lenders such as insurance companies and pension funds.[46] Thus, the term "lender" now includes virtually anyone who holds an interest in property to secure a debt, regardless of whether that person is a commercial lender.

Lender and Fiduciary Protections

As described above, the basic lender-liability protections of MERA were not affected by the MERA Amendments, except to make them applicable to "lenders" rather than "commercial lending institutions." A lender or other person acting as a fiduciary, who has not participated in the management of a facility, is exempted from liability as an owner or operator.[47] This exemption does not relieve a fiduciary from liability that the fiduciary has assumed or from liability for negligent, reckless, or willful misconduct,[48] nor does it prevent claims against the assets of an estate or trust that contains the facility.[49] These provisions also apply to lenders who assumed control of property in a fiduciary capacity as authorized by the banking code[50] or National Bank Act,[51] pursuant to an agreement entered on or before August 1, 1990.[52] If a lender assumes control of a facility in a fiduciary capacity pursuant to an agreement entered after August 1, 1990, the lender is subject to this exemption if the lender has served only in an administrative, custodial, or financial capacity and has not exercised sufficient involvement to control the owner or operator's hazardous-substance handling practices.[53]

A lender that conducts a BEA within 45 days of foreclosure on a facility and who is not responsible for an activity causing the contamination may take title to, or possession of, the property without liability for any existing contamination. If the lender chooses, it may then transfer the property to the state if the lender complies with the following requirements:[54] (1) lists the facility for sale for a period of at least 120 days within nine months following foreclosure, (2) takes reasonable care in maintaining the property and permanent fixtures, (3) provides all environmental information relating to the facility and available to the lender to MDEQ, (4) complies with any MDEQ order,[55] and (5) undertakes appropriate response activities to abate any threat of fire or explosion or an imminent hazard through direct contact with hazardous substances.[56]

Lenders before foreclosure, and state or local units of government, do not incur liability for damages resulting from activity taken in response to a release or threat of release.[57] This exemption does not protect state or local units of government from liability for grossly negligent, reckless, willful, wanton, or intentional misconduct.[58]

Other Special Liability Exemptions

Other exemptions are provided for state or local units of government that acquired ownership or control of a facility involuntarily by virtue of their governmental function, including through bankruptcy, tax delinquency, abandonment, seizure, receivership, forfeiture, or transfer from a lender.[59]

State and local units of government that hold or acquire an easement interest in a facility or an interest through dedication are also exempt.[60] A state or local unit of government, as lessor, is not liable for contamination at the property unless the state or local unit of government is otherwise responsible for the contamination.[61]

The MERA Amendments also include an "innocent purchaser" defense that exempts from liability a person who did not know, and had no reason to know, that the property was a facility at the time the person became an owner or operator of the property.[62] To qualify for this exemption, the owner or operator must have made all appropriate inquiry into the condition and history of the property before acquisition, consistent with good commercial or customary practice.[63] Any specialized knowledge or experience of the owner or operator, any difference between the purchase price and the value of the property if uncontaminated, commonly known information about the property, the obviousness of the presence of contamination, and the ability to detect a release by appropriate inspection are all relevant considerations in determining whether an owner or operator is entitled to the innocent-owner defense.[64]

Leaking Underground Storage Tank (LUST) Act

Recent amendments to Part 201 and Part 213 of the Environmental Code (the former Michigan LUST Act)[65] reconcile the liability provisions of both parts, matching the essential liability components of Part 213 to Part 201. Previously, the owner or operator of a LUST, or the property on which a LUST is located, was responsible for site cleanup, regardless of whether the person caused the leak.[66] Under the new definitions, only those "owners" or "operators" who control or hold an interest in underground storage tank (UST) systems and are liable for contamination under Part 201 of the Environmental Code are subject to responsibility for associated LUST contamination.[67]

Migration

The owner or operator of property onto which contamination migrates is not liable for that contamination unless he or she is responsible for an activity causing the contamination.[68] In addition, an owner or operator is exempt from the section 20107a due-care obligations (described below) if the contamination migrated onto the property and the owner or operator is not responsible for an activity causing contamination.[69]

Scope of Liability Protection

A person who is provided an affirmative determination by MDEQ of an exemption from liability in connection with a BEA, who has complied with the MERA Amendments, or who is exempt from liability under MERA is also exempt from liability under Part 17 (the former Michigan Environmental Protection Act), Part 31 (the former Water Resources Commission Act) and common law for claims related to response activity.[70] This exemption does not extend to liability arising from violation of a permit.[71] This exemption may also not apply to tort claims unrelated to the performance of response activities, tort claims for damages that result from response activities, or tort claims related to the person's responsibilities to exercise due care and take reasonable precautions.[72]

An additional innovative provision provided by the MERA Amendments is designed to provide limited protection for federal Comprehensive Environmental Response, Compensation, and Liability Act (CERCLA)/Superfund[73] claims and contribution protection under both CERCLA and MERA.[74] Under this provision, a person who is in compliance with MERA is considered to have resolved his or her liability to the state in an administratively approved settlement under CERCLA.[75] If upheld by the courts, this provision will provide contribution protection from any CERCLA claim that could be brought by the state, and prevent the state from suing under state law. This provision may not prevent the federal government from bringing federal claims under CERCLA.[76] Although the MERA Amendments substantially altered state law, liability and other aspects of federal law remain unaffected and must still be considered.[77]

Memorandum of Agreement

On July 10, 1996, Region V of the United States Environmental Protection Agency (USEPA) executed a Superfund Memorandum of Agreement Addendum I with MDEQ, in which USEPA promised not to "plan or anticipate any federal action against a covered party (owner, operator, generator or transporter) under Superfund" provided that the covered party is in compliance with MERA. This promise extends to properties in Michigan that are not National Priorities List (NPL) sites or otherwise subject to USEPA action. Although this agreement is not enforceable by a third party, it does offer some degree of assurance that USEPA will not initiate an enforcement action against a party who has, in good faith, addressed cleanup issues under state law.

Cleanup Criteria

One of the most significant innovations of the MERA Amendments is the restructuring of the cleanup criteria. The previous categories of cleanup criteria, Types A, B, and C were eliminated and replaced with an approach based on land use. Residential criteria will be applied to residential property, commercial criteria to commercial property, recreational criteria to recreational property, industrial criteria to industrial property, and so forth.[78] In addition, there are two categories of land-use criteria: limited and unlimited.[79] At an unlimited land-use site, exposure barriers are not necessary to prevent exposure to contamination on the property.[80] At a limited site, however, it may be necessary to impose an exposure barrier, such as a parking lot or other barrier, or limit access through some form of institutional control, such as a deed restriction, ordinance, or other related mechanism.[81] In any case, cleanup criteria cannot be more strict than the current Type A criteria, background, or the applicable detection limit.[82]

MDEQ cannot approve a remedial action plan using any cleanup category other than unlimited residential unless the person proposing the plan documents that (1) the current zoning of the property is consistent with the criteria category proposed, (2) the governing zoning authority intends to change the zoning designation so that the proposed criteria are consistent with the new zoning designation, or (3) the property use is a legal nonconforming use.[83] MDEQ may approve a remedial action plan that achieves categorical criteria based on greater exposure potential than the criteria applicable to current zoning.[84]

If a cleanup is based on criteria for the unlimited residential category, no land-use restrictions or monitoring are required once the standards are met.[85] If a remedial action plan uses one of the other unlimited land-use categories, a "Notice of Approved Environmental Remediation" must be recorded with the register of deeds by the owner of the property within twenty-one days after selection or approval by MDEQ of the remedial action, or after construction of the remedial action.[86] If MDEQ approves a remedial action using criteria for one of the limited land-use categories, a legally enforceable agreement with MDEQ must provide for (1) land-use or resource-use restrictions, (2) monitoring, (3) operation and maintenance, (4) permanent markers to describe restricted areas, and (5) financial assurance to pay for monitoring, operation, maintenance, oversight, and other costs.[87] If MDEQ concurs with an analysis provided in a remedial action plan that one or more of these requirements (other than the land-use or resource-use restrictions) is not necessary, that element may be omitted from the enforceable agreement.[88]

If a remedial action plan uses cleanup criteria for one of the limited land-use categories, the owner must file, with the register of deeds, a restrictive convenant that (1) describes generally the property uses that are consistent with the categorical criteria and limitations that are part of the remedial action plan, (2) restricts activities that may interfere with the remedial action or measures taken to assure the effectiveness of the remediation, (3) restricts activities that may result in exposures above the levels established in the remedial action plan, (4) requires notice to MDEQ fourteen days before any conveyance of the property, (5) grants to MDEQ the right to enter the property for the purpose of determining compliance with the remedial action plan, and (6) allows MDEQ to enforce the restrictive covenant by legal action in a court of appropriate jurisdiction.[89] If implementing the restrictions necessary for use of one of the limited land-use criteria through a restrictive covenant is impracticable, the necessary restrictions may be established through an institutional control, such as a local ordinance.[90]

Any land-use or resource-use restrictions relating to remedial action must be disclosed to the transferee before any interest in the property may be conveyed.[91] In addition, even if land-use or resource-use restrictions have not been imposed, any person who has information or knowledge that he or she owns a facility may not transfer an interest in that real property without providing to the transferee written notice of the general nature and extent of the contamination.[92] A person who implements an MDEQ-approved remedial action plan that relies on local zoning must, within thirty days of approval of the plan, provide notice of the land-use restrictions to the zoning authority for the local unit of government.[93]

Basis of Calculating Cleanup Criteria

When MERA was amended in 1990, the fundamental principles of risk assessment were adopted as the foundation for calculating the applicable cleanup criteria.[94] At that time, the risk basis for carcinogenic substances was established at a one-in-one-million risk of an additional cancer.[95] The MERA Amendments have revised the risk basis for carcinogens so that it is now "the 95% upper bound on the calculated risk of one additional cancer above the background cancer rate per 100,000 individuals."[96] For hazardous substances that pose a risk of an adverse health effect other than cancer, cleanup criteria for the noncarcinogenic effects shall be based on a hazard quotient of 1.0.[97]

The MERA Amendments establish a less conservative approach for establishing cleanup criteria than prior regulations, including specific provisions requiring that cleanup criteria be based on reasonable assumptions, using only relevant and reasonable exposure pathways.[98] In addition, for noncarcinogenic hazardous substances present in soil, the intake (amount consumed) is assumed to be 100 percent—the relative source contribution (RSC)—of the protective level rather than the 20 percent previously found in the statute. The 20 percent RSC, in effect, resulted in dividing all calculated cleanup criteria by five. A different RSC will be used if information is available that demonstrates that such an RSC is appropriate.[99]

The Michigan groundwater/surface-water interface values (GSI) remain viable and applicable to all remediation activities that may be associated with surface water.[100] Accordingly, any potential benefits in terms of less stringent cleanup criteria that may result from the change to a one-in-one-hundred-thousand risk level may be moot if a GSI applies to that site. This would be particularly true of most metals, which typically have a GSI that is already lower than even the former one-in-one-million risk level for human carcinogens.[101]

Remedial Actions

The rules in effect before the enactment of the MERA Amendments[102] limited the scope of acceptable groundwater remediation options by requiring that (1) the horizontal and vertical extent of contamination not increase after initiation of remediation and (2) all remedial actions addressing the remediation of an aquifer provide for the removal of contamination either through active remediation or as a result of naturally occurring "biological or chemical processes which can be documented to occur at the site."[102] Because these limitations are inconsistent with a risk-based approach to remediation, the MERA Amendments include a provision allowing approval of remedial action plans that do not meet these requirements if MDEQ makes a finding that the remedial action is protective of the public health, safety, and welfare, and the environment.[103] This exception does not apply if the contamination was caused by a release resulting from grossly negligent or intentional conduct, unless complying with these groundwater requirements would be technically infeasible.[104] Approval of a remedial action plan under this provision may be granted only if any one of several conditions is met: (1) compliance with the standards is technically impractical, (2) the remedial action will attain a standard of performance equivalent to that required in the Act 307 rules[105] in a reasonable amount of time, (3) the adverse environmental impact of attempting to comply with the rules would outweigh the benefit of complying, or (4) naturally occurring remediation is documented to occur and no adverse environmental impacts will occur from migration of the hazardous substances during the remedial action except as specified in an approved remedial action plan, the implementation of which includes enforceable land-use restrictions or other institutional controls to reduce any unacceptable risk from exposure to the hazardous substances.[106]

An aquifer monitoring plan is required to be included in all remedial action plans that address aquifer contamination, unless MDEQ determines that the vertical and horizontal extent of hazardous-substance concentrations above the residential criteria will not significantly increase in the absence of active removal of those hazardous substances from the aquifer.[107]

If a remedial action plan allows for venting of groundwater,[108] the discharge must comply with the requirements of Part 31 of the Environmental Code. A water-discharge permit would not be required for the venting of contaminated groundwater to the surface water, if it is part of an approved remedial action plan.[109] In addition, a change in Michigan law providing the authority to apply a "mixing zone" in the receiving surface water was also adopted concurrently with the MERA Amendments.[110]

If state funds will be used for a remedial action or there is a significant public interest, MDEQ must provide notice to the county and the local unit of government within thirty days after completion of the remedial investigation.[111] Among other things, the notice must provide an opportunity for people in the local unit of government to meet with MDEQ. The local unit of government (or any twenty-five citizens) may request that MDEQ hold a public meeting.[112] In addition, before a remedial action plan can be approved that involves (1) the use of state funds, (2) cleanup criteria for limited land-use categories, (3) institutional controls instead of restrictive covenants, or (4) significant public interest, MDEQ must provide public notice and an opportunity for comment, as well as an opportunity for a public meeting on the proposed plan.[113]

Coordination of Other Legal Requirements

Other Legal Claims

An important factor that historically posed the potential to undermine final resolution of site cleanups is whether there may be other parties who potentially have claims against someone who owns contaminated property. The MERA Amendments address, in several provisions, the general principal that once compliance with MERA has been attained, there should be no other claims under any other statute or common law. In addition, the MERA Amendments clarified the jurisdiction of MERA relative to Part 111 of the Environmental Code (the former Hazardous Waste Management Act) by providing that corrective action under Part 111 satisfies a person's remedial obligations under MERA and Part 31 of the Environmental Code.[114]

LUSTs

The relationship between the liability scheme for owners and operators of LUSTs and the provisions of Part 201 of the Environmental Code was clarified by legislation adopted in March 1996 (March 1996 Amendments).[115] Under Part 213 of the Environmental Code, the owner or operator of a LUST or the property on which a LUST is located was responsible for site cleanup, regardless of whether the party caused the leak.[116] The March 1996 Amendments modified Part 213 to limit the definition of "owner" and "operator" to those parties who control or hold an interest in UST systems *and* are liable for contamination under Part 201 of the Environmental Code.[117] As a result, the Part 201 liability scheme (causation-based liability) has been incorporated into Part 213 of the Environmental Code.

Provisions in the March 1996 Amendments also (1) address the removal of soil at LUST sites,[118] (2) provide that penalties under Part 213 do not begin to accrue until MDEQ has notified the party on whom the penalty is being imposed of the potential assessment of penalties, and (3) allow MDEQ to issue closure letters for sites meeting the Part 213 cleanup criteria.[119]

Contamination resulting solely from a LUST must be remediated according to Part 213 of the Environmental Code. At sites with contamination resulting from both a LUST and one or more other sources, the owner or operator may choose to remediate the LUST release under either Part 201 or Part 213 of the Environmental Code.[120] Part 213 previously required that MDEQ determine the part under which a site should be remediated.

By incorporating the Part 201 liability scheme into Part 213, the March 1996 Amendments permit purchasers of property contaminated by LUSTs to take advantage of the same liability protections as purchasers of other contaminated property under Part 201, such as limiting liability by preparing a BEA. This provision, which is retroactive,[121] should prove beneficial to purchasers of LUST sites who do not plan to operate USTs on the property. Purchasers who plan to operate USTs will, in most cases, need to conduct a Type C BEA or investigate alternative liability defenses on the premise that the same hazardous substances will be used on the property. If the USTs have been removed or properly closed in place, liability under Part 213 of the Environmental Code should be fulfilled and any remaining contamination problems would be subject to Part 201 and not Part 213.[122]

Due Care, Reasonable Precautions, and Exacerbation of Existing Contamination

Duty to Exercise Due Care and Take Reasonable Precautions

Section 20107a of the MERA Amendments imposes obligations (107a Obligations) on all owners and operators of property who are aware that it is contaminated (that is, they have knowledge it is a "facility") to exercise due care to avoid unacceptable exposure to the contamination on the property, to take reasonable precautions to avoid any inappropriate conduct by a third party, and to avoid exacerbating existing contamination on the site, regardless of whether that owner or operator caused the contamination.[123] These duties are in addition to any duty to perform response action under MERA.[124] Any party who fails to take due care to avoid unacceptable exposures, fails to take reasonable precautions to avoid foreseeable third-party actions, or exacerbates contamination will be subject to liability.[125] This liability extends to response costs and natural-resource damages attributable to the new or exacerbated contamination.[126] The liability does not extend to preexisting contamination unless the person is otherwise liable for that contamination.[127]

Exacerbation

"Exacerbation" refers to activities by the owner or operator that cause existing contamination to migrate beyond the boundaries of the source of the contamination at levels above the relevant cleanup criteria, or that cause increased response costs.[128]

Exemptions from 107a Obligations

The 107a Obligations do not apply to a local unit of government that is exempt from liability or that obtained the property before the effective date of the MERA Amendments.[129] These duties also do not apply to an owner or operator of property onto which contamination has migrated, unless the owner or operator is responsible for causing the release of the hazardous substance.[130] This section does not impose a duty on a person who holds a utility easement or severed subsurface-mineral rights, except in connection with that person's activities at the facility.[131]

The BEA

As noted earlier, the MERA Amendments retain the process whereby standard Phase I environmental assessments can be used to provide an innocent-purchaser defense if performed before a property is purchased.[132] If the property is known to be contaminated (that is, it is known to be a "facility" and the buyer cannot invoke the innocent-purchaser defense), a BEA and related documents may allow the buyer, nonetheless, to buy the property and qualify for an exemption from liability.[133]

Qualifying for the Exemption

To qualify for this exemption, a new owner or operator of property must perform a BEA within forty-five days after purchase, occupancy, or foreclosure, whichever is earliest.[134] A BEA is an evaluation of environmental conditions that exist at a facility at the time of purchase; it reasonably defines the existing conditions at the facility so that, in the event of a subsequent release caused by the new owner or operator, there is a means of distinguishing the new release from preexisting contamination.[135] The BEA may also be performed before acquiring the property.[136] If an initial assessment of the property confirms that the property is a facility and a BEA is prepared, the owner or operator must disclose the BEA to MDEQ and to any subsequent transferee, or potentially lose the benefit of this exemption.[137] If an initial assessment indicates that the property is not a facility, the owner or operator would need to qualify for a defense to liability as an innocent purchaser.[138] Under this provision, a person is not liable as an owner or operator of a facility if he or she did not know, and had no reason to know, the property was a facility, after making appropriate inquiry to determine whether the property was a facility at the time of acquisition. This defense assumes, of course, that the owner or operator is not otherwise responsible for an activity that caused a release at the facility.[139]

If the property is a facility, the owner or operator may, in addition to disclosing the BEA to MDEQ, petition MDEQ—within six months after completion of the BEA—for a written determination that the owner or operator qualifies for the exemption.[140] The petition must be accompanied by a $750 fee.[141] Along with the petition, a request may be submitted for a determination that the owner's or operator's proposed use of the facility will comply with the requirements of Part 201 to exercise due care, take reasonable precautions, and avoid exacerbation of contamination at the facility.[142]

MDEQ is required to grant or deny the requested determination within fifteen business days after receipt of a petition.[143] If MDEQ denies the petition, it must give the specific reasons for the denial and specify, if possible, how the applicant could qualify for the exemption.[144] MDEQ may condition approval of a petition on completion of response activities described in the petition.[145]

If MDEQ affirms that the new owner or operator qualifies for an exemption to liability under MERA, the owner or operator also qualifies for relief from liability, under Parts 17 and 31 of the Environmental Code and under the common law, resulting from the contamination described in the petition.[146] An affirmative determination in response to a petition also constitutes a "settlement" for purposes of CERCLA liability.[147] A positive response to a petition does not exempt the owner or operator from liability for a violation of any permit issued under state law, nor does it provide an exemption to the requirements of section 20107a for a use or activity on the property inconsistent with an MDEQ determination.[148]

If a BEA is performed in accordance with the requirements in the statute, the new buyer of the property will not incur liability for existing contamination. In effect, the state has taken the essential components of a covenant-not-to-sue and incorporated them into the simple concept of a BEA and the due care/reasonable precaution/exacerbation provisions found in section 20107a.

Affirmative Obligation to Remediate

Before the MERA Amendments, no affirmative obligation to remediate contamination existed under the statute. In other words, one could own contaminated property, know that it exceeded applicable cleanup standards, and yet not be in violation of the statute if no cleanup activities were initiated on that property without an order from MDEQ or a court.[149] This condition changed with the adoption of the MERA Amendments and with the incorporation of a new affirmative obligation to remediate.[150]

Except in the case of contamination caused solely by a permitted release,[151] the owner or operator of a facility who has knowledge that the property is a facility *and who is liable* under MERA must diligently pursue response activities.[152] This obligation is an independent affirmative obligation that does not require any direction or order from MDEQ.

For any release of hazardous substances that occurs after June 5, 1995, the owner or operator of the facility must immediately implement source control or removal measures to remove or contain such hazardous substances, if those measures are technically practical and cost effective, and provide protection to the environment.[153] If groundwater is likely to be affected by such a release, groundwater contamination must be prevented if it can be prevented, by measures that are technically practical and cost effective, and provide protection to the environment.[154]

A person who is liable can also be required by MDEQ to (1) plan and implement interim response activities, (2) plan and implement evaluation activities, (3) perform any technically sound response activity necessary to protect the environment, (4) submit a remedial action plan to achieve the appropriate cleanup criteria, and (5) implement a remedial action plan.[155] Prior approval for response activity is not required unless the response activity is being performed pursuant to a judicial decree or administrative order or agreement.[156]

A person who is liable under MERA for contamination at a facility may petition a court for access to the facility to conduct response activities that are approved by MDEQ.[157] If access is granted under this provision, the owner or operator of the facility is not liable for conditions associated with the response activity that may present a threat to public health or safety.[158] Of course, the owner or operator is also not liable for the contamination that is the subject of the response activity, unless the owner or operator is otherwise liable under section 20126.[159]

Within six months after receiving a response plan, MDEQ must review the plan and approve it or return it with suggested changes.[160]

Anyone who is not diligently pursuing response activities could potentially be subject to fines and penalties for the failure to do so. One of the issues that has been substantially debated is the definition of "diligently pursue." Given limited resources and time, many property owners may struggle with this term. Until further clarification of this issue occurs, it may be very difficult to develop a reasoned plan to address these potential obligations, particularly for those who own more than one parcel of potentially contaminated property. However, it is also possible that taking deliberate steps to address potentially contaminated sites will be adequate to demonstrate the owner has met its obligation to "diligently pursue." Anyone who manages a multitude of properties cannot be expected to respond to potential remediation issues at each site simultaneously. It certainly would not make sense for the state to interpret these provisions to impose an obligation on owners of many properties, which the state has demonstrated it cannot itself meet. That is, the state cannot manage and clean up all the sites for which it is responsible because it does not have enough money or manpower. How can the state reasonably expect a private party to perform under conditions the state has demonstrated it cannot perform? Everyone has a limit of time and resources to respond to remediation issues that must be considered in a reasonable interpretation and application of the MERA Amendments.

Conclusion

The MERA Amendments have dramatically changed the conditions and ground rules for real estate and other transactions in Michigan. Property that was previously considered untouchable can now be purchased and developed, without liability for existing contamination the developer did not cause. The substantial success in the first year after the MERA Amendments ($221 million in private investment and creation of over 2,300 jobs) is convincing evidence of the benefits created by the MERA Amendments.

The new financial incentives offer expanded opportunities to demonstrate the continued success of brownfields redevelopment in Michigan. As greater appreciation for these changes becomes better known, even more abandoned or underutilized properties should be reintroduced to productive uses.

Purchase Agreement

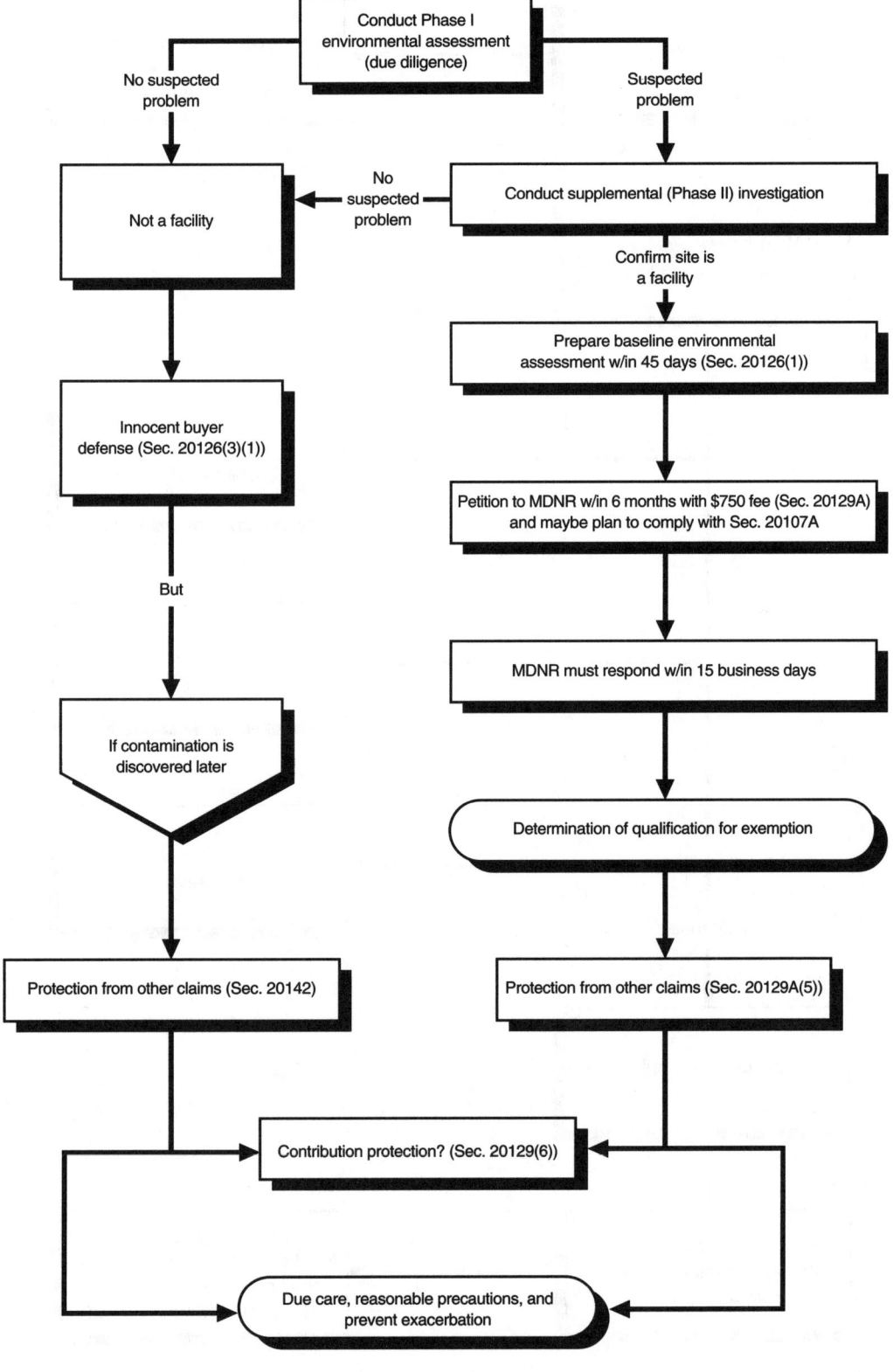

Section 107A Responsibilities

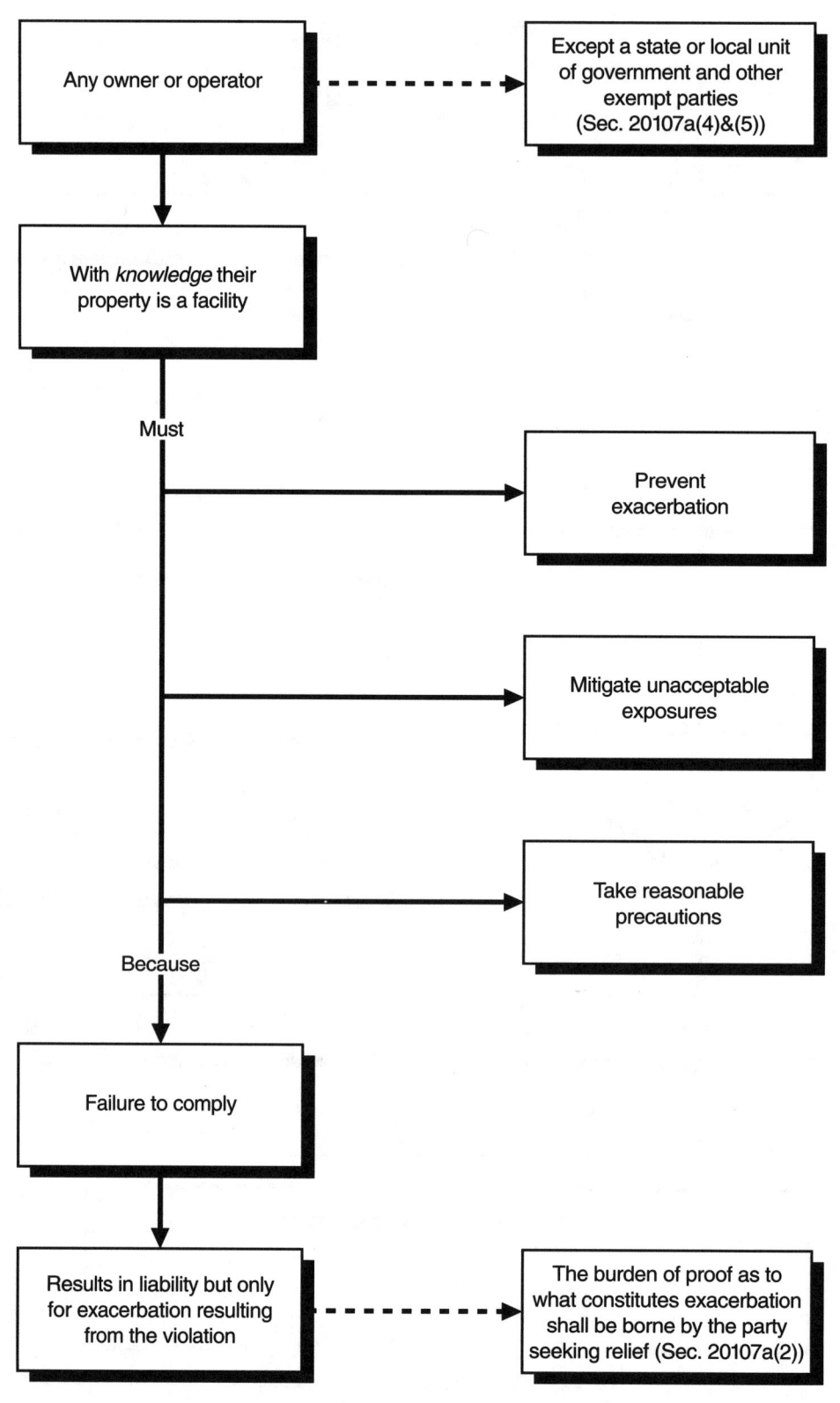

Section 114(1)(g) Affirmative Obligations

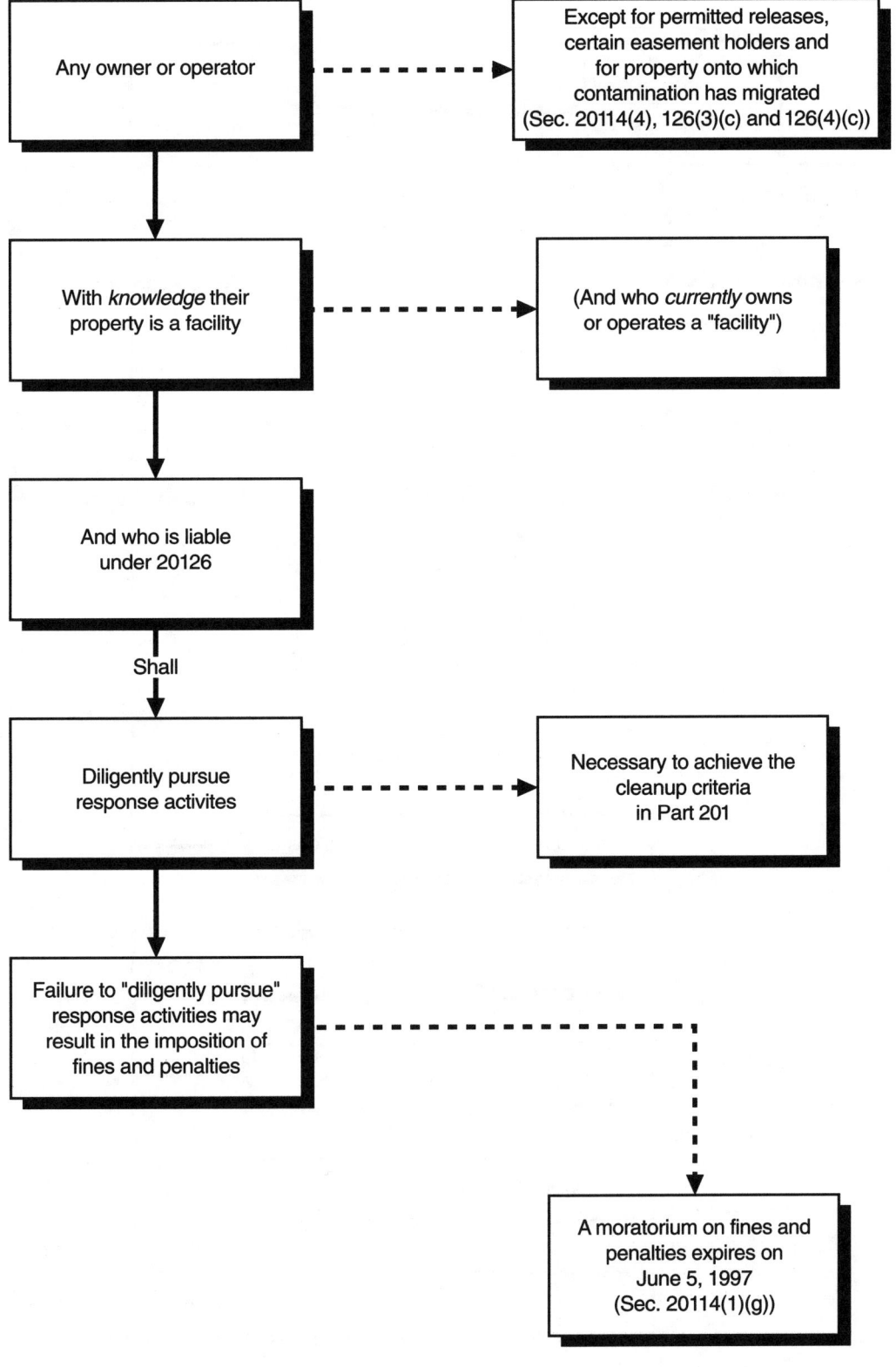

Land Use/Risk-Based Closures

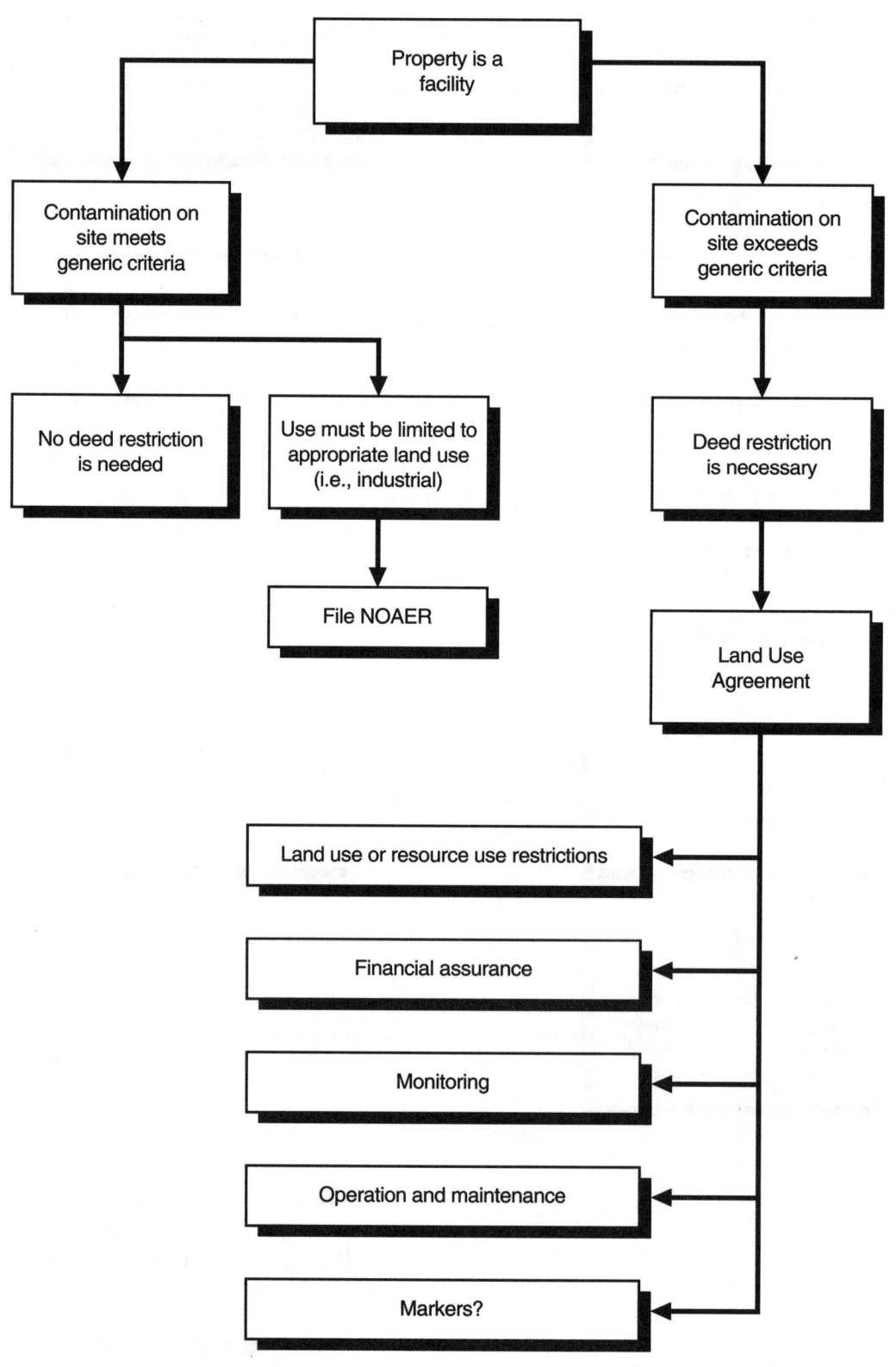

Notes

1. 1996 Mich. Pub. Acts 380–384.

2. The amendments became effective on June 5, 1995, and have been codified in part 201 of the Natural Resources and Environmental Protection Act (Public Act 451 of 1994, section 324.101 *et seq.* of Michigan Compiled Laws, also known as the "Environmental Code"). For consistency purposes, reference hereinafter will be to Part 201 of the Environmental Code.

3. MICH. COMP. LAWS §§ 208.38d(1)–(8) (1996).

4. The report was prepared by the Environmental Response Division of MDEQ and is entitled *The Part 201 Amendments: One year later, an interim evaluation of the effectiveness of the June 5, 1996 amendments to Part 201 of the Natural Resources and Environmental Protection Act, Act 451 of 1994, as amended.* Copies of the report may be obtained by contacting MDEQ at 517/373-9837.

5. *Id.* at i.

6. *Id.*

7. MICH. COMP. LAWS § 324.20104a (1996).

8. *Id.* § 324.20108b.

9. *Id.* § 324.20108a.

10. An applicant for these funds must be a county, city, township, village, or brownfields redevelopment authority (established under 1996 Mich. Public Acts 381) and the funds must be used for evaluation, demolition, and certain interim response activities at a contaminated or suspected contaminated site. To receive a loan, an applicant must enter an agreement with MDEQ that includes scope of services, timelines, and related provisions.

11. MICH. COMP. LAWS § 125.2651 *et seq.* (1996).

12. *Id.* § 125.2653.

13. *Id.* § 125.2657.

14. *Id.* § 125.2658.

15. *Id.* § 125.2655.

16. *Id.* § 125.2665(1).

17. An eligible property is defined to mean "a facility as that term is defined in section 20201 of part 201" of the Environmental Code, and "adjacent or contiguous parcels if the development of the adjacent and contiguous parcels is estimated to increase the captured taxable value of the facility for which eligible activities are proposed under a brownfield plan. Eligible property shall include, to the extent included in brownfield plan, personal property located on the facility." MICH. COMP. LAWS § 125.2652(1) (1996).

18. MICH. COMP. LAWS § 213.321 *et seq.* (1996).

19. *Id.* § 125.2663(1).

20. *Id.* § 125.2663(3).

21. *Id.* § 125.2663(10).

22. *Id.* § 125.2652(j).

23. *Id.* § 125.2652(a).

24. *See id.* § 125.2663(5).

25. *Id.* § 208.38d(1-8).

26. *See id.* § 324.20101(o).

27. *Id.* § 208.38d(2).

28. The amendments became effective on June 5, 1995, upon filing with the Secretary of State.

29. MICH. COMP. LAWS § 324.101 *et seq.* (1996). The MERA Amendments adopted through 1995 Mich. Public Acts 71 have been codified in Part 201 of the Environmental Code.

30. MICH. COMP. LAWS § 324.20114(1)(g) (1996).

31. *Id.* § 324.20107a.

32. *Id.* § 324.20126(1).

33. The definition of "facility" is found at section 324.20101(1)(o) of Michigan Compiled Laws. The MERA Amendments also deleted the term "stored" from the definition. As a result, arguably hazardous substances may now be "stored" at a site and the site would not be, by definition, a facility. The definition specifically notes that if response activities have been completed, consistent with residential criteria, the site will no longer be a facility. Also, the definition refers to section 20120a(17). Under this section, it is possible a site could become a facility even if all of the test results are below the general residential criteria, if MDEQ determines certain conditions are relevant at a site.

34. Anyone who becomes an owner or operator of a facility after June 5, 1995, and who does not prepare a BEA would not qualify for an exemption from liability under section 20126(1)(c) and would therefore be strictly liable for any existing contamination.

35. MICH. COMP. LAWS §§ 324.20126(1)(a)–(b) (1996).

36. *Id.* § 324.20126(1)(c).

37. *Id.* § 324.20126(2).

38. *Id.*

39. See former section 20126(1)(c), which was repealed by the MERA Amendments. Liability is established only for those parties listed in section 20126, which does not include "interim" owners or operators.

40. MICH. COMP. LAWS § 324.20126(1)(d) (1996).

41. *Id.* § 324.20126(1)(d)(i) (limits secondary material to "scrap metal, paper, plastic, glass, textiles, or rubber").

42. *Id.* § 324.20126(1)(d)(ii).

43. *Id.* §§ 324.20126(3)(f), 324.20126(3)(j).

44. *Id.* § 324.20126(6).

45. Representatives of the Attorney General's office suggested during the legislative debate that a rebuttable presumption should be established that would impose responsibility for an activity causing a release if (1) the release occurred during the period of the party's ownership or operation, and (2) the contamination found at the facility contained substances of the same character as those used by the party sought to be held liable. This presumption would have avoided MDEQ proving, in fact, what activity occurred that caused the contamination. The legislature did not incorporate this presumption.

46. MICH. COMP. LAWS §§ 324.20101(1)(s)(i)–(xii) (1996).

47. *Id.* § 324.20101b(1).

48. *Id.* § 324.20101b(1)(a).

49. *Id.* § 324.20101b(1)(b) (provides that protection from liability in section 20101b does not "[p]revent claims against the assets that are part of or all of the estate or trust that contains the facility; any other estate or trust of the decedent, grantor, ward, or other person whose estate or trust contains the facility that is administered by the lender or other person; or any other estate or trust of the decedent, grantor, ward, or other person whose estate or trust contains the facility. Such claims may be asserted against the fiduciary in its representative capacity, whether or not the fiduciary is personally liable.").

50. 1969 Mich. Pub. Acts 319 (sections 487.301 through 487.598 of Michigan Compiled Laws).

51. 12 U.S.C. § 1 *et seq.* (1996).

52. MICH. COMP. LAWS § 324.20101b(2) (1996).

53. *Id.* § 324.20101b(3).

54. Obviously, if the property has significant value and the lender can resell or otherwise realize a return on its investment, it would not simply turn the property over to the

state. On the other hand, if the lender is unable to sell or otherwise make use of the property, it may need the ability to transfer the property to the state to eliminate the property from its portfolio.

55. Section 324.20119 of Michigan Compiled Laws provides MDEQ with the authority (when there may be an imminent and substantial endangerment to the public health, safety, or welfare, or to the environment) to issue administrative orders to address such conditions.

56. MICH. COMP. LAWS §§ 324.20126(7)(a)–(e) (1996).

57. Id. § 324.20126(5).

58. Id. If a lender conducts response activities in a grossly negligent manner, does that lender have a defense to liability based on this section? If the limitation regarding grossly negligent conduct applies only to the local unit of government, could there be an argument that the limitation does not therefore apply to the lender? If so, a lender might end up with protection even greater than originally intended.

59. Id. § 324.20126(3)(a).

60. Id. § 324.20126(3)(b).

61. Id. § 324.20126(3)(e).

62. Id. § 324.20126(3)(h).

63. Id.

64. Id. In effect, there should be no substantive change in the way Phase I environmental assessments are prepared, because the essential standard that must be met has not changed in any material way. The scope of the BEA, however, is an entirely different matter, and may require additional review and strategic consideration of future-use expectations on the property to be purchased.

65. Part 213 refers to Part 213 of the Environmental Code, also commonly known as the former Leaking Underground Storage Tank Act. MICH. COMP. LAWS § 324.21301 et seq. (1996).

66. See the former definitions found at sections 324.21301(a) and (b) of Michigan Compiled Laws.

67. See the current definitions at sections 324.21301(a) and (b) of Michigan Compiled Laws, as amended effective March 6, 1996.

68. MICH. COMP. LAWS § 324.20126(4)(d) (1996).

69. Id. § 324.20107a(4).

70. Id. §§ 324.20129a(5), 324.20142.

71. Id. §§ 324.20126a(5), 324.20129a(5), 324.20142(1).

72. Id. § 324.20142(2). It is important to note, however, that section 20129a(5) does not contain specific language excluding tort claims from the exemption. The scope of the claims subject to the exemption is also drafted differently in the two sections. It is unclear how sections 20142 and 20129a will ultimately be reconciled by the courts.

73. 42 U.S.C. § 9601 et seq. (1996).

74. MICH. COMP. LAWS § 324.20129(6) (1996).

75. CERCLA provides that a "person who has resolved its liability to the United States or a State in an administrative or judicially approved settlement shall not be liable for claims for contribution regarding matters addressed in the settlement." 42 U.S.C. § 9613(f)(2) (1996).

76. The federal government has extensive authority beyond section 113 of CERCLA (42 U.S.C. § 9613), including section 122, which addresses the federal government's authority to provide contribution protection (42 U.S.C. § 9622(h)(4)).

77. Because of general resource limitations, if a person complies with state law, there is limited probability that the federal government would initiate an enforcement action contrary to state law. That possibility, however, must be recognized. In addition,

third parties may attempt an action under federal law and, under those circumstances, the scope and enforceability of contribution protection under section 20129(6) will be an important issue.

78. MICH. COMP. LAWS § 324.20120a (1996).
79. *Id.*
80. *Id.* § 324.20120b(2).
81. *Id.* §§ 324.20120b(3)–(5).
82. *Id.* § 324.20120a(11).
83. *Id.* § 324.20120a(6). MDEQ may not grant approval of a final remedial action plan that relies on a change in zoning until a final determination of that zoning change has been made by the local unit of government.
84. *Id.* Presumably, an appropriate exposure assessment would need to be prepared to demonstrate that the remedy was protective of human health and the environment. Even if the exposure *potential* is higher, that does not mean the *actual* exposure is above the applicable criteria.
85. *Id.* § 324.20120b(1).
86. *Id.* § 324.20120b(2).
87. *Id.* § 324.20120b(3).
88. *Id.*
89. *Id.* §§ 324.20120b(4)(a)–(f).
90. *Id.* § 324.20120b(5).
91. *Id.* § 324.20116(3).
92. *Id.* § 324.20116(1). This section arguably applies to *any interest* in the property, including a security interest or leasehold. Although the statute is not explicit, presumably the remedy for a violation of this provision would fall under the penalty provision in section 20137. Yet, only the Attorney General is empowered to bring an enforcement action under the penalty provision. Therefore, some form of common-law claim may be the only remedy for a private party to obtain any relief for the failure to file the required notice.
93. MICH. COMP. LAWS § 324.20120b(9) (1996).
94. Act 307 was substantially amended in 1990 (Public Act 233, effective July 1, 1991), establishing many provisions similar to CERCLA. The provisions creating the "Type C" remedy and the basis for calculating the "Type B" criteria were fundamental risk-based principles. Therefore, restructuring the cleanup criteria, though important, is not a significant variation from the risk principles previously adopted, which had served as the basis for the program since 1990.
95. MICH. ADMIN. CODE r. 299.5709(2)(a) (1990 AACS).
96. MICH. COMP. LAWS § 324.20120a(4) (1996).
97. The former method used for noncarcinogens was referred to as the Human Life Cycle Safe Concentration. *See* MICH. ADMIN. CODE r. 299.5709(2)(b), 299.5725 (1990 AACS).
98. MICH. COMP. LAWS §§ 324.20120a(2)–(3) (1996).
99. *Id.* § 324.20120a(4).
100. *Id.* § 324.20120a(15); MICH. ADMIN. CODE r. 299.5711(3) (1990 AACS).
101. *See* MERA Operational Memorandum No. 8, Revision No. 4 (June 5, 1995); MERA Operational Memorandum No. 14, Revision No. 2 (June 6, 1995).
102. MICH. ADMIN. CODE r. 299.5705(5)–(6) (1990 AACS).
103. MICH. COMP. LAWS §§ 324.20118(5)–(7) (1996).
104. *Id.* § 324.20118(5).
105. Although the MERA Amendments substantially altered many statutory provisions, most of the original Act 307 administrative rules, with a few exceptions, remain intact. As a result, and for consistency purposes, the rules will be referred to as the "Act 307 rules."

106. MICH. COMP. LAWS § 324.20118(6) (1996).

107. *Id.* §§ 324.20118(10)–(11).

108. Venting of groundwater is defined as "groundwater that is entering a surface water of the state from a facility." *Id.* § 324.20120a(15).

109. *Id.*

110. *Id.* § 324.3109a. Note, however, that the use of a mixing zone is prohibited in the Act 307 rules. MICH. ADMIN. CODE r. 299.5713(2) (1990 AACS). Although later legislative action (amending Part 31, via Public Act 70 of 1995, effective June 5, 1995) should be controlling, it may be necessary to delete the prohibition from the Act 307 rules to avoid any confusion.

111. MICH. COMP. LAWS § 324.20120d(1) (1996).

112. *Id.*

113. *Id.* § 324.20120d(3).

114. *Id.* § 324.11115b.

115. Public Acts 115 and 116 of 1996 became effective on March 6, 1996.

116. *See* the former definitions of "owner" and "operator" before the enactment of Public Act 116 of 1996, section 324.21303 of Michigan Compiled Laws.

117. MICH. COMP. LAWS §§ 324.21303(a)–(b) (1996).

118. *Id.* § 324.21304b.

119. *Id.* § 324.21315(2). Generally, as a result of these amendments, a purchaser or developer can obtain protection from liability related to USTs that may have been difficult to obtain before the amendments, thereby making it easier to acquire and redevelop such properties.

120. *Id.* § 324.21304a(5).

121. Under section 324.20126 of Michigan Compiled Laws, anyone who has acquired property since June 5, 1995, with a BEA is exempt from liability. Because anyone who is not liable under Part 201 of the Environmental Code is now not an owner or operator under Part 213, liability protection for USTs is available retroactively to at least June 5, 1995.

122. MICH. COMP. LAWS §§ 324.21303(a)–(b), 324.21304a(5), 324.21312a–324.21315 (1996).

123. *Id.* § 324.20107a.

124. *Id.* § 324.20107a(3).

125. *Id.* § 324.20107a(2).

126. *Id.*

127. *Id.*

128. *Id.* § 324.20101(1)(m).

129. *Id.* § 324.20107a(4).

130. *Id.* §§ 324.20107a(4), 324.20126(4)(d).

131. *Id.* §§ 324.20107a(5), 324.20126(3)(c)–(d).

132. *Id.* § 324.20126(3)(h).

133. *Id.* § 324.20126(1)(c).

134. *Id.*

135. *Id.* § 324.20101(1)(d).

136. *Id.* § 324.20129a(1).

137. *Id.* § 324.20126(1)(c)(ii). The implications from losing the exemption could be significant because the owner or operator would then become strictly liable for all the existing contamination, if no other defense is applicable.

138. *Id.* § 324.20126(3)(h). It is important to note that a BEA is prepared for a different purpose than meeting the requirements of the innocent-purchaser defense. If a site is not a facility, a BEA evaluation may not be sufficient to establish an innocent-purchaser defense.

139. *Id.* § 324.20126(3).

140. *Id.* § 324.20129a(1). *See also* the MDEQ BEA guidance documents released under cover of a January 20, 1996, memorandum from Al Howard, Chief of the Environmental Response Division, entitled "Revised Interim Instructions for the Preparation and Submittal of Baseline Environmental Assessments," as amended by a February 20, 1997, Addendum.

141. MICH. COMP. LAWS § 324.20129a(4) (1996).

142. *Id.* § 324.20129a(1).

143. *Id.* § 324.20129a(2).

144. *Id.* § 324.20129a(2)(b).

145. *Id.* § 324.20129a(3).

146. *Id.* § 324.20129a(5).

147. *Id.* § 324.20129(6).

148. *Id.* § 324.20129a(5).

149. There are other affirmative duties both under the prior act and after adoption of the MERA Amendments, including obligations to (1) determine the extent of a release (section 20114(1)(a)), (2) report the release (section 20114(1)(b)), (3) eliminate the threat of fire or explosion (section 20114(1)(e)), and (4) removal of "liquid phase" hazardous substances (section 20114(1)(f)).

150. MICH. COMP. LAWS § 324.20114(1)(g) (1996).

151. *Id.* §324.20114(4).

152. *Id.* § 324.20114(1)(g); see also the definition of "response activity" at section 20101(1)(bb).

153. MICH. COMP. LAWS § 324.20114(1)(h).

154. *Id.*

155. *Id.*

156. *Id.* § 324.20114(2).

157. *Id.* § 324.20135a(1).

158. *Id.* § 324.20135a(2)(b).

159. *Id.* § 324.20135a(2)(a).

160. *Id.* § 324.20114(8).

CHAPTER 34

Minnesota

PAUL S. MOE

The Minnesota Voluntary Investigation and Cleanup Program (VIC Program) provides incentives for property owners and others to volunteer to undertake investigation and remediation of environmentally impaired real property. The VIC Program establishes a framework for participants to select the type of liability protection that would be most beneficial, and to work with the Minnesota Pollution Control Agency (MPCA) in achieving the goals established by the participant.

Major Provisions of the VIC Program

The VIC Program has evolved over the past eight years into its present structure. In 1988, the Minnesota Environmental Response and Liability Act (the state "Superfund" act, known as MERLA)[1] was amended to enable MPCA to provide "technical assistance" to requesting parties for a fee determined by MPCA.[2]

In the ensuing years, MPCA offered requesting parties a number of administrative assurances, based upon its discretionary enforcement authority. In 1992, several of these assurances were enacted as the "land recycling act of 1992,"[3] and the policies established under the land recycling act became known as the VIC Program. Pursuant to the VIC Program, several different options are available to property owners, lenders, and others interested in voluntarily remediating real property in Minnesota.

A participant in the VIC Program can retain MPCA to provide technical assistance in connection with a proposed cleanup.[4] The statute allows MPCA to "assist a person in determining whether real property has been the site of a release or threatened release of a hazardous substance, pollutant, or contaminant," including "review and approval of a requester's investigation plans and reports and response action plans and implementation."[5] The requesting party is required to pay a fee to MPCA for such assistance, at rates determined by the MPCA Commissioner. MPCA staff will review environmental reports or cleanup plans prepared for a property, and provide technical commentary and other assistance related to such reports or plans. Participants may find this useful in determining, for example, the adequacy of an environmental report or cleanup plan prepared by another party.

MPCA also has authority to issue several different types of liability assurances to participants in the VIC Program. For example, MPCA has authority to issue "no association determinations" to VIC Program participants.[6] Under MERLA, the current owner of property may be responsible for environmental remediation costs (even if such owner did not own the property at the time of the release) if the owner has taken an action to "associate" itself with identified preexisting contamination on the site.[7] (Contrast this approach with the federal Comprehensive Environmental Response, Liability, and Compensation Act [CERCLA], which, on its face, imposes strict liability on the current owners of contaminated property, regardless of whether such current owner has "associated" itself with preexisting contamination on the property.[8]) A no-association determination is a finding by MPCA that specified conduct by a property owner will not "associate" that owner with the identified preexisting contamination, and therefore the owner will not have cleanup liability for such contamination under MERLA. If new information is discovered, however, MPCA is not precluded from taking other enforcement action.

In addition, MPCA has authority to issue "off-site source determinations."[9] An off-site source determination is a written finding from MPCA that it will take no action against a person who owns real property that has been affected by an identified release on adjacent or nearby real property.

MPCA also has authority to issue "no action" letters or agreements,[10] typically where contamination has been found, but at levels that MPCA considers not to be a threat to human health or welfare or the environment. A no-action letter is essentially an acknowledgment from MPCA that it will not exercise its discretionary authority to require a removal or remedial action under MERLA (or to take such action itself), because the data submitted to it does not indicate any unacceptable risks to human health or the environment. Similarly, MPCA may, in its discretion, provide a "limited no action letter" if, for example, only soil contamination at a site has been investigated.

If a party is interested in conducting voluntary remediation of real property, the party can take steps to receive a "certificate of completion."[11] A party seeking a certificate of completion must obtain MPCA approval of response action plan (RAP) describing the proposed cleanup.[12] An RAP is a detailed investigation of a release and a comprehensive description of a cleanup plan.[13] The VIC Program provides that MPCA may approve RAPs that do not require removal or remedy of all releases at a site, provided that certain criteria are met.[14] The RAP can be submitted to MPCA by a prospective purchaser before the acquisition of a parcel. By receiving MPCA approval of an RAP before a purchase, a prospective purchaser can become more confident of the potential financial obligations associated with remediation at a site.

Upon completion of the cleanup described in the RAP, a party can request that MPCA issue a certificate of completion for a property.[15] A certificate of completion from MPCA is, in effect, a covenant from MPCA not to sue the recipient of the certificate under MERLA.[16] A certificate of completion is the only liability assurance from MPCA that does not specifically permit MPCA to take subsequent action under MERLA if new information is discovered.[17] A certificate of completion cannot be "reopened" by MPCA because, for example, concentrations of metals in the soil at a site are found to be higher than originally thought. Nevertheless, the issuance of a certificate of completion does not preclude MPCA from taking action under MERLA against responsible persons.[18]

The VIC Program also specifically provides that investigation of a site (in accordance with an MPCA-approved investigation plan) will not constitute conduct "associating" the person conducting or authorizing the investigation with the contamination at the site.[19] Similarly, persons performing response actions pursuant to an MPCA-approved RAP will not become "associated" with preexisting releases.[20] In both cases, efforts to investigate or remediate contamination can be conducted in a manner that precludes additional liability under MERLA.

Eligibility for the VIC Program

Eligible Participants

Participation in the VIC Program is open both to persons not responsible for the release and to "responsible persons" under MERLA. Persons not otherwise responsible under MERLA for a release have a full range of options available under the VIC Program.[21] Participation is not limited to potential fee owners, but can include any party with an interest in real property subject to a release or threatened release, such as secured lenders or tenants. A responsible person under MERLA who enrolls in the VIC Program has fewer options. To receive a certificate of completion, the responsible person must receive approval from MPCA for an RAP that includes a plan for the remediation or removal of all releases and threatened releases at a particular site.[22] Moreover, the receipt of a certificate of completion by a responsible person does not preclude other official action by MPCA.[23] If, for example, after the issuance of a certificate of completion, an investigation of the parcel identified certain previously undetected contamination, MPCA would have the authority to take action against any person responsible for such release (even if such person was the recipient of the certificate of completion).

Participation Requirements

The various components of the VIC Program may have specific participation requirements. For example, to receive a "no-association determination," the proposed actions to which the determination will apply must be "response actions approved by [the MPCA], actions to improve or develop the real property, loans secured by the real property, or other similar actions."[24] Actions inconsistent with these statutory priorities may not be eligible for the benefit of a no-association determination. As noted above, MPCA may, under certain circumstances, approve voluntary response action plans that do not require the removal or remedy of all releases at a site, unless that RAP was submitted by a party responsible for the release.[25]

Eligible Properties

Although the VIC Program does not expressly exclude any categories of properties from its scope, MPCA will not allow a request to proceed if a site is currently receiving (or is expected to receive) federal funding of cleanup costs, or if MPCA determines that emergency action is required because of contamination of a drinking-water source.[26] Any other property with "known releases" or "threatened releases" of hazardous substances (and therefore within the scope of MERLA) can be enrolled in the VIC Program, including National Priorities List sites that are no longer receiving federal cleanup funds. Sites on MPCA's "Permanent List of Priorities" (the Minnesota "Superfund List") can also be enrolled in the VIC Program.

Enrolling a site in the VIC Program does not preclude MPCA from taking remedial action under MERLA to protect human health or the environment.[27] Also, MPCA can disqualify a site from participation in the VIC Program (and refer the matter to MPCA's Superfund program) if it believes that the site poses a significant health or environmental risk that requires state intervention. In addition, MPCA may terminate participation in the VIC Program if the participant is uncooperative with MPCA or unwilling or unable to complete investigation or remediation. In such cases, the site is typically referred to MPCA's Superfund program. Finally, a voluntary party can elect to withdraw from the VIC Program, which may result in a referral of the site to the state Superfund program.

Similarly, the federal government (pursuant to CERCLA, the Resource Conservation and Recovery Act, or other statutes) may have jurisdiction over cleanup at a site, thereby precluding its participation in the VIC Program. Because chapter 115B of the Minnesota Statutes defines "hazardous substances" as excluding petroleum products,[28] properties subject to releases of petroleum products are typically not enrolled in the VIC Program. Liabilities for releases of petroleum products, and standards for conducting remediation of petroleum releases, are generally addressed in the Minnesota Petroleum Tank Release Cleanup Act.[29] MPCA also has a Voluntary Petroleum Investigation and Cleanup Program (VPIC), which (among other available services) provides expedited review of voluntary cleanups of petroleum releases.[30]

Memorandum of Agreement

MPCA and the United States Environmental Protection Agency (USEPA), Region V, entered a Superfund Memorandum of Agreement on May 3, 1995, detailing USEPA's view of the liability assurances provided under the VIC

Program. Pursuant to the memorandum, if MPCA has issued a no-action determination or a certificate of completion for a site under the VIC Program, the Region V office "will not plan or anticipate any federal action . . . unless, in exceptional circumstances, the site poses an imminent and substantial endangerment or emergency situation."[31] If MPCA issues an "off-site source" determination, Region V "will not plan or anticipate any federal action with respect to the parties covered by the MPCA determination."[32] These protections are in addition to the fundamental agreement between USEPA and MPCA, pursuant to which USEPA has agreed not to interfere with ongoing state-supervised cleanups. It is important to note that no-association determinations (the most popular form of liability limitation requested from MPCA) are not recognized by USEPA as a protection against liability under CERCLA. Because CERCLA does not contain a requirement that a property owner become "associated" with preexisting contamination for liability to exist, USEPA has determined that it would be inconsistent with federal law to consider such determinations by MPCA to be binding on it.

Participants in the VIC Program also have the option of seeking "comfort letters" from the USEPA's Region V office, or negotiating for prospective-purchaser agreements with Region V, in an effort to minimize liability under CERCLA.[33]

Liability Protection

The VIC Program is a component of MERLA. Accordingly, the liability protections provided under the VIC Program can be described as simply a waiver of certain administrative claims that MPCA would have had under MERLA. The protections afforded by the VIC Program are available generally to all applicants with a real property interest in a site who are not otherwise responsible for cleanup under MERLA, including lenders, trustees, new owners, and current owners who are not responsible parties. In addition, current owners who are responsible persons under MERLA may receive MPCA approval of an RAP, provided that the RAP addresses all releases and threatened releases at a site.[34] Upon completion of the measures described in an approved RAP, a responsible person can receive a certificate of completion pursuant to the VIC Program.[35] As previously noted, however, receipt of a certificate of completion by a responsible person is not a guarantee that such person can have no future liability.

Many of the liability assurances provided by MPCA pursuant to the VIC Program will benefit the successors and assigns of the party that received the assurance. For example, off-site source determinations issued by MPCA "may extend to the successors and assigns of the person to whom it originally applies, if the successors and assigns are not otherwise responsible for the release and are bound by the conditions in the [off-site source] determination."[36] The same standard applies for no-association determinations issued by MPCA.[37] Certificates of completion issued by MPCA will also protect the successors and assigns of the voluntary party.[38] If the voluntary party is a responsible person, successors and assigns who are not otherwise responsible for the release will receive the benefits provided by the certificate of completion.[39]

MERLA also contains several protections from liability specifically intended for lenders, trustees, and personal representatives. MERLA includes a straightforward lender-liability rule that provides that the making of a loan will not associate a lender with preexisting contamination.[40] The statute also provides that trustees will not be personally liable for contamination on property[41] and personal representatives will have no personal liability for contaminants on property that they are administering.[42]

Financial Incentives

Neither the VIC Program nor MERLA generally contains explicit financial incentives to facilitate development of contaminated property. Other Minnesota laws, however, provide such incentives. For example, the Minnesota "contamination tax" provides for reduced aggregate taxes payable for a property that is owned by a person who is not responsible under MERLA for the cleanup of a release of hazardous substances on the property.[43]

The contamination tax was created by the Minnesota legislature after the owners of contaminated property demanded a reduction in their real estate taxes by arguing that the existence of contamination significantly lowered the fair market value of their property.[44] The legislature responded by imposing a contamination tax upon such properties, with the tax being based upon the amount of the reduction in market value resulting from the contamination.[45] The contamination tax and the traditional property taxes due for a property are paid together on a combined bill. The net effect is to eliminate the "windfall" of a reduction in property taxes owed because of contamination on a property.

The legislature created incentives, however, to enroll such sites in the VIC Program. In addition, the legislature provided a reduction in the contamination tax for landowners who are not "responsible persons" under MERLA: If the owner and the occupant are not responsible persons, the contamination tax is reduced to 25 percent of the regular contamination tax rate.[46] The combination of the incentives and reductions can dramatically reduce the taxes payable for a property. Upon the approval of an RAP for a contaminated site (pursuant to the VIC Program or otherwise), the contamination tax is reduced by 50 percent.[47] This reduction is available to responsible persons under MERLA and persons not responsible for the contamination. This reduction is combined with the reduction available for owners who are not responsible persons, with the result that the contamination tax is reduced to 12.5 percent of its normal amount if a nonresponsible party has its RAP approved.[48] Finally, once the RAP has been implemented, no contamination tax is owed,[49] based on the assumption that the market value of the property will rise upon cleanup, thereby driving up the amount of real estate tax due for the property.

In addition, the Minnesota legislature has provided for hazardous-waste tax increment subdistricts, pursuant to which tax revenue generated by a property may be used for cleanup costs.[50] Such tax increment financing programs are useful for significant redevelopment projects, in which the increase in land value attributable to the cleanup is sufficient to justify the transaction costs. The value of a contaminated parcel may be reduced by the estimated amount necessary to remove or remediate hazardous substances on the parcel, thereby increasing the tax increment and therefore the amount of funds available for cleanup.[51]

The state has established various loan and grant programs that may be useful to persons interested in participating in the VIC Program. MPCA administers a small-quantity-generator remediation loan program, pursuant to which entities that generate a small quantity of hazardous substances may obtain low-interest loans to help address contamination issues for which they are responsible. In addition, certain governmental development authorities may be entitled to contamination cleanup grants administered by the Minnesota Department of Trade and Economic Development,[52] and the Metropolitan Council has grant funds available for cleanups in the Twin Cities metropolitan area.[53]

Cleanup Standards

The cleanup standards imposed by MPCA for the VIC Program are, by statute, identical to the standards required for the cleanup of a Superfund site under MERLA.[54] As with state Superfund sites cleaned up under MERLA, the cleanup standards are tied to the intended use of the property.[55]

The standards themselves are a mixture of numeric standards and risk goals. MPCA generally takes a flexible approach to determining how "clean" a site must be before a certificate of completion will be issued. A variety of treatment technologies, as well as institutional controls (such as restrictive covenants filed against the property)[56] and engineering controls (such as nonpermeable caps over soil contamination)[57] are permitted, although not generally encouraged.[58]

The standards for soil cleanup are based upon federal risk-assessment techniques, and are generally described in MPCA Guidance Document Number 13 of the VIC Program, last revised in September 1994. The soil cleanup standards at a particular site are determined only after the VIC Program participant has conducted Phase I and Phase II investigations at a site and has submitted an RAP to MPCA for approval. In certain circumstances, however, specific standards are established by regulation.[59]

Groundwater standards are also based on site-specific risk assessments.[60] The goals of this process are to return groundwater to use as a potable water supply.[61] The standards applicable at a given site typically depend upon the condition of groundwater at the site. The highest priority of MPCA is to prevent the degradation of groundwater such that it exceeds the most stringent of the following standards: the health risk limits (HRLs) established by the Minnesota Department of Health; a one-in-one-hundred-thousand cumulative risk for carcinogens (as determined by the Minnesota Department of

Health); or the maximum containment levels (MCLs) established by USEPA.[62] For contaminated groundwater, remediation is expected to continue until the most stringent standard described above is achieved.[63] Under certain circumstances, however, MPCA may modify remediation standards on a site-by-site basis, provided that the standards do not present unacceptable risks to the public health or welfare or the environment.[64]

MPCA is considering new objective standards that may obviate the need for risk assessments in certain circumstances. The new standards are expected to recognize the concept that higher contaminant levels should be permitted when the risk of exposure is perceived to be lower.

MPCA has also described its expectations for Phase I and Phase II investigations in the VIC Program. Phase I investigations are expected to include broad historical information about the site; to determine whether a release of a hazardous substance, pollutant, or contaminant has occurred; and to identify any additional inquiry that should be conducted.[65] MPCA standards for a Phase I investigation are significantly higher than the traditional Phase I investigation conducted as part of a prepurchase investigation. MPCA's requirements for a Phase II investigation are also quite thorough.[66] MPCA recommends that a work plan be submitted to it for approval before undertaking a Phase II investigation.[67]

Confidentiality, Consultants, and Fees

Confidentiality and Privileges

The VIC Program does not contain any specific confidentiality provisions or other privileges. Minnesota has an environmental audit pilot program, which provides for certain penalty waivers for self-disclosed violations of environmental requirements.[68] The program is primarily intended to address violations of operating requirements and discharge limitations.

Certification of Consultants

Consultants providing services to a participant in the VIC Program are not required to be certified under any state certification programs.[69] As a practical matter, however, MPCA's thorough review of the work product of the consultant prevents unqualified consultants from pursuing unacceptable approaches to cleanup.

MPCA Fees; Review Process

MPCA must be retained at an hourly rate established by the MPCA Commissioner.[70] Typically these fees are approximately $85 per hour. As previously described, MPCA may be retained merely to provide technical assistance on a project. Other services include review of requests for no-action letters and off-site source determinations. If a voluntary party seeks a certificate of completion, MPCA's role will include review of a Phase I report, review of a Phase II work plan, review of a Phase II report, review of an RAP, and review of a response action implementation report.

Mechanics of Participating in the VIC Program

Parties interested in participating in the VIC Program must first execute and deliver to MPCA a program application (sometimes referred to as "request for assistance"), under which the party agrees to pay costs incurred by MPCA. Because the program application authorizes MPCA to begin calculating bills for the property owner, in many cases no help will be available from MPCA until the application has first been executed and submitted.

Once a program application has been completed and forwarded to MPCA, the VIC Program participant can then seek MPCA review of various reports. For technical assistance, a no-association determination, a no-action letter, or an off-site source determination, MPCA will review any and all Phase I and Phase II materials forwarded to it. In many cases, however, MPCA will request more background information about the property, and/or a new Phase I assessment describing the property and its history in greater detail. In addition, the effectiveness of a no-association determination or a no-action letter will be limited by inadequate investigation. For example, if the Phase II investigation identified a substance in the soil at the site, but no analysis of groundwater was conducted, the no-association determination will reference only the substances identified in the soil and state only that the requesting party is not associated with such substances in the soil, without addressing any contaminants that may be in the groundwater. It is in a participant's best interest to conduct as thorough an investigation as possible to get the maximum benefit from a no-association determination or no-action letter.

If the participant desires the issuance of a certificate of completion, several steps are required. To start the process, a VIC Program application must be submitted to MPCA. If not submitted with the application, an appropriate Phase I investigation is expected to be completed and submitted within ninety days. MPCA must also review and approve a Phase II work plan. This work plan can be submitted either with the Phase I report or separately. MPCA will approve the work plan if it considers the proposed investigation adequate to identify all the contaminants on a site. In addition, a party must prepare and submit to MPCA a site safety and contingency plan, which is designed to prepare the applicant for emergencies or for the discovery of hazardous materials during investigation or remediation.

After approval of the work plan and a site safety plan, the Phase I investigation can begin. A Phase II report is expected to identify the nature and scope of contamination of a site, the potential sources of the contamination, and the potential pathways and receptors of such contamination. If the Phase II investigation indicates that the identified contaminants are from an on-site source, a focused feasibility study (FFS) may be required by MPCA. The FFS is intended to include recommendations of response actions at a site.

Upon selection of an appropriate response action by the participant and MPCA, the participant is expected to prepare and submit to MPCA an RAP covering the site. As previously noted, an RAP submitted by a responsible person must address all sources of contamination at a site, but an RAP submitted by another party need not. Once the RAP has been completely implemented, a participant is eligible for a certificate of completion from MPCA.[71]

Because the VIC Program is, as indicated by its title, "voluntary," a party can withdraw at any time. Typically, however, a participant that withdraws from the program will find that its site has been referred to the state Superfund program. Participants in this situation often find MPCA less willing to allow the property owner to take the initiative in cleaning up the site, or to propose alternative solutions to contamination problems. Cleanup of a site pursuant to the state's site response section does not permit the flexible approaches permitted by the VIC Program.

Cost Recovery

The Minnesota land recycling act does not include a specific cost-recovery section for parties who participate in the VIC Program. However, such parties are not precluded from maintaining cost-recovery actions under MERLA, under other state common-law actions, or under federal cost-recovery actions. MERLA provides that responsible persons are strictly liable, jointly and severally, for "reasonable and necessary" removal costs incurred by another party.[72] A person who is not responsible for a release of hazardous substances at a site may recover removal costs from any responsible party. This right is not abrogated by participation in the VIC Program. Any responsible person held liable under MERLA for such removal costs may seek contribution for such loss from other responsible persons.[73]

Conclusion

In 1988, Minnesota became the first state to authorize by statute a Voluntary Cleanup Program.[74] Since that date, MPCA has remained a leader in encouraging the reuse and redevelopment of brownfields sites. Prospective purchasers that are not otherwise responsible for a release can receive an assurance that they will not be held liable by MPCA for remediation costs associated with the release. Alternatively, owners can remediate sites with limited oversight by MPCA, and receive a certification from MPCA that the site is clean. The VIC Program has been reviewed by USEPA, which has entered an addendum to its Memorandum of Agreement with MPCA confirming that USEPA will recognize the validity of certain decisions of MPCA. In brief, the Minnesota VIC Program was the pioneer in brownfields redevelopment, and remains today one of the most innovative and flexible governmental approaches to the brownfields cleanups.

Minnesota Voluntary Investigation and Cleanup Program

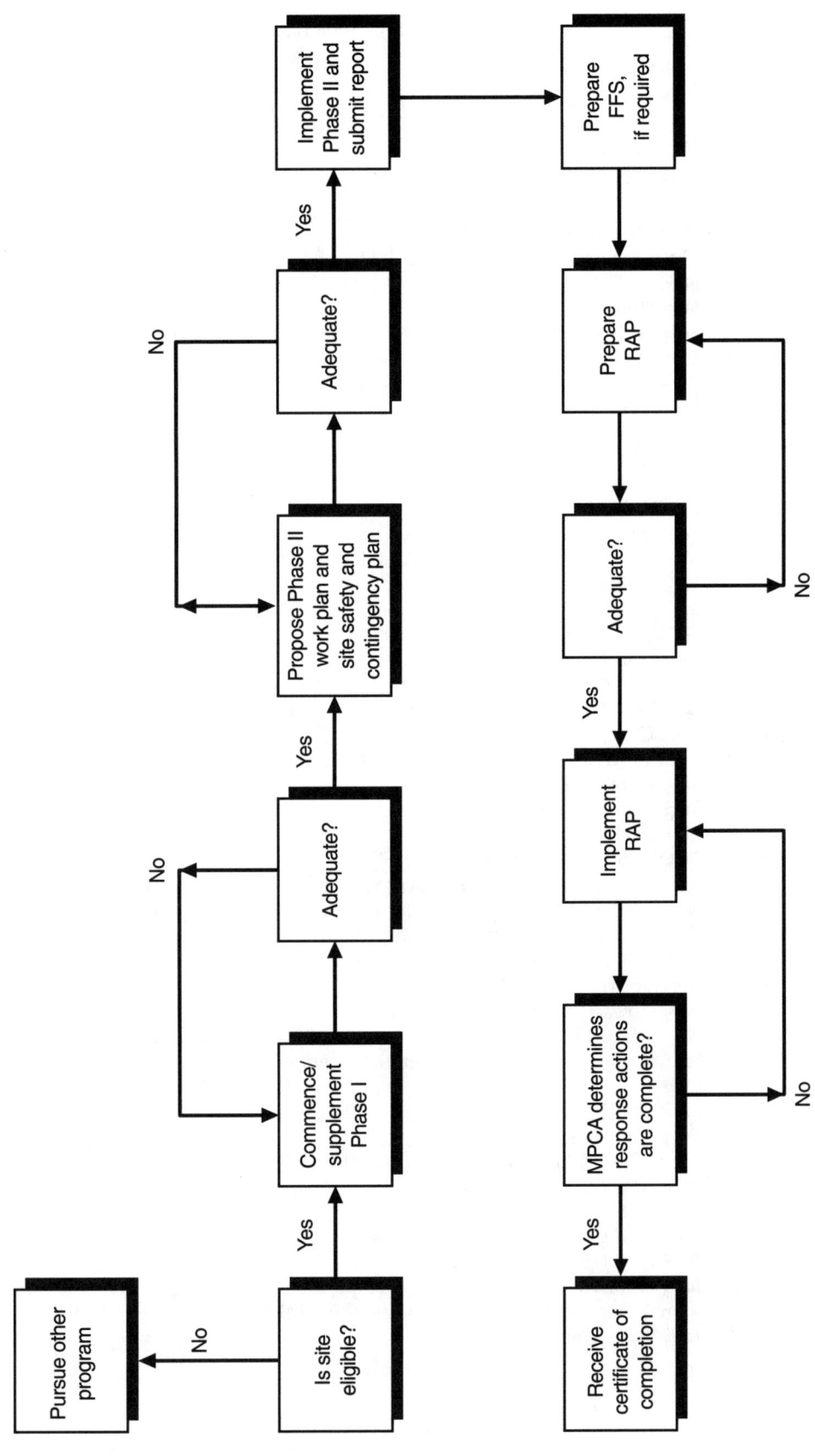

Notes

1. MINN. STAT. § 115B (1995).
2. 1988 Minn. Laws 685, *codified at* MINN. STAT. § 115B.17, subdivs. 14(a), 14(b) (1995).
3. 1992 Minn. Laws 512.
4. MINN. STAT. § 115B.7, subdiv. 14 (1995).
5. *Id.*
6. *Id.* § 115B.178.
7. *See id.* § 115B.03, subdiv. 3 (owner may be a responsible person); *id.* § 115B.04 (liability imposed on responsible persons).
8. *See* 42 U.S.C. § 9607 (1995).
9. MINN. STAT. § 115B.177 (1995).
10. *See id.* § 115B.17, subdiv. 1.
11. *Id.* § 115B.175.
12. *Id.* § 115B.175, subdiv. 3.
13. An RAP may also be used in connection with other assurances available under the VIC Program.
14. MINN. STAT. § 115B.175, subdiv. 2 (1995).
15. *Id.* § 115B.175, subdiv. 5.
16. *Id.* § 115B.175, subdiv. 1.
17. *See id.* § 115B.175, subdiv. 1(a).
18. *Id.* § 115B.175, subdiv. 8.
19. *Id.* § 115B.17, subdiv. 14(c).
20. *Id.* § 115B.175, subdiv. 4. *See* Musicland Group, Inc. v. Ceridian Corp., 508 N.W. 2d 524 (Minn. Ct. App. 1993).
21. *See, e.g.*, MINN. STAT. § 115B.175, subdiv. 1 (1995).
22. *Id.* § 115B.175, subdiv. 6a(b).
23. *Id.* § 115B.175, subdiv. 8.
24. *Id.* § 115B.178, subdiv. 1(a).
25. *Id.* §§ 115B.175, subdivs. 3, 6a.
26. MINN. POLLUTION CONTROL AGENCY, GUIDANCE DOCUMENT #1: INTRODUCTION TO THE VOLUNTARY INVESTIGATION AND CLEANUP PROGRAM (rev. Sept. 1994).
27. *See* MINN. STAT. §§ 115B.175, subdiv. 8, 115B.179 (1995).
28. *Id.* § 115B.02, subdiv. 8.
29. MINN. STAT. § 115C (1995).
30. *Id.* § 115C.03, subdiv. 9.
31. "Superfund Memorandum of Agreement" between the Minnesota Pollution Control Agency and the United States Environmental Protection Agency, Region V, at 2 (May 5, 1995).
32. *Id.*
33. *See* 60 Fed. Reg. 34,792 (1995).
34. MINN. STAT. § 115B.175, subdiv. 6a(b) (1995).
35. *Id.* § 115B.175, subdiv. 6a(a).
36. *Id.* § 115B.177, subdiv. 2(a).
37. *Id.* § 115B.178, subdiv. 2.
38. *Id.* § 115B.175, subdiv. 6(3).
39. *Id.* § 115B.175, subdiv. 6a(c)(3). A responsible party that receives a certificate of completion may still potentially have liability under MERLA.
40. *Id.* § 115B.03, subdiv. 6.

41. *Id.* § 115B.03, subdiv. 8.
42. *Id.* § 115B.03, subdiv. 9.
43. *See id.* §§ 270.91–270.98.
44. *See, e.g.,* Westling v. County of Mille Lacs, No. C5-92-341, C7-92-342, 1993 WL 35155 (Minn. Tax Feb. 10, 1993), *rev'd,* 512 N.W.2d 863 (Minn. 1994). The tax court concluded that the estimated market value of the property should be reduced from $974,200 to $100 because of the existence of contamination; the Minnesota Supreme Court reversed the decision.
45. MINN. STAT. § 270.91 (1995).
46. *Id.* § 270.91, subdiv. 3.
47. *Id.* § 270.91, subdiv. 4(c).
48. *Id.* § 270.91, subdiv. 4(d).
49. *Id.* § 270.94.
50. *Id.* §§ 469.174–469.176.
51. *Id.* § 469.175, subdiv. 7.
52. *Id.* § 116J.552, subdiv. 4.
53. *Id.* § 473F.08, subdiv. 3b.
54. *See id.* § 115B.175, subdiv. 2(a)(1) (partial cleanups); *id.* § 115B.175, subdiv. 3(c) (review, approval, and implementation of complete response action plans); *id.* § 115B.17, subdivs. 1–2 (general MERLA standards).
55. *Id.* § 115B.17, subdiv. 2(a).
56. *See, e.g., id.* § 115B.175, subdiv. 2 ("The [MPCA] may approve a voluntary response action plan . . . that does not require removal or remedy of all releases . . . at an identified area of real property" if certain criteria are met).
57. *Id.*
58. *See* MINN. POLLUTION CONTROL AGENCY, GUIDANCE DOCUMENT #15: REMEDY SELECTION/TREATMENT TECHNOLOGY (rev. Sept. 1994) ("The objectives of this document are . . . to promote the implementation of treatment technologies and discourage the request for approval of 'containment only' response actions, such as capping, vaults, and ground water pump and treatment systems.").
59. For example, Minnesota Rules part 4761.0300, subpart 4, establishes a bare soil standard for lead of 100 parts per million (mg/kg) for residential and playground areas.
60. *See* MINN. POLLUTION CONTROL AGENCY, GUIDANCE DOCUMENT #14: APPROACH TO GROUND WATER CLEANUP (rev. Sept. 1994).
61. MINN. R. 7060.0400 (1995); *see also* 40 C.F.R. § 300.430(a)(iii)(E) (1995).
62. MINN. STAT. § 103H.280 (1995).
63. *Id.*
64. *See id.* § 115B.17, subdiv. 1 (general standard for response actions). *See also* 40 C.F.R. § 300.430(e)(2)(i) (1995); Minn. Pollution Control Agency, Memorandum (on file with author, and entitled "Adjustments to Ground Water Cleanup Goals").
65. MINN. POLLUTION CONTROL AGENCY, GUIDANCE DOCUMENT #8: PHASE I INVESTIGATION (rev. Sept. 1994).
66. *See* MINN. POLLUTION CONTROL AGENCY, GUIDANCE DOCUMENT #12: PHASE II INVESTIGATION REPORT (rev. Sept. 1994).
67. *See* MINN. POLLUTION CONTROL AGENCY, GUIDANCE DOCUMENT #11: PHASE II INVESTIGATION WORK PLAN (rev. Sept. 1994).
68. 1995 Minn. Laws 168.
69. The MPCA has published recommendations on the process of selecting a consultant. *See* MINN. POLLUTION CONTROL AGENCY, GUIDANCE DOCUMENT #6: SELECTING A CONSULTANT (rev. Sept. 1994).

70. MINN. STAT. § 115B.17, subdiv. 14(b) (1995).

71. *See* MINN. POLLUTION CONTROL AGENCY, GUIDANCE DOCUMENT #7: SCHEDULING PHASES OF INVESTIGATIONS AND RESPONSE ACTION (rev. Sept. 1994) (general description of process necessary to receive certificate of completion).

72. MINN. STAT. § 115B.04, subdiv. 1 (1995).

73. *Id.* § 115B.08, subdiv. 1.

74. *See* Elizabeth Glass Geltman, *Recycling Land: Encouraging the Redevelopment of Contaminated Property*, NAT. RESOURCES & ENV'T, Spring 1996, at 3, 8.

CHAPTER 35

Missouri

JAMES T. PRICE
JENNIFER A. DOWNS

Missouri's legislative efforts to facilitate redevelopment of contaminated properties consist of two principal laws. In 1995, the legislature passed House Bill 414, the Abandoned Property Tax Credit Program or "Brownfields Initiative."[1] This program supplemented the existing Missouri Voluntary Remediation Program.[2] The stage now is set for local governmental agencies, private landowners, developers, lenders, and regulatory agencies to begin working together to bring about remediation, reclamation, and redevelopment in some of the state's traditional industrial and manufacturing centers.

Summary of Major Provisions

The Voluntary Remediation Program provides a way for persons with an interest in contaminated property to clean up that property according to the requirements of the Missouri Department of Natural Resources (MDNR) and receive a completion letter. The major provisions include the following:

- A party wishing to enter a site into the Voluntary Remediation Program must apply to MDNR by submitting copies of all reports, site assessments, investigations, and sample analyses requested by MDNR. At a minimum, these must include a Phase I noninvasive environmental site assessment.
- Upon application approval, MDNR and the participant enter a site-specific, environmental-remediation oversight agreement. The participant agrees to pay MDNR's costs, plus a multiplier, associated with the oversight.
- The participant then submits a remedial action plan to remediate contamination identified in the environmental site assessments. If the remedial action plan is determined by MDNR to be protective of human health and the environment, it is approved.
- If remediation pursuant to MDNR's approved plan is completed successfully, MDNR issues a completion letter stating that no further re-

medial action at the site is needed for contamination identified in the environmental assessments, provided the participant complies with all provisions of the Voluntary Remediation Program statute and regulations, takes remedial actions in accordance with the approved remedial action plan, and pays all applicable participation fees to MDNR.
- A participant may terminate participation at any time by providing MDNR with written notification by certified mail. According to the regulations, termination "does not affect the person's environmental liability."[3]
- If appeals are required, they are taken to the Missouri Hazardous Waste Management Commission.

The Brownfields Initiative, a separate but related program, is designed to encourage persons to remediate contaminated sites and provides economic incentives to do so. It also provides liability releases for qualifying projects. A significant question exists concerning whether the Brownfields Initiative provisions apply only to property owned by public entities or also to private property. The major provisions of the Brownfields Initiative are as follows:

- Eligible projects, if approved by the Departments of Economic Development, Natural Resources, and Revenue, can be entitled to receive a variety of economic-development incentives.
- Tort immunity is granted to the government agency owner and prospective purchaser arising out of performance of an eligible project, including any voluntary remediation.
- Governmental agencies are not liable in tort or under the state's environmental laws for ownership of property acquired by foreclosure or deed in lieu of foreclosure, provided the agency did not directly cause or contribute to new contamination.
- The purchaser of an eligible project is released from further liability based on the voluntary remediation work performed, subject to certain reopeners.

Eligibility for Programs

The requirements for participation in the Brownfields Initiative and the Voluntary Remediation Program vary somewhat. The Brownfields Initiative limits "eligible projects" to abandoned properties that meet certain requirements in connection with their intended future use. The Brownfields Initiative contains one provision that seems to limit its applicability to properties owned by government agencies. But another definition seems to broaden its applicability to cover any site, public or private, that goes through the Voluntary Remediation Program. The reach of this program will vary dramatically depending on which of these interpretations prevails.

An eligible project is an "abandoned property to be acquired, established, expanded, remodeled, rehabilitated or modernized[,] . . . the operation of which, alone or in conjunction with other facilities, will create new jobs or preserve existing jobs and employment opportunities, attract new businesses to the state, prevent existing businesses from leaving the state or improve the economic welfare of the people of the state."[4] Abandoned property is defined as "real property previously used for, or which has the potential to be used for, commercial or industrial purposes[,] which reverted to the ownership of the state, a county, or municipal government, or an agency thereof ... and has been vacant for a period of not less than three years."[5] But the term "eligible project" also includes voluntary remediation conducted pursuant to the Voluntary Remediation Program.[6] The inclusion of voluntary remediation projects within the definition of Brownfields Initiative "eligible projects" has the potential to expand significantly the pool of properties qualifying for assistance and liability limitations under the Brownfields Initiative, because the voluntary remediation project properties are not limited to properties abandoned or in the possession of a governmental agency.

Eligibility for the Voluntary Remediation Program is broad, covering all sites except those that are ineligible for the program.

The following sites are not eligible to participate in the Voluntary Remediation Program:[7]

1. Sites that warrant cleanup under the Resource Conservation and Recovery Act (RCRA),[8] the Comprehensive Environmental Response, Compensation, and Liability Act (CERCLA),[9] or the Missouri Hazardous Waste Management Law[10] and that fall within any of the following categories:
   - Site conditions constitute an imminent and substantial threat to public health or the environment
   - Site inspection is completed and the site is being evaluated for listing on the National Priorities List
   - Permitted or interim status RCRA facilities
   - Sites that warrant enforcement action for cleanup under RCRA, CERCLA, or the Missouri Hazardous Waste Management Law
2. Property not currently or previously used for industrial or commercial purposes[11]

The Brownfields Initiative excludes properties that have not been vacant for three years and are not owned by the state, a county, a municipality or one of their agencies. As discussed in the preceding section, the Brownfields Initiative may cover property that is the subject of a Voluntary Remediation Program cleanup.[12]

Memorandum of Agreement

MDNR and the United States Environmental Protection Agency (USEPA) Region VII have negotiated a Memorandum of Agreement in connection with the Missouri Voluntary Remediation Program. Sites that successfully complete remediation through this program will not be targets of federal cleanup action, subject to the conditions and reopeners in the Memorandum of Agreement.[13]

Liability Protection

Lenders, Trustees, and Fiduciaries

Under the Brownfields Initiative, private lenders are immune from liability, including potential liability from the incomplete or unsuccessful remediation of contaminated property or from indicia of ownership held to protect their security interests.[14] When a lender has foreclosed on property that is an eligible project and has held the abandoned property for at least two years, the lender may apply to the Property Reuse Fund established by the Brownfields Initiative for repayment of the unpaid amount of the defaulted loan.[15] This $10 million fund may or may not be funded by the state at any given time.

Program Participants: Brownfields Initiative

Liability protection for participants in the Missouri Brownfields Initiative, though as yet untested, appears quite comprehensive. Governmental agencies, including their officers and employees, and the prospective purchaser of an eligible project are immune from liability for any tort action arising from the performance of an eligible project or any voluntary remediation associated with that project while the project is in progress. The immunity is broad, extending to acts and failures to act by the governmental entity, regardless of whether the act or failure is required by law or is discretionary, or whether it is related to policymaking, planning, implementation, or enforcement duties.[16] The immunity does not include acts "[m]anifestly outside the scope of employment" or "[c]ommitted maliciously, in bad faith, or in a wanton or reckless manner."[17] The immunity of the governmental agency and the prospective purchaser includes acts and failures to act by contractors, subcontractors, and others performing the project. The tort immunity ceases after MDNR approves the completed voluntary remediation.[18]

Under the Brownfields Initiative, no local governmental entity or agency will bear tort or environmental liability for the ownership of real or personal property acquired by foreclosure pursuant to a defaulted loan. Further, no liability may attach to a governmental entity acquiring such property through repurchase or reversion under the provisions for payment of defaulted private loans from the Property Reuse Fund, provided that the governmental entity did not cause or contribute to the contamination.[19] In addition, when governmental entities and agencies act as lenders, they are also protected from environmental liability, in the same manner as private financial institutions are protected.[20]

Under the Brownfields Initiative, purchasers or new owners of eligible projects will be released from further liability to the extent of the remedial work performed and the level of risk to human health and the environment remaining after performance of the voluntary remediation, provided the release of hazardous substances on the property occurred before the acquisition by the purchaser.[21] A purchaser of an eligible project can obtain one of the following:

1. A letter from MDNR requiring no further action from the purchaser who performs a Phase I and Phase II environmental site assessment demonstrating that no remedial action is necessary to protect public health and welfare or the environment[22]
2. For a purchaser who performs a cleanup pursuant to the Voluntary Remediation Program, a no-further-action letter or a covenant-not-to-sue, depending upon the degree of cleanup
   - The purchaser receives a covenant-not-to-sue if the cleanup attains the approved cleanup goals of the voluntary remedial action *and* the remediation has treated all hazardous substances of concern to levels below regulatory action levels.[23] To receive a covenant-not-to-sue, the voluntary remediation plan must be submitted for public comment and public hearing.[24]
   - The purchaser receives a no-further-action letter if the goals of the voluntary remedial plan have been attained, but the remediation does not treat all hazardous substances present to levels below regulatory action levels and instead uses alternative cleanup goals, risk-reduction solutions, institutional controls, or other risk-based alternatives.[25] Such a letter will not be issued unless the remediation minimizes the harmful effects of the hazardous substance to "acceptable risk levels."[26]

Program Participants: Voluntary Remediation Program

Successful participants in Missouri's Voluntary Remediation Program who complete the MDNR-approved remedial action plan that addresses contamination identified in the environmental assessments receive from MDNR a letter stating that no remedial action and no further remedial action need be taken at the site related to any contamination identified in the environmental assessments. There is no other express provision governing protection against liability in this program.

Other Liability Considerations

Missouri statutes do not at present provide liability protection for current, as opposed to new, owners voluntarily disclosing contamination on their sites. There exists no audit privilege or other mechanism to protect owners who voluntarily disclose contamination for the purposes of a Voluntary Remediation Program application. In cases in which that contamination exceeds action levels under Missouri or federal law, the application will be rejected from the Voluntary Remediation Program and the property may be subject to corrective or enforcement actions.

Financial Incentives

The Brownfields Initiative is formally titled "The Abandoned Property Tax Credit Program" and provides an array of financial mechanisms and incentives for investment in the eligible properties. The Brownfields Initiative provides loans, loan guarantees, tax credits and abatements, income exemptions, and tax refunds to qualifying projects.[27] Loans and loan guarantees of up to $1 million each to acquire land, extend utilities, and construct new or renovate existing buildings are available under the program. The loaned and guaranteed funds may also be used to furnish and equip buildings and to pay for engineering, architectural, legal, remediation, and other costs necessary to complete a project. The statute provides for tax credits equal to 100 percent of the costs of voluntary remediation activities at an eligible site for up to twenty years. Tax credits are also available for jobs created or retained, hiring of previously unemployed persons, hiring of persons living near the facility, and total investment in the project. Abatements of at least 50 percent for at least ten years are offered for improvements to real property used for manufacturing. A tax refund of up to $75,000 over two years may be claimed for tax credits earned but not taken.

Tax credits under the Brownfields Initiative may be used to offset Missouri income-tax liability. Partnerships and S-corporations may pass through the tax credits to each partner or shareholder. The tax credits provided under the Brownfields Initiative may not be claimed in tandem with certain other Missouri tax credits.

Cleanup Standards

The cleanup standards for sites in the Voluntary Remediation Program are not explicitly risk-based and there is no language indicating that MDNR will consider the intended use of the property in approving the voluntary remedial action plan. The regulations state only that a remedial action plan will not be approved unless it is protective of human health and the environment.[28] However, MDNR has shown some willingness to negotiate the terms of remediation plans and cleanup standards in developing and approving site-specific remediation plans. The Missouri Department of Health, a separate agency, plays a major role in setting cleanup levels at Missouri sites.

The Brownfields Initiative specifically provides for site-specific, risk-based voluntary remediation plans. The statute names alternative cleanup goals, risk-reduction solutions, and institutional controls, such as deed restrictions, as examples of acceptable means of controlling risk and protecting human health and the environment.[29]

Participating in the Missouri Brownfields Programs

Voluntary Remediation Program[30]

Participants in the Missouri Voluntary Remediation Program must pay a $200 application fee, reimburse MDNR for all its costs, and pay project personnel their hourly rate times 2.5. When a site application is approved, participants must post a $5,000 deposit to cover anticipated initial oversight costs. The applicant and MDNR will then enter an environmental remediation oversight agreement. Also, upon approval of the application, the applicant must submit all site assessments and other reports concerning sampling or other investigations on the property, including, at the very least, a Phase I environmental assessment. The required Phase I assessment must be conducted in accord with the American Society for Testing and Materials (ASTM) Standard E.1527, and by a "technical consultant familiar with the nature of the operations and activities that occurred on the real property." [31]

MDNR must review the environmental reports and comment within 180 days regarding any additional assessment required. After MDNR review of any additional assessments, the approved applicant must submit a remedial action plan, including a work plan, a safety plan, testing protocols, and monitoring plans for contamination identified in the assessments. MDNR must approve the plan within 90 days if the plan satisfies the requirements of the statute. If at any time MDNR fails to meet the review deadlines, the applicant is not required to reimburse MDNR for the cost of the review.

Quarterly progress reports must be submitted to MDNR. When remediation is complete, the participant signs a final report documenting that all required work has been completed as required by the remedial action plan.

If an applicant's initial voluntary remediation application is not approved by MDNR, MDNR necessarily has determined that the property warrants an action under RCRA, CERCLA, or the Missouri Hazardous Waste Management Law. Currently, there is no audit privilege in Missouri or other mechanism to protect the applicant from further action by MDNR or USEPA. Although the regulations provide that withdrawal will not affect the person's liability for the remediation,[32] this is little comfort to property owners and others who may perceive themselves already to be liable. If the applicant's remedial action work plan is rejected later in the process, MDNR will advise the applicant of necessary revisions.

Brownfields Initiative

Under the Brownfields Initiative, a property owner may be eligible for a no-further-action letter if (1) it has completed Phase I and II environmental assessments, which demonstrate that no remedial action is required to protect public health and the environment, or (2) it completes a voluntary remediation action, verified by MDNR, even though the owner has not cleaned up all hazardous substances on the property to levels below regulatory action levels.[33] A no-further-action letter will not be issued unless the remediation activities "significantly restore, in whole or in part, the environment so as to minimize the harmful effects from a release of a hazardous substance to acceptable risk levels."[34]

In addition, after completion of a voluntary remediation project, a property owner may be eligible to receive a covenant-not-to-sue. The owner must certify, and MDNR verify, that the voluntary remediation action has "treated all hazardous substances of concern to levels below then existing regulatory action levels."[35] To qualify for a covenant-not-to-sue, the corrective action plan must be submitted for public notice and comment concerning the effectiveness of the remedy for the intended use of the property.[36]

Perhaps the most attractive incentive provided under the Brownfields Initiative is broad third-party liability protection.[37] The statute provides that upon completion of a voluntary remediation project, the property owner "shall be immune from liability in a civil action brought by any third party to recover clean-up costs, response costs or other legal or equitable damages, including costs of restitution."[38] This liability protection is limited to contamination identified and addressed in the voluntary remediation action and existing on the abandoned property before acquisition by the purchaser. It does not cover hazardous-substance contamination unknown at the time of the remedial activities.

Fees

Neither the Missouri Voluntary Remediation Program nor the Brownfields Initiative specifies fees for the acquisition of liability protection. The fees prescribed by the Voluntary Remediation Program, described above, are intended to cover costs incurred by MDNR in approving and overseeing the remediation activities.

If a person initially is accepted into the Voluntary Remediation Program or purchases property as part of a qualified project and later fails to remediate the property to the satisfaction of MDNR, the purchaser may be required immediately to repay monies granted or loaned from the Brownfields Initiative Property Reuse Fund, with interest and a 10 percent penalty on the total amount granted or loaned. Likewise, the purchaser failing to remediate would not be eligible for the tax exemptions, credits, and other incentives provided under the Brownfields Initiative.[39]

Cost-Recovery Actions

Neither the Missouri Voluntary Remediation Program nor the Brownfields Initiative provides for private cost recovery in connection with a voluntary cleanup. Parties may elect to pursue cost-recovery measures available under federal laws such as CERCLA, or under common law.

Conclusion

Missouri's Voluntary Remediation Program and its Brownfields Initiative provide the tools for property buyers, sellers, and lenders to proceed with contaminated-property transactions, especially on publicly owned property. It remains to be seen whether the economic-development incentives of the Brownfields Initiative will be funded to the extent necessary to spur such redevelopment. Privately owned property will receive benefits from the Voluntary Remediation Program, MDNR's no-further-action letters, and USEPA's Memorandum of Agreement with MDNR. Whether private property will receive benefits under the Brownfields Initiative will depend on how its eligibility is construed and on whether transactions can be structured to take advantage of it.

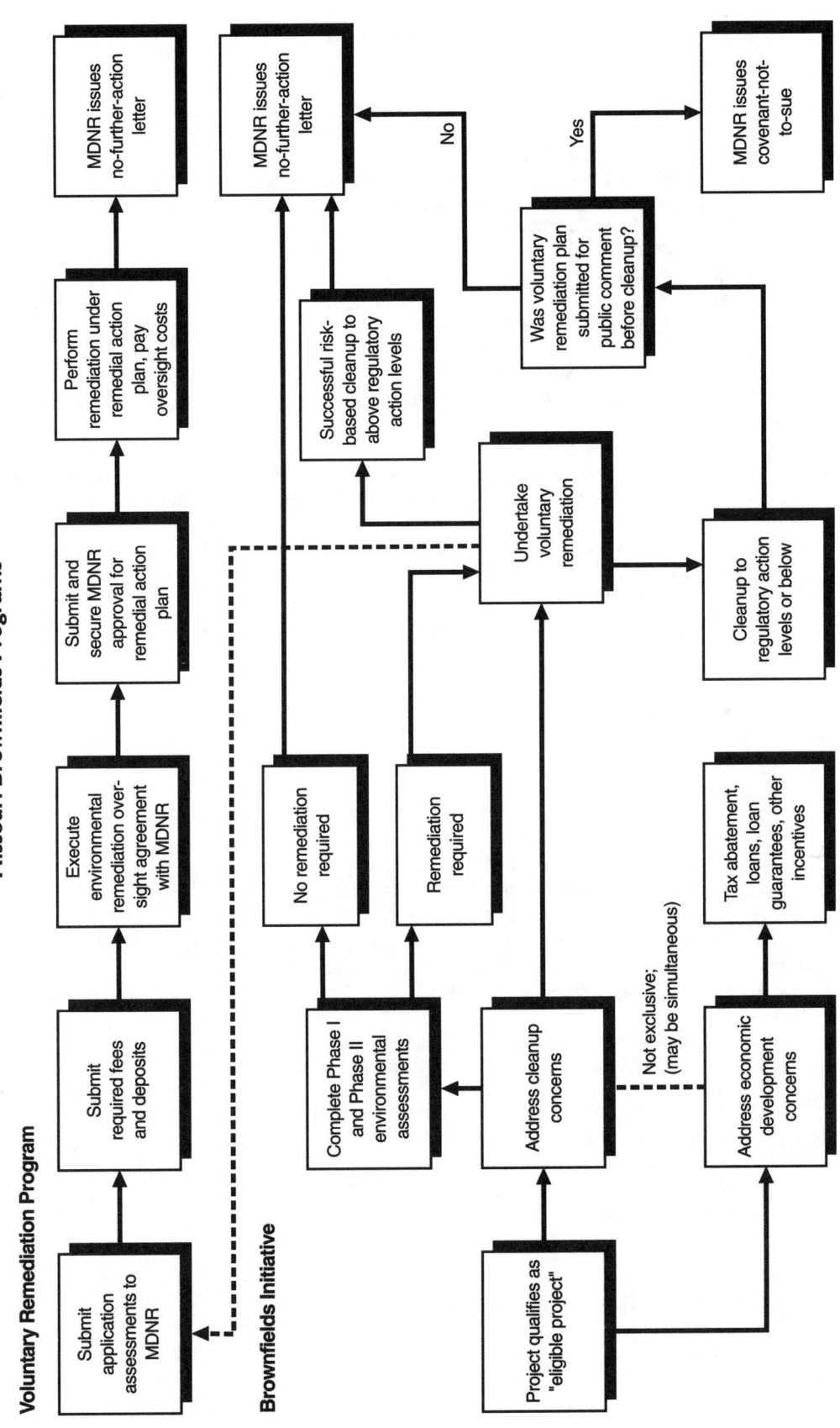

Notes

1. MO. REV. STAT. § 447.700 et seq. (1996).
2. Id. § 260.565–75; MO. CODE REGS. tit. 10, § 25-15.010 (1996).
3. MO. CODE REGS. tit.10, § 25-15.010(7)(A) (1996).
4. MO. REV. STAT. § 447.700(3) (1996).
5. Id. § 447.700(1).
6. Id. § 447.700(3).
7. MO. CODE REGS. tit. 10, § 25-15.010(3)D (1996).
8. 42 U.S.C. § 6901 et seq. (1996).
9. Id. § 9601 et seq.
10. MO. REV. STAT. § 260.350-430.
11. MO. REV. STAT. §§ 260.567, 260.565(4); MO. CODE REGS. tit. 10, § 25-15.010(2)(A)4 (1996).
12. MO. REV. STAT. § 447.700(3) (1996).
13. Memorandum of Agreement between the Missouri Department of Natural Resources and the United States Environmental Protection Agency, Region VII (Sept. 5, 1996).
14. MO. REV. STAT. § 447.704.6 (1996). See also id. §§ 427.011–427.041.
15. Id. § 447.710.
16. See id. § 447.712.2.
17. Id. § 447.712.3.
18. Id. § 447.712.2.
19. Id. § 447.712.4.
20. Id. §§ 427.011–427.041.
21. Id. § 447.714.1.
22. Id. § 447.714.2(1).
23. Id. § 447.714.2(3).
24. Id. § 447.714.2(4).
25. Id. § 447.714.2(2).
26. Id.
27. See MO. REV. STAT. §§ 447.702–447.708 (1996).
28. See MO. CODE REGS. tit. 10, § 25-15.010(5) (1996).
29. See MO. REV. STAT. § 447.714.2(2) (1996).
30. The specific requirements for participation in the Voluntary Remediation Program are set forth in section 260.567 of Missouri Revised Statutes.
31. MO. REV. STAT. § 260.565(3) (1996); MO. CODE REGS. tit. 10, § 25-15.010(2)(A)(7) (1996).
32. MO. CODE REGS. tit.10, 25-15.010(7)(A) (1996).
33. Id. § 25-15.010(2)(A)(7)–(8); MO. REV. STAT. § 447.714.2(2) (1996).
34. MO. REV. STAT. § 447.714.2(2).
35. Id. § 447.714.2(3).
36. Id. § 447.714.2(4).
37. Id. § 447.714.3.
38. Id.
39. Id. § 447.716.2.

CHAPTER 36

Nebraska

JAMES T. PRICE
ANNETTE KOVAR

Nebraska's Remedial Action Plan Monitoring Act[1] (the Act) provides a broad, flexible program for the voluntary cleanup of land or water pollution. The Act took effect January 1, 1995. It is designed to be available to any entity that voluntarily may choose to clean up land or water pollution. There are no statutory restrictions on who can apply to the program. The Nebraska Department of Environmental Quality (NDEQ) administers the program.

Under the program, NDEQ coordinates and monitors voluntary environmental cleanup activities. The costs of NDEQ oversight are paid by the program participants. When remedial activities at a site are complete, if the participant has met the terms required by NDEQ and paid all required fees, NDEQ issues a letter stating that no further action related to the contamination for which the remedial action was conducted is required at the site.

Summary of Major Provisions

The major provisions of the Remedial Action Plan Monitoring Program (the Program) include the following:

- There are no statutory restrictions on who can apply to the Program. The Program is available to any entity that voluntarily chooses to clean up a contaminated site. The program can include even National Priorities List sites and other large or difficult cleanup sites, if accepted by NDEQ.
- The participant, in consultation with NDEQ, develops a remedial action plan that is approved by NDEQ. The remedial action plan outlines the remedial activities to be conducted at the site.

The preparation of this chapter was a collaborative effort. The text does not necessarily, therefore, in every respect reflect the views of Mr. Price, Ms. Kovar, the Nebraska Department of Environmental Quality, or Spencer Fane Britt & Browne. It does not necessarily represent the view of any of their clients.

- The participant must reimburse NDEQ for all expenses incurred in monitoring the cleanup and reviewing the planning and other technical documents. The Program requires a $5,000 application fee and a $5,000 participation fee.
- When an applicant successfully completes the Program in compliance with the remedial action plan, NDEQ may issue a no-further-action letter. This letter states that no further action—relating to contamination for which remedial action had been taken in accordance with the approved remedial action plan—needs to be taken at the site. Certain reopeners may be required.

Participation in the Program

To participate in the Program, a person must be willing to conduct a remedial action plan for land or water pollution under NDEQ oversight, monitoring, and approval. A prospective participant must submit the following to NDEQ:[2]

1. A general description of the remedial action plan, on an application form provided by NDEQ, that conforms with procedures approved by NDEQ
   - NDEQ has not established formal written procedures. General directions are provided with the application.
2. Documentation regarding land or water pollution at the site, including information, when necessary, indicating that the applicant holds or can acquire title, easements, and rights-of-way for the cleanup
   - Typically, site investigation reports are submitted to fulfill this requirement.
3. A detailed remedial action plan for the proposed project, including project monitoring reports, engineering, scientific and financial feasibility data, and other data and information as may be required by NDEQ, such as additional site investigation work
   - NDEQ has accepted applications based on the general remedial action plan described in paragraph 1 above, and deferred submission of the detailed plan to a later date.
4. A payment schedule to reimburse NDEQ's expenses related to monitoring the progress of the remedial action plan, including expenses to review and evaluate the proposed plan
   - With large, complex sites, NDEQ may require a prepayment plan to ensure oversight activities are not interrupted by a lack of oversight funds.
5. A demonstration that the remedial action plan conforms to federal standards
   - Compliance with state standards, although not explicitly stated in the Act, is presumed.
6. Application and participation fees of $5,000 each
   - The application fee is used to offset direct NDEQ oversight costs for the site. The participation fee is used to fund general program expenses.[3] Both are nonrefundable.

For an applicant that successfully completes the Program, NDEQ may issue a letter stating that no further action—related to any contamination for which remedial action had been taken in accordance with the approved remedial action plan—needs to be taken at the site. The letter may provide, however, that NDEQ may require the person to conduct additional remedial action if monitoring indicates that (1) contamination is reoccurring, (2) additional contamination, which was not identified in the submissions to NDEQ, is present, or (3) additional contamination, for which remedial action was not taken according to the remedial action plan, is present.[4] As a general policy, NDEQ accepts sites for partial remediation, but does not issue a no-further-action letter unless all contamination at the site has been addressed.

Sites Ineligible for the Program

The Act does not specify any sites that are ineligible for the Program. NDEQ evaluates the applications to determine whether a site is eligible. Sites that may be subject to other statutory cleanup authorities, under either state or federal law, may participate in this Program, but are not relieved of these other statutory obligations. To the extent possible, however, NDEQ will coordinate these remedial activities. The authors are aware of at least one National Priorities List site at which NDEQ provides oversight under the Program to reduce the oversight of the United States Environmental Protection Agency (USEPA) in a private-party cleanup.

Memorandum of Agreement

Nebraska does not have a Memorandum of Agreement with USEPA for its Program. Therefore, a person participating in the Program theoretically could be subject to federal liability even after obtaining a no-further-action letter from NDEQ. Given USEPA's current enforcement position, it does not seem likely that USEPA would take an enforcement action in connection with a site participating in the Program. Private parties also would not be precluded from bringing a federal claim concerning a site participating in the Program. Participants can seek "comfort letters" or enter prospective-purchaser agreements with USEPA's Region VII office to minimize the threat of federal liability.[5] It is anticipated, however, that the vast majority of sites that will participate in the Program will not be sites for which USEPA believes there is a sufficient federal interest to issue such letters or enter prospective-purchaser agreements.[6]

Liability Protection

The Act provides that a person who completes a cleanup in accordance with Program requirements can receive from NDEQ a letter stating that no further cleanup of the site is required, in connection with the contamination for which remedial action has been taken. The Act does not, however, extend liability protection in any other fashion. Indeed, the final section of the Act provides that the Act supplements other Nebraska laws and does not amend or limit them.[7]

Financial Incentives under the Program

At this time, Nebraska does not have any financial incentives expressly directed to sites in the Program. However, Nebraska has limited financial incentives in the nature of a sales and use tax refund for equipment acquired for the purpose of industrial-pollution control or abatement.[8] NDEQ plans to assist participants in the Program by developing expertise and innovative remedial technologies that can be applied to sites in the Program.

Cleanup Standards

The Act does not explain how cleanup standards for sites accepted into the Program will be set. NDEQ sets such standards on a site-specific basis, using other applicable regulatory requirements. Principal among these is Title 118 of the Nebraska Administrative Code Rules and Regulations, Ground Water Quality Standards and Use Classification. Title 118 sets both narrative and numerical standards for groundwater protection. These standards are intended to protect beneficial uses of groundwater and they are set using national primary and secondary drinking-water regulations. Other contaminant levels may be set, either by pursuing a change in the groundwater classification or by seeking an alternate cleanup level. Appendix A to Title 118 describes a detailed groundwater remedial action protocol.

Mechanics of Participating in the Program

A participant submits the voluntary application containing the elements described above. NDEQ evaluates the proposed project, the proposed completion schedule, and the proposed reimbursement schedule to determine whether the site and plan are eligible. If so, NDEQ and the applicant execute an agreement under which NDEQ monitors the project and the applicant pays for NDEQ's time and expenses, and contractor expense (if any). As a policy matter, NDEQ intends to keep these agreements simple and basic, in contrast with federal Superfund negotiated agreements, by avoiding liability determinations, covenants-not-to-sue, and contribution protection issues. If the application does not meet the requirements, NDEQ returns it and may make recommendations concerning how it can be improved.

During the application process, a conference ordinarily will be held between NDEQ and the participant. Discussions may include future use, goals, cleanup levels, and expectations. These discussions may lead to development or refinement of the remedial action plan. NDEQ may suggest innovative technologies, which the agency encourages. NDEQ also will estimate its review costs. When all parties are in agreement on the elements of the remedial action plan and the required fees have been submitted, the remedial action plan may be implemented.

During remedial activities, the participant must keep NDEQ informed of the project's status. Depending on the project, NDEQ may review and monitor various reports and may visit the site for on-site inspections. Regular progress reports typically are required. NDEQ has designed the Program, however, to work simply. At most sites, NDEQ has eliminated many of the reporting requirements normally associated with environmental investigations and remedial actions. This allows participants to expedite their cleanup efforts in a cost-effective manner.

NDEQ submits regular billings for its review services when the fees for those services exceed the $5,000 initial application fee.

When the planned remedial activities are complete, NDEQ compares the remedial action plan with the final work. For sites where all contamination has been addressed, NDEQ will submit a letter to the participant stating that no further action needs to be taken in connection with the contamination for which the remedial action was conducted if (1) the remedial activities have been performed satisfactorily and according to the approved remedial action plan, (2) all applicable fees have been paid, and (3) all other provisions have been met. The letter will be tailored to fit each project. Sites where only partial remediation has occurred will not receive a no-further-action letter.

If the work is not satisfactory, NDEQ and the participant will confer to determine further actions necessary to achieve the outcome set forth in the remedial action plan and to receive the letter of completion from NDEQ.

Cost Recovery

The Act does not contain a cost-recovery provision, nor do other provisions of Nebraska law expressly contain cost-recovery provisions, with the exception of petroleum releases.[9] A participant who wants to seek cost recovery might turn to common law or federal law.

Conclusion

Unlike many states' Voluntary Cleanup Programs, the Nebraska Program is simple. It leaves significant discretion with NDEQ concerning the sites selected for inclusion. The statutory authorities are flexible enough, and NDEQ in its administration is flexible enough, to apply the Program to a wide range of environmental cleanups, from the simple to the most complex.

Nebraska Remedial Action Plan Monitoring Program

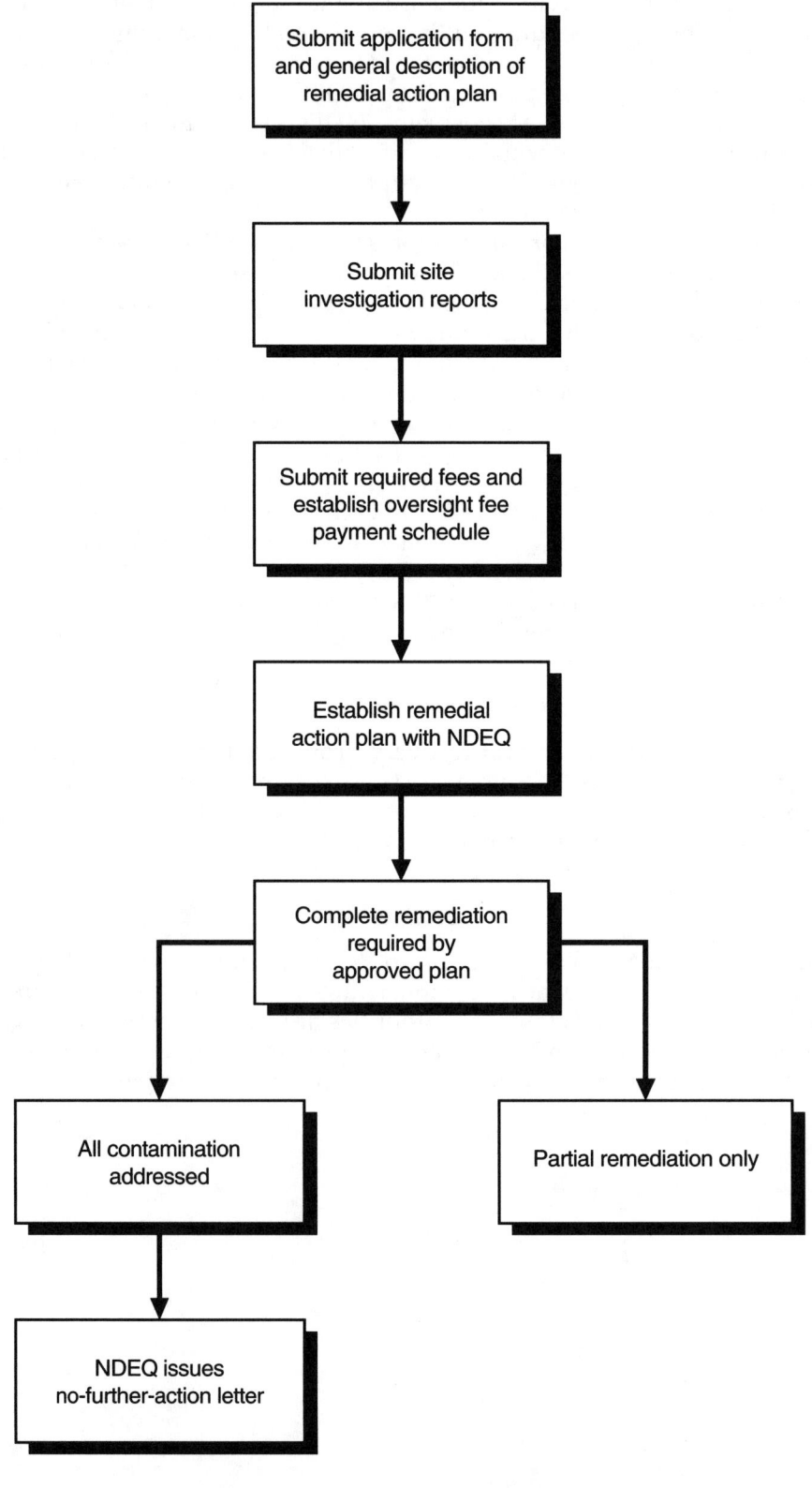

Notes

1. NEB. REV. STAT. §§ 81-15,181–188 (1996).
2. *Id.* § 81-15,184.
3. *Id.*
4. *Id.* § 81-15,186.
5. *See* 60 Fed. Reg. 34,792 (1995).
6. *Id.*
7. NEB. REV. STAT. § 81-51,188 (1996).
8. Air and Water Pollution Control Tax Refund Act, NEB. REV. STAT. §§ 77-27, 149–77-27,155 (1996).
9. *Id.* § 66-1529.02.

CHAPTER 37

New Jersey

I. LEO MOTIUK
SEAN T. MONAGHAN

The New Jersey Department of Environmental Protection (NJDEP) has traditionally used administrative consent orders (ACOs) (also known as remediation agreements) as its means for overseeing a party's participation in the cleanup of contaminated sites. However, in the early 1990s, it was increasingly clear that this approach did not provide NJDEP with an appropriate mechanism to respond to the needs of interested private parties to address sites that were not "high priority" sites. Many parties wanted a procedure allowing them to investigate and then consider whether to remediate a lesser-priority site, yet not face all the regulatory and penalty burdens common to an ACO. Accordingly, NJDEP instituted a Voluntary Cleanup Program (VCP). In early 1996, NJDEP reported that approximately 100 to 125 new voluntary cleanups occur every month.[1]

The VCP is set forth in NJDEP's "Procedures for Department Oversight of Contaminated Sites" (Oversight Rules), which are regulations promulgated under the New Jersey Administrative Code.[2] The Oversight Rules themselves were promulgated pursuant to the NJDEP rule-making authority granted under the New Jersey Spill Compensation and Control Act (Spill Act).[3]

Unlike some other states, New Jersey does not presently have a Memorandum of Agreement for its VCP with the United States Environmental Protection Agency (USEPA).

Eligibility

The VCP is open to parties without any responsibility for conditions at the site, as well as responsible parties.[4] Lenders and prospective purchasers, as well as municipalities, apply to participate in the VCP to develop more information before determining whether to foreclose or go ahead and take title. However, NJDEP will not permit sites for which there are existing enforcement agreements or that are identified as high priority sites to enter the VCP program.[5]

How Does the VCP Program Work?

The cornerstone of the VCP is a Memorandum of Agreement (MOA). NJDEP refers to the MOA as a contract established between the agency and the party electing to investigate and/or remediate the subject site. It is not an enforcement document. Under the VCP and pursuant to the terms of an MOA, all data developed during the remediation is required to be supplied to NJDEP; further, the private party's actions must not create any new environmental hazards at the site.[6] Finally, the private party must agree to pay all of NJDEP's oversight costs.[7] Unlike the ACO program, there are no stipulated penalties and no requirement to establish a remediation funding source.[8] In return for the undertakings of the private party, NJDEP attempts to commit to specific time frames to review and comment on specific submissions.[9]

A party interested in entering an MOA submits a completed application that details the work to be performed and the requested NJDEP involvement in those activities. Further, the private party should submit a schedule of the work to be performed, including the documents and reports to be submitted to NJDEP for review. The rules provide that NJDEP should review the document for administrative completeness and give notice of any deficiencies to the party conducting the investigation or cleanup within thirty days.[10] Upon a determination that the submission is administratively complete, NJDEP will notify the party in writing of the expected turnaround time for comments on the document. NJDEP will provide written comments on the submission, detailing the acceptance of any conclusions or recommendations and the need for any additional investigative or cleanup effort. Although the following does not always become reality, NJDEP has anticipated that the review process should take between thirty and sixty days, depending on the level of complexity of the submission.

NJDEP is very reluctant to alter the standard language of the MOA set forth in the Oversight Rules.[11] However, NJDEP recognizes the need for the MOA to be flexible enough to address different types of remedial activities at a variety of sites. Therefore, in its discretion, NJDEP allows additional information to be included in the findings section of the MOA.

What Happens If the VCP Is Not Completed?

The MOA specifically provides that a private party's entry into the VCP does not constitute an admission of fact or liability. Further, the MOA specifically provides that NJDEP does not surrender any rights or causes of action. As long as the private party pays the appropriate oversight fees and tenders all information to NJDEP, it can withdraw from the program at any time. When and if there is a withdrawal from the program, the site will remain on NJDEP's list of "known contaminated sites" if a discharge at the site has been identified. The site will be prioritized in accordance with NJDEP's priority scoring system. If NJDEP determines that the site is a high-priority site, it may issue an ACO to any responsible parties. Further, NJDEP retains the right to pursue all its statutory remedies against any party in any way responsible for the contamination.[12] If the party entering the MOA was not in any way responsible for the preexisting discharge, but may have generated new contamination as a result of its work, NJDEP would have the right to pursue that party under existing laws such as the Spill Act.

Confidentiality

At the present time, New Jersey has no statutory protection relating to information discovered during a voluntary investigation or cleanup. The MOA specifically requires that all information be turned over to NJDEP.[13]

Liability Protection

The VCP does not address liability protection, except for the representation in the MOA that entry into the MOA and performance thereunder does not in any way constitute any admission of fact or liability. However, the Spill Act does provide exemptions from liability for governmental entities acquiring title by virtue of bankruptcy, tax delinquency, abandonment, escheat, or eminent-domain proceedings.[14]

Furthermore, similar to the federal Comprehensive Environmental Response, Compensation, and Liability Act (CERCLA), the Spill Act has an exemption for the innocent purchaser who acquires real property after September 16, 1993. To qualify for this exemption, a party must establish, by a preponderance of the evidence, that (1) it acquired the property after the discharge occurred, (2) it either acquired the property by devise or succession or, at the time it acquired the property, it did not know and had no reason to know that any hazardous substance had been discharged at the property, (3) it did not discharge the hazardous substance and is not in any way responsible for the substance, and (4) it gave notice to NJDEP upon actual discovery of the discharge.[15] To establish that it had no reason to know that a hazardous substance had been discharged, the party must have undertaken, at the time of acquisition, all appropriate inquiry into the previous ownership and uses of the property.[16] Somewhat different from CERCLA, New Jersey law defines "all appropriate inquiry" as the performance of a preliminary assessment and site investigation (when necessary) in accordance with NJDEP regulations.[17] Further, if a person who owns real property obtains actual knowledge of a discharge during that period and subsequently transfers ownership without disclosing the knowledge, the transferor will be strictly liable for the cleanup and removal costs of the discharge and the innocent-purchaser defense will be lost.[18]

New Jersey law also exempts a secured creditor from liability for past releases of hazardous substances from a facility, as long as it did not actively participate in the management of the facility. A secured creditor can foreclose on its security interest without incurring liability for past environmental damage on the property. However, it is liable for any new discharge caused by its own negligence after the date of foreclosure. In such a case, its liability is limited to the cleanup costs related solely to postforeclosure discharge.[19]

As for trustee and other fiduciary liability, New Jersey law provides that for discharges from facilities that are all or part of the trust or estate, only the assets of the trust or estate shall be subject to the obligation to pay for the cleanup.[20]

Financial Incentives

In 1996, the New Jersey Environmental Opportunity Zone Act became law.[21] "Environmental opportunity zones" can now be established by municipalities for urban enterprise zones that are eligible under the New Jersey Special Municipal Aid Act.[22] The municipal governing body must pass an ordinance to identify one or more areas within such a municipality as an environmental opportunity zone and specifically find that one or more parcels of real property in that zone could be remediated and used for industrial or other commercial areas.[23] If so, the governing body can provide a property-tax exemption for a term of ten years.[24] Payments in lieu of property taxes to the municipality would be based on a sliding scale computation, which would start at 0 percent and increase by 10 percent each year.[25]

In 1995, New Jersey also enacted the Landfill Reclamation Act.[26] This law provides municipalities with new authority to help finance landfill closure and pollution abatement costs and to provide incentives for the redevelopment of areas where old landfills have not been properly closed. Similar legislation, the Large Site Landfill Reclamation and Improvement Law,[27] was signed into law on July 22, 1996.

In 1993, legislation commonly known as S-1070 was enacted. As part of the Hazardous Discharge Site Remediation Fund,[28] the S-1070 legislation includes funding for low-interest loans and grants for remediation of contaminated sites. Municipalities may apply for grants and loans for up to $2 million for investigations and cleanup activities. Grants are specifically provided for preliminary assessments and site investigations, while loans can be used for more comprehensive remedial investigations and cleanup actions. Consideration is presently being given to expanding the grant program for municipalities to cover the costs of remedial investigations, as well as reducing the interest rate on loans currently offered. It is believed that by using the data obtained from these remedial activities, local officials can develop accurate cost estimates of any cleanup work required at a site, thereby improving the marketability of the property.

Businesses and individuals, without responsibility for the contamination and who want to conduct such actions voluntarily, may qualify for loans of up to $1 million per year. Some property owners, who acquired their property before 1989, may also qualify for grants. Parties responsible for site contamination, but who do not have a funding source, can also qualify for such $1 million loans.[29]

NJDEP records reveal that as of September 1995, NJDEP and the New Jersey Economic Development Authority had approved $10.5 million in loans and $4.5 million in grants. These monies come from a $55 million fund created when the New Jersey legislature dedicated an unused portion of the State Hazardous Waste Bond Issue and New Jersey Economic Recovery Fund.

Two new funding sources have recently been added to the New Jersey cleanup program. First, on November 5, 1996, the New Jersey electorate passed a constitutional amendment that requires 4 percent of the money raised each year under the New Jersey Corporate Tax Act to go to a dedicated fund to be used for cleanup of hazardous-waste sites, as well as for the removal of underground storage tanks.[30] Second, on November 6, 1996, the governor of New Jersey signed into law a bill (S-294) that would give developers who build on abandoned landfills the opportunity to recoup up to 75 percent of their closure and remediation costs from sales-tax revenue generated from commercial activity at the site. Because there is very little, if any, public money for properly closing municipal landfills that were shut down many years ago, the state supported this bill as an incentive for developers to build on certain of these sites.[31]

Technical Regulations

New Jersey has specific regulations defining how to conduct remedial actions. Known as the Technical Requirements for Site Remediation (Technical Regulations), these rules establish the minimum criteria for performing preliminary assessments, site investigations, remedial investigation, remedial designs, and remedial actions at New Jersey sites.[32]

Other than in connection with underground storage tanks, New Jersey does not have any statutory or regulatory program governing environmental consultants, except to the extent that certain documents must be signed by a licensed professional engineer. However, in the definition section of the Technical Regulations, the term "person responsible for conducting the remediation" includes any person who is performing the remediation or has control over the person (that is, contractor or consultant) who is performing the remediation.[33] Therefore, depending on the authority of the consultant, he or she may have responsibilities such as submitting work plans and reports pursuant to a schedule contained in the oversight document.[34] In connection with the closure of regulated underground storage tanks, New Jersey licenses both consultants, who conduct investigations and assessments, and contractors, who perform the closure work.[35]

Cleanup Standards

As of yet, NJDEP has not formally adopted cleanup standards. However, in June 1994, NJDEP published "Guidance Document For The Remediation of Contaminated Soils." Until cleanup standards are adopted, NJDEP will rely on its "soil cleanup criteria" (SCC), which it publishes from time to time in *Site Remediation News*.[36]

There is no difference in treatment between sites in the VCP and other sites regarding cleanup standards. New Jersey's criteria for site cleanups include water quality criteria, maximum contaminant levels from the federal Safe Drinking Water Act, background levels, risk assessments, USEPA guidelines, and New Jersey's SCC guidelines. For soil cleanups, the state may use the SCC guidelines or case-specific levels derived from risk assessments. Although alternatives are being considered, at the present time a cancer risk level of one million is used.[37] If SCC levels are determined to be below background levels, then the cleanup level is background.[38]

The New Jersey groundwater quality standards and the New Jersey surface-water quality standards provide guidance that depends on the use and value of the resource. Likewise, NJDEP has developed SCC levels that accommodate various uses, both residential and nonresidential.[39]

Institutional and Engineering Controls

Though New Jersey law expresses a preference for permanent remedies, NJDEP recognizes that it does not always make sense to remove all or some contamination from a site. When a cleanup that limits workers' or residents' exposure to contamination is deemed to be protective of human health and the environment, NJDEP will approve nonpermanent remedial actions and/or engineering or institutional controls.[40] A condition of the approval of such nonpermanent remedial actions is the consent of the current property owner to the placement of an institutional control, commonly known as a "declaration of environmental restriction." Similarly, for groundwater contamination that exceeds the groundwater quality criteria, NJDEP may approve a "classification exemption area" under certain circumstances.[41]

No-Further-Action Letters

If a party completes investigative and remedial activities for an area of concern at a site in accordance with the Technical Regulations, or when further cleanup work is not required, a no-further-action (NFA) letter will be issued.[42] If an area of concern has received an NFA letter, NJDEP may compel additional remediation only if a cleanup standard that was applied has subsequently decreased by more than a factor of ten.[43] Further, this additional requirement can be imposed only on a party who is in any way responsible to clean up and remove contamination.[44] When a nonpermanent cleanup, including an engineering control, has been approved for a site, NJDEP may require additional remediation activities only if the control is no longer protective of human health and the environment.[45]

In New Jersey urban centers, there has historically been substantial construction on land that was created by filling in tidal and freshwater wetlands areas. New Jersey law establishes a presumption that encapsulation is the satisfactory remedy for sites with such "historic fill" that is contaminated.[46] However, it must be noted that there has been substantial controversy about whether the historic fill has been the cause of contamination. In addition, this presumption does not apply to (1) fill generated by the processing of metal or mineral ores, residues, slugs or tailings, or chemical-production waste, or (2) fill that was generated on the site where it was deposited.[47]

Developer Incentives

Before the New Jersey Environmental Opportunity Zone Act was passed, an earlier version of the bill contained a provision—later deleted—that would have insured that a person who acquired or operated any qualified real property and who remediated the property would not be liable (1) for the acts of the prior owner or operator for any preexisting contamination that was discovered post-closing, or (2) if the remediation standards subsequently changed. This provision was important because under current New Jersey law, a party who acquires or operates a site and who knew or should have known about preexisting contamination falls within the ambit of a "responsible party" and has all the liabilities of a current owner.[48] Such a provision in existing law discourages redevelopment, especially in formerly industrialized urban areas, by developers who would inherit such liability.

The New Jersey Urban Redevelopment Act, which was signed into law on July 12, 1996, attempts to remedy this problem.[49] Section 56 of this law provides that a prospective purchaser, who has no responsibility for the contamination except for its status as a current owner, could be eligible for a complete release from liability for any preexisting contamination. To be eligible for such a release, the prospective purchaser would need to propose a remedial action work plan for review by NJDEP. The remedial action work plan could be proposed for the entire site or part of the site and NJDEP would have the flexibility to approve the remedial action work plan for the entire site or part of the site. If the remedial action work plan was proposed and approved for the entire site, the approval would relieve the prospective purchaser and its successors-in-interest or successors-in-title from responsibility for any preexisting contamination, other than the contamination required to be addressed under the approved plan, as long as the prospective purchaser and its successors implement the approved remedial action work plan and comply with all institutional and engineering controls required by NJDEP. However, such a release would not be available to parties who were in any way responsible for the preexisting discharge.[50]

The New Jersey Urban Redevelopment Act protections referenced above are specifically limited to municipalities that are eligible to receive aid under the Special Municipal Aid Act or that have been classified as "special needs districts."[51] However, three bills (A-2250, S-39, and S-1539) that have recently been introduced would expand to all parts of the state the protections provided in the New Jersey Urban Redevelopment Act. Two of these bills (S-39 and S-1539) also (1) seek to reduce cleanup costs by restricting the ability of

NJDEP to assess "indirect" administrative costs on a developer, (2) eliminate the requirement for certain developers to undertake cleanups of contaminated groundwater, and (3) limit soil cleanups to situations involving imminent threats to the public.

NJDEP is also giving further consideration to covenants-not-to-sue as a means of providing some protection above and beyond the NFA letter.[52] As one possible course of action, it has entered a consent agreement with a prospective purchaser and then filed the consent agreement, along with a relevant complaint, in a United States district court for approval by the court. The action was brought pursuant to both federal and state law and sought to provide the broadest possible covenant-not-to-sue.[53]

Recovery from Third Parties

A party that performs a voluntary cleanup and that was not in any way responsible for the discharge in question may seek reimbursement from two sources. First, within one year of when it knew (or should have known) about the contamination, the party can make a claim against the New Jersey Spill Compensation Fund.[54] It is uncertain whether this fund would have sufficient resources to pay all the claims. Further, if it had eligible investigative and/or cleanup costs, the private party could seek reimbursement from parties in any way responsible for the discharge. With NJDEP approval, the private party may even be able to seek treble damages from the aforementioned responsible parties who have not complied with prior cleanup directives from NJDEP.[55] If the private party is successful in its treble-damage claim, it is entitled to retain two-thirds of the monies and one-third goes to the state of New Jersey.[56]

It should be noted that the Spill Act is broader than CERCLA in that petroleum is a covered "hazardous substance."[57] Further, there is no statute of limitations for bringing a contribution action under the Spill Act.[58] Liability under the Spill Act is joint and several.[59] To make out a claim under the Spill Act, a private party must demonstrate that it has incurred cleanup costs and that the other parties are "in any way responsible" for the discharge in question. The only statutory defenses to liability are acts of war, acts of God, and sabotage.[60]

Examples of Brownfields Projects in New Jersey

Over the last two years, there have been some interesting brownfields, non-Superfund site projects that are beginning to move forward in New Jersey. The Elizabeth Landfill (off exit 13A of the New Jersey Turnpike) may be developed into a major retail center, as may three old municipal landfills in Carteret. In the Newark central ward, developers completed a $200,000 cleanup and then moved ahead with the construction of fifty-one townhouses. At the former Volco Brass & Copper Company in Kenilworth, a private developer cleaned up a twelve-acre area of metal residues and moved ahead with the construction of a large strip mall. In Trenton, the old Roebling Wire Works passed NJDEP review and the developer plans to convert some of the historic buildings into a strip mall that would include a supermarket and a museum. In Camden, the New Jersey Economic Development Authority (EDA) purchased an abandoned factory building, razed two asbestos-ridden buildings, cleaned up the site, and put up two buildings that are now leased to a Fortune 500 company. A twelve-acre dumping ground in Wanaque will be the satellite campus of Passaic County College. A new state-of-the-art, public-works depot for Hackensack and a large, private, wholesale operation (The Price Club) now occupy the site of a former city public-works yard and a private, wholesale fuel-oil distribution facility. Harsimus Cove South is one of several sites along the Hudson River in Jersey City that have been and/or are being developed under NJDEP oversight. The site was previously used as a railroad yard. In Trenton, the former Roebling Works complex, which was the site where cables for suspension bridges were manufactured, has been developed into over two hundred thousand square feet of office and retail space.

Recommendations for Brownfields Redevelopment in New Jersey

Probably the most detailed look at brownfields issues in New Jersey has been conducted by the Regional Plan Association. In 1994, it issued a report, the goal of which was to analyze obstacles to redevelopment and to recommend and implement solutions that could be replicated. Most interesting are the sections dealing with recommendations for streamlining the permit process. These are as follows:

- Priority project designation by NJDEP
- NJDEP policy statement setting forth objectives and describing whether priority review process is adopted and how it works
- NJDEP Assistant Commissioner to oversee policy
- State case manager to serve as a permit coordinator with sufficient expertise and authority to expedite approvals
- Local coordinator who has the expertise and authority at the local level on engineering, environmental, and economic-development issues
- Project facilitator involvement
- Early NJDEP guidance and concept approval
- Monthly project team meetings
- Development of team approach for state and regional entities
- Gubernatorial executive order requiring quarterly reporting by NJDEP to the governor's office

The report made the following recommendations regarding the redevelopment process:

1. Incentives under the existing redevelopment statutes for Urban Aid communities should be used.
2. Environmental cleanup financing in an initial amount of $20 million should be provided to EDA to create an industrial and commercial redevelopment fund to leverage private and/or municipal cleanup of contaminated or environmentally damaged priority urban sites. EDA could provide revolving resource financing, as well as participate directly in redevelopment of the real estate.
3. Transportation development districts should be considered for areas in which a number of developers share an interest in funding area road and related infrastructure improvements.
4. Consideration should be given to tax increment financing to fund public investments in areas slated for redevelopment. This program would allow the recapturing of all, or a significant portion of, the increased tax revenue that might result if redevelopment stimulates private investment.[61]

New Jersey Brownfields Initiatives Flowchart

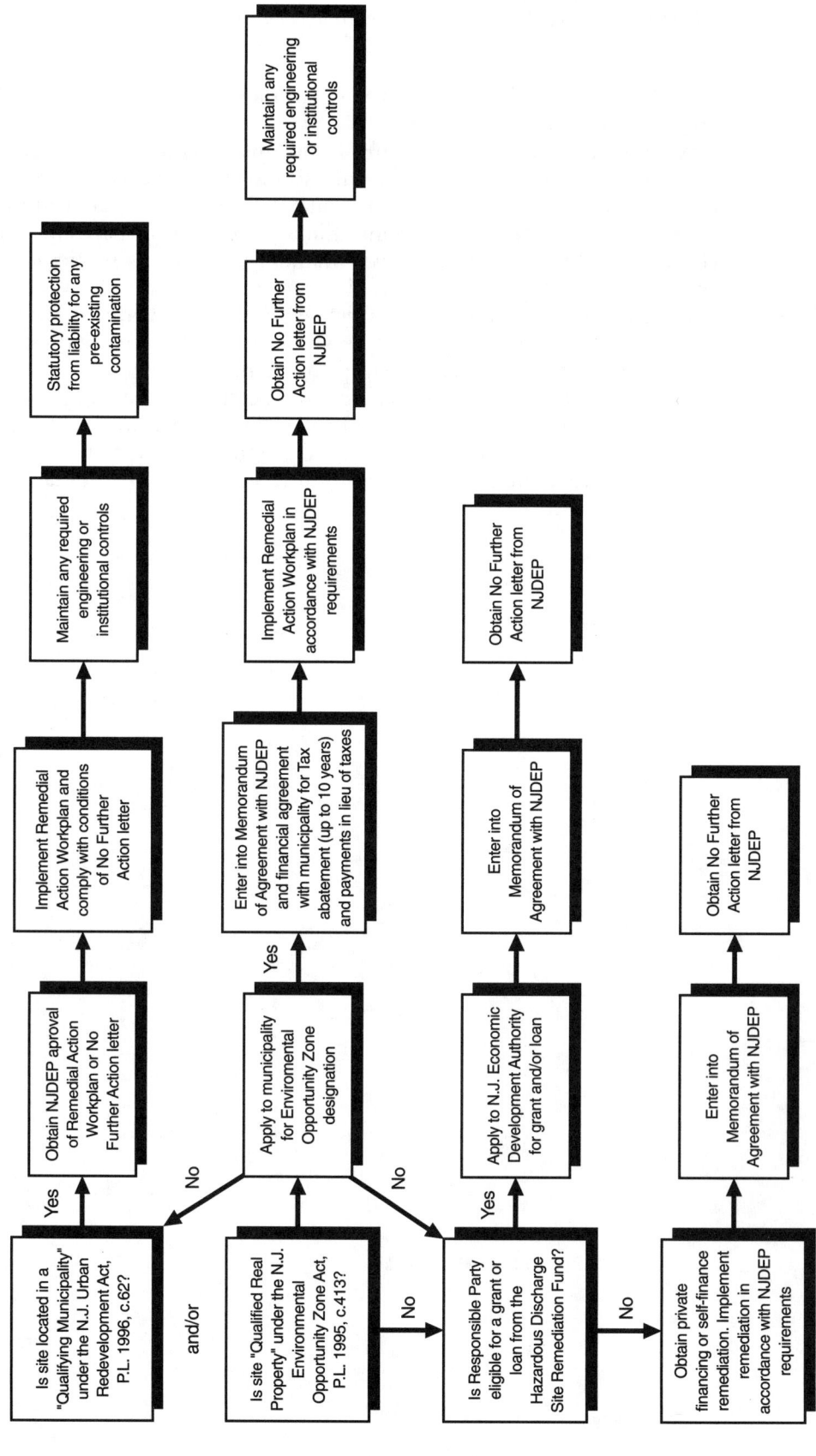

Notes

1. Jim Sinclair, *Industrial Site Remediation in the Post-ECRA Era*, N.J. Bus., May 1996, at 60. The Whitman Administration recently reported that in 1994, NJDEP reached 990 such agreements; 1,267 in 1995; and 1,436 in 1996 through November 19, 1996. *Whitman Defends Site Cleanup Program*, Star-Ledger, Nov. 20, 1996. Further, at a seminar on November 6, 1996, a representative of NJDEP reported that there are approximately 6,600 known contaminated sites in New Jersey. Pearson, McLaren Hart Seminar booklet, Nov. 1996.
2. N.J. Admin. Code tit. 7, § 26C *et seq.* (1997).
3. N.J. Stat. Ann. § 58:10-23.11 *et seq.* (West 1996).
4. N.J. 7:26C-2.2.
5. N.J. Admin. Code tit. 7, § 26C-3.1(b) (1996).
6. *Id.* § 26C, app. A.
7. *Id.*
8. *Id.*
9. *Id.*
10. *Id.* § 26C-3.1(d).
11. *Id.* § 26C-3.3.
12. *Id.* § 26C-2.1c.
13. For a limited period (specifically June 16, 1993, to July 16, 1994), New Jersey had an amnesty provision for violations of the Spill Act or the Environmental Cleanup Responsibility Act (ECRA) that occurred before June 16, 1993. More specifically, any party who, before that time, violated ECRA by closing or transferring operations without appropriate NJDEP approval, or who failed to report spills properly, had one year from June 16, 1993, to report such information and to enter an ACO or MOA with NJDEP to remediate the site. If the investigation and remediation were properly conducted, NJDEP could not use the aforementioned information to prosecute any violations against the party in question. Conversely, failure to comply with the terms of the ACO or the MOA could trigger penalties for violations that occurred both before and after June 16, 1993. N.J. Stat. Ann. §§ 13:1K-11.10, 58:10B-15 (West 1996).

The New Jersey legislature is presently considering a bill entitled "The Voluntary Environmental Auditing Promotion Act" (S384), which would establish an evidentiary privilege, for environmental audit reports, that makes audit information inadmissible in civil or administrative proceedings. Exceptions would include instances when the public interest outweighs the expectation of confidentiality and the information cannot be obtained otherwise in cases in which violations were not corrected. The bill would give businesses limited immunity from civil penalties and protection from citizen suits for violations discovered during an audit if the violations are disclosed promptly and corrected. This legislation is controversial and its outlook is uncertain.
14. N.J. Stat. Ann. § 58:10-23.11g(d)(4) (West 1996).
15. *Id.* § 58:10-23.11g(d)(2).
16. *Id.*
17. *Id.*
18. *Id.* § 58:10-23.11g(d)(3).
19. *Id.* §§ 58:10-23.11g4–8.
20. *Id.* § 58:10-23.11g9.
21. 1995 N.J. Laws 413.
22. N.J. Stat. Ann. § 54:4-3.152 (West 1996).
23. *Id.* § 54:4-3-154.

24. *Id.*
25. *Id.* § 54:4-3.156b.
26. 1995 N.J. Laws 173.
27. 1996 N.J. Laws 13.
28. N.J. STAT. ANN. § 58:10B-4 *et seq.* (West 1996). On November 5, 1996, the New Jersey electorate passed a general bond referendum that authorizes the state to issue $70 million of new bonds for hazardous-discharge sites.
29. *Id.* § 58:10B-6.
30. SCR 4160.
31. 1996 N.J. Laws 124 (commonly known as S-294 (Gormley)).
32. N.J. ADMIN. CODE tit. 7, § 26E (1995) (proposed revisions are now being considered).
33. *Id.* § 26E-1.8.
34. *See id.* § 26E-1.6(b).
35. N.J. STAT. ANN. §§ 58:10A-24.1–24.5 (West 1992).
36. Copies can be obtained by writing to NJDEP at the Program Support Element, Box CN 413, Trenton, New Jersey 08625-0413.
37. N.J. STAT. ANN. § 58:10B-12(d)(1) (West 1996).
38. *Id.* § 58:10B-12(g)(4).
39. *Id.* § 58:10B-12.
40. *Id.*
41. *Id.* §§ 58:10B-12, 58:10B-13.
42. *Id.* § 58:10B-13. In the seminar papers referenced in note 1, the NJDEP representative reported the following statistics: 7,200 Industrial Site Recovery Act (ISRA)/ECRA NFA letters were issued since 1984, 5,200 underground storage tank NFA letters since 1987, 4,400 voluntary cleanup and ACO NFA letters since 1981, and 120 Superfund subsite NFA letters since 1981, for a total of 16,920 NFA letters.
43. *Id.* § 58:10B-13e.
44. *Id.*
45. *Id.* § 58:10B-13f.
46. *Id.* § 58:10B-12h. Senate bill 39 attempts to facilitate the process for developers to obtain "historic fill" determinations.
47. *Id.*
48. *Id.* § 58:10-23.11g(d)(2).
49. 1996 N.J. Laws 62.
50. West's 1996 New Jersey Session Law Service No. 4 at 216 [§ 58:10-23.11g(e)] (West 1996).
51. *Id.* § 55:19-22.
52. "THE STARR REPORT" (Strategy To Advance Regulatory Reform), July, 1995 at II-21. Senate bills 39 and 1539 would require that covenants-not-to-sue be issued at the same time NFA letters are granted.
53. New Jersey Dep't of Envtl. Protection v. OENJ Corp., Inc. & N.J. Metromall, LLC, No. 96-3276 (D.N.J. July 9, 1996). NJDEP has indicated that there were unique circumstances in this matter and that it is unlikely that the agency will take the judicial consent decree approach in the future.
54. N.J. STAT. ANN. §§ 58:10-23.11g(a), 58:10-23.11k (West 1996).
55. *Id.* § 58:10-23.11f(a)(3).
56. *Id.*
57. *Id.* § 58:10-23.11b.

58. Pitney-Bowes v. Baker Indus., 277 N.J. Super. 484, 649 A.2d 1325 (N.J. Super. Ct. App. Div. 1994).

59. N.J. STAT. ANN. § 58:10-23.11g(c) (West 1996).

60. *Id.* § 58:10-23.11g(d). However, it should be noted that as a matter of practice, NJDEP will not require an owner or operator at a property to remediate conditions that have been caused by the actions of an off-site third party and that are not susceptible of correction without action being taken off-site. *See id.* 58:10B-12g(5).

61. N.J. REGIONAL PLAN ASS'N, UNION COUNTY MODEL SITE REDEVELOPMENT PROJECT (1994). *See also Easing the Re-Use of Once-Polluted Industrial Sites*, N.Y. TIMES, Oct. 27, 1996, at I7.

CHAPTER 38

New York

MARGARET MURPHY

Although the New York legislature has yet to enact any specific brownfields legislation, the New York Department of Environmental Conservation (NYDEC) has been encouraging the redevelopment of contaminated properties with its own Voluntary Cleanup Program.[1] Initiated in October 1994, NYDEC's Voluntary Cleanup Program is a departmental policy only and is not supported by formalized agency regulations or specific statutory authority. The Voluntary Cleanup Program is still undergoing development and revision. Most recently, for example, NYDEC expanded the Voluntary Cleanup Program to provide protection to certain potentially responsible parties (PRPs) when previously all PRPs were prohibited from benefiting from the Voluntary Cleanup Program.[2] In the future, NYDEC hopes to begin to formalize its policy by circulating the Voluntary Cleanup Program to interested parties for comment.[3]

New York Governor George Pataki supports codifying the Voluntary Cleanup Program and, in fact, drafted a brownfields bill in 1995.[4] In addition to formalizing the Voluntary Cleanup Program, the governor has proposed legislation that would provide environmental liability protection to fiduciaries, secured lenders, and state and public corporations that are nonmanagerial owners or operators of property.[5]

The New York legislature also has shown support for similar brownfields programs. In fact, the New York State Senate passed its own version of a voluntary program in the 1994–95 legislative session, although the bill did not reach consideration in the House.[6] The main issue of debate in the legislature over adopting a Voluntary Cleanup Program has been the extent to which PRPs should be allowed to participate and/or be relieved from liability for contamination.[7]

Summary of Major Provisions

The Voluntary Cleanup Program allows certain volunteers to enter agreements to clean up contaminated property and pay for NYDEC's oversight costs in exchange for a no-further-action letter and a release from liability for historical contamination, both of which are subject to certain "reopeners." If the property will be used for industrial or commercial purposes, the volunteer may use cleanup standards consistent with the safe use of the property. NYDEC may require the volunteer to place an environmental deed restriction on the property.

Eligibility for the New York Voluntary Cleanup Program

All sites over which NYDEC exercises enforcement jurisdiction are eligible for the Voluntary Cleanup Program.[8] This includes sites that are on the New York State Registry of Inactive Hazardous Waste Disposal Sites and those that are not.[9]

Anyone who would like to clean up a site that is under the jurisdiction of NYDEC is eligible to participate in the Voluntary Cleanup Program, unless the party is a PRP for the property sought to be cleaned up under the Voluntary Cleanup Program and the property is

1. a treatment, storage, or disposal facility (TSDF) subject to corrective action or closure under permit or order issued under NYDEC's hazardous-waste management regulatory (RCRA) program;
2. a TSDF that is operating under interim status under the RCRA program and is subject to enforcement action leading to the issuance of an order containing a corrective-action schedule; or
3. subject to any other "enforcement action" requiring the PRP to remove or remediate a hazardous substance.[10]

"Enforcement action," under the Voluntary Cleanup Program, means the issuance of a notice of violation, commencement of enforcement, or notification of PRP status under New York environmental law; the issuance of an accusatory instrument under the New York Criminal Procedural Law; or the issuance of any notification under federal law that requires the removal or remediation of hazardous substances.[11]

In addition to the above-listed PRPs that are ineligible for the Voluntary Cleanup Program, NYDEC is still evaluating "whether a PRP for a Class 1 or Class 2 site, as listed on the New York State Registry of Inactive Hazardous Waste Disposal Sites, may participate in the program."[12] A Class 1 site is a site that NYDEC has classified as needing immediate action and presenting an imminent danger to the public health. A Class 2 site is a site that NYDEC has classified as needing action and posing a significant threat to public health or the environment.[13]

Memorandum of Agreement

Currently, New York does not have a Memorandum of Agreement with the United States Environmental Protection Agency (USEPA) for its Voluntary Cleanup Program. Therefore, a person participating in the New York Voluntary Cleanup Program could be subject, theoretically, to federal liability even after obtaining a release from liability from the state. USEPA, however, is currently developing a formal policy under which it will not take action in connection with a site that is participating in a state brownfields program.[14] In addition, to minimize the threat of federal liability, parties may seek "comfort letters" from USEPA's Region II office.[15] There does not appear to be anything to preclude private parties from bringing a federal cause of action against a New York Voluntary Cleanup Program participant.

Liability Protection

Lenders, Trustees, and Fiduciaries

Neither the Voluntary Cleanup Program nor any New York statute or regulation provides lenders, trustees, or fiduciaries with any special protection against liability for hazardous-waste cleanup.[16] NYDEC's regulations define "responsible party" for purposes of New York's law governing inactive hazardous-waste disposal sites[17] to include the following: current owners of a site; owners and operators of a site at the time any hazardous-waste disposal occurred; generators and transporters of hazardous waste disposed of at the site; and persons determined to be responsible according to statutory or common law.[18]

Protection for Participants in the Voluntary Cleanup Program

A volunteer under the Voluntary Cleanup Program will receive a no-further-action letter from NYDEC if cleanup of the contaminated property to the agreed-upon levels is achieved.[19] A no-further-action letter from NYDEC includes a declaration that NYDEC "does not contemplate further action needing to be taken at the site" and a release from liability for past contamination, including liability for natural-resource damage.[20] The benefits of a no-further-action letter run to the volunteer's successors and assigns unless a successor or assignee is a PRP in connection with the site.[21] This prevents a PRP who is not qualified to enter the program from using a third party to obtain a no-further-action letter under the Voluntary Cleanup Program and subsequently assign the release to the PRP. It is unclear, however, whether a successor or assign who is a PRP based solely on his or her status as an owner of contaminated property is precluded from receiving the benefits of a no-further-action letter from a volunteer. This proposition, if true, would not conform to the rest of the Voluntary Cleanup Program, which allows no-fault/owner-only PRPs to benefit from the program.

It is clear, however, that NYDEC's no-further-action determination and release from liability can be reconsidered, or "reopened," after the no-further-action letter has been delivered. NYDEC will consider reopening a no-further-action letter if

1. the response action required of the volunteer was not sufficiently protective to allow the contemplated use of the site to proceed safely from a human health perspective;

2. the volunteer, or its successor, changes the site's use to a use requiring a lower level of residual contamination;
3. the volunteer fraudulently obtains the release; or
4. the environmental conditions present at the affected site at the time the voluntary agreement was executed by NYDEC were unknown to NYDEC at such time. This reopener affects only the volunteer, successor, or assign who owns or operates the property at the time of the reopening or thereafter. In other words, a volunteer who has subsequently sold, developed, or lent money for the purchase or redevelopment of a site will not be subject to liability based on the late discovery of historical contamination.[22]

NYDEC points out in its Voluntary Cleanup Program summary that because the program is subject to reopeners, the more comprehensive the remedial response devised in the volunteer's initial agreement, the less likely it is that the no-further-action letter will be reopened.[23] The Voluntary Cleanup Program does not state whether the benefits of the no-further-action letter run with the land.

Financial Incentives

New York does not currently provide any financial incentives for volunteers under the Voluntary Cleanup Program.

Cleanup Standards

NYDEC's policy under the Voluntary Cleanup Program is to require a volunteer to clean up the contaminated property to a level consistent with the safe use of the property.[24] The stated goal of NYDEC's current cleanup regulations is to restore "[a] site to pre-disposal conditions, to the extent feasible and authorized by law."[25] The regulations require a site's cleanup program to be

> designed so as to conform to standards and criteria that are generally applicable, consistently applied, and officially promulgated, that are either directly applicable, or that are not directly applicable but are relevant and appropriate, unless good cause exists why conformity should be dispensed with.[26]

The regulations further state that good cause for deviation from conformity exists if (1) conformity to a standard would result in a greater risk to the public or the environment or is technically impracticable or (2) the proposed deviation will attain an equivalent level of cleanup through use of another method.[27] New York has not promulgated cleanup standards any more specific than those discussed above. Although use-based cleanup standards under the Voluntary Cleanup Program may fit within NYDEC's regulations, such use-based standards also have the potential to be heavily scrutinized during the Voluntary Cleanup Program's public comment period for the lack of good cause in not requiring cleanup to predisposal conditions.

Mechanics of Participating in the Voluntary Cleanup Program

To participate in the Voluntary Cleanup Program, a volunteer must complete an application and submit it to NYDEC. Applications are available by calling NYDEC at 518/457-5861. The Voluntary Cleanup Program application assumes that a preliminary site assessment has already been performed and requires specific information about the property, including answers to the following questions:

- What is the site's location and how long have facilities at the site been operating?
- What hazardous substances are the subject of the voluntary cleanup, what containment measures have been taken, and what is the current accessibility of the potentially contaminated areas?
- What environmental permits have been issued to the site?
- What type of contamination is suspected?
- What is the character of the surrounding area—the population, the source of drinking water, and so forth?[28]

NYDEC will determine, within sixty days of receipt of an application, whether the prospective volunteer is eligible for the Voluntary Cleanup Program.[29] Most applications, however, are reviewed in a much shorter time.[30] If an application is approved, NYDEC must next determine whether the prospective volunteer is financially able to undertake the investigative and/or remedial plans proposed for the property.[31]

If the applicant is financially capable of participating in the Voluntary Cleanup Program, NYDEC and the volunteer may enter an agreement to clean up the contaminated property.[32] (If the volunteer is a PRP, he or she has the option of entering a consent order with NYDEC instead of an agreement; this option has rarely, if ever, been exercised under the Voluntary Cleanup Program.[33]) In entering an agreement with NYDEC, the volunteer may choose to agree to a phased plan, wherein the volunteer may commit first to a full site investigation before he or she commits to a full remediation agreement.[34] This allows the volunteer to opt out of the program if the volunteer decides that the necessary remediation would be too costly or time consuming to pursue.[35] If a volunteer withdraws from a cleanup, the volunteer "will be required to leave the site no worse, from an environmental and human health perspective, than when it began its activities."[36]

In formulating a work plan for the agreement, PRPs and non-PRPs are subject to different standards. PRP volunteers, not including volunteers who are PRPs based solely on their status as owners of the contaminated property, must remediate the off-site impacts of the contamination on the subject property, whereas non-PRP volunteers (and owner-only PRPs) must eliminate only sources of on-site contamination that cause off-site impacts; they do not need to remediate off-site contamination.[37]

When a volunteer and NYDEC have entered an agreement for a site that is not on the New York Registry of Sites, NYDEC will notify the public and the local governmental entities of the existence of the agreement and will afford the public and governmental entities a thirty-day opportunity to comment on the work plan in the agreement.[38] "Any new information received during this process may be used to reopen the agreement to revise the work plan appropriately."[39] If a site is listed in the Registry of Inactive Hazardous Waste Disposal Sites, public participation will be conducted pursuant to the requirements of the Inactive Hazardous Waste Disposal Site Remedial Program, the New York State Inactive Hazardous Waste Site Citizen Participation Plan, and the Citizen Participation Plan Interim Guidance.[40]

Once a work plan is completed to the satisfaction of NYDEC, it will issue a no-further-action letter and release the volunteer from liability for historical contamination, as discussed above.[41]

Recording, Admission of Liabilities, and Penalties

Recording

NYDEC may, in formulating its agreement with the volunteer, require the volunteer to place institutional restrictions on the subject property.[42] For example, the volunteer may be required to record a deed restriction that would run with the land and prevent the land from being used for any activity other than the type contemplated in the voluntary agreement with NYDEC.

Admission of Liability

Neither the Voluntary Cleanup Program nor New York statutes and regulations address the issue of the protection from disclosure of voluntary environmental audits. Although many other states have enacted bills that create a privilege for voluntary environmental audits, the New York legislature has yet to introduce a bill providing for an environmental audit privilege.[43] It is reported, however, that Governor Pataki actively supports the creation of a legislatively created privilege to protect environmental audit reports under certain circumstances.[44]

Under common law, a party may argue that an environmental audit is subject to protection from disclosure under the attorney-client privilege. This privilege may be waived, however, if the contents of the audit report are disclosed to third parties. Given the close interaction a volunteer under the Voluntary Cleanup Program has with NYDEC concerning the details of the environmental condition of the volunteer's property, it is likely that any privilege asserted for the protection of an environmental audit would be waived as having been disclosed to a third party.

Civil and Criminal Penalties

The Voluntary Cleanup Program does not provide for any specific penalties for violating the Voluntary Cleanup Program. Because the Voluntary Cleanup

Program is currently an informal, agency program, NYDEC relies mainly on the contractual obligations created by each voluntary agreement for its enforcement powers.[45]

Conclusion

Although the New York legislature has yet to enact a brownfields statute, NYDEC has taken the lead in developing its own brownfields initiative. NYDEC's Voluntary Cleanup Program allows certain volunteers, including some PRPs, to enter agreements to clean up contaminated property and pay for NYDEC's oversight costs in exchange for a no-further-action letter and a release from liability for historical contamination. NYDEC's no-further-action letters, however, are subject to certain "reopeners" that allow the state to require additional cleanup of a property upon the discovery of certain conditions. Parties interested in entering the Voluntary Cleanup Program should stay abreast of changes in NYDEC's still developing program and should watch for the potential creation by the New York legislature of a brownfields statute.

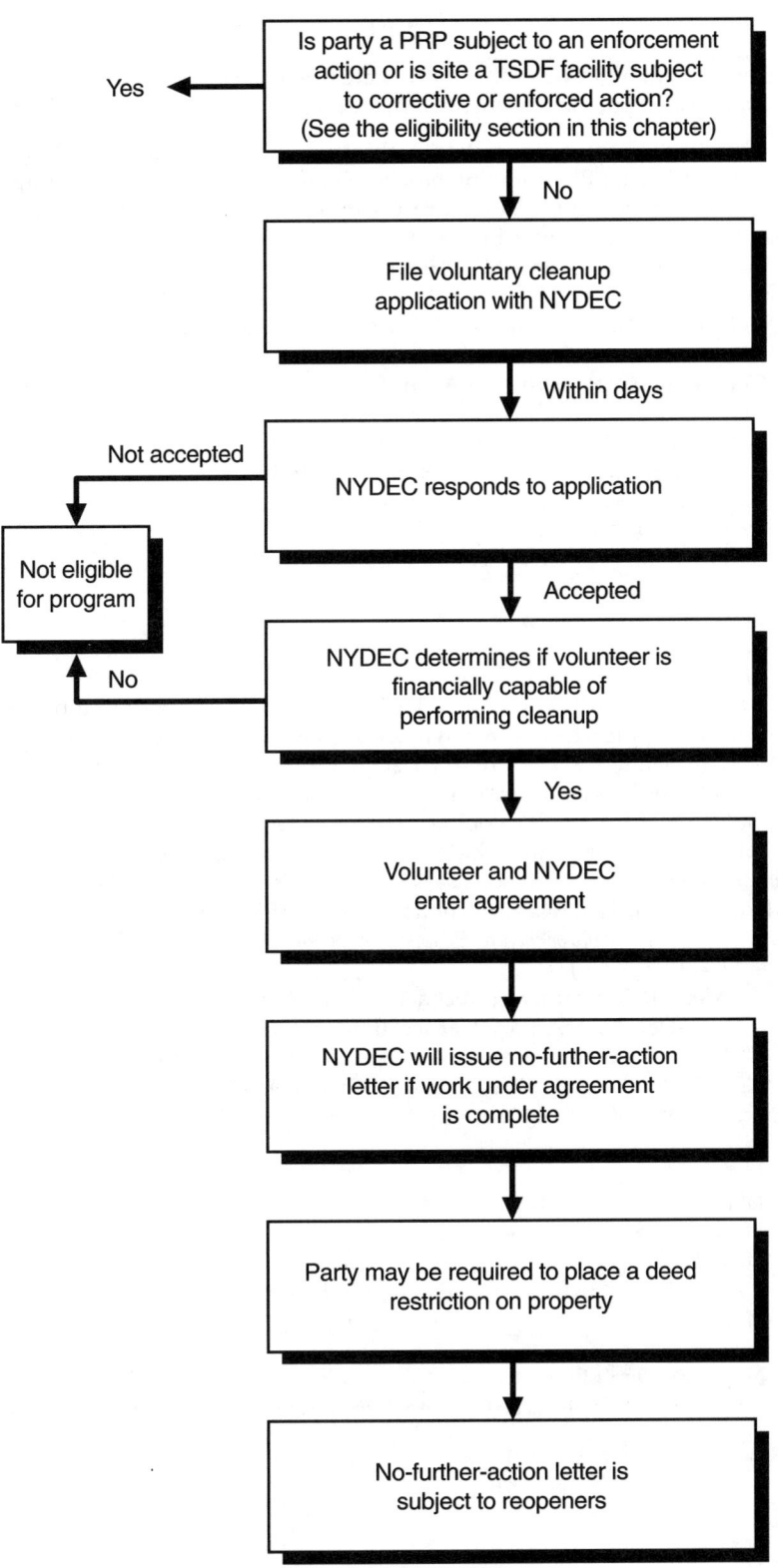

Notes

Note: Materials authored by the New York State Department of Environmental Conservation are unpublished but may be obtained by calling NYDEC at 518/457-5861.

1. N.Y. Dep't Envtl. Conservation, Voluntary Cleanup Program (undated) (summary of Voluntary Cleanup Program) [hereinafter Summary of Voluntary Cleanup Program]. This summary of New York's Voluntary Cleanup Program was current as of March 5, 1996. Telephone Interview with Christine Costopoulos, Remediation Program Coordinator for New York State Department of Environmental Conservation (Mar. 5, 1996) [hereinafter Telephone Interview with Christine Costopoulos].
2. Telephone Interview with Christine Costopoulos, *supra* note 1.
3. *Id.*
4. The Voluntary Remediation Act of 1995, Governor's Program Bill #46 (1995).
5. *Id.*
6. Beth Fitting, *Environmentalists, Developers, and the Regulators Finally Seem to be Pulling in the Same Direction*, CENTRAL N.Y. BUS. J., July 10, 1995, at 10.
7. *See* Charles Sullivan, Jr., *Voluntary Cleanup: The DEC's Solution*, Summer 1995 ALB. L. ENVTL. OUTLOOK 30, 31.
8. Summary of Voluntary Cleanup Program, *supra* note 1, at 1.
9. *Id.*
10. *Id.*
11. *Id.*
12. *Id.* at 2.
13. N.Y. ENVTL. CONSERV. LAW § 27-1305(4)(b) (McKinney 1984 & Supp. 1996).
14. Telephone Interview with David Evans, Office of Emergency and Remedial Response Division, United States Environmental Protection Agency (Mar. 18, 1996). The working title of USEPA's policy is "Guidance for State Voluntary Cleanup Programs."
15. USEPA is currently developing a policy under which it would advocate the issuance of "comfort letters" to brownfields volunteers (*see* Vicki Allen, *EPA to Ease City Waste Site Cleanups*, REUTERS N. AM. WIRE, Jan. 25, 1995 while some regional offices, like Region V, already make it a policy to issue "comfort letters" to brownfields volunteers (*see* David Fink, *Born Again: New Approaches to Redevelopment of Urban Industrial Sites*, BUS. DATELINE DETROITER, Mar. 1995, at 24).
16. *See* Meryl R. Lieberman & Michael Case, *Shelter from the Storm? EPA's Final Lender Liability Rules*, 65-Dec. N.Y. St. B.J. 32, at sec. III (1993).
17. N.Y. ENVTL. CONSERV. LAW § 27-1301 *et seq.* (McKinney 1984 & Supp. 1996).
18. N.Y. COMP. CODES R. & REGS. tit. 6, § 375-1.3(u) (1995).
19. Summary of Voluntary Cleanup Program, *supra* note 1, at 3.
20. *Id.*
21. *Id.*
22. *Id.* at 3–4.
23. *Id.* at 4.
24. *Id.* at 2.
25. N.Y. COMP. CODES R. & REGS. tit. 6, § 375-1.10(b) (1995).
26. *Id.* at (c)(1)(i).
27. *Id.*
28. N.Y. State Dep't Envtl. Conservation, Voluntary Cleanup Program, 1–2 (undated) (summary of Program Application) [hereinafter Summary of Program Application].
29. *Id.* at 3.

30. Telephone Interview with Christine Costopoulos, *supra* note 1.
31. Summary of Program Application, *supra* note 28, at 2.
32. Summary of Voluntary Cleanup Program, *supra* note 1.
33. Telephone Interview with Christine Costopoulos, *supra* note 1.
34. Summary of Voluntary Cleanup Program, *supra* note 1, at 2.
35. *Id.*
36. *Id.*
37. *Id.*
38. *Id.*
39. *Id.*
40. *Id.*; *see* N.Y. COMP. CODES R. & REGS. tit. 6, § 375 (1995).
41. Summary of Voluntary Cleanup Program, *supra* note 1, at 3.
42. *Id.* at 2.
43. *See Environmental Audits: State Privilege Legislation Multiplies in 1995; Predictions Differ about 1996,* Daily Env't Rep. (BNA), at AA-2 (Aug. 29, 1995).
44. John T. Kolaga, *New York Joins the Battle over Environmental Audit Reports: Will They Be Protected by Legal Privilege under the Pataki Administration?*, N.Y. St. B.J., July/Aug. 1995, at 38.
45. Telephone Interview with Christine Costopoulos, *supra* note 1.

CHAPTER 39

Ohio

TODD S. DAVIS
KEVIN D. MARGOLIS

Ohio's Voluntary Real Estate Reuse and Cleanup Program (Voluntary Action Program or Program) provides a tremendous benefit to the regulated community in Ohio by increasing the speed and predictability of voluntary environmental cleanups. The Voluntary Action Program, principally codified at chapter 3746 *et seq.* of the Ohio Revised Code, also creates an incentive to redevelop environmentally impaired sites in Ohio. The Voluntary Action Program is among the most comprehensive and progressive voluntary cleanup legislation enacted in any state to date.

Summary of Major Provisions

The major provisions of the Voluntary Action Program include the following:

1. Setting concrete, generic, numerical cleanup standards for soil, sediment and water
   - The Program also requires the development of risk-assessment procedures for deriving property-specific numerical standards.
2. Instituting state-certified environmental professionals and laboratories to conduct and approve much of the actual cleanup work that the Ohio Environmental Protection Agency (Ohio EPA) formerly coordinated, thereby shifting Ohio EPA's primary role to auditing completed cleanups.
3. Providing liability protection to volunteers in the form of a covenant-not-to-sue from Ohio EPA.
   - The covenant will release the party participating in the Program from all civil liability to the state of Ohio to perform additional investigation and remedial activities to address a release of hazardous substances or petroleum, with certain limited exceptions.
4. Allowing consolidated permitting to facilitate the cleanup.

- This consolidated permit is designed to decrease time delays in commencing a cleanup by obviating the need to obtain multiple permits from various Ohio EPA divisions.
5. Instituting strong penalties for those persons who abuse the system.
    - These sanctions include felony convictions resulting in two years of jail time and civil penalties of $10,000 per day, per violation.
6. Permitting private-party cost recovery through a system based on comparative fault.
7. Providing tax abatement, tax credits and low-interest loans as incentives for volunteers to remediate contaminated sites.

Participation in the Voluntary Action Program
Eligible Sites

To participate in the Program, a person must conduct a "voluntary action."[1] Voluntary actions may include, without limitation, a Phase I property assessment, a Phase II property assessment, a sampling plan, a remediation plan and remedial activities including such other activities the volunteer considers to be necessary or appropriate to address the contamination. These voluntary characterization and cleanup activities are followed by the issuance of a no further action (NFA) letter by a certified professional, indicating that the property meets "applicable standards."[2] The NFA letter signifies that either there is no information indicating there has been a release of hazardous substances or petroleum at the property, or there has been a release of hazardous substances or petroleum at the property and any releases were de minimis or applicable standards either were not exceeded, or have been (or will be) achieved in accordance with the Voluntary Action Program.[3]

Voluntary Action Program participants must use the services of certified laboratories and certified professionals.[4] The certified professional may issue a volunteer an NFA letter by demonstrating through a Phase I or Phase II property assessment that the property meets applicable cleanup standards. A property in the Program can meet applicable standards and be eligible for issuance of an NFA letter as a result of current conditions or through an environmental cleanup. If a cleanup relies on restrictions on the property's use to meet the cleanup standards, the restrictions must be recorded in the county recorder's office.[5] If a cleanup relies on engineering or institutional controls to meet the cleanup standards, a plan assuring the proper operation and maintenance of those controls must be developed.[6] The operation and maintenance plan describes the activities necessary to ensure that a remedy maintains compliance with applicable standards and is enforceable by Ohio EPA through an operation and maintenance agreement entered into by the volunteer with the state.[7] Unless specifically provided in the Program, participants in the Voluntary Action Program must comply with all other applicable laws.[8]

Sites Ineligible for the Program

Not all sites are eligible to participate in the Voluntary Action Program.[9] Specifically, the following properties are not eligible:

- Properties where the Voluntary Action Program is precluded by federal law or regulation, including without limitation, (1) the Water Pollution Control Act Amendments of 1972, as amended, (2) the Resource Conservation and Recovery Act of 1976, as amended, (3) the Toxic Substances Control Act, as amended, (4) the Comprehensive Environmental Response, Compensation, and Liability Act of 1980 (CERCLA), as amended, and (5) the Safe Drinking Water Act, as amended
- Those portions of properties where closure of a hazardous-waste facility or solid-waste facility is required under chapter 3734 of the Ohio Revised Code (Ohio's hazardous and solid waste laws)
- Properties subject to the remediation rules promulgated pursuant to chapters 3737.88, 3737.882, and 3737.889 of the Ohio Revised Code (Ohio's underground storage tank laws)
- Property that is subject to chapter 1509 of the Ohio Revised Code (Ohio's oil and gas laws)
- Any other property, if (1) the Director of Ohio EPA has issued a letter notifying the owner or operator of the property that the Director will issue an enforcement order under chapters 3704, 3734, or 6111 of the Ohio Revised Code, (2) a release or threatened release of a hazardous substance or petroleum from or at the property poses a substantial threat to public health or safety or the environment, *and* (3) the person subject to the order does not present "sufficient evidence" to the Director that he or she has entered the Voluntary Action Program and is proceeding expeditiously to address the environmental threat.[10]

Memorandum of Agreement

As of March, 1997, Ohio is negotiating, but does not have, a Memorandum of Agreement (MOA) with the United States Environmental Protection Agency (USEPA) for its Voluntary Action Program. The MOA would provide USEPA's specific approval of Ohio's Voluntary Action Program for cleanups or other voluntary actions. Therefore, until the MOA is final, a person participating in the Voluntary Action Program, theoretically, could be subject to a federal action brought by USEPA even after obtaining a covenant-not-to-sue under the Program. However, USEPA's current enforcement position does not contemplate pursuing an enforcement action in connection with a site that is participating in the Voluntary Action Program. Such participation, however, would not preclude private parties from bringing a federal cause of action in association with a site participating in the Voluntary Action Program. Participants may seek "comfort letters" from USEPA's Region V office or enter prospective-purchaser agreements with the federal government to minimize the threat of federal liability.[11]

Liability Protection

Lenders, Trustees, and Fiduciaries

The Ohio Voluntary Action Program has extended state law protection from civil environmental liability to lenders,[12] trustees, and fiduciaries.[13] A person who is holding property primarily to protect a security interest in the property will not be held liable for the costs of conducting a voluntary action under the Program, or for the costs of conducting any investigations or environmental remediation under any statutory or common-law action brought in Ohio.[14]

A person holding a security interest also may take title to environmentally distressed property and be protected by the secured-party "safe harbor." Adopting much of the structure of USEPA's 1992 Lender Liability Rule,[15] the security-interest exemption from environmental liability applies only if, after taking title to the property, the holder of a security interest conducts all activities occurring at the property in compliance with the applicable requirements of the security-interest exemption.[16]

Protection for Participants in the Program

Individuals, corporations and other "persons" may take advantage of the liability protection generally established throughout the Program, as long as they are eligible for participation. Prospective purchasers of contaminated real estate or current owners voluntarily addressing the presence of hazardous substances or petroleum at their sites can protect themselves from liability (1) by obtaining a covenant-not-to-sue, and (2) by cleaning up their property and removing or minimizing the environmental threat.

Financial Incentives under the Voluntary Action Program

To induce participation in the Voluntary Action Program, Ohio has established a number of financial incentives. These incentives include automatic tax abatement, low interest loans and tax credits.

Tax Abatement

After a covenant-not-to-sue has been issued by the Director of Ohio EPA, the increased assessed value of land and of buildings, fixtures, and improvements on the property due to a cleanup under the Program automatically will be exempted from real estate taxes for a ten-year period.[17] Any sale or other transfer of the affected property does not rescind or diminish the real estate tax abatement. Thus, tax abatement runs with the land. However, if the covenant-not-to-sue is revoked, the real estate tax abatement will be discontinued and the property owner must repay any taxes that would have been due if the property had not been exempted from such taxes.[18] Local governments also may afford tax abatement on real and personal property taxes for development projects for a period of up to ten years.[19]

Low-Interest Loans

Ohio has made low-interest loans available for Voluntary Action Program cleanups through the Ohio Water Pollution Control Loan Fund, the Ohio Pollution Prevention Loan Program, the Ohio Department of Development, and the Ohio Water Development Authority.[20] Loans may be available to finance both investigative and remedial activities; however, loans from the Ohio Water Pollution Control Loan Fund must assist activities associated with Voluntary Action Program projects that will have a water-quality benefit to either surface or groundwater.[21] Loan terms are negotiable and may be amortized on a schedule of up to twenty years.

Tax Credits

In the spring of 1996, the Ohio legislature passed (and the governor signed) legislation authorizing the Director of Development to grant corporation franchise and personal income-tax credits to entities and individuals that participate in the Voluntary Action Program. Application for these tax credits must be made after a covenant-not-to-sue has been issued by the Director of Ohio EPA. The taxpayer is entitled to a credit equal to a specified percentage of "eligible costs" of the cleanup, including site investigation and remedial costs, which must be claimed over a five-year period.[22]

Ohio is currently considering additional financial incentives for Voluntary Action Program cleanups including grants, additional loan funds, and tax increment financing.

Cleanup Standards

When the Voluntary Action Program became law in September of 1994, the Ohio EPA was required to adopt rules to establish both numerical cleanup standards for soil, sediment and groundwater and to establish risk assessment procedures for deriving property-specific cleanup standards.[23] On December 16, 1996, the final rule package was adopted and rules creating generic numerical cleanup standards and property-specific risk assessment procedures were promulgated.[24] In addition, this rule package also provided extensive regulations for groundwater classification and cleanup response requirements.[25]

The Voluntary Action Program cleanup standards are linked to the intended use of the property that is the subject of the cleanup and redevelopment. Three categories of generic cleanup standards have been established: (1) industrial land-use standards, (2) commercial land-use standards, and (3) residential land-use standards.[26] Generic standards tied to the intended use of the property are more lenient for industrial-use scenarios and more stringent for residential uses. The new generic numerical standards include soil cleanup standards for forty-three carcinogenic and noncarcinogenic chemicals as well as standards for polychlorinated biphenyls and lead.[27] The rules also provide over seventy cleanup standards for unrestricted potable groundwater.[28]

If a volunteer's site has contaminants for which there is no generic standard or if the volunteer's certified professional determines that a cleanup could be conducted more advantageously with site-specific cleanup standards, a risk assessment can be performed. The Voluntary Action Program provides detailed procedures for completing a property-specific risk assessment.[29] These procedures outline applicable risk goals that must be met, including a cumulative risk goal of 1×10^{-5} excess cancer risk for residential and commercial land uses and a cumulative risk goal of 1×10^{-4} excess cancer risk for industrial land uses.[30] Both institutional controls (such as deed restrictions) and engineering controls (such as caps) are acceptable approaches to meeting applicable remediation standards set through a risk assessment. [31]

The level of cleanup required for groundwater at a Voluntary Action Program site depends on (1) how the groundwater at the site has been classified and (2) from where the contamination at the site originates. The Program divides groundwater into classifications associated with the use (that is, potable or not potable) and the yield of the groundwater.[32] High-yield, critical resource, and potable groundwater suitable for unlimited use must meet higher standards than low-yield non-potable groundwater. The Program also affords the volunteer the opportunity to have his or her certified professional designate the groundwater at a site as within an "urban setting," essentially recognizing that the groundwater at the site is not used by the residents in the area for drinking water purposes and that the area is served by a community public water system.[33] Property designated as within an urban setting requires a lower level of groundwater cleanup.

Under certain conditions, a property or its contamination may possess characteristics that preclude the straightforward use of the established cleanup standards. In such a case, the volunteer may apply to Ohio EPA for a variance.[34]

Certified Professionals and Laboratories

The Voluntary Action Program provides certification procedures for environmental professionals and laboratories. Individuals who can demonstrate at least eight years of professional experience in the investigation or remediation of hazardous substances of petroleum, including at least three years of supervisory experience, may be deemed certified by Ohio EPA to issue NFA letters.[35] These individuals must, at a minimum, have earned a bachelor's degree in a science-related or appropriate engineering discipline from a recognized educational institution.[36] In addition, the environmental professional must demonstrate to Ohio EPA satisfactory professional competence and knowledge to perform tasks required of a certified professional.[37]

Laboratories authorized to participate in the Voluntary Action Program undergo a rigorous performance evaluation program and must demonstrate their ability to quantitate analyses within prescribed limits.[38] Routine compliance audits of certified laboratories by Ohio EPA are authorized.[39] Volunteers and their certified professionals must utilize certified laboratories in order to participate in the Program and support the issuance of a no further action letter.

Mechanics of Participating in the Ohio Voluntary Action Program

No-Further-Action (NFA) Letters

Once a volunteer has chosen to participate in the Program and engaged a certified professional, the following activities should take place. The certified professional will review information developed through participation in the Voluntary Action Program to verify whether the property meets applicable cleanup standards. If the certified professional determines that the property meets the applicable standards, he or she may prepare an NFA letter. The certified professional must have expertise in hydrogeology to issue NFA letters for properties with groundwater issues.[40] Any reliance on institutional or engineering controls must be reflected in the NFA letter.[41]

If the volunteer wishes to receive liability protection in the form of a covenant-not-to-sue from the Director of Ohio EPA, the certified professional must send the original NFA letter to the Ohio EPA.[42] If the certified professional finds that the property does not meet the applicable standards, he or she must notify the volunteer of his or her inability to issue the NFA letter.[43] With one limited qualification, there is no obligation to notify Ohio EPA that the property does not meet applicable cleanup standards. In fact, the investigation and remediation process remains confidential. Any information, documentation, reports, or data produced, or any samples collected as a result of entering and participating in the Voluntary Action Program, are neither admissible nor discoverable in any civil, criminal or administrative action brought against a volunteer.[44] (See the section below entitled, "Admission of Liability.") The only exception to the confidentiality provision is if an immediate threat to human health or the environment is discovered by the certified professional and he or she deems the situation "sufficiently important."[45]

Covenant-Not-to-Sue

The Director of Ohio EPA must issue a covenant-not-to-sue to an eligible volunteer who has received an NFA from a certified professional and who wishes to have a covenant.[46] The covenant releases the volunteer from all civil liability to the state except for claims for natural-resources damages (a mandatory state claim under CERCLA), and claims for cleanup costs if USEPA takes an action at the site and, as a result, the state of Ohio incurs costs. The covenants are conditioned upon (1) continuing the property under the same land use upon which the NFA was based and (2) the cleanup remedy operating properly. The covenant may be revoked only if the conditions at the property no longer meet applicable cleanup standards at the time the covenant was issued.[47]

The Director of Ohio EPA must issue or deny a covenant for a parcel of property, which does not involve a consolidated standards permit or institutional or engineering controls as a remedy, within thirty days of receiving an NFA letter.[48] For properties with a consolidated standards permit or that use institutional or engineering controls, the Director of Ohio EPA must issue or deny the covenant within ninety days of receiving the NFA letter.[49] The Director may deny a covenant if the NFA does not comply with the Voluntary Action Program statute (chapter 3746 of the Ohio Revised Code), if the remedy identified in the NFA does not protect public health and safety and the environment or if the NFA was fraudulently submitted.[50]

Recording of the NFA Letter and Covenant

Persons issued a covenant-not-to-sue must file both the NFA and the covenant in the appropriate county recorder's office so that subsequent purchasers of the property will enjoy the release of environmental liability and will be notified of any restrictions on uses of the land.[51] The covenant and any land-use restrictions will run with the property.[52]

Fees

Participation in the Voluntary Action Program comes with a price. The Program includes fees for submission of NFA letters in pursuit of covenants-not-to-sue ($950 with a Phase I, otherwise $4,950), variances ($18,500), use of operation and maintenance agreements to meet applicable standards ($2,950), and certification of professionals and laboratories ($2,000 and $3,000 per year, respectively).[53] In addition, the costs incurred by Ohio EPA to review and consider certain special voluntary actions (such as an urban setting designation) must be reimbursed by the volunteer.[54]

Auditing by Ohio EPA

As a statutorily mandated requirement of the Voluntary Action Program, the Director of Ohio EPA will conduct random audits on the condition of properties that received NFA letters to insure that the cleanup standards are appropriately met.[55] Certified professionals and certified laboratories also will be audited to ensure that the professionals and laboratories are maintaining their qualifications.[56]

If the Director finds that a certified professional or certified laboratory issued an NFA letter at a property where the applicable cleanup standards were not met, the Director may notify all persons who received an NFA letter from that certified professional of this finding and may audit all the relevant properties.[57] If the audit confirms that applicable standards were not met, the Director must give the holder of the relevant covenant an opportunity to bring the property into compliance with the mandated cleanup standards.[58] If the holder fails to do so within the allotted time frame, the Director may issue an order revoking the covenant.[59]

Admission of Liability

Participation in, or entry into, the Voluntary Action Program cannot be used as evidence in a civil or criminal prosecution to demonstrate that the volunteer is liable. Also, any information or data obtained by Ohio EPA from a person participating in the Program is inadmissible and not discoverable in any administrative or judicial action.[60]

Civil and Criminal Penalties

There are substantial penalties for violating certain provisions of the Voluntary Action Program. The Director of Ohio EPA may request in writing that the Attorney General, County Prosecutor, or local City Attorney criminally prosecute or seek injunctive relief, and civil penalties, for violating a consolidated standards permit[61] or for submitting to the Director information or data that is materially false.[62] The maximum civil penalty for violation of these sections is $10,000 per day, per violation.[63] A reckless violation of a consolidated standards permit constitutes a felony and requires a fine of at least $10,000 and/or imprisonment for at least two years. Further, the knowing submission of false information by a participant in the Voluntary Action Program also constitutes a felony and requires a fine of at least $10,000 and/or imprisonment for at least two years.[64]

Cost Recovery

Except for petroleum cleanups, persons who perform a voluntary action may bring a lawsuit against owners, operators, and any person who caused or contributed to a release of a hazardous substance (Liable Persons) for the costs incurred in conducting the voluntary action.[65] The court will apportion the shares of cleanup costs among Liable Persons based on their respective degrees of responsibility for causing the release. Factors for allocating such liability include (1) the nature and amount of hazardous substances stored, treated, disposed of, used, and released by each person, (2) the length of time that each person owned or operated the property, (3) each person's history of compliance with applicable federal and state environmental laws and rules in the use and operation of the property, and (4) any other factors that the jury or the court considers to be appropriate.[66]

Persons bringing such an action may recover investigation costs, remediation costs, the costs of the NFA letter, the costs paid to the Ohio EPA, attorney fees, court costs, and any other expenses incurred in bringing the lawsuit.[67] Persons may not recover any costs associated with improving the land uses of a property.[68] For instance, the costs associated with rendering a property suitable for residential use when the existing use of the site was industrial would not be recoverable. However, the costs of cleaning the site for industrial use in that scenario would be recoverable.

Prima Facie Case

A volunteer may establish a prima facie case to recover the costs of conducting a voluntary action as follows:

> Any person who, at the time when any of the hazardous substances identified and addressed by a Voluntary Action . . . were released at or upon the property . . . was the owner or operator of the property, and any other person who caused or contributed to a release of hazardous substances at or upon the property, is liable to the person who conducted the Voluntary Action for the costs of conducting the Voluntary Action.[69]

If the person who conducted the voluntary action did not cause or contribute to any release of hazardous substances at the property, he or she may recover the costs of conducting a voluntary action from (1) the owner or the operator of the property at the time when those releases occurred and (2) any other person who caused or contributed to the releases. A person bringing a private cost-recovery action, who is also liable, may recover the costs of conducting the Voluntary Action that are attributable to the release to which those owners, operators, and others caused or contributed.[70]

Liability under the Voluntary Action Program private cost-recovery action section is specifically made retroactive.[71] A person conducting the Voluntary Action may commence a private cost-recovery action any time after that person has commenced conducting the voluntary action.[72]

Statute of Limitations

A private cost-recovery action must be commenced within three years after the applicable NFA letter was submitted to the Director of Ohio EPA.[73]

Joinder of Parties

All owners and operators and other responsible persons must be joined as defendants in a private cost-recovery action.[74]

Contractual Allocation of Liability

Contractual allocation of liability for the costs of conducting a voluntary action specifically is permitted under the Voluntary Action Program.[75] Contractual allocations of those costs, however, do not affect the government's rights against the persons who allocated those costs.[76]

Defenses/Exemptions from Liability

The following parties are exempt from liability for the costs of conducting a Voluntary Action:[77]

1. a person who neither caused nor contributed to, in any material respect, a release of hazardous substances;
2. a landlord who did not know, and who could not reasonably have known, of tenant acts or omissions that caused or contributed to a release of hazardous substances;
3. the state or a political subdivision that involuntarily acquires ownership or control of property by virtue of its function as a sovereign;
4. the state or political subdivision if it voluntarily acquires ownership or control of property through purchase, appropriation under chapter 163 of the Ohio Revised Code, or other means;
5. an owner or operator or any other person who caused or contributed to a release of petroleum at or upon the property;[78]
6. a holder who is in compliance with the requirements of section 3746.26 of the Ohio Rev. Code (the lender-liability provisions); and
7. a fiduciary or trustee who is in compliance with the requirements of section 3734.27 of the Ohio Revised Code.

Conclusion

Since being enacted in 1994, Ohio's Voluntary Action Program has been considered one of the most progressive and complete state voluntary cleanup programs. The inclusion of specific cleanup standards, the provision of liability protection to participants in the form of a covenant-not-to-sue, and the ability of a volunteer to participate in the Program and implement a cleanup without the oversight of Ohio EPA create a new environment for potential redevelopers of contaminated property in Ohio. As a result, now that the complete set of administrative rules were promulgated at the end of 1996, it is expected that there will be a significant volume of sites taken through the Ohio Voluntary Action Program process.

Obtaining a Covenant-Not-to-Sue under the Ohio Voluntary Action Program

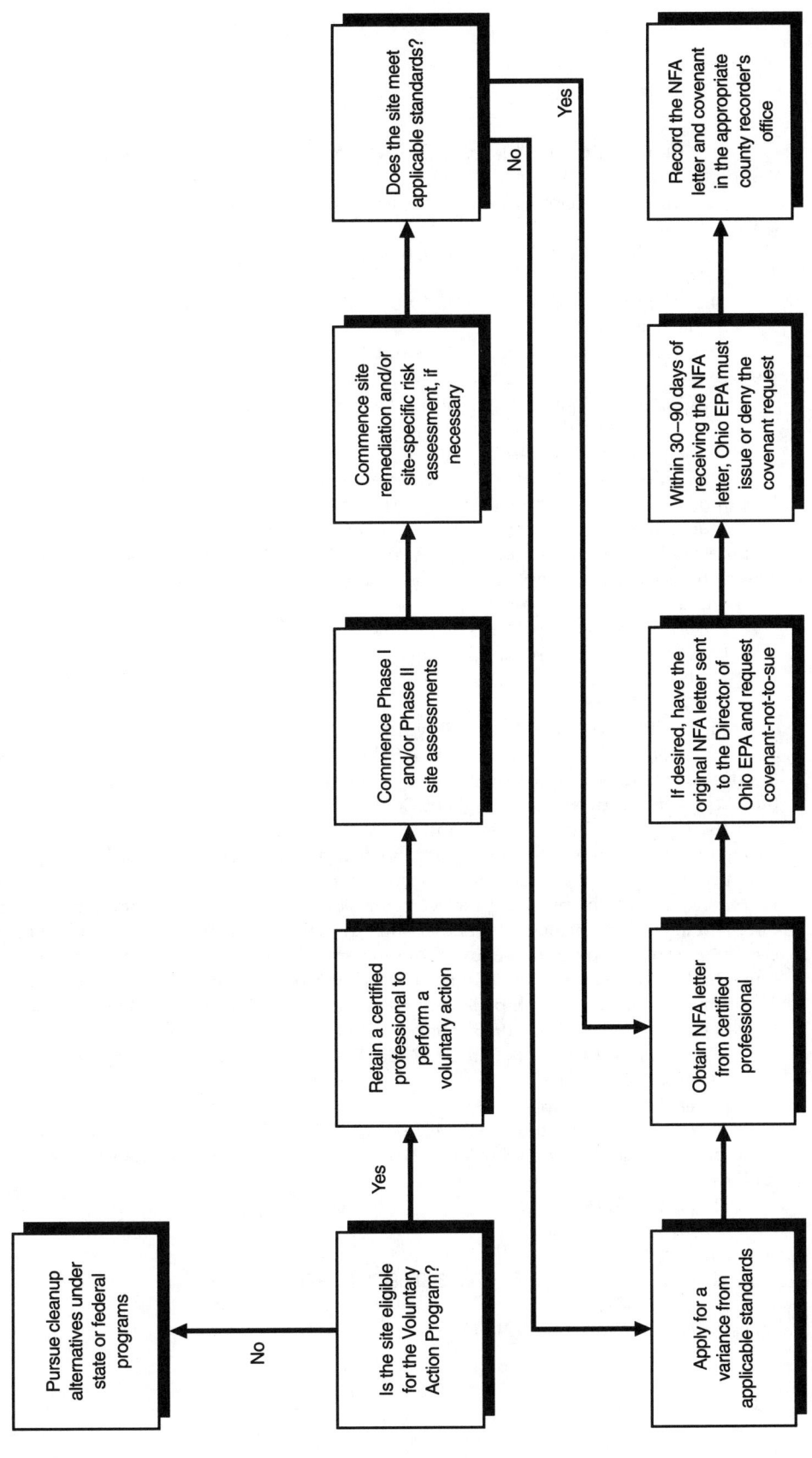

Notes

1. OHIO REV. CODE ANN. § 3746.01(O) (Banks-Baldwin 1995); OHIO ADMIN. CODE § 3745-300-01(50) (1997).
2. OHIO REV. CODE ANN. § 3746.01(B) (Banks-Baldwin 1995); OHIO ADMIN. CODE § 3745-300-01(5) (1997).
3. OHIO REV. CODE ANN. § 3746.01(O) (Banks-Baldwin 1995); OHIO ADMIN. CODE § 3745-300-13(A) (1997).
4. OHIO REV. CODE ANN. § 3746.10(B) (Banks-Baldwin 1995).
5. Id. § 3746.14.
6. OHIO REV. CODE ANN. § 3746.10(C)(4); OHIO ADMIN. CODE § 3745-300-15(A)(2) (1997).
7. OHIO REV. CODE ANN. § 3746-12(A)(2) (Banks-Baldwin 1995).
8. Id. § 3746.10D.
9. Id. §§ 3746.02(A)(1)–(5).
10. Evidence constituting "sufficient evidence" of entry into the Voluntary Action Program is defined in section 3745-300-02(D) of the Ohio Administrative Code and generally requires that within thirty days of receiving an enforcement letter that the prospective volunteer submit evidence to Ohio EPA that a Phase I Property Assessment and an initial statement of work for completing a Phase II Property Assessment were completed *prior* to receiving the enforcement letter.
11. *See* 60 Fed. Reg. 34,792 (1995).
12. OHIO REV. CODE ANN. § 3746.26 (Banks-Baldwin 1995).
13. Id. § 3746.27.
14. Id. § 3746.26.
15. USEPA's 1992 Lender Liability Rule was overturned by *Kelley v. United States Protection Agency*, 25 F.3d 1088 (D.C. Cir. 1994) but has now been codified by recent federal legislative action. *See* Asset Conservation, Lender Liability and Deposit Insurance Protection Act of 1996, P.L. No. 104–208 (Sept. 30, 1996).
16. The secured-creditor exemption provides that any person who, without participating in the management of a property, holds indicia of ownership in a property primarily to protect its security interest is protected from liability (1) for any civil action brought in connection with the property subject to the Voluntary Action Program, for investigative or remedial costs, regardless of whether such costs were incurred in connection with property subject to the Voluntary Action Program, and (2) for conducting or causing to be conducted all such nonremedial activity at the property in compliance with applicable law. The term "indicia of ownership" is defined at section 3746.26(B) of the Ohio Revised Code. *See also* U.S.C. § 9601(20)(A) (1988 & Supp. V 1993).
17. OHIO REV. CODE ANN. § 5709.87 (Banks-Baldwin 1995).
18. Id. § 5709.87(E).
19. Id. § 5709.88 *et seq.*
20. Id. §§ 6111.036, 6123.032, 6123.04, 6123.041. For further information regarding these programs contact the Ohio EPA, Division of Environmental Financial Assistance at (614) 644-2798.
21. Id. § 6111.036 *et seq.*
22. Id. § 122.19 *et seq.*
23. Id. § 3746.04.
24. OHIO ADMIN. CODE §§ 3745-300-08-09 (1997).
25. Id. § 3745-300-10.
26. Id. § 3745-300-08(B)(2)(c).

27. *Id.* § 3745-300-08(B)(3).
28. *Id.* § 3745-300-08(B)(4).
29. *Id.* § 3745-300-09.
30. *Id.* § 3745-300-09(C).
31. *Id.* §§ 3745-300-09(B)(2)(d), (D)(2)(d), (D)(3)(b)(ii), (I)(2), 300-15(D).
32. *Id.* § 3745-300-10(C).
33. *Id.* § 3745-300-10(D).
34. A variance from applicable cleanup standards may be based on technical infeasibility, cost/benefit analysis, incremental improvement to the environment, and the promotion of employment opportunities or property reuse. *See* OHIO REV. CODE ANN. § 3746.09 (Banks-Baldwin 1995); OHIO ADMIN. CODE § 3745-300-12 (1997).
35. OHIO ADMIN. CODE § 3745-300-05(B)(2)(b) (1997).
36. *Id.* § 3745-300-05(B)(2)(a).
37. *Id.* § 3745-300-05(B)(2)(c).
38. *Id.* § 3745-300-04(D).
39. *Id.* § 3745-300-04(J).
40. OHIO REV. CODE ANN. § 3746.071(B)(2) (Banks-Baldwin 1995).
41. *Id.* § 3746.10(C).
42. *Id.* § 3746.12(A). The specific requirements for the content and scope of NFA letters are described in section 3745-300-13 of the Ohio Administrative Code.
43. OHIO REV. CODE ANN. § 3746.11(B) (Banks-Baldwin 1995).
44. *Id.* § 3746.28(C).
45. *Id.* § 3746.071(B)(1)(c).
46. *Id.* § 3746.12(A).
47. *Id.* § 3746.12(E). As long as the property continues to meet standards in effect on the date the covenant was issued, the covenant may not be revoked due to modification in cleanup standards.
48. *Id.* § 3746.13(A). Persons who require multiple environmental permits to perform a cleanup may request and receive a consolidated standards permit under the Voluntary Action Program. The single consolidated permit will contain all standards and limitations for the cleanup activities requiring the permits. *See id.* § 3746.15(A). Consolidated permits cover all activities requiring permits under the Federal Water Pollution Control Act, as amended, 33 U.S.C.A. §§ 1251-1387 (West 1995), the Resource Conservation and Recovery Act, as amended, 42 U.S.C. §§ 6901-6992(k) (West 1995), and the Clean Air Act, as amended, 42 U.S.C.A. § 7401-7671 (West 1995).
49. OHIO REV. CODE ANN. § 3746.13(B) (Banks-Baldwin 1995).
50. *Id.* § 3746.12(C).
51. *Id.* § 3746.14.
52. *Id.* § 3746.12.
53. OHIO ADMIN. CODE § 3745-300-03(C) (1997).
54. *See, e.g., id.* § 3745-300-10(D)(F).
55. OHIO REV. CODE ANN. § 3746.17(A)(1) (Banks-Baldwin 1995); OHIO ADMIN. CODE § 3745-300-14 (1997).
56. OHIO REV. CODE ANN. §§ 3746.17(A)(2)-(3) (Banks-Baldwin 1995).
57. *Id.* § 3746.19.
58. *Id.* §§ 3746.12(B)(2)-(4).
59. *Id.* § 3746.12(B).
60. *Id.* § 3746.28.
61. *Id.* § 3746.15(B).
62. *Id.* § 3746-20.

63. *Id.* § 3746.22(B).
64. *Id.* § 3746.99.
65. *Id.* §3746.23.
66. *Id.* § 3746.23(D).
67. *Id.* § 3746.23(A).
68. *Id.*
69. *Id.* § 3746.23(B).
70. *Id.*
71. *Id.* §3746.23(E).
72. *Id.* § 3746.23(C).
73. *Id.*
74. § 3746.23(D). The statute does not define who must join all such parties.
75. *Id.* § 3746.23(F).
76. *Id.*
77. *Id.* § 3746.23(G).
78. However, if a petroleum release became mixed with a release of a hazardous substance at or upon the property, the owner or operator or other person who caused or contributed to the release of petroleum is not liable for that increment of the costs of conducting a voluntary action that is attributable to the presence of the petroleum release. *See id.* § 3746.23(G)(5).

CHAPTER 40

Oregon

RICHARD M. GLICK

The Oregon Recycled Lands Act of 1995 (the Act) facilitates the redevelopment of brownfields, or contaminated industrial property, by limiting the risk of environmental cleanup liability. The Act also institutionalizes risk-based corrective action as a means of controlling cleanup costs, while still protecting public health and the environment. The net effect will bring industrial lands in Oregon to full productive value and use, and reduce pressures leading to urban sprawl.

Summary of Major Provisions

The major provisions of the Act include the following:

- *Prospective-Purchaser Agreements:* A prospective buyer of industrial property may negotiate an agreement with the Oregon Department of Environmental Quality (DEQ) before committing to the property. Upon completion of the work, DEQ will execute a release from future environmental liability.
- *Probabilistic Risk Assessment:* Responsible parties are no longer required to clean up property to background or lowest feasible concentration levels. Instead, probabilistic risk assessment weighs not only the possibility of human exposure and harm due to contaminants, but also whether such exposure is likely to occur.
- *Remedy Selection:* Unless the contaminated property is a "hot spot" (discussed below), the Act removes the presumption that expensive removal or treatment is necessary. The new approach permits any remedy that reduces risk, including engineering or institutional controls. The Act requires DEQ to select the least costly remedy, unless more ex-

---

Mr. Glick wishes to acknowledge Christopher L. Blake for his able assistance in preparing this chapter.

pensive remedies are demonstrably more effective, reliable, or protective against immediate risk.

Participation in the Recycled Lands Act Program
Eligible Properties

The risk-based, cost-conscious approach to remediation under the Act applies to all contaminated property within the state. Prospective-purchaser agreements are equally available to all properties requiring remediation and meeting the criteria discussed below.

The Voluntary Cleanup Program

Before adopting the Act, DEQ had established a Voluntary Cleanup Program (VCP). The VCP was developed in response to the need for expedited agency oversight for cleanups incidental to fast-moving transactions.

Characteristics of the VCP include expedited cleanup and continual DEQ oversight and input during cleanup. DEQ assigns a project coordinator, who is readily available to the project, on a fee basis. A retainer deposit of $5,000 is customary and DEQ bills against it monthly. In certain cases, DEQ will waive the deposit or require less. Despite additional staff hired for the VCP, there still is a waiting list. DEQ estimates that most sites will complete the process within a year, unless a speedier cleanup is required. However, DEQ will generally provide timely response if informed of strict transactional deadlines.

The advantage of DEQ oversight is the certainty that comes with timely DEQ approval of the remedial action plan. The alternative is to clean the site without DEQ supervision and accept the attendant risk of nonacceptance of the results by DEQ, or to wait until DEQ can address the site in the normal course. Depending on how seriously contaminated the site is, absent the VCP, it may be years before DEQ can devote attention to a site. Such attention will likely be in the context of regulatory enforcement, as opposed to a voluntary compliance effort, which may make negotiations for cleanup standards more problematic.

The Act does not amend the VCP, except that the codification of prospective-purchaser agreements will likely add to the VCP's caseload as more sites are brought into the program. These agreements will most often require expedited agency review to meet transactional time frames.

Cleanup Standards

The Act requires DEQ to adopt implementing rules and a risk-assessment protocol within eighteen months after the Act's effective date (July 18, 1995).[1] A prospective purchaser may negotiate a range of remedies. The Act generally requires DEQ to select the least costly remedy. Essentially, the question used to be whether an expenditure would reduce concentrations of contaminants, and now the issue is whether the expenditure reduces risk. The risk assessment weighs not only the risk that someone could be exposed to, and harmed by, contaminants, but also whether such exposure is likely to occur.[2] DEQ may also consider current and reasonably likely land uses and zoning designations as a part of the risk assessment.[3]

Pending adoption of the implementing rules, DEQ will apply existing rules and policies, consistent with the intent of the new law "to the maximum extent practicable."[4]

Probabilistic Risk Assessment

The Act requires that the risk-assessment protocol consider the following factors:

1. existing and reasonably likely human exposure in the future, as well as significant adverse effects to the ecological receptor's health and viability (these must receive consideration in both a baseline risk assessment and in an assessment of residual risk after remedial action);[5]
2. reasonable estimates of plausible upperbound exposures (these must not grossly underestimate or grossly overestimate risks);[6]
3. consideration of the range of possibilities of risks actually occurring and the range of size of populations likely to receive exposure;[7]
4. current and reasonably likely future land uses, as well as the quantitative and qualitative descriptions of uncertainties;[8]
5. sources of toxicity information;[9]
6. definition of probabilistic modeling;[10]
7. identification of the criteria for the selection and application of fate and transport models;[11]
8. definition of the use of high-end and central-tendency exposure cases and assumptions;[12]
9. definition of population risk estimates in addition to individual risk estimates;[13]
10. definition of appropriate approaches for addressing cumulative risks posed by multiple contaminants or multiple exposure pathways;[14] and
11. establishment of appropriate sampling approaches and data quality requirements.[15]

Remedy Selection

After risk assessment, a prospective purchaser or current owner may negotiate a remedy with DEQ. DEQ must determine the protectiveness of a remedial action based on application of the following:

- In relation to protecting humans, the acceptable risk to individual carcinogens is a lifetime excess cancer risk of one per one million people

exposed. The acceptable risk for exposure to noncarcinogens is a hazard index number equal to, or less than, one.[16]
- For protection of ecological receptors, if a release of hazardous substances causes, or is reasonably likely to cause, significant adverse impacts to the health or viability of a species listed as threatened or endangered, the acceptable risk level should be the point before such impact occurs.[17]

When indicated, appropriate remedial actions (those protecting human and environmental health) consist of treatment that eliminates or reduces toxicity, mobility, or volume of hazardous substances. These actions include excavation, disposal, containment, and any other engineering or institutional controls. The actions may also consist of any combination of other methods providing protection, such as engineering techniques to halt migration of contaminants, and deed restrictions.[18]

The process of remedy selection balances the following: (1) effectiveness of the remedy,[19] (2) technical and practical implementation of the remedy,[20] (3) long-term reliability of the remedy,[21] (4) any short-term risk to the community, those engaged in the remedy's implementation, or the environment,[22] and (5) the reasonableness of the remedy's cost. A remedy's costs are not reasonable if they are disproportionate to benefits created through risk reduction or risk management.[23]

In an effort to facilitate remedial actions, the Act also requires DEQ to develop generic remedies for "common categories" of facilities, considering the balancing factors discussed immediately above.[24] Specifically, DEQ will consider demonstrated performance data from prior remedial actions. If a generic remedy is protective and satisfies the balancing factors discussed above, DEQ can select or approve the generic remedy for that site with limited evaluation of remedial alternatives.[25]

The Act maintains a specific preference for treatment of "hot spots," defined as discrete areas with the following types of hazardous substances:

- Hazardous substances, present in high concentrations, that are highly mobile or cannot be easily contained[26]
- Hazardous substances that would present a risk to human health and exceed the acceptable risk of exposure[27]
- Concentrations of hazardous substances in water, which have significant adverse effects on existing or likely future beneficial uses of the water, but for which treatment is reasonably likely to restore beneficial use within a reasonable time[28]

Prospective-Purchaser Agreements

The Act authorizes DEQ to provide a prospective purchaser with a written agreement that includes a release of liability by DEQ after the prospective purchaser completes the agreed-upon environmental remediation work.[29]

The party must not currently be liable for the release of a hazardous substance at the particular site and must satisfy the requirements discussed below under "Liability Protection."[30] DEQ must consult with affected land-use planning jurisdictions and consider reasonably anticipated future land uses at the site and surrounding properties.[31]

The agreement itself may be finalized in an administrative consent order, administrative agreement, or judicial consent decree.[32] The agreement must include all the following: (1) a commitment to undertake measures that result in a substantial public benefit,[33] (2) remedial actions performed pursuant to the agreement and under DEQ's oversight,[34] (3) a waiver by the party of any claim or cause of action against the state of Oregon and arising from contamination as of the date of acquisition,[35] (4) an irrevocable right of entry to DEQ for purposes of the agreement or for remedial actions authorized under this section,[36] (5) the reservation of rights of entities that are not parties to the agreement, and[37] (6) a legal description of the property.[38]

The agreement is available only if

1. the remediating party is not currently liable for an existing release of a hazardous substance;[39]
2. removal or remedial action is necessary to protect human health or the environment;[40]
3. proposed redevelopment or reuse of the property will not contribute to or exacerbate existing contamination, cause increased health risk, or interfere with necessary remedial measures;[41] and
4. a substantial public benefit results from the agreement, including, but not limited to,
   (a) the generation of substantial funding or other resources that facilitate remedial measures;
   (b) a commitment to perform substantial remedial measures;
   (c) productive reuse of vacant or abandoned industrial or commercial facilities; or
   (d) development by a governmental entity or nonprofit organization to address an important public purpose.[42]

Memorandum of Agreement

Currently, Oregon does not have a Memorandum of Agreement with the United States Environmental Protection Agency (USEPA) for its brownfields cleanup program under the Act. Therefore, a person obtaining a prospective-purchaser agreement or cleaning a site under the new approach, theoretically, could be subject to federal liability even after obtaining a release under the program. However, USEPA's current enforcement position does not contemplate taking an enforcement action in connection with a site that is participating in state supervised voluntary cleanups. Prospective-purchaser agreements are available from USEPA Region X, Seattle.

Liability Protection

If the party satisfactorily performs its obligations under the agreement, it will receive a release from DEQ for future cleanup liability.[43] The party asserting the release, however, bears the burden of proving that any contamination existed before its date of acquisition.[44] The agreement does not release the party from claims arising from

1. releases after acquisition;[45]
2. contribution or exacerbation of releases;[46]
3. interference or failure to cooperate with DEQ or persons conducting remedial measures;[47]
4. failure to exercise due care to take reasonable precautions at the facility; [48] or
5. any violation of federal, state, or local law.[49]

The Act requires the recordation of the agreement in the real-property records of the county where the property lies. All benefits and burdens of the agreement run with the land.[50] However, the release from liability affects only those who (1) are not liable for releases at the date of acquisition, and (2) assume, or are bound by, the terms of the agreement applicable to the site at the date of acquisition.[51]

Financial Incentives

The Act requires the Oregon Economic Development Department to establish a task force to explore funding strategies and financial incentives that are in existence or that may be established to facilitate use of the Act by prospective purchasers.[52] DEQ is to select the "least expensive remedial action," unless "the additional cost of a more expensive alternative is justified by proportionately greater benefits."[53]

Cost Recovery

A liable party can seek contribution from another liable party under Oregon law. If a claim for contribution is at trial, the court may determine the proportionate amount recoverable among the liable parties. The amounts are determined by equitable factors the court deems appropriate.[54] The factors may include, but are not limited to, the following:

- Amount of hazardous substances contributed[55]
- Degree of toxicity or hazard posed by the hazardous substances[56]
- Degree of involvement in release of the hazardous substance by the liable person[57]
- Relative culpability or negligence of the liable person[58]
- Degree of cooperation by liable persons with the government or persons who have a financial interest in the facility[59]
- Extent of participation by the liable person in remedial actions[60]

- Length of time the facility was owned or operated by the liable person during the time the release occurred[61]
- Whether the party's acts or omissions resulting in a release were in material compliance with the law[62]
- Economic benefit derived from the facility or from the acts or omissions that resulted in a release[63]
- Circumstances and conditions involved in the facility's conveyance, including the price paid and any discounts granted[64]
- Quality of the evidence concerning liability[65]

If a person is deceased or bankrupt at the time of trial, the court may, in its discretion, allocate that person's share to other responsible parties, taking into consideration any relationship between the orphan share's responsible party and any other party.[66]

Conclusion

The Recycled Lands Act has transformed the manner in which contaminated property is addressed in Oregon. Remediation costs will now be evaluated in terms of whether the expenditure reduces risk to human health and the environment, and not simply whether concentrations are reduced. The Act mandates that any remediation action be protective, but this does not necessarily mean expensive treatment or removal. "Least-cost" remedies are preferred and engineering or institutional controls are reasonable alternatives. Remedy selection will follow a balancing of common-sense factors, such as long-term effectiveness.

The new approach to site remediation in Oregon tracks national trends toward risk-based corrective action. This approach, coupled with the availability of prospective-purchaser agreements, should carry Oregon a long way toward redevelopment of its industrial lands.

Prospective-Purchaser Agreements under the Oregon Recycled Land Act

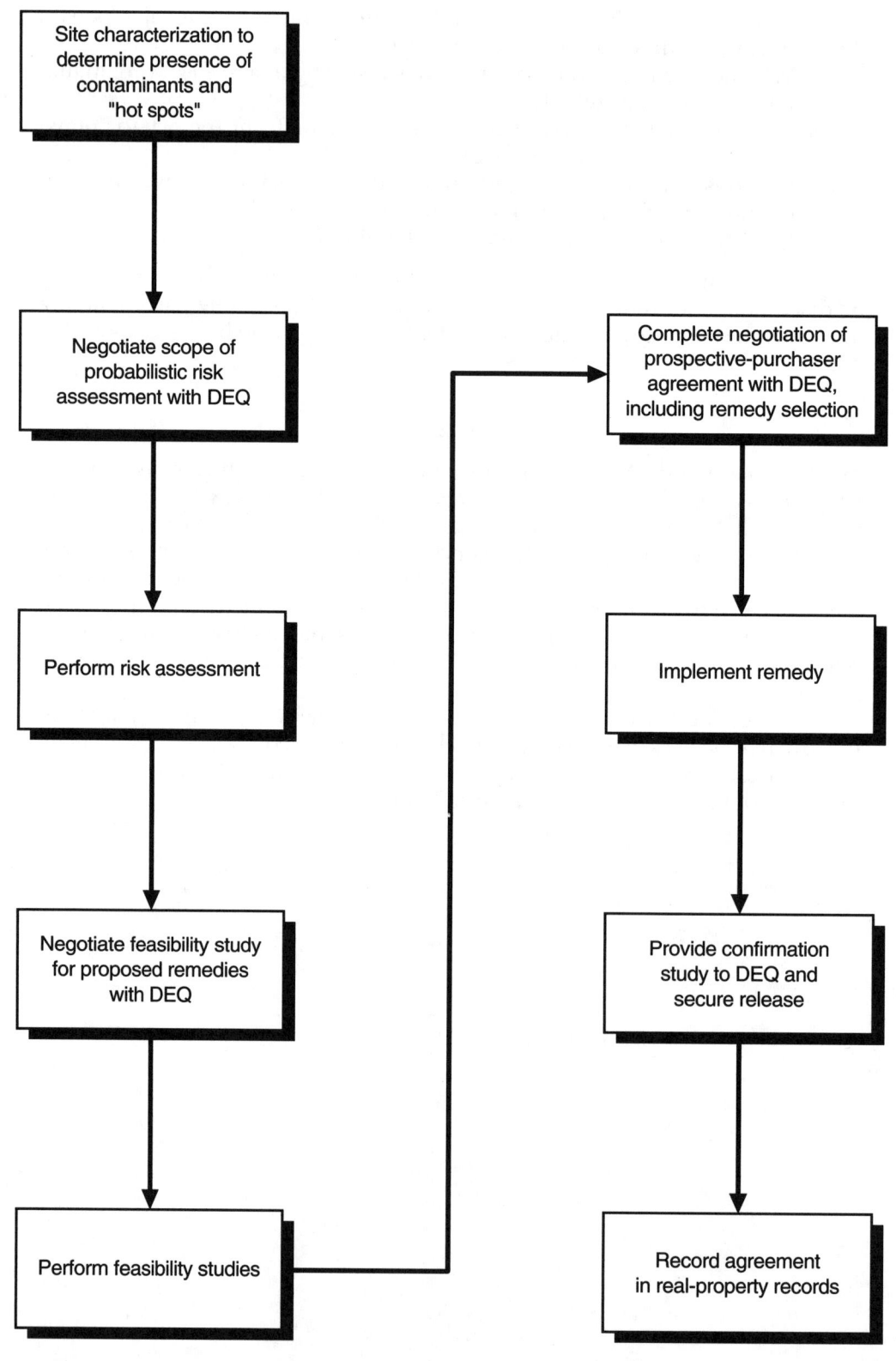

Notes

1. Or. Rev. Stat. § 465.315(2)(a) (1995).
2. *Id.* § 465.315(2)(C).
3. *Id.*
4. Or. Dep't Envtl. Quality, Summary of the 1995 Amendments to Oregon's Environmental Cleanup Law 2 (1995) (DEQ guidance document).
5. Or. Rev. Stat. § 465.315(2)(a)(A) (1995).
6. *Id.* § 465.315(2)(a)(B).
7. *Id.* § 465.315(a)(2)(C).
8. *Id.*
9. *Id.* § 465.315(2)(a)(D).
10. *Id.*
11. *Id.* § 465.315(2)(a)(F).
12. *Id.* § 465.315(2)(a)(G).
13. *Id.* § 465.315(2)(a)(H).
14. *Id.* § 465.315(2)(a)(I).
15. *Id.* § 465.315(2)(a)(J).
16. *Id.* § 465.315(1)(b)(A).
17. *Id.*
18. *Id.* § 465.315(1)(c)(A).
19. *Id.* § 465.315(1)(d)(A).
20. *Id.* § 465.315(1)(a)(B).
21. *Id.* § 465.315(1)(d)(C).
22. *Id.* § 465.315(1)(d)(D).
23. *Id.* § 465.315(1)(d)(E).
24. *Id.* § 465.315(1)(f).
25. *Id.*
26. *Id.* § 465.315(2)(b)(A).
27. *Id.*
28. *Id.* § 465.315(2)(b)(B).
29. *Id.* § 465.327.
30. *Id.* § 465.327(1).
31. *Id.* § 465.327(d)(2).
32. *Id.* § 465.327(d)(3).
33. *Id.* § 465.327(d)(3)(a).
34. *Id.* § 465.327(d)(3)(b).
35. *Id.* § 465.327(d)(3)(c).
36. *Id.* § 465.327(d)(3)(d).
37. *Id.* § 465.327(d)(3)(e).
38. *Id.* § 465.327(d)(3)(f).
39. *Id.* § 465.327(1)(a).
40. *Id.* § 465.327(1)(b).
41. *Id.* § 465.327(1)(c).
42. *Id.* § 465.327(1)(d).
43. *Id.* § 465.327(4).
44. *Id.*
45. *Id.* § 465.327(4)(a).
46. *Id.* § 465.327(4)(b).
47. *Id.* § 465.327(4)(c).

48. *Id.* § 465.327(4)(d).
49. *Id.* § 465.327(4)(e).
50. *Id.* § 465.327(5).
51. *Id.*
52. Oregon House Bill 3352, Section 6.
53. OR. REV. STAT. § 465.315(1)(d)(E) (1995).
54. *Id.* § 465.257(1).
55. *Id.* § 465.257(1)(a).
56. *Id.* § 465.257(1)(b).
57. *Id.* § 465.257(1)(c).
58. *Id.* § 465.257(1)(d).
59. *Id.* § 465.257(1)(e).
60. *Id.* § 465.257(1)(f).
61. *Id.* § 465.257(1)(g).
62. *Id.* § 465.257(1)(h).
63. *Id.* § 465.257(1)(i).
64. *Id.* § 465.257(1)(j).
65. *Id.* § 465.257(1)(k).
66. *Id.* § 465.257(2).

CHAPTER 41

Pennsylvania

ROBERT L. COLLINGS

Pennsylvania, like most states, is learning from its mistakes. In the past, Pennsylvania policies required contaminated properties to be cleaned up to pre-human, pristine conditions. The standards were often prohibitively expensive and impossible to meet. On top of that, the state was unwilling to let parties cut off liability for further cleanup at a site, even after making it safe. Lenders became entangled in the web of potential liability and refused to make loans for the purchase of contaminated sites. Property owners who wanted to clean up their property were faced with an unresponsive bureaucracy that emphasized process over results and sometimes took years to approve cleanup plans. The unintended consequence of this lack of realistic cleanup standards, never-ending liability, unavailability of credit, and an unresponsive bureaucracy was "brownfields"—old industrial sites that were warehoused, fenced off, or abandoned.[1]

Pennsylvania's legislature began to realize there was a brownfields problem in 1991. But it took four long years, and much discussion and debate, before Governor Ridge signed into law a three-bill package on May 19, 1995, creating the state's Land Recycling Program (the Program). Acts 2, 3, and 4 of 1995 (defined below) each provide some of the tools used by the Program.[2] The Program's goals are defined as follows:

- To make contaminated sites safe
- To return contaminated sites to productive use
- To preserve farmland and greenspace[3]

Various mottoes have been offered to describe the Program—"Rebuilding Pennsylvania's Industrial Heritage,"[4] "Putting Abandoned Industrial Sites

---

The descriptions and opinions are those of the author, who is deeply grateful for the careful review and many helpful comments of M. Joel Bolstein, Esquire, Deputy Secretary for Special Projects of the Pennsylvania Department of Environmental Protection. Mr. Bolstein is extensively involved in the implementation of this program.

Back to Work,"[5] or "A Timely Common Sense Solution to Cleaning Up Contaminated Sites."[6] All these mottoes convey the distinctly different spirit of this environmental cleanup program.

Summary of Major Provisions

The Land Recycling and Environmental Remediation Standards Act (Act 2)[7] establishes a realistic framework for setting cleanup standards based on health and environmental risks; it sets up a clear process for the Pennsylvania Department of Environmental Protection (DEP) to approve or reject cleanup plans with real deadlines; and it offers full releases of liability for identified contamination under state law once a site is made safe from those conditions. This release includes liability to the state, under citizen suits, or for contribution to other persons who are legally responsible for the contamination. Special treatment as an "innocent purchaser" is also available to individuals who clean up contamination at abandoned sites or in enterprise zones.[8] Act 2 also provides procedures and imposes a duty on the state to develop cleanup standards, and creates a fund to make grants and loans to government agencies and others for the assessment and cleanup of abandoned sites.

The Economic Development Agency, Fiduciary, and Lender Environmental Liability Protection Act (Act 3)[9] provides fiduciaries, lenders, and economic-development agencies protection from environmental cleanup liabilities if they did not cause the contamination at a subject site. This statutory defense includes liability under all state environmental statutes and common-law liability for environmental conditions.

The Industrial Sites Environmental Assessment Act (Act 4)[10] authorizes $2 million per year in grant money to finance environmental assessments at certain abandoned industrial sites in certain communities.[11] The Pennsylvania Department of Commerce administers this program.[12]

Elements of the Pennsylvania Program

Uniform Cleanup Standards

Ever since environmental programs expanded from setting treatment standards for ongoing discharges to cleaning up preexisting contamination, the basic problem has been a lack of governmental willingness to define cleanup standards for identified contamination problems. "How clean is clean?" became a question that was repeatedly asked, but never answered. Sites were investigated endlessly to identify risks, but few cleanups were completed.[13]

Act 2 ends this cycle at the state and local level by providing three categories of cleanup standards:

- Cleanup to background levels[14]
- Cleanup to statewide health standards, to be set by regulation[15]
- Cleanup to meet standards set through a site-specific exposure and risk-assessment process[16]

A special standard applies for innocent purchasers who agree to clean up imminent threats at abandoned sites and in enterprise zones.[17] The person performing the site remediation—not the state regulatory agency—selects the cleanup standard to be used, with knowledge of the degree of future legal liability[18] and site use restrictions[19] under each standard. Any of the three standards or a combination can be used to clean up different sources of contamination at a site.[20]

Cleanup to background levels, the only standard in widespread use before the state's issuance of the December 1993 "greenfields" cleanup standards,[21] continues to be offered, but under a clearer, reasonable definition. "Background" refers to area conditions measured by samples that do not reflect man-made conditions on the cleanup site.[22] Substances present in natural soils or from off-site releases are not included.[23]

The cleanup standards in Act 2 apply to all sites and cover both voluntary and government-ordered cleanup actions. There is a three-year phase-out period for existing standards that may be more (or less) stringent than the Act 2 standards.[24] Existing statewide standards may be used by DEP until July 18, 1998, but their use may be avoided by the selection and attainment of background or site-specific standard remediation. Some standards were superseded immediately by the Act because they were inconsistent with its provisions.[25]

Cleanup standards based on site risk must consider actual risks based on probable exposures.[26] In no case may a cleanup standard under any category require cleanup below the level at which the contamination may be accurately measured; that is, the practical quantitation limit (PQL).[27]

Standardized and Streamlined Administrative Cleanup Procedures

Act 2 creates a specific, uniform procedure for site remediation and release from liability.[28] Statutory deadlines for review and approval of cleanup reports include a deemed approval for failure of the state to identify deficiencies in a timely manner.[29] This allows greater planning of commercial transactions dependent on timely site cleanups to fix environmental liabilities. Greater opportunity for public input is also afforded by the process.[30]

Permit Waiver Provisions

Another important piece of administrative relief that is helpful to brownfields redevelopment in Pennsylvania is the provision for waiver of state and local permit requirements for remediation performed pursuant to Act 2's requirements.[31] Remediation programs conducted entirely on site may incorporate or waive permit requirements under other state environmental laws in any of the following circumstances:

- Permit standards will create greater risk than the remedial option selected

- Permit standards will substantially interfere with natural or artificial structures or features (for example, would render buildings or property unsafe)
- Remedial actions selected attain equivalent levels of performance
- Permit standards do not provide for cost-effective remedial action[32]

In no case may a waiver of permit standards waive compliance with the Act 2 remediation standards selected for the site,[33] nor may a waiver under Act 2 waive a permit standard in a state program when the standard is imposed to meet analogous federal requirements under which a state program has received federal authorization.[34]

Release from Liability

Act 2 contains a broad release of liability for contamination identified and remediated to Act 2 standards.[35] The release applies to all state statutory cleanup liability for the person doing the cleanup, and for current and future site owners, developers, occupants, successors and assigns, and public utilities.[36] Persons performing environmental site assessments are also protected from liability for site conditions not caused by their actions, as long as the site assessment is performed with "professional care."[37]

Act 3 provides even greater protections for economic-development agencies, fiduciaries, and lenders. The protection for such persons under Act 3 extends not only to all liability for cleanups, but to all claims, arising under state environmental laws and common law, related to environmental conditions on the property.[38] Very limited exceptions to this broad release exist when the lender, development authority, or fiduciary (1) actively participates in, or controls, activities that directly cause or aggravate an environmental problem, or (2) intentionally violates an environmental law.[39] Entities protected under Act 3 may own property without liability if the ownership is in the course of their protected relationship (through foreclosure, for example).[40]

Acts 2 and 3 are not exclusive vehicles for defense against liability. All other limits on liability, or defenses to liability under state environmental laws, continue to exist.[41]

Eligible Persons

Any person may apply for a liability release under the Act 2 remediation program.[42] Both voluntary cleanups and state-mandated cleanups qualify for Act 2 remediation.[43] The release from remedial liability does not include a release from penalty liability if the condition resulted from a violation of law.[44] Persons liable for a violation remain subject to penalties. However, remediation of a site may avoid penalty liability for future owners by eliminating what would otherwise be a continuing violation of law.

A liability release may be obtained only by the person performing or participating in the cleanup. Once obtained, the release applies to current and future site owners, developers, occupants, successors or assigns, and public utilities.[45]

Cleanup Standards

Background

A cleanup to background levels need not eliminate all man-made contamination.[46] Under Act 2, "background" is the concentration of regulated substances that is not associated with releases from the site.[47] Contaminants may be present, but the levels must reflect off-site conditions. Any elevated levels related to releases at a site must be removed. Background levels are set by either a background determination study, or through the use of background default values.[48] (These values are listed in the Background Default Table at Appendix A of the Technical Manual for the Program.)

Background must be achieved for each contaminant in each medium (that is, soil and groundwater).[49] Attainment of the standard is demonstrated by a statistical analysis of representative samples from each medium.[50]

The background standard test may not always be the most stringent test, particularly in areas of widespread, heavy industrial use. The background standard is generally more stringent than the site-risk standard, in that cleanup requirements may not be reduced through engineering controls such as fencing a site or imposing land-use restrictions to reduce exposure risks.[51]

State Health Standards

To meet state health standards,[52] DEP sets "medium specific concentrations,"[53] which are numerical limits for each residual contaminant. They are set by the state,[54] using existing federal and state health-based standards for residential and nonresidential use, and health advisory levels (HALs).[55] Risk-assessment procedures may be used to set additional limits as needed.[56] Statewide standards cannot be more stringent than federal standards.[57]

In addition to attaining soil and groundwater standards, cleanups must not cause or leave behind any source of air emissions or surface-water discharges in violation of state standards for those media.[58] The standards must also protect drinking water and agricultural uses of groundwater.[59] Statewide standards take into account residential[60] and nonresidential[61] exposure patterns, with appropriate differences in standards. Soils must be cleaned to acceptable standards, to a depth of fifteen feet to protect against direct exposure and to greater depths if necessary to protect groundwater standards.[62] Fencing and other use restrictions may not be used to remove any portion of the site from compliance with the statewide health standards under Act 2.[63] In addition, engineering controls—such as paving over contamination, building structures over contamination, and restricting site use—cannot be used to achieve statewide standards. However, institutional controls may be used to meet a site-specific risk-based cleanup standard in appropriate situations.

Site-Specific Standards

This is a "case-by-case" approach to setting cleanup standards, which is different from the familiar, ultraconservative government standards of the past.[64] Under Act 2, site-specific risk assessments must use reasonable assumptions to predict exposure and consequent health risks.[65] For instance, contamination buried at great depths may threaten groundwater, but it is unlikely that people will eat the contamination or even touch it, particularly if uses and site development are restricted. Appropriate statistical tests are used to measure contamination levels and exposure risks.[66]

A change in exposure-assessment approaches can produce a dramatic change in standards. For carcinogens, cleanup standards must assure that cancer rates do not increase more than one case in ten thousand exposed persons, but may not be set so conservatively that they prevent increased risks that are less than one in one million persons.[67]

Cleanup Standards for Special Industrial Sites[68]

When no financially viable responsible party is available to clean up an abandoned site or when a site is in a state-designated "enterprise zone," Act 2 provides special incentives for cleanup and reuse for such "special industrial sites." If the site qualifies as a special industrial site, then any person who acquires the site who did not cause or contribute to releases at the property has limited obligations. Section 502 of Act 2 states that the "person shall only be responsible for remediation of any immediate, direct or imminent threats to public health or the environment, such as drummed waste, which would prevent the property from being occupied for its intended purpose."[69] For example, if a person purchases an old industrial site in an enterprise zone where the groundwater is contaminated, but no one on the property will be using the groundwater, then under Act 2 all that person would be required to do is perform a baseline environmental investigation[70] and submit it to DEP. He or she would *not* need to clean up the groundwater because it does not pose a direct or imminent threat to people using the property for its intended purpose. DEP will enter an agreement[71] with the party acquiring such a property, outlining the purchaser's specific and limited cleanup obligations. The agreement will include a liability release that extends to all identified, existing contamination.[72] In the example above, the release would extend to liability for groundwater remediation, even though there was no cleanup of groundwater. In theory, the cleanup standards for special industrial sites may be less stringent than the three generally available cleanup standards.

Procedures For Performing Act 2 Cleanups

1. For cleanups attaining background or statewide standards, which will be completed within ninety days, the procedures are as follows:

- The party must perform the cleanup and document the site investigation (identification of contamination), remediation, and attainment of selected standard.[73]
2. For other cleanups (site specific, special sites, or background/statewide standard cleanups lasting more than ninety days), the procedures are as follows:
    - A notice of intent to remediate must be submitted to DEP and the local municipality, and must be published.[74]
    - For cleanups to site-specific standards and special industrial sites, a "public involvement plan" is required if requested by the municipality.[75]
    - For site-specific standard cleanups, a remedial investigation (RI) must be performed and an RI report submitted to DEP.[76] This identifies contamination and provides a basis for setting risk-based cleanup standards. A fate and transport analysis may be appropriate to show that no exposure paths exist.[77] If exposure pathways exist, a risk assessment (RA) must be performed to set appropriate cleanup standards.[78] An RA report may be required by DEP,[79] and the local municipality may be invited to comment before standards are set.[80] A "cleanup plan" discussing options to achieve risk-based standards must be submitted to DEP.[81] The RI, RA, and cleanup plan may be submitted simultaneously.[82] DEP must approve the standards and cleanup plan.
    - For special industrial sites, a baseline environmental report must be submitted to identify existing contamination and identify necessary remedial measures and proposed beneficial uses. DEP may balance the risks of the proposed residual contamination against the proposed benefits.[83]
    - Site remediation must be performed until attainment of cleanup standards is demonstrated.
    - For special industrial sites, the cleanup is specified without reference to standards, and only specified tasks need be performed.[84]
    - A final report, documenting all work and achievement of cleanup standards, must be submitted to DEP.
    - Public notice of the report submission is required for background/statewide standard cleanups lasting more than ninety days and for site-specific cleanups.[85]
3. The DEP review process is as follows:
    - For background and statewide-health-standard cleanups, DEP has sixty days to review final reports and note deficiencies.[86] If no response is made within sixty days, the report is deemed approved and cleanup liability ends.[87]
    - For site-specific standard and special industrial sites, DEP has ninety days to review the final report.[88] If deficiencies are not raised within ninety days, the reports are deemed approved.[89]

4. Deed notice requirements are as follows:
   - Act 2 requires notice in the deed of any contamination left in place above background or residential standard levels after a cleanup has been concluded.[90]
   - No deed notice is required for a cleanup achieving either background levels or the statewide health standard using residential exposure patterns.[91]
   - Deed notice is not required as part of the final report approval by DEP, but is made at the time property is conveyed.[92]

Liability Protection

Liability is released upon final-report approval by DEP.[93] In the case of special industrial sites, a prior agreement is needed to define the scope of liability release in relation to environmental benefits and intended use.[94]

Act 2 does contain some reopeners that were added in the last hours of the legislative process.[95] These reopeners are intended to address concerns over the release of liable persons in inappropriate circumstances. The liability release applies only to future cleanup of identified contamination addressed by the cleanup. As a result, a careful and thorough identification of contamination is very important. In particular, the site-specific cleanup process requires special care in documenting risk assessments and special site conditions.

Specifically, reopeners exist when

1. fraud was committed in demonstrating the attainment of standards, if such fraud resulted in avoiding further needed cleanup;[96]
2. newly identified areas of contamination exceed standards;[97]
3. the remedy failed to meet one, or a combination, of the three cleanup standards;[98] or
4. risks at the site increased above acceptable site-specific standards, due to changes in exposure conditions such as land-use changes, or new information about contaminant toxicity or exposure pathways (such as dermal absorption or inhalation).[99] When the increased risk results from a land-use change, the person who changes the use is responsible for new remediation. Otherwise, any liable person is presumably liable for remediation.[100]

Also, for sites not used for industrial activity before July 18, 1995, when releases occur after July 18, 1996, and institutional or engineering controls (limits on access to exposure) are used instead of treatment or removal to meet Act 2 standards, added treatment, removal, or destruction may be required if it becomes technically and economically feasible to do more treatment, removal, or destruction in the future.[101]

Financial Assistance

The Industrial Land Recycling Fund[102] is used by DEP to implement Act 2.[103] Funding is provided by state appropriation (supplemented by any federal funds that may be authorized), private contributions, and fines and penalties assessed under Act 2.[104]

Funded by legislative appropriation, the Industrial Sites Cleanup Fund[105] (ISCF) makes grants and loans for voluntary remediation of sites used for industrial activity before July 18, 1995. Administered by the Department of Commerce, the ISCF may fund up to 75 percent of the costs of environmental studies and cleanup.[106] Funding is not available to persons who caused or contributed to site contamination.[107]

Outright grants may be made to political subdivisions, instrumentalities, or economic-development authorities, if the grantee owns the site and oversees the cleanup. Up to 20 percent of the ISCF may be expended as grants in any fiscal year.[108] Political subdivisions, their instrumentalities, economic development agencies (EDAs), or other eligible persons may receive loans on favorable terms from the ISCF.[109]

Act 4 establishes a $2 million fund, administered by the Department of Commerce, to make grants to municipalities, local authorities, or nonprofit EDAs for environmental studies in "distressed communities," designated by the Department of Commerce under the Business Infrastructure Development Act.[110] Certain cities may also receive funds for remediation of sites contaminated before July 18, 1995.[111]

Guidance and Regulations/Buyer-Seller Agreements

DEP has produced a Technical Guidance Manual containing extensive information, tables, and forms for use under this program.[112] Regulations have been proposed,[113] which will be issued in 1997.

Conclusion

The Pennsylvania Land Recycling Program puts the Commonwealth at the forefront of brownfields redevelopment. Acts 2, 3, and 4 restore redevelopment relationships among developers, lenders, and the government, which had been sundered by environmental laws. The program needs to continue on a fast track, with model agreements and agency initiatives to identify suitable sites and actively promote their remediation and reuse. A public/private investment-fund initiative should also be explored. Finally, implementing policies that promote reasonable, consistent applications of agency discretion and limit the uncertainties of reopener provisions will help keep this program on the fast track where it has started.

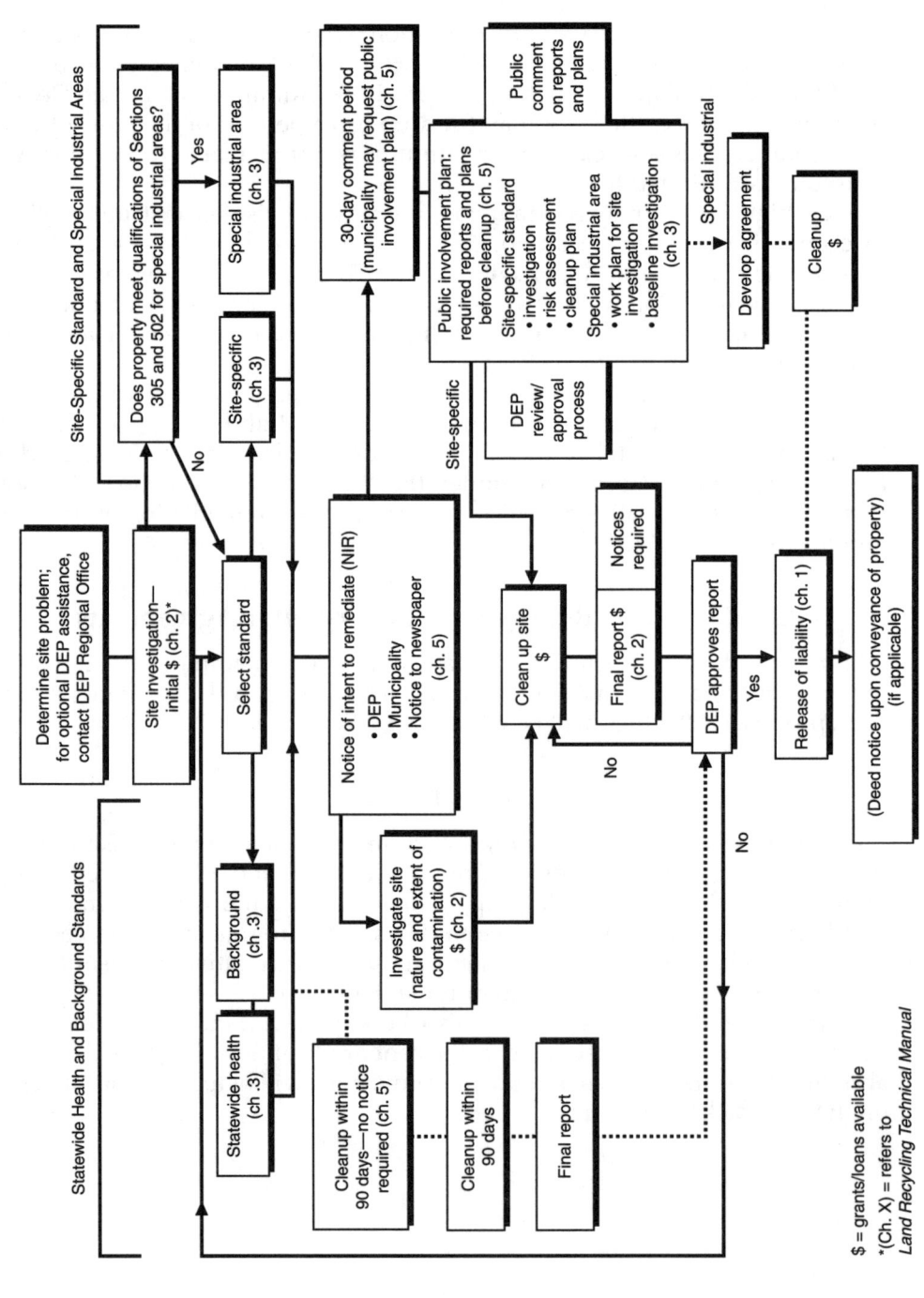

Special Industrial Areas (SIAs)

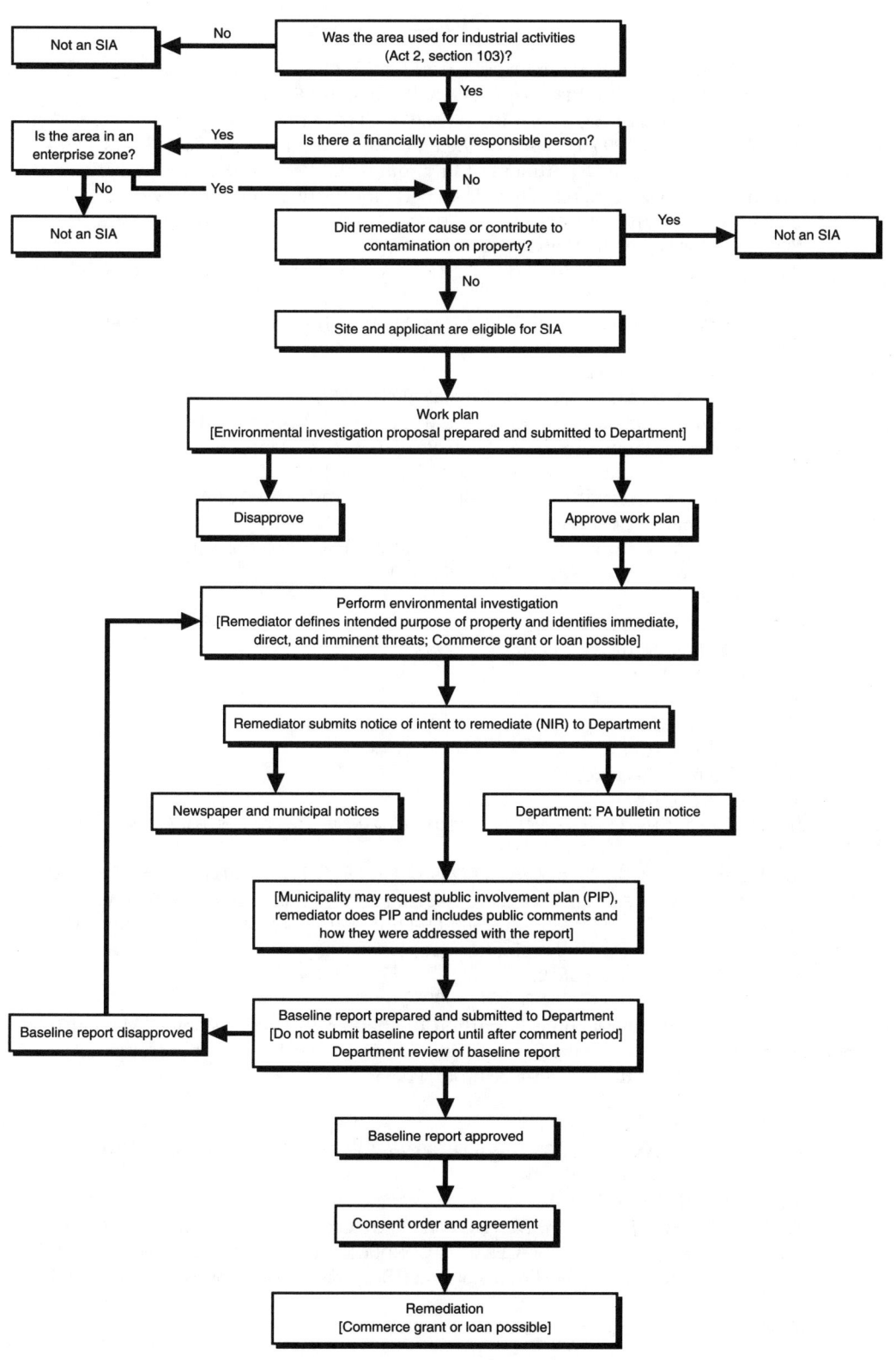

Notes

1. *See* PA. DEP'T ENVTL. PROTECTION, FACT SHEET 1: OVERVIEW OF THE LAND RECYCLING PROGRAM, THE NEED FOR LAND RECYCLING. (1995) [hereinafter FACT SHEET 1].

2. PA. STAT. ANN. tit. 35, § 6026.101 *et seq.* (1995) (Act 1995-2); *id.* § 6027.1 *et seq.* (Act 1995-3); *id.* § 6028.1 (Act 1995-4). Also see the Land Recycling Program Fact Sheets 2 and 3 issued by the Pennsylvania Department of Environmental Protection in 1995 (available at landrecycling@a1.dep.state.pa.us). These laws became effective on July 18, 1995.

3. PA. DEP'T ENVTL. PROTECTION, A CITIZEN'S GUIDE TO PENNSYLVANIA'S LAND RECYCLING PROGRAM (1995) [hereinafter CITIZEN'S GUIDE].

4. PA. DEP'T ENVTL. PROTECTION, PENNSYLVANIA'S LAND RECYCLING PROGRAM: SIX MONTH PROGRESS REPORT (1996).

5. PA. DEP'T ENVTL. PROTECTION, PROGRAM HIGHLIGHTS AND FREQUENTLY ASKED QUESTIONS (1995).

6. CITIZEN'S GUIDE, *supra* note 3.

7. PA. STAT. ANN. tit. 35, §§ 6026.101–6026.908 (1995).

8. A "cleanup" may include no action, when the DEP agrees that no action is needed because the site contamination already meets one or more standards.

9. PA. STAT. ANN. tit. 35, §§ 6027.1–6027.14 (1995); PA. DEP'T ENVTL. PROTECTION, FACT SHEET 3: LAND RECYCLING PROGRAM, SUMMARY OF ACT 3 AND ACT 4, NO. 2530 FS-DEP 1845, (1995) [hereinafter FACT SHEET 3].

10. PA. STAT. ANN. tit. 35, §§ 6028.1–6028.5 (1995); FACT SHEET 3, *supra* note 9.

11. PA. STAT. ANN. tit. 35, §§ 6028.3, 6028.4 (1995).

12. *Id.* § 6028.5.

13. FACT SHEET 1, *supra* note 1.

14. PA. STAT. ANN. tit. 35, § 6026.302 (1995).

15. *Id.* § 6026.303.

16. *Id.* § 6026.304.

17. *Id.* § 6026.305.

18. *Id.* § 6026.501(a).

19. *Id.* §§ 6026.302–6026.305.

20. *Id.* § 6026.301(b).

21. PA. DEP'T ENVTL. RESOURCES, INTERIM CLEANUP STANDARDS FOR CONTAMINATED SOILS (1993).

22. PA. STAT. ANN. tit. 35, § 6026.103 (1995); PA. DEP'T ENVTL. PROTECTION, TECHNICAL MANUAL, PENNSYLVANIA'S LAND RECYCLING PROGRAM (1995) [hereinafter TECHNICAL MANUAL]; PA. DEP'T ENVTL. PROTECTION, NO. 2500 FS-DEP 1846, FACT SHEET 4: CLEANUPS USING BACKGROUND STANDARDS (1995) [hereinafter FACT SHEET 4].

23. FACT SHEET 4, *supra* note 22.

24. PA. STAT. ANN. tit. 35, § 6026.107 (1995).

25. The groundwater-standards portion of chapter 3 of the Technical Manual (see note 22) appears to supersede the Pennsylvania Groundwater Policy, for example.

26. PA. STAT. ANN. tit. 35, § 6026.301(a)(3) (1995).

27. *Id.* § 6026.301(c).

28. FACT SHEET 1, *supra* note 1.

29. PA. STAT. ANN. tit. 35, §§ 6026.302(e)(3), 6026.303(h)(3), 6026.304(n)(2)(ii), 6026.304(n)(3), 6026.305(d) (1995).

30. *See, e.g., id.* § 6026.304(o).

31. *Id.* § 6026.902. *But see* PA. DEP'T ENVTL. PROTECTION, NO. 2530-FS-DEP 1852, FACT SHEET 10: PERMIT REQUIREMENTS FOR CLEANUP CONDUCTED UNDER ACT 2; *compare* 42 U.S.C. § 9621(e) (1995) (Comprehensive Environmental Response, Compensation, and Liability Act).

32. PA. STAT. ANN. tit. 35, § 6026.902(b) (1995).

33. *Id.*

34. *Id.*

35. *Id.* 6026.501(a).

36. *Id.*

37. *Id.* § 6026.501(b).

38. *Id.* §§ 6027.4–6027.6. The absence of a reference to economic-development agencies appears to be an oversight. But compare the inclusion of the reference concerning economic development agencies in section 6027.7.

39. *Id.*

40. *Id.* §§ 6027.4(2), 6027.5(a). Fiduciaries are protected from exposure of their personal or corporate assets. Assets held in trust may be reachable to satisfy cleanup obligations related to the assets.

41. *Id.* §§ 6026.503, 6027.7, 6027.8.

42. "Person" is broadly defined to include "an individual, firm, corporation, association, partnership, consortium, joint venture, commercial entity, authority, nonprofit corporation, interstate body or other legal entity which is recognized by law as the subject of rights and duties. The term includes the Federal Government, State Government, political subdivision and Commonwealth instrumentalities." *Id.* 6026.103.

43. *Id.* § 6026.106(A).

44. *Id.* § 6026.905(B).

45. *Id.* § 6026.501(a).

46. *Id.* § 6026.302. *See also* FACT SHEET 4, *supra* note 22; TECHNICAL MANUAL, *supra* note 22.

47. PA. STAT. ANN. tit. 35, § 6026.103 (1995).

48. *Id.* § 6026.302; TECHNICAL MANUAL, *supra* note 22.

49. PA. STAT. ANN. tit. 35, § 6026.302 (1995); TECHNICAL MANUAL, *supra* note 22.

50. PA. STAT. ANN. tit. 35, § 6026.302 (1995); TECHNICAL MANUAL, *supra* note 22.

51. PA. STAT. ANN. tit. 35, § 6026.302(b)(4) (1995).

52. *Id.* § 6026.303.

53. *Id.* § 6026.303(b).

54. Standards under section 6026.303 of Act 2 are adopted as regulations by the state Environmental Quality Board.

55. PA. STAT. ANN. tit. 35, § 6026.303(a) (1995).

56. *Id.* § 6026.303(c).

57. *Id.* § 6026.303(a).

58. *Id.* §§ 6026.303(b)(1), 6026.303(b)(2).

59. *Id.* § 6026.303(b)(3).

60. *Id.* § 6026.303(b)(4).

61. *Id.* §§ 6026.303(b)(5), 6026.303(b)(6).

62. *Id.* §§ 6026.303(b)(4), 6026.303(b)(5). *See also* PA. DEP'T ENVTL. PROTECTION, NO. 2350 FS-DEP 1847, FACT SHEET 5: LAND RECYCLING PROGRAM, CLEANUPS USING STATEWIDE HEALTH STANDARDS (1995) [hereinafter FACT SHEET 5].

63. PA. STAT. ANN. tit. 35, § 6026.303(e)(3) (1995); FACT SHEET 5, *supra* note 62.

64. PA. STAT. ANN. tit. 35, § 6026.304 (1995); PA. DEP'T ENVTL. PROTECTION, NO. 2530 FS-DEP 1848, FACT SHEET 6: LAND RECYCLING PROGRAM, CLEANUPS USING SITE-SPECIFIC STANDARDS (1995) [hereinafter FACT SHEET 6].

65. PA. STAT. ANN. tit. 35, § 6026.304 (1995); FACT SHEET 6, *supra* note 64.

66. For example, Monte Carlo simulations or other appropriate statistical techniques are sometimes used. PA. STAT. ANN. tit. 35, § 6026.304(f)(2) (1995).

67. *Id.* § 6026.304(b).

68. *Id.* § 305; PA. DEP'T ENVTL. PROTECTION, NO. 7530 FS-DEP 1849, FACT SHEET 7: LAND RECYCLING PROGRAM, SPECIAL INDUSTRIAL AREAS (1995) [hereinafter FACT SHEET 7].

69. PA. STAT. ANN. tit. 35, § 6026.502(b)(1) (1995).

70. *Id.* § 6026.305(b).

71. *Id.* §§ 6026.305(e), 6026.502(a).
72. *Id.* § 6026.502(a).
73. *Id.* §§ 6026.302(e)(4), 6026.303(h)(4).
74. *Id.* §§ 6026.302(e)(1), 6026.303(h)(1), 6026.304(n)(1), 6026.304(c)(1).
75. *Id.* §§ 6026.304(o), 6026.305(c)(2).
76. *Id.* § 6026.304(l)(1).
77. *Id.* § 6026.304(l)(1)(iv).
78. *Id.* § 6026.304(l)(2).
79. Section 6026.304(a), read in conjunction with sections 6026.304(l)(1) and 6026.304(l)(2), suggests this requirement.
80. PA. STAT. ANN. tit. 35, § 6026.304(l)(3) (1995). *See also* PA. DEP'T ENVTL. PROTECTION, NO. 2530 FS-DEP 1851, FACT SHEET 9: OPPORTUNITIES FOR PUBLIC PARTICIPATION IN THE LAND RECYCLING PROGRAM (1995) [hereinafter FACT SHEET 9].
81. PA. STAT. ANN. tit. 35, § 6026.304(l)(3) (1995); FACT SHEET 9, *supra* note 80.
82. PA. STAT. ANN. tit. 35, § 6026.304(l)(5) (1995).
83. *Id.* § 6026.305(b); FACT SHEET 7, *supra* note 68.
84. PA. STAT. ANN. tit. 35, § 6026.502(b)(1) (1995).
85. *Id.* §§ 6026.302(b)(2), 6026.303(e)(2), or 6026.304(l)(4).
86. *Id.* §§ 6026.302(e)(3), 6026.303(b)(3).
87. *Id.*
88. *Id.* §§ 6026.304(n)(2)(ii), 6026.305(d).
89. *Id.*
90. *Id.* § 6026.302(d).
91. *Id.*
92. Deed notice of certain environmental conditions is required when a deed issues. *See* PA. STAT. ANN. tit. 35, § 6018.405 (1995) (on-site disposal of hazardous waste under Solid Waste Management Act); *id.* § 6020.512 (on-site disposal of hazardous substances under Hazardous Sites Cleanup Act).
93. PA. STAT. ANN. tit. 35, § 6026.501 (1995).
94. *Id.* § 6026.502.
95. *Id.* § 6026.505.
96. *Id.* § 6026.505(1).
97. *Id.* § 6026.505(2).
98. *Id.* § 6026.505(3).
99. *Id.* § 6026.505(4).
100. *Id.*
101. *Id.* § 6026.505(5).
102. *Id.* § 6026.701.
103. *Id.* § 6026.701(b).
104. *Id.* § 6026.701(c).
105. *Id.* § 6026.702; PA. DEP'T ENVTL. PROTECTION, NO. 2530 FS-DEP 1850, FACT SHEET 8: LAND RECYCLING PROGRAM, FINANCIAL ASSISTANCE (1995).
106. PA. STAT. ANN. tit. 35, § 6026.702(b) (1995).
107. *Id.*
108. *Id.* § 6026.702(c).
109. *Id.* § 6026.702(d).
110. *Id.* § 6028.2(A)(1).
111. Philadelphia, Pittsburgh, and certain other large cities are eligible. *Id.* § 6028.702(A)(2).
112. The Technical Guidance Manual was issued on July 18, 1995, and supplemented on November 8, 1996. It is available, with other useful information, at http://www.dep.state.pa.us/dep/deputate/airwaste/wm/landrecy/default.htm.
113. 26 Pa. B. 3985-4065, No. 33 (August 17, 1996).

CHAPTER 42

Rhode Island

NED ABELSON
WILLIAM M. SEUCH
MAURA McCAFFERY

The Rhode Island brownfields program was enacted in July 1995 as part of the Rhode Island Industrial Property Remediation and Reuse Act (Reuse Act).[1] The Reuse Act was developed out of a joint effort by the Rhode Island Department of Environmental Management (DEM), the Division of Statewide Planning, the Department of Economic Development, now called the Economic Development Corporation (EDC), and various representatives from the private sector. Like so many other brownfields programs across the country, the Reuse Act was designed to encourage the cleanup of underutilized, contaminated sites—by means of numerical, risk-based cleanup standards tied directly to the intended use of the property—and bring these sites back into productive use. The Rhode Island brownfields program is available to those sites that are targeted for economic investment or redevelopment but are affected by environmental contamination that impedes that redevelopment.

The Rhode Island brownfields program is only one component of the Reuse Act. This recently enacted law includes a liability scheme that is similar to the Comprehensive Environmental Response, Compensation, and Liability Act (CERCLA), the federal Superfund statute, and is Rhode Island's first statute of this type. Before 1995, contamination caused by releases of hazardous material was regulated in Rhode Island by a variety of different statutes, including the Hazardous Waste Management Act,[2] the Water Pollution Act,[3] the Groundwater Protection Act,[4] and the Oil Pollution Control Act.[5]

Enforcement powers under these various acts are vested predominantly with DEM, with DEM's oversight being integral throughout the remediation process. Unlike a few other state programs, to date, environmental compliance in Rhode Island has not been privatized in any fashion, nor does DEM anticipate such a transition.

Similar to CERCLA, the Reuse Act makes the following types of parties liable under the statute: (1) current owners and operators, (2) owners or operators of the site at the time of disposal, (3) arrangers for transport or disposal, and (4) transporters for disposal.[6] It is important to note that according to some commentators, before the enactment of the Reuse Act, the state's authority to hold anyone other than the actual violating party liable for damage caused by environmental contamination was questionable. Under this new liability scheme, the scope of liability is much more clearly defined and is considerably more expansive.

Major Provisions of the Rhode Island Brownfields Program

Eligible Participants

The Rhode Island brownfields program provides opportunities primarily for two kinds of parties: (1) volunteers, defined as parties who do not currently own or operate the site and who come forward to conduct environmental assessment and/or remediation, and (2) bona fide prospective purchasers (BFPPs). To participate in the program, the site itself must have actual or suspected contamination and be targeted for redevelopment. However, no protections are afforded under the program for sites contaminated by virgin petroleum products or sites that pose an immediate and imminent threat to human health or the environment. These sites will continue to be handled by DEM on a case-by-case basis.

DEM envisions two categories of volunteers. The first is composed of those who are not responsible for the contamination, undertake a site assessment, and provide the results to DEM. The second category of volunteers is composed of those who are, again, not responsible for the contamination, and who undertake—and successfully complete—cleanup activity according to the terms of a remedial action plan that DEM has previously approved.

To obtain BFPP status, several criteria must be satisfied. Pursuant to section 23-19.14-3 of the Reuse Act, BFPP status is available only to those persons who (1) are not responsible for the contamination, (2) do not hold a 10 percent or greater interest in the ownership or operation of the site, (3) intend to buy the site and have documented their intent to buy in writing, and (4) have offered fair market value for the site in its contaminated condition. Although the Reuse Act does not specifically contemplate prospective tenants within the definition of BFPPs, DEM apparently views them as being eligible for the program. For example, DEM has already entered a settlement agreement with a tenant leasing contaminated property who agreed, with the approval of the property owner, to apply a percentage of the rent toward site investigation and remediation. Under the Reuse Act, parties potentially responsible for the contamination are not categorically excluded from entering a settlement agreement, but, according to DEM officials, it is unlikely that the state will grant such parties a covenant-not-to-sue.

Cleanup Standards

The steps that a settling party must take when investigating and remediating a site are contained in the *Rules and Regulations for the Investigation and Remedi-*

ation of Hazardous Materials Releases (the Site Remediation Regulations).[7] The level of remediation activity that a settling party must conduct will be driven by the present and foreseeable future use(s) of the property. For residential property, where the risk to human health and safety is potentially the greatest, the most stringent cleanup standards will apply. For industrial or commercial property, reduced cleanup levels in conjunction with engineering controls, such as a cap and/or a deed restriction, may suffice to protect human health, welfare, and the environment.[8]

To date, the relevant groundwater and soil cleanup standards have not been developed. As required by the Reuse Act, DEM must develop the cleanup standards by July 1996.[9] As currently envisioned by DEM, the soil standards will involve a three-tier process, each tier representing a different method for determining the cleanup objective for the site ranging from a quick and relatively easy chart of cleanup levels based upon the particular contaminant, the groundwater category, and the proposed use of the land, to protocols for a site-specific risk assessment. The remediating party will have the option of using any one of the three tiers.

Unlike the soil standards that DEM is drafting from inception, the groundwater standards will be modeled after the current groundwater policy published by DEM in May 1995.[10] This policy addresses urban locations where the groundwater is not a current or potential source for drinking water, and it sets remediation objectives based on the potential for volatile organic compounds found or suspected in these aquifers to volatize from the groundwater and migrate into indoor air in buildings located above the relevant groundwater. DEM currently intends to carry this type of risk-based analysis to nonurban locations and those where groundwater serves as a current or potential source of drinking water.

Liability Protections

The benefits to volunteers and BFPPs for participating in the brownfields program are conferred in the form of a settlement agreement with the state. These are the most significant protections afforded under the brownfields program, because they limit the state's ability to impose liability on the settling party for existing contamination, which would have otherwise attached as a result of property ownership or operation.[11] As part of the settlement agreement, the state of Rhode Island will issue a covenant-not-to-sue wherein the state agrees not to pursue enforcement actions against the settling party for all environmental conditions covered by the agreement, typically the contamination known at the time the parties enter the agreement. Essentially, the settlement agreement and covenant allow the settling party to assess and remediate property without automatically becoming a liable party.

To obtain a covenant-not-to-sue from the state, the settling party must agree to complete certain tasks. For example, the settling party must agree to investigate and assess the environmental condition of the property, if it has not already been fully investigated, and to undertake and complete all remedial action required by DEM.[12] Through each step in the investigation and remediation process, the settling party must also submit for DEM's approval all work plans and final reports. As a practical matter for BFPPs, as parties entering into the chain of title, a complete site assessment should be conducted before entering the agreement in any event, so that all essential environmental conditions are known and so these conditions can be covered by the covenant-not-to-sue.

Once the settling party has completed all work required under a DEM-approved work plan, the state will issue a letter of compliance for the property, pursuant to the Site Remediation Regulations.[13] In addition, the state will use its best efforts to obtain a comfort letter from the United States Environmental Protection Agency (USEPA) for the settling party, stating that USEPA will not pursue listing the property on the National Priorities List for matters covered in the settlement agreement.[14] To date, Rhode Island and USEPA have not entered a Memorandum of Agreement providing an official USEPA sign-off on the Rhode Island brownfields program.

As required in the covenant, copies of the settlement agreement, the state's letter of compliance, and USEPA's comfort letter, if any, must be recorded in the land evidence records for the city or town in which the property is located. In most circumstances, all duties and protections afforded by the settlement agreement may be transferred or assigned to successors in the chain of title, as long as they are not also responsible parties under the Reuse Act.[15]

The covenant-not-to-sue does not cover liability under five notable circumstances. They are as follows: (1) claims by the state based on the settling party's failure to comply with the terms of the settlement agreement, (2) any liability arising from contamination after the effective date of the settlement agreement, (3) any liability from contamination not addressed in the settlement agreement, (4) criminal liability, and (5) liability for violations of local, state, or federal law or regulations concerning the response actions to be performed.[16]

The settlement agreement also confers contribution protection. The protection covers "all matters addressed in the agreement," defined as all response actions taken (or to be taken) and response costs incurred (or to be incurred) by the state or any other person in connection with any hazardous substances, pollutants, or contaminants present or existing on the property as of the date of the agreement.[17] Essentially, the settling party is protected against suits by other potentially responsible parties for any additional remediation costs sought by the state.

The settlement agreement, incorporating the contribution-protection provision, is predicated upon a fourteen-day public comment period.[18] Tied into the comment period is an environmental equity and public-participation requirement, wherein abutting residents and other interested parties have an opportunity to provide input throughout the cleanup process.[19]

It is important to note that the contribution protection afforded by section 23-19.14-12 of the Reuse Act does not protect the settling party from third-party claims brought by adjacent property owners for damages to their property caused by contaminant migration, for example, or personal-injury damages. It bars claims only of other responsible parties or potentially responsible parties for contribution of costs that have been incurred by the state.

Liability Reopeners

If a volunteer exacerbates the environmental condition of a site while conducting remediation activity, or a BFPP either fails to enter a settlement agreement before purchasing the property or violates the agreement, the volunteer or BFPP will become a responsible party and liable to the state just like any other current owner or operator. Under section 23-19.14-6 of the Reuse Act, responsible parties have strict, joint, and several liability for the following:

(i) all removal or remedial actions necessary to rectify the effects of a release of hazardous material so that it does not cause a substantial danger to present or future public health or welfare or the environment; (ii) all costs of removal or remedial action incurred by the state including direct costs, indirect costs and the cost of overseeing response actions conducted by private parties; (iii) any other necessary costs of removal or remedial action incurred by any other person, and (iv) damages for injury to, destruction of, or loss of natural resources, including the reasonable costs of assessing such injury, destruction, or loss resulting from such a release.[20]

In addition, the Reuse Act includes a punitive damages section, wherein the state can collect from any responsible party additional money, above and beyond that which the state has already recovered from that responsible party, if that person—without good cause—failed to remediate pursuant to a final order from DEM.[21] The responsible party is potentially liable for up to three times the amount of any costs incurred by the state because of that party's failure to remediate.

A settling party, once an owner or operator for liability purposes, may also be saddled by the state with a "windfall" lien. Pursuant to section 23-19.14-14 of the Reuse Act, "[a]t any site where there are unrecovered response costs and/or additional remedial actions required, the state may place a lien upon the site for such unrecovered costs or outstanding actions."[22] The lien will remain until all response costs incurred by the state are paid and the performance of all outstanding remedial actions is completed. The amount of the lien may not exceed the increase in fair market value of the site that is attributable to the response action or redevelopment activities at the time of the subsequent sale or other disposition of the property. What appears clear is that a settling party in violation of an agreement and covenant-not-to-sue would be subject to the windfall lien for any additional remedial actions that are required. What is less clear is whether a settling party who is in compliance with the agreement, and thus presumably conducting all required response actions and reimbursing DEM for its oversight costs, is still subject to the windfall lien.

Fees and Private Cost Recovery

The only fees associated with obtaining the settlement agreement and covenant-not-to-sue are those that DEM has incurred in overseeing the settling party's response actions addressed in the agreement.[23] Currently, there are no administrative fees associated with entering the agreement itself.

The Reuse Act, along with the Hazardous Waste Management Act and the Water Pollution Act, create a right of cost recovery for private parties.[24] However, based on the dearth of reported cases, it appears that few claims by private parties have been brought under these statutes. Most private cost-recovery actions have been brought under CERCLA and common-law theories of negligence, nuisance, trespass, and strict liability for conducting ultrahazardous activities.

The relevant language of the Reuse Act creating a right of private cost recovery is found in section 23-19.14-6:

> Responsible parties are strictly, jointly and severally liable for the following: . . . (i) all removal or remedial actions necessary to rectify the effects of a release of hazardous material so that it does not cause a substantial danger to present or future public health or welfare or the environment [and] . . . (iii) any other necessary costs of removal or remedial action incurred by any other person.[25]

Whether this provision of the act provides an effective avenue of recourse in private cost-recovery actions remains to be seen, as, to date, no cases have been reported using this section of the act. It is also important to reiterate here that the Reuse Act excludes from its definitions of hazardous materials all virgin petroleum products.[26] Thus, it appears that the Reuse Act does not afford protections against releases of virgin petroleum products.

Attempts by private parties to recover for environmental contamination under common-law theories in Rhode Island have achieved mixed results. Success is based largely upon whether the contamination was caused by on-site activity of a prior owner or operator or whether the contamination migrated on-site from an off-site source. In general, Rhode Island courts have held that current owners do not have a cause of action against prior owners or operators of that property for contamination arising on-site under theories of nuisance, trespass, or strict liability for conducting an abnormally dangerous activity. Such causes of action require contamination migrating from an off-site source.

The shortcomings of common-law theories of recovery against prior owners for contamination arising from on-site activity is evident in the decisions of *Wilson Auto Enterprises, Inc. v. Mobil*,[27] and *Hydro-Manufacturing, Inc. v. Kayser-Roth Corp*.[28] In *Wilson*, the plaintiff, as the current landowner, brought common-law claims of negligence, trespass, nuisance, and strict liability for abnormally dangerous activity, *inter alia*, against a prior tenant of the property who allegedly caused a release of chemicals on the property. When the defendant filed a motion to dismiss for failure to state a claim, the United States District Court for the District of Rhode Island dismissed all the plaintiff's claims except trespass, as it related solely to the air stripper remediation equipment that the tenant left on the site.

As for the plaintiff's claim for negligence, the court found that prior owners and tenants owe no duty to successive owners to maintain their property in any certain condition, even if it would adversely affect the property into the future. Absent misrepresentation or contractual privity, the purchaser bears the risk of existing defects. In this respect, the court gave great weight to the fact that the plaintiff knew the tenant had operated an automobile service station on the property, but had not inspected the property for contamination before purchase.

The court responded to the plaintiff's trespass claim with a similar analysis. Upon leasing the land, the tenant's right of possession permitted it to occupy and use the land within the parameters of the lease. According to the court, none of the tenant's activities during the leasehold could constitute trespass against the plaintiff. Neither the plaintiff nor the court addressed a continuing-trespass theory.

The nuisance claim failed because, under Rhode Island nuisance law, the offending activity or condition must originate from outside the plaintiff's land.

As for strict liability for abnormally dangerous activity, the court undertook a two-part analysis. As a basic premise, the court cited precedent for rejecting the doctrine of *Rylands v. Fletcher*.[29] Then, under a basic tort analysis, the court concluded that, similar to negligence, to find liability, the defendant must first owe a duty to the plaintiff. And thus, relying on the doctrine of caveat emptor, the tenant could not be held responsible for its activities before plaintiff's ownership.

The Rhode Island Supreme Court reached similar conclusions in *Hydro-Manufacturing*. The plaintiff, again a current owner, brought an action against the successor-in-interest to a corporation that had previously owned the land. The court refused to extend the common-law theories of negligence, nuisance, or strict liability to attach liability to a prior owner.

Similar to the *Wilson* decision, the doctrine of caveat emptor was preeminent in the court's analysis. However, unlike *Wilson*, the court recognized an exception to the doctrine: when hidden defects, which are known or should have been known to the seller, create an unreasonable risk of harm to others. The duty arising from this exception applies only to immediate parties to the transaction, not successive remote buyers, as was the case here. It appeared that the court was not reluctant to reach its decision, given plaintiff's ability to pursue a cause of action under CERCLA based on the alleged facts.

In light of *Wilson* and *Hydro-Manufacturing*, the case law in Rhode Island sets difficult precedent for an owner of contaminated property to collect damages from a prior owner or operator. However, read together, these cases leave open the question of whether, in the narrow instance of a current purchaser who conducted assessment but nonetheless did not discover an imminent hazard known to the seller, the owner could prevail. Neither case prohibits such a result.

Private causes of action under common-law theories have been more successful in Rhode Island when the contamination migrated from off-site sources. In these cases, property owners have prevailed under theories of continuing trespass and nuisance. Although less clear, case law exists suggesting that a cause of action for strict liability for abnormally dangerous activity also exists for contamination caused by off-site sources.

The leading case is *Wood v. Picillo*,[30] wherein adjoining property owners brought suit to enjoin the operators and owners of a nearby landfill from continuing operations and for cleanup costs. The case was brought on theories of public and private nuisance.

The court rejected the defendant's argument that the plaintiff had not established any negligent conduct on the defendant's behalf. "Distinguished from negligence liability, liability in nuisance is predicated upon unreasonable injury rather than upon unreasonable conduct."[31] For the court, actionable nuisance arises when persons suffer harm or are threatened with injuries that they ought not bear. The plaintiffs prevailed on both the private and public nuisance counts.

The *Wood* decision also leaves the door open to strict liability claims. As dictum, footnote 7 states as follows:

> It could well be argued that one who utilizes his land for abnormally dangerous activities or for storage of abnormally dangerous substances may be strictly liable for resulting injuries, even in the absence of a finding of nuisance or negligence. *Rylands v. Fletcher*, (1868) L.R. 3 E&I App. 330. In the case at bar, the finding of both private and public nuisance makes it unnecessary to consider the doctrine of strict liability.[32]

The Rhode Island courts also recognize continuing trespass as a means for property owners to recover from abutting violators. In *Regan v. Cherry Corp.*,[33] the court denied a motion to dismiss by the defendant corporations, the Cherry Corporation and its predecessor-in-interest, for the plaintiff's claim of continuing trespass for dumping hazardous material on his property. The plaintiff had a viable claim even though the dumping occurred before plaintiff's interest in the land.

Lender-Liability Protection

Second to the threat of liability for existing contamination, the difficulties in obtaining financing for contaminated property have been a significant impediment to remediating and redeveloping brownfields. To help overcome this barrier and induce the availability of capital for these types of projects, the Reuse Act has carved out liability protections for secured lenders and receivers.

The liability carve-outs exempt certain parties from the liability provisions of section 23-19.14-6 discussed above. Lenders are exempt when they maintain an indicia of ownership only to protect a secured interest in land and do not act as operators. Also, according to DEM, Rhode Island is considering adopting safe-haven provisions modeled after the USEPA regulations on lender liability, which were overturned by the United States Court of Appeals for the District of Columbia in 1994 in *Kelley v. USEPA*.[34]

In addition, the Reuse Act exempts from liability custodial receivers who do not act as operators. Section 23-19.14-6 gives the Rhode Island Superior Court jurisdiction to supervise the receiver to determine the nature and extent of the receiver's compliance obligations under the provisions of the act. This exemption for receivers provides an important avenue for redeveloping a distinct class of brownfields sites—for example, those properties that, because of market-driven forces, require court intervention to be redeveloped.

Financial Incentives

Section 23-9.14-19 of the Reuse Act has created a fund to be used as a financing tool for BFPPs by drawing from the existing Tire Site Remediation Account of the Vehicle Tire Storage and Remediation Act.[35] By an amendment to the Vehicle Tire Storage and Remediation Act, the legislature has created a funding source for low-interest loans to assist in facilitating brownfields redevelopment.[36] The Rhode Island Port Authority and EDC are authorized to use up to 50 percent of the monies in the Tire Site Remediation Account as a revolving fund for low-interest loans to finance site assessments.[37] This money is currently available only for properties that the EDC has identified as a critical economic concern.

Conclusion

The Rhode Island brownfields program was designed to allow properties to be cleaned up using standards commensurate with the properties' intended use, to reduce the liability risks of redeveloping contaminated property, and to reduce the stigma of environmental contamination. Whether the program will meet these goals remains an open question, because the program is still within its first year and various changes are anticipated as DEM's practical application of the program progresses. Much of the success of the program will depend on the soil and groundwater standards that DEM is currently developing.

Rhode Island Brownfields Program

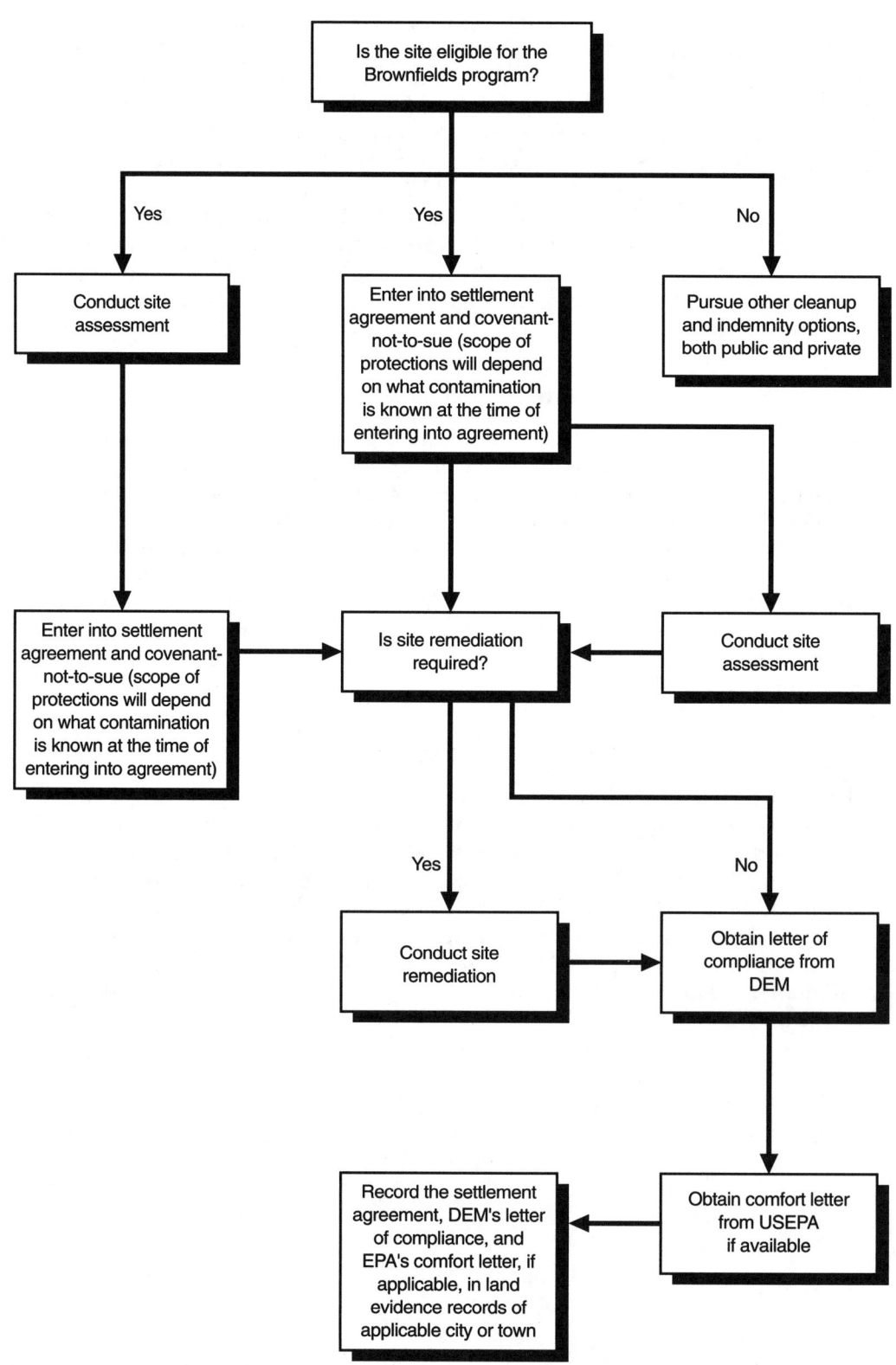

Notes

1. R.I. GEN. LAWS § 23-19.14-1 *et seq.* (1996).
2. *Id.* § 23-19.1-1 *et seq.*
3. *Id.* § 46-12-1 *et seq.*
4. *Id.* § 46-13.1-1 *et seq.*
5. *Id.* § 46-12.5-1 *et seq.*
6. *Id.* § 23-19.14-6.
7. R.I. DEP'T ENVTL. MANAGEMENT, NO. DSR-01-93, RULES AND REGULATIONS FOR THE INVESTIGATION AND REMEDIATION OF HAZARDOUS MATERIALS RELEASES (1996).
8. *Id.*
9. R.I. GEN. LAWS § 23-19.14.4 (1996).
10. R.I. Dep't Envtl. Management, Policy Memo 95-02, Guidelines on the Management and Restoration of Groundwater in Urban Areas (1991).
11. *See* R.I. Dep't Envtl. Management, Model Settlement Agreement, art. VII (1996) [hereinafter Model Settlement Agreement].
12. *See id.* art. III.
13. *See id.*
14. *Id.*
15. R.I. GEN. LAWS § 23-19.14-10 (1996).
16. *See* Model Settlement Agreement, *supra* note 11, art. VIII.
17. R.I. GEN. LAWS § 23-19.14-12 (1996); *see* Model Settlement Agreement, *supra* note 11, art. XIII.
18. R.I. GEN. LAWS § 23-19.14-11 (1996).
19. *Id.* § 23-19.14-5.
20. *Id.* § 23-19.14-6.
21. *Id.* § 23-10.14-16.
22. *Id.* § 23-19.14-14.
23. *See* Model Settlement Agreement, *supra* note 11, art. III.
24. R.I. GEN. LAWS §§ 23-19.14-6, 23-19.1-22, 46-12-21 (1996).
25. *Id.* § 23-19.14-6.
26. *Id.* § 23-19.14-3.
27. Wilson Auto Enters., Inc. v. Mobil, 778 F. Supp. 101 (D.R.I. 1991).
28. Hydro-Mfg., Inc. v. Kayser-Roth Corp., 640 A.2d 950 (R.I. 1994).
29. Rylands v. Fletcher, L.R. 3 E&I App. 330 (1868).
30. Wood v. Picillo, 443 A.2d 1244 (R.I. 1982).
31. *Wood*, 443 A.2d at 1247.
32. *Wood*, 443 A.2d at 1249.
33. Regan v. Cherry Corp., 706 F. Supp. 145 (D.R.I. 1989).
34. Kelley v. U.S. EPA, 15 F.3d 1100 (D.C. Cir. 1994), *reh'g denied*, 25 F.3d 1088 (D.C. Cir. 1994).
35. R.I. GEN. LAWS § 23-63-4.2 (Supp. 1996).
36. *Id.*
37. *Id.* § 23-19.14-9.

CHAPTER 43

Tennessee

GREER TIDWELL

Tennessee was brownfields when brownfields wasn't cool.[1] Before the word "brownfields" gained popularity for addressing reuse of contaminated properties, Tennessee joined in an effort with the United States Environmental Protection Agency (USEPA) Region IV and Boliden Intertrade A.G. to establish one of the nation's first prospective-purchaser agreements to allow continued commercial use of a facility and gain substantial, incremental improvements in the severely impacted environment at the Copper Hill site in southeast Tennessee. The Copper Hill site was widely known as one of the few man-made features visible from satellites orbiting the earth due to the enormous area of deforestation resulting from decades of mining operations. This precursor to the current brownfields movement was driven by recognition that with proper controls and incremental environmental improvements built into an agreement, there would be significant benefits to bringing active commercial operations back to the site. Although this first of Tennessee's prospective-purchaser agreements required intensive multiparty negotiations, the ultimate resolution and incremental environmental improvements resulting from this arrangement set the stage for Tennessee to enter the era of brownfields redevelopment as a significant component of its state environmental programs.

Currently, Tennessee has adopted a Voluntary Cleanup Oversight and Assistance Program[2] (sometimes referred to by the acronym VOAP or as the Voluntary Action Program), which became effective May 9, 1994, with minor amendments in 1995. This program is one of the tools available for addressing contaminated sites. As a practical matter, the VOAP should be viewed in the context of two other significant and encouraging changes in Tennessee law:

- Apportioned liability[3]
- Lender-liability protection[4]

Before addressing the legal framework for brownfields redevelopment and the Voluntary Action Program, the details of three particular Tennessee projects should be noted, as they will impact the development of Tennessee's regulatory approach for redevelopment of contaminated properties. The city of Knoxville has been selected as one of USEPA's national Brownfields Pilot Projects. Knoxville is addressing redevelopment of its central core business district.[5] Along with Knoxville's brownfields program, reindustrialization projects at the Department of Energy's Oak Ridge Reservation in Oak Ridge, Tennessee, and the Department of Defense's Volunteer Army Ammunition Plant in Chattanooga, Tennessee, are aggressively addressing how to develop commercially viable uses for property with identified areas of contamination. United States Congressman Zack Wamp has spurred the movement toward re-commercializing these two properties by establishing a series of summits addressing economic development in the "Tennessee Technology Corridor," stretching from Chattanooga through Oak Ridge, Knoxville, and up to the Tri-Cities of northeast Tennessee. Although there will certainly be many opportunities for cleanup and reuse of properties in a much more simplified context than these enormous facilities, the procedural developments and the experience that Tennessee regulators gain in supporting the safe redevelopment of these areas will likely be important in the long term.[6]

Summary of Major Provisions

The recently established Voluntary Action Program is not a separate program, but rather, it fits into the Tennessee Division of Superfund operations and is structured to incorporate the other two primary statutory tools for brownfields development—liability apportionment and lender-liability protection. The major provisions of Tennessee's Voluntary Action Program include the following:

- Parties seeking to use the Voluntary Action Program must be willing and able to conduct an investigation and cleanup.
- The Tennessee Department of Environment and Conservation (TDEC) will enter a consent order and agreement with the party, outlining steps to be taken for investigation, cleanup, monitoring, and maintenance. This consent order and agreement can be performed in phases, and the extent of—and time frame for—work is based on mutual agreement.
- Upon completion of the work described in the consent order and agreement, a letter is provided by TDEC, indicating that obligations have been completed and, if appropriate, no further action will be required of the participating party.
- To enter the Voluntary Action Program, a party must pay a $5,000 participation fee. Also, the consent order and agreement will address re-

imbursement of monies TDEC may have already spent in prior response activities at the site.

Eligibility for the Voluntary Action Program

For admission into Tennessee's Voluntary Action Program, a potential participant must submit a written request to the Division of Superfund. Currently, the key contact person is Mr. Andy Shivas (at 615/532-0912). Three major factors for determining eligibility in the program are (1) the party's willingness and ability to conduct the investigation and cleanup, (2) the complexity of the site, and (3) the availability of TDEC staff to provide oversight. Also, upon acceptance into the program, the party must submit the initial $5,000 participation fee and obligate itself to the terms of the consent order and agreement.

The Voluntary Action Program applies to inactive hazardous-substance sites, which by definition do not include sites where petroleum is the only substance released.[7] Sites on USEPA's National Priorities List, or those proposed for listing, are also excluded from the program.

Memorandum of Agreement

Tennessee is currently working very closely with USEPA on development of a Memorandum of Agreement (MOA) that recognizes Tennessee's Voluntary Action Program. Federal recognition may provide some certainty and security to a liable party.

Liability Protection

Perhaps the most important legal tool in addressing brownfields issues in Tennessee is the statutory scheme of apportioned liability, pursuant to Tennessee's Hazardous Waste Management Act of 1983 (the state Superfund statute).[8] Tennessee's law, modeled on the federal law, defines the categories of "liable parties" as

1. the owner or operator of an inactive hazardous-substance site;
2. any person who, at the time of disposal, was the owner or operator of an inactive hazardous-substance site;
3. any generator of hazardous substance who, at the time of disposal, caused such substance to be disposed of at an inactive hazardous-substance site; or
4. any transporter of hazardous substance, which is disposed of at an inactive hazardous-substance site, who, at the time of disposal, selected the site of disposal of such substance.[9]

Tennessee deviates significantly from the federal model by providing apportioned liability. A party's apportioned or allocated share is assessed on the basis of "equitable factors," including, but not limited to, the following:

- Any monetary or other benefit accruing to each liable party from the disposal of hazardous substances upon the sites
- The culpability of each liable party in placing hazardous substances upon the site
- Efforts of each liable party to restore the land, water, air, and all other aspects of the site to their natural conditions.
- The party's portion of the total volume of hazardous substances at the hazardous substance site
- The monetary benefit accruing to an owner as a result of the cleanup of the site if, at the time of acquisition of the site, such owner knew or should have known that hazardous substances were previously disposed of at the site
- The monetary benefit accruing to an owner as a result of the cleanup of the site if such owner was the owner at the time hazardous substances were disposed of on the property and knew or should have known of such disposal[10]

The Voluntary Action Program incorporates the apportioned-liability concept and establishes the authority to set allocation of liability among the known liable parties. Any expenditures required by the law and made by a liable party will be credited toward that party's share of the cost. Tennessee has established a hazardous-waste remedial action fund, or state Superfund, based primarily on annual hazardous-waste generator fees. The state may "expend monies from the remedial action fund to pay that portion of the investigation, cleanup, monitoring, maintenance and oversight of an inactive hazardous substance site to the extent such expenditures are not allocated under the Consent Order to the potentially liable party."[11] Pursuant to statutory amendments in 1995, this discretionary allocation authority was a significant change in the Voluntary Action Program. The details of working with such an allocation scheme are still being developed. Nevertheless, the specific limitation on liability through an allocated-liability scheme, as opposed to joint and several liability, is significant in the context of property redevelopment, to the extent that it eliminates one arm of the daunting redevelopment question, "How much will the cleanup cost and what will be my share?"

Lender-Liability Protection

In 1995, Tennessee added a new part to its state Superfund statute, entitled "Indicia of Ownership."[12] In general, this statute provides that persons who are maintaining indicia of ownership primarily to protect their security interests and who do not participate in the management of the site or are not owners/operators of the site will not be deemed responsible parties and will not be liable for cleanup costs or damages resulting from discharge from a site.[13] Of particular interest in the protections afforded to lenders is the extensive definition of "active participation in management," which "does not include the mere capacity, or the ability to influence, or the unexercised right to control a site, vessel or facility operations."[14] This provision should significantly reduce the uncertainties perceived by lending institutions in transactions generally and specifically involving contaminated properties.

Benefits for Participants in the Program

As a practical matter, certain benefits go along with participation in Tennessee's Voluntary Action Program, including the ability to address the contamination of the site with state regulatory involvement but without the property being listed on the state's "Superfund Site List." Participation also may project an image in the community that a participant is living up to its responsibilities as a good corporate citizen. Furthermore, by using the consent order and agreement tool, a party avoids being subjected to a direct commissioner's order. This allows for mutually agreed-upon schedules, with input from the participating party throughout the investigation and cleanup process, as well as opportunities through mutual agreements to amend the consent order and agreement. A participant also forgoes the burdens of a deed notice identifying the property as an inactive hazardous-substance site and a potential lien against the property for monies expended by the Department.[15] Most importantly, at the conclusion of participating in the Voluntary Action Program, after the terms of the consent order and agreement are fulfilled, the Commissioner issues a letter to the participating party indicating that such party's obligations under the consent order have been completed and that no further action will be required of the participating party.[16]

Cleanup Standards

Tennessee's Voluntary Action Program uses the same criteria for selecting containment and cleanup actions, including monitoring and maintenance, as are used for non-VOAP sites.[17] In selecting containment and cleanup actions, including monitoring and maintenance, reasonable alternatives are evaluated, and actions necessary to protect public health, safety, and the environment are to be selected. The goal of any such action is the cleanup and containment of the site through the elimination of the threat to public health, safety, and the environment posed by the hazardous substance. In choosing necessary actions at each site, the following factors must be considered:

- The technological feasibility of each alternative
- The cost-effectiveness of each alternative
- The nature of the danger to the public health, safety, and the environment posed by the hazardous substance at the site
- The extent to which each alternative would achieve the goal of the statute[18]

Regulations have been developed regarding site assessment and cleanup criteria, and generally provide for consistency with the National Contingency Plan.[19] These regulations allow for remedy selection that is similar to the federal "applicable or relevant and appropriate requirements" approach as well as a risk-based criteria selection approach.[20] Also, the TDEC is drafting more detailed guidance about participation in the Voluntary Action Program, which was to be finalized during the summer of 1996.

Participating in the Voluntary Action Program

TDEC considers the Voluntary Action Program as a tool for the official administration of cleanup activities, with the goal of encouraging private enterprise to redevelop contaminated properties. The discretion to allow a particular site or proposal into this program rests with the Department. Anyone seeking to use this tool should contact the Division of Superfund within TDEC and initiate discussions regarding what is already known about a site and the plans for addressing contamination.

When the private parties and TDEC staff concur on using the Voluntary Action Program, an agreement in the form of a consent order and agreement will be entered. Upon payment of the $5,000 initial participation fee, as well as reimbursement of prior Department expenditures regarding the site, work on addressing the contamination will go forward according to a schedule established in the consent order and agreement.

Admission of Liability

If the site being addressed under the Voluntary Action Program involves an apportionment of liability to allocate a certain percentage of the cost to the party participating in the VOAP, such an allocation "is not admissible, in any suit, hearing, or other proceeding that involves a person not a party to such Consent Order."[21] However, information submitted by a participant, including records, reports, and test results, are not considered confidential and are open to public review.

Cost Recovery

The Voluntary Action Program does not directly establish any new cause of action for a participating party to seek cost reimbursement from other potentially responsible parties. Nevertheless, the VOAP does not specifically prohibit such cost-recovery actions. The Voluntary Action Program does recognize TDEC's authority to seek recovery from other liable parties in the full amount of their respective allocated share of liability.[22]

If a participant in the Voluntary Action Program does not fulfill its requirements as established in the consent order and agreement, TDEC may revert to its ordinary tools for addressing such sites outside this program.[23] These tools include the authority to request a liable or potentially liable party to investigate possible inactive hazardous-substance sites; the authority to issue an order requiring an investigation or containment, cleanup, monitoring, and maintenance.[24] If such an order is not followed or a liable party cannot reasonably be identified, TDEC may conduct such actions itself, with the capacity to seek cost recovery for expenditures, plus punitive damages, in an amount equal to 150 percent of the amount of any costs incurred.[25]

Conclusion

In Tennessee, the Department of Environment and Conservation is leading the way to provide appropriate legal tools and programs to foster safe reuse of brownfields. Using the tools of fair-share liability, lender protection, and voluntary cleanup actions, the Department can safely address hazardous-substance sites while not unnecessarily discouraging beneficial reuse of the sites. The Department is committed to staying on the cutting edge of ideas, legal tools, and programs to do what is good for the environment and good for the communities.

Tennessee Voluntary Action Program

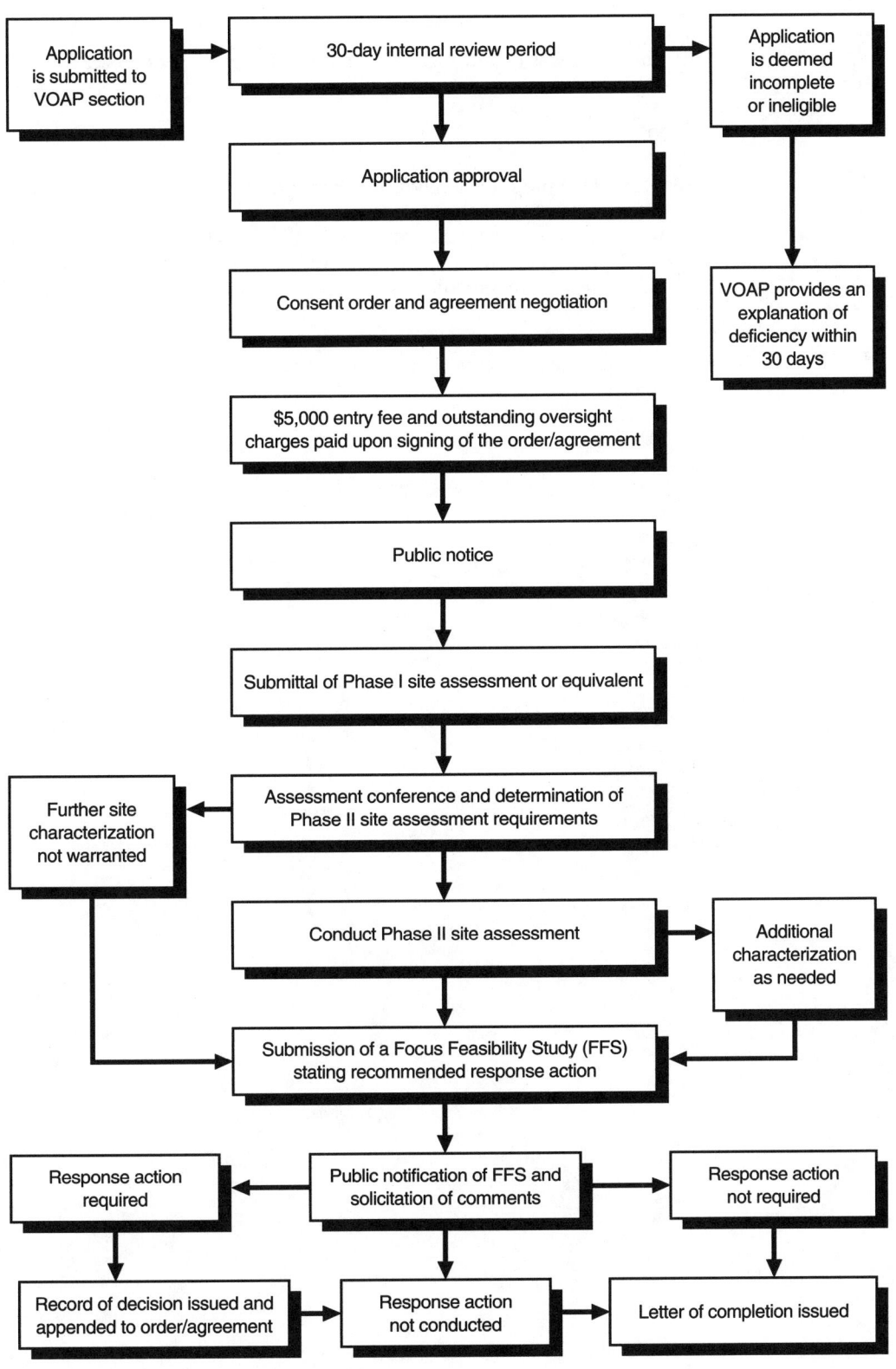

Notes

1. Sung to the tune of, "I was country when country wasn't cool," by Barbara Mandrell.
2. TENN. CODE ANN. § 68-212-224 (1996).
3. *Id.* § 68-212-207.
4. *Id.* § 68-212-204 *et seq.*
5. Details of the Knoxville program can be addressed with Mr. Charles Barker, Economic Development Manager, at 423/215-4545.
6. For further information regarding the Oak Ridge site, contact the state's office of the Department of Energy Oversight, Division Director, Mr. Earl Leming, at 423/481-0995; for the Department of Defense's Volunteer Site in Chattanooga, contact Mr. Sid Saunders, of ICI Americas, at 423/855-7256.
7. TENN. CODE ANN. § 68-212-202(2) (1996) (defining "hazardous substance" as defined in the Comprehensive Environmental Response, Compensation, and Liability Act with the "petroleum exclusions").
8. *Id.* § 68-212-201 *et seq.*
9. *Id.* § 68-212-202(4).
10. *Id.* § 68-212-207(b)(1).
11. *Id.* § 68-212-224(d)(2).
12. *Id.* § 68-212-401 *et seq.*
13. *Id.* § 68-212-402.
14. *Id.* § 68-212-401(1).
15. *Id.* § 68-212-209.
16. *Id.* § 68-212-224(g).
17. *Id.* § 68-212-224(e).
18. *Id.* § 68-212-206(d).
19. *Id.* § 68-212-206(d).
20. TENN. COMP. R. & REGS. tit. 1200, ch. 13 (1996).
21. TENN. CODE ANN. § 68-212-224(h)(2).
22. *Id.* § 68-212-224(d)(2).
23. *Id.* § 68-212-224(g).
24. *Id.* § 68-212-206(a).
25. *Id.* § 68-212-2007(c).

CHAPTER 44

Texas

JOHN SLAVICH

The Texas legislature amended the Texas Health and Safety Code to create a Voluntary Cleanup Program (VCP, or the Program), effective September 1, 1995. The stated purpose of the VCP, codified at sections 361.601 through 361.613 of the Texas Health and Safety Code[1] (the Statute), is to provide incentives to remediate property by removing the environmental liability of future lenders and future landowners.[2] The Program is not intended to replace other remedial programs and is restricted to voluntary actions. The Program is administered by the Voluntary Cleanup Section of the Pollution Cleanup Division of the Texas Natural Resource Conservation Commission (TNRCC). TNRCC issued proposed rules for the Program (Proposed Rules) for public comment in November 1995.[3]

Summary of Major Provisions

A person electing to participate in the Program must enter a voluntary cleanup agreement with TNRCC and pay all TNRCC's costs of oversight of the cleanup. The participant must submit work plans and reports that reflect remediation goals and proposed methods of remediation acceptable to TNRCC under applicable TNRCC technical standards. Following TNRCC's determination of a successful cleanup, TNRCC will issue a certificate of completion. The certificate acknowledges the release from cleanup liability to the state of Texas for nonresponsible persons specified in the Statute. The certificate also indicates the permitted future land use of the site.

Eligibility for the Program

The Statute provides that any site in the state of Texas is eligible for participation in the VCP, except the portion of a site that is subject to a TNRCC permit or order.[4] Sites at which a notice of violation has been issued by the state are eligible until an order is issued. Additionally, the Statute excludes from the definition of a "response action" under the VCP the cleanup of substances regulated by, or that result from activities under the jurisdiction of, the Railroad Commission of Texas under specified sections of the Natural Resources Code and the Water Code. TNRCC takes the position that sites may be rejected from participating in the VCP if they are subject to any other administrative, state, or federal enforcement action, or when a federal grant requires an enforcement action be taken.

To participate in the VCP, a person must enter a voluntary cleanup agreement with TNRCC[5] and pay all costs of TNRCC oversight of the voluntary cleanup. The agreement designates rules and regulations directly applicable to the site and sets forth the submittals required, along with a time schedule for compliance. The form of agreement specifies that it is not to be construed as an admission of liability under the Texas Solid Waste Disposal Act or any other law, or a waiver of any defense to such liability.[6] The form of agreement also provides that TNRCC and the applicant each reserve all rights and defenses unless specifically waived. TNRCC specifically reserves the right to bring an action against the applicant for any violations of applicable laws, except for the specific violations or releases to be remediated under the agreement. TNRCC will not bring an enforcement action against the applicant in connection with the specific violations or releases being remediated, unless the applicant withdraws from the agreement before completion of the cleanup.

TNRCC or the participant may terminate the agreement by giving fifteen days' advance written notice to the other. TNRCC may recover costs it incurred or obligated before it received the agreement termination notice.[7]

Memorandum of Agreement

TNRCC and the United States Environmental Protection Agency, Region VI, have entered a Memorandum of Agreement (MOA). Pursuant to the MOA, under certain conditions, participants who complete a cleanup of contaminated property under the VCP will also not be subject to federal enforcement action.[8]

Liability Protection

A person who is not otherwise a responsible party under pertinent enforcement provisions of subchapter I of the Health and Safety Code at the time the person applies to perform a voluntary cleanup will be released from liability to the state of Texas for cleanup of areas covered by the certificate of completion.[9] The release becomes effective at the time the state issues the certificate.[10] The release from liability is not effective if a certificate of completion is acquired by fraud, misrepresentation, or knowing failure to disclose material information.[11]

The Statute provides that following the issuance of the certificate for a site, a person who becomes the owner of the property on which the site is located, or a lender who makes a loan secured by that property after the date of the issuance of the certificate, is released from all liability for cleanup of contamination released before the date of the certificate for the areas covered by the certificate.[12] Such liability protection does not apply if the owner or lender was originally included as a responsible party.[13] Additionally, the release does not apply to a person who changes land use from that specified in the certificate, if the new use may result in increased risks to human health or the environment.[14]

Financial Incentives

The Statute does not provide for financial incentives to facilitate redevelopment of brownfields. Other provisions of Texas law do, however, provide for tax abatement for certain environmental improvements.[15] It is anticipated that the VCP will facilitate private-sector financing by lenders that would not otherwise consider financing due to the presence of contaminants on the property that would serve as the collateral for such loan.

Cleanup Standards

The Statute specifies that the voluntary cleanup agreement is to "state the technical standards to be applied in evaluating the work plans and reports [submitted to TNRCC under the agreement] with reference to the proposed future land use to be achieved."[16] The preamble to the Proposed Rules indicates that section 333.7 (voluntary cleanup work plans and reports) and section 333.8 (response action standards) of the Proposed Rules will allow added flexibility in the use of site-specific information in directing investigations and response actions. TNRCC currently uses two primary technical standards that are applied under different programs: (1) industrial solid waste and municipal hazardous waste risk-reduction standards,[17] and (2) petroleum storage tank target concentration criteria.[18] These programs allow cleanup under either numeric standards for contaminants or risk-based standards, and apply different technical standards for residential and nonresidential land use. The preamble states that the goal of TNRCC is to have one set of technical requirements for all sites that currently operate under these different standards. TNRCC plans to operate under the current standard that would be appropriate for the particular site being considered under the VCP, with such standard modified as contemplated in the Proposed Rules, as discussed below.

Section 333.7 of the Proposed Rules states that the VCP work plans and reports must include an investigation of the full nature and extent of contamination in all media at a site, unless site-specific conditions allow otherwise. Applicants may seek TNRCC's authorization to perform a focused site investigation, rather than a full investigation. TNRCC's consideration of such a request will require the applicant to submit a conceptual exposure assessment model (CEAM) for the site. The CEAM will include a determination of the current and reasonably anticipated use of the property and its resources, and an evaluation of human health and environmental exposure to the contaminated media of concern, including exposure pathways. TNRCC is to develop a guidance document that addresses the CEAM process and how the CEAM may be used with the currently applicable technical standards.[19]

Institutional controls (such as property-use restrictions) and engineering controls (such as site caps) are currently used under the Texas risk-reduction standards and would therefore be applicable to the VCP.

The Program does not require any special certification for the environmental professionals used by a Program participant.

Participation in the Program

Application

A completed Program application must be submitted for a site to be included in the Program. The application calls for, among other things, information about the site, the applicant, and the current property owner, if different from the applicant. Persons who have, or may in the future have, a property interest other than property owner (such as a mortgage holder, prospective purchaser, or lessee) may be applicants under the Program. The application includes an "intent to participate" statement in which the applicant requests TNRCC oversight for investigation and cleanup activities and indicates the applicant's interest to negotiate in good faith a voluntary cleanup agreement with TNRCC. The applicant must also acknowledge that it has the financial capability to perform a voluntary cleanup.

Liability Protection

The Statute provides for a release of a person performing voluntary cleanup, on issuance of a certificate of completion, from all liability to the state for cleanup of areas of the site covered by the certificate, except for releases and consequences that the person causes, provided that the applicant is not a responsible party under sections 361.271[20] or 361.275(g)[21] of the Texas Solid Waste Disposal Act.[22] The Statute provides more specifically that if a certificate of completion for a site is issued by TNRCC, an owner who acquires the property on which the site is located, or a lender who makes a loan secured by that property, after the date of issuance of the certificate is released from all liability for cleanup of contamination released before the date of the certificate for the areas covered by the certificate, unless the owner or lender was originally included as a responsible party under sections 361.271[23] or 361.275(g).[24] A release of liability does not apply to a person who changes land use from the use specified in the certificate of completion, if the new use may result in increased risks to human health or the environment.[25] There is no instrument that evidences the release of liability other than the certificate of completion. The release from liability is not effective if a certificate of completion is acquired by fraud, misrepresentation, or knowing failure to disclose material information.[26]

Fees

A fee of $1,000 must accompany the application to the Program. Initial costs incurred by TNRCC for review and oversight are applied against the application fee. The applicant is then invoiced for all additional review and oversight costs. TNRCC costs are to include all direct, indirect, overhead, salaries, equipment, utilities, legal, management, and support costs fairly attributable to the VCP.

VCP Approval Process

The technical standards to be applied by TNRCC in evaluating the work plans and reports are established in the process of negotiating the voluntary cleanup agreement and are detailed in the agreement.[27]

TNRCC is to review and evaluate the work plans and reports submitted by a participant for accuracy, quality, and completeness. TNRCC may approve a voluntary cleanup work plan or report or, if a work plan or report is not approved, notify the participant concerning additional information or actions needed to obtain approval.[28] After considering future land use, TNRCC may approve work plans and reports submitted that do not require removal or remedy of all discharges, releases, and threatened releases at a site, if the partial response actions for the property (1) will be completed in a manner that protects human health and the environment, (2) will not cause, contribute to, or exacerbate discharges, releases, or threatened releases that are not required to be removed or remedied under the work plan, and (3) will not interfere with, or substantially increase, the cost of response actions to address the remaining discharges, releases, or threatened releases.[29]

Unless the reports regarding cleanup work performed demonstrate, to the satisfaction of TNRCC, that no further action is required to protect human health and the environment, TNRCC will not issue a certificate of completion.[30] If TNRCC determines that an approved cleanup has not been successfully completed, TNRCC must notify the person who undertakes the cleanup and the site owner.[31]

Cost Recovery

The form of agreement to be entered between TNRCC and an applicant specifies that the applicant expressly reserves all rights against third parties, other than TNRCC. The VCP does not, however, provide for cost recovery. The general statutory requirements for cost recovery under Texas law are set forth in section 361.344 of the Texas Health & Safety Code. It is not yet clear whether a cleanup performed under a VCP agreement, rather than under a TNRCC order, would qualify for cost recovery under section 361.344.

Audit Privilege

Audit-privilege issues are not addressed in the Statute, but are covered in the Environmental, Health and Safety Audit Privilege Act (Privilege Act).[32] The Privilege Act was adopted in the same legislative session as the VCP.

An audit report, as defined in the Privilege Act,[33] includes not only the actual audit report (along with exhibits and appendices), information gained in the audit, findings, conclusions, and recommendations, but also all analyses of the foregoing information. Also included within the definition of an audit report are implementation plans or tracking systems to correct past noncompliance, improve current compliance, or prevent future noncompliance.[34] An audit must be completed within a reasonable time, which must not exceed six months, without an extension approval by the appropriate governmental entity.[35]

Subject to various exceptions provided in the Privilege Act, an audit report is considered to be privileged and is not admissible as evidence or subject to discovery in civil, criminal, or administrative proceedings.[36] A person who participates in the preparation of an audit may testify about any physical events of violation actually observed, but may not be compelled to testify about, or produce documents related to, the privileged part of the audit or the audit report.[37]

Other persons involved in the audit, or to whom audit results are disclosed in a manner that does not waive the privilege under the Privilege Act, may not be compelled to testify or produce a document related to the audit if that testimony or document would disclose privileged material.[38] Additionally, an employee of a state agency may not request, review, or otherwise use an audit report during an inspection of a regulated facility or operation.[39]

The privilege may be expressly waived by the owner or operator who prepared the audit.[40] The Privilege Act provides for specified disclosures that do not waive the privilege. Permissible disclosure includes specified disclosures made in the corrective action process addressing audit results, as well as disclosures made under the terms of a confidentiality agreement. Additionally, disclosure may be made under a claim of confidentiality to a governmental official or agency,[41] in which case the information is not subject to disclosure under the Texas Public Information Act.[42]

A court or administrative hearings official may require disclosure of an audit report in a civil, criminal, or administrative proceeding if (1) it is determined, after an *in camera* review, that the privilege is asserted for a fraudulent purpose, (2) if the materials sought are, in fact, nonprivileged, and (3) if the audit report shows noncompliance without corresponding efforts to initiate and pursue achievement of compliance.[43] The Privilege Act also provides for a process whereby the state may obtain an audit report subject to privilege if there is reasonable cause to believe a criminal offense has been committed under an environmental or health and safety (EHS) law.[44]

The privilege does not apply to information required to be developed or reported under applicable federal or state laws. Also excluded is information obtained by observation, sampling, or monitoring by a regulatory agency, or information obtained from a source not involved in the preparation of the audit report.

The Privilege Act provides that a person who makes voluntary disclosure of a violation of an EHS law is, under specific circumstances, immune from an administrative, civil, or criminal penalty for the violation disclosed. To qualify for the immunity, the disclosure must arise out of a voluntary audit, independent of any governmental actions, and must be made promptly after knowledge of the violation. Additionally, the noncompliance must be addressed and corrected, and the disclosing party must cooperate with the appropriate agency in investigating the issues identified in the disclosure. Finally, the violation cannot have resulted in injury to persons at the site or created substantial off-site harm. A report to a regulatory agency required solely by a specific condition of an enforcement order or decree does not qualify as voluntary disclosure.

To receive immunity under the Privilege Act, a facility conducting an audit must have given notice to the appropriate regulatory agency of the fact that it is planning to commence the audit, along with information concerning the facility affected, the anticipated commencement of the audit, and the general scope of the audit.

The immunity does not apply in the case of specified intentional, knowing, or reckless conduct. In these instances, the Privilege Act provides for certain factors that can be considered in mitigation of a penalty. The immunity also does not apply if a court or administrative law judge finds that the person claiming the immunity has a history of serious violations and has not attempted to bring the facility into compliance, thereby constituting a pattern of disregarding EHS laws. TNRCC takes the position that even though a company may be immune from a penalty, the company is not immune from TNRCC enforcement that imposes requirements and schedules generally referred to by TNRCC as "technical recommendations."

Developments

The Texas VCP is in its early stages, effective only since September 1, 1995. Final rules under the Statute were adopted in April of 1996. TNRCC has provided guidance with respect to technical issues relating to VCP cleanups in "Guidance for Initiating and Reporting Response Actions Conducted Under TNRCC's Voluntary Cleanup Program." The Guidance also contains a copy of the Statute and final Rules, along with the preamble from the Texas Register with respect to adoption of the final Rules.

A TNRCC home page for the VCP may be accessed through the Internet at http://www.tnrcc.state.tx.us/waste/pcd/vcp/index.html. The web site has various materials available for downloading. Those materials include a spreadsheet that provides details with respect to all properties that have been entered into the VCP.

To the extent that the VCP provides sufficient comfort for sellers, buyers, and lenders, it will serve as a tool to assist in the potential redevelopment of environmentally impacted properties in Texas and the return of those properties to productive use.

Texas Voluntary Cleanup Program (VCP)

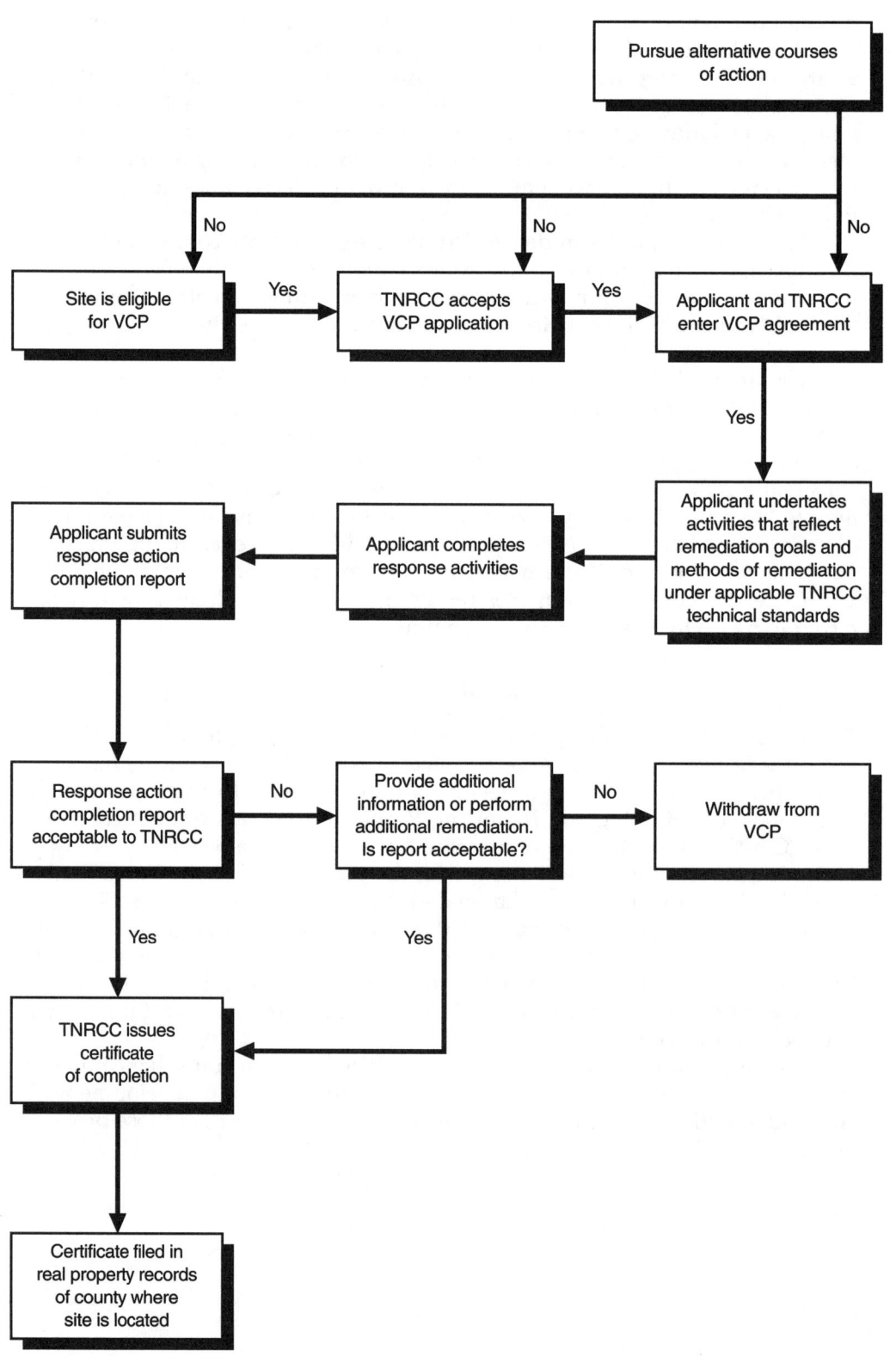

Notes

1. TEX. HEALTH & SAFETY CODE ANN. §§ 361.601–361.613 (West 1992 & Supp. 1996).
2. *Id.* § 361.602.
3. 20 Tex. Reg. 9250–58 (1995).
4. TEX. HEALTH & SAFETY CODE ANN. § 361.603 (West Supp. 1997).
5. *Id.* § 361.606.
6. *Id.* § 361.001 *et seq.*
7. *Id.* § 361.607(a).
8. *EPA Agrees to Relieve Landowners of Liability under Brownfields Program*, DAILY ENV'T REP., May 2, 1996, at A3.
9. TEX. HEALTH & SAFETY CODE ANN. § 361.610(a) (West Supp. 1997).
10. *Id.*
11. *Id.* § 361.610(b).
12. *Id.* § 361.610(c).
13. *Id.*
14. *Id.*
15. *See, e.g.,* TEX. TAX CODE ANN. § 11.31 (West Supp. 1996).
16. TEX. HEALTH & SAFETY CODE ANN. § 361.606(c)(4) (West Supp. 1997).
17. TEX. ADMIN. CODE tit. 30, §§ 335.551–68 (1994).
18. 20 Tex. Reg. 8808–12 (1995) (codified at TEX. ADMIN. CODE tit. 30, §§ 334.201–08 (1996)).
19. 20 Tex. Reg. 9255–56 (1995).
20. TEX. HEALTH & SAFETY CODE ANN. § 361.271 (West Supp. 1996).
21. TEX. HEALTH & SAFETY CODE ANN. § 361.275(g) (West 1992).
22. *Id.* § 361.610(a).
23. *Id.* § 361.271.
24. *Id.* § 361.275(g).
25. *Id.* § 361.610(c).
26. *Id.* § 361.610(b).
27. *Id.* § 361.606(c)(4).
28. *Id.* § 361.608(b).
29. *Id.* § 361.609(d).
30. 20 Tex. Reg. 9258 (1995).
31. TEX. HEALTH & SAFETY CODE ANN. § 361.609(d) (West Supp. 1997).
32. Texas Environmental, Health and Safety Audit Privilege Act, 74th Leg. R.S., ch. 219, § 1 *et seq.*, 1995 Tex. Gen. Laws 1963–69 [hereinafter Audit Privilege Act].
33. Audit Privilege Act, *supra* note 32, § 4.
34. *Id.* § 4(b).
35. *Id.* § 4(e).
36. *Id.* §§ 5(a)–(b).
37. *Id.* § 5(d).
38. *Id.* § 5(c).
39. *Id.* § 5(e).
40. *Id.* § 6(a).
41. *Id.* § 6(b)(3).
42. TEX. GOV'T CODE ANN. § 552.001 *et seq.* (West 1994 & Supp. 1996).
43. Audit Privilege Act, *supra* note 32, § 7(a).
44. *Id.* § 9.

CHAPTER 45

Utah

H. MICHAEL KELLER

Utah law provides encouragement for brownfields redevelopment. The Utah Hazardous Substances Mitigation Act (HSMA)[1] contains a provision authorizing the Utah Department of Environmental Quality (DEQ) to enter voluntary agreements with responsible parties to investigate or clean up sites that contain hazardous materials.[2] HSMA was enacted in 1989 to accelerate remedial investigations and feasibility studies at Utah sites listed or eligible for listing on the National Priority List (NPL) under the federal Comprehensive Environmental Response, Compensation, and Liability Act (CERCLA)[3] and to provide authority to DEQ to address hazardous-material sites and releases not adequately addressed by Utah's existing water quality, underground storage tank, and hazardous-waste management programs. HSMA also provides DEQ with emergency response authority and funds for responding to releases of hazardous materials presenting a direct and immediate threat to public health or the environment.

HSMA's provision for voluntary agreements (Voluntary Agreement Program),[4] however, lacks the breadth and detail of "brownfields" legislation recently adopted in other states. The limitations of the Utah program generated increasing interest in the adoption of more comprehensive legislation. In March of 1997, the Utah legislature passed and submitted to the Governor new legislation to establish the Utah "Voluntary Cleanup Program," providing more comprehensive procedures and stronger incentives for conducting voluntary cleanups in Utah, including procedures for certifying cleanups conducted prior to its enactment. Nevertheless, the current voluntary agreement provisions of HSMA, which are discussed below, remain in effect.

Summary of Major Provisions

The major provisions of the Voluntary Agreement Program under HSMA include the following:

- Authorizing the DEQ Executive Director to enter a voluntary agreement with a responsible party for the responsible party to conduct an

investigation or a cleanup action on a site that contains hazardous materials
- Requiring notification of all known potentially responsible parties and providing them an opportunity to comment on the proposed agreement
- Authorizing the DEQ Executive Director to receive funds from any responsible party that signs a voluntary agreement allowing the Executive Director to review and oversee the proposed investigation or cleanup action
- Authorizing the DEQ Executive Director to take action and seek penalties to enforce a voluntary agreement
- Precluding the DEQ Executive Director from using the cost-recovery provisions of HSMA to recover costs received or expended pursuant to a voluntary agreement from any person not a party to that agreement
- Authorizing any party under a voluntary agreement who incurs costs in excess of its liability to seek contribution from any other party who is or may be liable under HSMA for the excess costs
- Requiring the court in such contribution actions to allocate costs in proportion to each responsible party's respective contribution to the release
- Protecting lenders, trustees, and fiduciaries from such contribution actions

Eligibility for the Utah Voluntary Agreement Program

The voluntary agreement provision of HSMA authorizes the DEQ Executive Director to enter a voluntary agreement with a "responsible party," providing for that party to conduct an investigation or a cleanup action on a site containing hazardous materials.[6] Thus, under the express language of the statute, a party must be a "responsible party" to participate in the Voluntary Agreement Program. The term "responsible party" is defined elsewhere in HSMA.[7] It includes the same cast of owners, operators, arrangers, and transporters who constitute potentially responsible parties under section 107 of CERCLA.[8] Security-interest holders and fiduciaries, however, are excluded from the definition of "responsible party."[9]

It is unclear whether DEQ may enter a voluntary agreement with parties, such as lenders or fiduciaries, who do not constitute "responsible parties" under HSMA. DEQ has traditionally taken the position that notwithstanding the definitional limitations in the voluntary cleanup provision of HSMA, DEQ, under its various statutory authorities, may enter voluntary agreements with a party to undertake action to achieve environmental compliance.

The Voluntary Agreement Program applies to investigations or cleanup actions on sites that contain "hazardous materials." This term is defined to include hazardous waste as defined under the Utah Hazardous Waste Management Regulations, polychlorinated biphenyls, dioxin, asbestos, or "a substance regulated under 42 U.S.C. Section 6991(2)."[10] The cited federal statute[11] defines a "regulated substance" to include petroleum and any hazardous substance defined in section 101(14) of CERCLA[12] (except hazardous waste). The reference to substances "regulated under" the federal underground storage tank (UST) law raises the question whether hazardous materials under HSMA may be limited to only those CERCLA hazardous substances and petroleum released from a UST. DEQ, however, has taken the position that the definition of "hazardous materials" in HSMA broadly includes CERCLA hazardous substances and petroleum, regardless of whether they originated from a UST.

A voluntary agreement may not be entered by the DEQ Executive Director and a responsible party unless all known potentially responsible parties have been notified of the proposed agreement and given an opportunity to comment on the proposed agreement before its consummation.

Subject to the constraints posed by the definitions of "responsible party" and "hazardous material" under HSMA, there is no express limitation on the eligibility of sites for the Voluntary Agreement Program. It has been the practice of DEQ, however, not to enter such agreements for NPL-qualifying sites or sites where the cleanup action is governed wholly by the Utah Underground Storage Tank Act,[13] the Utah Solid and Hazardous Waste Act,[14] or other state environmental regulatory programs.

Memorandum of Agreement

Utah has a Memorandum of Agreement with the United States Environmental Protection Agency (USEPA) for cooperative actions on the federal NPL qualifying sites but has no such memorandum regarding its Voluntary Agreement Program. Therefore, a responsible party entering a voluntary agreement under HSMA could arguably be subject to federal liability, even after completing a cleanup in accordance with a voluntary agreement with DEQ. The party would need to seek a comfort letter or other assurance directly from USEPA to minimize the risk of such liability.

Liability Protection

Lenders and Fiduciaries

The liability protections provided by HSMA do not extend to any person who does not participate in the management of a facility and who holds indicia of ownership primarily to protect a security interest in a facility.[15] Thus, lenders or other persons who hold indicia of ownership primarily to protect a security interest in a facility and who do not participate in the management of the facility cannot be held liable in a contribution action by a responsible party for investigation or cleanup costs incurred pursuant to a voluntary agreement under HSMA.

Similarly, HSMA does not extend to any person who does not participate in the management of a facility and who holds indicia of ownership as a fiduciary or custodian under the Uniform Probate Code or an employee benefit plan.[16] The scope of the exemption for fiduciaries and custodians is more limited than that for holders of security interests. Fiduciaries and custodians are not exempt from actions taken by the state or its officials or agencies.[17] Thus, fiduciaries and custodians are protected from cost-recovery actions by private parties but are not protected from such actions by the state. As noted above, however, DEQ is precluded from seeking cost recovery in connection with action taken by a responsible party under a voluntary agreement.

The scope of the exemptions provided for security-interest holders, fiduciaries, and custodians is defined by reference to key definitions in "40 C.F.R. 300.1100, National Contingency Plan,"[18] an express cross-reference to USEPA's 1992 Lender Liability Rule, which was invalidated by court decision but generally reinstated by Congress through provisions of the Omnibus Appropriations Act of 1997.[19]

Governmental Entities

HSMA does not extend liability protection to governmental ownership or control of property arising from involuntary transfers as provided in section 101(20)(D)[20] of CERCLA.[21]

Participants in the Program

There is no express liability protection granted to participants under a voluntary agreement. The Voluntary Agreement Program under HSMA contains no express authorization for, or prohibition against, DEQ granting contribution protection or a covenant-not-to-sue to a party who enters a voluntary agreement. Typically, such provisions have not been included in voluntary agreements entered under HSMA. The allowance for cost recovery in HSMA, however, implicitly recognizes that a party who incurs costs under a voluntary agreement "in excess of" that party's liability may seek contribution from any other party who is, or may be, liable for the excess cost.

Financial Incentives

Other than the right of contribution, no financial incentives are provided to participants under a voluntary agreement under the Utah HSMA.

Cleanup Standards

Nothing in the voluntary agreement provision of HSMA defines cleanup standards or explains how they will be established for a cleanup to be undertaken pursuant to a voluntary agreement. HSMA simply provides for DEQ and the contracting responsible party to enter an agreement for conducting an investigation or cleanup action. The statute does not articulate how the agreed-upon action will be taken. DEQ looks to standards and protocols established under other regulatory programs for setting cleanup standards.

Surface and groundwater quality are closely regulated in Utah. Under the Utah Water Quality Act,[22] DEQ, through the Division of Water Quality and Water Quality Board, has established groundwater classifications based on quality and corresponding groundwater quality protection levels.[23]

Sites contaminated by hazardous waste or its residues are subject to the Utah Hazardous Waste Management Program.[24] Under that program, DEQ traditionally required that sites be cleaned to background levels (or nondetect in the case of organic contaminants) to achieve a clean closure. In 1994, DEQ, through its Division of Solid and Hazardous Waste, issued a rule establishing cleanup action and risk-based closure standards.[25] These closure standards are applicable to any party involved in management of a site contaminated with hazardous waste or hazardous constituents. They allow a clean closure to be achieved even if background or nondetect levels will not be achieved. The new standards establish information requirements to support risk-based cleanup and closure standards at sites for which remediation or removal of hazardous constituents to background levels will not be achieved. These standards permit clean closure to be confirmed if the level of risk present at a site at which hazardous waste or hazardous constituents remain at or above background is less than one in one million for carcinogens and poses a hazard index of less than one for noncarcinogens, based on a risk assessment conducted in accordance with the rule.

For sites contaminated by petroleum constituents, DEQ looks to soil and water cleanup action levels developed and used under the Utah UST program.[26]

Certified Professionals and Laboratories

The Voluntary Agreement Program in HSMA does not expressly require the use of certified professionals or laboratories. However, in negotiating investigation or cleanup plans under any environmental program, including the Voluntary Agreement Program, it is the practice of DEQ and its divisions to require specific protocols for the collection and analysis of samples in accordance with accepted protocols of the USEPA. Moreover, actions taken under Utah's UST program require the use of certified professionals.[27]

Mechanics of Participating in the Voluntary Agreement Program

The mechanics for participating in the Voluntary Agreement Program are relatively straightforward. A responsible party who desires to enter a voluntary agreement for the investigation or cleanup of a site containing hazardous materials must contact the DEQ Executive Director and negotiate a proposed voluntary agreement. Either the DEQ Executive Director or the responsible party must then ensure that all known potentially responsible parties receive notice of the proposed agreement and are given an opportunity to comment on the proposed agreement before the agreement becomes effective.[28] Following the comment period, DEQ and the contracting party may finalize and sign the agreement.[29]

There is no prescribed format for voluntary agreements under HSMA. Agreements are negotiated uniquely for each situation. DEQ tends to rely on formats used for consent agreements under other programs. Some voluntary agreements specify the work that will be performed, while others provide for the contracting responsible party to submit for DEQ's approval a work plan detailing how the proposed investigation or cleanup will proceed, including cleanup levels and protocols for sampling and analysis. In the latter situation, a question of inadequate notice could arise if only the proposed agreement is sent to other responsible parties. To avoid any question of adequate notice under HSMA, it would be prudent for DEQ and the contracting party to ensure that the subsequently submitted work plan is also provided to all such parties.

The DEQ Executive Director and the contracting responsible party may also negotiate provisions providing for the responsible party to provide funds to the DEQ Executive Director to undertake the review of the work plan and to oversee the investigation or cleanup action.[30] The DEQ Executive Director also has authority to negotiate enforcement provisions that would impose penalties upon the contracting responsible party in the event it fails to perform as required by the voluntary agreement.[31]

Upon completion of the work required under a voluntary agreement, the contracting responsible party submits a report of the investigation or cleanup action to DEQ and requests written confirmation from DEQ that the work has been completed in accordance with the agreement and that no further action is required.

Cost Recovery

The voluntary agreement provision of HSMA expressly authorizes any party who incurs costs under a voluntary agreement in excess of its liability to seek contribution from any other party who is, or may be, liable under HSMA for the "excess costs" by filing an action in a Utah district court. DEQ, however, is expressly precluded from recovering costs received or expended pursuant to a voluntary agreement from any person not a party to that agreement.

In resolving contribution claims by a contracting responsible party, the court is required to allocate costs using standards set forth in section 19-6-310(2) of the Utah Code. The burden of proving proportionate contribution must be borne by each responsible party. Liability must be apportioned in accordance with each responsible party's respective contribution to the release. Apportionment must be based on equitable factors, including the quantity, mobility, persistence, and toxicity of hazardous materials contributed by a responsible party, and the comparative behavior of a responsible party in contributing to the release, relative to other responsible parties. A responsible party may be considered to have contributed to the release and may be liable for a proportionate share of costs either by affirmatively causing a release or by failing to take action to prevent or abate a release that has originated at or from the facility. A person whose property is contaminated by migration from an off-site release is not considered to have contributed to the release, unless that person takes actions that exacerbate the release.

HSMA does not define the term "costs," and there is no express limitation on which costs may or may not be recoverable under the Voluntary Agreement Program; it simply refers to costs incurred under a voluntary agreement. The liability apportionment provisions of HSMA, however, refer only to the "costs of the investigation and abatement."[32] In contrast, the Utah UST Act is quite explicit about which costs are recoverable under that statute.[33]

Defenses/Exemptions from Liability

The following parties are exempt from liability for costs incurred by any responsible party under a voluntary agreement:

- Any person who does not participate in the management of a facility and holds indicia of ownership primarily to protect a security interest in a facility or as a fiduciary or custodian under the Uniform Probate Code or under an employee benefit plan
- A unit of state or local government owning or controlling property acquired through involuntary transfer, as provided in section 101(20)(D) of CERCLA
- A current or previous owner or operator (1) who acquired or became the operator of the facility before March 18, 1985, (2) who did not know that any hazardous material that is the subject of a release was on, in, or at the facility before acquisition or operation of the facility, and (3) whose act or omission did not result in the release.
- A current or previous owner or operator (1) who acquired or became the operator of the facility on or after March 18, 1985, (2) who did not know and had no reason to know, after having taken all appropriate inquiry into the previous ownership and uses of the facility, consistent with good commercial or customary practice at the time of the purchase, that any hazardous material that is the subject of a release was on, in, or at the facility before acquisition or operation of the facility, and (3) whose act or omission did not result in the release.[34]

Conclusion

The Voluntary Agreement Program under HSMA provides limited incentives for acquisition and development of environmentally impaired property. In most cases, the risks of brownfields development in Utah outweigh the benefits, and the regulatory approval process remains time consuming and uncertain. The enactment of broader voluntary cleanup legislation in 1997 will provide improved procedures and stronger incentives for conducting voluntary cleanups in Utah.

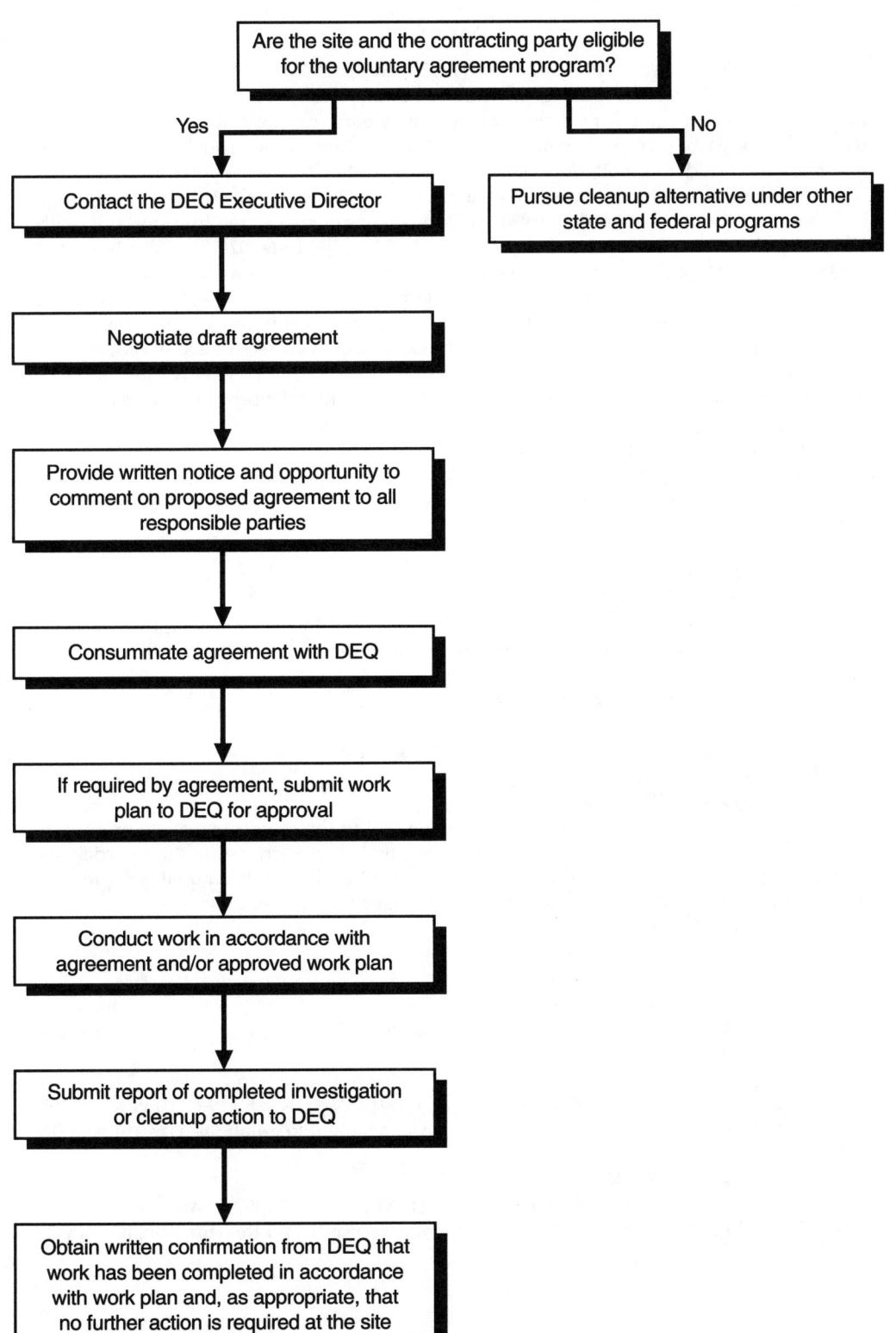

Notes

1. UTAH CODE ANN. § 19-6-301 et seq. (1995 & Supp. 1996).

2. Id. § 19-6-325. The Utah Underground Storage Tank Act also contains authority for DEQ to enter an agreement with any responsible party (as defined under that statute) regarding that party's proportionate liability or "any action" to be taken by that party. Id. § 19-6-424.5(4). Under this act, only "owners" or "operators," as strictly defined under the federal underground storage tank (UST) law, 42 U.S.C.A. § 6991(3)–(4) (West 1995), may be subjected to compliance orders from DEQ to take corrective action in connection with USTs or petroleum releases therefrom. UTAH CODE ANN. §§ 19-6-402(17)–(18), 19-6-420(2) (1995 & Supp. 1996). Other parties, however, who do not strictly constitute owners or operators, may still be "responsible parties," as defined in section 19-6-402(24) of the Utah Code, and be subject to apportioned liability for costs expended in connection with actions taken for leaking USTs. Such parties, although not subject to corrective action orders, may "voluntarily" step forward and enter an agreement with DEQ to resolve their liability and/or undertake action in connection with a UST site, under the provisions of section 19-6-424.5 of the Utah Code.

3. 42 U.S.C.A. § 9601, et seq. (West 1995).

4. UTAH CODE ANN. § 19-6-325.

5. S.B. 190, 1997 General Session (to be codified at UTAH CODE ANN. §§ 19-8-10110-118.

6. UTAH CODE ANN. § 19-6-325 (1995).

7. Id. § 19-6-302(18).

8. 42 U.S.C.A. § 9601 (West 1995).

9. UTAH CODE ANN. § 19-6-302(18)(c) (1995 & Supp. 1996). The exemption for fiduciaries does not apply to actions by the state or its agencies. Id. § 19-6-302(18)(d).

10. Id. § 19-6-302(7).

11. 42 U.S.C.A. §§ 6991-6991h (West 1995).

12. Id. § 9601(14).

13. UTAH CODE ANN. § 19-6-401, et seq. (1995 & Supp. 1996).

14. Id. § 19-6-101 et seq.

15. Id. § 19-6-302(18)(c)(i)(A).

16. Id. § 19-6-302(18)(c)(i)(B). Although the intent of this provision was to exempt fiduciaries and custodians from liability, the prerequisite that such parties "not participate in management" would effectively preclude a trustee of property from qualifying for the exemption. A trustee's legal responsibility is to manage the trust and its assets.

17. Id. § 19-6-302(18)(d).

18. Id. §§ 19-6-302(18)(e), 19-6-302(18)(f).

19. Pub. L. No. 104-208, § 2504 (1996).

20. 42 U.S.C.A. § 9601(20)(D) (West 1995).

21. UTAH CODE ANN. § 19-6-302(18)(c)(ii) (Supp. 1996).

22. Id. § 19-5-101 et seq.

23. UTAH ADMIN. R. 317-6 (1996).

24. Utah Solid and Hazardous Waste Act, UTAH CODE ANN. §§ 19-6-101 et seq. (1995 & Supp. 1996), and Solid and Hazardous Waste Management Regulations, UTAH ADMIN. R. 315-1 et seq. (1996)

25. UTAH ADMIN. R. 315-101 (1996).

26. See UTAH CODE ANN. § 19-6-401 et seq. (1995 & Supp. 1996); UTAH ADMIN. R. 311 (1996). DEQ, through its Division of Environmental Response and Remediation, has devel-

oped "tier 1" screening levels for petroleum-contaminated soil and groundwater at UST sites, and is developing guidelines for preparing risk assessments to establish alternative risk-based corrective action levels at such sites.

27. UST consultants, inspectors, testers, installers, and removers, and groundwater and soil samplers, must be certified. UTAH ADMIN. R. 311-201 (1996).

28. UTAH CODE ANN. § 19-6-325(1)(b) (1995).

29. *Id.*

30. *Id.* § 19-6-325(2).

31. *Id.* § 19-6-325(3).

32. *Id.* § 19-6-310(2).

33. *Id.* § 19-6-402(9).

34. The limitations on liability of owners and operators were added by the legislature in 1995, in response to the decision in *Utah Department of Environmental Quality v. Wind River Petroleum*, 881 P.2d 869 (Utah 1994), which held that a current owner or operator who did not contribute to the release could be liable for DEQ's costs in responding to a petroleum release under the emergency response provisions of HSMA.

CHAPTER 46

Vermont

A. JAY KENLAN
W. ANDREW HAZELTON

The Vermont Redevelopment of Contaminated Properties Program (RCPP, or Program), codified at title 10, section 6615a of the Vermont Statutes, is a property redevelopment program conceived by the Vermont Department of Economic Development, but created and implemented by the Vermont Agency of Natural Resources (ANR, or Agency). Predictably, the RCPP emphasizes the goals and attitudes of ANR over those of economic development. As a result, the Program ultimately fails to fulfill its central purpose of restoring economic vitality to environmentally impaired properties.

According to ANR, the RCPP is designed to alleviate the major impediments to voluntary cleanup of environmentally impaired property.[1] In actuality, the Program gives persons looking for a vehicle to redevelop contaminated sites in this state little to applaud; the Program provides little meaningful liability protection to attract participants to the RCPP and contains no tax or other financial incentives. The only benefit bought by participation in the RCPP is limited protection from liability to the state under Vermont's Hazardous Waste Management Act (HWMA)[2] as a present "owner or operator" of the contaminated property. Because this benefit will seldom outweigh the financial costs of Program participation, it is unlikely that the RCPP will accomplish its purpose to a significant degree.

Summary of Major Provisions

The major benefits and limitations of the Program can be capsulized as follows. Major benefits:

- Successful participants in the RCPP will not be liable to the state as owners or operators under the HWMA.
- The Program does not require that a participant reimburse the state for its response costs incurred before the participant's purchase of the property. As a result, participation may be attractive at those sites

where the state has already invested time and money toward remediation, and the anticipated remaining cleanup costs are relatively small.

Major limitations:

- The RCPP offers participants no protection from liability under state environmental laws other than the HWMA (for example, the Vermont Water Pollution Control statute and the Vermont Groundwater Protection statute). Moreover, the RCPP does not provide Program participants with protection from arranger or transporter liability under the HWMA.
- Because Vermont does not currently have a Memorandum of Agreement with the United States Environmental Protection Agency (USEPA), participation in the RCPP does not alleviate the threat of federal environmental liability.
- The Program contains no express protection from third-party liability. Further, the enabling statute provides that "[a]ny determination of nonliability under [the RCPP] shall not have any collateral effect in other proceedings."[3] As a result, participants remain open to contribution suits by other parties held responsible for contamination on the property.
- The Program provides no independent authority for participants to seek contribution from responsible parties under state law. The HWMA, as enacted in 1985, does not contain any language allowing for the recovery of privately incurred response costs.[4] However, the statute was amended in 1993 to permit contribution and indemnification actions between responsible parties.[5] Whether a Program participant qualifies as a "responsible party" for purposes of maintaining a recovery action under the HWMA is unclear, and a participant may ultimately be left to his or her remedies under the federal Comprehensive Environmental Response, Compensation, and Liability Act (CERCLA) to recoup response costs incurred under the Program.
- The RCPP does not provide participants with liability protection when the participant transfers the property before completing the Program and the successor fails to complete the Program or withdraws from the Program.
- The RCPP does not provide less stringent remediation standards, or other variance mechanisms, for participating properties.
- The RCPP contains no financial incentives for participation (such as low-interest loans or tax abatement).

Application and Eligibility Determination

The RCPP may be divided into three parts: (1) application/eligibility determination, (2) site investigation/remediation, and (3) postcompletion/liability protection. Each of these components will be discussed in turn.

Application Procedure

Participation in the Program begins with the submission of an application and a determination by the Agency regarding the applicant's eligibility for the RCPP. The Program application requires the following:

- A nonrefundable application fee of $500
- A preliminary environmental assessment of the property, including information on the operational history of the property and information concerning the nature and extent of the contamination on the property and the health and environmental risks posed by that contamination
- A description of the proposed redevelopment and use of the property
- Eligibility information, including certification that the Program applicant has not had any involvement with the subject property that could subject the applicant to liability for contamination at the property[6]

Eligibility for the Program

Within thirty days of submitting the application, ANR must make a determination of the applicant's eligibility. That decision is unappealable. To qualify for participation in the RCPP, both the purchaser and the property must meet mandated eligibility requirements.

Persons Eligible for Participation in the Program

By statute, all persons[7] are eligible for the RCPP, including secured lenders who hold indicia of ownership in the property primarily to ensure repayment of a financial obligation,[8] provided the person or lender is not potentially liable for a release or threatened release on the subject property as a prior owner or operator, generator, or transporter.[9] However, the fact sheet issued by ANR—setting out its interpretation of the Program's mechanics—excludes any applicant that "has *in any way* been previously involved with the property."[10] Although the Agency fact sheet is designed to be an informational tool, and does not have the effect of law, it makes uncertain the extent to which the Agency will recognize, at a given property, holders of security interests on the property, officers and employees of previous property owners, and regulating governmental agencies, as persons eligible for the Program.

Properties Eligible for Participation in the Program

A property will be eligible for the Program if (1) there has been a release or threatened release of a hazardous material[11] at the property, and (2) the property is vacant, abandoned, or "substantially underutilized," or is to be acquired by a municipality. According to the ANR fact sheet, "a 'substantially underutilized' property is one which upon [sic] the proposed redevelopment will significantly reverse a sustained or substantial economic decline in the number of jobs or overall economic activity at the property."[12]

Properties Ineligible for Participation in the Program
Not all properties are eligible for participation in the RCPP. The following properties are ineligible:

- Facilities listed on the National Priorities List established under CERCLA[13]
- Hazardous-waste facilities certified under section 6606 of the Vermont Statutes and subject to a corrective action under the Resource, Conservation, and Recovery Act[14]
- Properties at which the only known release or threatened release relates to an underground storage tank subject to enforcement under section 1941 of the Vermont Statutes[15]
- Properties that are not vacant or abandoned, or do not otherwise qualify as "substantially underutilized," unless the property is to be acquired by a municipality

Site Investigation and Site Remediation

If ANR determines, in its sole discretion, that an applicant is an eligible person for the Program and accepts the submitted application as complete, the person is made a participant of the RCPP. Upon entering the Program, the participant submits a site investigation work plan, and a fee of $5,000 to be applied to the Agency's costs for review and oversight of the performance of the site investigation and any corrective action plan.[16] The $5,000 fee is not a one-time expense, but must be replenished periodically as the Agency's review and oversight costs exceed the initially funded amount.[17] There is no statutorily imposed cap on the Agency's review and oversight costs.

Contents of Site Investigation Work Plan

The site investigation work plan builds on the information contained in the preliminary environmental assessment submitted with the Program application. As contemplated under the RCPP, the preliminary environmental assessment is roughly comparable to the preliminary assessment/site investigation process under the federal National Contingency Plan (NCP),[18] and the work plan is roughly comparable to the NCP's remedial investigation/feasibility study provisions.

According to ANR personnel, the liability protections offered to participants in the Program necessitate closer scrutiny by the Agency of submitted information and will likely require participants to provide more complete and more detailed information in the work plan than would be needed for a site investigation and characterization conducted outside the Program. Persons interested in participating in the RCPP should be mindful that the heightened scrutiny applied by ANR to Program work plans may substantially *increase* the costs of site investigation, modeling, and remedial action planning.

Site Investigation Work Plan Approval

A participant must receive ANR approval of the proposed work plan before its implementation. The Agency may approve, disapprove, or approve with modifications the submitted work plan.[19] If a work plan is not approved, the participant must resubmit a revised work plan for approval or withdraw from the Program.[20] The potential consequences for withdrawing from the Program before completion of the remedial action plan are discussed below. Only work plans approved by the Agency may be implemented under the Program.

Site Investigation Report

Upon completion of the work plan, the participant must submit a "site investigation report" detailing the results of the work plan to the Agency. ANR may approve the report or may revise the work plan to require further site investigation and characterization.[21] Once the report is approved, the participant must develop and submit a corrective action plan (CAP) to address the contamination on the property.

If information gathered during the site investigation process indicates that no further removal, remediation, or monitoring activities are required to protect human health and the environment adequately, the Program participant may request a no-further-action determination by ANR.[22] The Agency may grant such a determination if it is concluded that

1. redevelopment and reuse of the property will not cause, allow, contribute to, or worsen any release or threatened release of hazardous materials at the property;
2. releases or threatened releases that are not abated, removed, or remediated do not pose an unacceptable risk to human health and the environment; and
3. the applicant will provide access to both the state and responsible parties for purposes of undertaking additional investigation, abatement, removal, remediation, or monitoring activities on the subject property.[23]

According to ANR personnel, a release of a hazardous material is deemed not to pose "an unacceptable risk to human health and the environment" only when concentrations do not exceed state enforcement standards.[24] Given the requirements for property eligibility in the Program—vacant, abandoned, or substantially underutilized property from which there has been a release or threatened release of a hazardous material—it is unlikely that a participating property could secure a no-further-action determination from ANR without some corrective action.

CAP

If ANR finds that based upon the approved site investigation report, remediation or monitoring is required "to adequately protect human health and the environment and to meet all applicable cleanup standards," the participant must submit a CAP detailing a remediation strategy for the property.[25] Again, the Agency may approve, approve with conditions, or disapprove the submitted CAP.[26] Likewise, if the CAP is disapproved, the applicant must submit a revised CAP for approval or must withdraw from the Program.[27] Before the Agency's approval of the CAP, the CAP is put out for public comment, and any public comment must be considered by the Agency before its approval of the CAP.

By statute, "the decision of the secretary [of ANR] as to whether a corrective action plan should be approved is within the secretary's sole discretion and is final."[28] A Program participant must be prepared to implement an Agency-approved CAP, and must be mindful that given the liability protections provided by the Program, the Agency-approved CAP may be far more extensive than the plan developed by the participant.

The CAP, once approved, may be modified and amended by the Agency at any time during its implementation, before the issuance of the certificate of completion. The approved CAP is binding on the participant and any eligible successor of the participant.[29]

Liability Protection

Certificate of Completion

Upon completion of all the activities required under the approved CAP and any additional work required by modification or amendment of the CAP, the Agency will issue a certificate of completion that contains the following statements:

- Certification that the work is completed in accordance with the CAP and any amendments
- No further investigation, abatement, removal, remediation, or monitoring activities are required at the subject property
- The participant is protected from further liability under the HWMA[30]

HWMA Liability Protection

A participant in the RCPP is not liable under the HWMA for releases or threatened releases of hazardous materials on the subject property as an owner or operator of the property, provided the participant completes the Program and receives a certificate of completion.[31] This protection extends to successors-in-interest, as well.[32] Specifically, the successful participant is not liable for the following:

- Releases of hazardous materials discovered after approval of the CAP that could not have been discovered before that time by "recognized standard methods" (As noted above, a likely consequence of this protection is a demand by ANR for rigorous site investigation.)
- Releases of substances that were not regulated as a hazardous material until after approval of the CAP.

- Additional cleanup required by the adoption of more stringent cleanup standards after approval of the CAP.[33]

The Program's liability protection is available only as long as the participant "is implementing, in good faith, the approved site investigative work plan or corrective action plan."[34] Thus, if the participant withdraws from the Program before the issuance of a certificate of completion, the state may seek to hold the participant liable as a responsible party at the site under the HWMA. Although this liability is not explicitly stated within the Program's enabling legislation, the statute does provide that a person will not be liable as an owner/operator of the property "if the eligible person complies with [the Program requirements] and obtains a certificate of completion."[35]

This potential liability is significant, as both the site investigation work plan and the CAP must be approved by ANR before their implementation, and may be modified by the Agency at any time before the issuance of a certificate of completion. If the submissions are disapproved, the participant faces the uncomfortable choice of accepting the costs of a work plan and CAP dictated by ANR, or withdrawing from the Program and risking unlimited liability as a responsible party under the HWMA. Either choice may be financially ruinous to the participant.

Similarly, the protections offered by the Program do not apply to instances when the participant, before the issuance of the certificate of completion, transfers the property and the successor owner fails to complete the Program.[36] As a result, participants must seek financial assurances from prospective buyers, or remain prepared to return and complete the Program, should the purchaser default on Program obligations. Participants also may be subject to liability when the property is transferred before the issuance of the certificate of completion and the successor is deemed by ANR to be a responsible party under the HWMA, and therefore ineligible for the Program. Again, the participant must be prepared to return and complete the Program to secure its liability protections.

The liability protection granted by the RCPP relates only to contamination present on the property at the time of its acquisition by the participant, and contamination—caused by the participant during completion of the Program—that is fully remediated before the issuance of a certificate of completion. The enabling legislation specifically withholds liability protection for the following:

- Releases or threatened releases that occur after submission of an application and are not fully remediated by the participant
- Releases resulting from the reckless actions of the participant
- Releases existing on the property that were not included in information provided to ANR by the applicant (even though existing at the time of the application)
- Liability based upon the participant's actions as a generator or transporter
- When the application approval was secured by fraud, knowing nondisclosure or intentional misrepresentation

- When the participant worsens an existing release or threatened release that is not fully remediated pursuant to an approved CAP[37]

Finally, as noted above, the Program does not protect the participant from liability under other Vermont environmental laws or from third-party liability, a point specifically recognized and detailed by ANR in its Program fact sheet, which states that

> [t]he waiver of liability applies only to owner/operator liability under 10 V.S.A. Section 6615(a)(1). It does not release a party from liability under any other state or federal statute, including liability under CERCLA or RCRA. The protection from liability also does not extend to suits brought by third parties.[38]

Additional Obligations of the Program Participant

In exchange for the limited liability protection offered by the state under the RCPP, a participant in the Program agrees to perform all the activities in the approved site investigation work plan and the approved CAP, and all amendments to either plan. In addition to fulfilling the work plan and CAP requirements, a participant covenants further that he or she

1. will indemnify and hold the state harmless from claims arising from the implementation of the work plan and CAP;
2. will not sue the state for claims arising from the approval or implementation of the work plan and CAP, or seek any other cost damages or attorney fees from the state;
3. will not sue the state for claims arising from hazardous-material contamination on the property, except to the extent the state may be liable as a responsible party under the HWMA;
4. will provide the state with all documents and information relating to the participant's activities on the subject property;
5. will disclose all information known to the participant relating to the releases or threatened releases of hazardous materials on the property;
6. will provide site access to ANR and any responsible parties for the purpose of undertaking investigation, abatement, removal, remediation, or monitoring activities on the property; and
7. will not leave the property in a condition that presents a greater threat to human health and the environment than existed before entry into the Program, if he or she withdraws from the RCPP before obtaining a certificate of completion.[39]

Cleanup Standards

The state's cleanup standards are contained in ANR's Ground Water Protection Rules.[40] The rules establish primary groundwater quality standards for approximately 110 substances. The enforcement standard for each listed substance is based upon the USEPA maximum contaminant levels for drinking water.[41] It should be noted that the enforcement standards relate solely to groundwater contamination; no standards have been established for soil and air contamination. It is not known whether the state would apply the ground water enforcement standards to situations that do not involve groundwater pollution, or whether it would seek to apply some different standard.

Conclusion

The RCPP lacks many of the tools necessary for a successful property redevelopment program. It offers participants only limited liability protection from the state, while exacting significant financial commitments from the participant. Given ANR's absolute discretion over remedy selection within the Program, many of the costs associated with participation in the Program are not quantifiable by the prospective applicant. Indeed, the two most appealing features of the RCPP are that it is voluntary—a quality unique among Vermont's host of environmental programs—and that it is a pilot program, set to expire on July 1, 2000.[42] With luck, that will be enough time to retool the Program to alleviate the major impediments to voluntary cleanup of environmentally impaired property in this state.

Vermont Redevelopment of Contaminated Properties Program

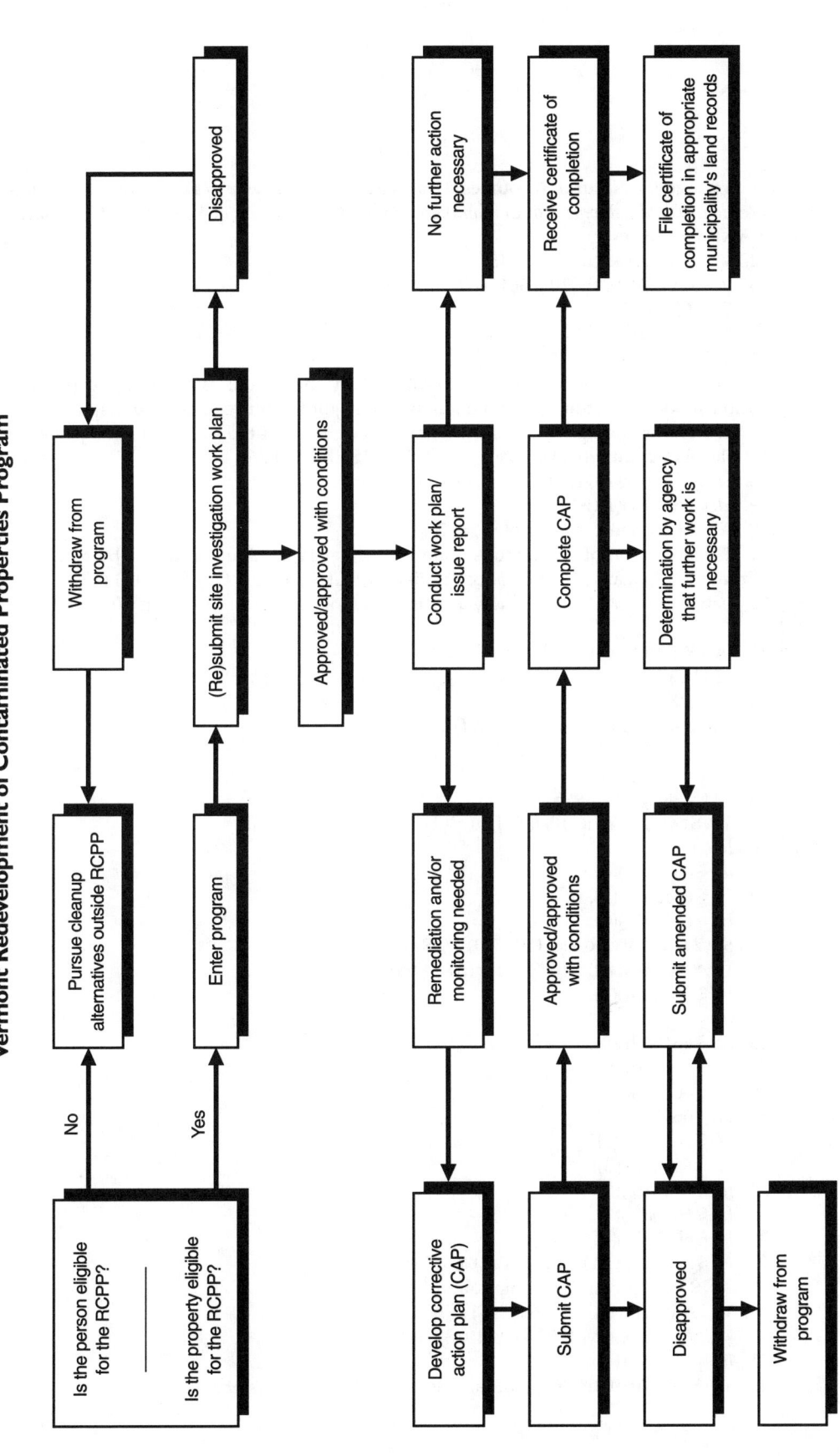

Notes

1. Vt. Agency Natural Resources, Draft Fact Sheet, *Cleanup of Hazardous Waste Sites under Vermont's Redevelopment of Contaminated Properties Program*, at 1 (Sept. 1, 1995) [hereinafter ANR Fact Sheet].
2. Vt. Stat. Ann. tit. 10, ch. 159.
3. Vt. Stat. Ann. tit. 10 § 6615a(f)(1)(B) (1995).
4. *See id.* § 6615(a).
5. *Id.* § 6615(i).
6. *Id.* § 6615a(e).
7. The term "person" is defined to mean "any individual, partnership, company, corporation, association, unincorporated association, joint venture, trust, municipality, the state of Vermont or any agency, department or subdivision of the state, federal agency, or any other legal or commercial entity." *See id.* §§ 6615a(b), 6602(6).
8. *See id.* § 6602(23) (defining "secured lender").
9. *Id.* § 6615a(f)(1)(B).
10. ANR Fact Sheet, *supra* note 1, at 2.
11. The definition of "hazardous material" under HWMA incorporates by reference the definition of "hazardous substance" under CERCLA, 42 U.S.C. § 9601(14) (1994), and includes petroleum, crude oil, or any fraction thereof. Vt. Stat. Ann. tit. 10, § 6602(16) (1993).
12. ANR Fact Sheet, *supra* note 1, at 2.
13. 42 U.S.C. § 9601 *et seq.* (1994).
14. *Id.* § 6901 *et seq.*
15. Vt. Stat. Ann. tit. 10, § 6615a(f)(2) (1995).
16. *Id.* §§ 6615a(e)(2), 6615a(g)(1).
17. *Id.* § 6615a(j)(1).
18. 40 C.F.R. § 300 (1995).
19. Vt. Stat. Ann. tit. 10, § 6615a(g)(2) (1995).
20. *Id.*
21. *Id.* § 6615a(g)(4).
22. *Id.* § 6615a(g)(5).
23. *Id.* § 6615a(g)(6).
24. *See* 7 Code Vt. Rules 12037001 (1988).
25. Vt. Stat. Ann. tit. 10, § 6615a(h) (1995).
26. *Id.* § 6615a(h)(2).
27. *Id.*
28. *Id.* § 6615a(h)(5).
29. *Id.* § 6615a(j)(2).
30. *Id.* § 6615a(k).
31. *Id.* § 6615a(c)(1).
32. *Id.* § 6615a(c)(2).
33. *Id.* § 6615a(c)(3).
34. *Id.* § 6615a(c)(5).
35. *Id.* § 6615a(c)(1).
36. *Id.* § 6615a(d)(4).
37. *Id.* §§ 6615a(d)(1), 6615a(d)(2), 6615a(d)(5).
38. ANR Fact Sheet, *supra* note 1, at 4.
39. Vt. Stat. Ann. tit. 10, 6615a(j) (1995).
40. 7 Code Vt. Rules 12037001 (1988).
41. *Id.* § 12037001-26.
42. Vt. Stat. Ann. tit. 10, § 6615a, at Reporter's Notes: History.

CHAPTER 47

Virginia

MARK D. ANDERSON

The Virginia voluntary remediation program (VRP, or Chapter 622) was enacted in 1995. The law was enacted to encourage hazardous-substance cleanups by streamlining mechanisms and providing incentives for voluntary participation. The law does this by requiring the Virginia Department of Environmental Quality (VADEQ) to take into account future land use when determining cleanup standards and providing liability releases upon cleanup completion. Chapter 622 is not a very descriptive law. Instead, it mandates that VADEQ promulgate regulations to administer the VRP consistent with the law by July 1, 1997.[1] Until promulgation of regulations, VADEQ will administer the VRP on a case-by-case basis, consistent with criteria established in the new law for promulgation of the regulations.[2]

Summary of Major Provisions of Chapter 622

There are three sections to Chapter 622. The first section requires VADEQ to promulgate regulations in five categories:

1. methodologies to determine site-specific, risk-based remediation standards;[3]
2. procedures that minimize delay and expense of the remediation, to be followed by a volunteer and by VADEQ in processing submissions and overseeing remediation;[4]
3. certifications of satisfactory completion of remediation, based on then-present conditions and available information, when a voluntary cleanup achieves applicable cleanup standards or when VADEQ determines that no further action is required;[5]
4. procedures to waive or expedite issuance of any permits required to initiate and complete a voluntary cleanup consistent with applicable federal law;[6] and
5. registration fees to be collected from persons conducting voluntary remediation, to defray the actual and reasonable costs of the VRP.[7]

The second section of Chapter 622 addresses releases from liability.[8] The third section addresses the ability of volunteers to gain access to neighboring properties to conduct a cleanup.[9]

Eligibility for the VRP

The VRP is an extensive law, with broad applicability. A wide range of parties and sites are eligible to participate in the program. Any person who owns, operates, has a security interest in, or enters a contract for the purchase of contaminated property can participate in the program.[10] Potential responsibility for the contamination is not relevant to the issue of eligibility for the VRP.

The types of sites that may be remediated under the VRP are also broadly defined by the law. Sites contaminated with hazardous substances, hazardous waste, solid waste, or petroleum may participate in the program.[11] However, several types of sites cannot participate in the program. A site is ineligible if remediation of the release is "clearly mandated" under one of the following laws:

- The Comprehensive Environmental Response, Compensation, and Liability Act (CERCLA)[12]
- The Resource Conservation and Recovery Act (RCRA)[13]
- The Virginia Waste Management Act[14]
- The Virginia Water Control Law[15]
- Other applicable statutory or common law with jurisdiction to require environmental remediation

There has been considerable discussion in Virginia about the "clear mandate" standard. VADEQ has taken the approach that "clearly mandated" is defined as any site that is subject to remediation. Other stakeholders in Virginia have argued that "clearly mandated" arises only after a governmental order for cleanup has been issued or a permit entered. It is anticipated that VADEQ will address the issue in updated internal guidance for the program.

Petroleum Releases

Petroleum releases, which are regulated under the Virginia Water Control Law, may be eligible for participation in the VRP if they are not clearly subject to cleanup under articles 9 and 11 of the law. However, if petroleum contamination is cleaned up as part of an action taken under the VRP, the cleanup is not eligible for reimbursement through the Virginia Petroleum Storage Tank Fund remediation reimbursement program.

Memorandum of Agreement

Virginia has not entered a Memorandum of Agreement with the United States Environmental Protection Agency (USEPA) for its VRP. However,

USEPA Region III is developing a Memorandum of Agreement with Delaware and Pennsylvania, which VADEQ is watching closely. Until such time as a Memorandum of Agreement is entered by the state, liability protection under the VRP does not extend to potential USEPA enforcement actions. Participants who require federal liability assurances may seek comfort letters from USEPA Region III. If the participant is a prospective purchaser without current federal liability for the contamination, prospective-purchaser agreements also may be entered with USEPA Region III.[16] Under either scenario, the participant should strongly weigh the advantages and disadvantages of including USEPA as a stakeholder at a site.

Liability Protection

Lenders, Trustees, and Fiduciaries

There are no provisions in Virginia to provide liability protection for lenders, trustees, and fiduciaries. However, Virginia does not have a strict-liability scheme.[17] As such, lenders and others obtaining a security interest in the property are not subject to the same potential liability that a strict-liability scheme creates. However, lenders and security holders remain subject to the federal liability and must take federal liability concerns into account when securing an interest in contaminated property.

Protection for Participants in the Program

Chapter 622 provides liability protection for parties receiving certifications of satisfactory completion of remediation from VADEQ. The completion certification provides immunity from enforcement actions under Virginia environmental laws. However, the immunity does not provide protection against third-party contribution liability or toxic-tort liability.

Cleanup Standards

The Virginia VRP anticipates cleanup standards on a site-specific basis. Cleanup standards under the program must not be more stringent than applicable or appropriate federal standards for soil, groundwater, and sediments.[18] According to Chapter 622, cleanup standards must take the following into account:

1. protection of public health and the environment;
2. future industrial, commercial, residential, or other use of the property to be remediated and of surrounding properties;
3. reasonably available and effective remediation technology and analytical quantitative technology;
4. availability of institutional or engineering controls that are protective of human health or the environment; and
5. natural background levels for hazardous constituents.[19]

Draft internal VADEQ guidance indicates that there will be two types of cleanup standards; unrestricted land uses or restricted land uses.[20] An unrestricted land-use cleanup takes into account a greater exposure scenario and, as such, is more difficult to achieve. However, once achieved, there will be no zoning restrictions on the site. Cleanup to a restricted land use permits participants to take future land use into account. However, in a restricted land-use cleanup, the restriction will be attached to the deed for the site after the cleanup has been completed.

Unrestricted Use Cleanups

When determining an unrestricted-use cleanup level, participants have the option of cleaning up to background levels or developing a site-specific risk assessment. When developing a site-specific risk assessment, sites are evaluated based on a potential residential use and with regard to whether there is an aquifer as a drinking-water source. The resulting cleanup standard of a risk assessment must be the lower of the two levels.[21]

When determining the impact of carcinogens to human health, participants in the VRP must use a one-in-one-million cancer risk for a single contaminant and a one-in-ten-thousand combined cancer risk for multiple contaminants.[22] For sites with noncarcinogenic contaminants, levels at which a human population could be exposed on a daily basis without a deleterious risk during a lifetime will govern as the applicable standard.[23] For the determination of effects upon the environment, VADEQ has promulgated ecological standards.[24] In cases with an existing standard, that standard will be applied as the cleanup level. For detected contaminants without promulgated standards, cleanup levels will be established on a case-by-case basis.[25]

Restricted Use Cleanups

Like unrestricted-use cleanups, cleanup levels for risk assessments developed for restricted-use cleanups must consider the proposed restricted use and the existence of aquifers as a potable or nonpotable water source. The resulting cleanup standard will be the lower of the two levels.[26]

Cleanup standards for restricted-use cleanups are based on the same levels for human health and environmental assessment as nonrestricted use. However, concentration exposure points are determined differently. Unrestricted-use cleanups assume the receptor point concentration and the source concentration to be equal. No transport modeling of the contaminants outside the area of contamination is allowed. In a restricted-use cleanup, the point of exposure is determined by considering site-specific conditions. Cleanup concentrations at the source will be determined by back-calculating acceptable concentrations at the point of exposure and possible dilution and attenuation by using fate and transport modeling.

Deed Restrictions

Guidance governing deed restrictions in Virginia is now being developed. The forthcoming regulations will provide requirements for deed recordation of completion certifications when restricted cleanup levels are attained. For unrestricted cleanups, deed recordation requirements are now being discussed. Some stakeholders have recommended that deed recordation be required at any cleanup, while others contend that deed recordation at unrestricted voluntary cleanups is unnecessary and will cause a blemish upon the title.

Mechanics of Participating in the VRP

Certification of Eligibility

The first step in determining eligibility for the VRP is the submission of a "certification of eligibility." The certification of eligibility is the form on which VADEQ will determine an applicant's eligibility for participation in the program. VADEQ has drafted internal guidance to assist a volunteer in providing the necessary information.[27] The information required to complete the certification of eligibility is quite extensive, as it is the basis for VADEQ's determination of site eligibility. Structurally, the certification of eligibility is a letter to VADEQ, which includes the following:

1. a declaration of intent by the volunteer to participate in the VRP;
2. a certification that, to the best of the applicant's knowledge, the site is not subject to remediation under any other environmental program or applicable statutory or common law; and
3. site background information to facilitate VADEQ's verification of eligibility.[28]

Because the purpose of the certification of eligibility is to provide VADEQ a basis on which to determine site eligibility, there should be comprehensive information in the certification about the subject site. The site background portion of the certification of eligibility should provide a thorough summary of conditions at the site and potential jurisdiction of other existing environmental programs. The summary should include the following elements:

- Basic information about the site location, including the name and address of the owner or operator, the length of time of operation of facilities on the site, a listing of the products that were manufactured on the site, and a narrative about how contamination was first identified. Information about whether a site investigation or remediation work has been performed should be included, as well as a contact person for information about the site.[29]

- Information about the nature of the contaminants of concern should also be included. This should include the physical state, estimated quantities, hazardous characteristics, hazardous constituents expected to be present, and the environmental media that has likely been affected. The sources of this information should also be included for VADEQ to determine the credibility of the analysis.[30]
- Information identifying any waste management or environmental permits issued for the site, and information about any waste treatment, storage, and disposal activities that have occurred on the site. Information regarding any past and current enforcement actions should also be included.[31]
- An assessment of the potential statutory jurisdiction that may be asserted for the contamination released at the site should be prepared, including each program, regulation, or statute that may have jurisdiction and could require remediation of the release.[32]

The information above should be summarized in brief narrative sections. Each section should include the sources of information used to complete each portion. Following submission, VADEQ will review the letter and either enroll the site in the VRP or classify the site as ineligible.

Voluntary Remediation Agreement

Once VADEQ has enrolled a site in the VRP, the next step is entering a voluntary remediation agreement. VADEQ prepares the agreement as a contract between the volunteer and VADEQ. The voluntary remediation agreement provides that VADEQ staff will review and provide oversight of the cleanup. Under the agreement, the volunteer commits to activities that will be expected to be completed as part of the environmental remediation of the site. The agreement also provides a demonstration of the intent of the volunteer to remediate the site until completion.[33]

Although a template serves as the basis for all voluntary remediation agreements, site-specific conditions may warrant reconsideration of certain portions of the form agreement after the cleanup has begun. In these cases, the agreement may be modified by amendment by mutual consent of VADEQ and the volunteer.

The agreement formally establishes site eligibility, and defines the work to be performed, registration fees to be paid, and termination provisions.[34] There are five situations provided in the agreement for termination:

1. the Director of VADEQ determines that no remedial action is necessary at the site,
2. remediation at the site has been successfully achieved,
3. the terms of the voluntary remediation agreement have been materially breached,
4. new information relative to site eligibility has been disclosed, or
5. either party elects to terminate the voluntary remediation agreement upon thirty days notice.[35]

Registration Fee

As part of the voluntary remediation agreement, the volunteer must pay a registration fee. This fee is required for VADEQ to perform review services.[36] The amount of the fee is the lesser of $5,000 or 1 percent of the cost of remediation.[37]

The volunteer has two options regarding the registration fee: pay the $5,000 or calculate an estimate of cleanup costs to determine the fee. To determine the registration fee, the volunteer may provide a generalized written estimate of the total cost of remediation. This includes costs for site investigation, report development, remedial system installation, and operation and maintenance. Before completion of the cleanup, the volunteer must provide the actual costs of cleanup for VADEQ to calculate balance adjustments. A balance owed to VADEQ must be paid before issuance of a completion certification by VADEQ.

Voluntary Remediation Report

Before completing work under the VRP, the volunteer must prepare a voluntary remediation report. The report is a compilation of all documentation pertaining to remedial activities at the site. The voluntary remediation report must contain three sections: (1) site characterization, (2) remedial action work plan, and (3) demonstration of completion.[38]

The site characterization is the investigative component of the report. The site characterization should include those things typical in a Phase II site investigation, such as a delineation of the nature and extent of releases to all media and an evaluation of the risks to human health and the environment posed by the release. Additionally, a proposed set of remedial standards that are protective of human health and the environment, as well as a recommended remedial action, should be described.[39]

The remedial action work plan is the design component of the report. This section should detail the activities, schedule, and specific design plans for implementing the remedial action established in the site characterization component.[40]

The demonstration of completion is the closure component of the report. It should include a detailed summary of the performance of the remedial action implemented at the site. Complete cost information should be included in this component. Finally, sampling results that confirm that the site-specific remedial standards have been met must be included.[41]

Certification of Satisfactory Completion of Remediation

The certification of satisfactory completion is the final step in the remediation process. This completion certification, which is issued by VADEQ, contains the liability-release provisions described earlier in this chapter.[42] The completion certification is issued after VADEQ examines the demonstration of completion portion of the voluntary remediation report. VADEQ will issue a completion certification if the report shows that the objectives of the remedial action work plan have been met and the established standards have been achieved.

Site Access

Neighboring properties often must be disrupted while conducting a voluntary remediation. Chapter 622 provides for access to property, not owned by the cleanup participant, to complete a voluntary remediation under the program. To invoke the provision, a request must be made to VADEQ, including a demonstration that the person requesting access has used reasonable efforts to obtain access.[43] VADEQ will then request access on behalf of the participant. Denial of access by a property owner creates a rebuttable presumption that the owner waives all claims against the participant performing the remediation.[44] Because the neighboring property could be the recipient of off-site migration, this rebuttable presumption could mean the loss of a claim against the party responsible for the contamination.

Cost Recovery

There are no cost-recovery provisions in Chapter 622. Further, there are no statutory causes of action for cost recovery between private parties in Virginia. As such, parties seeking reimbursement for cleanup costs attributable to another party must proceed through a common-law action or use the provisions of sections 107 or 113 of CERCLA.[45]

Conclusion

Despite the lack of regulatory guidance governing the Virginia VRP, in the first year of the VRP, VADEQ entered into agreements for twenty-three voluntary cleanups. That number is expected to increase dramatically when regulations governing the program are finalized in 1997. VADEQ has reviewed an internal draft regarding the program and a final draft for public comment was expected in the fall of 1996. The regulations are expected to codify the existing procedures governing the Voluntary Cleanup Program (with additional detail), so the program procedures should not change significantly when the rules are finalized. For now, voluntary cleanups must be negotiated on a case-by-case basis, consistent with the law as described in this chapter.

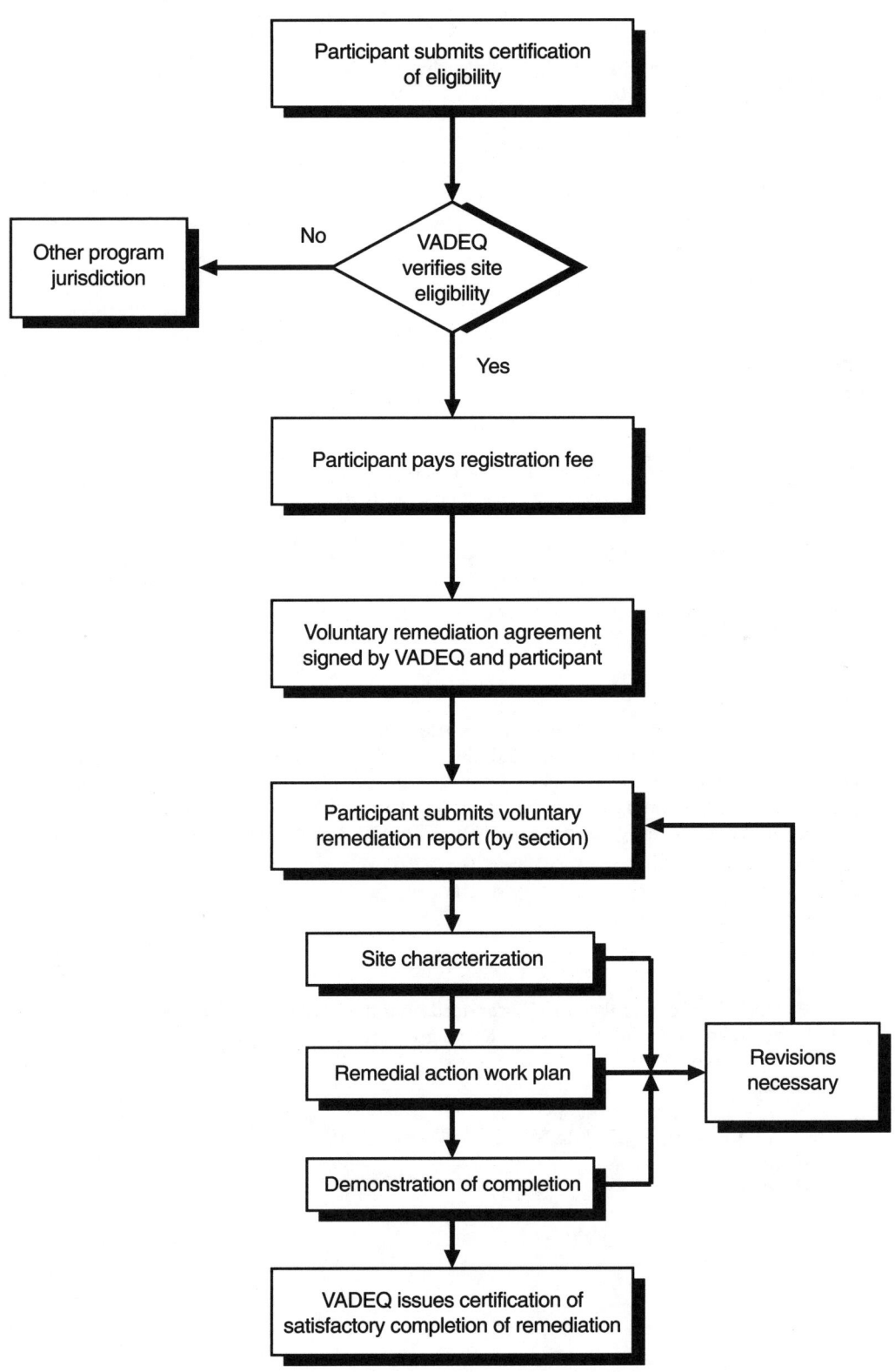

Notes

1. VA. CODE ANN. § 10.1-1429.1(A) (Michie 1995).
2. *Id.* § 10.1-1429.1(B).
3. *Id.*
4. *Id.*
5. *Id.*
6. *Id.*
7. *Id.*
8. *Id.* § 10.1-1429.2.
9. *Id.* § 10.1-1429.3.
10. *Id.* § 10.1-1429.1(A).
11. *Id.*
12. 42 U.S.C. § 9601 *et seq.* (1995).
13. *Id.* § 6901 *et seq.*
14. VA. CODE ANN. § 10.1-1400 *et seq.* (Michie 1995).
15. *Id.* § 62.1-44.2 *et seq.*
16. *See* 60 Fed. Reg. 34,792 (1995).
17. *See generally* VA. CODE ANN. ch. 14 (Michie 1995).
18. VA. CODE ANN. § 10.1-1429.1(A)(1) (Michie 1995).
19. *Id.*
20. Va. Dep't. of Envtl. Quality, Guidance for Conducting Risk Assessment for Voluntary Remediation Sites, § 2.1.1 (July 18, 1995) (draft internal guidance document).
21. *Id.* § III.
22. *Id.* § III.1.
23. *Id.* § III.3.
24. *Id.* App. V.
25. *Id.* § III B.
26. Va. Dep't. Envtl. Quality, *supra* note 20, § 2.1.2.
27. Va. Dep't. Envtl. Quality, Guidelines for the Application of Statutory Criteria for Voluntary Remediation (Sept. 1, 1995) (working draft). This is internal guidance that is being used as a discussion document for purposes of rule promulgation. The draft is likely to be altered. As such, the guidance should not be relied upon without consultation with VADEQ.
28. *Id.*
29. *Id.*
30. *Id.*
31. *Id.*
32. *Id.*
33. VADEQ is currently using a standard voluntary remediation agreement. However, the Agreement is considered to be internal guidance and may vary. *See* Va. Dep't Envtl. Quality, *supra* note 27, app. II.
34. *Id.*
35. *Id.*
36. VA. CODE ANN. § 10.1-1429.1(A)(5) (Michie 1995).
37. *Id.*
38. *Id.* § 5.
39. *Id.*
40. *Id.*

41. *Id.*
42. *See id.* § 10.1-1429.2.
43. *Id.* § 10.1-1429.3.
44. *Id.*
45. 42 U.S.C. §§ 9607, 9613 (1988).

CHAPTER 48

Washington

BRADLEY M. MARTEN
ROBIN K. ROCK

Washington has implemented a number of programs intended to spur the redevelopment of brownfields. Foremost among these is the Model Toxics Control Act (MTCA), Washington's Superfund, which allows prospective purchasers and lessors of contaminated property to enter agreements with the Washington State Department of Ecology (Ecology),[1] limiting their liability for cleanup costs. Washington's MTCA also includes a Voluntary Cleanup Program, known as the Independent Remedial Action Program (IRAP),[2] through which owners of contaminated land may obtain letters stating Ecology's intention to take no further action at a contaminated site. The Washington legislature also has adopted provisions limiting a lender's liability under MTCA.[3] These programs have improved the incentives, and removed barriers, to redevelopment of contaminated property in Washington. As of May 1996, Ecology had completed seven prospective-purchaser agreements and issued more than twenty-eight no-further-action (NFA) letters.[4]

Summary of Major Provisions

Prospective-Purchaser Agreements

Washington's prospective-purchaser legislation allows a potential purchaser or long-term lessee of contaminated property to resolve its potential liability for remedial action costs before purchase or lease, subject to certain reopeners, and it extends liability protection to successors-in-interest.[5] Such an agreement is documented in a consent decree issued by a state court and is available to parties who (1) are not already liable at a site, (2) can show that their project will provide a substantial public benefit, (3) yield new resources to facilitate cleanup, and (4) meet and expedite remedial state cleanup standards.[6]

IRAP

Ecology will review independent cleanups for consistency with state cleanup regulations if a fee is paid to cover the state's costs.[7] Participation in the IRAP is strictly voluntary, although all independent cleanups must be reported to Ecology.[8] After reviewing an independent remedial action report, Ecology provides a written determination that indicates whether the cleanup is complete. If so, Ecology will issue an NFA letter; if not, the state will specify the particular actions needed to meet cleanup requirements.[9]

Eligibility for the Prospective-Purchaser Program

Ecology has adopted an "interim" policy on prospective-purchaser agreements.[10] Generally, the state follows the steps in the interim guidance in deciding whether to enter an agreement. The decision to enter a prospective-purchaser agreement is up to Ecology and the Washington State Attorney General.[11] Both must approve the agreement, and the Attorney General acts as an independent decision maker, not solely as the lawyer for Ecology, its client. The applicant must have a legal commitment to purchase, redevelop, or reuse the site, in the form of a long-term lease or purchase option.[12] This commitment must include a schedule for starting the project within a "reasonable time frame."[13]

Prospective-purchaser agreements are available only when they would yield substantial new resources "to facilitate site cleanup."[14] Ordinarily, they will not be available when there is a party who already has a legal obligation to clean up the site. The applicant must demonstrate that it will expedite a cleanup that would not otherwise occur or would occur much later without the agreement. In deciding whether an applicant for a prospective-purchaser agreement has met this threshold, Ecology and the Attorney General's office consider

1. the adequacy of the existing environmental information about the site (the more, the better);
2. the certainty and the completeness of the proposed cleanup (the more, the better);
3. the availability of other potentially liable parties (PLPs) or other funds to complete the cleanup process (the fewer, the better);
4. whether private investment in the cleanup will preserve public funds that would otherwise need to be used; and
5. the site's priority (higher-priority sites are more favored).[15]

The state normally will not accept proposed agreements for sites where a remedial action is already underway under an order or decree.[16] Likewise, the state may not enter a prospective-purchaser agreement with a person already named as a PLP at the site.[17]

Memorandum of Agreement

There is no Memorandum of Agreement between the United States Environmental Protection Agency (USEPA) and the state of Washington governing the state's prospective-purchaser program. Therefore, an agreement with the state does not protect a prospective purchaser from federal liability. The practice in Washington is for a prospective purchaser to either negotiate a separate (or joint) agreement with USEPA, or obtain a "comfort letter" from USEPA Region X. There have also been instances in which a purchaser entered a prospective-purchaser agreement with USEPA and obtained an NFA letter from the state.

Liability Protection
Lenders, Trustees, and Fiduciaries

Washington's prospective-purchaser agreement program does not extend liability protection to lenders, trustees, and fiduciaries. The state has, however, enacted a lender-liability law.

Under MTCA, like the federal Comprehensive Environmental Response, Compensation, and Liability Act of 1980 (CERCLA), the term "owner or operator" is defined to exclude any person who, without participating in the management of a facility, holds indicia of ownership primarily to protect the person's security interest in the facility.[18] Legislation enacted in 1995 extends this exemption to include lenders after foreclosure and lenders who engage in policing activities primarily to protect the holder's security interest in the facility.[19] MTCA defines "policing activities" as those taken to ensure that the borrower complies with the terms of a loan or security interest, or actions the holder takes—or requires the borrower to take—to maintain the value of its security.[20] Examples include requiring the borrower to conduct remedial actions at the facility; requiring the borrower to comply, or come into compliance, with applicable federal, state, and local environmental laws; or securing or exercising authority to monitor or inspect the facility. A lender who wishes to take advantage of this exemption must

1. properly maintain the environmental compliance measures already in place at the facility;
2. comply with the reporting requirements in the MTCA regulations;
3. comply with any order issued to the holder by Ecology to abate an imminent or substantial endangerment;
4. allow access to Ecology or potentially liable persons under an order, agreed order, or settlement agreement to conduct remedial actions, and not to impede the conduct of such action;
5. comply with any preexisting requirements identified by Ecology during the implementation of any remedial action;
6. not exacerbate an existing release.[21]

There is no protection afforded lenders who cause or contribute to a new release or threatened release or who are otherwise liable under MTCA.[22] However, if the lender can establish that any new release has been remediated according to the requirements of MTCA, and that any hazardous substances remaining at the facility after remediation of the new release are divisible from the new release, liability protection will be preserved.[23]

Protection for Participants in the Prospective-Purchaser Program

Any settlement under MTCA must be entered as a consent decree in state court.[24] The statute specifies some of the provisions that must be included in a consent decree. For example, a covenant-not-to-sue must contain a reopener clause allowing Ecology to take or order additional remedial actions if factors not known at the time of entry of settlement are discovered and present a previously unknown threat to human health or the environment.[25] As with any MTCA settlement, a prospective purchaser who enters a consent decree with Ecology obtains protection from contribution claims brought under state law.[26] Such protection does not extend, however, to claims brought under federal law, or to claims other than for cleanup costs (such as personal injury or property damage claims).[27]

Ecology may, and routinely does, allow the benefits of the decree to be transferred with the property to future site owners. For successors to obtain the same protection, the following conditions must be met:

- The future site owner cannot be a PLP.
- The same or comparable public benefits must be maintained after the transfer.
- The subsequent purchase/development must comply with MTCA's substantive requirements, including cleanup requirements, and must not interfere with any remedial action or contribute to the existing release, or cause health problems.
- The successor must adhere to conditions in the decree.
- The state must be notified of the transfer of interest at least sixty days in advance of the transfer.
- The person to whom the agreement is being transferred must agree to become a party to the decree and to be bound by it.
- All other "site-specific factors" must be met.[28]

At the time of publication, legislation was proposed that would automatically transfer the consent decree protections, including contribution protection and the covenant-not-to-sue, to any subsequent owner of the property[29] who is not otherwise liable at the facility. This transfer will not apply if the consent decree was based on circumstances unique to the settling party that do not exist with regard to the successor, such as financial hardship.

The IRAP—No-Further-Action Letter

There are many situations in which a prospective purchaser of contaminated property will not qualify for, or will not wish to pursue, a prospective-purchaser agreement. An alternative is to pursue an NFA letter through the state's Voluntary Cleanup Program or IRAP.[30] The IRAP provides a prospective purchaser the option of having Ecology review its environmental report and provide a determination about the adequacy of the remedial action conducted on the property that it is considering acquiring. Alternatively, the seller may wish to pursue an NFA letter to improve the marketability of its property.

The IRAP does not provide binding legal protection from liability. The NFA letter is, in essence, an "advisory" opinion, which Ecology may change.[31] However, many existing property owners, prospective purchasers, and lenders feel that this level of review sufficiently meets their needs, and the IRAP has become quite popular in Washington.

To participate in the IRAP, an applicant must submit a report describing the nature of the hazardous-substance release and the remedial action taken. The report must be prepared in accordance with Ecology guidance, and a fee must be paid to cover the costs of Ecology's review.[32] The report may address the entire site or an operable unit at the site, or be used to confirm that no release ever occurred at the site.[33] After reviewing the report, Ecology will provide the applicant with a written determination that indicates whether further remedial actions are required at the site.

Ecology's initial review of an application for an NFA letter must be completed within ninety days of submission.[34] Ecology may issue an "interim" status letter if it feels that monitoring data is required before it can issue an NFA letter.[35] If Ecology believes additional action is required, it will issue a deficiency letter stating that the remedial action is insufficient or the report is inadequate. An applicant may reapply once the deficiencies are addressed.[36]

Financial Incentives

Washington does not have a formal program of loans, tax credits, grants, or other financial incentives to purchasers/developers of contaminated parcels. However, on a case-by-case basis, developers of contaminated lands may qualify for tax credits or other incentives under economic-development programs not directly tied to environmental programs.

Cleanup Standards

Washington's regulations provide for a certain amount of flexibility in determining what cleanup standards will be applicable to a given site. MTCA regulations generally provide two ranges of cleanup options: one for "residential" property, and the other for "industrial" property.[37] The goal is to restore property to a condition that poses an acceptable risk to human health and the environment, while not unduly restricting the property's future use. Something less than full removal of contaminants and complete restoration may be acceptable for many reasons, including past uses of the property and adjacent property, lack of effective cleanup technologies, and situations in which only marginal environmental benefits would be achieved by a more stringent cleanup. These issues should be negotiated with Ecology during the prospective-purchaser agreement application process.

Ecology uses a two-step selection process in selecting a remedy. The first step is selecting appropriate "cleanup levels" for impacted media (such as soil and groundwater), including the "points of compliance" for measuring these levels.[38] These cleanup levels must be in compliance with all "applicable or relevant and appropriate requirements" of federal, state, and local laws.[39] The second step in establishing cleanup requirements is selecting a cleanup action.[40]

The commonly understood concept of setting cleanup standards actually includes two distinct decisions under MTCA—"cleanup levels" and "points of compliance." A cleanup level is the concentration of a hazardous substance that Ecology deems safe for exposure to people and biota. Although many PLPs focus almost exclusively on these cleanup levels, it is equally important to know *where* the cleanup levels will be measured. The locations on a site where these levels must be achieved are called the points of compliance.

Cleanup Levels

MTCA differs significantly from CERCLA in its methods for setting cleanup levels. Unlike CERCLA, MTCA uses well-defined formulas to derive cleanup levels, without the extensive site-specific analysis used under CERCLA. Fortunately, MTCA does allow the consideration of some site-specific characteristics, particularly the subject site's future land use. For example, if property is designated as "residential," MTCA requires Ecology to use the most stringent levels. If, on the other hand, property is classified for future industrial use, Ecology may use a less stringent method to set cleanup levels. In general, the regulations provide three options for establishing site-specific cleanup levels: Methods A, B, and C.[41]

Method A is used only for straightforward, routine cleanups at small sites.[42] Method A is simply a chart with predetermined cleanup levels for twenty-five of the most common hazardous substances.[43] Because Method A is designed to be used at a wide range of sites, the cleanup levels are set using very conservative risk assumptions. Method A does prescribe different cleanup levels for industrial[44] and residential sites.[45]

Method B is the most common method for determining cleanup levels. Unless a site specifically qualifies for Methods A or C, MTCA assumes that the site is potentially subject to residential use, which requires Ecology to set protective cleanup levels under Method B.[46] Under Method B, cleanup levels are set using a risk-based algorithm, with most of the variables specified in the MTCA regulations.[47]

Use of the less stringent Method C is allowed in limited circumstances.[48] Exposure assumptions are still predetermined but are less conservative than under Method B. For example, the risk level for carcinogens is set at equal to or less than one in one hundred thousand (1×10^{-5}) under Method C.[49] The circumstances under which Method C is allowed include (1) when Method A or B levels are higher than background levels at the site, (2) when attainment of these Method A or B levels would create a greater overall threat to the environment, or (3) when the site is a qualifying industrial site under section 173-340-745 of the Washington Administrative Code.[50]

Point of Compliance

The point of compliance is the point on the site where cleanup levels are measured and must be met.[51] In most instances, cleanup levels must be met throughout the site. In certain instances, however, the point of compliance may be set at the property boundary or, in the case of soils, beyond the property boundary.[52] Establishing a "conditional point of compliance" can significantly reduce the cost of a remedy, while still providing the requisite protection of human health and the environment.

Compliance with ARARs

After Ecology has determined the cleanup levels and selected the point of compliance, it must also ensure compliance with "applicable or relevant and appropriate requirements" of other laws (also known as ARARs).[53] Compliance with ARARs involves a two-part analysis. First, Ecology will decide if a given requirement is applicable; that is, if it "specifically addresses a hazardous substance, cleanup action, location or other circumstances" at the subject site.[54] If it deems the requirement not applicable, Ecology will then look to see if the requirement is nevertheless both relevant and appropriate. A requirement is relevant and appropriate if, notwithstanding the fact that it is not legally applicable, it addresses problems or situations "sufficiently similar to those at the site that [its] use is well suited to the particular site."[55]

Cleanup Action

After determining the cleanup levels for each substance, Ecology will then choose the appropriate methods to remediate the site to those standards. MTCA requires Ecology to use "permanent" cleanup methods "to the maximum extent practicable."[56] Ecology may take cost into account in selecting cleanup actions.[57] It may also take into account technological limitations.

Ecology will then select "cleanup action levels" to implement the remedy.[58] These cleanup action levels must be achieved within a "reasonable restoration time frame." As an example, Ecology might set a cleanup action level of 1,000 parts per million (ppm) of total petroleum hydrocarbons in soil at a particular site, thus requiring that all soil above this level be remediated through excavation and thermal desorption or some other technology. Soil below the 1,000 ppm "cleanup action level" would not be remediated, even though it is above the Method A cleanup level of 200 ppm. In other words, Ecology may allow cleanup action levels that exceed cleanup levels at portions of a site, provided the more stringent cleanup level is met at the "point of compliance."

Risk Assessments

The preparation of a remedial investigation and feasibility study and the selection of cleanup standards may include preparation of a risk assessment.[59] The risk assessment must characterize the current and potential threats to human health and the environment that may be posed by hazardous substances. This assessment may not be required when Ecology determines that proposed cleanup standards are obvious and undisputed and allow an adequate margin of safety for protection of human health and the environment.

Institutional Controls

Institutional controls are a part of the MTCA remedy selection process. Under MTCA regulations, institutional controls are one of the specified technologies to be considered as part of the cleanup action plan.[60] Legislation enacted in 1995 clarified and strengthened Ecology's authority to use institutional controls. Regulations implementing the legislation also clarified that institutional controls do not automatically mean deed restrictions, but may instead include other forms of land-use control.

MTCA differentiates between the types of institutional controls that are appropriate for land owned by PLPs, and those for adjacent property. For land owned by PLPs, appropriate institutional controls generally include deed restrictions (restrictive covenants).[61] The covenant must be executed by the property owner and recorded with the register of deeds for the county in which the property is located. The covenant runs with the land and is binding on the owner's successors and assigns. However, for properties containing hazardous substances when the land owner is not a PLP, restrictive covenants are not necessary.[62] Other forms of institutional controls include physical measures (such as fences and signs) and legal and administrative measures (such as educational mailings, zoning overlays, and placing notices in local zoning or building department records and state land records).[63]

Mechanics of Participating in Washington's Prospective-Purchaser Program

Generally, the prospective-purchaser agreement process contains three distinct phases: the initial submittal, the detailed submittal, and the consent decree. The applicant begins the process by preparing an initial submittal to Ecology. The level of detail in this submission will vary depending upon the availability of information. However, Ecology will generally look more favorably upon applicants who have a well-characterized site and a well-developed remedial action plan. At a minimum, the following information must be included in the initial submittal:

- A description of the facility and a description of the proposed remedial action:

This should include information that demonstrates that the settlement will lead to a more expeditious cleanup and resolution of the environmental problems to be addressed at the facility. It should also include sufficient information to rank the site. It should include any special scheduling consideration for implementing the remedial actions, including the current proposed schedule for purchase, redevelopment, or reuse of the site.
- A summary of the relevant historical-use conditions at the facility:
   This should include names of other persons who the applicant has reason to believe may be PLPs at the facility.
- A proposed public-participation plan
- Identification of the applicant, including relevant corporate information
- A general description of the proposed development and the public benefits of the development and the settlement
- The date on which the person proposing the settlement will be ready to submit a detailed proposal[64]

The state will review the initial application for a prospective-purchaser agreement at no charge and respond to the applicant within sixty days.[65] If the initial application is accepted, Ecology will ask the applicant to "prepay" the state's costs of negotiations, and ask for a detailed application. The detailed application must include the following:

- A remedial investigation/feasibility study that complies with MTCA
- A proposed cleanup plan for the site
- Identification of the applicant's proposed share of the cleanup costs and information describing the applicant's ability to conduct or finance the proposed remedial actions and the purchase and redevelopment:
   For proposals indicating that the current owners are expected to pay for part of the cleanup, a notarized letter from the current owner accepting status as a PLP and indicating a willingness to negotiate cleanup is necessary.
- A detailed description of, and an updated schedule for, the purchase and redevelopment of the site:
   The applicant should include information demonstrating that use of the property will not contribute to the existing release or threatened release or interfere with the remedial actions that may be needed, and information demonstrating that the proposed use for the site will not likely increase human health risks to persons at, or in the vicinity of, the site.
- Notarized letters from the current owners and operators of the site authorizing Ecology to access the site:
   Notarized letters from the applicant and the seller, indicating that they have fully disclosed all relevant information, should also be included.

- A detailed description of the substantial public benefits of the project
- A proposed schedule for negotiating the prospective-purchaser agreement
- Any additional information requested by Ecology or thought necessary by the applicant[66]

Ecology and the Attorney General's office jointly review the detailed application and make a discretionary decision whether to accept it, generally within sixty days of submission.[67] If the application is accepted, the applicant enters consent decree negotiations with the agencies. During this phase, the applicant and the agencies negotiate particular terms of the consent decree, including the covenant-not-to-sue. Once finalized, the decree is issued for public review and comment, generally for thirty days.[68] After this period has ended, Ecology must complete a responsiveness survey, responding to any comments received. After the comment period, Ecology will lodge the decree in state court and move for entry.

Cost Recovery

MTCA allows private rights of action and contribution actions for cost recovery.[69] A private right of action is available only if the cleanup is the "substantial equivalent" of an Ecology-conducted or supervised cleanup.[70] The meaning of "substantial equivalent" is outlined in the regulations, providing guidance for those conducting independent actions.[71] Generally, such cases must be brought within three years from the date that remedial action confirms that cleanup standards have been met.[72]

Unlike CERCLA, MTCA authorizes prevailing parties to recover their reasonable attorneys' fees and expenses.[73] The recovery of attorneys' fees, in combination with the inclusion of petroleum and petroleum products as hazardous substances, represents a substantial incentive to bring recovery actions under MTCA rather than CERCLA.

Conclusion

There are a number of programs available to aid the redevelopment of brownfields in Washington. Depending on the size and timing of the project, a developer, owner, or potential owner may choose to pursue either a prospective-purchaser agreement or a no-further-action letter from the Washington State Department of Ecology. In addition, the new MTCA lender-liability provisions make Washington an attractive state for lenders. These programs have already spurred the redevelopment of property in Washington, a trend that can be expected to continue in the coming years.

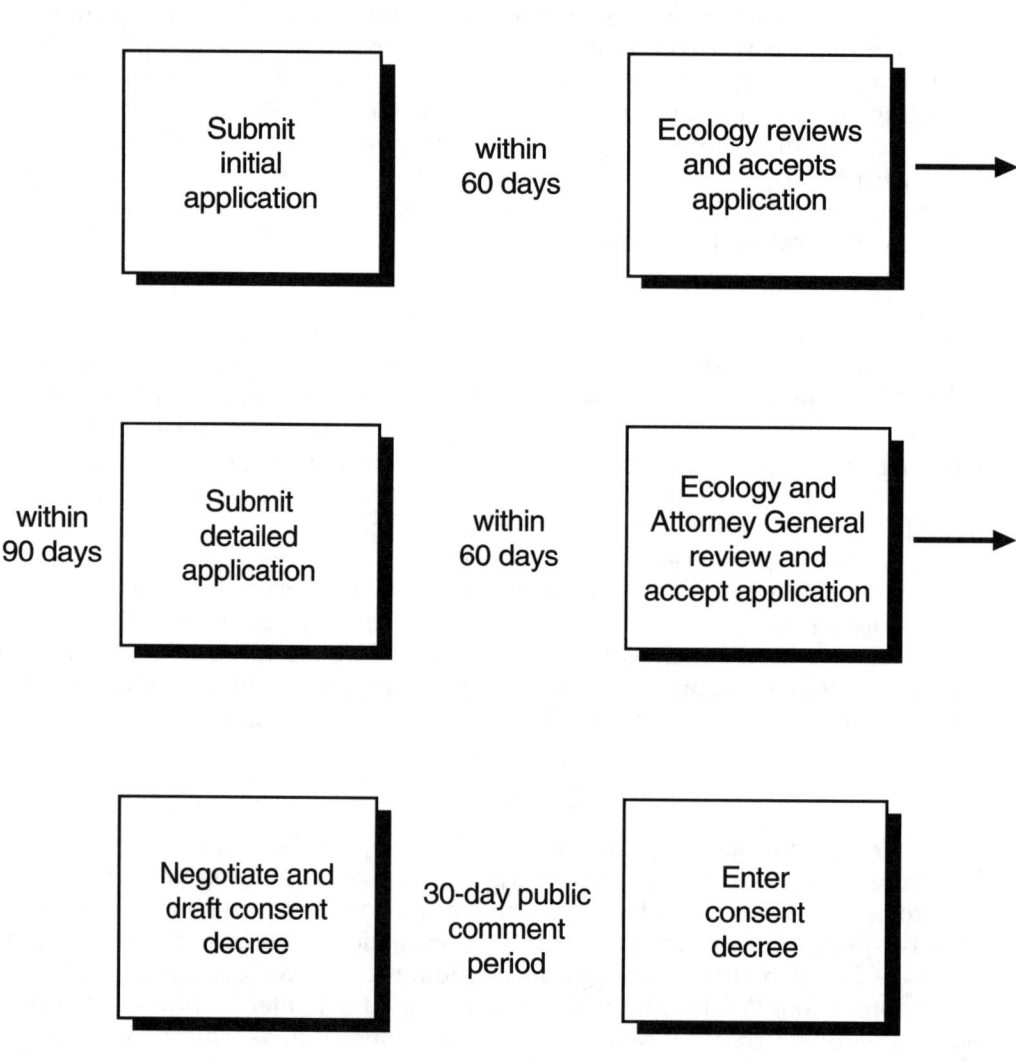

Notes

1. WASH. REV. CODE § 70.105D.040(5) (1990 & Supp. 1995).
2. *See generally* Wash. State Dep't Ecology, Independent Remedial Action Program, Questions and Answers (Jan. 1995) [hereinafter IRAP Questions and Answers].
3. WASH. REV. CODE § 70.105D.020(11) (1990 & Supp. 1995).
4. *See* Wash. State Dep't Ecology, Focus Sheet, State Review of Independent Cleanups (Jan. 1994, revised Aug. 1995).
5. *See* Wash. State Dep't Ecology, Prospective Purchaser Agreements—Interim Policy (Aug. 29, 1994) [hereinafter Interim Policy].
6. WASH. REV. CODE § 70.105D.040(5) (1990 & Supp. 1995).
7. WASH. ADMIN. CODE § 173-340-310 (1995); *see also* IRAP Questions and Answers, *supra* note 2.
8. WASH. ADMIN. CODE § 173-340-300 (1996).
9. *Id.* § 173-340-310(4); *see also* IRAP Questions and Answers, *supra* note 2.
10. *See* Interim Policy, *supra* note 5.
11. *Id.*
12. *Id.*
13. *Id.*
14. WASH. REV. CODE § 70.105D.040(5)(b) (1990 & Supp. 1995).
15. *See* Interim Policy, *supra* note 5.
16. *Id.*
17. WASH. REV. CODE §§ 70.105D.040(4), 70.105D.040(5) (1990 & Supp. 1995).
18. *Id.* § 70.105D.020(11).
19. *Id.*
20. *Id.* § 70.105D.020(14).
21. *Id.* § 70.105D.020(11).
22. *Id.*
23. *Id.*
24. *Id.* § 70.105D.040(4)(b).
25. *Id.* § 70.105D.040(4)(c).
26. *Id.* § 70.105D.040(4)(d).
27. *Id.*
28. *See* Interim Policy, *supra* note 5.
29. S.B. 7900.
30. *See* IRAP Questions and Answers, *supra* note 2.
31. *Id.*
32. *Id.*
33. *Id.*
34. WASH. ADMIN. CODE § 173-340-300(5) (1995).
35. *Id.* § 173-340-310(4)(c).
36. *See* IRAP Questions and Answers, *supra* note 2.
37. *See, e.g.,* WASH. ADMIN. CODE §§ 173-340-740, 173-340-745 (1995).
38. *Id.* § 173-340-700.
39. *Id.* §§ 173-340-700(4)(A), 173-340-710.
40. *Id.* § 173-340-700(2)(a).
41. *Id.*
42. *Id.* § 173-340-704.
43. *Id.*
44. *Id.* § 173-340-745.

45. *Id.* § 173-340-740.
46. *Id.* § 173-340-705.
47. *Id.*
48. *Id.* § 173-340-706.
49. *Id.*
50. *Id.*
51. *Id.* §§ 173-340-720(6), 173-340-730(6).
52. The point of compliance for groundwater, however, may not extend beyond the property boundary. *Id.*
53. WASH. REV. CODE § 70.105D.030(2)(d) (1990 & Supp. 1995); WASH. ADMIN. CODE § 173-340-170 (1995).
54. WASH. ADMIN. CODE § 173-340-710(2) (1995).
55. *Id.* § 173-340-710(3).
56. *Id.* § 173-340-360.
57. *Id.*
58. *Id.*
59. *Id.* § 173-340-350.
60. *Id.* § 173-340-360(4).
61. *Id.* § 173-340-440(4)(a).
62. *Id.* § 173-340-440(4)(b).
63. *Id.*
64. *See* Interim Policy, *supra* note 5.
65. *Id.*
66. *Id.*
67. *Id.*
68. *Id.*; WASH. ADMIN. CODE § 173-340-600 (1995).
69. WASH. REV. CODE § 70.105D.080 (1995).
70. *Id.*
71. WASH. ADMIN. CODE § 173-340-550(5) (1995).
72. WASH. REV. CODE § 105D.080 (1990 & Supp. 1995).
73. *Id.*

CHAPTER 49

West Virginia

BARBARA D. LITTLE

West Virginia's Voluntary Remediation and Redevelopment Act[1], effective July 1, 1996 (the Act), encourages redevelopment of former industrial sites by attempting to provide legal certainty as an inducement to development. The Act affects all properties that are or may be contaminated, and includes special provisions for brownfields sites, defined as industrial or commercial properties that are abandoned or not being actively used by the owners. Rules for implementation of the Act have been proposed (60 CSR 3) by the West Virginia Director of Environmental Protection and should be effective, assuming legislative enactment and the Governor's approval, by June of 1997. The Act offers limited financial incentives to a development authority or person who did not cause or contribute to the contamination of the property. The West Virginia Legislature's emphasis on voluntary compliance is a shift from traditional enforcement action, and the Act encourages brownfields development by increasing the number of tools available for achieving compliance.

Summary of Major Provisions of the Act

The Act includes the following major provisions:

- Limited financial incentives for brownfields sites, including funding for site assessment loans and remediation loans
- Liability limitations for persons who remediate in compliance with applicable standards
- Procedures for licensure of remediation specialists, approves the use of such specialists for compliance monitoring activities, and mandates sanctions for failure to perform
- Community notification procedures and community input in formulating a brownfields remediation plan
- Criminal penalties for persons associated with a remediation project who engage in fraudulent acts or representation

- Certificate of completion, issued by the Director of Environmental Protection, which provides a limitation on liability to site remediation participants
- Reopener clause in particular circumstances
- Limited protection of further remediation liability to current or future owners or operators of the site, site developers, successors or assigns, public utilities, remediation contractors, licensed remediation specialists, lenders, the parties requesting the site assessment, and the party performing the site assessment
- Affirmative defenses
- Engineering controls (such as capping or treatment) or institutional measures (such as deed restriction covenants) as tools for environmental compliance

Eligibility for the Voluntary Action Program

Eligible Parties

Only certain parties are eligible for the voluntary action program.[2] A person or development authority[3] who did not cause or contribute to the contamination of the property may apply for a site assessment loan and remediation loans under the Act.

Eligible Sites

The Act affects two classes of property: (1) brownfields—those properties that are derelict (that is, abandoned or not being used by the owner at the time the Act goes into effect), and (2) all other properties to be investigated or remediated under the Act.[4] Brownfields sites are eligible for money from one or both of two funds established under the Act. Nonbrownfields sites would not qualify for such funds. To be eligible as a brownfields site, the industrial or commercial property must be abandoned or not actively used as of July 1, 1996.

Ineligible Sites

No property qualifies for the voluntary remediation program if a unilateral final order has been issued under any state or federal environmental program or if the willful negligence or gross misconduct of the person wishing to perform the remediation caused the contamination of the site.[5] The term "unilateral enforcement order" means a written final order issued by a federal or state agency, compelling the fulfillment of an obligation imposed by law or rule against a person without the person's voluntary consent.[6] Unilateral enforcement orders do not include notices of violation.

Liability Protection

The Act provides liability protection for site owners or operators, development authorities, lenders, developers, remediation contractors, and fiduciaries for environmental contamination at a brownfield or voluntary remediation site.[7] Once the participant completes the remediation of the subject site, the Act contains no provisions for postremediation audits. Furthermore, certificates of completion may be sought even when a site assessment demonstrates that remediation is unnecessary. A certificate of completion provides a clean bill of health for uncontaminated sites thus making site marketing efforts easier. Within one year following enactment of the Act, the Director of Environmental Protection must establish cleanup levels, risk-assessment procedures, and voluntary agreement guidelines, and must promulgate any other rules needed to implement the Act.

Participants, Successors, and Assigns

Remediators may apply to the Department of Environmental Protection (DEP) for a certificate of completion once the work contemplated in a remediation agreement is completed and either a Licensed Remediation Specialist (LRS) submits a final report to the remediator certifying that the property meets all applicable standards and that all work has been completed, or a site assessment shows that all applicable standards are being met. The certificate of completion will include a release relieving the remediator and subsequent successors and assigns from all liability to the state of West Virginia for further remediation of the property. This release remains effective as long as the property complies with applicable standards in effect at the time that DEP issued the certificate of completion.

A certificate of completion may require a land-use covenant[8] if institutional and engineering controls are used, in whole or in part, to achieve a remediation standard. The land-use covenant must contain a release relieving remediators and their subsequent successors and assigns from all civil liability to the state for further remediation, and remains effective as long as the property complies with applicable standards in effect when the covenant was issued. The land-use covenant must appear in the chain of title by deed properly recorded in the office of the county clerk where the remediation site is located. DEP's Director will establish rules for recording land-use covenants and all necessary deed restrictions.

In addition to the release from liability for further remediation, the Act provides that contamination identified in a remediation agreement submitted to, and approved by, DEP will not be subject to citizens' suits or contribution actions.[9]

Other Parties Protected from Liability

The following persons are protected from further remediation liability by the Act: (1) current or future owners or operators of the site, including development authorities and fiduciaries who participated in the remediation of the site, (2) persons who develop or otherwise occupy the site, (3) successors or any assigns of any person protected from liability, (4) public utilities and utilities engaged in the storage and transportation of natural gas, to the extent the public utility performs activities on the site, (5) remediation contractors, (6) LRSs, and (7) lenders or developers who engage in routine commercial lending, including the provision of financial services, holding of security interests, work-out practices, foreclosure, or the recovery of funds from the sale of a site. Persons do not receive liability protection simply by conducting or having a site assessment conducted (a certificate of completion must be obtained), nor does the Act relieve persons from liability for failing to exercise due diligence in performing a site assessment.

Special Liability Protection for Remediation Contractors

The Act provides additional liability protection for remediation contractors.[10] The Act relieves remediation contractors from liability for any harm, damage, or injury caused by a release of a contaminant that occurred before beginning work at a site. In addition, remediation contractors are not liable for a release or a threatened release of contaminants resulting from work properly performed pursuant to a remediation agreement. These protections for remediation contractors do not apply if the remediation contractor's gross negligence or willful misconduct causes a release or a threatened release. Remediation contractors are not required to obtain permits for remediation activities, although the owner or operator of the site may be required to obtain a permit, when applicable.

Reopening Remediation

Further remediation of a property may be ordered, despite issuance of a certificate of completion, in certain limited circumstances.[11] The DEP Director may order further remediation of a site if (1) the certificate of completion was issued as a result of fraud, (2) unremediated areas are discovered at the site, or (3) new information is obtained about a substance at the site that revises exposure assumptions beyond acceptable ranges. DEP can also, in limited situations, require additional remediation when preferred remediation efforts have become technically and economically practicable. Although the DEP Director bears the burden of demonstrating that the conditions for reopening and requiring additional remediation have been met, the Act provides no process for appealing a decision to require additional remediation.

Affirmative Defenses

The Act provides specific affirmative defenses available to any person alleged to have violated an environmental law or the common-law equivalent while conducting remediation activities.[12] These affirmative defenses include (1) an act of God, (2) an intervening act of a public agency, (3) migration from property owned by a third party, (4) actions taken or omitted in the course of rendering care, assistance, or advice in accordance with environmental laws or at DEP's direction, (5) an act of a third party not an agent or employee of a lender, fiduciary, developer, remediation contractor, or development authority, or (6) situations in which the alleged liability occurred after foreclosure and the lender, fiduciary, developer, or development authority exercised due care concerning its knowledge about the contaminants and, based upon such knowledge, took reasonable precautions against foreseeable actions of third parties and the consequences arising therefrom. A lender, fiduciary, developer, remediation contractor, or development authority may avoid liability by proving any other defenses that may be available to it.

Criminal Penalties

The Act provides for criminal sanctions as described below:

- Any person who divulges or discloses any information that the DEP Director certifies to be confidential may be guilty of a misdemeanor and, if convicted, can be fined not more than five thousand dollars, imprisoned in a county jail for not more than one year, or both fined and imprisoned.[13]
- If an LRS has violated provisions of the Act or any rules promulgated by the DEP Director, the DEP Director may issue an order suspending or revoking licenses or requiring the LRS to take remedial action, or issue a cease and desist order and request the prosecuting attorney of the county in which the alleged violation occurred to bring a criminal action. Any person issued an order may file a reconsideration request with the DEP Director within seven days of receipt of the order. Within ten days after the filing, the DEP Director must conduct a hearing, but filing does not stay or suspend execution or enforcement of the order. Furthermore, if an LRS fraudulently misrepresents that the work has been completed and such action results in unjustified and inexcusable disregard for the safety of others, resulting in imminent danger or contributing to ongoing harm to the environment, the LRS may be guilty of a felony. If convicted, the LRS can be fined not more than fifty thousand dollars, imprisoned not less than one nor more than two years, or both fined and imprisoned. If any person associated with remediation of a brownfield or voluntary remediation site engages in fraudulent acts or representations, that person may be guilty of a felony.[14]
- Whoever violates a land-use covenant by converting nonresidential property to residential property is guilty of a felony, and if convicted can be fined not more than twenty-five thousand dollars, imprisoned not more than five years, or both.[15]

Incentives under the Act

The Legislature identified one purpose of the Act as providing financial incentives to entice investment in brownfields sites.[16] Adding the remediation of contaminated property as a project eligible for tax increment financing is identified as one of the objectives of the Act.[17] The Act created two special revenue funds, including the "Voluntary Remediation Administrative Fund" and the "Brownfields Revolving Fund." The first fund consists of fees collected by the DEP Director plus interest earned. This fund will be expended primarily for administrative, licensing, enforcement, inspection, monitoring, planning, research, and other activities required by the Act.[18] The "Brownfields Revolving Fund" will be used to make loans to persons for site assessments of eligible brownfields sites, and other authorized activities including costs incurred in administering this fund. This fund consists principally of money expressly allocated to the state by the federal government for establishing and maintaining a state brownfields redevelopment revolving fund, and moneys appropriated by the Legislature.[19] Persons who did not cause or contribute to the contamination of the site will be considered for remediation loans.[20] The Act provides that development authorities, but not other government entities, may be given financial and other assistance from DEP in remediating property.[21] Additionally, access to information in the possession of the DEP Director is available to persons undertaking brownfields remediation who did not cause or contribute to the contamination of the brownfields site.[22]

Mechanics of Participating in the West Virginia Voluntary Action Program

Applying for Coverage under the Act

Regardless of whether a person wishes to remediate a brownfields site or another type of property, when applying for coverage under the Act[23] such person must complete a site assessment and file an application to remediate with the DEP Director.[24] All information about a remediation is within the public domain, except information that is expressly marked or identified as a trade secret.[25] The DEP Director must negotiate a written agreement with the remediator, containing terms and conditions necessary to carry out the remediation once the application is approved.[26] If the DEP Director rejects an application, applicants may appeal to the state Environmental Quality Board.[27] Under the Act, the DEP Director may reject or return an application only because (1) a federal requirement precludes the site's eligibility for participation, (2) the application is not complete or accurate, or (3) the site is ineligible under the provisions of the Act. An application fee, to be established through rule making, must accompany applications for participation in the voluntary remediation program. The remediation agreement must provide for DEP's recovery of all reasonable costs—incurred in reviewing the remediator's work plan and reports—that are in excess of the fees submitted by the applicant. One important component of the Act is that DEP may not take enforcement action against a person for the contamination that is the subject of a remediation agreement, or for activity that resulted in the contamination, unless there is an imminent threat to the public.

Cleanup Standards

The DEP Director has proposed rules (60 CSR 3) that establish (1) risk-based standards for remediation, (2) standards for the remediation of property, (3) a risk protocol for conducting risk assessments and establishing risk-based standards, and (4) criteria to evaluate and approve measurement methods of contaminants using the practical quantitation level and related laboratory standards and practices to be used by certified laboratories.[28] The Act specifically allows municipalities and local governments to participate in the remedy selection process.[29]

LRSs

Persons carrying out a voluntary remediation must use an LRS to oversee work approved by the DEP Director.[30] The DEP Director must establish a written test that all LRSs must pass to be qualified to oversee remediation. In addition, LRSs must demonstrate a practical knowledge of remediation activities. LRSs must renew their licenses every two years.

The LRS is responsible for any release of contaminants during remediation activities taken under the approved remediation agreement, work plans, or reports generated by the remediation activity. If the LRS is unable to protect the safety, health, and welfare of the public, the LRS may either cease his/her relationship with the remediator or refuse professional responsibility for the work plan, report, or design. LRSs also must notify DEP if there is a threat to the environment or public health, safety, or welfare. LRSs may be subject to civil or criminal penalties if they violate the Act.

Special Provisions under the Act Applicable to Brownfields Sites

Several special provisions in the Act apply only to brownfields sites being voluntarily remediated.[31] Persons remediating brownfields sites must comply with public notice and public review requirements. These requirements include submitting a notice of intent to remediate to DEP, which is then published by DEP in a publication of general circulation. At the same time, DEP must provide a copy of the notice of intent to the municipality and the county in which the site is located, and a summary of the notice of intent must be published in a newspaper of general circulation serving the area in which the site is located. The public notice must include a thirty-day comment period, during which the public, county, and municipality can request involvement in the development of the remediation and reuse plans for the brownfields site. If requested by such persons, the remediator must develop and implement a public involvement plan, meeting requirements set forth by DEP. Persons remediating nonbrownfields sites are not subject to any public participation requirements.

Conclusion

The creation of West Virginia's voluntary remediation program by passage of the Act is a big step toward defining the liability of individuals and entities interested in remediating brownfields and other contaminated sites. In addition, once the rules have been promulgated, a clearer definition of acceptable remediation standards and methods will be available as an encouragement to undertaking a cleanup. Finally, the financial incentives now available as a result of the creation of the two special revenue funds should, in the private development community, ignite interest in brownfields redevelopment in West Virginia.

West Virginia Voluntary Cleanup Program

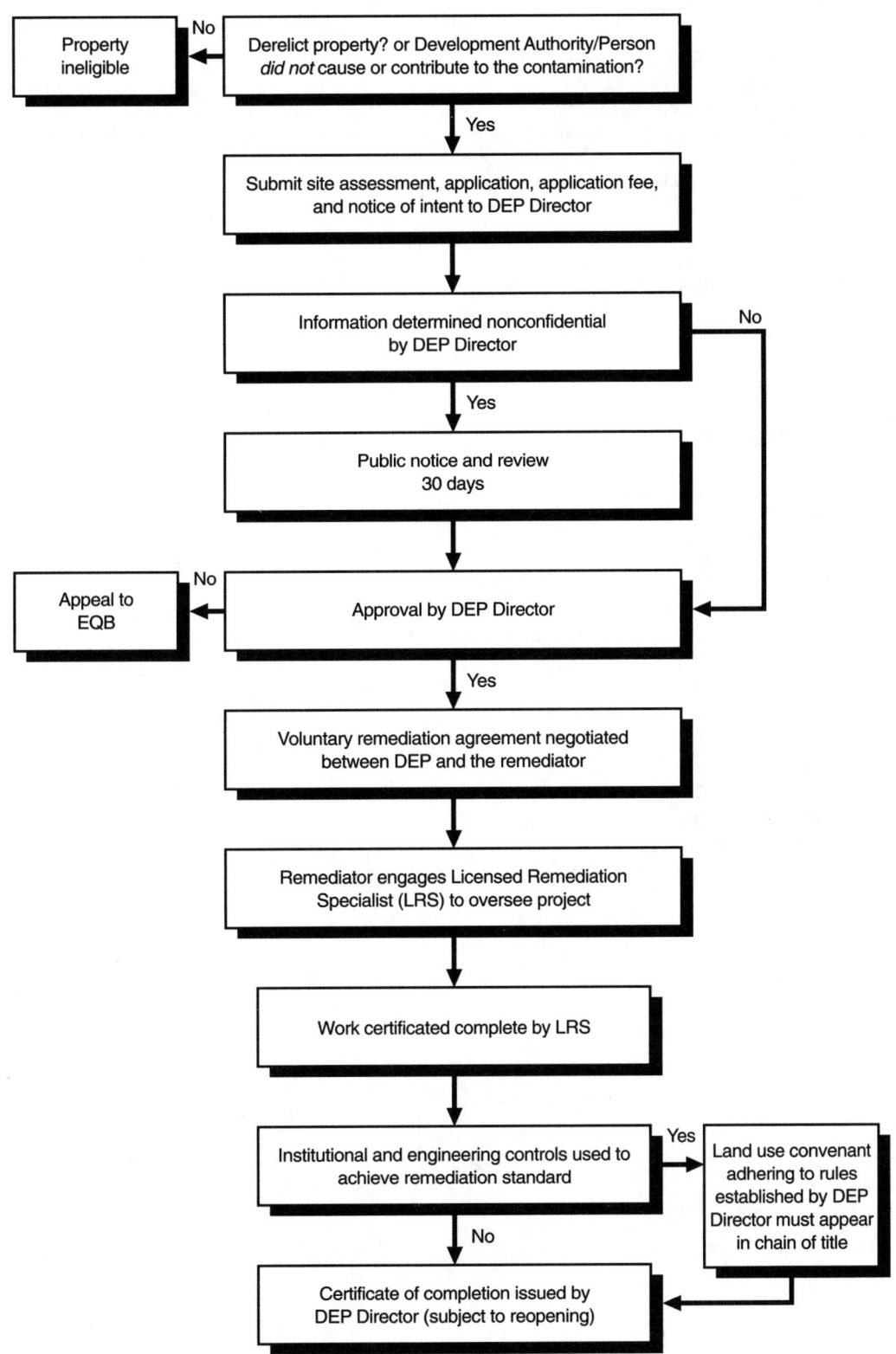

Notes

1. W. VA. CODE § 22-22 *et seq.* (1996).
2. *See generally id.* § 22-22-4.
3. "Development authority" is defined in sections 5B-2-5 (West Virginia Development Office) and 7-12 (County and Municipal Development Authorities of the West Virginia Code).
4. *See generally* W. VA. CODE § 22-22-5 (1996).
5. *Id.* § 22-22-4.
6. *Id.* § 22-22-2(ee).
7. *See generally id.* § 22-22-18.
8. *Id.* § 22-22-14.
9. *Id.* § 22-22-18.
10. *See generally id.* § 22-22-19.
11. *Id.* § 22-22-15.
12. *Id.* § 22-22-20.
13. *Id.* § 22-22-4(d).
14. *Id.* § 22-22-12.
15. *Id.* § 22-22-14.
16. *Id.* § 22-22-1(d)(2).
17. 1996 W. Va. Acts 4605.
18. W. VA. CODE § 22-22-6(a).
19. *Id.* § 22-22-6(b).
20. *Id.* § 22-22-5(c).
21. *Id.* § 22-22-5.
22. *Id.*
23. *See generally id.* §§ 22-22-4, 22-22-7(f).
24. *Id.* § 22-22-4.
25. *Id.*
26. *Id.* § 22-22-7.
27. *Id.* § 22-22-4.
28. *Id.* §§ 22-22-3(d)–(h).
29. *See generally id.* § 22-22-2(t).
30. *Id.* § 22-22-11.
31. *See generally id.* § 22-22-17.

CHAPTER 50

Wisconsin

ARTHUR J. HARRINGTON

Wisconsin's new Land Recycling Act, codified at sections 292.11–.21 of the Wisconsin Statutes (Land Recycling Act, or Act), provides an exemption from liability for certain qualified categories of parties as an incentive to develop contaminated property in Wisconsin. The Land Recycling Act, which became effective on May 13, 1994,[1] is the first attempt by the Wisconsin legislature to provide cleanup exemptions as an incentive to develop contaminated property. In addition, the Wisconsin Department of Natural Resources (DNR) has adopted new soil cleanup standards, which for the first time codify cleanup levels for certain identified contaminants found in the soil. Furthermore, DNR is in the process of adopting a less stringent groundwater cleanup program, specifically designed to save an estimated $800 million on groundwater cleanup in Wisconsin. All these developments provide significant incentives for parties to develop contaminated property located in urban settings in the state of Wisconsin.

Summary of Major Provisions

The major provisions of the Land Recycling Act and other brownfields redevelopment initiatives in Wisconsin include the following:

- The establishment of exemptions from cleanup liability under the Wisconsin Spill Statute (defined below) for certain qualified parties, such as purchasers, lenders, fiduciaries, and environmental taxing authorities, who take title to contaminated, tax-delinquent property
- New soil cleanup standards, which provide not only objective criteria for soil cleanups, but also the opportunity to formulate site-specific cleanup standards based on a risk/benefit analysis
- An expansion of Wisconsin's tax incremental-financing law, to encourage use of innovative financing for environmental cleanups in urban areas

- The development of a new, less stringent cleanup program for groundwater contamination, which is estimated to save over $800 million in costs for the cleanup of groundwater that otherwise would be required under existing law
- A Memorandum of Understanding with the United States Environmental Protection Agency (USEPA), which provides Wisconsin with formal, lead cleanup authority for hazardous-substance remediations in the state

Cleanup Liability—the Wisconsin Spill Statute

The primary statutory authority used by DNR for environmental cleanups is the Wisconsin Spill Statute (the Spill Statute).[2] The Spill Statute gives DNR broad authority (1) to require parties responsible for environmental contamination to take actions to prevent adverse impacts to human health or the environment, and (2) to require cleanup after the discharge has occurred.[3] In addition, the law generally requires responsible parties to notify DNR immediately when a discharge has occurred.[4]

Persons who are subject to the Spill Statute must be aware of the law's requirements, as failure to adhere to these requirements can subject the responsible party to fines or penalties of not less than $10 nor more than $5,000 for each violation.[5] Each day that the violation continues is viewed as a separate offense for purposes of determining penalty calculations.[6] The broad coverage of the Spill Statute makes it the preferred regulatory authority of choice by DNR in cleanup actions in Wisconsin.

"Possession or Control" Cleanup Liability

The obligations of notification and cleanup under the Spill Statute attach to a person who "possesses or controls a hazardous substance" in addition to the person who "causes the discharge of a hazardous substance."[7] The only case to interpret the breadth of the phrase "possess or control" is *State v. Mauthe*.[8] In *Mauthe*, the Wisconsin Supreme Court ruled that the current property owner whose property was contaminated with hazardous substances was responsible for cleaning up contamination that was migrating from his property, even though he did not originally cause the contamination on his property. In *Mauthe*, the court interpreted the continuing migration of contamination to the groundwater as a "discharge" that the current property owner "possessed or controlled" within the meaning of the cleanup responsibility provisions under the Spill Statute.[9]

It is important to note that the facts in *State v. Mauthe* involved migration of contamination from the property that was the source of the original contamination to an adjoining property. To date, no appellate court in Wisconsin has decided a case that involves a present property owner's responsibilities for cleanup of hazardous substances that he or she did not cause, without evidence of migration to an adjoining property owner.

A number of courts in other states have interpreted similar language and concluded that an innocent property owner is not responsible for cleanup of hazardous substances located on his or her property if he or she did not cause the original release. These cases have interpreted the term "discharge" as requiring some human activity of a current property owner that resulted in the contamination on the owner's property.

Nonetheless, the prudent course of action in Wisconsin, until further clarification is provided by the appellate courts, is to assume that cleanup liability may attach based upon the mere existence of hazardous substances on an owner's property, even if the current owner did not cause the original release of the hazardous substances. This broad interpretation of cleanup responsibility in Wisconsin places a premium on knowing as much as possible about the status of contamination on property before purchase.

Notification Requirements

The duty to notify the state of the existence of hazardous substances under the Spill Statute is interpreted very broadly by DNR. First, DNR interprets the notification requirements to include discharges not only to land but also to the air and waters of the state. Second, as indicated in the previous section, DNR takes a broad interpretation of the term "discharge." Under this interpretation, DNR takes the position that a property owner's discovery of the existence of contamination in the soil and/or groundwater, without actual evidence of the owner causing the original discharge, obligates the owner to report the existence of the contamination to DNR immediately.[10]

Third, unlike the analogous federal laws, the Spill Statute does not specifically identify materials that constitute hazardous substances, nor does the law provide any reportable quantities of any such substances that trigger a "reporting obligation." In particular, the definition of "hazardous substances" is very broad and, in general, will include any material that may create potential harm to human health or the environment because of its characteristics upon discharge.[11]

Under current law, DNR is required to adopt rules that further define the characteristics of materials that constitute hazardous substances, and the quantities that trigger notification obligations under the Spill Statute. Until further clarification is provided by DNR in its rule-making authority, it is prudent for a current owner who discovers contamination on his or her property through the existence of monitoring information or other methods to report such conditions to DNR immediately. Failure to do so could expose the responsible party to fines or penalties.[12]

Liability Protection

Purchaser Liability Exemption

The Wisconsin Spill Statute[13] requires a person to restore the environment, to the extent practicable, if he or she possessed or controlled the hazardous substance discharged or otherwise caused a discharge. Such a person is liable for reimbursement for investigation, cleanup, and related expenses incurred by the state for the discharged substances.[14]

The Land Recycling Act exempts certain purchasers, lenders, and municipalities from the Spill Statute liability. However, it is important to note that the Act does not affect potential liability under the federal Superfund law.

Under the new purchaser-liability exemption, a qualified purchaser is exempt from certain attributes of cleanup liability under the Spill Statute. A "purchaser" is defined in the Act as a "person who acquires property in an arm's length, good faith transaction."[15] A "person" is broadly defined to include individuals, corporations, municipalities, and state and federal agencies.[16] To qualify as a purchaser for the liability exemptions, the person must also meet the following requirements: (1) did not participate in the management, and was not the owner, of the business or entity that caused the release, (2) did not own the property at the time of the release, and (3) did not otherwise cause the release.[17]

The scope of the purchaser-liability exemption applies to preacquisition releases. In particular, the exemption relates to (1) the purchaser's obligation to restore the environment,[18] (2) DNR's authority to order preventive measures,[19] and (3) DNR's right to reimbursement.[20]

The purchaser-liability exemption contained in the Spill Statute is provided not only to the qualified purchaser, but also to the purchaser's successors or assignees who comply with the monitoring and maintenance requirements of the Act, unless the successor or assignee knows certification was wrongfully obtained.[21]

The Mechanics of Obtaining a Purchaser-Liability Exemption

For the purchaser to obtain the liability exemption available under the Land Recycling Act, all the following requirements must be met:

1. The purchaser (or the seller pursuant to a contract) must conduct a thorough environmental investigation of the property, which is approved by DNR.[22] The scope of the environmental assessment is further defined in applicable rules adopted by DNR.[23]
2. The purchaser must restore the environment and minimize harmful effects from the releases, in accordance with DNR rules or a contract with DNR.[24] It is important to note that the seller cannot perform the necessary cleanup. Rather, the actual cleanup must be performed by the buyer to meet the purchaser exemption requirements under the Act.
3. The purchaser must obtain "certification" from DNR that the property has been satisfactorily restored and that the harmful effects from the release of the hazardous substance have been minimized.[25]
4. The purchaser must maintain and monitor the property as required by applicable DNR rules or the remediation contract.[26]
5. The purchaser must not disrupt the property by engaging in activities that are inconsistent with such maintenance and monitoring.[27]
6. The purchaser must not have obtained DNR certification by fraudulent methods. A fraudulent certification would include an instance

when a purchaser knew or should have known about pollution beyond that revealed in the environmental investigation.[28]

If a purchaser meets these requirements, the purchaser is exempt from cleanup requirements after the date of the DNR certification, notwithstanding the occurrence of any of the following:

- Future statutes, rules, or regulations imposing greater responsibilities on property owners[29]
- Unsuccessful remediation of a site when the remediation was performed in accordance with DNR's cleanup rules (The lack of success in performing a cleanup could result from the provision in the Spill Statute that a party clean up the environment only to the extent "practicable"[30])
- Contamination from a hazardous substance that is the subject of the remediation is discovered to be more extensive than anticipated by the purchaser and DNR at the time the environmental assessment was conducted[31]

To take advantage of the purchaser-liability exemption contained in the Act, an applicant must submit a completed application form along with the requisite application fee to DNR. A successful applicant will then receive a letter from DNR indicating the purchaser's eligibility for the program. After an eligibility letter is received, the applicant must submit to DNR a Phase I environmental assessment report as well as a Phase II environmental assessment scope-of-work report. In addition, the purchaser must immediately notify DNR of any hazardous substance discharges required to be reported under applicable law.[32]

After submitting the Phase I and Phase II environmental reports, the applicant will receive a letter from DNR with a projected date for DNR review to begin. Upon approval of the Phase I and Phase II documents by DNR, the applicant must complete a thorough environmental investigation of the property in accordance with applicable rules.[33] After completing the Phase I report and submitting the findings to DNR, the applicant must receive DNR approval of the thoroughness of the environmental investigation. The applicant must then conduct a cleanup of the property, and any contamination that migrated off the property, in accordance with applicable cleanup rules.[34] Upon the completion of the cleanup, the applicant must request and receive DNR closeout.[35] In addition, the applicant must maintain and monitor the property as required by DNR, in accordance with the approved closeout conditions.

Liability Protections for Governmental Entities Holding Contaminated, Tax-Delinquent Property

There are special liability exemptions for cleanup liability under the Spill Statute for qualified municipalities.[36] In general, a qualified municipality is exempt from specifically identified cleanup obligations under the Spill Statute if the municipality acquired the property through tax-delinquency proceedings or as a result of an order issued by a bankruptcy court.[37] However, it is important to note that a municipality cannot qualify for this exemption if the discharge of hazardous substances was caused by any of the following:

- Action taken by the municipality
- A failure of the municipality to take appropriate action to restrict access to the property, to minimize costs or damages that may result from unauthorized persons entering the property
- A failure of the municipality to sample and analyze unidentified substances in containers stored aboveground on the property
- A failure of the municipality to remove and properly dispose of, or to place in a different container and properly store, any hazardous substances that are stored aboveground on the property in a container that is leaking or is likely to leak[38]

A municipality that acquires contaminated property possesses another important, transferable benefit that could provide an increased potential for subsequent development of the contaminated parcel. In particular, any purchaser who obtains contaminated property from a qualified municipality has, in turn, certain important limitations of cleanup liability under the Spill Statute. A qualified purchaser who acquires contaminated property from a municipality that qualified for a liability exemption under the Spill Statute may obtain the right to limit the cost of any subsequent cleanup the purchaser conducts. If the qualified purchaser agrees to conduct a cleanup and executes an agreement with DNR for such a cleanup, the purchaser may end the cleanup when the cost of the cleanup equals 125 percent of the DNR-approved estimated expenses of the cleanup.[39] Under this special provision, in the event the actual costs of the cleanup exceed this 125 percent cap, the purchaser may discontinue the cleanup and the purchaser will continue to receive the benefits of the liability exemption after the cessation of the cleanup, provided the purchaser (1) maintains and operates the property pursuant to certain requirements, and (2) uses "reasonable efforts" to sell the property in a manner equivalent to comparable federal rules.[40]

Lender and "Representative" (Fiduciary) Liability Protection

There is a liability exemption provided under the Spill Statute for a lender engaged in certain enumerated lending activities.[41] To qualify for this exemption, the lender must engage in lending activities as its primary business, or be an insurance company, pension fund, or governmental agency engaged in secured lending.[42] If the entity is deemed a lender under the Spill Statute, various lending activities in which the lender is engaged are generally exempt from liability under the Spill Statute.[43] However, a lender engaged in qualified lending activities cannot take advantage of the exemption from Spill Statute liability if the lender actually caused the discharge or, through tortious conduct in connection with lending activities, causes or exacerbates a discharge.[44]

Normally, lenders are concerned that merely requiring compliance and environmental assessment audits could subject the lender to liability for cleanup under environmental laws. However, the Spill Statute creates a special exemption for lenders that inspect real property for compliance with environmental laws, conduct any portion of an environmental assessment, conduct an investigation to determine the degree and extent of contamination, or perform remedial action, if the lender meets all the following requirements:

- The required environmental investigation and remediation activities occur before the date the lender acquires title to, or possession or control of, real property through the enforcement of a security interest.
- The lender notifies DNR of any discharge of a hazardous substance identified as a result of such assessment/remediation activities.
- The lender conducts an investigation or performs remedial action in accordance with DNR rules.
- The lender does not physically cause a discharge or exacerbate an existing discharge.[45]

There are also special liability exemptions provided for lenders under the Spill Statute when the lender acquires title to, or possession or control of, real property through enforcement of a security interest. This liability exemption applies to lenders that obtain title to contaminated property if they meet all the following requirements:

- The lender does not intentionally or negligently cause a new discharge or exacerbate an existing discharge.
- The discharge of a hazardous substance was not from an underground storage tank regulated under federal underground storage tank laws.
- The lender notifies DNR of any known discharge.
- When the releases of hazardous substances occur on or after the date a lender obtains title, possession, or control, the lender must not be operating a business at the property, completing work in progress, or otherwise conducting the borrower's business, and the lender must implement an emergency response action for the postacquisition discharge of hazardous substances.[46]

In addition, to be exempt from Spill Statute liability for the acquisition of real property through the enforcement of a security interest, a lender must conduct an environmental assessment of the property not more than 90 days after the date the lender obtains title, possession, or control of the property, and must file a complete copy of the environmental assessment with DNR not more than 180 days after the date the lender obtains title, possession, or control of the property.[47]

There is also a special liability exemption under the Spill Statute for lenders that enforce a security interest in personal property or fixtures at a particular location.[48] To maintain this liability exemption, the lender must not obtain title, possession, or control of the real property at that location for any purpose other than to protect and remove personal property or fixtures.[49]

It is important to note that these liability exemptions for lending activities apply only to Spill Statute liability and do not apply to other statutory cleanup liabilities under federal and Wisconsin law, including the Wisconsin environmental response and repair law,[50] the Wisconsin abandoned-container law,[51] and the Wisconsin hazardous-waste law.[52] However, because the Spill Statute is an important basis for cleanup liability in Wisconsin, this exemption should provide a significant degree of comfort to lenders who are considering lending activities for industrial property that may be contaminated.

There is also an important new exemption to Spill Statute liability that applies to a broad category of "representatives."[53] Under certain circumstances, a qualified representative who acquires title to, or possession or control of, real or personal property is not personally liable for contaminated property under the Spill Statute. A qualified representative means any person acting in the capacity of a conservator, guardian, court-appointed receiver, personal representative, executor, administrator, testamentary trustee of a deceased person, trustee of a living trust, or fiduciary of real or personal property.[54] To gain the benefits of this exemption, a person must be acting in the capacity of a representative and he or she must not

1. knowingly, willfully, or recklessly physically cause a discharge;
2. have a beneficial interest in a trust, estate, or similar entity that owns, possesses, or controls the real or personal property; or
3. knowingly, willfully, or recklessly fail to provide DNR a Spill Statute notice of a discharge.[55]

The Spill Statute limitation on personal liability for a representative does not apply if (1) the representative knew or should have known that the subject trust, estate, or similar entity was established, or that assets were transferred to trust, estate, or similar entity, to avoid responsibility for a discharge; or (2) the representative fails to act in good faith to cause the subject trust, estate, or similar entity to restore the environment or reimburse DNR for its restoration expenses.[56]

Super-Priority Liens for DNR Cleanup Expenses

The Spill Statute grants DNR authority to file a "super lien" with the register of deeds, in the amount of expenditures made by DNR for a response action required under the Spill Statute for contaminated property.[57] The priority lien authorized under this section is superior to all other liens, except valid prior liens on residential property.[58] To take advantage of the super-priority lien provision, DNR must provide the current owner of the property and any mortgagees-of-record a notice containing all the following information:

- A brief description of the property for which DNR expects to incur cleanup expenses
- A brief description of the types of assessment/remedial activities that DNR expects may be conducted on the property

- A statement that the property owner could be liable for expenses incurred by DNR
- A statement that DNR could file a lien against the property to recover the expenses
- An explanation of whom to contact at DNR to discuss the matter[59]

Any expenditures made by DNR after providing the requisite notice will constitute a lien upon the property for which expenses are incurred, provided DNR files the lien with the register of deeds in the county in which the property is located.[60]

This imposition of a super-priority lien on behalf of DNR is a new development that should be watched closely by potential purchasers, lenders, and issuers of title policies on contaminated property in Wisconsin. The potential for a super-priority lien creates the obvious risk of displacing prior existing liens on all but residential real estate, even though prior existing liens may have limited value if the property is contaminated and not cleaned up by DNR or some other entity.

Soil Cleanup Standards

Wisconsin has established a new framework for determining the level of cleanup required for contaminated soils.[61] DNR has developed these new cleanup standards to govern environmental cleanups conducted by responsible parties or DNR. The new cleanup standards apply to a broad array of cleanups under various Wisconsin solid-waste laws, including the abandoned container law,[62] the Spill Statute,[63] and the environmental response and repair law.[64]

The initial focus of the Wisconsin soil cleanup standards was the cleanup of contaminated soil associated with underground storage tanks and the Petroleum Environmental Cleanup Fund Act (PECFA).[65] The focus on petroleum-related contamination was partly due to concerns raised by the Wisconsin legislature about the financial viability of the PECFA fund, resulting from the case-by-case cleanup standards used by DNR on underground storage tank cleanups. In particular, the legislature was concerned that DNR had too much discretion and, accordingly, was using standards that were too stringent. This had the effect of financially depleting the PECFA fund. In part, the establishment of soil cleanup standards and the initial focus on petroleum contamination were designed to create uniformity, thereby assisting in preserving the financial viability of the PECFA fund.

The new soil cleanup standards establish generic cleanup standards that apply to sites that qualify as "simple sites." In contrast, "complex sites" are those sites that will have "site-specific determinations" for cleanups.[66] For sites classified as "simple sites," numeric soil cleanup standards apply to the site and such sites are designed to be cleaned up with minimal DNR oversight.

Under the new cleanup standards, "simple sites" are those sites that meet all the following requirements:

- *Petroleum sites:* The site is contaminated by gasoline-related constituents[67] or various identified metals.[68] The simple-site process also applies to sites where gasoline-range organics or diesel-range organics are the only other contaminants of concern.[69]
- *Minimal surface-water impacts:* There is no residual soil contamination at the site or facility that will adversely affect surface water.[70]
- *No sensitive-site characteristics:* There is no residual soil contamination at the site or facility that will adversely affect a "sensitive" environment.[71]
- *No plant uptake concerns:* There is no residual soil contamination at the site that will concentrate through plant uptake.[72]

If the responsible party can meet the simple-site requirements and chooses to proceed to clean up under this process, DNR oversight is minimal. However, if simple-site standards are selected, generic numerical cleanup standards apply and there are no opportunities to seek exemptions from these numeric standards. Cleanup under these standards is designed to be conducted primarily under the direction of the responsible party and its environmental consultant. The simple-site process requires progress reports to be submitted to DNR at six-month intervals, beginning six months after the responsible party notifies DNR of the discharge.[73] The six-month status reports must be filed until the responsible party files a letter of compliance stating that the cleanup has been completed in accordance with the numeric cleanup standards. The letter of compliance required at the conclusion of the cleanup under the simple-site process must include a certification and documentation that applicable cleanup standards have been followed and no further action is necessary for the site.[74]

Generally speaking, under the simple-site process, DNR does not approve or disapprove these reports or letters of compliance. Rather, DNR's options generally are limited to determining whether it should take enforcement actions against parties who are responsible for the site or the facility.[75] DNR is required to provide written acknowledgment of the receipt of a letter of compliance submitted by the responsible party, and the accompanying final report, within thirty days of receipt.[76]

DNR does retain the option of auditing facilities for compliance with the simple-site cleanup process. Under its audit authority, DNR has authority to (1) concur with the letter of compliance, (2) determine that the letter of compliance is not complete and require the responsible party to submit additional information to determine whether the response action was inadequate, or (3) require the responsible party to take additional response actions that DNR determines are necessary.[77]

The "complex site" cleanup standard applies to any site that does not meet the simple-site criteria. A responsible party retains the option of using the complex-site process even if the site meets all the simple-site characteristics.[78] For complex-site cleanups, responsible parties must submit to DNR a site investigation report and a draft remedial options report,[79] as well as a final report for response actions, including a letter of compliance documenting that the response action has been completed in accordance with applicable cleanup requirements.[80]

Under the complex-site process, DNR is required to provide written acknowledgment of the receipt of the site investigation report and draft remedial options report within thirty days, and estimate the likely date for completion of review of the documents.[81] DNR also has the option under the complex-site procedure to require additional submittals for more extensive review in any of the following circumstances:

- If the site is eligible for PECFA reimbursement[82]
- If a purchaser is seeking a purchaser-liability exemption and certification under the Spill Statute[83]
- If DNR determines that environmental standards are not achieved and additional response action is required before the issuance of a no-further-action determination[84]

Although the complex-site process is more time intensive and involves a greater amount of DNR oversight than the simple-site process, it provides the responsible party with the option to seek site-specific cleanup standards that may be less restrictive than the generic numerical cleanup standards provided under the simple-site soil cleanup program.

Procedures for Establishing Soil Cleanup Standards

Wisconsin has identified the procedures for establishing soil cleanup standards at a contaminated site or facility. Responsible parties must establish the residual contaminant level, which is the lowest concentration determined in accordance with a target goal procedure designed to protect groundwater.[85] In the alternative, a responsible party can establish soil cleanup standards in accordance with performance standards.[86]

Under these new soil standards, a responsible party may use generic residual contaminant levels (that is, numeric standards identified in tables) as the cleanup goals if all the following requirements for the site are met:

- An investigation has been conducted and completed in accordance with applicable site investigation requirements[87] and the contaminants of concern are the subject of numeric cleanup values.[88]
- The appropriate investigation has been conducted to determine the vertical and horizontal extent of contamination.[89]
- At least one meter of vertical distance exists between bedrock and the base of the soil contamination.[90]
- Only six meters or less of the vertical thickness of residual soil contamination exists,[91] and the residual contaminants do not contribute to facilitated transport or cosolvent effects.[92]

A methodology also authorized for determining generic residual contaminant levels involves the responsible party establishing that the dilution effect of groundwater is greater than that assumed by DNR in developing the generic site values in the soil cleanup standards.[93]

For sites involving petroleum contamination, the generic residual contaminant levels for gasoline-range organics or diesel-range organics is 250 mg/kg where soils below the contaminant soils meet specified hydraulic conductivity requirements.[94] When the requisite hydraulic conductivity is not met, the generic residual contaminant levels for gasoline-range organics and diesel-range organics is 100 mg/kg.[95]

The soil cleanup standards vary depending on the land-use classification for the site considered for cleanup. More restrictive generic cleanup standards apply for nonindustrial sites than for those contaminated sites that are classified as industrial sites.[96]

If any of the foregoing criteria are not satisfied, the responsible parties may use an alternative site-specific procedure for establishing cleanup standards.[97] In addition, if background concentrations for substances in the soil at a site are higher than the residual contaminant levels for that substance, the background concentrations in the soils may be used as the residual contaminant cleanup goals for that substance.[98]

Memorandum of Agreement

In October 1995, DNR and USEPA Region V executed a brownfields Memorandum of Agreement (MOA). Pursuant to the MOA, the parties agreed to certain operating principles that will preclude federal action under the Superfund law, except for sites that pose an imminent threat to public health or the environment. Generally, the MOA provides that when DNR is proceeding in accordance with its own cleanup standards, USEPA will not seek to impose its authority under federal law to require cleanup, but will defer to DNR under Wisconsin's own cleanup authority.

Conclusion

These innovative Wisconsin brownfields redevelopment initiatives have all been enacted since 1994 and are designed to provide incentives for new brownfields redevelopment. It is likely that the provisions of these programs will be modified in the future to make them more "user friendly." Nonetheless, the Wisconsin programs are an important first step in the direction of promoting brownfields redevelopment.

Wisconsin Land Recyling Act Program

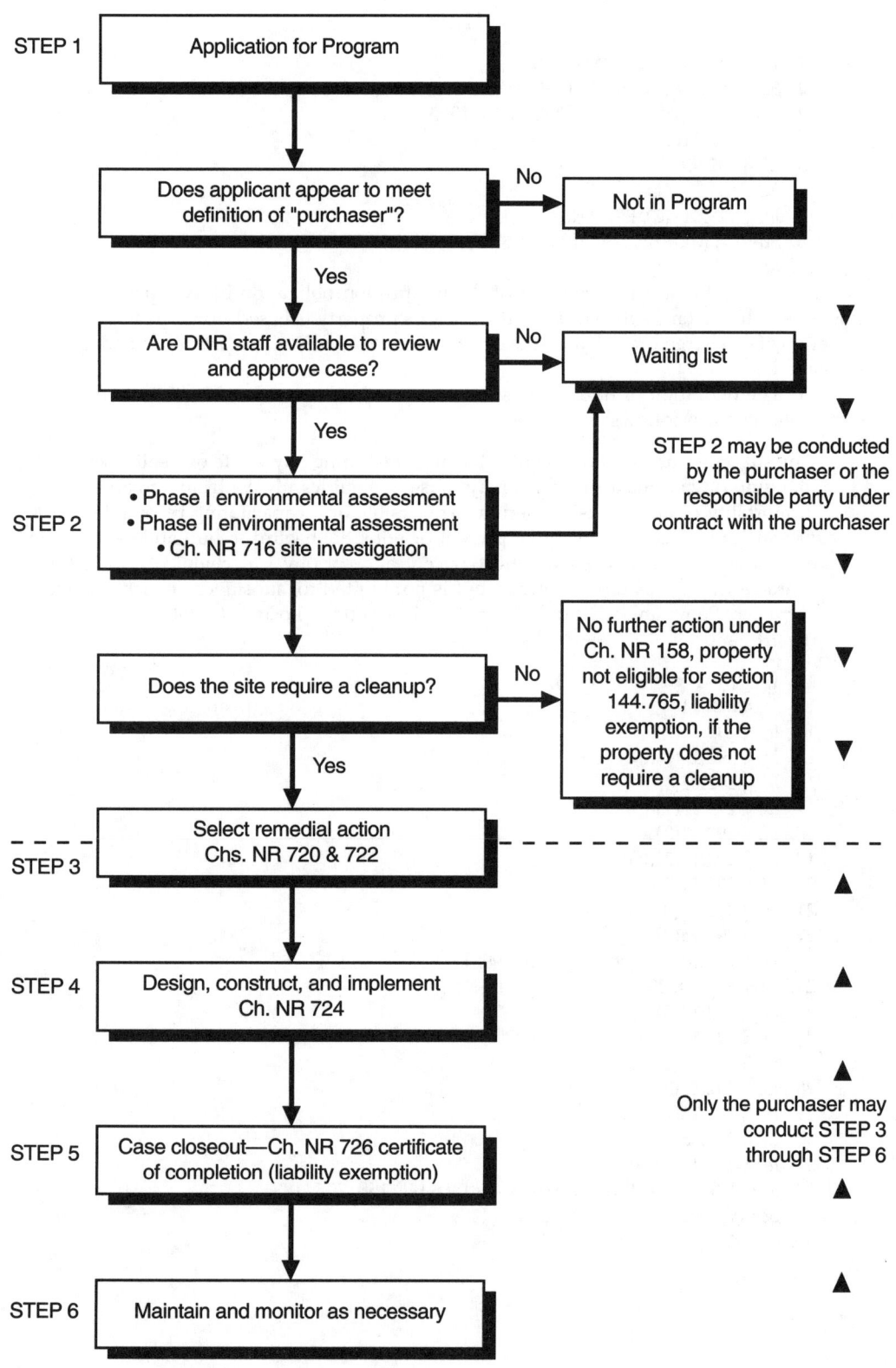

Notes

1. 1993 Wis. Laws 45 (SB 462).
2. *See* WIS. STAT. § 292.11 (1995); WIS. ADMIN. CODE § NR 158 (1995).
3. WIS. STAT. §§ 292.11(3), 292.11(4) (1995).
4. *Id.* § 292.11(2).
5. *Id.* § 292.99.
6. *Id.*
7. *Id.* §§ 292.11(2), 292.11(3).
8. State v. Mauthe, 366 N.W.2d 871 (Wis. 1985).
9. *Id.*
10. DNR's broad interpretation of the notification obligation based upon the mere existence of hazardous substances on the owner's property is based upon the broad interpretation of discharge provided by the court in *State v. Mauthe. See supra* note 8 and accompanying text.
11. The definition of hazardous substance provided in section 292.01(5) of the Wisconsin Statutes is as follows:

> Any substance or combination of substances including any waste of a solid, semi-solid, liquid or gaseous form which may cause or significantly contribute to an increase in mortality or an increase in serious irreversible or incapacitating reversible illness or which may pose a substantial present or potential hazard to human health or the environment because of its quantity, concentration or physical, chemical or infectious characteristics. This term includes, but is not limited to, substances which are toxic, corrosive, flammable, irritants, strong sensitizers or explosives as determined by the department.

12. WIS. STAT. § 292.99 (1995).
13. *Id.* § 292.11
14. *Id.* § 292.11(7).
15. *Id.* § 292.15(1)(c).
16. *Id.* § 292.01(13).
17. *Id.* § 292.15(1)(c).
18. *Id.* § 292.11(3).
19. *Id.* § 292.11(7)(b).
20. *Id.* § 292.11(7)(c).
21. *Id.* § 292.15(3).
22. *Id.* § 292.15(2)(a).
23. WIS. ADMIN. CODE § NR 716 (1995).
24. WIS. STAT. § 292.15(2)(a) (1995).
25. *Id.* § 292.15(2)(a)3.
26. *Id.* § 292.15(2)(a)4.
27. *Id.* § 292.15(2)(a)5.
28. *Id.* § 292.15(2)(a)6.
29. *Id.* § 292.15(2)(b)1.
30. *Id.* § 292.15(2)(b)2.
31. *Id.* § 292.15(2)(b)3.
32. *See id.* § 292.11(2); WIS. ADMIN. CODE § NR 158, 705 (1995).
33. *See* WIS. ADMIN. CODE § NR 716 (1995).
34. *See id.* § NR 700 *et seq.*

35. The closeout rules are contained in section NR 726 of the Wisconsin Administrative Code.

36. "Municipality" includes a redevelopment authority under section 66.431 of the Wisconsin Statutes or a public body designated by a municipality under section 66.435(4) of the Wisconsin Statutes. *See* WIS. STAT. § 292.11(9)(e) (1995).

37. *See id.* § 292.11(9)(e)1. The specific cleanup liability obligations for which qualified municipalities are exempt under this section include the cleanup obligations under sections 292.11(3), 292.11(4), 292.11(7)(b), and 292.11(7)(c).

38. *Id.* § 292.11(9)(e)2.

39. *Id.* § 292.15(4).

40. *See* 40 C.F.R. § 300.1100(d)(2)(i) (1996) (defining the comparable "reasonable efforts" rules to sell the property). The general purchaser requirements that are applicable to the 125 percent cap are contained in sections 292.15(4)(b) and 292.15(4)(c) of the Wisconsin Statutes.

41. WIS. STAT. § 292.21(1) (1995).

42. The term "lender" is defined as "a bank, credit union, savings bank, savings and loan association, mortgage banker or similar financial institution, the primary business of which is to engage in lending activities or an insurance company, pension fund or government agency engaged in secured lending." *Id.* § 144.76(1)(b).

43. The lending activities qualified for the exemption are defined in section 202.01(9)(m) of the Wisconsin Statutes as follows:

> "Lending activities" means advancing funds or credit to and collecting funds from another person; entering into security agreements, including executing mortgages, liens, factoring agreements, accounts receivable financing agreements, conditional sales, sale and lease back arrangements and installment sales contracts; conducting inspections of or monitoring a borrower's business and collateral; providing financial assistance; restructuring or renegotiating the terms of a loan obligation; requiring payment of additional interest; extending the payment period of a loan obligation; initiating foreclosure or other proceedings to enforce a security interest in property before obtaining title; requesting and obtaining the appointment of a receiver; and making decisions related to extending or refusing to extend credit.

44. WIS. STAT. § 292.21(a)2 (1995).

45. *Id.* § 292.21(b).

46. *Id.* § 292.21(c)1.

47. *Id.* § 292.21(c)1d.

48. *Id.* §§ 292.21(d), 292.21(1)(c)2.

49. *Id.*

50. *Id.* § 292.31; WIS. ADMIN. CODE § NR 550 (1995).

51. WIS. STAT. § 292.41 (1995); WIS. ADMIN. CODE § NR 551 (1995).

52. WIS. STAT. Chapter 291 (1995); WIS. ADMIN. CODE § NR 600 *et seq.* (1995).

53. WIS. STAT. § 292.21(2) (1995).

54. *Id.* § 292.01(16).

55. *Id.* § 292.21(2)(a).

56. *Id.* §§ 292.21(2)(b).

57. *Id.* § 292.81(3).

58. *Id.*

59. *Id.* § 292.81(2).

60. *Id.* § 292.81(3).

61. WIS. ADMIN. CODE § NR 700–36 (1995).
62. WIS. STAT. § 292.41 (1995).
63. *Id.* § 292.11.
64. *Id.* § 292.31.
65. *Id.* § 101.143.
66. WIS. ADMIN. CODE § NR 720.09(1)–(2) (1995).
67. These contaminants are listed in Table 1 of section NR 720 of the Wisconsin Administrative Code, and include benzene, 1,2-dichloroethane, ethyl benzene, toluene, and xylenes.
68. These metals are listed in Table 2 of section NR 720 of the Wisconsin Administrative Code, and include arsenic, cadmium, hexavalent chromium, trivalent chromium, and lead.
69. The rule provides the following example of contaminants falling within this category:

> For example, if polynuclear aromatic hydrocarbon (PAH) compounds are present, they would be considered contaminants of concern. With the exception of naphthalene, PAH compounds are generally only of concern for direct contact due to their relatively low migration potential. PAH compounds include, but are not limited to: acenaphthene, acenaphthylene, anthracene, benzo(a)anthracene, benzo(a)pyrene, benzo(b)fluoranthene, benzo(ghi)perylene, benzo(k)fluoranthene, chrysene, dibenzo(a,h)anthracene, fluoranthene, fluorene, indeno(1,2,3-cd)pyrene, 1-methyl naphthalene, 2-methyl naphthalene, naphthalene, phenanthrene, and pyrene. *See* WIS. ADMIN. CODE § NR 700.09(1).

70. WIS. ADMIN. CODE § NR 700.09(1)(a)(d) (1995).
71. *Id.* § NR 700.09(1)(a)(c).
72. *Id.* § NR 700.09(1)(a)(d).
73. *Id.* § NR 700.11(1)(a).
74. *Id.* § NR 700.11(1)(b).
75. *Id.* § NR 700.11(1)(c).
76. *Id.* § NR 700.11(1)(d).
77. *Id.* § NR 700.11(1)(e).
78. *Id.* § NR 700.09(2).
79. The site investigation report must be prepared in accordance with the requirements contained in section NR 716 of the Wisconsin Administrative Code, and the draft remedial options report must meet the requirements of section NR 722 of the Wisconsin Administrative Code. *See* WIS. ADMIN. CODE § NR 700.11(2)(b) (1995).
80. *Id.* § NR 700.11(2)(a).
81. *Id.* § NR 700.11(2)(c).
82. There is a more extensive review procedure for PECFA-eligible sites provided in section ILHR 47 of the Wisconsin Administrative Code. Generally speaking, a DNR review of PECFA-eligible sites will be more frequent than described in the simple-site review process under section NR 700.11(1) of the Wisconsin Administrative Code.
83. The purchaser certification exemption is provided under section 292.765 of the Wisconsin Statutes.
84. WIS. ADMIN. CODE § NR 700.11(3)(c) (1995).
85. These procedures are detailed in sections NR 720.09 through 720.19(3) of the Wisconsin Administrative Code.
86. The procedure for establishing performance standards is contained in section NR 720.19(2) of the Wisconsin Administrative Code.

87. These site investigation requirements are contained in section NR 720.05(1) of the Wisconsin Administrative Code.

88. The contaminants of concern are listed in Table 1 as follows:

BASELINE CONCENTRATIONS, DILUTION, ATTENUATION FACTORS AND RESIDUAL CONTAMINANT LEVELS BASED ON PROTECTION OF GROUNDWATER

Substance	Baseline Concentration ($\mu g/kg$)	Dilution Attenuation Factor	Residual Contaminant Level
Benzene	0.093	59	5.5
1,2-dichlorethane	0.041	120	4.9
ethylbenzene	42	70	2900
toluene	18	81	1500
xylenes (total)	47	87	4100

At sites where petroleum contamination in the form of gasoline-range organics or diesel-range organics or both are present, and are the only contaminants of concern present other than the contaminants listed in Table 1, standards listed in section NR 720.09(4)(a) of the Wisconsin Administrative Code may be used for nonspecific gasoline-range organic or diesel-range organic contaminants.

89. WIS. ADMIN. CODE § NR 720.09(1)(c) (1995).
90. *Id.* § NR 720.09(1)(d).
91. *Id.* § NR 720.09(1)(e).
92. *See id.* § NR 720.09(1)(f). The rules provide examples of facilitated transport, which include polychlorinated biphenyls in the presence of an oily phase. An example of cosolvency, provided by DNR, includes petroleum aromatic hydrocarbons (PAHs) in the presence of alcohols when the alcohol acts to increase solubility of the PAHs. *See id.,* note following § NR 720.09(1)(f).
93. The methodology for determining an alternative dilution value for establishing alternative generic residual contaminant levels is detailed in section NR 720.09(3)(b) of the Wisconsin Administrative Code.
94. In this instance, the soils below the contaminated soil for a depth of three meters must have hydraulic conductivity of $1\text{-}10_6$ cm/s or less.
95. WIS. ADMIN. CODE § NR 720.11(4) (1995).
96. Industrial site classification may apply to sites that otherwise do not meet the individual site requirement if the area is subject to restricted access. *See id.* § NR 720.11(1).
97. *Id.* § NR 720.11(4).
98. *Id.* § NR 720.11(5).

ABOUT THE EDITORS

Todd S. Davis is President of Hemisphere Corporation, located in Cleveland, Ohio, a company that acquires and redevelops environmentally distressed real estate and serves as national brownfields disposition advisor for a large public company. Mr. Davis also is a partner in the law firm of Benesch, Friedlander, Coplan & Aronoff, LLP, and is co-chair of the firm's Environmental Practice Group.

Mr. Davis is a vice chair of the ABA's Brownfields Task Force. He was invited to advise the USEPA on structuring the federal Brownfields Program. He also has been appointed to the Ohio Environmental Protection Agency's Phase I/Phase II rulemaking subcommittee for Ohio's Voluntary Real Estate Reuse and Cleanup law, and is a member of the Ohio Brownfields Finance Partnership. He has participated in the Brownfields Working Group, sponsored by the Cuyahoga County, Ohio Board of Commissioners, which was charged with developing creative strategies to facilitate the redevelopment of contaminated properties.

He also has written and contributed to numerous books and articles on environmental issues, including *The Underground Storage Tank Manual,* a comprehensive book on Ohio UST law. Mr. Davis speaks frequently both nationally and locally on various environmental topics, including redeveloping contaminated real estate, financing contaminated property, lender liability, and understanding environmental remediation. Recent speaking engagements include presentations to the National Governors Association and Brownfields '96, a national conference sponsored by the USEPA and the ABA. Mr. Davis received a bachelor's degree in business administration from the University of Michigan, with distinction, and a law degree from George Washington University, with honors.

Kevin D. Margolis is Vice President of Hemisphere Corporation, located in Cleveland, Ohio, a company that acquires and redevelops environmentally distressed real estate and serves as national brownfields disposition advisor for a large public company. Mr. Margolis also is a partner in the Environmental Practice Group of the law firm of Benesch, Friedlander, Coplan & Aronoff, LLP.

Mr. Margolis has been appointed to the Ohio Environmental Protection Agency's rulemaking committee for Generic Numeric Cleanup Standards and Procedures for Risk Assessment for Ohio's Voluntary Real Estate Reuse and Cleanup law, is a member of the ABA's Brownfields Task Force, and is a member of the Ohio Brownfield's Finance Partnership.

Mr. Margolis is the coauthor of *The Underground Storage Tank Manual*, a comprehensive book on Ohio underground storage tank law. His published articles on environmental law topics include "Banking Bogs" (an article discussing wetland mitigation banking), published in *Affinity*. Mr. Margolis also was a significant contributor to the preparation of an article published in 1990 by the American College of Real Estate Lawyers, "Lawyer's Criminal and Civil Liability in Workouts," and has spoken frequently on business, real estate, and environmental issues, including presentations at Brownfields '96, a national conference sponsored by the United States Environmental Protection Agency and the American Bar Association. Mr. Margolis received a bachelor of arts degree from Northwestern University and a law degree from Case Western Reserve University School of Law.

To send or request information about recent brownfields issues, please write to either Todd S. Davis or Kevin D. Margolis at 2300 BP America Building, 200 Public Square, Cleveland, OH 44114-2378 or phone (216) 771-5900.

ABOUT THE CONTRIBUTORS

Ned Abelson is a director at Goulston & Storrs in Boston, Massachusetts, where he concentrates his practice on environmental law. A considerable amount of his practice involves property subject to the Massachusetts Contingency Plan and sites involved with brownfields issues. Mr. Abelson is a member of the ABA's Brownfields Task Force and the Liability Subcommittee of the Massachusetts Brownfields Advisory Group. He is also a member of the Chicago Brownfields Forum Model Loan Package Group and the Toxic Waste Subcommittee of The International Council of Shopping Centers. Mr. Abelson regularly writes articles and speaks about hazardous-waste issues. He graduated *magna cum laude* from Brown University and received his law degree from the University of Pennsylvania. He is a member of Phi Beta Kappa.

B. Robert Amjad is an associate at Cleveland Real Estate Partners, working primarily with the Consulting Services Group. Mr. Amjad recently served as the project manager for the Regional Transit Authority (RTA) Cleveland State University Transit Center site acquisition. His responsibilities included coordinating the appraisal, surveying, tenant relocation, and environmental studies on the eleven targeted properties and the thirty tenants of those properties. He also supported the negotiations with the owners of the parcels before cancellation of the project by the RTA. He is a graduate of Case Western Reserve University's School of Law and Weatherhead School of Management.

Mark D. Anderson is a lawyer at Stateside Associates in Arlington, Virginia. Mr. Anderson also acts as counsel to The Greenfields Group, an industry coalition promoting the establishment of state voluntary cleanup programs. Mr. Anderson is a graduate of the George Washington University National Law Center.

Anne Slaughter Andrew is a partner with the law firm of Baker & Daniels, in its Indianapolis office. Ms. Andrew concentrates her practice in the representation of clients in environmental matters. She has represented clients in federal, state, and local administrative forums on environmental rulemaking, permitting, and enforcement matters. Ms. Andrew has also represented clients in civil litigation regarding alleged violations of statutory law and common-law claims, including clients facing liability under the Comprehensive Environmental Response, Compensation, and Liability Act (Superfund), toxic-tort liability, or citizen suits. In conjunction with the firm's corporate real estate lawyers, Ms. Andrew counsels clients regarding environmental liability in real estate purchase transactions.

Ms. Andrew received a bachelor of arts degree from Georgetown University and a juris doctor degree, *cum laude,* from Indiana University. She was chair of the Environmental Law Section of the Indiana State Bar Association from 1992–1993. Additionally, she has authored several articles and participated in seminars on environmental law topics.

Diane R. Archangeli is Claims Manager and Counsel and Marketing Manager at Risk International in Akron, Ohio. Ms. Archangeli graduated from the State University of New York in 1983 with a bachelor of arts degree in political science. She received her juris doctor degree from the University of Baltimore School of Law in 1987 and is a member of the Pennsylvania, Maryland, and Connecticut Bar Associations. She joined The Travelers Companies in 1988 as an account executive in the District Environmental Claim office in Baltimore, and was promoted to an account manager in the Law Department in Hartford, Connecticut in 1989. While working in The Travelers Companies Law Department, Ms. Archangeli was responsible for handling all aspects of coverage determinations and liability issues related to hazardous waste and pollution claims. Most recently, she worked with the Reinsurance Department of Aetna Insurance Company's Reserves and Financial Services Group as liaison between Aetna's Environmental Claims Department, Reinsurance Department, and domestic and foreign reinsurers and brokers. In her position with Risk International, Ms. Archangeli is responsible for a full array of environmental insurance claim recovery efforts, as well as marketing projects.

Judith A. Armold is of counsel to Venable, Baetjer and Howard, L.L.P., where she practices environmental and land-use law in the firm's Baltimore, Maryland office. She received her undergraduate degree from Western Maryland College and her law degree from the University of Maryland. From 1977 to 1987, Ms. Armold served as an Assistant Attorney General of Maryland and as general counsel to the Maryland Department of State Planning.

Dan B. Brown is a certified professional in the state of Ohio for the Voluntary Action Program, which is a part of the Ohio Real Estate Reuse Act (Brownfields Act), and is a member of the Brownfields Multi-Disciplinary Board, Administrative Procedures Subcommittee. He is also a certified professional geologist and received a bachelor of science degree in geology from Fort

Lewis College. Mr. Brown is currently pursuing a master's degree in business administration from John Carroll University. His experience includes conducting all types of environmental investigations, ranging from Phase I environmental site assessments (ESAs) to large-scale Phase II ESAs and subsurface investigations. His experience also includes managing the installation of large-scale remediation projects. Mr. Brown has managed multiple subsurface investigation projects that required fate and transport modeling, source delineation, and remediation system design. He is also experienced in conducting site investigations and site assessments, and designing and installing remediation systems relating to underground storage tanks. Mr. Brown has also supervised the construction of monitoring and recovery wells, including soil sample description, well logging, the collection of samples for laboratory analysis, and preparation of reports.

Juliet T. Browne is a lawyer with Verrill & Dana's Environmental Practice Group in Portland, Maine. Previously, she was an Assistant Attorney General for the Republic of Palau, a former United States Trust Territory located in the Western Pacific, where she represented the government in civil trials and appeals and counseled administrative agencies. Before joining the Palau Attorney General's office, Ms. Browne practiced with the San Francisco law firm of Skadden, Arps, Slate, Meagher & Flom, where her practice involved primarily environmental litigation. Ms. Browne received her bachelor of arts degree from the University of Michigan and her juris doctor degree from Boalt Hall School of Law at the University of California in Berkeley. She is currently a member of the California and Maine bars.

Anthony M. Carey is a partner and the head of the Energy and Environmental Practice Group at the law firm of Venable, Baetjer and Howard, L.L.P., resident in the firm's Baltimore office. He is a graduate of Princeton University and Harvard Law School, and a former Assistant Attorney General of the state of Maryland. He also served as a special assistant for energy matters to the United States Department of Housing and Urban Development.

John C. Chambers is a partner with the firm of McKenna & Cuneo in Washington, D.C., where he specializes in environmental counseling and litigation, with emphasis on hazardous-waste regulation under the Resource Conservation and Recovery Act (RCRA) and the Comprehensive Environmental Response, Compensation, and Liability Act. He has written extensively on related topics. Before joining the firm, Mr. Chambers served as in-house counsel for CONOCO in Houston, Texas. From 1981 to 1984, he was principal RCRA attorney for the American Petroleum Institute and was elected chair of its RCRA task force in 1985. Mr. Chambers is also the founder of the Brownfields Business Information Network. He received a bachelor of arts degree from the University of Pennsylvania and a juris doctor degree from the Washington College of Law, the American University.

Edward J. Cichon is Senior Vice President at Metcalf & Eddy, Inc., a worldwide environmental services company specializing in water, wastewater, and industrial-waste treatment, remediation of soil and groundwater, and pollution prevention. In his sixteen years at Metcalf & Eddy, Dr. Cichon has held numerous positions covering the areas of research and development, operations, and business development. His work has encompassed all aspects of environmental management, from wastewater treatment to potable water supply to remediation of soils and groundwater. Currently, he is responsible for Metcalf & Eddy's industrial and remediation services on a worldwide basis, and recently established Metcalf & Eddy's brownfields practice.

Robert L. Collings is a partner in the Government Regulation Section in the Philadelphia office of Morgan, Lewis & Bockius L.L.P. His practice includes handling a variety of environmental litigation matters, including the defense of government and private Superfund claims, United States Environmental Protection Agency (USEPA) and state lawsuits under environmental laws, and administrative proceedings challenging enforcement actions, permits, and other actions. A former USEPA Associate Regional Counsel and Branch Chief for water and waste programs, Mr. Collings also provides regulatory advice regarding proceedings, permit applications, and compliance issues under federal and state environmental laws, and transactional advice, including evaluating environmental liabilities in business transactions and drafting agreement language to protect client interests. He is the author of several articles on environmental law issues, including Superfund, infectious waste, and lender liability, and is the editor of the *Environmental Spill Reporting Handbook* and the editor of the stormwater and solid waste chapters in the *Municipal Solicitors Handbook*. He is a member of the executive board of the Water Resources Association of the Delaware River Basin and recipient of its 1996 Achievement Award. He is also a former chair (1986) of the Philadelphia Bar Association's Environmental Law Committee. He has lectured for the Pennsylvania Bar Institute, the American Bar Association, and other organizations. He was named one of the best lawyers in the Delaware Valley (*Corporate Monthly*, April 1989) for his work in the field of environmental law. Mr. Collings received his law degree from Boston College Law School. He is a graduate of Harvard College.

Richard L. Cristea, M.D., is Vice President and Director of Environmental Toxicology of Hemisphere Corporation, a company that acquires and redevelops environmentally distressed real estate. Dr. Cristea received his medical degree from the Chicago Medical School and completed his neurology residency and training in electromyography at the Cleveland Clinic Foundation. Dr. Cristea is certified by the American Board of Psychiatry & Neurology, with Added Qualifications in Clinical Neurophysiology, the American Board of Electrodiagnostic Medicine, and the American Society of Neurorehabilitation. He is a primary shareholder in the Northern Indiana Neurological Institute, where he serves as Director of Electromyography. He is also an adjunct staff member of the Cleveland Clinic Foundation's Department of Neurology. Dr. Cristea has lectured both nationally and internationally, including presentations in Europe, China, and Vietnam. He has published articles in numerous professional publications, including *Surgical Neurology, The Journal of General Internal Medicine,* and *The Journal of Cerebrovascular Disease and Stroke.*

Christine DiCato-Thaxton is the Laboratory Manager and the Director of Marketing for GeoAnalytical, Inc., an environmental laboratory dedicated to the analysis of soil and groundwater. Ms. DiCato-Thaxton monitors and controls the daily operations of the laboratory, coordinates incoming laboratory submissions, and tracks the laboratory's performance and turnaround times. She also schedules and coordinates project work and negotiates contractual agreements. Ms. DiCato-Thaxton has spent the last year involved with the Ohio Environmental Protection Agency, participating on the laboratory subcommittee for the voluntary action program, "Ohio's Answer to Brownfields." The subcommittee had input on the certification of laboratories for the program. She is a graduate of Kent State University, with a bachelor of science degree in geology and comprehensive science.

Alan S. Doris is a partner with the firm of Benesch, Friedlander, Coplan & Aronoff LLP in Cleveland, Ohio, and serves as chair of the firm's Tax Practice Group. He focuses his tax practice on corporate and partnership taxation, tax controversies, and capital equipment financing. He has additional expertise in the area of leasing and has extensive experience in the wholesale and electrical-supply industry. Mr. Doris also is a certified public accountant. Mr. Doris is the chair of the Capital Recovery and Leasing Committee and the former chair of the Amortization of Intangibles Subcommittee for the Section of Taxation of the American Bar Association. He also serves the ABA's Section of Taxation as the co-chair of the Task Force on Section 197 Regulations. He is the former chair of the Cleveland Tax Institute. Mr. Doris placed first in the Ohio 1969 certified public accountant exam after graduating *cum laude* from Miami University and earning the distinctions of Phi Beta Kappa and Beta Gamma Sigma. He earned his juris doctor degree in 1972 from Harvard University.

Jennifer A. Downs is an associate in the Kansas City, Missouri law firm of Spencer Fane Britt & Browne, where she is a member of the firm's Environmental Practice Group. She is active in the Kansas City Metropolitan Bar Association's Environmental Law Committee.

From 1993 to 1994, she was a law clerk for the Honorable Pasco M. Bowman II, United States Court of Appeals for the Eighth Circuit. She is a 1989 honors graduate of Dartmouth College and graduated in 1993, with honors, from the Duke University School of Law.

David J. Engel is a partner at Sidley & Austin, where he focuses his practice on environmental litigation, client counseling, and transactions. He concentrates on regulatory, permitting, and enforcement issues under the Clean Water Act, the Clean Air Act, and the Resource Conservation and Recovery Act. He also represents clients in connection with liability, allocation, and insurance issues relating to state and federal Superfund. Mr. Engel is a member of the Chicago Brownfields Forum and was recently appointed to a committee established by a consortium of municipalities that received a United States Environmental Protection Agency Brownfields Pilot Project grant. From 1992 through 1994, Mr. Engel served as the Associate Director of the Illinois Environmental Protection Agency. He assisted the director and agency senior management with the development of legislation, approved all Agency enforcement referrals and submissions to the Illinois Pollution Control Board, and directly managed major legal cases that required policy coordination and attention of the director. Mr. Engel received his undergraduate degree in biology from Clarion University of Pennsylvania, his master of science degree in limnology from Michigan State University, and his law degree from Wayne State University, where he served as editor-in-chief of the *Wayne Law Review*. After law school, Mr. Engel clerked for Judge Cornelia Kennedy on the United States Court of Appeals for the Sixth Circuit.

Adam Fishman has spent the last twelve years in northeastern Ohio pursuing a commercial real estate career in brokerage, development, and advisory capacities. Mr. Fishman joined the Victor S. Voinovich Company in office-leasing brokerage in 1985. During his tenure with the company, Mr. Fishman earned numerous awards in addition to handling prestigious developments, including Cleveland's landmark office building, the Terminal Tower. In 1987, he joined Trammell Crow Company and was actively involved in the successful development, marketing, leasing, and management of all the Trammell Crow Company assets in northeastern Ohio. Mr. Fishman's other responsibilities included construction management, municipal relations, and lender relations. In early 1992, Mr. Fishman joined Cleveland Real Estate Partners as Vice President. He has been engaged in assignments including office and industrial tenant representation/consulting for both local and national firms, construction management services for large-scale new construction and renovation projects, and varied aspects of the corporate real estate consulting practice. In 1994, Mr. Fishman became a principal of the company and assumed day-to-day responsibilities of firm management as Managing Principal in January 1995. A 1984 business graduate of Miami University, Oxford, Ohio, Mr. Fishman holds degrees in business finance and economics.

Michael L. Gargas is a principal health scientist with the ChemRisk Division of McLaren/Hart, which has eighteen offices throughout the United States and abroad. Dr. Gargas is the Operations Manager of the Cleveland, Ohio office, with over forty staff members, representing risk assessment (ChemRisk), regulatory compliance management, and remedial investigation disciplines. Technically, Dr. Gargas oversees and prepares human health risk assessments, conducts toxic-tort support investigations, and addresses various toxicological issues through applied research on behalf of clients in the private sector. Dr. Gargas has eighteen years of experience in human health risk assessment and biochemical toxicology research, with over fifty publications in the peer-reviewed literature on these topics. He also serves on the editorial board of *Toxicology and Applied Pharmacology* and is a member of the Society of Toxicology and the Society for Risk Analysis.

Scott D. Garson is the Manager of Planning and Development at MidTown Corridor, Inc., in Cleveland, Ohio. Mr. Garson has worked for two community development corporations during the last eight years, after a three-year career in the private sector. He has extensive experience in brownfields issues from a community development perspective. A graduate of Kenyon College, he is also working toward a graduate degree in urban planning.

Allan Gates is a partner in the law firm of Mitchell, Williams, Selig, Gates & Woodyard in Little Rock, Arkansas. Mr. Gates served as law clerk to Chief Judge Pat Mehaffy of the Eighth Circuit Court of Appeals and to Associate Justice Harry A. Blackmun. Since then, Mr. Gates has engaged in the private practice of law in Washington, D.C. and in Arkansas. His practice has focused primarily on environmental law and litigation. Mr. Gates has authored several articles and book chapters on subjects relating to environmental law and litigation. He is the editor of the *Arkansas Environmental Law Handbook* series. Mr. Gates teaches environmental law as an adjunct faculty member at the University of Arkansas at Little Rock School of Law. Mr. Gates is a member of the American, Arkansas, District of Columbia, and Pulaski County Bar Associations. He is a member of the American Law Institute and the Environmental Law Institute. He is a graduate of Yale University and Vanderbilt University School of Law.

Richard M. Glick, a partner in the Portland office of Davis Wright Tremaine, serves as chair of the firm's Environmental & Natural Resources Department. His practice emphasizes hazardous substances, water, and energy law. He has represented his clients before state and federal agencies in matters such as water rights, cleanup actions, and permit proceedings. Mr. Glick also represents his business and real estate clients in assessment and management of environmental risk.

Before joining the firm, Mr. Glick was Staff Counsel to the State Water Resources Control Board in Sacramento, and Deputy City Attorney for the city of Portland.

He is past chair of the Oregon State Bar Association's Environmental and Natural Resources Section. Mr. Glick is the author of the Oregon State Bar Professional Liability Fund's *The Environmental Law Handbook: A Malpractice Avoidance Guide for Every Oregon Lawyer* (August 1990). He is co-author of

"Water," a chapter in the *Oregon Law Institute Treatise on Environmental Law for Oregon Practitioners.*

David F. Goossen is an associate with the law firm of Isaacson, Rosenbaum, Woods & Levy, P.C., in Denver, Colorado. He is a member of the firm's Environmental Practice Group and has experience in environmental litigation, hazardous-waste remediation, environmental compliance, environmental audits, and contractual allocation of environmental liabilities. Mr. Goossen is actively involved with brownfields in Colorado, working with federal, state, and local agencies to facilitate the purchase and sale, remediation, and reuse of contaminated properties in the state. He is the author of several articles on environmental law, including, on the subject of brownfields, "Contractual Allocation of Environmental Liabilities in Real Estate Transactions," 25 *The Colorado Lawyer* 79 (March 1996).

Arthur J. Harrington has practiced environmental law for nearly two decades and is well versed in all major state and federal environmental programs. He has authored numerous articles on environmental law and has spoken at more than seventy seminars on the subject.

After graduating from the University of Wisconsin in 1972, Mr. Harrington received his juris doctor degree the University of Wisconsin Law School in 1975. He is a member of Phi Beta Kappa and the Order of the Coif, and was the note and comment editor for the Wisconsin Law Review. Since 1988, Mr. Harrington has been selected for inclusion in every edition of *The Best Lawyers in America* in the "Environmental Law" category. He is the inaugural past chair of the Environmental Law Section of the State Bar of Wisconsin and a past president of the Milwaukee Bar Association.

W. Andrew Hazelton is an associate with the firm of Reiber, Kenlan, Schwiebert, Hall & Facey, P.C., in Rutland, Vermont. His practice focuses on local and state land-use planning and permitting, and environmental defense. He received a bachelor of arts degree from Dartmouth College in 1986 and a juris doctor degree from Albany Law School in 1994.

Joel L. Herz of the Law Offices of Joel L. Herz in Tucson, Arizona, specializes in complex environmental and related litigation matters. From 1987 to 1994, Mr. Herz practiced in the environmental and litigation practice groups with the New York office of the international law firm of Fried, Frank, Harris, Shriver and Jacobsen. His firm is currently handling more than forty major Superfund and state enforcement cleanup proceedings for clients including Fruit of the Loom, Inc.; Farley Industries, Inc.; Salant Corporation; and DeSoto, Inc. He graduated from the University of Michigan with honors in 1985, and the University of Michigan Law School in 1987.

Donald T. Iannone directs three centers at Cleveland State University in the areas of economic development and environmental finance. Over his twenty-year career, he has become nationally recognized for his work in both areas. As the Executive Director of the Great Lakes Environmental Finance Center, Mr. Iannone assists communities and states in a six-state region to prepare and implement innovative and cost-effective strategies to fund brownfields cleanup and redevelopment. He is an advisor to the United States Environmental Protection Agency's prestigious Environmental Finance Advisory Board, and a former board member of the American Economic Development Council and the National Council for Urban Economic Development.

Bernard J. Jerlstrom is a tax lawyer with the law firm of Benesch, Friedlander, Coplan & Aronoff LLP in Cleveland, where he concentrates his practice on corporate tax matters, corporate reorganizations, and international tax matters. He is also a member of the adjunct faculty of Case Western Reserve University School of Law's Graduate Tax Program. Mr. Jerlstrom received a bachelor's degree from Duke University, a juris doctor degree from the University of Miami, and a master of laws degree in taxation from George Washington University.

H. Michael Keller practices environmental and natural-resources law as a shareholder with the Salt Lake City law firm of Van Cott, Bagley, Cornwall & McCarthy, where he chairs the firm's Natural Resources and Environmental Law Section. He received his bachelor's and master's degrees in geology from Dartmouth College and his juris doctor degree in 1978 from Duke University. His practice focuses on environmental permitting, compliance, and enforcement. Mr. Keller is a past chair of the Energy, Natural Resources, and Environmental Law Section of the Utah State Bar and currently serves as a vice chair of the Hard Minerals Committee of the ABA Section of Natural Resources, Energy, and Environmental Law.

A. Jay Kenlan is a lawyer with Reiber, Kenlan, Schwiebert, Hall & Facey. His areas of specialization include environmental, commercial, and general business law.

The principal focus of Mr. Kenlan's practice has been local and state land-use planning and permit processes, state and federal environmental regulations, and the legal and constitutional implication of land use and environmental regulations. In pursuit and defense of his clients' property rights, he regularly appears before municipal zoning boards and planning commissions, district environmental commissions, and the Vermont environmental board, state agencies that regulate land use and development, and Vermont and federal courts. Mr. Kenlan's professional career has been devoted to understanding and defending the legitimate rights of business and property owners to use, manage, and enjoy their property free of unwarranted governmental regulation or interference.

Mr. Kenlan received a bachelor of science degree in government from Norwich University, attended the University of Maine School of Law, and received a juris doctor degree from the University of North Carolina School of Law. He is admitted to practice in Massachusetts and Vermont, as well as the United States District Court, Vermont, the Second Circuit Court of Appeals, and the United States Supreme Court. Mr. Kenlan has made several seminar presentations on environmental and real estate topics.

James T. Kilbreth heads Verrill & Dana's Environmental Practice Group in Portland, Maine. He represents numerous industrial, utility, financial institution, and state agency clients on a range of environmental matters, including compliance, audits, permitting, enforcement, and related litigation. Mr. Kilbreth drafted the legislation creating Maine's Voluntary Response Action Program on behalf of a major financial institution. Mr. Kilbreth was formerly Chief Deputy Attorney General of the state of Maine, where he was involved in all major environmental permitting and civil and criminal enforcement activities within the Attorney General's office. For over two years during his tenure at the Attorney General's Office, he was the principal lawyer representing the Department of Environmental Protection and advising all bureaus of the Department on interpretation and implementation of the environmental statutes. Mr. Kilbreth negotiated the first Superfund site cleanup in the state and brought the first actions on behalf of the state under the Uncontrolled Hazardous Substance Site Law.

Annette Kovar has been legal counsel with the Nebraska Department of Environmental Quality since 1987. From 1980 to 1987, she was a member of the Nebraska Natural Resources Commission. She is a former chair of the Natural Resources and Environmental Law Section. In 1985 she received the United States German Marshall Fund Fellow. She was a contributing author of *Environmental Law Practice Guide*.

Ms. Kovar received her juris doctor degree from the University of Nebraska College of Law and is accredited with the Nebraska State Bar Association.

Jane B. Kroesche is an environmental lawyer with the law firm of Skadden, Arps, Slate, Meagher & Flom in San Francisco, California. Ms. Kroesche spearheads the firm's west coast environmental transaction practice. In this role, Ms. Kroesche has advised and represented clients on all aspects of environmental law in connection with complex real estate, corporate, and securities transactions. Ms. Kroesche has supervised environmental site reviews (including environmental assessments and compliance audits), and represented clients in administrative and judicial proceedings in connection with site cleanup and environmental compliance involving, among other things, state and federal land use, air, water, and hazardous-waste management and cleanup matters. Ms. Kroesche is a member of the ABA's Section of Natural Resources, Energy, and Environmental Law (SONREEL) Brownfields Task Force, SONREEL's Coordinating Group on Alternate Dispute Resolution, and the Executive Committee of the Bar Association of San Francisco's Environmental Law Section. Ms. Kroesche received her juris doctor degree in 1984 from the University of California, Los Angeles.

Howard C. Landau is President of Wyse-Landau Public Relations, a leading communications consulting firm headquartered in northeast Ohio. Mr. Landau has provided public relations counsel to major real estate developers in cities around the world, and has worked with some of the country's leading architects, engineers, and construction firms.

Barbara D. Little is a partner in the Charleston, West Virginia office of Jackson & Kelly and has, for the past twenty years, specialized in the practice of environmental law. From 1981 through 1985, before joining Jackson & Kelly, she was chief environmental counsel for Browning-Ferris Industries, Inc. (Houston, Texas) and its subsidiary CECOS International, Inc. (Buffalo, New York), where she was responsible for all environmental matters, including Superfund claims, for the multinational waste-management concern.

From 1977 to 1981, Ms. Little served as the principal environmental lawyer for Shell Oil Company (Houston, Texas), handling all environmental matters for several Shell Oil divisions. From 1974 to 1977, she was an adjunct professor of law at the University of Houston Law School, teaching environmental and administrative law. Ms. Little received her bachelor of arts and juris doctor degrees, *summa cum laude*, from the University of Texas at Austin.

Admitted to practice law in West Virginia and Texas, Ms. Little is an expert in all areas of environmental law, has authored numerous environmental law articles, and is a frequent seminar lecturer. Ms. Little has been active in the American Petroleum Institute, the Chemical Manufacturers Association, the National Solid Waste Management Association, the American Institute of Chemical Engineers, the American Law Institute/ABA, the Section of National Resources, Energy, and Environmental Law of the ABA, and the Texas and West Virginia state bars.

Thomas F. Long is a senior health scientist with the ChemRisk Division of McLaren/Hart Environmental Engineering, located in Cleveland, Ohio. He received his undergraduate degree in biology from Wabash College and graduate degree in pharmacology and toxicology from Purdue University. Before his association with ChemRisk, he served for over a decade as Senior Toxicologist for the Illinois Department of Public Health, conducting risk assessments, setting health-based standards and guidelines, and conducting investigations of the effects of exposure to hazardous substances on human health. He served as adjunct faculty at four state universities, where he taught undergraduate and graduate courses in toxicology, industrial hygiene, risk assessment and management, risk communication, environmental health, and public policy.

Bradley M. Marten is a partner at Marten & Brown. He advises clients on environmental issues in the solid and hazardous waste, mining, chemical, manufacturing, chemical distribution, energy, railroad, pulp and paper, gasoline storage, transit, telecommunications, real estate, lending, food, aviation, scrap metal, and shipping industries. His special areas of expertise are in hazardous and solid waste, natural-resource damages, and water quality. He has obtained permits for numerous solid and hazardous-waste facilities, and has extensive experience in the remediation and restoration of contaminated sites. He represented the state of Alaska on natural-resource damage issues in the Exxon Valdez litigation. He has successfully represented industrial clients in administrative proceedings and regulatory negotiations before federal and state environmental agencies. Mr. Marten is a frequent lecturer on environmental issues. His publications include the leading treatise on Washington environmental law, the *Washington Environmental Law Handbook* (Government Institutes, 1990). He is also the coauthor of the *MTCA Handbook* (1991), and author of an article entitled "Litigating CERCLA Natural Resource Damage Claims," 22 *Environmental Reporter* 670 (BNA, July 19, 1991). He received a bachelor's degree in 1975 from Cornell University, a master's degree in 1977 from Yale University, and a juris doctor degree from Harvard Law School in 1981. From 1981 to 1983, Mr. Marten clerked for Judge Donald Voorhees, United States District Court, Western District of Washington.

Maura McCaffery is a lawyer with the law firm of Goulston & Storrs in Boston, Massachusetts, where she focuses on environmental compliance counseling associated with industrial and commercial clients, as well as environmental due diligence for real estate transactions. Ms. McCaffery regularly participates in the efforts of the Massachusetts Brownfields Advisory Group and has published several articles on brownfields and other environmental topics. Before joining Goulston & Storrs, Ms. McCaffery represented corporate and industrial clients in environmental litigation. Ms. McCaffery received her bachelor of arts degree from the University of Pennsylvania and her juris doctor degree from Boston University School of Law.

William McElroy is a program and product-line manager in the Environmental Liability Department of the Zurich-American Insurance Group, based in New York. His expertise includes environmental risk evaluation, particularly in the area of commercial real estate and underground storage tanks. He is a frequent speaker on environmental issues, and has served as a member of the United States Department of Housing and Urban Development Title X Task Force on Lead-Based Paint Hazard Reduction and Financing.

Before joining Zurich-American, Mr. McElroy was with the American International Group (AIG). While at AIG, he held a number of positions, including that of program manager and senior staff technical underwriter in the environmental liability group. Mr. McElroy holds a bachelor of science degree in biology from Rutgers University, New Brunswick, New Jersey.

Michelle A. Meertens was a 1996 summer associate at McKenna & Cuneo, L.L.P. She attends New York University School of Law and expects to receive her juris doctor degree in May 1997. She received her bachelor of arts degree from Yale University.

Paul S. Moe is a lawyer with Faegre & Benson. His practice includes a wide variety of environmental and real estate matters, with a focus on remediation, development, financing, and the sale and leasing of contaminated commercial property. He is a real property law specialist, certified by the Real Property Section of the Minnesota State Bar Association. He has conducted seminars and published articles on a variety of topics related to environmental liability and real property, including underground storage tanks, asbestos, lead-based paint, and voluntary remediation. Mr. Moe received a bachelor of arts degree, *magna cum laude*, from St. Olaf College, and a juris doctor degree, *magna cum laude*, from the University of Minnesota Law School. He is a member of Phi Beta Kappa. He was a research editor for the University of Minnesota Law Review, and was admitted to the Order of the Coif.

Sean T. Monaghan is counsel at Shanley & Fisher and a member of the firm's Environmental Group. His practice of environmental law includes environmental aspects of business transactions, regulatory compliance, and environmental litigation. His real estate, commercial, and corporate law background has prepared him to handle environmental aspects of real estate and commercial transactions, including compliance with New Jersey's Industrial Site Recovery Act (ISRA). Mr. Monaghan is a member of the New Jersey Bar Association's Real Property, Probate and Trust Law Section, Environmental Law Section, and Corporate and Business Law Section. He is a member of the ISRA Committee of the Corporate and Business Law Section and the Transactions Committee of the Environmental Law Section. Mr. Monaghan has authored papers on lender liability under the federal Superfund law and environmental issues for borrowers. He has coauthored papers with other members of the Environmental Law Group, on environmental transfer laws including the New Jersey Environmental Cleanup Responsibility Act and incentives and protections for prospective purchasers and operators of contaminated property, the latter of which is included in the treatises *Contamination Procedures and Techniques* and *Nichols on Eminent Domain*, published by Matthew Bender and Company, Inc. Mr. Monaghan has lectured to lawyer and business groups on site remediation and environmental liability issues. Mr. Monaghan is a graduate of the University of Pennsylvania and Seton Hall University School of Law.

I. Leo Motiuk is a partner at Shanley & Fisher and co-chair of the Environmental Group. His activities range from involvement with federal and state agencies in seeking permits, handling Superfund and Spill Act claims, and defending against enforcement actions, to engaging in transactional, "due diligence," and cost-recovery work for sellers, buyers, and lenders, and the development and implementation of corporate environmental policies. He has been actively involved with strategic planning for meeting the impact of ever-increasing regulation in such areas as air pollution, hazardous-waste minimization, and occupational safety and health. Mr. Motiuk has served as chair of the Environmental Committee of the ABA Section on Real Property, Probate and Trust Law, and as chair and a member of the Board of Directors of the Environmental Section of the New Jersey State Bar Association. He presently serves as a member of the Board of Advisors of the BNA *Toxics Law Reporter.* He is a frequent lecturer around the country on issues such as brownfields/ redevelopment of contaminated properties, the development and implementation of corporate environmental policies, lender liability, defending against enforcement actions, air pollution, and occupational safety and health. He previously served as Deputy Attorney General and Assistant Counsel to the Governor of New Jersey at a time when landmark environmental legislation was being developed in New Jersey. Mr. Motiuk is a graduate of Dickinson College (bachelor of arts), Duquesne University School of Law (juris doctor), and New York University School of Law (master of laws). He has been an adjunct professor at the Rutgers University Institute of Labor-Management Relations and the Rider University School of Business Administration.

Bill Mundy, Ph.D., MAI, CRE, is President and Senior Analyst at Mundy & Associates, Seattle, Washington, a firm specializing in the valuation of conservation, preservation, and contaminated properties. Mr. Mundy has published numerous articles in various professional and academic journals. He also contributed to, and reviewed, *The Appraisal of Real Estate* (9th, 10th, 11th eds.), published by the Appraisal Institute.

Margaret Murphy is a partner of Shearman & Sterling, an international law firm headquartered in New York with offices in London, Paris, Frankfurt, Budapest, and other European, Middle Eastern, and Asian business centers. Ms. Murphy is the head of the Shearman & Sterling Environmental Practice Group. She has extensive experience in all aspects of environmental law. Her clients include multinational corporations, financial institutions, and governments. Most recently, she was chief legal advisor to the government of Kuwait in the preparation of claims for environmental damages resulting from the 1990 Gulf War. She advised the government of Mexico on environmental aspects of the North American Free Trade Agreement, and currently is environmental counsel to the Government of Kazakhstan in the development of hydrocarbon resources in the Caspian Sea. More generally, her practice includes transactional work in the corporate, merger and acquisition, finance, and real estate areas in the United States and around the world. She has served as lead environmental counsel to Bankers Trust Co., British Steel, Chemical Bank, Citibank, Dresdner Bank, Goldman Sachs & Co., Maraven, Merrill Lynch, Morgan Stanley, Petroleos De Venezuela, and Union Bank of Switzerland, among others, in numerous United States and international transactions involving environmental matters.

James T. Price is a partner in the Kansas City, Missouri, law firm of Spencer Fane Britt & Browne, where he chairs the firm's Environmental Practice Group. Mr. Price served as chair, from 1990 to 1992, of the Solid and Hazardous Waste Committee of the ABA Section of Natural Resources, Energy, and Environmental Law. From 1985 to 1990, he served as vice chair. From 1992 to 1995, he served as a member of the Section Council. He has chaired the section's Brownfields Task Force since 1995.

Mr. Price is a graduate of the University of Missouri School of Journalism and a graduate of Harvard Law School.

James R. Rocco is a civil engineer with over twenty years of experience in emergency response, corrective action, and environmental compliance. He is Manager of Environmental Remediation for BP Oil Company, and has been employed by BP for the past twenty-two years in engineering and management positions related to construction, maintenance, and environmental activities. In his current position, he is responsible for corrective-action activities at BP Oil's marketing, terminal, pipeline, and refinery facilities, and for related environmental legislative and regulatory activities. He has extensive experience and involvement with environmental regulatory development and legislative issues on the state and federal level, and technical issues related to corrective action, and environmental compliance. Mr. Rocco is currently the chair of the Ohio Petroleum Underground Storage Tank Release Compensation Board. He is a founding member of the USEPA/Industry Partnership in RBCA Implementation (PIRI). He is active in the American Petroleum Institute Marketing Operations and Engineering Committee and participates in various committees and task forces addressing marketing engineering and environmental issues. He is active as a member of the National Fire Protection Association Technical Committee on Tank Leakage and Repair Safeguards and is chair of the American Society for Testing and Materials E50.01 Task Force on Corrective Action, which has developed the ASTM Standards for Corrective Action at Petroleum Release Sites, Accelerated Site Assessment, and Remediation by Natural Attenuation. He is an active member of the ASTM Task Force that developed the ASTM Risk-Based Corrective Action (RBCA) standard and chair of an RBCA work group that developed and coordinates the training program for the RBCA standard.

Robin K. Rock is an associate at Marten & Brown, specializing in environmental law. Ms. Rock advises clients on compliance matters involving federal and state environmental issues and legislation. Her experience includes addressing issues related to Superfund, the Clean Air Act, the Resource Conservation and Recovery Act, natural-resource damages, cross-border waste, and environmental audits. She has also advised clients on the environmental issues arising out of a variety of business and real estate transactions. Ms. Rock's experience includes negotiating operating permits for a veneer mill in Washington state, implementing an environmental compliance program for a real estate development company, and acting as a temporary environmental in-house counsel for a timber-industry client. She is a coeditor and author of the *Washington Environmental Law Deskbook*. Ms. Rock received her bachelor of arts degree from Smith College, and her law degree, *cum laude*, from Georgetown University Law Center.

R. Judson Scaggs, Jr. is a partner in the Wilmington law firm of Morris, Nichols, Arsht & Tunnell, where his practice is concentrated in environmental aspects of land use and development and environmental and corporate litigation. A graduate of Washington & Lee University, Mr. Scaggs received his juris doctor degree from the College of William and Mary, Order of the Coif, where he was a member of the William and Mary Law Review. Mr. Scaggs is a member of the Delaware, Virginia, and American Bar Associations. He is the vice chair of the Environmental Law Section of the Delaware Bar Association and is a frequent speaker and author on environmental law topics.

William M. Seuch is a lawyer with the firm of Goulston & Storrs in Boston, Massachusetts, where he concentrates his practice on environmental law, including brownfields issues associated with the development of previously contaminated properties. He counsels clients on the environmental issues associated with real estate transactions, industrial operations, corporate mergers and acquisitions, international business transactions, and financings regarding all these matters. Mr. Seuch also has significant experience in environmental litigation, including insurance coverage disputes, and he frequently writes articles and speaks about environmental issues. Mr. Seuch graduated *cum laude* from the University of Connecticut and received his law degree from the Georgetown University Law Center.

John Slavich, a shareholder in the firm of Guida, Slavich & Flores, has a diversified environmental law practice focusing on environmental issues arising in corporate, real estate, and lending transactions, and on counseling for environmental compliance issues involving a broad range of industries. His environmental practice is complemented by more than a decade of experience in corporate, finance, securities, and real estate law. Mr. Slavich previously was a partner in the Environmental Section of Andrews & Kurth, L.L.P. Mr. Slavich received his law degree from Washington University School of Law, where he was a member of the board of editors of the *Washington University Law Quarterly*. He earned his master of business administration degree at Southern Methodist University and his undergraduate degree at Earlham College. Mr. Slavich is admitted to the bar in Texas. He has regularly lectured on environmental law topics at the University of Texas at Dallas and at various continuing legal education programs. Recent publications include "Financial Disclosure of Environmental Liabilities," in *Advanced Environmental Law Course* (State Bar of Texas, 1994); "Environmental Issues Affecting Emerging Growth Companies," in *Emerging Growth Companies* (State Bar of Texas, 1993); "Beyond Permitting: Environmental Compliance Management Challenges for the 1990s," 2 *Journal of Environmental Permitting* 525

(1993); and "Environmental Issues in the Acquisition of a Going Concern," in *What Every Business Lawyer Needs to Know about Environmental Law* (State Bar of Texas, 1991). He is also the coauthor of "Contractual Efforts to Allocate the Risk of Environmental Liability: Is There a Way to Make Indemnities Worth More Than the Paper They Are Written On?," 44 *Southwestern Law Journal* 1349 (1991). He is also a member of the Dallas Brownfields Forum. Mr. Slavich additionally is a senior editor of *Environmental Regulation and Permitting*.

Henry L. Stephens, Jr. is of counsel to Greenebaum, Doll & McDonald in the Environmental Law Section and is a resident of the Covington office. He is also former dean and professor of law at Salmon P. Chase College of Law, Northern Kentucky University. Dean Stephens is the founder and executive director of the Ohio Valley Environmental Law & Natural Resources Law Institute affiliated with the College of Law. A former prosecutor with the Kentucky Natural Resources & Environmental Protection Cabinet, Dean Stephens has been on the faculty of Chase College of Law since 1979, where he teaches courses in environmental and mining law. In addition, Dean Stephens has been a consultant to business and industry, exclusively in the environmental area, for more than twenty years. He is a past chair of the Natural Resources Section of the Kentucky Bar Association and a past member of the Commonwealth of Kentucky's Environmental Quality Commission. He is presently a trustee-at-large of the Eastern Mineral Law Foundation and is president-elect of the Kentucky Bar Foundation. Dean Stephens is a frequent lecturer and writer on environmental topics in Kentucky.

Greer Tidwell is General Counsel of the Tennessee Department of Environment and Conservation. His office is responsible for administrative enforcement cases and legal counseling for Tennessee's environmental and natural-resource programs, including the state parks. Before serving as General Counsel, Mr. Tidwell was Environmental Counsel with Bridgestone/Firestone Corporation. Also, he coordinated development of a Clean Air Act Title V business plan and federal legislative advocacy. Before his in-house position, Mr. Tidwell was with the Baker, Donelson, Bearman & Caldwell law firm, where he represented industrial and municipal clients before state and federal environmental protection agencies in cases under the Resource Conservation and Recovery Act, the Comprehensive Environmental Response, Compensation, and Liability Act, the Clean Water Act, and the New Jersey Environmental Cleanup and Responsibility Act. Mr. Tidwell graduated from Vanderbilt Law School, where he was on the Moot Court Board. He is a graduate of Baylor University and Tennessee Technological University in Civil Engineering. Before law school, he prepared waste-minimization plans and conducted compliance auditing with a consulting engineering firm. Mr. Tidwell is a licensed patent attorney. He serves as chair of the Tennessee Environmental Law Section and, before his public position, was vice president of the Tennessee Conservation League.

Grant R. Trigger is a partner in the Environmental Law Department of Honigman Miller Schwartz and Cohn. Mr. Trigger received a bachelor of science degree in environmental sciences engineering from the University of Michigan in 1975, and is a 1985 *cum laude* graduate of the University of Detroit Law School. He is a registered professional engineer in Michigan and has consulting experience in wastewater treatment, environmental permitting, hydrogeological and landfill analysis, and general civil engineering. Mr. Trigger has been actively involved in the development of environmental law and policy in Michigan for almost twenty years, and he served on the Michigan Toxic Substance Control Commission from 1983 to 1989. He participated as a representative of business interests on the Michigan Department of Natural Resources (MDNR) workgroup that developed the Act 307 rules. He is currently a member of the Act 307 Program Advisory Group, established to provide input to MDNR on Act 307 implementation issues. In December 1992, he was appointed by Governor Engler to the Natural Resources Management and Environmental Code Commission. In addition, as counsel to the Michigan Chemical Council, he participated extensively in the negotiation of the recent Act 307 amendments, now known as Part 201 of the Environmental Code. He currently serves as chair of the Environmental Committee of the Michigan Economic Developers Association, which focuses on brownfields redevelopment methods.

Wendy E. Wagner graduated *summa cum laude* from Hanover College and went on to receive a master's degree in environmental studies from the Yale School of Forestry and Environmental Studies and a juris doctor degree from Yale Law School. After a clerkship with Albert Engel of the United States Court of Appeals for the Sixth Circuit, Ms. Wagner worked as a trial lawyer with the Environment and Natural Resources Division of the United States Department of Justice, and then served as the Pollution Control Coordinator for one year in the Office of General Counsel, United States Department of Agriculture. Ms. Wagner joined the Case Western Reserve University School of Law faculty as an assistant professor in 1992. She teaches in the areas of environmental law, torts, and law and science. Her recent publications can be found in the *Columbia Law Review* and the *Texas Law Review*.

Jonathan D. Weiss works on economic redevelopment issues in the Office of the Vice President at the White House. He is serving on special assignment from the United States Environmental Protection Agency, where he was Senior Brownfields Counsel. He has a juris doctor degree from the University of Virginia and master's and bachelor's degrees from the University of Michigan, where he graduated Phi Beta Kappa.

Lesley Hay Wilson is a chemical engineer with over ten years of experience in corrective action, environmental compliance, and emergency response. She is the Northeast Environmental Remediation Team Leader for BP Oil Company and has held several different environmental compliance and corrective-action positions during her tenure at BP Oil. In her current position, she is responsible for corrective-action projects for current and former BP Oil refining, terminal, pipeline, and retail facilities in a nine-state region. She is currently the BP Oil Remediation Team Facilitator for risk-based corrective action (RBCA) for all oil company corrective-action sites, including development of a company RBCA program and the accompanying staff and consultant training. Mrs. Wilson is currently the Work Group Leader for Site Characterization and Remediation Technologies for the American Petroleum Institute Soil and Groundwater Technical Task Force. She is active in the American Society for Testing and Materials E50.01 Task Force on Corrective Action and a contributor to the ASTM RBCA Standard (E1739). She was active in the development of the RBCA state regulatory personnel training and the ASTM user group training. She is actively involved in the development of the voluntary corrective-action RBCA standard under ASTM E50.04.

PUBLICATIONS
SECTION OF NATURAL RESOURCES, ENERGY, and ENVIRONMENTAL LAW
ORDER FORM

____ I am a member of the Section of Natural Resources, Energy, and Environmental Law and am eligible for the reduced member prices listed on this form.

____ I wish to join the Section of Natural Resources, Energy, and Environmental Law and take advantage of the member price now. I have enclosed a separate check for $35 (payable to American Bar Association—SONREEL). Please send the publications I have indicated at the member prices. I am an ABA member.

SONREEL Books

Qty	Title	SONREEL price	Nonmember price	Amount
	A Complete Guide to Environmental Audits (5350057)	$75.00	$85.00	$
	Brownfields: A Comprehensive Guide to Redeveloping Contaminated Property (5350056)	$139.95	$149.95	$
	CERCLA Enforcement (5350053)	$110.00	$120.00	$
	Environmental Aspects of Real Estate Transactions (5350052)	$94.00	$104.00	$
	The Natural Resources Law Manual (5350051)	$75.00	$80.00	$
	Water Law (5350050)	$70.00	$75.00	$
	Underground Storage Tanks: A Primer on the Federal Regulatory Program, Second Ed. (5350048)	$55.00	$60.00	$
	The Environmental Law Manual (5350035)	$75.00	$80.00	$
	What to Do When the Environmental Client Calls (5350043)	$55.00	$60.00	$
	The Clean Water Act Handbook (5350045)	$75.00	$80.00	$
	The Clean Air Act Operating Permit Program: A Handbook for Counsel, Environmental Managers & Plant Managers (5350040)	$75.00	$75.00	$
	The RCRA Practice Manual (5350044)	$84.95	$94.95	$
	The RCRA Policy Documents: Finding Your Way Through the Maze of EPA Guidance on Solid & Hazardous Waste (5350037)	$55.00	$60.00	$

SONREEL Monographs

On oil & gas issues

Qty	Title	SONREEL price	Nonmember price	Amount
	Gas Dispatch, Allocation, and Balancing Agreement—A Proposed Form #22 (5350047)	$29.95	$39.95	$
	Drafting Indemnity Provisions in Oil and Gas Contracts #21 (5350046)	$29.95	$39.95	$
	Legal Research Checklists for International Petroleum Operations #20 (5350042)	$29.95	$39.95	$
	Federal Oil and Gas Appeals in the Department of the Interior #18 (5350036)	$29.95	$39.95	$
	Legal Aspects of the Purchase & Sale of Oil & Gas Properties #17 (5350034)	$29.95	$39.95	$
	International Oil & Gas Joint Ventures #16 (5350033)	$29.95	$39.95	$
	Improved 1989 Joint Operating Agreement #15 (5350030)	$29.95	$39.95	$
	Drafting Oil & Gas Exploration Agreements #14 (5350029)	$19.95	$29.95	$
	Package: 1989 Joint Operating Agreement and Drafting Oil & Gas Exploration Agreements (5350031)	$39.95	$49.95	$
	An Analysis & Evaluation of Rules and Policies Governing OCS Operations #13 (5350027)	$29.95	$39.95	$
	Drafting Natural Gas Contracts After Order 436 #11 (5350021)	$19.95	$29.95	$
	Federal Oil & Gas Leasing Reform Act of 1987 #9 (5350018)	$29.95	$39.95	$
	Protecting Oil & Gas Lien and Security Agreements #6 (5350014)	$19.95	$29.95	$

Qty	Title	SONREEL price	Nonmember price	Amount
	Offshore Bidding Agreements #3 (5350010)	$19.95	$29.95	$
	1982 Joint Operating Agreement #2 (5350008)	$19.95	$29.95	$
	Standard Form Farmout Agreements #1 (5350007)	$19.95	$29.95	$
	Package: 1982 Joint Operating Agreement and Farmout Agreement (5350009)	$29.95	$39.95	$

On land patents

Qty	Title	SONREEL price	Nonmember price	Amount
	Railroad Land Grants #4 (5350011)	$29.95	$39.95	$

Environmental Issues

Qty	Title	SONREEL price	Nonmember price	Amount
	The Application of RCRA and CERCLA to Wastes Produced by Oil & Gas Operations #19 (5350039)	$29.95	$39.95	$
	Identification of RCRA-Related Substances #8 (5350016)	$19.95	$29.95	$
	Checklists for Preparing NEPA Documents #5 (5350013)	$19.95	$29.95	$

The Year in Review. *Information on pertinent legislation, administrative, and regulatory actions in such areas as oil; natural gas; electric power; alternate energy sources; water, marine, and forest resources; air, water, and environmental quality; and solid and hazardous wastes.*

Qty	Title	SONREEL price	Nonmember price	Amount
	1996 Year in Review 5350055	$29.95	$39.95	$
	1996 Year in Review on Disk 5350055D	$25.00	$30.00	$
	1996 Year in Review Package 5350055P	$44.95	$54.95	$
	1995 Year in Review 5350054	$29.95	$39.95	$
	1995 Year in Review on Disk 5350054D	$25.00	$30.00	$
	1995 Year in Review Package 5350054P	$29.95	$39.95	$
	1994 Year in Review 5350049	$29.95	$39.95	$
	1994 Year in Review on Disk 5350049D	$25.00	$30.00	$
	1994 Year in Review Package 5350049P	$44.95	$54.95	$
	1993 Year in Review (5350041)	$29.95	$39.95	$
	1992 Year in Review (5350038)	$29.95	$39.95	$

* IL residents add 8.75%; DC residents add 5.75%; MD residents add 5%
** For orders of $10-$24.99, add $3.95; orders of $25-$49.95, add $4.95; orders $50+, add $5.95

Subtotal $
Tax* $
Handling** $
Total $

Payment

☐ Check enclosed, payable to American Bar Association
☐ VISA   ☐ MasterCard   ☐ American Express

CARD NUMBER _____ EXPIRATION DATE _____

CARDHOLDER SIGNATURE _____

NAME _____

FIRM/ORGANIZATION _____

STREET ADDRESS (NO BOX NUMBERS PLEASE) _____

CITY/STATE/ZIP _____ DAY PHONE NUMBER _____

Mail to: ABA Publication Orders
P.O. Box 10892
Chicago, IL, 60610-0892
Phone: 1-800-285-2221 (ABA1)
Fax: (312) 988-5568
E-mail: abasvcctr@abanet.org
WWW: http://www.abanet.org

If, for any reason, you are not satisfied with your purchase, return it to us within 30 days of receipt for a full refund of the price of the book. No questions asked!

Allow 1-2 weeks for delivery.
Thank you for your order.

SC: BFB97